Foodborne Pathogens and Antibiotic Resistance

Foodborne Pathogens and Antibiotic Resistance

Edited by Om V. Singh

WILEY Blackwell

Published by John Wiley & Sons, Inc., Hoboken, New Jersey
Published simultaneously in Canada

Library of Congress Cataloging-in-Publication Data:

Names: Singh, Om V., editor.
Title: Foodborne pathogens and antibiotic resistance / [edited by] Om V. Singh.
Description: Hoboken, New Jersey : John Wiley & Sons, 2017. | Includes bibliographical references and index.
Identifiers: LCCN 2016028176 (print) | LCCN 2016042961 (ebook) | ISBN 9781119139157 (cloth) |
ISBN 9781119139164 (pdf) | ISBN 9781119139171 (epub)
Subjects: LCSH: Foodborne diseases. | Drug resistance in microorganisms.
Classification: LCC QR201.F62 F665 2017 (print) | LCC QR201.F62 (ebook) | DDC 615.9/54–dc23
LC record available at https://lccn.loc.gov/2016028176

Set in 10/12pt Warnock by SPi Global, Pondicherry, India

Printed in Singapore by C.O.S. Printers Pte Ltd

10 9 8 7 6 5 4 3 2 1

The editor gratefully dedicates this book to Daisaku Ikeda, Uday V. Singh, and Indu Bala in appreciation for their encouragement.

Contents

List of Contributors

Akram R. Alaboudi
Department of Veterinary Pathology and Public Health
Jordan University of Science and Technology
Irbid
Jordan

Heather K. Allen
Food Safety and Enteric Pathogens Research Unit
National Animal Disease Center
ARS, USDA
Ames, Iowa
USA

F.N. Arroyo-López
Food Biotechnology Department
Instituto de la Grasa (CSIC)
Campus Universitario Pablo de Olavide
Seville, Spain

Sharla R. Barr
Department of Biology
Eastern New Mexico University
Portales, New Mexico
USA

I. Bascón
Department of Food Science and Technology
University of Cordoba. Campus de Rabanales
Córdoba
Spain

Bradley L. Bearson
Agroecosystems Management Research Unit
National Laboratory for Agriculture and the Environment
ARS, USDA
Ames, Iowa, USA

Bailey C. Benham
Department of Biology
Eastern New Mexico University
Portales, New Mexico
USA

Ashim Kumar Biswas
Division of Post-Harvest Technology
ICAR-Central Avian Research Institute
Uttar Pradesh
India

Debabrata Biswas
Department of Animal and Avian Sciences; Center for Food Safety and Security Systems
University of Maryland
Maryland
USA

A. Bolivar
Department of Food Science and Technology
University of Cordoba. Campus de Rabanales
Córdoba
Spain

Blaise Pascal Bougnom
University of Yaounde I
Department of Microbiology
Laboratory of Microbiology
Yaounde
Cameroon

Brian W. Brunelle
Food Safety and Enteric Pathogens Research Unit
National Animal Disease Center
ARS, USDA
Ames, Iowa, USA

Conrado Carrascosa
Department of Animal Pathology and Production
Bromatology and Food Technology
Faculty of Veterinary
Universidad de Las Palmas de Gran Canaria
Arucas, Spain

Michael Carroll
Department of Pathobiology & Veterinary Science
University of Connecticut
Connecticut
USA

Kok-Gan Chan
Division of Genetics and Molecular Biology
Institute of Biological Sciences
Faculty of Science
University of Malaya
Kuala Lumpur
Malaysia

Mariah L. Cole
Division of Biological and Health Sciences
University of Pittsburgh
Bradford, Pennsylvania
USA

J.C.C.P. Costa
Department of Food Science and Technology
University of Cordoba. Campus de Rabanales
Córdoba, Spain

Georgios Delis
Laboratory of Pharmacology
School of Veterinary Medicine
Faculty of Health Sciences
Aristotle University of Thessaloniki
Thessaloniki
Greece

Fernanda Domingues
CICS-UBI – Health Sciences Research Centre
Chemistry Department
University of Beira Interior
Covilhã, Portugal

Catarina Tinoco de Faria
Centro de Investigação Interdisciplinar Egas
Moniz (CiiEM)
Instituto Superior de Ciências da Saúde Egas Moniz (ISCSEM)
Caparica
Portugal
Department of Animal Pathology and Production
Bromatology and Food Technology
Faculty of Veterinary
Universidad de Las Palmas de Gran Canaria
Arucas
Spain

Vangelis Economou
Laboratory of Animal Food Products Hygiene – Veterinary Public Health
School of Veterinary Medicine
Faculty of Health Sciences
Aristotle University of Thessaloniki
Thessaloniki
Greece

Francois-Xavier Etoa
University of Yaounde I
Department of Microbiology
Laboratory of Microbiology
Yaounde
Cameroon

Susana Ferreira
CICS-UBI - Health Sciences Research Centre
University of Beira Interior
Covilhã
Portugal

María Antonia Ferrús
Biotechnology Department
Centro Avanzado de Microbiología de Alimentos
Universitat Politècnica de Valencia
Valencia, Spain

Bey-Hing Goh
Novel Bacteria and Drug Discovery Research Group
School of Pharmacy
Monash University Malaysia
Selangor Darul Ehsan
Malaysia
Center of Health Outcomes Research and Therapeutic Safety (Cohorts)
School of Pharmaceutical Sciences
University of Phayao, Phayao
Thailand

Panagiota Gousia
Food-Water Microbiology Group
Department of Microbiology
School of Medicine
Faculty of Health Sciences
University of Ioannina
Ioannina
Greece

Alberto J. Hernandez
Department of Biology
Eastern New Mexico University
Portales, New Mexico
USA

Prathusha Kakarla
Department of Biology
Eastern New Mexico University
Portales, New Mexico
USA

Ranjana K.C.
Department of Biology
Eastern New Mexico University
Portales, New Mexico
USA

Sanath Kumar
QC Laboratory
Harvest and Post Harvest Technology Division
Central Institute of Fisheries Education (CIFE)
Mumbai, India

Prabhat Kumar Mandal
Department Livestock Products Technology
Rajiv Gandhi Institute of Veterinary Education & Research
Pondicherry, India

Learn-Han Lee
Novel Bacteria and Drug Discovery Research Group
School of Pharmacy, Monash University Malaysia
Malaysia
Center of Health Outcomes Research and Therapeutic Safety (Cohorts)
School of Pharmaceutical Sciences
University of Phayao, Phayao
Thailand

Manjusha Lekshmi
QC Laboratory
Harvest and Post Harvest Technology Division
ICAR-Central Institute of Fisheries Education (CIFE)
Mumbai, India

E. Medina
Food Biotechnology Department
Instituto de la Grasa (CSIC)
Campus Universitario Pablo de Olavide
Seville
Spain

Shirley A. Micallef
Department of Plant Science and Landscape Architecture
Center for Food Safety and Security Systems
University of Maryland
Maryland
USA

Mun Mun Mukherjee
Department of Biology
Eastern New Mexico University
Portales, New Mexico
USA

Sujay Y. Naik
Department of Biotechnology
Dayananda Sagar College of Engineering
Bangalore
Visvesvaraya Technological University
Karnataka
India

Meera Surendran Nair
Department of Animal Science
University of Connecticut
Connecticut
USA

Binaya Bhusan Nayak
QC Laboratory
Harvest and Post Harvest Technology Division
ICAR-Central Institute of Fisheries Education (CIFE)
Mumbai, India

Maximilienne Nyegue
University of Yaounde I
Department of Microbiology, Laboratory of Microbiology
Départment of Biochemistry
Laboratory of Phytobiochemistry and Medicinal Plant Study
Yaounde
Cameroon

Mónica Oleastro
National Reference Laboratory for Gastrointestinal Infections
Department of Infectious Diseases
National Institute of Health Dr. Ricardo Jorge
Lisboa, Portugal

Ammini Parvathi
CSIR-National Institute of Oceanography (NIO) Regional Centre
Kochi, India

Esteban Pérez
Department of Animal Pathology and Production
Bromatology and Food Technology
Faculty of Veterinary
Universidad de Las Palmas de Gran Canaria
Arucas
Spain

F. Perez-Rodríguez
Department of Food Science and Technology
University of Cordoba. Campus de Rabanales
Córdoba
Spain

António Raposo
Centro de Investigação Interdisciplinar Egas Moniz (CiiEM)
Instituto Superior de Ciências da Saúde Egas Moniz (ISCSEM)
Caparica
Portugal

Indrika Ranaweera
Department of Biology
Eastern New Mexico University
Portales, New Mexico
USA

Mary Anne Roshni Amalaradjou
Department of Animal Science
University of Connecticut
Connecticut
USA

Hercules Sakkas
Food-Water Microbiology Group
Department of Microbiology
School of Medicine
Faculty of Health Sciences
University of Ioannina
Ioannina
Greece

Anagha S. Setlur
Department of Biotechnology
Dayananda Sagar College of Engineering
Bangalore
Visvesvaraya Technological University
Karnataka
India

Om V. Singh
Division of Biological and Health Sciences
University of Pittsburgh
Bradford
Pennsylvania
USA

Sinosh Skariyachan
Department of Biotechnology
Dayananda Sagar College of
Engineering
Bangalore
Visvesvaraya Technological University
Karnataka
India

Adeyanju Gladys Taiwo
CAHS (COCIN Animal Health Services) Vwang
Nigeria

Indu Upadhyaya
Department of Animal Science
University of Connecticut
Connecticut
USA

A. Valero
Department of Food Science and Technology
University of Cordoba. Campus de Rabanales
Córdoba, Spain

Manuel F. Varela
Department of Biology
Eastern New Mexico University
Portales, New Mexico
USA

Kumar Venkitanarayanan
Department of Animal Science
University of Connecticut
Connecticut
USA

Stève Olugu Voundi
University of Yaounde I
Department of Microbiology
Laboratory of Microbiology
Yaounde
Cameroon

T. Mark Willmon
Department of Biology
Eastern New Mexico University
Portales, New Mexico
USA

Jodi Woan-Fei Law
Novel Bacteria and Drug Discovery
Research Group
School of Pharmacy
Monash University Malaysia Selangor
Darul Ehsan
Malaysia

Anna Zadernowska
University of Warmia and Mazury
in Olsztyn
Poland

Preface

Food is an essential means for humans and other animals to acquire the necessary elements needed for survival. However, it is also a transport vehicle for foodborne pathogens, such as *Salmonella* and *Escherichia coli*, which can pose great threats to human health. Use of antibiotics (e.g., penicillin, kanomycin, streptomycin) has been enhanced in the human health system for multiple generations; however, selective pressure among bacteria allows the development for antibiotic resistance. Recent technological advances have opened the door to explore alternative solutions to antibiotics resistance that might prove useful toward food safety in food industry at large.

Foodborne Pathogens and Antibiotic Resistance features outstanding articles by expert scientists who shed light on broad-ranging areas of progress in the development of food safety interpreting antibiotic resistance. It bridges technological gaps, focusing on critical aspects of foodborne pathogen detection and mechanisms regulating antibiotic resistance that are relevant to human health and foodborne illnesses.

This ground-breaking guide:

- Introduces the microbial presence on variety of food items for human and animal consumption.
- Provides the detection strategies to screen and identify the variety of food pathogens.
- Reviews the literature on diversity of foodborne pathogens on varying food matrices.
- Provides microbial molecular mechanism of food spoilage.
- Discusses molecular mechanism of microorganisms acquiring antibiotic resistance in food.
- Discusses systems biology of foodborne pathogens in terms of detection and food spoilage.
- Discusses FDA's regulations and Hazard Analysis and Critical Control Point (HACCP) toward challenges and possibilities of developing global food safety.

Foodborne Pathogens and Antibiotic Resistance is an immensely useful resource for graduate students and researchers in the food science, food microbiology, microbiology, and industrial biotechnology.

Introduction

Food is the way that humans and other animals acquire the necessary elements needed for survival. However, it is also a transport vehicle for foodborne pathogens, such as *Salmonella* and *Escherichia coli*, which can pose great threats to human health. Raw food, ready-to-eat vegetables, dairy products, pork, beef, and poultry have been shown to harbor antibiotic-resistant pathogens, as well as multi-drug-resistant pathogens. Food spoilage can be a visual sign of pathogenic bacteria. Spoilage due to enzymatic oxidation and the reduction mechanisms of pathogenic bacteria can be prevented if appropriate measures are taken to preserve the food: heating, cold preservation, fermentation, moisture reduction, chemical preservation, ultra-high pressure, or irradiation.

In the United States, foods are regulated under the Federal Food, Drug, and Cosmetic Act (FD&C Act; 21 U.S.C. Sec 321). The classification of food products ("i. articles used for food or drink for man or other animals, ii. chewing gum, and iii. articles used for components of any other such article") defines how rigorously the food or related product is regulated, or if the product is even legal to consume as a food. Food adulteration is defined in section 342 of the United States Code of Federal Regulations (CFR), title 21, chapter 9 (FD&C Act), subchapter IV (food).[1] According to the FD&C Act, "food is [also] adulterated if it has been prepared, packed, or held under unsanitary conditions; may become contaminated with filth; or has been rendered injurious to human health." To make sure food does not become adulterated and human illness does not occur, levels of microorganisms need to be kept low or nonexistent. In general, federal and state law prohibits the selling of adulterated foods.

Worldwide, significant efforts are being made to improve food safety with new policies and recommendations. Antibiotics (e.g., penicillin, kanomycin, and streptomycin) have been used in food safety and healthcare for multiple generations; however, selective pressure among bacteria has allowed the development of antibiotic resistance. Among these bacteria are foodborne pathogens, which have acquired antibiotic resistance genes through horizontal gene transfer and mobile genetic elements (e.g., transposons, plasmids). Due to the overusage of antibiotics in the agricultural system, antibiotic resistance genes have been transferred through agricultural waste, soil, meat, vegetables, and water systems. There are a variety of alternatives to antibiotics (i.e., phage therapy, bacterial vaccinations, probiotics, and prebiotics) that are undergoing evaluation, but are not as reliable due to many shortcomings such as cost-effectiveness, specificity, and likelihood of bacteria becoming resistant. Human society must learn from ongoing microbial outbreaks to deal better with emerging antibiotic resistance and multiple-drug resistance.

This book continues to bridge the gaps on technology and focuses on exploring the diversity of food pathogens in varying food matrices, in addition to the inescapable question of antibiotic resistance among foodborne pathogens and its impact on human society. Biswas and

Foodborne Pathogens and Antibiotic Resistance, First Edition. Edited by Om V. Singh.

Micallef in Chapter 1 present the prevalence of common as well as under-researched bacterial pathogens and parasites along with the limitations of current detection techniques. They lay out a future direction toward improving detection techniques for under-researched foodborne pathogens and parasites in produce and animal products.

The connection between seafood and human health is undeniable; the National Marine Fisheries Services (NMFS) yearbook of fishery statistics for the United States for 20142 reported that the estimated U.S. per capita consumption of fish and shellfish was 14.6 pounds (edible meat) in 2014. The United States is the second largest consumer of seafood in the world, after China and before Japan.[2] Bolivar *et al.* in Chapter 2 explain and characterize foodborne pathogens and spoilage bacteria in Mediterranean fish species and seafood products. The same chapter provides insights into European seafood regulations and microbiological standards for fishery and aquaculture products. Notably, the authors also present the global legislative requirements for controlling the safety and quality of fish and seafood products defined under Hazard Analysis and Critical Control Point (HACCP).

The *Pseudomonas* genus is one of the most diverse ecologically significant groups of known bacteria, and includes species that have been isolated worldwide in all types of environments. This genus contaminates foods from many sources and is able to utilize a wide range of materials, such as red meat, fish, milk and dairy, etc., as substrates for growth. In Chapter 3, Raposo *et al.* give an overview of the mechanistic views of food spoilage by Pseudomonas spp.

The technological implementation of pathogen detection is an undeniable aspect of food safety. The microorganism *Arcobacter* spp. is a food- and waterborne pathogen distributed throughout the food chain. In fact, consumption of *Arcobacter*-contaminated food or water is the most common cause of infection that poses a serious hazard to human health.[3] Ferreira *et al.* in Chapter 4 explain the specifics of *Arcobacter* and human and animal infections associated with this genus. The chapter also describes unique "-omics" molecular approaches to detect *Arcobacter* in varying food matrices.

Vegetables are important for a healthy diet, as they are sources of phytosterols, dietary fiber, phytochemicals, minerals, and water-soluble vitamins, among other important compounds. Table olives are a traditional fermented vegetable with many centuries of history; they are a good source of antioxidants (hydroxytyrosol, oleuropein, methyloleuropein, etc.) and triterpenic acids. The global consumption of olive oil has almost doubled over the last 25 years, with a jump of 73%.[4] Based on the importance of the health benefits from olives, Valero-Diaz *et al.* in Chapter 5 describe varieties of olives, microbial hazards, and their implications in the production of table olives.

Under stress, many microorganisms can temporarily change their genetic expression to form spores or endospores. Bacterial species can thrive under adverse environmental conditions such as food sterilization in this way, potentially causing foodborne diseases. Voundi *et al.* in Chapter 6 describe the phenomenon by which a vegetative cell becomes a spore and goes through the germination process under favorable conditions. The chapter also highlights the problem of controlling spore-forming bacteria in food.

Due to challenges in timely detection of foodborne pathogens, the Centers for Disease Control and Prevention (CDC) reports, a staggering number of people die annually due to foodborne diseases,[5] even though the United States has the safest food supplies in the world. In Chapter 7, Law *et al.* examine rapid detection methods and their applications for four major foodborne bacterial pathogens, that is, *Listeria monocytogenes*, *Vibrio parahaemolyticus*, *Escherichia coli*, and *Salmonella*, along with their advantages and limitations.

Salmonella sp. has been recognized as the main culprit in food poisoning around the world. Zadernowska and Chajecka-Wierzchowska in Chapter 8 describe alternative rapid detection methods for Salmonella in food.

Many different measures have been used to prevent food spoilage. Genetic engineering is one new measure. In Chapter 9, Carroll and Zhou describe the use of CRISPR (clustered, regularly interspaced, short palindromic repeats) in bacterial genome editing for food safety and industrial purposes. Biswas and Mandal in Chapter 10 propose other natural modes of food safety, such as the use of natural antimicrobials against meatborne pathogens for food safety.

Antibiotics have long been important for treating infectious diseases in human and animals. Their use in the food industry has risen rapidly in the past few decades, and the overuse of antibiotics has caused bacteria to develop a means of protection against them, that is, resistance genes. Cole and Singh in Chapter 11 discuss the apparent linkage of foodborne pathogens with antibiotic resistance, describing the biochemistry of food spoilage, food preservation techniques, and systems biology approaches as effective ways to detect foodborne pathogens. Additionally, the chapter discusses alternative strategies for targeting foodborne pathogens. In continuation, Nair *et al.* in Chapter 12 discuss common preservatives and disinfectants used in the food industry, their modes of action, bacterial responses to these antimicrobials, and resistance mechanisms.

Many foodborne pathogens cause infectious diseases. Chemotherapy is being considered as a prominent mode of treatment, but bacterial variants continually evolve resistance to clinically relevant antimicrobial chemotherapeutic agents, potentially confounding effective treatments, especially when the pathogens are multi-drug-resistant. In Chapter 13, Ranjana *et al.* discuss recent developments in key multidrug efflux pumps of the major facilitator superfamily from key foodborne bacterial pathogens.

As stated earlier, the foodborne pathogen *Salmonella* sp. is ubiquitous in agricultural, environmental, and human reservoirs, causing human gastrointestinal illness worldwide. The National Antimicrobial Resistance Monitoring System (NARMS) recognized the development of antibiotic-resistant *Salmonella*, which has been a public health concern for over 40 years and continues to persist in nontyphoidal *Salmonella* (CDC, 2015).[6] In Chapter 14, Brunelle *et al.* explore the prevalence, evolution, and dissemination of antibiotic resistance in *Salmonella*.

The genus *Staphylococcus* is a broad group of microorganisms containing coagulase-positive staphylococci (CPS) and coagulase-negative staphylococci (CNS). Chakecka-Wierzchowska and Zadernowska in Chapter 15 describe antibiotic resistance in both CPS and CNS isolated from food.

Enterococcus spp. microorganisms have been used for decades in food fermentation and preservation. Organisms in this genus have also emerged as nosocomial- and community-acquired pathogens. Due to the nature of pathogenicity, the *Enterococcus* spp. are regarded as reservoirs of antimicrobial resistance genes and indicators of antibiotic resistance. In Chapter 16, Vangelis *et al.* describe antibiotic resistance in *Enterococcus* spp.

Seafood remains on the top tier of the global food market that has led to food mobility across continents, but also increases the chances of food contamination with a number of pathogenic microorganisms. Kumar *et al.* in Chapter 17 describe antibiotic resistance in seafood-borne pathogens.

Campylobacter spp. are part of the normal intestinal flora of wild and domestic animals, including birds, and the most frequently recognized bacterial cause of human gastroenteritis. Poultry and their meat products are the main source of *Campylobacter* spp. contamination that affects humans. In Chapter 18, Osaili and

Alaboudi describe occurrences of *Campylobacter* sp. and its antimicrobial resistance in animals and humans.

Advancements in technology have greatly improved the lifestyles of people living in developed countries, but developing nations are still struggling with the basic necessities. Foodborne diseases due to pathogenic organisms are one common issue that has not received much attention from local and international authorities. In Chapter 19, Adeyanju discusses the prevalence and antibiograms of foodborne pathogens in developing African countries. Globally, the misuse of antibiotics is another issue to be raised, as it promotes bacterial growth and resistance against multiple antibiotics. In Chapter 20, Skariyachan *et al.* discuss the evolution and prevalence of multi-drug resistance among foodborne pathogens.

This book, *Foodborne Pathogens and Antibiotic Resistance*, is a collection of outstanding articles elucidating several broad-ranging areas of progress and challenges related to foodborne pathogens. This book will contribute to research efforts in the scientific community and commercially significant work for corporate businesses, with the goal of establishing a long-term safe and sustainable food supply with minimum impact on the necessary elements needed for survival.

We hope readers will find these articles interesting and informative for their research pursuits. It has been my pleasure to put together this book with Wiley-Blackwell Press. I would like to thank all of the contributing authors for sharing their quality research and ideas with the scientific community through this book.

References

1. Title 21, Food and Drugs, Chapter 9 – Federal Food, Drug, and Cosmetic Act, Subchapter IV – Food, Sec. 342 – Adulterated food. Available at: http://www.gpo.gov/fdsys/pkg/USCODE-2010-title21/html/USCODE-2010-title21-chap9-subchapIV-sec342.htm [Accessed March 1, 2016].
2. Fisheries of the United States. (2014), Current Fishery Statistics No. 2014. National Marine Fisheries Service Office of Science and Technology. Available at: http://www.st.nmfs.noaa.gov/Assets/commercial/fus/fus14/documents/FUS2014.pdf (Last visited: March 2016).
3. International Commission on Microbiological Specifications for Foods (ICMSF). (2002). *Microorganisms in Food 7 - Microbiological Testing in Food Safety Management*. New York: Springer Science & Business Media.
4. Coldiretti Forza Amica Del Paese (2015), Avaliable at: http://www.coldiretti.it/news/Pagine/143—26-Febbraio-2016.aspx [Accessed March 1, 2016].
5. Centers for Disease Control and Prevention (CDC). (2011). CDC estimates of foodborne illness in the United States. Available at: http://www.cdc.gov/foodborneburden/PDFs/FACTSHEET_A_FINDINGS_updated4-13.pdf. [Accessed March 1, 2016].
6. Centers for Disease Control and Prevention (CDC). (2015). National Antimicrobial Resistance Monitoring System for Enteric Bacteria (NARMS): Human Isolates Final Report, 2013. Atlanta, Georgia: U.S. Department of Health and Human Services.

1

Diversity of Foodborne Bacterial Pathogens and Parasites in Produce and Animal Products and Limitations of Current Detection Practices

Debabrata Biswas[1,3] *and Shirley A. Micallef*[2,3]

[1] *Department of Animal and Avian Sciences*
[2] *Department of Plant Science and Landscape Architecture*
[3] *Center for Food Safety and Security Systems, University of Maryland, Maryland, USA*

1.1 Introduction

The Centers for Disease Control and Prevention (CDC) estimates that financial losses from foodborne illnesses, including medical costs and losses in productivity, range from $500 million to $2.3 billion annually. More than 250 foodborne diseases are recognized and the major causative agents associated with foodborne illness are bacteria, virus, and parasites. *Salmonella, Campylobacter,* pathogenic *E. coli, Listeria, Shigella,* and *Vibrio are* the most common foodborne pathogens associated with meat and animal products (Mead, 2004; Hutchison *et al.*, 2005). A decade ago, animal origin products used to cause known cases of foodborne disease, but now the whole scenario has been changed due to current health food habits in the United States. Green vegetables and fruits are highly recommended for a healthy life, and in the United States, consumption of produce has increased significantly in the last decade (Smith *et al.*, 2013; Sivapalasingam *et al.*, 2004; Scallan *et al.*, 2011). Further, organic produce has been shown to be of superior nutritional value, compared to conventional produce (Lester and Saftner, 2011; Hallmann, 2012; Hallmann and Rembiałkowska, 2012; Vinha *et al.*, 2014). Many consumers opt for organic products due to their nutritional value, and because the produce is free of synthetic pesticides and antibiotic residues. Simultaneously, the CDC reported that the proportion of foodborne bacterial disease outbreaks associated with fruits and vegetables have increased significantly (Johnston *et al.*, 2005; Berger *et al.*, 2010; CDC, 2011; Gould *et al.*, 2013). More specifically, the microbiological quality of organic versus conventional produce has only been compared for some produce types (Pagadala *et al.*, 2015; Marine *et al.*, 2015). According to CDC outbreak data, plant products, including fruits, vegetables, spices, and grains, are responsible for >51% (4,924,877 recorded cases) of foodborne illness in the United States (CDC, 2013). Among plant products, produce-only origin attributed to more than 45% (4,423,310) of recorded cases (CDC, 2013). A wide spectrum of pathogens has been documented in produce-associated outbreaks and a significant number of the infectious agents (>20%) that were responsible for the produce-borne infections are unknown (Scharfe, 2011; CDC, 2013). Further, these numbers do not represent the actual number of cases of foodborne infection in the United States, because sporadic cases remain largely unreported and/or undiagnosed.

Foodborne Pathogens and Antibiotic Resistance, First Edition. Edited by Om V. Singh.

On the other hand, animal food products including meat, egg, and milk are commonly known as a major contributor to zoonotic infections (Mead, 2004; Hutchison *et al.*, 2005), as many of the zoonoses are commonly found in farm animal gut as normal flora. In addition, with commonly known bacterial foodborne pathogens, farm animals also harbor varieties of under-researched microbial pathogens, specifically, parasite and viruses. Food processing and packaging facilities struggle with limitations in resources, and often lack in the latest scientific information and techniques to detect the possible contaminants that exist in these type of food products. According to the diverse sources and causative agents of foodborne infection, the quality control practices in food processing, storage, and transportation facilities in the United States need to make further progress to meet the current demand in reducing/controlling foodborne infections.

1.2 Common Bacterial Pathogens and Parasites Found in Produce and Animal Products

Most studies have investigated only major and mostly known foodborne pathogen prevalence, such as *Salmonella, Campylobacter*, pathogenic *E. coli, Shigella, Vibrio*, and *Listeria* (Berger *et al.*, 2010; Kozak *et al.*, 2013; Bolton *et al.*, 2012; Cartwright *et al.*, 2013; Johnston *et al.*, 2005; Mukherjee *et al.*, 2006; Yokoyama *et al.*, 1998; Peralta *et al.*, 1994; CDC, 2008; Scallan *et al.*, 2011; Zhao *et al.*, 2014). *Salmonella enterica* subspecies *enterica* serovar Typhimurium (*S.* Typhimurium) and serovar Enteritidis (*S.* Enteritidis) are the most common serovars, and can cause disease syndromes, such as gastroenteritis and systemic infections, in a wide range of animal species and humans (Yokoyama *et al.*, 1998; Peralta *et al.*, 1994). Major produce-associated serotypes include *S.* Javiana and *S.* Newport, with fruits and nuts most commonly associated with the former, whereas vine and stalk vegetables are associated with *S.* Newport (Painter *et al.*, 2013). It has been reported that approximately 10–20% of the poultry meat at the retail level is positive for many different serotypes. The most frequently reported serotypes in layer flocks in 2002 were *S.* Enteritidis (57.7%), *S.* Typhimurium (9.6%), and *S.* Infantis (6.9%) (Mead, 2004).

Another major foodborne pathogen, *Campylobacter jejuni*, is a microaerophilic, spiral-shaped, Gram-negative bacterium and causes bacterial gastroenteritis worldwide. The CDC estimated that *C. jejuni* causes 2.4 million cases in the United States each year (CDC, 2008) and is the causative agent for 5–14% of all diarrheal diseases worldwide (CDC, 2008). Campylobacteriosis, gastroenteritis with *C. jejuni*, is characterized by the rapid onset of fever, abdominal cramps, and bloody diarrhea (Skirrow and Blaser, 1992). Sporadic cases are most common and are often associated with handling and consumption of undercooked poultry, as *C. jejuni* is part of the normal intestinal flora in chicken (Shane, 1992; Skirrow and Blaser, 1992; Deming *et al.*, 1987; Tauxe, 1992; Biswas *et al.*, 2007). The presence of *C. jejuni* in processed chicken carcasses offered for retail sale was reported to range from 7% to 32% during the winter months and from 87% to 97% during the summer months (Willis *et al.*, 2000). Campylobacteriosis is less commonly attributed to fresh produce, although a major *C. jejuni* outbreak in Alaska associated with raw peas contaminated with bird feces demonstrated the risk of fresh crop contamination posed by wildlife (Gardner *et al.*, 2011).

Pathogenic foodborne *E. coli*, specifically enterohemorrhagic *E. coli* (EHEC), enteropathogenic *E. coli* (EPEC), and enterotoxigenic *E. coli* (ETEC) are also involved in thousand of foodborne infections in the United States (CDC, 2014). Foodborne pathogenic *E. coli* O157:H7 is commonly discussed in the media in association with foodborne illness outbreaks, because of the severity of the disease. The major reservoir of this bacterial pathogen is cattle, and eating raw or undercooked ground beef or drinking unpasteurized beverages or dairy products are mostly associated with the bacterial infections (CDC, 2014). Major produce-associated outbreaks have demonstrated the adaptability of these pathogens to the plant niche, especially leafy greens (Grant *et al.*, 2008; CDC, 2006; Marder *et al.*, 2014) and sprouts (Scheutz *et al.*, 2011).

Listeria spp., including *L. monocytogenes*, are commonly found in soil, water, and decaying plant material (Weis and Seeliger, 1975; Linke *et al.*, 2012). As such, there are many potential routes for contamination of foods with this organism. One characteristic that makes *L. monocytogenes* particularly difficult to control is its ability to grow in foods at refrigeration temperatures. Although *L. monocytogenes* has been known as a human pathogen since the early nineteenth century, it has only recently been recognized as a foodborne pathogen (Pradhan *et al.*, 2009). Several large outbreaks of listeriosis were reported due to consumption of contaminated foods, specifically refrigerated, ready-to-eat foods, such as hot dogs and deli meats, unpasteurized milk and dairy products, and raw and undercooked meat, poultry and seafood, salad, and fruits (CDC, 2011). The largest outbreak associated with fruit was caused by contaminated cantaloupe in 2011, sickening 146 people and causing 30 fatalities and one miscarriage (CDC, 2011).

Seafood consumption in the United States has been associated with a number of foodborne bacterial infectious agents. Specifically, *Vibrio parahaemolyticus* has been associated with sporadic infections and outbreaks of gastroenteritis, whereas *V. vulnificus* infections occur almost exclusively as sporadic cases. Clinical symptoms most often associated with *V. parahaemolyticus* infection include watery diarrhea, abdominal cramps, nausea, and vomiting; wound infections and septicemia occur less commonly (Iwamoto *et al.*, 2010, Daniels *et al.*, 2000). *V. vulnificus* is particularly virulent, especially among patients with liver disease and iron storage disorders, which are at increased risk of invasive infection such as sepsis and bacteremia (Iwamoto *et al.*, 2010; Levine and Griffin, 1993).

In addition to bacterial pathogens and viruses, the risk of contamination of animal and plant food products with parasites exists. Parasites remain understudied because of the complexity of methods of isolation and identification. Therefore, minimizing the risks and enhancing intervention strategies to prevent cross-contamination of organic and conventional animal products and produce with parasites is a priority. Prevalence of parasites in various food products is crucial for the development of effective control strategies against identified risk factors and management of foodborne infections with under-researched pathogens. The possibility of contamination of produce grown in organic or integrated crop-livestock farms with parasites such as *Cryptosporidium parvum/hominis, Cyclospora cayetanensis*, and *Giardia duodenalis* are potentially high (Putignan and Menichella, 2010; Pullin, 1987). Recent *C. cayetanensis* outbreaks have proved difficult to trace back and control, including ones associated with cilantro and other unidentified products (CDC, 2013; Nichols *et al.*, 2015). Another parasite commonly found in various farm animals including pig and chicken is *Toxoplasma gondii* (Dubey and Hill, 2002). Toxoplasmosis caused by *T. gondii* is an emerging public health problem in individuals who are at high risk for foodborne illness—pregnant women, infants, older adults, and people with weakened immune systems. Animals raised in unconfined conditions are at higher risk of being contaminated at all levels from farm to retail (Guo *et al.*, 2015).

1.3 Unusual Bacterial Pathogens and Parasites in Produce and Animal Products

The CDC has estimated that less than a fifth of estimated foodborne illnesses per year are attributed to a known agent, with over 38 million remaining unknown (CDC, 2011; Painter *et al.*, 2013). Further, the prevalence of lesser-known or under-researched zoonotic pathogens and their roles in cross-contamination of produce and etiology of human gastrointestinal infections have not been investigated in depth. This paucity of data is attributable to difficulties in identifying cases and the lack of reliable methods for detecting certain bacterial pathogens and parasites in animal and plant food products (Jolly and Lewis, 2005). Likely, under-researched zoonotic pathogens enter the food chain through direct contamination with fecal matter from farm animal to animal and plant food products, or indirectly via contaminated soil or water contaminated with fecal matter. Table 1.1 summarizes several unusual/under researched bacterial pathogens and parasites, and their sources caused outbreak or sporadic cases of foodborne infection.

In a study in the United States' Upper Midwest region, Mukherjee *et al.* (2006) investigated the contamination level of organic produce with common zoonotic bacterial pathogens at pre- and post-harvest levels and concluded that some of the conventionally produced fruit and vegetables had significantly lower coliform counts than did semi-organic (uncertified) or organic produce. In another study in Canada, Kozak *et al.* (2013) found that in addition to bacterial pathogens, several parasites were also often associated with produce-borne infections. This may vary depending on commodity, but to our knowledge, no data appears to exist on on-farm cross-contamination for under-researched microbial pathogens at any production scale. However, due to the proximity of animal and crop cultivation areas on smaller farms, it is possible that risks of pathogen dissemination onto produce are higher in small- and medium-scale mixed or integrated crop-livestock farm environments. Vehicular and human traffic, prevailing winds, rain run-off, and wildlife could all contribute to dispersal of human pathogens from animal rearing and manure composting areas to pre-harvest produce production areas (Salaheen *et al.*, 2015).

The common livestock grown in integrated crop-livestock farms are pig, goat, sheep, cattle, and poultry (Hoffman, 2010; Strawn *et al.*, 2013). These livestock are known major reservoirs for zoonotic pathogens, including under-researched foodborne pathogens, the prevalence of *Staphylococcus aureus* and *Yersinia enterocolitica* in produce are quite high, and these pathogens are not studied yet at all in the United States. Integrated crop-livestock farm products such as fresh fruits and vegetables (spinach, carrots, lettuce, tomatoes, cucumber, apples, and strawberries) are high-risk foods with respect to contamination with these bacterial pathogens. In addition, the possibility of contamination of the produce grown in integrated crop-livestock farm with parasites such as *Cryptosporidium parvum/hominis* and *Giardia duodenalis* are potentially high and these parasites are mostly unknown because of the complex methods of isolation and identification. Therefore, minimizing the risks and intervention strategies of cross-contamination of organic or conventional produce with these under-researched bacterial pathogens and parasites are required. Such data are crucial for the development of effective control strategies against identified risk factors and management of foodborne infections with under- researched pathogens.

1.4 Farming Systems and Mixed (Integrated) Crop-Livestock Farming

In a European study, it was found that the level of contamination with foodborne pathogens was higher in produce samples cultivated under organic practices on integrated farms compared to those grown in produce-only farms in the absence of livestock (Bolton *et al.*, 2012). Parasites such as *Giardia, Cryptosporidium*, and many bacterial pathogens including *Salmonella, E. coli* O157:H7, *Staphylococcus*, and *Yersinia*, could be introduced to integrated or mixed crop-livestock farms (MCLF) and its products at the pre-harvest level through contaminated water, dirt, insects, animal waste fertilizer, shared/commonly used instruments, and/or farm animals, birds, and wild animals (Natvig *et al.*, 2002). It appears, however, that

Table 1.1 List of Unusual/Underresearched Bacterial Pathogens and Parasites, and Their Sources Associated With Outbreak or Sporadic Cases.

Causative agent	Food sources	Outbreak/sporadic cases	State/Country (Year)	Reference
Brucella	Beef	Outbreak	South Korea	Yoo *et al.*, 2015
	Food served in Cafeteria	Out break	Peru	Roman *et al.*, 2013
	milk, cheese, etc.	Outbreak/sporadic	California	CDC, 2013
Pasteurella	Various	Outbreak/sporadic	United States	CDC, 2015
	Various	Sporadic	Canada	Glass-Kaastra *et al.*, 2014
Staphylococcus	Dairy products and other	Outbreak	Italy	Gallina *et al.*, 2013
	Fried chicken	Outbreak	South Korea	Ji-Yeon *et al.*, 2013
	Various	Outbreak/sporadic	United States	MMWR, 2013
Strptococcus	Egg	Outbreak	Australia	Levy *et al.*, 2003
	Corn	Outbreak	Israel	Linhart *et al.*, 2008
	Various	Outbreak	Israel	Kaluski *et al.*, 2006
Yersinia	Dairy products	Outbreak	Vermont and New Hampshire	Ackers *et al.*, 2000
	Salad	Outbreak	Norway	McDonald *et al.*, 2012
Cryptospordium	Various	Outbreak/sporadic	United States	Drinkard *et al.*, 2015
	Various	Outbreak	Finland	Ponka *et al.*, 2009
Toxoplasma	Meat products	Outbreak/sporadic	United States	Jones and Dubey, 2012
	Various	Outbreak/sporadic	Brazil	Ekman *et al.*, 2012
Giardia	Chicken, salads, etc.	Outbreak/sporadic	United States	Foodborne Illness outbreak, 2007
	Various	Outbreak/sporadic	United States	Foodborne Illness outbreak, 2008

elevated risk might be associated with the close proximity to animal operations and not the implementation of organic farming practices. A large body of evidence suggests that the safety of produce cultivated under organic farming practices is equivalent to that produced conventionally and not to harbor higher levels of foodborne pathogens. Bohaychuk *et al.* (2009) did not isolate any foodborne bacterial pathogens or parasites in produce grown organically or conventionally in Alberta, Canada, and levels of generic *E. coli* were not statistically different between the two farming systems. *Salmonella enterica* was also not isolated from conventional or organic tomato farms in the mid-Atlantic region of the United States (Micallef *et al.*, 2012; 2013; Pagadala *et al.*, 2015). Investigations of leafy greens farms in the same region did measure a 2.2% *Salmonella* positive rate for leafy greens, however, contaminated produce originated equally from organic and conventional farms (Marine *et al.*, 2015). The latter study also reported no statistical difference in generic *E. coli* population levels or number of positive samples collected from conventional and organic farms. Other studies, however, have reported higher levels of indicator *E. coli* on farms managed under organic practices (Mukherjee *et al.*, 2006), and the use of manure or young composts was associated with a higher prevalence of generic *E. coli* (Mukherjee *et al.*, 2006). Manure application to a field within a year has also been identified as a risk factor for *Samonella* presence in fields (Strawn *et al.*, 2013), whereas irrigation and soil moisture appear to increase the likelihood of *L. monocytogenes* prevalence in agricultural soils (Chapin *et al.*, 2014; Weller *et al.*, 2015).

1.5 Major Sources of Unusual/ Under-Researched Bacterial Pathogens and Parasites in Food

In spite of a dearth of data on sources of under-researched pathogens, reports on such pathogens point to fresh fruits and vegetables as common vehicles. Plant food source attributions to foodborne illness outbreaks occurring between 1998 and 2008 in the United States identify fruits and nuts, leafy greens, vine and stalked vegetables, and root vegetables as important vehicles for *C. cayatensis*, with nine, five, four, and three outbreaks, respectively (Painter *et al.*, 2013). A study investigating the prevalence of parasites in ready-to-eat packaged leafy greens grown in the United States, Canada, and Mexico revealed a high prevalence of *Cryptosporidium* (5.9%), *Cyclospora* (1.7%), and *Giardia* (1.8%) uses polymerase chain reaction (PCR)-based detection methods. *Cryptosporidium* oocysts, *Cyclospora*-like oocysts, and *Giardia* cysts were confirmed in leafy greens samples using microscopy (Dixon *et al.*, 2013). In another study in Canada, Kozak *et al.* (2013) found that many uncommon microbial pathogens were linked to produce-related outbreaks. In Norway, MacDonald *et al.* (2011) found that an outbreak of *Y. enterocolitica* was linked to ready-to-eat salad mixes, an unusual vehicle for this pathogen, since pigs are the major reservoir (Laukkanen *et al.*, 2008). Raw carrots were also the implicated vehicle for a *Y. pseudotuberculosis* O:1 outbreak in Finland in 2004, traced back to the farm where spoiled carrots, fluid from spoiled carrots, and intestinal samples of the common shrew all yielded the outbreak strain (Kangas *et al.*, 2008). *Yersinia* spp. are under-researched in the United States, apparently causing few illnesses (Painter *et al.*, 2013), but *Y. enterocolitica* is a major gastrointestinal pathogen worldwide (Rahman *et al.*, 2011). In Europe, *Y. enterocolitica* is listed in the annual reports of the European Food Safety Authority as the third most common enteropathogen (Zadernowska *et al.*, 2013), but often this serious pathogen is not recorded in the United States. The most recent yersiniosis outbreak in the United States was linked to inadequately pasteurized dairy products, causing 22 infections in 2011 (Longenberger *et al.*, 2014).

The source of contamination of fresh produce with enteric pathogens can frequently be traced back to environmental reservoirs associated with farm and wild animals (Brinton *et al.*, 2009; Park *et al.*, 2012). The small (goat and sheep) and large ruminants (cattle and buffalo) are potential reservoirs for several unusual bacterial pathogens including *Bacillus*, *Clostridium* (Salaheen *et al.*, 2015) and parasites including *Giardia* and *Cryptosporidium* (Iwamoto *et al.*, 2010; Salaheen *et al.*, 2015). *Bacillus* does not appear to cross over to produce crops, but *C. perfringens* and *C. botulinum* are important pathogens of food plants. Between 1998 and 2008, *C. perfringens* caused 16 outbreaks in grains/beans, 8 outbreaks in vined and stalked vegetables, and 1 outbreak in leafy greens, with a total of over 1,600 reported illnesses (Gould *et al.*, 2013). Poultry can also serve as a potential source for some unusual bacterial pathogens in food such as avian pathogenic *E. coli* (APEC), which one common pathogen causes urinary tract infections in humans (Rodriguez-Siek *et al.*, 2005). In MCLFs, produce and livestock, such as poultry, cattle, swine, goat, and sheep, co-exist in a single facility, and feral animals, birds, and rodents commonly co-exist. This increases the possibility of introducing pathogenic microbes to crop production environments. The survival/multiplication ability during recycling of animal waste as a sole source of fertilizer is also high if manure is not fully composted. Mixed crop-livestock/back yard farmers may not follow all proper guidelines, and in some cases, the facilities are open to visitors (pick-your-own), or subject to wildlife intrusion (Jolly and Lewis, 2005).

1.6 Diversity of Farming and Processing Practices and Possible Risks

A recent report from Economic Research Service (ERS) estimates that the organic and naturally grown food market is the fastest growing sector in the U.S. food industry. This rapid growth is due to an increase in consumer concerns combined with the evaluation of new organic production and marketing systems (USDA-ERS, 2014). The Organic Trade Association reported that the organic food industry has grown from $1.0 billion dollars in 1990 to $28 billion in 2012 (USDA-ERS, 2013). Organic products are currently sold in >73% of all conventional grocery stores (Jolly and Lewis, 2005). Another important location for locally grown, small farm-produced and organic food marketing is farmers markets. The USDA promotes farmers markets across the country, and currently more than 8,000 farmers markets are now listed in the National Farmers Market Directory (USDA, 2014; Johnson *et al.*, 2013). A significant share of the produce sold at farmers markets and/or farm/road side markets is grown on small- and medium-scale farms and MCLFs. In the United States, specifically in the Mid-Atlantic and Corn Belt regions, a large number of farming practices are organic MCLF or in organic transition and contributing a significant amount of food products specifically produce to the United States organic food supply chain (Abler and Shortle, 2000; Sulc and Tracy, 2007; Luna *et al.*, 1994).

Typically, the products of small- to medium-sized MCLFs are sold on a local or regional scale, thus the chances of any contamination of produce causing large or widespread outbreaks are low. However, locally sold produce that has been contaminated with foodborne pathogens may play an important role in sporadic cases or localized outbreaks. Due to the nature of MCLF systems, cross-contamination between animal and fresh crop produce may occur, since animals may serve as reservoirs for pathogens (van den Berg *et al.*, 2007; Hoffman, 2010; Strawn *et al.*, 2013) that could also colonize farm crops. Identifying the prevalence, identity, and antibiotic resistance patterns of microbial pathogens in MCLS products—specifically fresh produce mostly available at farmers markets, roadside stands, local grocery stores and their production facilities—are crucial to fully assess the risks of sporadic cases or localized outbreaks. Such information could significantly contribute to enhancing the safety and biosecurity of these products.

Products from mixed farming, including meat, egg, and fresh produce, are at greater risk of cross-contamination as they are grown in the same facility and are currently considered to be high-risk foods (Adl *et al.*, 2011). On the mixed farm, growers typically compost animal bedding and waste and use it to grow specialty crops such as lettuce, spinach, tomatoes, green pepper, and cantaloupe (Nascimbene *et al.*, 2012). These crops can be vehicles for a variety of enteric bacterial pathogens including *Salmonella* and *E. coli* O157:H7. The source of contamination of fresh produce with enteric pathogens can frequently be traced back to environmental reservoirs associated with farm animals such as poultry, cattle, swine, goat, and sheep (Brinton *et al.*, 2009; Park *et al.*, 2012). It appears that environmental factors and farming practices can affect *Salmonella* and *E. coli* O157:H7 transmission from animal reservoirs to leafy greens via soil, water, and vectors.

Most pasture animals and poultry are raised with access to the outdoors for at least one-third of their entire life cycle. The lack of proper biosecurity measures potentially increases the contact livestock and flocks may have with sources of pests and pathogens including wild birds, rodents, insects, and other wild animals (Berg, 2001). Proper biosecurity measures can reduce pathogen-shedding rates of housed farm animals (Heyndrickx *et al.*, 2001), but the increasing popularity of organic and pasture-raised meat and other animal food products raises the question of whether the welfare benefits for pasture-raised and organic agricultural animals can occur in conjunction with animal health. More species of helminthes and heavier worm burdens have been found in hens reared organically (Thamsborg *et al.*, 1999). In a study in Belgium, Permin *et al.* (1999) reported that *Capillaria anatis* and *Capillaria caudinflata* were present only in organic/free-ranging flocks.

1.7 Current Hygienic Practices and Their Effects on These Under-Researched Pathogens

Sanitizers, primarily hypochlorite and chlorine, are commonly used in fruit and vegetable processing water. Often, the produce industry uses the sanitizers to treat the fruit or vegetable to prevent cross-contamination, but all sanitizers do not act against all microbial pathogens at the same concentrations. Many spore-forming bacteria are resistant to hypochlorite or chlorine and cysts or oocysts of parasites are also resistant to these sanitizers. Brackett (1994) compared the disinfecting effects of sodium hypochlorite on the survival of several bacterial pathogens in Brussels sprouts. He found that 200 mg/mL of hypochlorite reduced populations by several logs in water but not in all types of produce. Moreover, he also noted that washing Brussels sprouts in water alone reduced populations by 1 log. Typically, sanitizer acts to prevent cross-contamination, and not to sanitize produce per se. Zhuang *et al.* (1995) found that chlorine was of minimal value in reducing populations of microbial pathogens on produce, and Beuchat *et al.* (2001) reported that chlorine had little effect on reducing microbial loads on tomatoes.

Treatment of lettuce with 20 ppm chlorine at either 20 or 50 °C did not result in significant reductions in populations of *E. coli* O157:H7 compared with treatments in water without chlorine (Li *et al.*, 2001). *Cryptosporidium* and some round worm eggs can also survive in the presence of 5.25% sodium hypochlorite (Fayer, 1995). Further, once pathogens internalize into plant cells, they cannot be removed by sanitizers and normal washing, and can only be inactivated through cooking. However, because most produce is consumed raw, any internalized pathogens are likely threats to consumers. Hence, although hypochlorite and chlorine may help reduce cross-contamination, they cannot guarantee complete elimination of pathogens from already contaminated food, particularly internalized pathogens. Additional and novel decontamination methods are earnestly needed.

Organic producers must use sanitizers as stipulated in the National Organic Standards (NOS). This includes a number of products, such as peroxyacetic acid, that are generally more expensive than chlorine, but very effective. Unfortunately, barriers to sanitizer use in leafy greens wash water among small farms using organic practices have been identified (Xu *et al.*, 2015). In this same study, washed leafy greens (mostly in water to which no sanitizer was added) were found to carry higher levels of some microbial indicators, possibly attributed to the lack of sanitizer use in wash water (Xu *et al.*, 2015). This increases the risk of cross-contamination not only for well-known pathogens, but also lesser-studied ones.

1.8 Current Detection Methods and Their Limitations

The currently most common used conventional methods for detecting the foodborne bacterial pathogens and parasites present in various types of food are based on culturing the microorganisms on agar plates followed by standard biochemical identifications (Mandal *et al.*, 2011). The major advantages of these conventional detection methods are usually inexpensive and simple for common pathogens but these methods can be time consuming and have low sensitivity (Lee *et al.*, 2014) as they depend on the ability of the microorganisms to grow in different culture media such as enrichment and/or selective enrichment media. Even when culturing is possible, these methods generally require more than two days for isolation and presumptive identification and more than a week for confirmation of the pathogen species (Zhao *et al.*, 2014). Additionally, conventional detection methods can be expensive if more differential and selective chromogenic media are used, and are limited in detecting parasites and viruses as these agents do not grow in *in vitro* culture conditions without mammalian cells or in vivo animal models. Conventional detection methods are also laborious, as they require the preparation of culture media, inoculation of plates, and colony counting (Mandal *et al.*, 2011), as well as specialists, such as bacteriologist or parasitologist. Another important but potentially underappreciated disadvantages of standard culturing methods is the possibility of false-negative results due to bacteria existing in environmental or food samples in injured, or in viable but non-culturable states (Dreux *et al.*, 2007; Dinu and Bach, 2011), thus impeding the ability to successfully culture these microorganisms in the laboratory. The failure to detect foodborne pathogens impacts food safety and could contribute to foodborne infections as well as recall of products.

A variety of advanced and rapid detection techniques for bacterial pathogens and parasites may be used as alternatives or in conjunction with culture methods. Nucleic acid sequence-based amplification (NASBA), using PCR, multiplex PCR, and real-time PCR may be used to detect and quantify target genes specific to the pathogen of interest. Samples positive by PCR may be successfully used to narrow down the number of samples from which isolation by culture is attempted (Marine *et al.*, 2015). Other nucleic acid-based methods include loop mediated isothermal amplification (LAMP) and oligonucleotide DNA microarray. Amplification-based methods offer an advantage over culture-based methods by being cheaper, faster, and more sensitive. However, amplification-based methods do not yield live organisms needed for downstream serotyping, genotyping, and whole genome sequencing, crucial tools for epidemiological studies and trace back investigations. Other detection methods include optical, electrochemical, and mass-based biosensors, which are classified as biosensor-based methods, and enzyme-linked immunosorbent assays (ELISA). Detection of parasites in animal and plant food samples requires *in vivo* mice or any other susceptible and reliable model.

1.9 Recommendation to Improve the Detection Level

Recently, several advanced and reliable methods have been developed for the rapid detection of foodborne pathogens but most of them still require improvement in sensitivity, selectivity, or accuracy to be of any practical use. Nucleic-acid-based methods have high sensitivity and require a shorter time than conventional culture-based techniques for detection of foodborne bacterial pathogens and parasites, but need expensive instruments and special training to operate the instruments. The development of immunological methods helped improve the time required to yield results but the specificity and the sensitivity of immunological techniques are still questionable as well as this methods need to adapt with the sample types, including type of food products such as produce, egg, milk, meat or meat products, and other interfering factors such as other non-target cells, DNA, and proteins.

Biosensors-based methods are easy to perform and it does not require longer training and produce results in a short period of time but sensitivity and sample (food matrixes) type need to adapt selectivity comparable to the culture-based methods. For immunological detection techniques, further studies are essential to improve the detections of bacterial pathogens or parasites in the food products by concentration prior to detection and more specific monoclonal antibody development, by which enhancing the sensitivity and reducing cost. Therefore, an appropriate detection method need to develop that is reliable, accurate, rapid, simple, sensitive, selective, and cost-effective and such methods would offer both bacterial pathogens and parasites detection in commercial food industrial practices.

1.10 Conclusion

As the possibility of contaminants, including bacterial pathogens and parasites in the produce grown in various environments, vary widely, and the complex methods of isolation and identification of parasites are also another consideration for regular practices in the small- and mid-size food processing plants, prevalence data for minimizing the risks and intervention strategies of cross-contamination of bacterial pathogens and parasites are required. Such data are crucial for the development of effective control strategies against identified risk factors and management of foodborne infections with under-researched pathogens. To improve the detection methodology for uncommon bacterial pathogens and parasites in various food products are also critical to control foodborne infections and identify the real causative agents. In addition, recent growing interests of using natural antimicrobials for food processing, specifically post-harvest level to eliminate microbial pathogens from produce to improve safety in a consumer friendly manner, is also important.

References

Abler, D.G. and Shortle, J.S. (2000). Climate change and agriculture in the Mid-Atlantic Region. *Climate Res.*, *14*, pp. 185–194.

Ackers, M.L., et al. (2000). An outbreak of *Yersinia enterocolitica* O:8 infection associated with pasteurized milk. *J. Infect. Dis.*, *181*, pp. 1834–1837.

Adl, S., Iron, D., and Kolokolnikov, T. (2011). A threshold area ratio of organic to conventional agriculture causes recurrent pathogen outbreaks in organic agriculture. *Sci. Total Environ.*, *409*, pp. 2192–2197.

Berg C. (2001). Health and welfare in organic poultry production. *Acta. Vet. Scand.* Supp *95*(Supl), pp. 37–45.

Berger, C.N., Sodha, S.V., Shaw, R.K., Griffin, P.M., Pink, D., Hand, P. and Frankel, G. (2010). Fresh fruit and vegetables as vehicles for the transmission of human pathogens. *Env. Microbiol.*, *12*(9), pp. 2385–2397.

Beuchat, L.R., Harris, L.J., Ward, T.E. and Kajs, T.M. (2001). Development of a proposed standard method for assessing the efficacy of fresh produce sanitizers. *J. Food Prot.*, *64*(8), pp. 1103–1109.

Biswas, D., et al. (2007). Correlation between in vitro secretion of virulence-associated proteins of *Campylobacter jejuni* and colonization of chickens. *Current Microbiol.*, *54*, pp. 207–212.

Brinton, W.F., Storms, P. and Blewett, T.C. (2009). Occurrence and Levels of Fecal Indicators and Pathogenic Bacteria in Market-Ready Recycled Organic Matter Composts. *J. Food Prot.*, *72*(2), pp. 332–339.

Bohaychuk, V.M., Gensel, G.E., King, R.K., Manninen, K.I., Sorensen, O. and Wu, J.T. (2009). A microbiological survey of selected Alberta-grown fresh produce from farmers markets in Alberta, Canada. *J. Food Prot.*, *72*(2), pp. 415–420.

Bolton, D.J., O'Neill, C.J. and Fanning, S. (2012). A preliminary study of *Salmonella*, verocytotoxigenic *Escherichia coli*/*Escherichia coli* O157 and *Campylobacter* on four mixed farms. *Zoonoses and Public Health*, *59*(3), pp. 217–228.

Brackett, R.E. (1994). Microbiological spoilage and pathogens in minimally processed refrigerated fruits and vegetables. In: R.C. *Wiley, ed., Minimally Processed Refrigerated Fruits and Vegetables*. New York: Chapman and Hall, pp. 269–312.

Cartwright, E.J., Jackson, K.A., Johnson, S.D., Graves, L.M., Silk, B.J. and Mahon, B.E. (2013). Listeriosis outbreaks and associated food vehicles, United States, 1998–2008. *Emerging Infect. Dis.*, *19*(1), pp. 1–9.

CDC (Centers for Disease Control and Prevention). (2002). *Yersinia enterocolitica* gastroenteritis among infants exposed to chitterlings—Chicago, Illinois, 2002. *MMWR Morbidity Mortality Weekly Report.*, *52*, pp. 956–958.

CDC (Centers for Disease Control and Prevention). (2006). Ongoing Multistate Outbreak of *Escherichia coli* serotype O157:H7 Infections Associated with Consumption of Fresh Spinach — United States, September 2006. *MMWR Morb. Mortal. Wkly. Rep.*, *55*(38), pp. 1045–1046.

CDC (Centers for Disease Control and Prevention). (2008). Outbreak of *salmonella* serotype Saintpaul infections associated with multiple produce items-United States 2008. *MMWR Morb. Mortal. Wkly. Rep.*, *60*, pp. 929–934.

CDC (Centers for Disease Control and Prevention). (2011). Multistate Outbreak of Listeriosis Associated with Jensen Farms Cantaloupe — United States, August–September 2011. *MMWR Morb. Mortal. Wkly. Rep.*, *60*(39), pp. 1357–58.

CDC (Centers for Disease Control and Prevention). (2013). Surveillance for foodborne disease outbreaks-United States, 1998–2008. Surveillance summaries. *MMWR Morb. Mortal. Wkly. Rep.*, *62*(2), pp. 1–34.

CDC (Centers for Disease Control and Prevention). (2014). Available at: http://www.cdc.gov/foodborneburden/2011-foodborne-estimates.html [Accessed April 25, 2014].

CDC (Centers for Disease Control and Prevention). (2015). Pasteurellosis deaths in the United States, 1993 to 2006. Data are based on CDC general mortality tables. Available at: http://www.cdc.gov/nchs/nvss/mortality_tables.htm [Accessed April 25, 2014].

CDC (Centers for Disease Control and Prevention)-DCDC. 2015. Epidemiological summary of human brucellosis in California, 2009–2012. Available at: https://www.cdph.ca.gov/programs/sss/Documents/BrucellosisEpiSummary2009-2012.pdf [Accessed April 25, 2014].

Chapin, T.K., Nightingale, K.K., Worobo, R.W., Weidman, M. and Strawn, L.K. (2014). Geographical and meteorological factors associated with isolation of Listeria species in New York state produce production and natural environments. *J. Food Prot.*, *77*, pp. 1919–1928.

Daniels, N.A., et al. (2000). Emergence of new *Vibrio parahaemolyticus* serotype in raw oysters: a prevention quandary. *JAMA*, *284*, pp. 1541–1545.

Deming, M.S., et al. (1987). *Campylobacter entertis* at a university: transmission from eating chicken and from cats. *American J. Epidimol.*, *126*, pp. 526–534.

Dinu, L. and Bach, S. (2011). Induction of viable but nonculturable E. coli O157:H7 in the phyllosphere of lettuce: a food safety risk factors. *Appl. Environ. Microbiol.*, *77*(23), pp. 8295–8302.

Dixon, B., Parrington, L., Cook, A., Pollari, F. and Farber, J. (2013). Detection of *Cyclospora, Cryptosporidium,* and *Giardia* in ready-to-eat packaged leafy greens in Ontario, Canada. *J. Food Prot.*, *76*(2), pp. 307–313.

Dreux, N., Albagnac, C., Federighi, M., Carlin, C.E. and Nguyen, C. (2007). Viable but non-culturable *Listeria monocytogenes* on parsley leaves and absence of recovery to a culturable state. *J. Appl. Microbiol.*, *103*(4), pp. 1272–1281.

Drinkard, L.N., et al. (2015). Outbreak of Cryptosporidiosis among veterinary medicine students-Philadelphia, Pennsylvania. MMWR. *Morb. Mortal. Wkly. Rep.*, *64*(28), p. 773.

Dubey, J.P. and Hill, D. (2002). *Toxoplasma gondaii*: transmission, diagnosis and prevention. *Clin. Microbiol. Infect.*, *8*, pp. 634–640.

Ekman, C.C.J., et al. (2012). Case-control study of an outbreak of acute Toxoplasmosis in an industrial plant in the state of Sao Paulo, Brazil. *Rev. Inst. Med. Trop. Sao Paulo.*, *54*(5), pp. 239–244.

Fayer, R. (1995). Effect of sodium hypochlorite exposure on infectivity *of Cryptosporidium parvum* oocysts for neonatal BALB/c mice. *Appl. Env. Microbiol.*, *61*(2), pp. 844–846.

Foodborne Illness Outbreak. (2007). Available at: http://outbreakdatabase.com/details/wyoming-office-setting-unknown-2007/? [Accessed June 14, 2016].

Foodborne Illness Outbreak. 2008. Available at: http://outbreakdatabase.com/details/wyoming-office-setting-unknown-2008/? [Accessed June 14, 2016].

Gallina, S., et al. (2013). Staphylococcal poisoning foodborne outbreak: Epidemiological investigation and strain genotyping. *J. Food Prot.*, *76*(12), pp. 2093–2098.

Gardner, T.J., et al. (2011). Outbreak of Campylobacteriosis associated with consumption of raw peas. *Clin. Infect. Dis.*, *53*(1), pp. 26–32.

Glass-Kaastra, S.K. and Pearl, D.L. (2013). Multiple-class antimicrobial resistance surveillance in swine Escherichia coli F4, *Pasteurella multocida* and *Streptococcus suis* isolates from Ontario and the impact of the 2004-2006 Porcine Circovirus type-2 Associated Disease outbreak. *Prev. Vet. Med.*, *113*(2), pp. 159–64.

Grant, J., et al. (2008). Spinach-associated *Escherichia coli* O157:H7 outbreak, Utah and New Mexico, 2006. *Emerging Infect. Dis.*, *14*(10), pp. 1633–1636.

Gould, L.H., Walsh, K.A., Viera, A.R., Herman, K., William, I.T., Hall, A.J. and Cole, D. (2013). Surveillance for foodborne disease outbreaks-United States, 1998-2008. Morbidity and mortality weekly report. Surveillance Summaries, *62*(2), pp. 1–34.

Guo, M., Buchanan, R.L., Dubey, J.P., Hill, D. E., Lambertini, E., Ying, Y., Gamble, H.R., Jones, J.L. and Pradhan, A.K. (2015). Qualitative assessment for *Toxoplasma gondii* exposure risk associated with meat products in the United States. *J. Food Prot.*, *12*, pp. 2108–2307.

Hallmann, E. (2012). The influence of organic and conventional cultivation systems on the nutritional value and content of bioactive compounds in selected tomato types. *J. Sc. Food Agri.*, *92*(14), pp. 2840–2848.

Hallmann, E. and Rembiałkowska, E. (2012). Characterisation of antioxidant compounds in sweet bell pepper (*Capsicum annuum* L.) under organic and conventional growing systems. *J. Sc. Food Agri.*, *92*(12), pp. 2409–2415.

Heyndrickx, M., Vandekerchove, D., Herman, L., Rollier, I., Grijspeerdt, K. and De Zutter, L. (2001). Routes for Salmonella contamination of poultry meat: epidemiological study from hatchery to slaughterhouse. *Epi. Infect.*, *1290*, pp. 253–265.

Hoffman, I. (2010). Climate change and the characterization, breeding and conservation of animal genetic resources. *41*(Suppl. 1), pp. 32–46.

Hutchison, M.L., Walters, L.D., Mead, G.C., Howell, M. and Allen, V.M. (2005). An assessment of sampling methods and microbiological hygiene indicators for process verification in poultry slaughterhouses. *J. Food Prot.*, *69*, pp. 145–153.

Iwamoto, M., Ayers, T., Mahon, B.E. and Swerdlow, D.V. (2010). Epidemiology of seafood-associated infections in the United States. *Clin. Microbiol. Rev.*, *23*(2), pp. 399–411.

Ji-Yeon, H., et al. (2013). A foodborne outbreak of *Staphylococcus aureus* associated with fried chicken in Republic of Korea. *J. Microbiol. Biotechol.*, *23*(1), pp. 85–87.

Jolly, D. and Lewis, C. (2005). Food Safety at Farmers Markets and Agritourism Venues. UC Small Farm Center Report. Available at: http://ucfoodsafety.ucdavis.edu/files/26480.pdf [Accessed DATEJune 14, 2016].

Johnson, R., Aussenberg, R.A. and Cowan, T. (2013). The Role of Local Food Systems in U.S. Farm Policy. Congressional Research Service March 7-5700. Available at: http://www.crs.gov [Accessed June 14, 2016].

Johnston, L.M., Jaykus, L., Moli, D., Martinez, M.C., Anciso, J., Mora, B. and Moe, C.L. (2005). A field study of the microbiological quality of fresh produces. *J Food Prot.*, *68*(9), pp. 1840–1847.

Jones, J.L. and Dubey, J.P. (2012). Foodborne Toxoplasmosis. *Clin. Infect. Dis.*, *55*(5), pp. 845–851.

Kaluski, D.N., et al. (2006). A large foodborne outbreak of group A streptococcal pharyngitis in an industrial plant: Potential for deliberate contamination. *IMAJ.*, *8*, pp. 618–621.

Kangas, S., Takkinen, J., Hakkinen, M., Nakari, U., Johansson, T., Henttomen, H., Vitaluoto, L., Siitonen, A., Ollgren, J. and Kuusi, M. (2008). *Yersinia pseudotuberculosis* O:1 Traced to Raw Carrots, Finland. *Emerging Infect. Dis.*, *14*(2), pp. 1959–1961.

Kozak, G.A., MacDonald, D., Landry, L. and Fabber, J.M. (2013). Foodborne outbreaks in Canada linked to produce: 2001 through 2009. *J. Food Prot.*, *76*(1), pp. 173–183.

Laukkanen, R., Martinez, P.O., Siekkinen, K.M., Ranta, J., Maijala, R. and Korkeala, H. (2009). Contamination of Carcasses with Human Pathogenic *Yersinia enterocolitica* 4/O:3 Originates from Pigs Infected on Farms. *Foodborne Path. Dis.*, *6*(6), pp. 681–688.

Lee, N., Kwon, K.Y., Oh, S.K., Chang, H.J., Chun, H.S. and Choi, S. W. (2014). A multiplex PCR assay for simultaneous detection of *Escherichia coli* O157:H7, *Bacillus cereus, Vibrio parahaemolyticus, Salmonella* spp., *Listeria monocytogenes*, and *Staphylococcus aureus* in Korea ready-to-eat food. *Foodborne Pathog. Dis.*, *11*, pp. 574–580.

Lester, G.E. and Saftner, R.A. (2011). Organically versus conventionally grown produce: common production inputs, nutritional quality, and nitrogen delivery between the two systems. *J. Agri. Food Chem.*, *59*(19), pp. 10401–10406.

Levine, W.C. and Griffin, P.M. (1993). *Vibrio* infections on the Gulf cost: results of first year of regional surveillance. *J. Infect. Dis.*, *167*, pp. 479–493.

Levy, M., Johnson, C.G. and Kraa, E. (2003). Tonsillopharyngitis caused by foodborne Group a Streptococcus: A prison based outbreak. *Clin. Infect. Dis.*, *36*, pp. 175–182.

Li, Y., Brackett, R.E., Chen, J. and Beuchat, L.R. (2001). Survival and growth of *Escherichia coli* O157:H7 inoculated onto cut lettuce before or after heating in chlorinated water, followed by storage at 5 °C or 15 °C. *J. Food Prot.*, *64*(3), pp. 305–309.

Linhart, Y., et al. (2008). A foodborne outbreak of streptococcal pharyngitis. *Isr. Med. Assoc. J.*, *10*(8-9), pp. 617–620.

Linke, K., Ruckerl, I., Brugger, K., Karpiskova, R., Walland, J., Muri-Klinger, S., Tichy, A., Wagner, M. and Stessl, B. (2012). Reservoirs of *Listeria* species in three environmental ecosystems. *Appl. Env. Microbiol.*, *80*, pp. 5583–5592.

Longenberger, A.H., et al. (2014). *Yersinia enterocolitica* infections associated with improperly pasteurized milk products: southwest Pennsylvania, March-August, 2011. *Epidemiol. Infect.*, *142*, p. 1640.

Luna, J., Allen, V., Fontenot, J., Daniels, L., Vaughan, D., Hagood, S., Taylor, D. and Laub, C. (1994). Whole farm systems research: An integrated crop and livestock systems comparison study. *Am. J. Alternative Agriculture*, *9*(1–2), pp. 57–63.

MacDonald, E., Heier, B.T., Nygård, K., Stalheim, T., Coudjoe, K.S., Skjerdal, T., Wester, A.L., Lindstedt, B.A., Stavnes, T.L. and Vold, L. (2012). *Yersinia Enterocolitica* Outbreak Associated with Ready-to-Eat Salad Mix, Norway, 2011. *Emerging Infect. Dis.*, *18*(9), pp. 1496–1499.

Mandal, P.K., Biswas, A.K., Choi, K. and Pal, U.K. (2011). Methods for rapid detection of foodborne pathogens: an overview. *Am. J. Food. Technol.*, *6*, pp. 87–102.

Marder, E.P., Garman, K.N., Ingram L.A. and Dunn, J.R. (2014). *Foodborne Path. Dis.*, *11*(8), pp. 593–595.

Marine, S.C., Pagadala, S., Wang, F., Pahl, D.M., Melendez, M.V., Kline, W.L., Oni, R., Walsh, C., Everts, K.L., Buchanan, R. and Micallef, S.A. (2015). The growing season, but not the farming system, is a food safety risk determinant for leafy greens in the Mid-Atlantic region of the United States. *Appl. Env. Microbiol.*, *81*(7), pp. 2395–2407.

McDonald, E., et al. (2012). *Yersinia enterocolitica* outbreak associated with ready to eat salad mix, Norway, 2011. *Emerg. Infect. Dis.*, *18*(9), pp. 1496–1501.

Mead, G.C. (2004). Poultry Meat Processing and Quality, England: CRCPress, Woodhead Publishing Limited.

Micallef, S.A., Rosenberg, G.R.E., George, A., Kleinfelter, L., Boyer, M.S., McLaughlin, C.R., Estrin, A., Ewing, L., Jean-Gilles Beaubrun, J., Hanes, D.E., Kothary, M.H., Tall, B.D., Razeq, J.H., Joseph, S.W. and Sapkota, A.R. (2012). Occurrence and antibiotic resistance of multiple *Salmonella* serotypes recovered from water, sediment and soil on mid-Atlantic tomato farms. *Env. Res.*, *114*, pp. 31–39.

Micallef, S.A., Rosenberg, G.R.E., George, A., Ewing, L., Tall, B.D., Boyer, M.S., Joseph, S.W. and Sapkota, A.R. (2013). Diversity, distribution and antibiotic resistance of *Enterococcus* spp. recovered from tomatoes, leaves, water and soil on U.S. Mid-Atlantic farms. *Food Microbiol.*, *36*(2), pp. 465–474.

MMWR. (2013). Outbreak of staphylococcal dood poising from a military unit lunch party-United States, July 2012. *MMWR.*, *62*(50), p. 1026.

Mukherjee, A., Speh, D., Jones, A.T., Buesing, K.M. and Diez-Gonzalez, F. (2006). Longitudinal microbiological survey of fresh produce grown by farmers in the upper Midwest. *J. Food Protection*, *69*(8), pp. 1928–1936.

Nascimbene, J., Marini, L. and Paoletti, M.G. (2012). Organic Farming Benefits Local Plant Diversity in Vineyard Farms Located in Intensive Agricultural Landscapes'. *Envir. Management.*, *49*(5), pp. 1054–1060.

Natvig, E.E., Ingham, S.C., Ingham, B.H., Cooperband, L.R. and Roper, T.R. (2002). *Salmonella enterica* serovar Typhimurium and *Escherichia coli* contamination of root and leaf vegetables grown in soils with incorporated bovine manure. *Appl. Env. Microbiol.*, *68*(6), pp. 2737–2744.

Nichols, G.N., et al. (2015). Cyclospora infection linked to travel to Mexico, June to September 2015. *Euro Surveill.*, *20*(43), doi: 10.2807/1560-7917.ES.2015.20.43.30048

Pagadala, S., Marine, S.C., Micallef, S.A., Wang, F., Pahl, D.M., Melendez, M.V., Kline, W.L., Oni, R.A., Walsh, C.S., Everts, K.L. and Buchanan, R.L. (2015). Assessment of region, farming system, irrigation source and sampling time as food safety risk factors for tomatoes. *Int. J. Food Microbiol.*, *196*, pp. 98–108.

Painter, J.A., Hoekstra, R.M., Ayers, T., Tauxe, R.V., Braden, C.R., Angulo, F.J. and Griffin, P.M. (2013). Attribution of foodborne illnesses, hospitalizations, and deaths to food commodities by using outbreak data, United States, 1998–2008. *Emerging Infect. Dis.*, *19*(3), pp. 407–415.

Park, S., Szonyi, B., Gautam, R., Nightingale, K., Anciso, J. and Ivanek, R. (2012). Risk factors for microbial contamination in fruits and vegetables at the preharvest level: a systematic review. *J. Food Prot.*, *75*(11), pp. 2055–2081.

Peralta, R.C., Yokoyama, H., Ikemori, Y., Kuroki, M. and Kodama, Y. (1994). Passive immunisation against experimental salmonellosis in mice by orally administered hen egg-yolk antibodies specific for 14-kDa fimbriae of *Salmonella* Enteritidis. *J. Med. Microbiol., 41*, pp. 29–35.

Permin, A., Bisgaard, M., Frandsen, F., Pearman, M., Kold, J. and Nansen, P. (1999). Prevalence of gastrointestinal helminths in different poultry production systems. *Brt. Poultry Sc., 40*(4), pp. 439–443.

Ponka, A., et al. 2009. A foodborne outbreak due to *Cryptsporidium parvum* in Helsinki, November 2008. *EuroSurvillance., 14*(28), pp. 1–3.

Pradhan, A.K., Ivanek, R., Gröhn, Y.T., Geomaras, I., Sofos, J.N. and Widemann, M. (2009). Quantitative risk assessment for *Listeria monocytogenes* in selected categories of deli meats: impact of lactate and diacetate on listeriosis cases and deaths. *J. Food Prot., 72*, pp. 978–989.

Pullin, R.S.C.V. (1987). Third-world aquaculture and the environment. AIT (Asian Institute of Technology) Research Report No. 205, Page 324.

Putignan, L. and Menichella, D. (2010). Global distribution, public health and clinical impact of protozoan pathogen Cryptosporidium. *Interdisciplinary Perspective Infect. Dis., 2010*. pii: 753512. doi: 10.1155/2010/753512.

Rahman, A., Bonny, T.S., Stonsaovapak, S. and Ananchaipattana, C. (2011). Yersinia enterocolitica: Epidemiological studies and outbreaks. *J. Pathogens., 2011*:239391. doi: 10.4061/2011/239391.

Rodriguez-Siek, K.E., Giddings, W.C., Doetkott, C., Johson, T.J., Fakhr, M.K. and Nolan, L.K. (2005). Comparison of *Escherichia coli* isolates implicated in human urinary tract infection and avian colibacillosis. *Microbiol., 151*, pp. 2097–2110.

Roman, K., et al. (2013). A foodborne outbreak of brucellosis at a police cafeteria, Lima, Peru. *Am. J. Trop. Med. Hyg., 88*(3), pp. 552–558.

Salaheen, S., Chowdhury, N., Hanning, I. and Biswas, D. (2015). Zoonotic bacterial pathogens and mixed crop-livestock farming. *Poultry Sc., 94*(6), pp. 1398–1410.

Scallan, E., Hoekstra, R.M., Angulo, F.A., Tauxe, R.V., Widdowson, M.A., Roy, S.L., Jones, J.L. and Griffin, P.M. (2011). Foodborne Illness Acquired in the United States-Major Pathogens. *Emerging Infect. Dis., 17*(1), pp. 7–15.

Scharfe, R.L. (2011). Health-related costs from foodborne illness in the United States. Available at: http://www.producesafetyproject.org/admin/assets/files/Health-Related-Foodborne-Illness-Costs- Report.pdf-1.pdf [Accessed August 29, 2011].

Scheutz, F., et al. (2011). Characteristics of the enteroaggregative Shiga toxin/verotoxin-producing *Escherichia coli* O104:H4 strain causing the outbreak of haemolytic uraemic syndrome in Germany, May to June 2011. *Euro. Surveill., 16*(24), pii: 19889.

Shane, S.M. (1992). The significance of *Campylobacter jejuni* infection in poultry: A review. *Avian Pathol., 21*(2), pp. 189–213.

Sivapalasingam, S., Friedman, C.R., Cohen, L. and Tauxe, R.V. (2004). Fresh produce: a growing cause of outbreaks of foodborne illness in the United States, 1973 through 1997. *J. Food Prot., 67*(10), pp. 2342–2353.

Skirrow, M.B. and Blaser, M.J. (1992). Clinical and epidemiologic considerations. In: Nachamkin I, Blaser MJ, Tompkins LS, eds., Campylobacter jejuni. Current status and future trends. Washington, DC: American Society Microbiol, pp. 3–8.

Smith, L.P., Ng, S.W. and Popkin, B.M. (2013). Trends in US home food preparation and consumption: analysis of national nutrition surveys and time use studies from 1965–1966 to 2007–2008. *Nutrition J., 12* pp. 45–54.

Strawn, L.K., Fortes, E.D., Bihn, E.A., Nightingale, K.K., Grohrn, Y.T., Worobo, R.W., Wiedmann, M. and Bergholz, P.W. (2013). Landscape and metrological factors affecting prevalence of three food-borne pathogens in fruit and vegetable farms. *Appl. Env. Microbiol., 79*(2), pp. 588–600.

Sulc, R.M. and Tracy, B.F. (2007). Integrated crop-livestock systems in the US Corn Belt. *Agronomy J., 99*, pp. 335–345.

Tauxe, R.V. (1992). Epidemiology of *Campylobacter jejuni* infections in the United States and Other industrialized nations, pp. 9–19. In: I. Nachamkin, M.J. Blaser, and L.S. Tompkins, eds., *Campylobacter jejuni*: current status and future trends. Washington, DC: American Society for Microbiology, 1992.

Thamsborg, S.M., Roepstoff, A. and Larsen, M. (1999). Integrated and biological control of parasites in organic and conventional production systems. *Vet. Parasitol.*, *84*, pp. 169–186.

USDA (United States Department of Agriculture). (2014). National Farmers Market Directory. Available at: http://search.ams.usda.gov/farmersmarkets [Accessed June 14, 2016].

USDA-ERS (United States Department of Agriculture Economic Research Service). (2013). Growth Patterns in the U.S. Organic Industry. Available at: http://www.ers.usda.gov/amber-waves/2013-october/growth-patterns-in-the-us-organic-industry.aspx#.VRhMjGctHIU [Accessed June 14, 2016].

USDA-ERS (United States Department of Agriculture Economic Research Service). (2014). Available at: http://www.ers.usda.gov/topics/natural-resources-environment/organic-agriculture/organic-market-overview.aspx [Accessed June 14, 2016].

van den Berg, M.M., Hengsdijk, H., Wolf, J., Van Ittersum, M.K., Guanghuo, W. and Roetter, R.P. (2007). The impact of increasing farm size and mechanization on rural income and rice production in Zhejiang province, China. *Agriculture System*, *94*(3), pp. 841–850.

Vinha, A.F., Barreira, S.V., Costa, A.S., Alves, R.C. and Oliveira, M.B. (2014). Organic versus conventional tomatoes: Influence on physicochemical parameters, bioactive compounds and sensorial attributes. *Food Chemical Toxicol.*, *67*, pp. 139–144.

Weis J and Seeliger HPR. (1975). Incidence of *Listeria monocytogenes* in nature. *Appl. Microbiol.*, *30*, pp. 29–32.

Weller, D., Andrus, A., Wiedmann, M. and den Bakker, H.C. (2015). *Listeria booriae* sp. nov. and *Listeria newyorkensis* sp. nov., from food processing environments in the USA. *Int. J. Syst. Evol. Microbiol.*, *65*(Pt 1), pp. 286–292.

Willis, W.L., Murray, C. and Talbott, C. (2000). Effect of delayed placement on the incidence of *Campylobacter jejuni* in broiler chickens. *Poultry Sci.*, *79*, pp. 1392–1395.

Xu, A., Pahl, D.M., Buchanan, R.L. and Micallef, S.A. (2015). Comparing the Microbiological Status of Pre- and Postharvest Produce from Small Organic Production. *Journal of Food Protection*, *78*(6), pp. 1072–1080.

Yokoyama, H., Umeda, K., Peralta, R.C., Hashi, T., Kuroki, M., Ikemori, Y. and Kodama, Y. (1998). Oral passive immunization against experimental salmonellosis in mice using chicken egg yolk antibodies specific for *Salmonella* Enteritidis and *S.* Typhimurium. *Vaccine*, *16*, pp. 388–393.

Yoo, J.R., Heo, S.T., Lee, K.H., Kim, Y.R. and Yoo, S.J. (2015). Foodborne outbreak of human brucellosis caused by ingested raw materials of fetal calf on Jeju Island. *Am. J. Trop. Hyg.*, *92*(2), pp. 267–269.

Zadernowska, A., Chajęcka-Wierzchowska, W. and Łaniewska-Trokenheim, L. (2013). *Yersinia enterocolitica*: A Dangerous, But Often Ignored, Foodborne Pathogen. *Foodborne Pathogen, Food Reviews Int.*, *30*, pp. 53–70.

Zhao, X., Lin, C., Wang, J. and Oh, D.H. (2014). Advances in rapid detection methods for foodborne pathogens. *J. Microbiol. Biotechnol.*, *24*(3), pp. 297–312.

Zhuang, R.Y., Beuchat, L.R. and Angulo, F.J. (1995). Fate of *Salmonella* Montevideo on and in raw tomatoes as affected by temperature and treatment with chlorine. *Appl. Env. Microbiol.*, *61*(6), pp. 2127–2131.

2

Characterization of Foodborne Pathogens and Spoilage Bacteria in Mediterranean Fish Species and Seafood Products

A. Bolivar, J.C.C.P. Costa, G.D. Posada-Izquierdo, F. Pérez-Rodríguez, I. Bascón, G. Zurera, and A. Valero

Department of Food Science and Technology, University of Cordoba, Campus de Rabanales, Córdoba, Spain

2.1 Fish Quality Assurance

Traditionally, fish processors have regarded quality assurance as the responsibility of the regulatory governmental agency, and the means used by these agencies have been the formulation of food laws and regulations, inspection of facilities, and processes and final product testing. The processors' own efforts have in many cases been based entirely on final product testing. Such a system is costly and ineffective; it provides no guarantee of quality but merely a false sense of safety. At this point, quality assurance (QA), according to International Standards (ISO 8402), can be defined by "all those planned and systematic actions necessary to provide adequate confidence that a product or service will satisfy given requirements for quality." In other words, QA is a strategic management function, which establishes policies, adapts programs to meet established goals, and provides confidence that these measures are being effectively applied.

2.2 Microbiological Standards To Be Accomplished

Microbiological hazards in foodstuffs form a major source of foodborne diseases in humans. From January 1, 2006, the European Union (EU) health conditions for the production and placing on the market of fishery products are laid down in the consolidated Hygiene Regulations 852/2004, 853/2004, and 854/2004 and Official Feed and Food Control Regulation 882/2004. Seafood production in third countries (not EU member states) has to match EU standards in terms of hygiene and food safety. That means that it is covered by the same general principles of food law and food safety as exist in the EU. These general principles are laid out in Council Regulation 178/2002. The EU hygiene legislation also applies Regulation (EC) No. 2073/2005, which establishes microbiological criteria for a range of foods. The aim of this legislation is to complement food hygiene requirements, ensuring that foods being placed on the market do not pose a risk to human health, and the legislation applies to all businesses involved in food production and handling. Chapter 1 of the regulation focuses on food safety criteria that cover foods such as ready to eat foods, fishery products, and live bivalve molluscs. Chapter 2 focuses on process hygiene criteria, with Chapter 2.4 referring to fishery products.

Rules relating to visual inspections for the detection of parasites in fishery products are stated in Annex II (obligations on the competent authorities) of Regulation (EC) No. 2074/2005.

Foodborne Pathogens and Antibiotic Resistance, First Edition. Edited by Om V. Singh.

This covers total volatile basic nitrogen (TVBN) limits and how to determine them.

A number of microbiological tests in fishery products are used by authorities to check their microbiological status. The purpose of these tests is to detect pathogenic bacteria (*Salmonella, Staphylococcus aureus*, and *E. coli*), indicator organisms of fecal pollution or other types of general contamination or poor handling practices (coliform bacteria, faecal streptococci, and total viable count) (Huss, 1995a). The most widely accepted microbiological criteria for chilled and frozen raw fish are those set for aerobic plate counts (APC) at 25 °C and *E. coli* proposed by the International Commission on Microbiological Specifications for Foods (ICMSF).

The following regulations about fish designations need to be considered:

- Regulation (EC) No 104/2000 on the common organisation of the markets in fishery and aquaculture products.
- Regulation (EC) No 2065/2001 laying down detailed rules for the application of Council Regulation (EC) No 104/2000 as regards informing consumers about fishery and aquaculture products

International requirements are based upon having a risk-based, preventive management system in place at all stages of the supply chain. The requirements for this preventive approach is described as follows:

- Fish and fishery products should be prepared in plants certified by the local competent authority. All certified plants should comply with the good hygienic practices (GHP).
- The fisheries industry should take responsibility, implement, and maintain safety management systems based upon HACCP.
- The national competent authority is responsible for the certification of fish processing and manufacturing plants, verification of effective systems, and issuing of certificates of compliance for export products. This includes auditing and inspection programs.
- National surveillance and monitoring programs should be in place to demonstrate that all identified hazards are under control, for example, biotoxins, and to identify potentially emerging hazards.

The (International) Codex Committee has published a Code of Practice for Fish and Fishery Products. In the United States, the National Advisory Committee on Microbiological Criteria for Foods (NACMCF) has published a number of recommendations on the safety of modified-atmosphere packaging and vacuum packaging refrigerated raw fishery products.

The U.S. Food and Drug Administration (FDA) operates a mandatory safety program for all fish and fishery products under the provisions of the Federal Food, Drug and Cosmetic (FD&C) Act, the Public Health Service Act, and related regulations. The FDA program includes research, inspection, compliance, enforcement, outreach, and the development of regulations and guidance. As a cornerstone of that program, FDA publishes the Fish and Fisheries Products Hazards and Controls Guidance, an extensive compilation of the most up-to-date science and policy on the hazards that affect fish and fishery products and effective controls to prevent their occurrence. The fourth edition of this guidance document, which has become the foundation of fish and fishery product regulatory programs around the world, is now available.

2.3 Hazard Analysis and Critical Control Points (HACCP) Implemented in the Fishery Industry

The global legislative requirement for controlling the safety (and quality) of fish and seafood products is to apply a preventive, risk-based system based upon the seven principles of Hazard Analysis and Critical Control Point (HACCP) supported by pre-requisite or good practices programs at all stages of the food chain, from farm to fork. Therefore any microbiological criteria need to be part of this HACCP-based system to verify that the system is under control. One reason for this development is that a number of national food legislations today are placing full responsibility for food quality on the producer (e.g., EEC Council Directive no. 91/493/EEC) and the use of the HACCP system is required.

The HACCP team should ideally have access to expertise on the appropriate fisheries, including practices on board fishing vessels, or in aquaculture operations. These factors are likely to have a significant effect on some potential hazards, for example, contamination by foodborne pathogens as a result of time/temperature abuse or unsatisfactory handling practice. In addition, there may be species-specific hazards related to certain fish and shellfish, such as enteric viruses in shellfish, marine toxins associated with particular fish and shellfish species, aquaculture drug residues, and potentially pathogenic *Vibrio* spp. naturally present on fish from warm waters (>15 °C). Expertise on such inherent hazards, challenge testing, and inoculation studies for evaluation of safety aspects are therefore essential for an effective HACCP team.

The hazard analysis for seafood products is fairly straightforward and uncomplicated. The live animals are caught in the sea, handled and processed without any use of additives or chemical preservatives, and finally distributed with icing or freezing as the only means of preservation. Contamination with pathogenic bacteria from the human/animal reservoir can occur when the landing place is unhygienic or when the fish is washed with contaminated water. Certain products may be contaminated or carry pathogenic organisms as a part of the natural flora. If the processing does not include a kill step, the only critical control point (CCP) that can render the product safe from pathogenic organisms is adequate heat treatment during preparation. However, it must be noted here that certain toxins (algal and shellfish) are not destroyed by heat treatments; contaminated fish and shellfish should not be used.

The term "fish and seafood" includes an extremely varied group of products, and there are equally varied ranges of potential hazards associated with them. For example, there are particular hazards associated with contamination of shellfish by microorganisms from human sewage, and the potential growth of *Listeria monocytogenes* on smoked fish.

Many of the microbiological hazards associated with fish and seafood products are derived from the raw materials. Pathogens may be part of the resident microflora of the living animal (e.g., *Vibrio* spp.), or may originate from polluted water or from post-capture contamination (e.g., *Salmonella* spp. and viruses). The incidence of psychrotrophic, non-proteolytic *Clostridium botulinum* on fresh fish is also sufficiently high that its presence may be assumed. There may also be inherent hazards from parasites such as the roundworm *Anisakis simplex* and from marine toxins such as ciguatera in reef fish (derived from microalgae) and scombrotoxin (biogenic amines) development in fish containing high levels of histidine.

Hazards introduced during processing of fish products depend very much on the characteristics of the process. For example, modified atmosphere packed of chilled raw fish may provide conditions suitable for the growth of psychrotrophic *C. botulinum*, or poor control of batter mixes in frozen battered fish portions may allow growth of *Staphylococcus aureus* and production of enterotoxin. Therefore, it is not possible, or desirable, to generalize about expected hazards and the reader is once again referred to the appropriate product chapter for additional advice on specific hazards.

It can be said that effective control measures are likely to include the following:

- Careful selection of sources for raw materials
- Adequate temperature control
- Effective sanitation
- Food handler hygiene
- Prevention of cross-contamination

Critical limits separate acceptable from unacceptable products. Where possible, critical limits should be specified and validated for each CCP. More than one critical limit may be defined for a single step. For example, it is necessary to specify both time and temperature for a thermal process. Criteria used to set critical limits must be measurable and may include physical, chemical, biological, or sensory parameters. Some examples relevant to fish products are:

- Time and temperature limits for pasteurized and cooked products
- Temperature limits for chilled fish
- Brine concentration in cured fish
- Water activity (aw) values for dried fish
- pH values in fermented fish
- Microbiological quality of water in shellfish production areas

2.4 Microbial Ecology of Mediterranean Fishery Products

2.4.1 Chilled and Frozen Fish

Fresh fish is a food that undergoes biochemical and microbiological changes over storage from capture until consumption. It is a product highly susceptible to rapid spoilage that leads to a short shelf life. Fish spoilage is mainly caused by oxidation of lipids, denaturation of proteins by enzymatic activity of fish enzymes, as well as microbial enzymes, and production of off-odor metabolites.

The microbial population of fish is a mixture of subpopulations present on the live animal and those microorganisms contaminating the product during processing and handling. Fish microbiota largely varies with the pollution level and temperature of the water (Gram, 2009). Bacteria from many groups, as well as viruses, parasites, and protozoa, can be present in the raw materials. Each processing operation has its own unique microflora reflecting the raw materials and the preservation parameters used (Bagge-Ravn *et al.*, 2003). Among these microbial groups, only a few parts will be able to tolerate the product specific conditions (i.e., temperature) and proliferate during subsequent storage.

Fish muscle is initially sterile, but scales, gills, and intestines harbor high microbial loads. Fresh or warm-water fish tend to have a biota that is composed of more mesophilic gram-positive bacteria than cold-water fish, which tends to be largely gram-negative (Jay *et al.*, 2005). Fish can have around 10^3 to 10^8 bacterial cells/g. In general, the microorganisms that are part of the microbial ecology of Mediterranean fish can highlight genera as total mesophilic aerobic, psychrotrophic, sulphite-reducing clostridia (SRC), *Aeromonas, Enterobacter, Escherichia coli, Lactobacillus, Listeria, Salmonella, Pseudomonas, Photobacterium, Shewanella, Vibrio*, yeasts, and some molds (Carrascosa *et al.*, 2015; Esteve *et al.*, 2012; Koutsoumanis and Nychas, 2000).

The effects of chilling and freezing on fish are well documented (Yin *et al.*, 2014; Ladip *et al.*, 2013; Viegas *et al.*, 2013). Bacteriological spoilage largely depends on type of fish microflora, thus influencing on product shelf life. The effect of biochemical changes induced by the bacterial growth on fish spoilage is not pronounced until the specific spoilage organisms have increased to a certain level (Gram and Huss, 1996). In addition, as microbial counts increase, enzymes secreted by microorganisms may also cause additional softening of the fish (Nielsen *et al.*, 2001).

During storage at cooling temperatures, death and sub-lethal injury is initially more pronounced (about 7 days) and decrease during storage afterward. Temperature fluctuations are the most significant factors affecting the quality of both chilled and frozen fish (Flemming *et al.*, 2014). The effect is more pronounced with gram-negative organisms such as *E. coli*, *Pseudomonas aeruginosa*, and *Salmonella* Typhimurium. The lethal effect of freezing on microorganisms in fishery products varies depending on storage duration, rate of cooling and thawing, and storage temperatures (IIR, 1986). Incidence of pathogens, including *Salmonella*, *Vibrio*, and *Listeria* spp. in frozen fishery products has been of great concern in the international trade of the commodities.

During freezing, as water temperature is being reduced, a large proportion of water becomes frozen, thus growth of most microbial species is inhibited, apart from some psychrophilic bacteria, yeasts, and molds. At −20 °C most cells have lethal or sub-lethal injury. The freezing rate dictates the extent of microbial damage resulting from the formation of ice crystals. Bacteria viability of often lost at abrupt temperature decreases (i.e., from 37 to 0 °C) (Venugopal, 2006).

2.4.2 Molluscan/Crustacean Shellfish

The microflora of molluscan/crustacean shellfish is more variable than that of fish. The number and type of microorganisms present in the aquatic habitat, from which the animals were harvested, depend various factors such as salinity, environmental conditions, water temperature, feeding regime, fish capturing, and chilling conditions (Vernocchi *et al.*, 2007; Cao *et al.*, 2009). The fundamental differences in spoilage of these foods are referred, generally, to the way in which they are handled, and to their specific chemical composition (Jay *et al.*, 2005).

Molluscan shellfish are filter feeders and can concentrate toxic substances and microorganisms. Several studies have reported that contaminants, such as heavy metals (Copat *et al.*, 2013), viruses such as human noroviruses or hepatitis A virus (HAV) (Campos and Lees, 2014; Suffredini *et al.*, 2012), bacteria (Iwamoto *et al.*, 2010), and marine toxins produced by algae (USFDA, 2012) can be accumulated in shellfish. Therefore, bivalves can act as carriers of food contaminants and under certain circumstances, they can produce human diseases.

Molluscs usually have a resident bacterial population that in the case of oysters fluctuates between 10^4 and 10^6 cfu/g of tissue, being the higher counts present when water temperatures are high (ICMSF, 2005). The dominant groups of bacteria found in shellfish are gram-negative of the genera *Vibrio, Pseudomonas, Acinetobacter, Moraxella, Aeromonas, Flavobacterium,* and *Alcaligenes*. Lower numbers of gram-positive bacteria may also be present such as species of *Bacillus, Corynebacterium,* and *Micrococcus*. Bacterial genera such as *Escherichia, Enterobacter, and Lactobacillus* were isolated in spoiled oysters (Jay *et al.*, 2005).

Regarding crustacean species, microbial spoilage of shrimps is more prevalent than that of crabs and lobsters. Whereas crabs and lobsters remain alive until they are processed, shrimps die during harvest. The flesh of crustacean is rich in NPN compounds (amino acids, especially arginine and trimethylamine oxide), contains ca 0.5% glycogen, and has a pH above 6.0 (Ray, 2005). These factors allow growth of spoilage microorganisms such as *Morganella* spp., *Proteus* spp., or *Pseudomonas* spp. (Matches, 1982).

The bacterial profile of freshly caught crustaceans should be expected to reflect the waters from which these foods are caught, and contaminants from the deck, handlers, and washing waters. Some representative species causing spoilage in crustaceans are *Pseudomonas, Acinetobacter, Moraxella*, and yeasts (Jay *et al.*, 2005). *Pseudomonas fragi* and *Shewanella putrefaciens* have been identified as the primary spoilage agents of chill-stored shrimps, with *P. fragi* spoiling iced-stored shrimp and *S. putrefaciens* being the dominant microorganism in shrimp stored in ice slurry (Chinivasagam *et al.*, 1996). Microbial spoilage of shrimp is dominated by odor changes due to production of volatile metabolites of NPN compounds (from decay and putrefaction), slime production, and loss of texture (soft) and color. If the shrimps are processed and frozen rapidly, spoilage can be minimized. Spoilage in other crustacean species such as prawns, are characterized by the formation of amines, sulphides, and esters associated to the growth of *P. fragi* and *S. putrefaciens* (Chinivasagam *et al.*, 1998). Lobsters are frozen following processing or sold live and thus are not generally exposed to spoilage conditions. Crabs, lobsters, and shrimps are also cooked to extend their shelf life. Blue crabs are steamed under pressure, and the meat is picked and marketed as fresh crabmeat. To extend shelf life (and safety), the meat is also heat processed (85 °C for 1 min) and stored at refrigerated temperature. Under refrigerated conditions, they have a limited shelf life because of growth of surviving bacteria and post heat contaminants (Ray, 2005).

2.4.3 Cured, Smoked, Dried, and Fermented Seafood

Preservation methods like salt-curing and drying have been used for centuries to obtain fully preserved products and access to good, safe, and nutritious food in all seasons and areas where the availability of fresh food is limited (Leroi *et al.*, 2006). Processing methods include drying, salting, smoking, pickling, and marinating of fish. There can be also different combinations of these methods and preservation of fish by fermentation (Food and Agriculture Organization (FAO), 1983; Jarvis, 1988).

The fish curing industry has flourished through the ages and has not been affected to any great extent by modern fish preservation and processing techniques. This is because cured fish is a highly appreciated and traditional product in many countries, mainly due to its excellent storage stability, special organoleptic characteristics, and nutritional value (Lauritzsen *et al.*, 2004).

Traditionally, ground fish species (cod, haddock, ling, blue ling, and tusk) are used for salting processes, including light salting and heavy salting, mainly because fish muscle has a low lipid content. Pelagic species (herring, sardine, capelin, blue whiting, and mackerel) and salmonids (salmon, trout, and arctic char) with a higher lipid content are more suited to other curing processes (smoking, marinating). Salt-cured cod, the precursor to klipfish, and known as the traditional product *bacalao* in Spain and *bachalau* in Portugal, has had this position for centuries, but today salt-cured cod is popular due to its sensory properties rather than lack of availability of other foods (Leroi *et al.*, 2006).

Salt-cured and dried fish products are generally regarded as safe, even though they are produced in relatively open houses with limited possibilities to regulate temperature and maintain good hygienic conditions. It is considered that salt curing is an effective barrier against bacteria. However, rehydrated salt-cured cod spoils rapidly, and it is found that this is due to growth of *Psychrobacter* spp. These bacteria are present on the skin of fresh fish, survive in a non-growing mode during salt curing, but recover and grow during and after rehydration (Bjorkevoll *et al.*, 2003). A number of other bacteria have also been found to survive the salt-curing step (Barat *et al.*, 2006). *Listeria* spp. and *Staphylococcus* spp. are occasionally found in salt-cured cod products but it has not been clear whether these bacteria survive in the fish if introduced to the fish prior to salt curing or only when they are introduced directly to the salt-cured cod shortly before the sample is taken (Pedro *et al.*, 2004).

Smoking is one of the oldest methods used to process and preserve fish and meat. Smoking introduces flavor, taste, and preservative ingredients into fish muscle by exposing fish to the smoke produced by burning or smoldering plant materials, most often wood. The volatile compounds in smoke penetrate into the fish muscle. Smoking usually extends the shelf-life of fish due to the combination effects of (a) salting, which lowers water activity resulting microbial growth, (b) elevated temperature drying, which provides a physical surface barrier to the passage of microorganisms, and c) deposition of antimicrobial and antioxidant compounds, such as aldehydes, carboxylic acid, and phenols, which delays microbial growth and rancidity development (Efiuvwevwere and Ajiboye, 1996; Leroi and Joffraud, 2000).

Moreover, fish muscle exposed to smoke in combination with a high temperature can effectively limit harmful enzymatic reactions (FAO, 1992). The smoking process is usually characterized by an integrated combination of salting, drying, heating, and smoking steps in smoking chamber/smokehouses (Alcicek and Atar, 2010). Although the general operations in all smoked fish processing plants are similar, the specific processing procedures can vary considerably. This variability relates to differences in equipment, regional and ethnic consumer preferences, raw materials, and tradition (Flick and Kuhn, 2012). Smoked fish is a perishable food, so to maintain its good quality and to prevent foodborne illnesses it must be preserved after smoking by processing techniques. Bacteria, yeasts, and molds are microorganisms associated with smoked fish. The microflora isolated from smoked fish include *Lactobacillus curvatus*, *L. sakei*, *L. plantarum*, *Carnobacterium* spp., *Leuconostoc* spp., *Serratia liquefaciens*, *S. grimesii*, *Enterobacter agglomerans*, *Hafnia alvei*, *Photobacterium phosphoreum*, *Brochothrix thermosphacta*, *Aeromonas* spp., *Micrococcus luteus*, *Pseudomonas* spp., *Alcaligenes* spp., *Staphylococcus sciuri*, *S. xylosus*, *Bacillus cereus*, *Raoultella ornithinolytica*, *B. thuringiensis*, *Citrobacter freundii*, *Klebsiella pneumoniae*, *E. aerogenes*, *E. cloacae*, *Psychrobacter immobilis*, and *Shewanella putrefaciens* (Lakshmanan *et al.*, 2002; Bjorkevoll *et al.*, 2003; Hsu *et al.*, 2009).

Drying is one of the thermal treatments applied in many food industries, and the use of dried foods is expanding rapidly (Senadeera *et al.*, 2005). Drying processes can be broadly classified as (Nguyen *et al.*, 2014):

- Thermal drying: a gaseous or void medium is use to remove water from the material, thus thermal drying can be divided into three types: (a) air drying, (b) low air environment drying, and (c) modified atmosphere drying (Rahman and Perera, 1999).
- Osmotic dehydration: solvent or solution is applied to remove water, whereas in mechanical dewatering, physical force is use to remove water. Consideration should be given to many factors before selecting a drying. These factors are (a) the type of product to be dried, (b) properties of the finished product desired, (c) the products susceptibility to head, (d) pretreatments required, (e) capital and processing cost, and (f) environmental factors. There is no one best technique for all products (Rahman and Perera, 1999).
- Mechanical and chemical dewatering. Drying is the removal of water content to safe levels that can slow down the actions of the enzymes, bacteria, yeasts, and molds. The effects of heating on the activity of microorganisms and enzymes are also important when food is dried (Rahman, 2006).

Fishes are prone to rapid microbial spoilage, thus adequate care must be taken in drying these products. Microbial standards are usually based on the total number of indicator organisms or number of pathogens. The microbial load and its change during drying and storage are important information for establishing a standard that will ensure food safety. Poor processing, handling, and storage practices often results in a limited storage life of the dried salted fish (Rahman, 2006).

Fermentation plays an important role in food technology. It is one of the oldest and most widely used food preservation methods in households and small-scale food industries as well as in large companies. Fermented foods are those in which the action of microorganisms or enzymes is involved, causing desirable biochemical changes and significant modifications in comparison to the fresh product (Tanasupawat and Visessanguan, 2012).

The main components of the fermentation ecosystem include microbes (yeasts, molds, and bacteria), organic materials to be fermented, a solution in which the fermentation takes place, a vessel with a controlled gate, and various tools that may be used to develop and monitor the fermentation (Scott and Sullivan, 2008).

The highest production of fermented food occurs in Europe, North America, and Africa. In Europe, traditional pickled fish products such as rakeorret in Norway, maatjes in Netherlands, and gravad fish in Scandinavia and Scotland are made with a low salt content, whereas surstromming in Sweden is produced with a medium salt content. Traditional Mediterranean products are fermented anchovies, where the fish is dry-salted and fermented at 15–20 °C. After a few days, further salt is added.

After 6 months of storage, the fish is chilled, washed in brine, filleted by hand, and the fillets are blotted dry. The product is canned in vegetal oil (Venugopal, 2006).

The term "fermented fish products" is used to describe the products of freshwater and marine finfish, shellfish, and crustaceans that are processed by the combined action of fish enzymes and bacterial enzymes with salt to cause fermentation and thereby to prevent putrefaction (Ruddle and Ishige, 2010). Fish fermentation produces three different types of products: those that retain the fish in their original form, those that reduce fish to paste or structured products, and those that reduce fish to a liquid sauce (Joshi and Petricorena, 2012). Fermentation is an energy-neutral process that extends the shelf life of fish and fish contents products and increases their palatability and nutritional value. In the fermentation in fish products the muscle proteins are degraded into smaller peptides and amino acids that are nutrients for microorganisms. Hence, this process is often combined with the addition of salt or drying to reduce water activity and eliminate proteolytic and putrefying microorganisms (Salampessy *et al.*, 2010). The basic aim of the fermentation process is to transform the highly perishable substrate muscle into a stable and safe product maintaining optimum nutritive values and sensory quality. The nature of the raw material and activity of the microorganism affects the process of fermentation (Peredes-Lopez and Harry, 1988).

In fermentation, raw materials are converted into metabolites through the activity of endogenous enzymes or microorganisms (bacteria, yeasts, and molds). The increase of bacteria with time corresponds to a decrease of pH that can be attributed to the production of acid by lactic acid bacteria (LAB) (Taira *et al.*, 2007). Such behavior occurs in the production of both fish sauce and fish paste (Thapa *et al.*, 2004).

The microbial species used for fermentation generally belong to *Lactobacillus*, *Streptococcus*, *Pediococcus*, or *Leuconostoc* spp. (Owens and Mendoza, 1985). Generally, the microbial population experiences a slight reduction in the first 10 days, and the first high initial microbial increase is probably due to fermentable sugar that promotes the growth of acid-forming bacteria (Kasankala *et al.*, 2011). Xu *et al.* (2010) reported an increase of lactic acid bacteria (LAB) within 24 hours of fermentation and a decrease in other microbes, which can be explained by the rapid drop in pH and an increased stress against the growth of other microorganisms such as gram-negative bacteria (Xu *et al.*, 2008).

2.5 Fish and Seafood Spoilage: Characterization of Spoilage Microorganisms During Capture, Manufacture, and Distribution of Fishery Products

The fisheries and aquaculture sector has achieved to blur the outlines of the geographical borders evolving to a global market. In this way, the manufacturing and distribution chain is by road, rail, air, and sea. Because of during these stages there are a big number of intermediaries who handle these products, it is necessary to identify where and which are the microbiological agents responsible for the fishery product's damage.

For this reason, spoilage microorganisms can occur during capture, manufacture, and distribution steps.

During capture, fishery products are very susceptible to changes caused by microorganisms and to autolytic and oxidative processes. Natural spoilage processes that take place post-mortem, include not only microbiological changes, but physical, chemical, and biochemical changes affecting the initial characteristics of the product from the time of capture (Ólafsdóttir *et al.*, 1997). These types of products, which have high non-protein nitrogen content and a high pH (>6), allow the proliferation of many microorganisms (Gram and Huss, 1996).

In order to reduce the incidence of deterioration factors and also to ensure a proper conservation of organoleptic properties, it is very important to take into account the time between the moment of capture and the reduction of temperature in the distribution chain (Ólafsdóttir *et al.*, 2004).

There are both extrinsic and intrinsic factors that influence the intensity and rate of fish deterioration; like the species and fish physiology, age, sex, seasonality, and capture area (Huss, 1995b; Nazrul and Razzaq, 2005).

Temperature fluctuations can favor the loss of quality and especially the safety of fishery products, allowing the proliferation of spoilage microorganisms (Koutsoumanis *et al.*, 2002; Mejlholm *et al.*, 2010).

During fish manufacture good handling practices are essential to prevent tissue damage, such as cuts and wounds, which facilitate bacteria internalization and in turn, lead to cross-contamination in gutting and filleting operations, for example. By contrast, in the packaging step the deterioration rate can be reduced by using modified atmosphere or vacuum packaging, which hinder the ratio of initial microbial load and final product.

Finally, in the distribution step, autolytic processes and microbial growth on fish products can occur if abuse storage conditions are applied. The control of temperature is the most important factor to ensure the preservation of the product (Simpson *et al.*, 2003). The multiplication of bacterial load is responsible for the loss of quality due to the progressive deterioration during storage (Huss, 1995).

The most representative spoilage microorganisms in fishery products and their associated effects are represented in Table 2.1.

Table 2.1 Representative Microbial Species of Spoilage Bacteria in Fishery Products.

Spoilage microorganism	Characteristics and growth requirements	Effects on fish products
Pseudomonas spp.	Strict aerobic bacilli, gram negative, catalase and oxidase positive, able to grow in ice storage	Natural microbiota, spoilage during storage, responsible for unpleasant odors, mucus, and tissue degradation. Present in packaged products
Shewanella spp.	Facultative anaerobic bacilli, gram negative, catalase/oxydase positive, able to grow on ice storage	Responsible for the production of unpleasant odors from (L-cisteine)
Photobacterium phosphoreum	Facultative anaerobic bacilli, gram negative, it is in the intestinal tract, able to grow at low temperatures	Responsible for the deterioration of packaged fillets, through the production of trimethylamine
Lactic acid bacteria	Able to grow in a wide range of temperature, generally psychrotrophic	Responsible for the deterioration in freshwater fish in vacuum packed, brine, and modified atmosphere products
Enterobacteriaceae	Facultative anaerobic bacilli, gram negative, mesophilic and psychrotrophic	Responsible for the deterioration of packaged products. Production of unpleasant odours and flavors
Brochothrix thermosphacta	Gram-positive bacilli, part of the dominant flora from the Mediterranean sea	Responsible for the deterioration of packaged products in modified atmosphere and smoked vacuum-packed products.

2.6 Foodborne Pathogens in Mediterranean Fishery Products

Fishery products can be contaminated with foodborne pathogens that could be present in the water, at the processing factories or within the intestinal tract of living species.

The most representative species are summarized below:

2.6.1 *Aeromonas* spp.

It is a gram-negative, non-sporulated rod and facultative anaerobic bacterium. *Aeromonas* spp. have many similarities to the family Enterobacteriaceae. The genus is divided into two groups: psychrophilic Aeromonas (*A. salmonicida*) and mesophilic *Aeromonas* (single polar flagellum), considered potentially hazardous to human health (*A. hydrophila, A. caviae, A. veronii* subsp. *sober, A. jandaei, A. veronii* subsp. *Veronii,* and *A. schubertii*).

This bacterium is ubiquitous in all freshwater environments and in brackish water. Some strains of *A. hydrophila* are able of causing gastroenteritis and other infections in humans. It is believed that disease-causing strains are only a fraction of the diversity of strains present in the environment.

Traditional techniques for its detection in foods consist on a pre-enrichment in alkaline peptone water (incubation at 30 °C for 24 h). Then, Base Phenol Red Agar is used adding 10% of starch and ampicillin (10 mg / L); incubating at 30 °C for 24 h. Presumptive colonies appear with a yellow halo after addition of lugol solution.

Further biochemical confirmation includes tests such as Gram stain, cytochrome c oxidase activity, catalase activity, and oxidation-fermentation of glucose. Differentiation from *Plesiomonas* spp. or Vibrio spp. can be achieved through the resistance to vibriostatic compound O-129. Further confirmation tests include Sulphide-Indole-Mobility (SIM), hydrolysis of aesculin, gas production from D-glucose, methyl-red test Voges-Proskauer, decarboxylation of L-lysine and L-ornithine and hydrolysis of L-arginine, acid production from D-mannitol, salicin, L-arabinose, inositol, and sucrose using traditional protocols in each case.

2.6.2 *Clostridium perfringens*

It is an anaerobic, but aero tolerant, bacterium. It is a gram-positive and spore-forming rod that produces enterotoxin. This bacterium is relatively psychrophilic, and its spores are heat-resistant. *C. perfringens* has been traditionally isolated from freshwater fish products, being frequently found in the intestinal tract of animal species. This is also found in improperly sterilized canned foods. *C. perfringens* has many isotopes, including the isotope A, which contains CPE gene associated to exterotoxin production. Also the isotopes B, C, D, and E might contain this gene. The infective dose is $>10^6$ vegetative cells or spores/g of food. Toxin production in the digestive tract is associated with sporulation.

The horizontal method for the enumeration of *C. perfringens* is referred to the ISO 7937: 2004, using TSC medium (Tryptose Sulfite Cycloserine) and incubating at 37 °C for 20 h under anaerobic conditions. Black colonies could be presumptive of *C. perfringens.* Biochemical confirmation is carried out in Usp Thioglycollate medium, incubated for 18-24 h at 37 °C under anaerobic conditions. Final confirmation is achieved in half Lactose Thioglycolate Sulphite Broth at 46 °C for 18–24 h. The presence of $>1/4$ gas production together with black colonies is confirmatory of *C. perfringens.*

2.6.3 *Clostridium botulinum*

C. botulinum is a gram-positive rod-shaped, obligate anaerobe, spore-forming, and toxin-producing bacterium (botulinum toxin, the causative agent of botulism). *C. botulinum* is classified into proteolytic (protein digestion performed and H_2S production) and non-proteolytic. There are eight types of botulinum toxins, namely A, B, C, D, E, F, G, and H. The spores can survive in most environments and they are difficult to be destroyed even at the boiling point of water. They have a special importance in canning, both animal and vegetable origin.

The bacteria growth can be prevented with acid pH (<4.5), high sugar concentration, high levels of oxygen, and low humidity. The high toxicity of botulinum toxins requires detection methods capable of toxin measurement in the low to sub pg/mL range. Polymerase chain reaction (PCR) methods can readily detect the presence of low levels of *C. botulinum* DNA but do not detect the presence or absence of the toxin. Current methods for toxin detection rely on 1) the gold standard mouse bioassay, or 2) in vitro tests such as molecular tests, immunoassays, and/or activity-based assays. Activity-based assays can detect active toxin but generally have poorer detection limits than immunoassays. New assays must also be carefully validated in individual food matrices or for as many toxin subtypes as possible, in order to establish assay performance standards.

2.6.4 *Listeria monocytogenes*

It is a gram-positive, rod-shaped, facultative bacterium, motile by means of flagella, that is among the leading causes of death from foodborne illness. It has 13 serotypes, including 1/2a, 1/2b, 1/2c, 3a, 3b, 3c, 4a, 4ab, 4b, 4c, 4d, 4e, and 7. Among them, serotypes 1/2a, 1/2b, and 4b have been associated with the vast majority of foodborne infections. *L. monocytogenes* is recognized as salt-tolerant, surviving and growing at refrigeration temperatures, unlike many other pathogens. It has been isolated in fresh fish, processed fish, and seafood products. Fatality rates associated to *L. monocytogenes* infection is around 15 to 30%. The infective dose of *L. monocytogenes* is undetermined, but is believed to vary with the strain and susceptibility of the host, and the food matrix involved. The enumeration and detection method for *L. monocytogenes* is based on the ISO 11290-1: 1996 and ISO 11290-2: 1998. These methods are based on the use of Half-Fraser broth (incubation at 30 °C for 24 h) and Fraser broth (incubation at 37 °C fir 48 h). Then, the final culture is streaked on Oxford, Palcam and chromogenic (ALOA) selective media. Presumptive positives are finally submitted to biochemical confirmation using haemolysis, acid production (rhamnose and xylose), and cAMP assays (control strains: *S. aureus* and *Rhodococcus equi*).

2.6.5 *Salmonella*

Salmonella is a genus of rod shaped, gram-positive, non-spore-forming, and facultative anaerobic bacteria included in *Enterobacteriacea* family. *Salmonella* possesses peritrichous flagellas. There are two species of *Salmonella*, *Salmonella bongori*, and *Salmonella enterica*. *Salmonella enterica* comprises six subspecies and around 2500 serovars (Wingstrand and Aabo, 2014). This microorganism exists in the intestinal tract of warm-blooded animals, though most of which are subclinical excretors. Besides that, *Salmonella* can survive in other environments such as moist soil, water, fecal material, animal feeds, and foods. Infection by *Salmonella enterica* is a zoonosis that can be transmitted by foods, this microorganism being of the most relevant foodborne pathogens.

For enumeration purposes, when detectable levels are expected, homogenized samples are directly cultured onto selective agar media, such as Xylose-Lysine-Desoxycholate (XLD) agar, and incubated at 37 °C for 18–24 hours. The microbiological method most widely used by food industry is that proposed by ISO corresponding to the standard ISO 6579: 2002. It is an agar-based method for detection of *Salmonella* spp. in products intended for human consumption and feeding of animals and in environmental samples in the area of food production and food handling.

Basically, this method, in a first stage, develops a pre-enrichment process by using buffered peptone water incubated at 37 °C for 18 h. Then, pre-enriched culture is subcultured into two selective enrichment media, that is, Rappaport Vassiliadis Soy Broth (RVS and Muller-Kauffmann Tetrathionate-Novobiocin, and incubated for 24 h at 45.5 °C and at 37 °C, respectively. Then, selective enrichment culture showing growth are inoculated onto two selective agar media and incubated at 37 °C for 24 h. At least, one of the agars should correspond to Xylose-Lysine-Desoxycholate (XLD) and other optional selective agar. There are several alternatives such as Bismuth Sulphite agar, Brilliant Green agar, and Hektoen Enteric agar. A number of selective chromogenic agar media specifically designed for the differentiation of *Salmonella* colonies are commercially available. Typical *Salmonella* colonies on selective agar are subcultured onto non-selective media prior to confirmatory testing (Dykes, 2016).

2.6.6 *Vibrio*

Vibrio is a genus of curve rod-shaped, facultative anaerobic, non-spore forming and gram-negative bacteria included in the group of Proteobacteria where *Salmonella* spp and *Escherichia coli* are also included. *Vibrio* spp. are important components of the coastal microbial communities, thus, most members of this genus are halophilic but sensitive to acid pH and tolerant to basic pH values. These microorganisms are oxidase positive and motile due to the presence of a flagellum. Some species are opportunistic human pathogens associated with the ingestion of contaminated water or contaminated (undercooked) seafood products and also can cause septicemia through open wounds. The most known pathogenic vibrio species are *V. chloreare*, *V. parahaemolyticus*, and *V. vulnificus*.

Conventional methods to detect and enumerate pathogenic *Vibrio* spp. are based on the use of selective agar media and Most Probable Number (MPN). The most widely selective agar media for *Vibrio* isolation and enumeration is Thiosulfate-citrate-bile salts sucrose (TCBS) (Mandal and Mandal, 2014). The basis of this selective agar is the formulation of sucrose and the presence of a pH indicator. Generally, *Vibrio* spp. are not able to ferment sucrose yielding green colonies, while those fermenting sucrose results in areas of reduced pH with yellow colonies. However, the capacity of discrimination of TCBS is not adequate for some bacterial species. For example, *V. cholerae* is sucrose positive and other non-vibrio species growing on TCBS such as *Aeromonas* and *Shewanella* are sucrose negative. The medium CHROMagar Vibrio(CaV) allows to select *Vibrio* spp. using an alkaline pH = 9 and based on the formulation of different chromogenic substrates, it can discriminate between *V. parahaemolyticus* and *V. vulnificus* but not between *V. vulnificus* and *V. cholera* (Dick *et al.*, 2012). Conversely, TCBS can differentiate between *V. vulnificus* and *V. cholerae*. Based on the thin-agar layer method, some researchers have developed a double-layer agar plate (DLAP) method for direct enumeration of *V. parahaemolyticus* (Wang *et al.*, 2015). This method is prepared using a chromogenic selective agar, such as Bo-chrome Vibrio medium (BCVM), as bottom layer overlaid with a nonselective medium (e.g., tryptone soy agar [TSA]) supplemented with 3% NaCl. Results demonstrated that DLAP was as efficient as NMP method, showing better recovery of heat-injured *V. parahaemolyticus* (Wang *et al.*, 2015).

The Most Probable Number (MPN) technique described in Bacteriological Analytical Manual Online is based on using ASPW, followed by plating onto selective agar, generally TCBS. In this method, results can be obtained after 5–7 days.

The method ISO/TS 21872–1 is intended to detect enteropathogenic *Vibrio* spp. particularly, *V. parahaemolyticus* and *V. cholerae*. To this end, a first selective enrichment with Alkaline Saline Peptone Water (ASPW) is used to perform the initial suspension, which should be incubated at 37 °C for 6 h for deep-frozen products or at 41.5 °C for 6 h for fresh, dried, or salted products. The second selective enrichment is also performed with ASPW with incubation at 41.5 °C for 18 h. After, incubation cultures from ASPW are used to inoculate TCBS agar plate. Likewise, a second optional selective agar medium is required such as CHROMagar Vibrio. Typical colonies will be isolated to test presumptive identification by oxidase test, microscopic examination, halotolerance test, and biochemical tests (detection of ornithine decarboxylase, L-Lysine decarboxylase, arginine dihydrolase, B-galactosidase, indole). To confirm pathogenicity, serology test, and presence of toxin production or toxigenicity genes for *V. cholerae* as well as thermostable direct haemolysin or TDH-related haemolysin gene for *V. parahaemolyticus* should be performed.

2.7 Molecular Methods for Pathogen Detection in Fishery Products

Development of efficient and accurate molecular methods for the detection of pathogenic species in fishery products is deemed essential to ensure safe food for consumers. Inasmuch fish and seafood are perishable products, the use of molecular methods have evolved substantially in the last decade, providing a rapid detection and identification of human pathogens. Fast detection of pathogens is important not only in infected fish (clinically and sub-clinically), but also in the environment, for example, between harvesting and re-stocking, and as an "early warning system." The application of antibody probes and DNA primers/probes in pathogen detection and identification has made a significant impact on the development of such rapid diagnostic methods. Standardization and validation of these methods, however, has scarcely been addressed.

Among the diverse methods employed, Polymerase-Chain-Reaction (PCR)-amplification, DNA, or oligonucleotide probe-based hybridisation methods, or immunological methods have been mainly oriented for pathogenic species of *Vibrio* spp. and *Listeria* spp.

2.7.1 Polymerase Chain Reactions

In Europe, there is possibility of using alternative methods, if they are properly validated. One of the major challenges still lies on the use of real-time quantitative polymerase chain reaction (RT-qPCR). The main advantages of using qPCR are related to the differentiation of viable and non-culturable cells and rapid detection of pathogens (*Listeria* can be detected in 48 h, as shown by Boyer and Combrisson, 2013). In fishery products, qPCR can used for simultaneously detection of *Salmonella* and *Listeria* in salmon (Amagliani *et al.*, 2010).

A number of PCR assays have been used to amplify *Vibrio* spp. from food, blood, or environmental samples. Pathogenicity of *V. parahaemolyticus* is strongly linked to production of the thermostable direct hemolysin (*tdh*) and/or the thermostable direct-related hemolysin (*trh*), although other virulence factors are likely involved (Yeung and Boor, 2004). Therefore, many molecular detection assays have targeted the genes responsible for production of the *tdh* and *trh* hemolysins (Bej *et al.*, 1999; Nemoto *et al.*, 2009; Nordstrom *et al.*, 2007; Tada *et al.*, 1992). Multiplex PCR has also been widely used for the detection of *Vibrio* spp. (*V. parahaemolyticus*, *V. cholerae*, and *V. vulnificus*) in raw and cooked seafood and fish samples (Messelhäusser *et al.*, 2010). Espiñeira *et al.* (2010) validated a method for the detection of five *Vibrio* species using multiplex PCR in seafood. They concluded that the method is sufficiently robust and adaptable to identify the serogroup, biotype, and /or virulence factors.

2.7.2 DNA Microarrays and Biosensors

One limitation of traditional PCR-based methods is that they are normally limited to the analysis of a single pathogen, a small group of related pathogens, or a small number of relevant genes. Microarray technology enables a significant expansion of the capability of DNA-based methods in terms of the number of DNA sequences that can be analyzed simultaneously. This enables molecular identification and characterization of multiple pathogens and genes in a single array assay. Microarray analysis of microbial pathogens has potential uses in research, food safety, medical, agricultural, regulatory, public health, and industrial settings. DNA microarrays have been used to detect fish pathogens including *A. hydrophila, V. alginolyticus, V. anguillarum*, or *V. harveyi*, corroborating the results of parallel molecular methods (plate culturing and 16 S rNA gene sequencing) (Shi *et al*., 2012).

Biosensor-based methods rely on the recognition of antigen targets or receptors by antibodies, aptamers, or high-affinity ligands. The captured antigens may be then directly or indirectly detected through an antibody or high-affinity and high-specificity recognition molecule. The possibility to use antigens in combination with a transducer such as a piezoelectric crystal detector offers the ability to develop a group-specific piezo-immunosensor for the detection of pathogenic bacteria in fish products (Guilbault and Luong, 1991).

2.7.3 Antibody-Based Tests

A variety of antibody-based tests and molecular tests have been developed to detect mainly bacterial and viral fish pathogens, although tests have also recently been reported for parasites and fungal agents. The antibody-based tests include slide agglutination, co-agglutination/latex agglutination, immunodiffusion, direct and indirect fluorescent antibody tests (FAT and IFAT), immunohistochemistry (IHC), and enzyme-linked immunosorbent assay (ELISA), dot blot/dip stick, and western blot (WB) (reviewed by Adams, 1999). The antibody-based test selected for the identification of pathogens depends on a variety of factors since each method has its merits and disadvantages. Although such methods are useful for the detection of pathogens in pure culture or/and in infected fish tissue, their sensitivity thresholds limit use in environmental samples, especially where pathogen levels are extremely low. Studies performed in seafood samples used monoclonal antibodies (mAbs) for detection of *V. parahaemolyticus*. Normally, *trh* recombinant proteins of pathogenic *V. parahaemolyticus* can be applied with sandwich enzyme-linked immunosorbent assays (ELISA). Advantages include performance simplicity, high throughput, speed, and cost (Krishna Kumar *et al*., 2011). Alternatives are the use of monoclonal antibody based immunochromatographic strip tests for direct detection of *V. cholerae* O1 (Chaivisuthangkura *et al*., 2013). However, the detection sensitivities were also depending on the content of the other bacteria that might inhibit the growth of *V. cholerae* during pre-enrichment step.

References

Adams, A. (1999). Application of antibody probes in the diagnosis and control of fish diseases. Silver Jubilee Lecture, Mangalore, India.

Alcicek, Z. and Atar, H.H. (2010). The effects of salting on chemical quality of vacuum packed liquid smoked and traditional smoked rainbow trout *(Oncorhyncus mykiss)* fillets during chilled storage. *Journal of Animal and Veterinary Advances, 9*, pp. 2778–2783.

Amagliani, G., Omiccioli, E., Brandi, G., Bruce, I.J. and Magnani, M. (2010). A multiplex magnetic capture hybridisation and multiplex Real-Time PCR protocol for pathogen detection in seafood. *Food Microbiology, 27*, pp. 580–585.

Bagge-Raven, D., Yin, N., Hjelm, M., Chiristiansen, J.N., Johansen, C. and Gram, L. (2003). The microbial ecology of processing equipment in different fish industries – analysis of the microflora during processing and following cleaning and disinfection. *International Journal of Food Microbiology, 87*, pp. 239–250.

Barat, J.M., Gallart-Jornet, L., André, A., Aske, L., Carlehoe, G.M. and Skjerdal, O.T. (2006). Influence of cod freshness on the salting, drying and desalting stages. *Journal Food Engineering, 73*, pp. 6–19.

Bej, A.K., Patterson, D.P., Brasher, C.W., Vickery, M.C.L., Jones, D.D. and Kaysner, C.A. (1999). Detection of total and hemolysin-producing

Vibrio parahaemolyticus in shellfish using multiplex PCR amplification of *tl, tdh* and *trh. Journal of Microbiological Methods, 36,* pp. 215–225.

Bjorkevoll, I., Olsen, R.L. and Skjerdal, O.T. (2003). Origin and spoilage potential of the microbiota dominating genus *Psychrobacter* in sterile rehydrated salt-cured and dried salt-cured cod *(Gadus morhua). International Journal of Food Microbiology, 84,* pp. 175–187.

Boyer M. and Combrisson J. (2013). Analytical opportunities of quantitative polymerase chain reaction in dairy microbiology. *International Dairy Journal, 30,* pp. 45–52.

Campos, C.J. and Lees, D.N. (2014). Environmental transmission of human noroviruses in shellfish waters. *Applied and Environmental Microbiology 80*(12), pp. 3552–3561.

Cao, R., Xue, C., Liu., Q. and Xue, Y. (2009). Microbiological, chemical, and sensory assessment of Pacific Oysters *(Crassostrea gigas)* stored at different temperatures. *Czech Journal of Food Sciences, 27*(2), pp. 102–108.

Carrascosa, C., Millán, R., Saavedra, P., Jaber, J.R., Raposo, A., Pérez, E., Montenegro, T. and Sanjuán, E. (2015). Microbiological evolution of gilthead sea bream (*Sparus aurata*) in Canary Islands during ice storage. *International Journal of Food Science and Technology, 52*(3), pp. 1586–1593.

Chaivisuthangkura, P., Pengsuk, C., Longyant, S., Sithigorngul, P. (2013). Evaluation of monoclonal antibody based immunochromatographic strip test for direct detection of *Vibrio cholerae* O1 contamination in seafood samples. *Journal of Microbiological Methods, 95,* pp. 304–311.

Chinivasagam, H.N., Bremner, H.A., Thrower, S.J. and Nottingham, S.M. (1996). Spoilage pattern of five species of Australian prawns: Deterioration is influenced by environment of capture and mode of storage. *Journal of Aquatic Food Product Technology, 5,* pp. 25–50.

Chinivasagam, H.N., Bremner, H.A., Wood, A.F. and Nottingham, S.M. (1998). Volatile components associated with bacterial spoilage of tropical prawns. *International Journal of Food Microbiology, 42,* pp. 45–55.

Copat, C., Arena, G., Fiore, M., Ledda, C. Fallico, R., Sciacca, S. and Ferrante, M. (2013). Heavy metals concentrations in fish and shellfish from eastern Mediterranean Sea: Consumption advisories. *Food and Chemical Toxicology, 53,* pp. 33–37.

Dick, M.H., Guillerm, M., Moussy, F. and Chaignat, C.L. (2012). Review of two decades of cholera diagnostics--how far have we really come? *PLoS Neglected Tropical Diseases, 6*(10), p. e1845.

Dykes, G.A. (2016). *Encyclopedia of Food and Health. Encyclopedia of Food and Health.* Elsevier.Available at: http://doi.org/10.1016/B978-0-12-384947-2.00606-1 [Accessed 1 December 2015].

Efiuvwevwere, B.J. and Ajiboye, M.O. (1996). Control of microbiological quality and shelf-life of catfish (*Clarias gariepinus*) by chemical preservatives and smoking. *Journal of Applied Bacteriology, 80,* pp. 465–470.

Espiñeira, M., Atanassova, M., Vieites, J.M. and Santaclara, F.J. (2010). Validation of a method for the detection of five species, serogroups, biotypes and virulence factors of *Vibrio* by multiplex PCR in fish and seafood. *Food Microbiology, 27,* pp. 122–131.

Esteve, C., Alcaide, E. and Blasco, M.D. (2012). *Aeromonas hydrophila* subsp. *dhakensis* isolated from feces, water and fish in Mediterranean Spain. *Microbes and Environments, 27*(4), pp. 367–373.

FAO (Food Agricultural Organization). (1983). Cured fish: market patterns and prospects. *Fisheries Technical Papers, 233,* pp. 1–10.

FAO (Food Agricultural Organization). (1992). Small-scale food processing – a guide for appropriate equipment. London: Intermediate Technology. Available at: http://www.fao.org/wairdocs/x5434e/x5434e00.htm [Accessed 30 November 2015].

Flemming, J., Nielsen, J. and Larsen, E. (2014). Chilling and freezing of fish. In: I. Boziaris, ed., *Seafood Processing: Technology, quality and safety*. United Kingdom: Wiley Blackwell, pp. 33–37.

Flick, G.,J. and Kuhn, D. (2012). Smoked, cured and dried fish. In: L.A. Granada, G.J. Jr. Flick and R.E. Martin, eds., The seafood industry 2 ed. New York: Wiley-Blackwell, pp. 404–426.

Gram, L. (2009). Microbiological spoilage of fish and seafood products. In: W.H. Sperber and M.P. Doyle, eds., *Compendium of the microbiological spoilage of foods and beverages.* New York: Springer, pp. 87–119.

Gram, L. and Huss, H.H. (1996). Microbiological spoilage of fish and fish products. *International Journal of Food Microbiology, 33*, pp. 121–137.

Guilbault, G.G. and Luong, J.H.T. (1991). Piezoelectric immunosensors and their applications in food analysis. In: G. Wagner and G.G. Guilbault, eds., *Food Biosensor Analysis.* New York: Marcel Dekker, pp. 151–172.

Hsu, H.H., Chuang, T.C., Loin, H.C., Huang, Y.R., Lin, C.M., Kung, H.F. and Tsai, Y.H. (2009). Histamine content and histamine-forming bacteria in dried milkfish *(Chanos chanos). Food Chemistry, 114*, pp. 933–936.

Huss, H. (1995). Quality and quality changes in fresh fish. In: H.H. Huss Ed., FAO Fisheries Technical Paper no. 348. FAO, Rome, Italy.

ICMSF. (2005). Microorganisms in foods, vol 6: Microbial ecology of food commodities. 2 ed, pp. 174–236.

IIIR, International Institute of Refrigeration, (1996). General principles for freezing, storage and thawing of foodstuffs. In: Recommendation for the processing and handling of frozen foods, 3rd ed. Paris, France, p. 42.

Iwamoto, M., Ayers, T., Mahon, B. E. and Swerdlow, D.L. (2010). Epidemiology of seafood-associated infections in the United States. *Clinical Microbiology Reviews, 23*(2), pp. 399–411.

Jarvis, N.D. (1988). Curing and Canning of Fishery Products: A History. *Marine Fisheries Review, 50*, pp. 180–185.

Jay, J.M., Loessner, M.J. and Golden, D.A. (2005). Modern food microbiology, 7th. New York: Springer, pp. 101–124.

Joshi, N.H. and Petricorena, Z.C. (2012). Fermented seafood products. In: B.M. Metha, A. Kamal-Eldin and R.Z. Iwanski, eds., *Fermentation effects on food properties.* Boca Raton: CRC Press, pp. 285–301.

Kasankala, L.M., Xiong, Y.L. and Chen, J. (2011). The influence of douche starter cultures on the composition of extractive components, microbiological activity, and sensory properties of fermented fish pastes. *Journal of Food Science, 76*(1), pp. C154–C161.

Koutsoumanis, K., Giannakourou, M., Taoukis, P. and Nychas, G. (2002). Application of shelf life decision system (SLDS) to marine cultured fish quality. *International Journal of Food Microbiology, 73*, pp. 375–382.

Koutsoumanis, K. and Nychas, G.J. (2000). Application of a systematic procedure to develop a microbial model for rapid fish shelf life predictions. *International Journal of Food Microbiology, 60*, pp. 171–184.

Krishna Kumar, B., Raghunath, P., Devegowda, D., Deekshit, V.K., Venugopal, M.N., Karunasagar, I. and Karunasagar, I. (2011). Development of monoclonal antibody based sandwich ELISA for the rapid detection of pathogenic *Vibrio parahaemolyticus* in seafood. *International Journal of Food Microbiology, 145*, pp. 244–249.

Ladip, L.D., Zzaman, W. and Yang, T.A. (2013). Effect of chilled-frozen storage on the physico-chemical, microbial and sensory quality of farmed Bighead Carp (*Hypophthalmichthys nobilis*). *Journal of Fisheries and Aquatic Science, 8*(6), pp. 686–696.

Lakshmanan, R., Shakila, R.J. and Jeyasekaran, G. (2002). Changes in halophilic amine forming bacterial flora during salt-drying of sardines (*Sardinella gibbosa*). *Food Research International, 35*, pp. 541–546.

Lauritzsen, K., Akse, L., Gundersen, B. and Olsen, R.L. (2004). Effects of calcium, magnesium and pH during salting of cod *(Gadus morhua L.). Journal of the Science of Food and Agriculture, 84*, pp. 683–692.

Leroi, F., Amarita, F., Arboleya, J.C., Bjørkevoll, I., Cruz, Z., Dousset, X., Izurieta, E., Joffraud, J.J., Lasagabaster, A., Lauzon, H.L., Lorentzen, G., Martínez de Marañón, I., Matamoros, S., Miranda, I., Nuin, M., Olabarrieta, I., Olsen, R., Pilet, M.F., Prévost, H. and Skjerdal, T. (2006).

Hurdle technology to ensure the safety of seafood products. In: T. Borresen, ed., *Improving seafood products for the consumer*. Boca Raton: CRC Press, pp. 399–425.

Leroi, F. and Joffraud, J.J. (2000). Salt and smoke simultaneously affect chemical and sensory quality of cold-smoked salmon during 5 °C storage predicted using factorial design. *Journal of Food Protection, 63*, pp. 1222–1227.

Mandal, S. and Mandal, M. (2014). *Encyclopedia of Food Microbiology*. Elsevier. Available at: http://doi.org/10.1016/B978-0-12-384730-0.00346-3 [accessed 1 December 2015].

Matches, J.R. (1982). Effects of temperature on the decomposition of Pacific coast shrimp (*Pandalus jordani*). *Journal of Food Science, 47*(4), pp. 1044–1047.

Mejlholm, O., Gunvig, A., Borggaard, C., Blom-Hanssen, J., Mellefont, L., Ross, T., Leroi, F., Else, T., Visser, D. and Dalgaarg, P. (2010). Predicting growth rates and growth boundary of *Listeria monocytogenes*-An international validation study with focus on processed and ready-to-eat meat and seafood. *International Journal of Food Microbiology, 141*, pp. 137–150.

Messelhäusser, U., Colditz, J., Thärigen, D., Kleih, W., Höller, C. and Busch, U. (2010). Detection and differentiation of *Vibrio* spp. in seafood and fish samples with cultural and molecular methods. *International Journal of Food Microbiology, 42*, pp. 360–364.

Nazrul, I. and Razzaq, A. (2005). Seasonal variation of the proximate composition of freshwater gobi, Glossogius giuris (Hamilton) from the River Padma. *Pakistan Journal of Biological Science, 8*, pp. 532–536.

Nemoto J., Sugawara C., Akahane K., Hashimoto K., Kojima T., Ikedo M., Konuma H. and Hara-Kudo Y. (2009). Rapid and specific detection of the thermostable direct hemolysin gene in *Vibrio parahaemolyticus* by loop-mediated isothermal amplification. *Journal of Food Protection, 72*, pp. 748–754.

Nguyen, M.V., Arason, S. and Eikevik, T.M. (2014). Drying of fish. In: I. Boziaris, ed., *Seafood Processing: Technology, quality and safety*. United Kingdom: Wiley Blackwell, pp. 161–175.

Nielsen, J., Larsen, E. and Jessen, F. (2001). Chilling and freezing of fish and fishery products. In: D.W. Sun, ed., *Advances in Food Refrigeration*. Leatherhead Publishing.

Nordstrom, J.L., Vickery, M.C.L., Blackstone, G.M., Murray, S.L. and Depaola, A. (2007). Development of a multiplex real-time PCR assay with an internal amplification control for the detection of total and pathogenic *Vibrio parahaemolyticus* bacteria in oysters. *Applied and Environmental Microbiology, 73*, pp. 5840–5847.

Olafsdóttir, G., Martinsdóttir, E., Oehlenschäger, J., Dalgaard, P., Jensen, B., Undeland, I., Mackie, I. Henehan, G., Nielsen, J. and Nielsen, H. (1997). Methods to evaluate fish freshness in research and industry. *Trends in Food Science & Technology, 8*, pp. 258–265.

Olafsdóttir, G., Nesvadba, P., Di Natale, C., Careche, M., Oehlenschläger, J., Tryggvadóttir, S., Schubring, R., Kroeger, M., Heia, K., Esaiassen, M., Macagnano, A. and JØrgensen, B. (2004). Multisensor for fish quality determination. *Trends in Food Science & Technology, 15*, pp. 86–93.

Owens, J.D. and Mendoza, L.S. (1985). Enzymatically hydrolyzed and bacterially fermented fishery products. *Journal Food Technology, 20*(2), pp. 273–293.

Pedro, S., Albuquercque, M., Nunes, L. and Bernando, F. (2004). Pathogenic bacteria and indicators and salted cod *(Gadus morhua)* and desalted products at low and high temperatures. *Journal Aquaculture Food Prod Technology, 13*, pp. 39–49.

Peredes-Lopez, O. and Harry, G.I. (1988). Food biotechnology review: Traditional solid state fermentations of plant raw materials: Application, nutritional significance, and future prospects. *Critical Review in Food Science and Nutrition, 27*, pp. 159–187.

Rahman, M.S. (2006). Drying of fish and seafood. In: A.S. Mujumdar, ed., *Handbook of Industrial Drying*. Boca Raton: FL, CRC Press, pp. 547–559.

Rahman, M.S. and Perera, C.O. (1999). Drying and food preservation. In: M.S. Rahman, ed., *Handbook of Food Preservation*. New York, Marcel Dekker, pp. 173–216.

Ray, B. (2005). *Fundamental food microbiology*. Boca Raton: CRC Press, p. 275.

Ruddle, K. and Ishige, N. (2010). On the origins, diffusion and cultural context of fermented fish products in Southeast Asia, in Globalization, Food and Social Identities in the Asia Pacific Region. Ed. (J. Farrer), Sophia University Institute of Comparative Culture, Tokyo. Available at: http://icc.fla.sophia.ac.jp/global%20food%20papers/ [Accessed 01 December 2015].

Salampessy, J., Kailasapathy, K. and Thapa, N. (2010). Fermented fish products. In: J.P. Tamanq and K. Kailasapathy, eds., *Fermented foods and beverages of the world*. Boca Raton, FL: CRC Press, pp. 289–307.

Scott, R. and Sullivan W.C. (2008). Ecology of fermented foods. *Human Ecology Review, 15*, pp. 25–31.

Senadeera, W., Bhandari, B.R., Young, G. and Wijesinghe, B. (2005). Modelling dimension shrinkage of shaped foods in fluidised bed drying. *Journal of Food Processing and Preservation, 29*, pp. 109–119.

Shi, Y.H., Chen, J., Li, C.H., Lu, X.J., Zhang, D.M., Li, H.Y., Zhao, Z.X. and Li, M.Y. (2012). Detection of bacterial pathogens in aquaculture samples by DNA microarray analysis. *Aquaculture, 338–341*, pp. 29–35.

Simpson, R., Almonacid, S., Acevedo, C. and Cortés, C. (2003). Mathematical model to predict effect of temperature abuse in MAP systems applied to pacific hake (*Merluccius australis*). *Journal of Food Process Engineering, 26*, pp. 413–434.

Suffredini, W., Magnabosco, C., Civettini, M., Rossetti, E., Arcangeli, G. and Croci, L. (2012). Norovirus contamination inf different shellfish species harvested in the same production areas. *Journal Applied Microbiology, 133*(3), pp. 686–692.

Tada J., Ohashi T., Nishimura N., Shirasaki Y., Ozaki H., Fukushima S., Takano J., Nishibuchi M. and Takeda Y. (1992). Detection of the thermostable direct hemolysin gene (*tdh*) and the thermostable direct hemolysin-related hemolysin gene (*trh*) of *Vibrio parahaemolyticus* by polymerase chain reaction. *Molecular and Cellular Probes, 6*, pp. 477–487.

Taira, W., Funatsu, T., Satomi, M., Takano, T. and Abe, H. (2007). Changes in extractive components acid microbial proliferation during fermentation of fish sauce from underutilized fish species and quality of final products. *Fisheries Science, 73*, pp. 913–923.

Tanasupawat, S. and Visessanguan, W. (2012). Fish fermentation. In: I. Boziaris, ed., *Seafood Processing: Technology, quality and safety*. United Kingdom: Wiley Blackwell, pp. 177–207.

Thapa, N., Pal. J., Tamang, P. (2004). Microbial diversity in ngari, hentak, and tungkap, fermented fish products of North-East India. *World Journal of Microbiology and Biotechnology, 20*, pp. 599–607.

USFDA (United States Food and Drug Administration). (2012). Various Shellfish-Associated Toxins. Foodborne Pathogenic Microorganisms and Natural Toxins Handbook. Available at: http://www.fda.gov/Food/FoodborneIllnessContaminants/CausesOfIllnessBadBugBook/ucm070795.htm [Accessed 29 November 2015].

Venugopal, V. (2006). *Seafood processing: adding value through quick freezing, retortable packaging, and cook-chilling*. Boca Raton: CRC Press, pp. 95–139.

Vernocchi, P., Maffei, M., Lanciotti, R., Suzzi, G. and Gardini, F. (2007). Characterization of Mediterranean mussels *(Mytilus galloprovincialies)* harvested in Adriatic Sea (Italy). *Food Control, 18*, pp. 1575–1583.

Viegas, E.M.M., Barbieri de Carvalho, M.R, Campagnoli de Oliveira Filho, P.R. and Kirschnik, P.G. (2013). Changes during chilled storage of whole tilapia and short-term frozen storage of tilapia fillets. *Journal of Aquatic Food Product Technology, 22*(2), pp. 192–200.

Wang, Y., Shen, X., Gu, R.R., Shi, Y.F. and Tian, L.L. (2015). Application of a rapid method for detecting *Vibrio parahaemolyticus* in seafood. *Journal of Food Safety, 35*(1), pp. 26–31.

Wingstrand, A. and Aabo, S. (2014). *Encyclopedia of Meat Sciences.* Elsevier. Available at: http://doi.org/10.1016/B978-0-12-384731-7.00034-9 [Accessed 1 December 2015].

Xu, W., Yu, G., Xue, C., Xue, Y., and Ren, Y. (2008). Biochemical changes associated with fast fermentation of squid processing by-products for low salt fish sauce. *Food Chemistry, 107,* pp. 1597–1606.

Xu, Y., Xu, W., Yang, F., Kim, J.M. and Nie, X. (2010). Effect of fermentation temperature on the microbial and physicochemical properties of silver carp sausages inoculated with *Pediococcus pentosaceus. Food Chemistry, 118*(5), pp. 12–518.

Yamasaki S., Shirai H., Takeda Y. and Nicshibuchi M. (1991). Analysis of the gene of *Vibrio hollisae* encoding the haemolysin similar to the thermostable direct hemolysin of *Vibrio parahaemolyticus. FEMS Microbiology Letters, 80,* pp. 259–264.

Yeung, P.S. and Boor, K.J. (2004). Epidemiology, pathogenesis, and prevention of foodborne *Vibrio parahaemolyticus* infections. *Foodborne Pathogen and Diseases, 1,* pp. 74–88.

Yin, X., Luo, Y., Fan, H., Wu, H. and Feng, L. (2014). Effect of previous frozen storage on quality changes of grass carp (*Ctenopharyngodon idellus*) fillets during short-term chilled storage. *International Journal of Food Science and Technology, 49*(6), pp. 1449–1460.

3

Food Spoilage by *Pseudomonas* spp.—An Overview

António Raposo[1], Esteban Pérez[2], Catarina Tinoco de Faria[1,2], María Antonia Ferrús[3], and Conrado Carrasosa[2]

[1] *Centro de Investigação Interdisciplinar Egas Moniz, CiiEM, Instituto Superior de Ciências da Saúde Egas Moniz, ISCSEM, Caparica, Portugal*
[2] *Department of Animal Pathology and Production, Bromatology and Food Technology, Faculty of Veterinary, Universidad de Las Palmas de Gran Canaria, Arucas, Spain*
[3] *Biotechnology Department, Centro Avanzado de Microbiología de Alimentos, Universitat Politècnica de Valencia, Valencia, Spain*

3.1 Introduction

Food spoilage is a serious global problem, especially in developing countries, because of inadequate processing and refrigeration facilities (Arslan *et al.*, 2011). It can be considered to be any change that renders a product unacceptable for human consumption (Jos and In't Veld, 1996; Gram *et al.*, 2002). During harvesting, processing, and handling operations, food may become contaminated with a high range of microorganisms (Jos and In't Veld, 1996). Microbial spoilage is by far the commonest cause of spoilage, and it leads to significant economic loss for the food industry (Gram *et al.*, 2002; Dogan and Boor, 2003). The point of spoilage may be defined by a certain maximum acceptable bacterial level, an unacceptable off-odor/off-flavor, or appearance (Borch *et al.*, 1996). The microbial flora that colonizes a particular food or beverage depends strongly on the characteristics of the product, and the way it is processed and stored (Jos and In't Veld, 1996). Each and every food product harbors its own specific and characteristic microflora at any given point in time during production and storage. This microflora is a function of raw material flora, processing, preservation, and storage conditions. It can be predicted according to knowledge of the food's origin, the substrate base, and a few central preservation parameters, such as temperature, atmosphere, and pH (Gram *et al.*, 2002).

Pseudomonas spp. are the commonest spoilage psychrotrophic organisms, particularly in aerobically stored foods with a high water content and natural pH, such as red meat (Dainty and Mackey, 1992; Borch *et al.*, 1996), fish (Gram and Huss, 1996), poultry (Gallo *et al.*, 1988), and milk and dairy products (Craven and Macauly, 1992). The *Pseudomonas* genus is one of the most diverse ecologically significant groups of known bacteria (Molina *et al.*, 2013; Franzetti and Scarpellini, 2007), and includes species that have been isolated worldwide in all types of environments (Molina *et al.*, 2013; Jos and In't Veld, 1996; Silby *et al.*, 2011). They are found in natural habitats, such as soil, fresh water, and marine environments, but have also been isolated from clinical instruments, contaminated aseptic solutions, cosmetics, and medical products (Molina *et al.*, 2013).

Foodborne Pathogens and Antibiotic Resistance, First Edition. Edited by Om V. Singh.

This genus may also contaminate foods from many sources and are able to utilize a wide range of materials as substrates for growth (Jos and In't Veld, 1996; Blackburn, 2006). The basic morphological characters common to pseudomonads are rod-shaped cells (Arslan *et al.*, 2011; Molina *et al.*, 2013; Franzetti and Scarpellini, 2007; Blackburn, 2006), which are either straight or slightly curved on one plane, presence of polar flagella, a gram-negative reaction, and absence of spores (Blackburn, 2006). They are also characterized by good metabolic versatility due to the presence of a complex enzymatic system (Franzetti and Scarpellini, 2007; Silby *et al.*, 2011). Some species are medically relevant because they are considered opportunistic pathogens for humans and animals, whereas others, like phytopathogens, are very important in the agricultural sector (Ridgway and Safarik, 1990). Whereas microorganisms from the *Pseudomonas* genus are frequently associated with lack of quality in food products, and also with pathogenic infections in plants and humans, the majority of *Pseudomonas* species are non-pathogenic (Molina *et al.*, 2013). Palleroni *et al.* (1973) began a complex study on the genus, and recognized five rRNA similarity groups. Nowadays, only rRNA similarity group 1 is considered the *Pseudomonas "sensu stricto"* (Moore *et al.*, 1996). Most members of group 1 are psychrotrophic, can be fluorescent or non-fluorescent, and have long since been known to be responsible for chilled food spoilage. The most important fluorescent species are *Pseudomonas fluorescens*, *Pseudomonas aeruginosa*, *Pseudomonas putida*, *Pseudomonas chlororaphis*, and plant pathogenic species such as *Pseudomonas syringae* and *Pseudomonas chichorii* (Scarpellini *et al.*, 2004). *Pseudomonas fluorescens*, *Pseudomonas putida*, and *Pseudomonas fragi* are the most frequently found species. However, the species distribution within the food ecosystem remains relatively unknown (Arnaut-Rollier *et al.*, 1999a).

Food spoilage due to pseudomonads may occur in a number of ways. In foods of animal origin, the non-protein nitrogen fraction is first metabolized. Subsequently, the production of lipases or proteases will release fatty and amino acids, which can also result in off-odors, off-flavors and rancidity, which alters metabolism. At later stages, the production of extracellular slime and the development of often pigmented growth become visible (Dainty, 1996). Pectolytic strains of *Pseudomonas flourescens*, *Pseudomonas viridiflava*, and *Xanthomonas campestris* are associated with spoilage of fresh fruits and vegetables (Blackburn, 2006). They increase in number during storage and processing, and they play an important role in the browning phenomenon in minimally processed vegetables because of their pectinolytic activity (Riva *et al.*, 2001). *Pseudomonas* spp. also plays an important role in milk spoilage, where mainly post-pasteurization contaminants gain access (Schokker and Van Boekel, 1999). While raw milk is stored, they produce many thermo-tolerant lipolytic and proteolytic enzymes that reduce both the quality and shelf life of processed milk (Dogan and Boor, 2003; Wiedmann *et al.*, 2000). Some *Pseudomonas* cause disease in diverse host organisms. The most widely studied and economically important species are the opportunistic pathogen *P. aeruginosa* and the plant pathogen *P. syringae*. The economic and health costs associated with *Pseudomonas* infections have attracted intense interest (Silby *et al.*, 2011).

P. aeruginosa is an aquatic and soil bacterium that can infect a range of organisms, including plants (Rahme *et al.*, 2000), nematodes (Tan *et al.*, 1999), fruits, and various mammals (Snouwaert *et al.*, 1992; Colledge *et al.*, 1995; Potvin *et al.*, 2003). Despite being a natural soil inhabitant, *P. aeruginosa* is versatile in its metabolic potential, which allows it to survive in a variety of natural and hospital environments. It appears that the combination of environmental persistence, versatility in virulence mechanisms, and multiple virulence factors enables *P. aeruginosa* to be effective as both an opportunistic human pathogen and a plant pathogen (Rahme *et al.*, 2000).

In humans, *P. aeruginosa* is capable of infecting a range of tissues and sites (Lyczak *et al.*, 2000), and is one of the top six infectious disease threats that cause serious infections in immunocompromised patients or whose natural defenses are otherwise breached. The pathophysiology of infections caused by *P. aeruginosa* is complex, as shown by the clinical diversity of diseases associated with this organism and the multiplicity of the virulence factors it produces (Rahme *et al.*, 2000). These virulence factors facilitate tissue invasion and systemic spread, and include pili and flagella, endotoxins, exotoxins, vascular permeability factors, and a variety of excreted enzymes (Fick, 1993). *P. aeruginosa* is also the major pathogen that contributes to the morbidity and mortality associated with cystic fibrosis, causes chronic lung infection, and considerably contributes to the burden of hospital-acquired infections (Govan and Deretic, 1996). In hospitals, it is a frequent cause of respiratory and urinary tract infections, and is a common cause of hospital-acquired pneumonia, health care-associated pneumonia and ventilator-associated pneumonia (American Thoracic Society, 2005; Gaynes and Edwards, 2005; Kolleff *et al.*, 2005) and pediatric patients in intensive care units (Richards *et al.*, 1999). Increasing resistance rates among *P. aeruginosa*, associated with reports of isolates that are resistant to all clinically available and/or only susceptible compounds to polymyxins (Livermore, 2009; Bonomo and Szabo, 2006), highlight the importance of understanding *P. aeruginosa* epidemiology and resistance mechanisms. *P. aeruginosa* poses a serious therapeutic challenge for the treatment of both community-acquired and nosocomial infections, and the selection of the appropriate antibiotic to initiate therapy is essential for optimizing clinical outcome (Bisbe *et al.*, 1988; Micek *et al.*, 2005). Unfortunately, the selection of the most appropriate antibiotic is complicated by *P. aeruginosa's* ability to develop resistance to multiple classes of antibacterial agents, even during treatment and infection (Lister *et al.*, 2009). A multidrug-resistant phenotype can arise in *P. aeruginosa* through the acquisition of multiple imported resistance mechanisms on mobile genetic elements, a combination of imported and chromosomally encoded resistance mechanisms, the accumulation of multiple chromosomal changes over time, and/or a single mutational event that leads to the overexpression of a multidrug resistance mechanism, such as an efflux pump (Lister *et al.*, 2009). *P. aeruginosa* infections are particularly challenging because of the organisms' broad intrinsic antimicrobial resistance (Lister *et al.*, 2009). *P. aeruginosa* is often resistant to multiple antibiotics, and has consequently joined the ranks of "superbugs." Compared with other pathogens, *P. aeruginosa* is very difficult to eradicate as it displays high intrinsic resistance to a wide variety of antibiotics, including aminoglycosides, fluoroquinolones, and β-lactams (Breidenstein *et al.*, 2011). This is largely because of the low permeability of its outer membrane (Nicas and Hancock, 1983), which limits the rate of penetration of antibiotic molecules into cells due to the inefficient porins in this bacterium (Breidenstein *et al.*, 2011). Therefore, the high intrinsic resistance of this pathogen is absolutely dependent on other intrinsic and adaptive secondary mechanisms such as rapid efflux (Li *et al.*, 1995; Li *et al.*, 2000) due to the intrinsic or induced expression of efflux pumps. In addition to its high intrinsic resistance, *P. aeruginosa* can become even less susceptible owing to the acquisition of inheritable traits. The two types of acquired resistance involve the horizontal transfer of genetic elements and mutational resistance. DNA elements, including plasmids, transposons, integrons, prophages, and resistance islands, can harbor antibiotic resistance genes, and can be acquired by conjugation, transformation, and transduction (Breidenstein *et al.*, 2011). This can increase antibiotic resistance and even multidrug resistance as plasmids contain multiple resistance cassettes. Carbapenems are important antimicrobial agents used to treat *P. aeruginosa* infections (Giamerellou and Poulakou, 2009; Riera *et al.*, 2011), and increasing resistance to these compounds has limited the use of carbapenems in certain geographic areas (Livermore, 2009; Giamerellou and Poulakou, 2009; Bonomo and Szabo, 2006). Carbapenem resistance to *P. aeruginosa* can be caused by the acquisition of acquired carbapenamases, mainly Ambler class B metallo-β-lactamases. More recently, however, KPC serine carbapenemases have been detected in this species (Villegas *et al.*, 2007).

A variety of selective and differential media has been proposed, particularly for the detection of *Pseudomonas* spp. from clinical, environmental, and food samples. A selective medium for their detection may be very valuable for assessing the potential shelf life of a product. *Pseudomonas* spp. grow well on non-selective media (e.g., plate count agar and blood agar) and on routine primary isolation media (e.g., MacConkey agar and eosin methylene blue agar) when incubated at a suitable temperature for their growth (Jeppesen, 1995). One problem associated with using the media developed for isolating *Pseudomonas* spp. from foods is considerable interference from non-pseudomonads (Brant, 1975), so they lack specificity (Solberg *et al.*, 1972).

Mead and Adams (1977) developed a new selective medium with cephaloridine, fucidin, and cetrimide as selective agents (CFC), which permits the enumeration of both pigmented and non-pigmented *Pseudomonas* spp. This medium can be used for the enumeration of *Pseudomonas* spp. in various food types (Mead, 1985), and is one of the most successful media for their detection. One major disadvantage of traditional culture methods is the time required to produce results (Vasavada and White, 1993), and also their failure to detect viable, but non-culturable, organisms (Davey and Kell, 1996; Gunasekera *et al.*, 2002).

Several factors need to be considered when selecting a molecular typing strategy. These include discriminatory power, reproducibility, and typeability, as well as the biological basis for grouping similar strains, cost, and logistics (Clinical and Laboratory Standards Institute, 2007; Tenover *et al.*, 1997). Fluorescent oligonucleotide probe hybridization is becoming a major tool in food microbiology to identify specific microorganisms in mixed communities without the need for isolation in pure cultures (Gunasekera *et al.*, 2003). The fluorescence *in situ* hybridisation (FISH) technique utilizes fluorescently labeled DNA oligonucleotide probes to detect specific sequences of ribosomal RNA (Wallner *et al.*, 1993; Amann *et al.*, 1995). The *Pseudomonas*-specific FISH probe has been tested on *Pseudomonas* spp. isolated from milk and various other bacteria commonly encountered in milk. The probe´s specificity has also been validated by a polymerase chain reaction (PCR) method using the *Pseudomonas*-specific sequence as a target for PCR primers (Gunesekera *et al.*, 2003).

Therefore, the current chapter aims to review food spoilage by *Pseudomonas* spp. in specific food groups (milk and dairy products, meat, fish, water, fruit, and vegetables) and of the detection and confirmation techniques implicated, including biochemical and molecular methods.

3.2 *Pseudomonas* spp. in Milk and Dairy Products

In milk and dairy products, psychrotrophic bacteria are very diverse, including both gram-positive species (*Bacillus, Clostridium, Corynebacterium, Microbacterium, Micrococcus,* Streptococcus, *Staphylococcus,* and *Lactobacillus*) and gram-negative species (*Pseudomonas,* Aeromonas, *Serratia, Acinetobacter, Alcaligenes, Achromobacter, Enterobacter, Flavobacterium, Burkholderia, Sphingomonas, Stenotrophomonas* (Cousin and Marth, 1977; Sorhaug and Stepaniak, 1997; Munsch-Alatossava and Alatossava, 2006; Hantsis-Zacharov and Halpern, 2007). Of these, *Pseudomonas* spp. are considered predominant, whose counts range from <10% to >75% of the total bacterial count of milk (Ternstrom *et al.*, 1993; Jaspe *et al.*, 1995; Wiedmann *et al.*, 2000; Dogan and Boor, 2003; Gunasekera *et al.*, 2003).

Microorganisms of the *Pseudomonas* genus are possibly the psychrotrophs that have been more widely studied in recent years as microbiota that alter foodstuff, especially milk and dairy products. A huge number of psychrotolerant isolates has been identified as members of the genus *Pseudomonas* (Martins *et al.*, 2006; Hantsis-Zacharov and Halpern, 2007), which make it one of the most important genera in the dairy sector (von Neubeck *et al.*, 2015).

These undesirable microorganisms are responsible for spoilage of dairy products, and pathogens being introduced into milk and milk products through unsatisfactory milk production practices, failures in processing systems, or unhygienic practices are of primary concern (Richter and Vedamuthu, 2001). Given the high nutritional value, water content, and near neutral pH of milk, many spoilage and pathogenic microorganisms can grow in it (Ray, 2004).

Contamination by *Pseudomonas* spp. plays an important role in milk spoilage because *Pseudomonas* spp. produce many thermo-tolerant lipolytic and proteolytic enzymes, which reduce both the quality and shelf life of processed milk (Wiedmann *et al.*, 2000; Dogan and Boor, 2003) as they develop during prolonged refrigerated storage of raw milk on farms and in dairy plants (De Jonghe *et al.*, 2011).

Many of these enzymes survive pasteurisation and ultra-high temperature (UHT) treatments (Ray, 2004; Bhunia, 2008). These can also be detrimental to cheese quality by causing bitter or rancid flavors, and by impairing milk's coagulation properties (Richter and Vedamuthu, 2001). Proteases degrade casein, which results in bitterness of milk, gelation of UHT-sterilized milk, and smaller yields of soft cheese. Lipases catalyse the hydrolysis of triglycerides to free fatty acids and glycerol, which produce flavor defects (Arslan *et al.*, 2011). Lecithinases are able to disrupt milk fat globule membranes by increasing milk fat's susceptibility to the action of lipases (Herrera, 2001; Ray, 2004). Thus storage of fresh raw milk at low temperature neither prevents the proliferation of psychrotrophic nor reduces the shelf life of dairy products.

The combination of longer storage time and lower temperature is a selective advantage for psychrotolerant bacteria, especially *Pseudomonas* members that enter raw milk via biofilms in milk tanks, contaminated water or soil (Simões *et al.*, 2009). The use of stringent agent cleaning, hygiene programs, and legal pasteurization regimes (e.g., High Temperature, Short Time) effectively minimizes the gram-negative spoilage bacteria present in raw milk (Van Tassell *et al.*, 2012). Therefore, reducing the presence of these spoilage organisms (*Pseudomonas* spp.) in dairy processing environments and finished products is critical for extending the shelf life of pasteurized dairy products (Van Tassell *et al.*, 2012). However, the heat treatment used to destroy vegetative bacterial cells can also activate bacterial spores, which actually enhance spore germination and outgrowth (Meer *et al.*, 1991). Ranieri *et al.* (2009) showed that high-temperature, short-time (HTST) pasteurization temperatures inversely affected the post-processing aerobic plate count (APC) of fluid milk products after storage at 6 °C for 21 d. The lower APC for milk pasteurized at 72.9 °C could be due to the reduced activation of the spore population found in fluid milk in relation to spore activation, which is achieved at higher temperatures. This could be consistent with a previous report on optimal spore germination after an 80 °C heat treatment (Morn *et al.*, 1990). At day 21 post-pasteurization, the mean APC was higherr for the milk treated at 85 °C. These results suggest that thermosensitive factors are intrinsic to milk (lactoperoxidase system), to spores, or that both influence spore germination and outgrowth in fluid milk systems.

Currently, *Pseudomonas* spp. represent the most predominant psychrotolerant gram-negative bacteria that contributee to milk spoilage, but these microbes are heat-sensitive and do not usually survive pasteurization (Cousin, 1982), unless present in raw milk in very large numbers (Griffiths and Phillips, 1984). Typically, such bacteria enter milk through post-pasteurization contamination in the filling machine (Ralyea *et al.*, 1998). In recent years, many commercial fluid milk processors have reduced post-pasteurization contamination after implementing improved filler and packaging technology to help produce milk that does not result in consumer complaints at 15 to 25 d post-pasteurization (Carey *et al.*, 2005).

Yet these are not the only forms of damage caused by these microorganisms since a major feature associated with *Pseudomonas* spp. (e.g. *P. aeruginosa, P. fluorescens, P. putida*) is secretion of a fluorescent yellow-green pigment called pyoverdin (fluorescein) (Meyer *et al.*, 2002). Certain *Pseudomonas* spp., such as *P. aeruginosa* (Lau *et al.*, 2004) and *P. fluorescens*, produce additional pigments, for example, pyocyanin (blue pigment). *P. fluorescens* biovar I (Cantoni *et al.*, 2001) may also cause green coloration due to pyoverdin production, and *P. fluorescens* biovar IV generates blue-green coloration due to the production of pigments pyocyanin and pyoverdin (Cantoni *et al.*, 2001).

In 2010 many cases of blue discoloration in mozzarella cheese—popularly termed blue mozzarella—were reported to the public opinion. The strong media impact of such cases confused consumers and experts. On the one hand, isolation of the bacteria of the genus *Pseudomonas*, already widely known for such alterations, highlighted the lack of danger for consumer health. On the other hand, repeated seizures of products with blue discoloration and the unpredictability of the alteration have become a serious problem for producers in the dairy sector (Chiesa *et al.*, 2012).

Pseudomonas spp. often dominates the microbiota at the end of shelf life (Baruzzi *et al.*, 2012). In fact, Pseudomonas counts that exceed 10^6 cfu/g have been associated with less product acceptability (Lucera *et al.*, 2014). Physico-chemical phenomena due to salt diffusion contribute to loss of quality during storage (Faccia *et al.*, 2013).

P. fluorescens contamination in this processing facility can be ultimately controlled by using a combination of improved plant cleaning and hygiene procedures, the use of starter cultures, and the addition of potassium sorbate (0.15%) in the grinding step of processing. The addition of starter culture as a type of hurdle technique has been recommended for reducing the risk of pathogen growth (Mendoza-Yepes *et al.*, 1999), but has also proved useful for controlling food spoilage organisms. However, other authors (Carrascosa *et al.*, 2015a) have suggested a new cleaning and disinfection program that includes an enzymatic detergent (Enzivix 30, Proquimia S.A., Spain), and peracetic acid (Asep 150, Proquimia S.A., Spain), which allowed them to control the *Pseudomonas* spp. biofilm until blue cheese coloration had been completely eliminated.

The control of the dairy water sources and water provided to animals is essential to avoid cheese contamination by *Pseudomona fluorescens*. Therefore, cleaning and disinfection programs, and the performance of biofilm detection tests, are necessary to prevent such contamination (Carrascosa *et al.*, 2015a).

The short shelf life of traditional mozzarella cheese has been attributed to microbiological spoilage. This spoilage is often caused by growth of coliforms, *Pseudomonas* spp. and/or by psychrotrophic bacteria that grow on cheese surfaces, which mostly come from the water used during manufacturing (Cantoni *et al.*, 2003a; Cantoni *et al.*, 2003b; Parisi, 2003). Na2-EDTA and lysozyme have proven effective for inhibiting the growth of spoilage microorganisms, such as coliforms and *Pseudomonas* spp., without affecting lactic acid bacteria (Sinigaglia *et al.*, 2008).

Pseudomonas spp. strains can be identified by different methods, such as traditional microbiology, phenotypical testing, fluorescence spectroscopy (Leblanc and Dufour, 2002) and genotypic identification (16s rDNA Sequences or Ribotyping) (Martin *et al.*, 2011). Phenotypic test systems can potentially entail inherent problems since not all strains within a given species exhibit a particular characteristic (Brisse *et al.*, 2002). The same strain may yield different results upon repeated testing (Brisse *et al.*, 2002) and the corresponding databases may have certain limitations (Bosshard *et al.*, 2006). However, this method is becoming an alternative or a complementary method to the established phenotypical identification procedures, and is a widely accepted tool for molecular identification purposes (Patel, 2001).

3.3 Meat Spoilage by *Pseudomonas* spp.

The microbiological quality of meat depends on the physiological status of the animal upon slaughter, spread of contamination during slaughter and processing, and temperature and other storage and distribution conditions (Nychas *et al.*, 2008). In fact, some microorganisms originate from animal intestinal tracts and from the environment that animals have been in for some time before or during their slaughter (Koutsoumanis and Sofos, 2004). This is evident from studies conducted into the origin of contaminants that have associated work surfaces with presence of Enterobacteriaceae in meat.

Microorganisms that can colonize fresh meat depend heavily on meat characteristics, and on the way meat is processed and stored (Huis in't Veld, 1996). During low-temperature aerobic storage, several *Pseudomonas* species are isolated from spoilt meat (Stanbridge and Davies, 1998; Labadie, 1999; Liao, 2006; Ercolini *et al.*, 2007, 2010a) and *Pseudomonas fragi* is the most frequently found species, followed by *Pseudomonas lundensis* and *Pseudomonas fluorescens* (Erichsen and Molin, 1981; Banks and Board, 1983; Molin *et al.*, 1986; Dainty and Mackey, 1992).

Meat spoilage is not always evident, and consumers would agree that gross discoloration, strong off-odors, and development of slime would be the main qualitative criteria for rejecting meat. In general, spoilage is a subjective judgment made by consumers, which may be influenced by cultural and economic considerations and background, by an individual's sensory acuity, and by intensity of change (Nychas *et al.*, 2008).

Apart from physical damage, oxidation and color change, among other spoilage symptoms, are ascribable to undesired microorganism growth at unacceptable levels, and also to the activity of endogenous enzymes. In meat, microbial spoilage leads to the development of off-odors, and often to slime formation, which make the product undesirable for human consumption (Jackson *et al.*, 1997).

Pseudomonas spoils a wide range of foodstuff, and meat (from fresh meat to meat products) is no exception. Unlike other environments, the meat ecosystem has been rarely studied at the strain level. Therefore, how homogeneous or diverse the biotypes of the same species often remains unknown. *P. fragi* is recognized as being one of the principal agents of meat spoilage (Labadie, 1999), and is frequently isolated from fresh and spoiled meat products (Ercolini *et al.*, 2007, 2009; Arnaut-Rollier *et al.*, 1999b). The fresh meat environment is a particularly suitable substrate for *P. fragi* growth. Beyond the nutritional value of meat components, this psychrotrophic species is favored by the strict chill chain applied to fresh meat production, which goes from slaughtering through to portioning, to distribution.

Organoleptic spoilage development is related to the microbial consumption of meat nutrients, such as sugars and free amino acids, and to the release of undesired volatile metabolites. Therefore, deterioration depends on contaminating microbial populations and storage conditions. The microbial populations associated with the meat environment are known to belong to pseudomonas and other microorganisms like *Enterobacteriaceae*, lactic acid bacteria, and *Brochothrix thermosphacta* (Borch *et al.*, 1996; Labadie, 1999; Ercolini *et al.*, 2006).

Among pseudomonads, *P. fragi* is by far one of the most threatening species for meat spoilage. The study of its diversity, ecology and spoilage behavior in meat at low temperatures is of utmost importance for us to understand its actual role in the meat environment as a first step toward a well thought out control of spoilage dynamics. Currently, available studies

Table 3.1 Common Defects in Meat Products and Causal Bacteria.

Defect	Meat product	Bacteria
Slime	Meats	*Pseudomonas, Lactobacillus, Enterococcus, Weissella, Brochothrix*
H_2O_2 greening	Meats	*Weisella, Leuconostoc, Enterococcus, Lactobacillus*
H_2S greening	Vacuum-packaged meats	*Shewanella*
H_2S production	Cured meats	*Vibrio, Enterobacteriaceae*
Sulfide odor	Vacuum-packaged meats	*Clostridium, Hafnia*
Cabbage odor	Bacon	*Providencia*
Putrefaction	Ham	*Enterobacteriaceae, Proteus*
Bone taint	Whole meats	*Clostridium, Enterococcus*
Souring	Ham	Lactic acid bacteria, *Enterococcus, Micrococcus, Bacillus, Clostridium*

Based on Nychas *et al.*, 2007; Skandamis and Nychas, 2002; Nychas *et al.*, 2006.

on the development of microbial populations in meat are either based merely on viable counts or focus on one species or a few, and their occurrence and dynamics.

Information on the diversity- and spoilage-related activities of the *Pseudomonas* populations that usually develop in meat is completely lacking, which include a considerable number of strains of the same species. Do different strains of the same species display the same spoilage behavior while they grow in meat? How heterogeneously can a population of *P. fragi* biotypes behave in a meat environment? In addition, better knowledge on spoilage development by certain species needs to be investigated in situ.

A vast number of studies in meat microbiology have established that spoilage is caused only by the fraction of the initial microbial association that dominates (Nychas *et al.*, 2007). Although similar genera have been reported to dominate in this fraction in the literature, findings at the species level are diverse (Nychas *et al.*, 2007).

A consortium of bacteria, commonly dominated by *Pseudomonas* spp., is mostly responsible for spoilage of meat stored aerobically and at various temperatures (1 °C to 25 °C) (Koutsoumanis *et al.*, 2006; Stanbridge and Davis, 1998). It has been well-established that in aerobic storage, three species of *Pseudomonas* (*P. fragi, P. fluorescens* and *P. lundensis*) are the most important. The pseudomonads population has been attributed to slime and off-odor formation (Table 3.1).

However in practice, both these characteristics become evident when pseudomonas have exhausted the glucose and lactate present in meat, and begin to metabolize nitrogenous compounds such as amino acids (Table 3.2).

Glucose in raw meat is the first substrate to be utilized preferably by most bacteria that develop in meat during chill storage under any packaging conditions (Gill, 1982; Nychas *et al.*, 1988; Nychas and Arkoudelos, 1990; Borch and Agerhem, 1992; Drosinos and Board, 1995a), and has been found to be the precursor of many off-odors that occur during meat storage such as: acetate, acetoin, diacetyl, acetic acid, iso-butyric acid, iso-valeric acid, 2-methylbutyric acid, 3methylbutanol, 2 methylpropanol, and ethanol (Nychas *et al.*, 1998; Tsigarida *et al.*, 2003; Nychas *et al.*, 2007). Changes in glucose

Table 3.2 Order (1 = first) of Substrate Utilization During Growth of Major Muscle Spoilage Bacteria. (A: *Pseudomonas* spp; B: *Shewanella putrefaciens*; C: *Brochothrix thermosphacta*; D: *Enterobacter* spp.; E: Lactic acid bacteria.)

Substrate	Aerobic					Anaerobic				
	A	B	C	D	E	A	B	C	D	E
Glucose/glucose-6-Ph	1	1	1	1	1	1	1	1	1	1
Lactate	2	2		2						
Pyruvate	3	3				2[c]				
Gluconate(Gl-6-Ph)	4	4				2[c]				
Proprionate		5								
Formate							1[c]			
Ethanol		6								
Acetate		7				2[c]				
Amino acids	5	8	2	3		2[c]	1[c]		2	2
Ribose			3							
Glycerol			4							

Modified from Nychas *et al.*, 2007.

and lactate, and their products, could be used to describe or predict the degree of spoilage (Nychas *et al.*, 1988; Boers *et al.*, 1994; Seymour *et al.*, 1994). In fact, it has been demonstrated that their concentration can affect the type and the rate of spoilage, and they seem to be the principal precursors of the microbial metabolites responsible for the olfactory detection of spoilage (Koutsoumanis and Nychas, 2000; Nychas *et al.*, 1998; Tsigarida and Nychas, 2001; Skandamis and Nychas, 2002). This is particularly evident in meat that is stored under aerobic conditions where *Pseudomonas* are major spoilage microorganisms.

Pseudomonas sequentially catabolize D-glucose and L and D-lactic acid, along with oxidation of glucose and glucose-6-P, via the extracellular pathway, which leads to transient D-gluconate accumulation and to higher gluconate-6-P concentrations (Nychas and Arkoudelos, 1990). Subsequently, *Pseudomonas* spp. utilize lactate, pyruvate, and gluconate under aerobic conditions, and pyruvate, gluconate, acetate, and amino acids under anaerobic conditions (Nychas *et al.*, 2007, 2008). When glucose is depleted, *Pseudomonas* spp. can attack amino acids, with the consequent production of volatiles, malodorous sulphides, esters, and amines (Dainty *et al.*, 1985; Edwards *et al.*, 1987; Samelis, 2006).

We have to also bear in mind the key role that *Pseudomonas* spp. play in the production of volatile compounds, such alcohols, aldehydes, esters and sulphur compounds. *Pseudomonas* spp., with *Carnobacterium* spp., are able to produce alcohols during refrigerated meat storage, and the commonest alcohols are 3 methyl-1-butanol, 1-octen-3-ol, 2-ethyl-1hexanol, 2,3-butanediol, butanol, 1-heptanol, 1-hexanol, and 3phenoxy-1-propanol.

Aldehydes are reported as particularly important components in imparting fatty flavor; short-chain aldehydes tend to be rather sharp or acid, and fattiness increases with chain length and degree of unsaturation (Insausti *et al.*, 2002; Soncin *et al.*, 2007). Aldehydes can derive from the hydrolysis of triglycerides and fatty acid metabolism, such as b-oxidation of unsaturated fatty acids or lipid auto-oxidation. Aldehydes can also arise from amino acids by transamination, which leads to an intermediate imide (Montel *et al.*, 1998; Moretti *et al.*, 2004). The aldehydes associated with spoilage microorganisms during the storage of fresh meat in air, VP and MAP, are reported in Table 3.2. *Pseudomonas* spp., *Carnobacterium* spp. and *Enterobacteriaceae* are microorganisms capable of producing the largest number of aldehydes. The commonest aldehydes found in naturally spoilt meat and in inoculated model meat systems are hexanal, nonanal, heptanal, benzaldehyde, and 3-methylbutanal (Table 3.2). These aldehydes are important contributors to meat flavor profiles.

A range of esters is produced by spoilage bacteria during meat storage under different conditions. These are among the components that dominate in inoculated meat stored aerobically, and their production is related to presence of *P. fragi*, addressed as the major ester producer in meat (Nychas *et al.*, 1988; Drosinos and Board, 1994; Ercolini *et al.*, 2009, 2010b).

The major esters found in naturally spoiled meat and inoculated model meat systems are ethyl acetate, ethyl butanoate, ethyl-3-methylbutanoate, ethyloctanoate, ethyl hexenoate, and ethyl decanoate. As far as sulphur compounds are concerned, the commonest sulphur compound in meat is dimethylsulphide, followed by dimethyldisulphide, dimethyltrisulphide, and methyl thioacetate, which are associated mostly with presence of *Pseudomonas* spp.

Despite us having focused mainly on *Pseudomonas*, we ought not to forget that food spoilage is the result of different kinds of bacteria that interact with the substrate and also interact (Gram *et al.*, 2002). These interactions may be classified according to their effects being detrimental or beneficial (Fredrickson, 1977).

3.4 Fish Spoilage by *Pseudomonas* spp.

It is well-known that fish spoilage is primarily due to autolysis, bacterial growth, and metabolism, which result in the formation of off-flavor compounds and chemical oxidation of lipids. Of all these reasons, microbiological activity is by far the most important factor to influence fish quality (Aberoumand, 2010; Ahne *et al.*, 1982; Chapman and Hall, 1995). However, not all microorganisms in seafood are equally important for changes in quality. Fish feeding habits, geographical location, season, sea temperature, type of fish, place where fish were harvested, and storage conditions, including the temperature and composition of packaging atmospheres, all determine the spoilage domains of specific spoilage organisms (SSO) (Comai *et al.*, 1982; Drosinos *et al.*, 1995b). However, only *Shewanella putrefaciens* is the SSO of the marine cold-water fish stored in ice, whereas *Photobacterium phosphoreum* is the SSO of the fish stored under modified-atmosphere conditions. *Pseudomonas* spp. and *Shewanella* spp. have been reported as the SSO in the fish obtained from Mediterranean Sea temperate waters, and stored in ice aerobically (Herald and Zottola, 1989; Curiale, 1996; Carrascosa *et al.*, 2014; Carrascosa *et al.*, 2015b; Carrascosa *et al.*, 2016).

Nevertheless, it should be noted that the determination of microbial growth and fish shelf life by traditional microbiological challenge tests is expensive and time-consuming (Bruckner *et al.*, 2013). One alternative is the predictive microbiology concept, which uses mathematical models to predict microbial growth and to, thus, estimate shelf life (McMeekin *et al.*, 1993; Whiting, 1995). Predictive microbiology involves the development of mathematical models to help describe the effect of the most important environmental factors on the biology of microorganisms in foods such as fish SSO (Ross *et al.*, 2000). The majority of these models are based on experimental data that derive from laboratory media, and information available on the assessment of individual species' variability in foods is limited (François *et al.*, 2006; Manios *et al.*, 2011). Models have immediate practical applications to improve microbial food safety and quality, and can also provide quantitative data on the microbial ecology of foods (Ross *et al.*, 2000). The implementation of these predictive models contributes to the improved control of food safety and spoilage by quantifying the effect of storage and distribution on microbial proliferation via the hazard analysis and critical control points (HACCP) system (Van Impe *et al.*, 2013).

Within the *Pseudomonas* species that cause the most economically serious diseases in marine cultures, we find *P. chlororaphis*, *P. anguilliseptica*, *P. fluorescens*, *P. putida*, and *P. plecoglossicida* (Toranzo *et al.*, 2005), and *P. anguilliseptica* is considered the most significant pathogen for cultured fish (Austin and Austin, 1999; Toranzo *et al.*, 2005, Toranzo and Barja, 1993).

P. anguilliseptica was originally described in 1972 as the aetiological agent of "Sekiten-bio," or "red spot disease," which caused very high mortality in pond-cultured Japanese eels in Japan (Wakabayashi and Egusa, 1972). Since then, this bacterium has been recorded in European eels reared in Taiwan, Scotland, and Denmark (Kuo and Kou, 1978; Stewart *et al.*, 1983). The pathogen has been subsequently isolated from other fish species, such as black sea bream and ayu in Japan (Nakai *et al.*, 1985), salmonids in Finland (Wiklund and Bylund, 1990), wild herrings (*Clupea harengus membras*) in the Baltic sea (Lönnström *et al.*, 1994), and has been considered the responsible agent of the "winter disease syndrome" since 1995, which is characteristic of the gilthead sea bream cultured in the Mediterranean region (Berthe *et al.*, 1995; Doménech *et al.*, 1997). *P. anguilliseptica* has also been recovered recently as an emerging pathogen of turbot and black spot sea bream (*Pagellus bogaraveo*) cultured in Spain (López-Romalde *et al.*, 2003a,c).

Disease occurs at low temperature (below 16 °C) in winter months. The main clinical signs of fish affected by this septicaemia are abdominal distension and hemorrhagic petechia on skin and in internal organs, but lesions in eels are always more severe than those observed in gilthead sea bream.

P. anguilliseptica grows very slowly and weakly on conventional media, but shows enhanced growth on blood agar. *P. anguilliseptica* seems to be a biochemically homogeneous pathogen, regardless of the isolation source (Berthe *et al.*, 1995; Doménech *et al.*, 1997; López-Romalde *et al.*, 2003a). However, the study of serological characteristics has indicated the existence of two major O serotypes related to the fish host: one is characteristic of eel isolates and the other is typical of gilthead sea bream, turbot, and black spot sea bream isolates (López-Romalde *et al.*, 2003b,c). Genetic characterization studies, conducted by random amplified polymorphic DNA (RAPD) techniques, have revealed the presence of two genetic groups, which were coincident with the two serological groups (López-Romalde *et al.*, 2003a). For eels, Japanese workers have described that virulence of strains is related to the presence of a capsular (K) antigen, which confers resistance to the bactericidal action of fish serum (Nakai, 1985). This information is useful for developing an adequate vaccine against this disease.

Two (polymerase chain reaction) PCR protocols, based on the amplification of the 16S rRNA gene, have been described for the rapid identification of *P. anguilliseptica* (Blanco *et al.*, 2002; Romalde *et al.*, 2004). However, only one (Romalde *et al.*, 2004) has shown sufficient sensitivity to directly detect the pathogen in fish tissues, and can hence become a powerful tool for diagnosing fish pseudomonadiasis under field conditions.

In the last decade, some research efforts have led to the development of aqueous and non-mineral oil-adjuvanted bacterins (including both the major serotypes detected), which have proved effective in experimental trials in gilthead sea bream and turbot (Romalde *et al.*, 2005).

3.5 Water Contamination by *Pseudomonas* spp.

When we consider the prototrophic character, metabolic versatility the plasticity of the genome and the ability to cope with different forms (physical, chemical, and antibacterial compounds) of stress (Palleroni, 2010) of the genus *Pseudomonas* to be remarkable characteristics, we can state that such characteristics are the driving forces for adaptability, which justifies the presence of pseudomonads in all major natural terrestrial and environments (Mena and Gerba, 2009; Palleroni, 2010). In particular, drinking water is considered a relevant habitat of *Pseudomonas* spp. (Reitler and Seligmann, 1957; Manaia *et al.*, 1990; Rusin *et al.*, 1997; Jayasekara *et al.*, 1998; Palleroni, 2010; Vaz-Moreira *et al.*, 2012). Although its presence in such habitats may result from aquifer colonization, the ability to grow in final drinking water is also assumed to occur. This expectation led Ribas *et al.* (2000) to propose using *Pseudomonas* spp. as an indicator of potential regrowth in water distribution systems.

Nevertheless, the widespread distribution of *Pseudomonas* spp. may pose some public health concerns. Indeed, this ubiquitous genus includes species that are considered opportunistic pathogens that can colonize animals and humans, mostly due to inadvertent contamination via water, for instance (Mena and Gerba, 2009). Regarding opportunistic pathogens, the species *P. aeruginosa* is by far the most extensively studied in this genus. Members of this species can cause severe infections in either healthy or immunocompromised individuals (Mena and Gerba, 2009).

If we bear in mind that water is the natural reservoir for *P. aeruginosa* (Molina *et al.*, 1991), some authors have considered the presence of *P. aeruginosa* to be an indicator of contamination from surface runoff, domestic and agricultural effluents, or human fecal matter (Warburton *et al.*, 1994). The major source of *P. aeruginosa* in surface water is thought, by some authors, to be domestic sewage (de Vicente *et al.*, 1988), and *P. aeruginosa* is found in 90% of sewage samples (Geldreich, 1996). Its concentration in surface waters that receive waste and storm water discharge ranges from 1 to 10,000 cells/100 mL; in storm water, concentrations are 1,000–100,000 mL^{-1} (Geldreich, 1996). However, the intestinal carriage rate for *P. aeruginosa* in humans is low, which suggests that its presence in water does not necessarily result from sewage contamination. Other sources of the organism are thought to be water leaching from agricultural soils, barnyard drainage, and urban run-off.

Although considered by some to be environmentally ubiquitous, others believe the presence of *P. aeruginosa* is linked directly to human activities (Relai and Rosati, 1994). These latter authors have reported up to 2,400 P. aeruginosa/100 mL in surface waters heavily contaminated by fecal discharge while it was absent in less polluted water. They have also found that *P. aeruginosa* can always be isolated when sampled from distribution plants (reservoirs and pipes), but not from wells or springs. Isolates from environmental waters tend to be less resistant to antibiotics than those from clinical specimens (Relai and Rosati, 1994).

P. aeruginosa is considered the most important pseudomonad in drinking water (Geldreich, 1996; Stoler *et al.*, 2015). Allen and Geldreich (1975) found that 3% of drinking water samples contained *P. aeruginosa*, with counts ranging between 1 and 2,300 organisms per millilitre. Geldreich (1996) later expressed concern for *P. aeruginosa* being present in the water supply because of its potential to increase exposure to senior citizens, AIDS patients, and infants. He believed ingestion was the route responsible for transmitting this organism. He suggested that amplification may occur in static water areas and in sediment-laden water pipes during warm-water periods.

Gambassini *et al.* (1990) found *P. aeruginosa* in only 1 of 66 drinking water samples; single detection emerged from a sample at the terminal end of a distribution system. Carbon filter point-of-use devices may also become colonized with *P. aeruginosa*. Therefore, Geldreich (1996) has suggested that these devices should not be used on untreated water supplies of questionable quality, unless they are fitted with an associated microbial barrier.

A survey conducted in southern Greece has shown that *P. aeruginosa* occurs in 9% of tap water samples at a mean concentration of 7 CFU/100 mL, and in 18.8% of bottled

Table 3.3 Occurrence of *P. aeruginosa* in Drinking Water.

Type of water	Percentage of positive samples or concentration found	Reference
Tap water	<2.0	Levesque *et al.*, 1994
Tap water	9, 7 CFU/100 mL	Papapetropoulou *et al.*, 1994
Tap water	<10 (none detected)	Payment *et al.*, 1988
Tap water	(0–0.25 CFU/L)	Payment *et al.*, 1988
Tap water	0.01–0.6	Clark *et al.*, 1982
Tap water	3, 1–2,300 mL^{-1}	Allen and Geldreich, 1975
Tap water	24.1, 0.01–0.38 mL^{-1}	Reitler and Seligmann, 1957
Treated drinking water	1.5	Gambassini *et al.*, 1990
Mineral water	8.3	Manaia *et al.*, 1990
Water cooler water	<2.0	Levesque *et al.*, 1994
Water vending machine	16.1, 0.002–0.16 mL^{-1}	Chaidez *et al.*, 1999
Bottled water	18.8, 1,000 CFU/100 mL	Papapetropoulou *et al.*, 1994
Bottled water	0.5	Warburton *et al.*, 1992
Bottled water	<0.4 (0 of 256)	Duquino and Rosenberg, 1986

water samples at a mean concentration of 1,000 CFU/100 mL (Papapetropoulou *et al.*, 1994). These authors have also noted that *P. aeruginosa* isolates are resistant to certain antibiotics, such as tetracycline, streptomycin, and naladixic acid. A synopsis of *P. aeruginosa* occurrence in drinking water is shown in Table 3.3.

It would appear that major *P. aeruginosa* exposure via tap water results from the colonization of tap fixtures (i.e., faucets, showerheads, sinks) and its introduction into water when these devices are used. In one study conducted at a children's hospital in Germany, 81% of all sinks were contaminated with *P. aeruginosa* (Doring *et al.*, 1991). Of the 484 tap water samples collected from 39 faucets in a medical intensive care unit, more than 11% contained *P. aeruginosa* (Rogues *et al.*, 2007). Other exposures via tap water may result from the growth of *P. aeruginosa* in vending machines (Schillinger and Du Vall Knorr, 2004), drinking water dispensers (coolers) (Baumgartner and Grand, 2006), and water softeners (Hambsch *et al.*, 2004).

In order to remove *P. aeruginosa* from drinking water, one of the following treatment methods is commonly applied: chlorine, chloramines, ozone, iodine, and UV disinfection. Laboratory testing has not shown *P. aeruginosa* to be unusually resistant to chlorine, chloramine, ozone, or iodine. When iodine-resistant *P. aeruginosa*, isolated from a hospital povidone-iodine solution, was tested against iodine, it was found to be more sensitive than several other tested bacteria: *P. fluorescens*, *P. cepacia*, *Bacillus* spp., and staphylococci (Pyle and McFeters, 1989). Determining the sensitivity of *P. aeruginosa* to disinfectants is complicated by the fact that the sensitivity of cells is affected by temperature, previous growth conditions, and the growth phase at the time of testing (Carson *et al.*, 1972; Pyle and McFeters, 1989). Cells are more rapidly killed when incubated at higher temperatures than at lower ones. They are also more sensitive to disinfection after being subcultured on laboratory media when tested in a natural water environment. The die-off rate may be higher when log phase populations are tested than when stationary phase cells are used.

In a study by Aspinall and Graham (1989), the levels of free chlorine >1 ppm were sufficient to kill an initial count of 3.8×10^4 CFU/mL of *P. aeruginosa* after a 10-second contact. However, these bacteria were isolated from a whirlpool and then grown in laboratory media before testing (Aspinall and Graham, 1989). This required 1–3 h to obtain a 99.9% kill in swimming pools with 0.5 mg/L chlorine at 25 °C (Fitzgerald and Der Vartanian, 1969). These authors also demonstrated that natural waters required 0.5 mg of chlorine per liter to kill 99.9% of *P. aeruginosa* cells in 1 h at 25 °C.

Price and Ahearn (1988) demonstrated that *P. aeruginosa* cells regrew when the chlorine concentration decreased in whirlpools. *P. aeruginosa* has been recovered from whirlpool water that contained 2 mg/L free chlorine. However, Highsmith *et al.* (1985) showed that *P. aeruginosa* test strains from whirlpool water were just as sensitive to low chlorine concentrations (0.2–0.6 ppm) as control strains. This implied that chlorine resistance *per se* does not appear to be a characteristic of certain *P. aeruginosa* strains. These bacteria can excrete glycocalyx material, which is capable of shielding them from chlorine. There is no evidence available o indicate that they are innately resistant to chlorine.

Vess *et al.* (1993) allowed *P. aeruginosa* to colonize PVC pipe material for 8 weeks. The biofilm of pure *P. aeruginosa* was then exposed to 10-15 ppm free chlorine for 7 days. During this time samples were taken at intervals to test for bacterial survival. After 7 days, the pipe was refilled with sterile distilled water and sampled at 7-day intervals for 6 weeks for recovery and regrowth. Although no *P. aeruginosa* was detected at 7 days after chlorine treatment, large numbers were recovered after 14 days. Scanning electron microscope pictures of the PVC wall surface showed cells embedded in heavy deposits of extracellular material, which probably accounted for survival in the presence of chlorine.

The sensitivity of *P. aeruginosa* to chloramine has been compared with *Klebsiella pneumoniae, Salmonella typhimurium,* and *E. coli* (Ward *et al.*, 1984). *P. aeruginosa* was just as sensitive to chloramine, or even more so, as were the other bacteria. The time for 99.9% inactivation at 1 mg/L chloramine ranged from 4 to 9.5 min, with more rapid activation at a pH of 6 than at one of 8. The time for 99.9% inactivation at 3 mg/L ranged from 1.5 to 4.5 min with the same influence of pH.

When *P. aeruginosa* was tested against *N*-chloramine (3-chloro-4,4-dimethyl-2-oxazolidnone), it was more sensitive than *Staphylococcus aureus* or *Shigella boydii* (Williams *et al.*, 1985). At 2.5–5 ppm chlorine, no *P. aeruginosa* was recovered after 5–10 min at pH 7 and 22 °C. At the same temperature, but at a pH of 4.5, *P. aeruginosa* was not detectable after exposure to 2.5 ppm chlorine for 5–10 min, or after exposure to 5.0 ppm for 2–5 min.

P. aeruginosa has been shown to be no more resistant to ozonation than *Salmonella* or *Yersinia*. *P. aeruginosa* underwent 6-log reduction in deionized water within 1 min using 0.64–0.188 ppm ozone (Restaino *et al.*, 1995). According to Box *et al.* (1982), *P. aeruginosa* also appeared to be no more resistant to UV light than other bacteria. *Bacillus subtilis* was most resistant to UV light at wavelengths from 250 to 270 nm, followed in order of decreasing resistance by *E. coli*, *P. aeruginosa*, and *S. aureus*. Fernandez and Pizarro (1996) also reported that *P. aeruginosa* was not resistant to UV light; in fact, they reported that it was more sensitive than *E. coli*. However, a comparison of the sensitivity of *P. aeruginosa* to several other bacteria has been described by Wolfe (1990), who found that *P. aeruginosa* is more resistant than the other tested bacteria (Table 3.4).

In recent years, copper and silver ions have been used to control *Legionella* and other water-based pathogens in water (Silvestry-Rodriquez *et al.*, 2007a). *P. aeruginosa* appears susceptible to silver ions, but mutated forms of the organism may be more silver-tolerant.

Table 3.4 Approximate Doses of UV Light Needed to Inactivate 90% of Selected Bacteria.

Bacterium	Dose (μW-s)/cm^2
E. coli	3,000
S. typhi	2,500
P. aeruginosa	5,500
S. enteritidis	4,000
S. dysenteriae	2,200
S. paradysenteriae	1,700
S. flexneri	1,700
S. sonnei	3,000
S. aureus	4,500
L. pneumophila	380
Vibrio cholerae	3,400

Silver requires longer contact to disinfect than oxidizing disinfectants do, but may enhance the action of other disinfectants, where preventing colonization is important (e.g., faucets). Silver ions, at a concentration of 80 mg/L, have been reported to produce a reduction of 99.999% in *P. aeruginosa* after 6 h of contact (Huang *et al.*, 2008). The advantage of silver is that it is effective in the presence of organic loads, which render chlorine ineffective (Silvestry-Rodriquez *et al.*, 2007b).

3.6 *Pseudomonas* spp. in Fruit and Vegetables

The final quality of fresh or processed foods of plant origin depends on several factors: preharvest aspects and conditions that occur during processing, collection methods, and postharvest handling systems (Kader and Barrett, 1996). Postharvest handling systems have changed in the last decade with grade IV fruit and vegetables, which are processed and made ready for consumption. The sensory, nutritional, and organoleptic quality of fresh products begins to decline immediately after collection owing to unbalanced metabolism products and microbial growth. Microbial contamination of fruit and vegetables during growth, processing, and postharvest is a serious problem in agricultural sectors. The relative importance and individual factors related to transpiration, senescence and fruit ripening, as well as external factors (temperature, packaging, etc.) and microbiological contamination, are useful life-limiting factors for each product (Rahman, 2002).

Fresh fruit and vegetables always contain a complex microbiota alterative from being in continuous intense contact with various types of microorganisms during growth and postharvest handling. The high acidity of many fruits (pH <4.6) limits the disruption caused by molds, yeasts, and lactic acid-tolerant bacteria. Vegetables generally have a pH of between 6.0 and 7.0, and lack intrinsic protection. Healthy microbial spoilage of products, without injury, can only be due to microorganisms capable of penetrating skin, which requires the presence of specific enzymatic systems. In vegetables, gram-negative pectinolytic bacteria of genera *Pseudomonas* and *Enterobacter* are frequently involved in spoilage. In vegetables packaged in protective atmospheres, CO_2 effectively inhibits the growth of aerobic microorganisms, such as *Pseudomonas*, *Micrococcus*, and *Bacillus*, and most derive from spoilage microorganisms of aerobic fruit and vegetable products. Most fruits have an excessively low pH to allow pathogenic bacteria to grow. To this inhibitory effect, cold fruit preservation should be added (Rahman, 2002).

Outbreaks of human infections associated with consumption of raw fruits and vegetables have occurred and their frequency has increased in the past decade. The factors that contribute to this increase may include changes in agronomic and processing practices, an increase in per capita consumption of raw or minimally processed fruits and vegetables, increased international trade and distribution, and more immuno-compromised consumers. A general lack of efficacy of sanitizers in removing or killing pathogens in raw fruits and vegetables is lacking, which has been partly attributed to their inaccessibility to locations within structures and tissues that may harbor pathogens (Beuchat, 2002). The pathogen *Pseudomonas* should not be thought to cause outbreaks, but it should be seen as one of the most important spoilage microorganisms that can affect fruits and vegetables.

Before new packaging techniques dominated the vegetable market, bacterial soft rot was already known as the leading cause of decaying perishable vegetables in retail and wholesale markets (Smith *et al.*, 1966). Although pectolytic *Erwinia* species are commonly assumed to be the chief cause of the problem, extensive efforts have not yet been made to identify and to compare the relative importance of pectolytic bacteria of other genera (*Pseudomonas*, *Bacillus*, *Clostridium*, and *Flavobacterium*) (Lund, 1983). Pectolytic fluorescent pseudomonads have been previously reported to cause postharvest rot in celery (Surico and Lacobellis, 1978), chicory (Cuppels and Kelman, 1980), lettuce (Beraha and Kivolek, 1975), Chinese chard (Burton, 1971), cabbage (Brocklehurst and Lund, 1981), and potato (Garrard, 1945). Although frequently named *P. marginalis* (Misaghi and Donndelinger, 1983), organisms probably consist in a nutritionally and physiologically heterogeneous group. Sands and Hankin (1975) examined 140 strains of pectolytic fluorescent pseudomonads from plant tissues, soil, and sewage. They found that the strains that caused maceration of potato disks fell within the phenotype from *P. fluorescens* to *P. putida*. Wang and Kelman (1982) detected pectolytic fluorescent pseudomonads with phenotypes that were representative of various biovars of *P. fluorescens* in soil at different depths. It is presently unknown whether pectolytic fluorescent pseudomonads associated with rot of fresh vegetables on markets are as diverse as those detected from other sources. Pectolytic fluorescent pseudomonads may not be as destructive as soft-rotting erwinias in the field, but are undoubtedly a threat to fresh produce stored at low temperature. However, during cold storage of grade IV leafy vegetables pectinolytic strains are responsible for bacterial soft rot (Liao and Wells, 1986).

In reference, microbial decay of minimally processed lettuce is largely due to the growth of microorganisms that originate from preharvest environments (Boyette *et al.*, 1993). Gram-negative bacteria and smaller numbers of yeasts make up the bulk of the initial microbial flora (Carlin *et al.*, 1995). Numbers of mesophilic aerobes and psychrotrophic microorganisms tend to predominate in stored shredded lettuce, whereas mold and lactic acid bacteria generally remain low (Marchetti *et al.*, 1992). Bacterial populations are often dominated by *P. fluorescens* (Zagory, 1999).

3.7 Biochemical and Molecular Techniques for *Pseudomonas* spp. Detection

There is no doubt about the role of microorganisms, specifically *Pseudomonas* spp., in food spoilage. For this reason, fast complete identification is very important.

In order to not go beyond the scope of this chapter or to dwell on things, this section presents some of the most recent studies conducted on this issue. Traditional microbiological techniques for bacterial identification, such as observation of growth patterns on selective and differentiative media, biochemical reactions and microscopy, often lack discriminatory power and are usually ineffective at establishing a causal relationship between the contamination of a finished product and the environmental source (Dogan and Boor, 2003). Several bacteriological media are available for selectively detecting gram-negative bacteria in dairy products. Crystal violet tetrazolium agar (CVTA; Frank and Yousef, 2004), MacConkey agar (MAC), violet red bile agar (VRBA; Frank and Yousef, 2004), and 3 M Coliform Petrifilm (CPETRI, 3 M, St. Paul, MN) are examples of gram-negative selective media.

For vegetables, Xu *et al.* (2015) developed an immunochromatographic strip for the rapid detection of *Pseudomonas syringae.* Rapid methods for detection in bottled water have been developed by Zhang *et al.* (2013), who have also developed rapid detection methods for bottled water. Meat and milk can also be analysed for Pseudomonas by methods that range from colorimetric assays (Gutierrez *et al.*, 1997a) to infrared spectroscopy (Cámara-Martos *et al.*, 2015) and monoclonal antibodies (Gutierrez *et al.*, 1997b).

In the particular case of meat, thanks to the evolution of bioinformatics (e.g., principal components analyses [PCA], hierarchical cluster analyses [HCA] and artificial neural networks), new approaches have been applied in parallel with mathematical models to evaluate meat spoilage (Nychas and Skandamis, 2005; Nychas *et al.*, 2007). These approaches can rapidly provide information on the contribution of ephemeral spoilage organisms or to categorize meat in spoilage terms.

Several authors, such as Martin *et al.* (2011) and Carrascosa *et al.* (2015a), have employed several phenotypic and genotypic methods to identify different species of *Pseudomonas* in cheese after isolating them in specific culture media for *Pseudomonas*. Isolates can be taken from storage, streaked on BHI agar and incubated at 32 °C, unless otherwise noted. Afterward, the colonies that are likely to belong to some of the species of gender *Pseudomonas* can be identified by the methods indicated below:

3.7.1 16S rDNA Sequencing

Sequencing the 3′ end of a 616-nucleotide 16S rDNA fragment, as described previously by Huck *et al.* (2007), is used to characterize each isolate. 16S sequence data can be employed to characterize isolates to the genus and species levels with the Ribosomal Database Project (RDP) database (http://rdp.cme.msu.edu/index.jsp).

3.7.2 Biolog

The Biolog GN2 MicroPlate system (Biolog Inc., Hayward, CA) can be used following the manufacturer's Gram-Negative Non-Enteric (GN-NENT) protocol. Cultures can be streaked onto BHI agar and incubated for 18 h at 32 °C. The uniform suspension (transmittance level of 52%) of each isolate can be prepared in gram-negative/gram-positive-inoculating fluid (GN/GP-IF). Suspensions can be inoculated into MicroPlates, incubated at 32 °C for 18 h, and read by the MicroLog 3 system.

3.7.3 Analytical Profile Index

The Analytical Profile Index (API) 20NE test kit (BioMerieux Vitek Inc., Hazelwood, MO) can be used following the manufacturer's instructions. Strips should be examined after 24 and 48 h. Identification of isolates is performed according to the API 20NE identification instructions online (http://apiweb.biomerieux.com).

3.7.4 Pulsed Field Gel Electrophoresis

Pulsed field gel electrophoresis (PFGE) of isolates is performed using the standard Centers for Disease Control and Prevention (CDC) PulseNet protocol for enteric pathogens (Ribot *et al.*, 2006). Briefly, isolates are grown on BHI agar plates at 32 °C for 18 h. Bacterial cultures are embedded in 1% agarose plugs (SeaKem Gold agarose, Cambrex, Rockland, ME), lysed, washed, and digested with restriction enzyme *Xba*I for 2 h at 37 °C. Size separation of restricted DNA fragments is carried out for 20–22 h in 1% agarose gels with a contour clamped homogeneous electrical field Mapper XA (Bio-Rad Laboratories, Hercules, CA); voltage set at 6 V/cm with switch times of 4 s to 40.01 s. The *Xba*I-digested *Salmonella enterica* serotype Braenderup (CDCH9812) DNA is used as a reference size standard. Pattern images are captured with a Bio-Rad Gel Doc and the Multi Analyst software, version 1.1 (Bio-Rad Laboratories). The PFGE patterns can be analyzed and compared using the BioNumerics version 3.5 software (Applied Maths, Sint-Martens- Latem, Belgium).

3.7.5 Ribotyping

Automated ribotyping can be performed by the RiboPrinter Microbial Characterization System (DuPont Qualicon, Wilmington, DE), and by the procedure described by Bruce (1996) and reported by Wiedmann *et al.* (1997). Briefly, this automated process includes bacterial cell lysis and DNA cleavage using restriction enzyme *Eco*RI. Then DNA fragments are separated according to size by electrophoresis and are analysed by a modified Southern hybridization blotting technique. DNA is hybridized with a labeled rRNA operon probe derived from *Escherichia coli,* and the bands are detected using a chemluminescent substrate. An image, captured with a customized charge-coupled device camera, is electronically transferred to the system's computer. Each sample lane of data is normalized to a standard marker set, characterized and identified by similarity measurements to previously run isolates and reference patterns.

The results can be difficult to reproduce. Genotypic methods such as ribotyping offer enhanced discriminatory power (van der Vossen and Hofstra, 1996). Ribotyping has been demonstrated to rapidly and reproducibly type bacterial isolates to the genus, species, and strain levels (Bruce *et al.*, 1995; Bruce, 1996). Ribotyping is a useful discriminatory typing method for differentiating Pseudomonas spp. capable of cutting shelf lives of fluid milk products. Ribotyping reveals high genomic diversity among the Pseudomonas spp. isolated from these dairy plants (Dogan and Boor, 2003). However, for Martin *et al.* (2011) and Carrascosa *et al.* (2015), pulsed field gel electrophoresis is the only method applied capable of distinguishing between isolates that produce pigment and those that do not in cheese samples that contain *Pseudomonas fluorescent.*

3.8 Conclusions

Pseudomonas spp. do not constitute a serious risk to public health. However, the most pathogenic *P. aeruginosa* belongs to risk group II (WHO, Laboratory Biosafety Manual, 3rd edition, 2005), and is an opportunistic pathogen and one of the main bacteria that cause nosocomial infections in hospitals that affects immunocompromised persons. The most frequent infections involve the cornea, skin, urinary tract, and respiratory tract, but can occur at essentially all anatomical sites (Yahr and Parsek, 2006).

Notwithstanding, we should consider *Pseudomonas* an opportunistic microorganism that are resistant to treatment with antibiotics, have the remarkable ability to form biofilm on surfaces, and easily adapt to different environments to survive. While food is usually conveyed by water and they subsequently recontaminate food, we must not forget that they can infect animals, act as asymptomatic carriers and be excreted through milk or part of the microbiota of meat. Although *Pseudomonas* spp. are a major concern as to the proteolytic degradation of milk, other proteolytic organisms, such as *Achromobacter, Aeromonas, Flavobacterium,* and *Xanthomonas* spp., can be found in milk and can cause proteolytic degradation and milk coagulation. Their isolation can be simple and involves specific culture media. However, species identification can be more costly due to the phenotypic and genotypic similarity of the different species that make up the genus *Pseudomonas.*

References

Aberoumand, A. (2010). A Review on *Pseudomonas* in Marine Fish. *World Journal of Fish and Marine Sciences,* 2(4), pp. 291–296.

Ahne, W., Popp, W. and Hoffman, R. (1982). Pseudomonas fluorescens as a pathogen for tench. *Bulletin of the European Association of Fish Pathologists,* 2, pp. 56–57.

Allen, M.J. and Geldreich, E.E. (1975). Bacteriological criteria for groundwater quality. *Ground Water, 13,* pp. 45–52.

Amann, I., Ludwig, W. and Schleifer, H. (1995). Phylogenetic identification and *in situ* detection of individual microbial cells without cultivation. *Microbiology Review, 59,* pp. 143–169.

American Thoracic Society. (2005). Guidelines for the management of adults with hospital-acquired, ventilator-associated, and healthcare-associated pneumonia. *Am. J. Resp. Crit. Care., 171,* pp. 388–416.

Arnaut-Rollier, I., Vauterin, L., De Vos, P., Massart, L., Devriese, A., De Zutter, L. and Van Hoof, J. (1999a). A numerical taxonomic study of the *Pseudomonas* flora isolated from poultry meat. *J. Appl. Microbiol., 87,* 15–28.

Arnaut-Rollier, L., De Zutter, and J. Van Hoof. (1999b). Identities of the Pseudomonas spp. in flora from chilled chicken. *International Journal of Food Microbiology, 48,* pp. 87–96.

Arslan, S., Eyi, A. and Ozdemir, F. (2011). Spoilage potentials and antimicrobial resistance of *Pseudomonas spp.* isolated from cheeses. *Journal of Dairy Science, 94,* pp. 5851–5856.

Aspinall, S.T. and Graham, R. (1989). Two sources of contamination of a hydrotherapy pool by environmental organisms. *Journal of Hospital Infection, 14,* pp. 285–292.

Austin, B. and Austin, D.A. (1999). *Bacterial fish pathogens. diseases of farmed and wild fish.* United Kingdom: Springer-Praxis Publishing, Ltd.

Banks, J.G. and Board, R.G. (1983). The classification of pseudomonads and other obligately aerobic Gram-negative bacteria from British pork sausage and ingredients. *Systematic and Applied Microbiology, 4,* pp. 424–438.

Baruzzi, F., Lagonigro, R., Quintieri, L., Morea, M. and Caputo, L. (2012). Occurrence of non-lactic acid bacteria populations involved in protein hydrolysis of cold stored high moisture Mozzarella cheese. *Food Microbiology, 30,* pp. 37–44.

Baumgartner, A. and Grand, M. (2006). Bacteriological quality of drinking water from dispensers (coolers) and possible control measures. *Journal of Food Protection, 69,* pp. 3043–3046.

Bisbe, J., Gatell, J. and Puig, J. (1988). *Pseudomonas aeruginosa* bacteremia: univariate and multivariate analyses of factors influencing the prognosis. *Rev. Infect. Dis., 10,* pp. 629–635.

Blackburn, C. (2006). *Food spoilage microorganisms.* Woodhead Publishing in Food Science and Nutrition.

Blanco, M.M., Gibello, A., Vela, A.I., Moreno, M.A., Domínguez, L. and Fernández-Garayzábal, J.F. (2002). PCR detection and PFGE DNA macrorestriction analysis of clinical isolates of *Pseudomonas anguilliseptica* from winter disease outbreaks in sea bream *Sparus aurata. Diseases of Aquatic Organisms, 50,* pp. 19–27.

Beraha, L. and Kivolek, W.F. (1975). Prevalence and extent of eight market disorders of Western-grown head lettuce during 1973 and 1974 in the greater Chicago, Illinois Area. *Plant Dis. Rep., 59,* pp. 1001–1004.

Berthe, F.C.J., Michel, C. and Bernardet, J.F. (1995). Identification of *Pseudomonas anguilliseptica* isolated from several fish species in France. *Diseases of Aquatic Organisms, 21,* pp. 151–155.

Beuchat, L.R. (2002). Ecological factors influencing survival and growth of human pathogens on raw fruits and vegetables. *Microbes and Infection, 4*(4), pp. 413–423.

Bhunia, A.K. (2008). *Foodborne microbial Pathogens: mechanisms and pathogenesis.* West Lafayette, IN: Springer.

Boers, R.H., Dijkmann, K.E., and Wijngaards, G. (1994). Shelf-life of vacuum-package wild boar meat in relation to that of vacuum-packaged pork: relevance of intrinsic factors. *Meat Science, 37,* pp. 91–102.

Bonomo, A. and Szabo, D. (2006). Mechanisms of multidrug resistance in *Acinetobacter* species and *Pseudomonas aeruginosa. Clin. Infect. Dis., 43,* pp. 49–56.

Borch, E. and Agerhem, H. (1992). Chemical, microbiological and sensory changes during the anaerobic cold storage of beef inoculated with a homofermentative Lactobacillus sp. or a Leuconostoc sp. *International Journal of Food Microbiology, 15*, pp. 99–108.

Borch, E., Kant-Muermans, M. and Blixt, Y. (1996). Bacterial spoilage of meat and cured meat products. *International Journal of Food Microbiology, 33*, pp. 103–120.

Bosshard, P.P., Zbinden, R., Abels, S., Böddinghaus, B. Altwegg, M. and Böttger, E.C. (2006). 16S rRNA Gene Sequencing versus the API 20 NE System and the VITEK 2 ID-GNB card for identification of nonfermenting gram–negative bacteria in the clinical laboratory. *Journal of Clinical Microbiology, 44*(4), pp. 1359–1366.

Box, J.A., Sugden, J.K. and Younis, N.M.T. (1982). The use of ultraviolet light to sterilize water. *Pharmaceutica Acta Helvetiae, 57*, pp. 330–333.

Brant, W. (1975). Selective media for psychrotrophic spoilage bacteria of fresh poultry. *2nd Eur. Symp. Poult. Mear. Oosterbeek*, p. 31.

Boyette, M.D., Ritchie, D.F., Carballo, S.J., Blankenship, S.M. and Sanders, D.C. (1993). Chlorination and postharvest disease control. *HortTechnolgy, 3*, pp. 395–400.

Breidenstein, E., De la Fuente-Núnez, C. and Hancock, R. (2011). *Pseudomonas aeruginosa:* all roads lead to resistance. *Trends in Microbiology, 19*, pp. 419–426.

Brisse, S., Stefani, S., Verhoef, J., Van Belkum, A., Vandamme, P. and Goessens, W. (2002). Comparative evaluation of the BD Phoenix and VITEK 2 automated instruments for identification of isolates of the *Burkholderia cepacia* complex. *Journal of Clinical Microbiology, 435*(40), pp. 1743–1748.

Brocklehurst, T.F. and Lund, B.M. (1981). Properties of pseudomonads causing spoilage of vegetables stored at low temperature. *J. Appl. Bacteriol. 50*, pp. 259–266.

Bruce, J. (1996). Automated system rapidly identifies and characterizes microorganisms in food. *Food Technol., 50*, pp. 77–81.

Bruce, J.L., Hubner, R.J., Cole, E.M., McDowell, C.I. and Webster, J.A. (1995). Sets of *Eco*EcoRI fragments containing ribosomal RNA sequences are conserved among different strains of *Listeria monocytogenes. Proc. Natl. Acad. Sci. USA, 92*, pp. 5229–5233.

Bruckner, S., Albrecht, A., Petersen, B. and Kreyenschmidt, J. (2013). A predictive shelf life model as a tool for the improvement of quality management in pork and poultry chains. *Food Control, 29*, pp. 451–460.

Burton, C.L. (1971). Bacterial soft rot and black spot disease of Bok. Choy (Chinese chard). *Plant Dis. Rep., 55*, pp. 1037–1039.

Cámara-Martos, F., Lopes, J.A., Moreno-Rojas, R. and Pérez-Rodríguez, F. (2015). Detection and quantification of Escherichia coli and Pseudomonas aeruginosa in cow milk by near-infrared spectroscopy. *International Journal of Dairy Technology, 68*, pp. 357–365.

Cantoni, C., Stella, S., Ripamonti, B., and Marchese, R. (2001). Anomalous colouration of mozzarella cheese. *Industrie Alimentari, 40*(399), pp. 33–35.

Cantoni, C., Iacumin, L., and Comi, G. (2003a). Alterazione giallo-arancio di mozzarella. *Industrie Alimentari, 422*, pp. 134–136.

Cantoni, C., Stella, S., Cozzi, M., Iacumin, L., and Comi, G. (2003b). Colorazione blu di mozzarelle. *Industrie Alimentari, 428*, pp. 840–843.

Carey, N.R., Murphy S.C., Zadoks, R.N. and Boor, K.J. (2005). Shelf lives of pasteurized fluid milk products in New York State: A tenyear study. *Food Prot. Trends, 25*, pp. 102–113.

Carlin, F., Nguyen-the, C. and Silva, A.A.D. (1995). Factors a¡ecting the growth of Listeria monocyte genes on minimally processed fresh endive. *J. Appl. Bacteriol., 78*, pp. 636–646.

Carrascosa, C., Millán, R., Saavedra, P., Jaber, J.R., Montenegro, T., Raposo, A., Pérez, E. and Sanjuán, E. (2014). Predictive models for bacterial growth in sea bass (*Dicentrarchus labrax*) stored in ice. *International Journal of Food Science & Technology, 49*, pp. 354–363.

Carrascosa, C., Millán, R., Raduán Jaber, J., Lupiola, P., del Rosario-Quintana, C., Mauricio, C.

and Sanjuán, E. (2015a). Blue pigment in fresh cheese produced by *Pseudomonas fluorescens*. *Food Control, 54*, pp. 95–102.

Carrascosa, C., Millán, R., Saavedra, P., Jaber, J.R., Raposo, A., Pérez, E., Montenegro, T. and Sanjuán, E. (2015b). Microbiological evolution of gilthead sea bream (*Sparus aurata*) in Canary Islands during ice storage. *Journal of Food Science & Technology, 52*(3), pp. 1586–1593.

Carrascosa, C., Saavedra, P., Millán, R., Jaber, J.R., Montenegro, T., Raposo, A. and Sanjuán, E. (2016). Microbial Growth Models in Gilthead Sea Bream (Sparus aurata) Stored in Ice. *Journal of Aquatic Food Product Technology, 25*(3), pp. 307–322.

Carson, L.A., Favero, M.S., Bond, W.W. and Petersen, N.J. (1972). Factors affecting comparative resistance of naturally occurring and subcultured *Pseudomonas aeruginosa* to disinfectants. *Applied Microbiology, 23*, pp. 863–869.

Chaidez, C., Rusia, P., Naranjo, J., Gerba, C.P. (1999). Microbiological quality of water vending machines. *International Journal of Environmental Health Research, 9*, pp. 197–206.

Chapman and Hall. (1995). *Dictionary of Natural Products*. CD-ROM, Release 4, pp. 1. London, UK.

Chiesa, F., Lomonaco, S., Nucera, D., Garoglio, D., Dalmasso, A. and Civera, T. (2014). Distribution of *Pseudomonas* species in a dairy plant affected by occasional blue discoloration. *Italian Journal of Food Safety, 3*, p. 1722.

Clark, J.A., Burger, C.A., Sabatinos, L.E. (1982). Characterization of indicator bacteria in municipal raw water, drinking water, and new main water samples. *Canadian Journal of Microbiology, 28*, pp. 1002–1013.

Clinical and Laboratory Standards Institute. (2007). Molecular methods for bacterial strain typing approved guideline. CLSI document MM11-A. Wayne, PA: Clinical and Laboratory Standards Institute.

Colledge, H., et al. (1995). Generation and characterization of a DeltaF508 cystic fibrosis mousse model. *Nat. Genet., 10*, pp. 445–452.

Comai, L., Surico, G. and Kosuge, T. (1982). Relation of plasmid DNA to indolacetic acid production in different strains of *Pseudomonas syringae* pv. *Savasatoni*. *Journal of General Microbiology, 128*, pp. 2157–2163.

Cousin, M.A. and Marth, E.A. (1977). Psychrotrophic bacteria cause changes in stability of milk to coagulation by rennet or heat. *J. Dairy Sci., 60*, pp. 1042–1047.

Cousin, M.A. (1982). Presence and activity of psychrotrophic microorganisms in milk and dairy products: A review. *J. Food Prot., 45*, pp. 172–207.

Craven, M. and Macauly, J. (1992). Microorganisms in pasteurised milk after refrigerated storage I. Identification of types. *J. Dairy Technol., 47*, pp. 38–45.

Cuppels, D.A. and Kelman, A. (1980). Isolation of pectolytic fluorescent psuedomonads from soil and potatoes, *Phytopathology, 70*, pp. 1110–1115.

Curiale, M.S., Lepper, N. and Robison, B. (1996). ELISA for detection of *Listeria monocytogenes* in dairy products, seafoods and meats: Collaborative study. *Journal of AOAC International, 77*, pp. 1472–1489.

Dainty, R.H. (1996). Chemical/biochemical detection of spoilage. *International Journal of Food Microbiology, 33*, pp. 19–34.

Dainty, R.H., Edwards, R.A. and Hibbard, C.M. (1985). Time course of volatile compounds formation during refrigerated storage of naturally contaminated beef in air. *Journal of Applied Bacteriology, 59*, pp. 303–309.

Dainty, H. and Mackey, M. (1992). The relationship between the phenotypic properties of bacteria from chill-stored meat and spoilage processes. *J. Appl. Bacteriol., 73*, pp. 113S–114S.

Davey, M. and Kell, B. (1996). Flow cytometry and cell sorting of heterogeneous microbial populations: the importance of single cell analyses. *Microbiology Review, 60*, pp. 641–696.

De Jonghe, V., Coorevits, A., Van Hoorde, K., Messens, W., Van Landschoot, A., De Vos, P. and Heyndrickx, M. (2010). Influence of Storage Conditions on the Growth of Pseudomonas Species in Refrigerated Raw Milk. *Applied and Environmental Microbiology, 77*(2), pp. 460–470.

de Vicente, A., Aviles, M., Borrego, J.J. and Romero, P. (1988). Die-off and survival of *Pseudomonas aeruginosa* in freshwater. *Zbl Bakt Hyg B., 185*, pp. 534–547.

Dogan, B. and Boor, J. (2003). Genetic diversity and spoilage potentials among *Pseudomonas* from fluid milk products and dairy processing plants. *Appl. Environ. Microbiol., 69*, pp. 130–138.

Doménech, A., Fernéndez-Garayzabal, J.F., Lawson, P., García, J.A., Cutúli, M.T., Blanco, M., Gibello, A., Moreno, M.A., Collins, M.D. and Domínguez, L. (1997). Winter disease outbreak in sea bream (*Sparus aurata*) associated with *Pseudomonas anguilliseptica* infection. *Aquaculture, 156*, pp. 317–326.

Doring, G., Ulrich, M. and Muller, W. (1991). Generation of *Pseudomonas aeruginosa* aerosols during hand washing from contaminated sink drains, transmission to hands of hospital personnel, and its prevention by use of a new heating device. *Zbl Hyg., 191*, pp. 494–505.

Drosinos, E.H. and Board, R.G. (1994). Metabolic activities of pseudomonads in bach cultures in extract of minced lamb. *J. Appl. Bacteriol., 77*, pp. 613–620.

Drosinos, E.H. and Board, R.G. (1995a). Attributes of microbial associations of meat growing as xenic batch cultures in a meat juice at 4 C. *International Journal of Food Microbiology, 26*, pp. 279–293.

Drosinos, E.H. and Board, R.G. (1995b). Microbial and physicochemical attributes of minced lamb: Sources of contamination with pseudomonads. *Food Microbiology, 12*, pp. 189–197.

Duquino, H.H. and Rosenberg, F.A. (1986) Antibiotic-resistant Pseudomonas in bottled drinking water. *Canadian Journal of Microbiology, 33*, pp. 286–289.

Edwards, R.A., Dainty, R.H. and Hibbard, C.M. (1987). Volatile compounds produced by meat pseudomonads and related reference strains during growth on beef stored in air at chill temperatures. *J. Appl. Bacteriol., 62*, pp. 403–412.

Ercolini, D., Ferrocino, I., La Storia, A., Mauriello, G., Gigli, S., Masi, P. and Villani, F., (2010a). Development of spoilage microbiota in beef stored in nisin activated packaging. *Food Microbiology, 27*, pp. 137–143.

Ercolini, D., Casaburi, A., Nasi, A., Ferrocino, I., Di Monaco, R., Ferranti, P., Mauriello, G. and Villani, F. (2010b). Different molecular types of Pseudomonas fragi have the overall behaviour as meat spoilers. *Int. J. Food Microbiol., 142*, pp. 120–131.

Ercolini, D., Russo, F., Blaiotta, G., Pepe, I., Mauriello G. and Villani, F. (2007). Simultaneous detection of Pseudomonas fragi, P. lundensis, and P. putida from meat by use of a multiplex PCR assay targeting the carA gene. *Applied and Environmental Microbiology, 73*, pp. 2354–2359.

Ercolini, D., Russo, F., Nasi, A., Ferranti, P. and Villani, F. (2009). Mesophilic and psychrotrophic bacteria from meat and their spoilage potential in vitro and in beef. *Applied and Environmental Microbiology, 75*, pp. 1990–2001.

Ercolini, D., Russo, F., Torrieri, E., Masi, P. and Villani, F. (2006). Changes in the spoilage-related microbiota of beef during refrigerated storage under different packaging conditions. *Applied and Environmental Microbiology, 72*, pp. 4663–4671.

Erichsen, I. and Molin, G. (1981). Microbial flora of normal and high pH beef stored at 4 °C in different gas environments. *Journal of Food Protection, 44*, pp. 866–869.

Faccia, M., Luisa, A., Marianna, M., Amalia, C., and Del Nobile, M.A. (2013). The effect of incorporating calcium lactate in the saline solution on improving the shelf life of Fiordilatte cheese. *International Journal of Dairy Technology, 66*, pp. 373–381.

Fernandez, R.O. and Pizarro, R.A. (1996). Lethal effect induced in *Pseudomonas aeruginosa* exposed to ultraviolet-A radiation. *Photochemistry and Photobiology., 64*, pp. 334–339.

Fick, R.B.Jr. (1993). *Pseudomonas aeruginosa* - the microbial hyena and its role in disease: An introduction. In: Fica, R.B., Jr. (ed.). *Pseudomonas aeruginosa* the opportunist: Pathogenesis and disease (pp. 1–5). Boca Raton, FL: CRC Press.

Fitzgerald, G. and Der Vartanian, M. (1969). *Pseudomonas aeruginosa* for the evaluation of swimming pool chlorination and algicides. *Applied and Environmental Microbiology., 17,* pp. 415–421.

François, K., Devlieghere, F., Uyttendaele, M., Standaert, A.R., Geeraerd, A.H., Nadal, P., Van Impe, J.F. and Debevere, J. (2006). Single cell variability of *L. monocytogenes* grown on liver pâté and cooked ham at 7 °C: comparing challenge test data to predictive simulations. *Journal of Applied Microbiology., 100,* pp. 800–812.

Frank, J.F. and Yousef, A.E. (2004). Tests for groups of microorganisms. In: H.M. Wehr and J.F. Frank, eds., *Standard Methods for the Examination of Dairy Products,* 17th ed. Washington, DC: American Public Health Association, pp. 227–248.

Franzetti, L., and Scarpellini, M. (2007). Characterisation of *Pseudomonas spp.* isolated from foods. *Annals of Microbiology, 57*(1), pp. 39–47.

Fredrickson, A.G. (1977). Behavior of mixed cultures of microorganisms. *Ann. Rev. Microbiol., 31,* pp. 63–87.

Gallo, L., Schmitt, E. and Schmidt-Lorenz, W. (1988). Microbial spoilage of refrigerated fresh broilers. J. Bacterial flora and growth during storage. *Lebensm. Wiss. U-Technol., 21,* pp. 216–223.

Gambassini, L., Sacco, C. and Lanciotti, E. (1990). Microbial quality of the water in the distribution system of Florence. *Aqua., 39,* pp. 258–264.

Garrard, E.H. (1945). A storage rot of potatoes caused by a fluorescent organism resembling *Pseudomonas fluorescens* (Flügge) Migula. *Canadian Journal of Research, 23,* pp. 79–84.

Gaynes, R. and Edwards, J. (2005). Overview of nosocomial infections caused by gram-negative bacilli. *Clin. Infect. Dis. 41,* pp. 848–854.

Geldreich, E.E. (1996). Microbial quality of water supply in distribution systems. Boca Raton, FL: CRC, p. 293.

Giamarellou, H. and Poulakou, G. (2009). Multidrug-resistant Gram-negative infections: what are the treatment options? *Drugs, 69,* pp. 1879–1901.

Gill, C.O. (1982). Microbial interaction with meats. In: M.H. Brown, ed., *Meat Microbiology.* London: Applied Science Publishers, pp. 225–264.

Govan, R. and Deretic, V. (1996). Microbial pathogenesis in cystic fibrosis: mucoid *Pseudomonas aeruginosa* and *Burkholderia cepacia. Microbial Rev, 60,* pp. 539–574.

Gram, L. and Huss, H. (1996). Microbial spoilage of fish and fish products. *Int. J. of Food Microbiol. 33*(1), pp. 121–137.

Gram, L., Ravn, L., Rasch M., Bruhn, J., Christensen A. and Givskov, M. (2002). Food spoilage-interactions between food spoilage bacteria. *International Journal of Food Microbiology, 78,* pp. 79–97.

Griffiths, M.W., and Phillips, J.D. (1984). Effect of aeration on extracellular enzyme synthesis by psychrotrophs growing in milk during refrigerated storage. *J. Food Prot., 47,* pp. 697–702.

Gunasekera, T., Dorsch, M., Slade, M. and Veal, D. (2003). Specific detection of *Pseudomonas spp.* in milk by fluorescence in situ hybridization using ribosomal RNA directed probes. *Journal of Applied Microbiology, 94,* pp. 936–945.

Gunasekera, T., Sorensen, A., Attfield, V., Sorensen, S. and Veal, A. (2002). Inducible gene expression by non-culturable bacteria in milk after pasteurisation. *Applied and Environmental Microbiology, 68,* pp. 1987–1993.

Gutierrez, R., García, T., Gonzalez, I., Sanz, B., Hernandez, P. E. and Martin, R. (1997a). Immunostick colorimetric assay for detection of Pseudomonas spp. in refrigerated meat and milk. *Journal of Food Protection, 60,* pp. 908–911.

Gutierrez, R., García, T., González, I., Sanz, B., Hernández, P. E. and Martín, R. (1997b).

Monoclonal antibody detection of Pseudomonas spp. in refrigerated meat by an indirect ELISA. *Letters in Applied Microbiology*, *24*, pp. 5–8.

Haas, D. and Defago, G. (2005). Biological control of soil-borne pathogens by fluorescent pseudomonads. *Nat. Rev. Microbial.*, *3*, pp. 307–319.

Hambsch, B., Sacre, C. and Wagner, I. (2004). Heterotrophic plate count and consumer's health under special consideration of water softeners. *International Journal of Food Microbiology.*, *92*, pp. 365–373.

Hantsis-Zacharov, E. and Halpern, M., (2007). Culturable psychrotrophic bacterial communities in raw milk and their proteolytic and lipolytic traits. *Appl. Environ. Microbiol.* *73*, pp. 7162–7168.

Herald, P.J. and Zottola, E.A. (1989). Effect of various agents upon the attachment of *Pseudomonas fragi* to stainless steel. *Journal of Food Science*, *54*(2), pp. 461–464.

Herrera, A.G. (2001). Psychrotrophic microorganisms. In: J.F.T. Spencer and de A.L.R. Spencer, ed., *Food Microbiology Protocols*. Totowa, NJ: Humana Press Inc., pp. 3–11.

Highsmith, A.K., Le, P.N., Khabbaz, R.F. and Munn, V.P. (1985). Characteristics of *Pseudomonas aeruginosa* isolated form whirlpools and bathers. *Infection Control.*, *6*, pp. 407–412.

Huang, H.I., Shih, H.Y., Lee, C.M., Yang, T.C., Lay, J.J. and Lin, Y.E. (2008). In vitro efficacy of copper and silver ions in eradicating *Pseudomonas aeruginosa*, *Stenotrophomonas maltophilat*, and *Acinetobacter baumannii*: Implications for on-site disinfection for hospital infection control. *Water Research.*, *42*, pp. 73–80.

Huck, J.R., Woodcock, N.H., Ralyea, R.D. and Boor, K.J. (2007). Molecular subtyping and characterization of psychrotolerant endospore forming bacteria in two New York State fluid milk processing systems. *J. Food Prot. 70*, pp. 2354–2364.

Huis in't Veld, J.H.J. (1996). Microbial and biochemical spoilage of foods: An overview. *International Journal of Food Microbiology*, *33*, pp. 1–18.

Insausti, K., Beriain, M.J., Gorraiz, C. and Purroy, A. (2002). Volatile compounds of raw beef from 5 local Spanish cattle breeds stored under modified atmosphere. *J. Food Sci.*, *67*, pp. 1580–1589.

Jackson, T.C., Acuff, G.R. and J.S. Dickson. (1997). Meat, poultry and seafood. In: M.P. Doyle and T.J. Beuchat, eds., *Food Microbiology: Fundamentals and Frontiers*, Washington, DC: ASM Press, pp. 83–100.

Jaspe, A., Oviedo, P., Fernandez, L., Palacios, P., and Sanjose, C. (1995). Cooling raw milk: change in the spoilage potential of contaminating *Pseudomonas*. *Journal of Food Protection*, *58*, pp. 915–921.

Jayasekara, N.Y., Heard, G.M., Cox, J.M. and Fleet, G.H. (1998). Populations of pseudomonads and related bacteria associated with bottled non-carbonated mineral water. *Food Microbiology.*, *15*, pp. 167–176.

Jeppesen, C. (1995). Media for *Aeromonas spp.*, *Plesiomonas shigelloides* and *Pseudomonas spp.* from foof and environment. *Int. J. of Food Microbiol. 26*, pp. 25–41.

Jos, J. and In't Veld, H. (1996). Microbial and biochemical spoilage of foods: an overview. *International Journal of Food Microbiology*, *33*, pp. 1–18.

Kader, A. and Barrett, D.M. (1996). Classification, composition of fruits, and posharvest maintenance of quality. In: L.P. Somogyi, H.S. Ramaswami, and Y.H. Hui, eds., *Processing Fruits: Science and technology*, vol *1*, Lancaster, PA: Tchnomic Publishing Company, p. 1.

Kolleff, H., Shorr, A., Tabak, Y., Gupta, V., Liu, L. and Johannes, R. (2005). Epidemiology and outcomes of health-care associated pneumonia: results from a large US database of culture-positive pneumonia. *Chest*, *128*, pp. 3854–3862.

Koutsoumanis, K. and Nychas, G.J.E. (2000). Application of a systematic experimental procedure to develop a microbial model for rapid fish shelf life predictions. *International Journal of Food Microbiology, 60*, pp. 171–184.

Koutsoumanis, K., Stamatiou, A., Skandamis, P. and Nychas, G.-J.E. (2006). Development of a microbial model for the combined effect of temperature and pH on spoilage of ground meat and validation of the model under dynamic temperature conditions. *Applied Environmental Microbiology, 72*, pp. 124–134.

Koutsoumanis, K.P. and Sofos, J.N. (2004). Microbial contamination of carcasses and cuts. In: W.K. Jensens, ed., *Encyclopedia of Meat Sciences*, Amsterdam: Elsevier Academic Press, pp. 727–737.

Kuo, S.C. and Kou, G.H. (1978). *Pseudomonas anguilliseptica* isolated from red spot disease of pond-cultured eel, *Anguilla japonica. Rep. Inst. Fish. Biol., Min. Econ. Aff. Nat. Taiwan Univ., 3*, pp. 19–23.

Labadie, J. (1999). Consequences of packaging on bacterial growth. Meat is an ecological niche. *Meat Science, 52*, pp. 299–305.

Lau, G.W., Hassett, D.J., Ran, H. and Kong, F. (2004). The role of pyocyanin in *Pseudomonas aeruginosa* infection. *Trends in molecular medicine, 10*, pp. 599–606.

Leblanc, L. and Dufour, É. (2002). Monitoring the identity of bacteria using their intrinsic fluorescence. *FEMS Microbiology Letters, 211*, pp. 147–153.

Levesque, B., Simard, P. and Gauvin, D. (1994). Comparison of the microbiological quality of water coolers and that of municipal water systems. *Applied and Environmental Microbiology., 60*, pp. 1174–1178.

Li, Z., Nikaido, H. and Poole, K. (1995). Role of MexA-MexB-OprM in antibiotic efflux in *Pseudomonas aeruginosa. Antimicrob. Agents Chemother. 39*, pp. 1948–1953.

Li, Z., Zhang, L. and Poole, K. (2000). Interplay between the MexA-MexB-OprM multirug efflux system and the outer membrane barrier in the multiple antibiotic resistance of *Pseudomonas aeruginosa. J. Antimicrob. Chemother. 45*, pp. 433–436.

Liao, C.H., (2006). Pseudomonas and related genera. In: C.W. Blackburn, ed., *Food Spoilage Microorganisms*. Cambridge, UK: Woodhead Publishing Limited, pp. 213–286.

Liao, C.H. and Wells, J.M. (1986). Diversity of pectolytic, *Fluorescent Pseudomonads* causing soft rots of fresh vegetables at produce markets. *Postharvest Pathology and Mycotoxins, 77*(5), pp. 673–677.

Lister, D., Wolter, J. and Hanson, D. (2009). Antibacterial-resistant *Pseudomonas aeruginosa*: clinical impact and complex regulation of chromosomally encoded resistance mechanisms. *Clin. Microbial. Rev., 22*, pp. 582–610.

Livermore, D.M. (2009). Has the era of untreatable infections arrived? *J. Antimicrob. Chemother, 64*, pp. 29–36.

Lönnström, L., Wiklund, T. and Bylund, G. (1994). *Pseudomonas anguilliseptica* isolated from Baltic herring *Clupea harengus* membras with eye lesions. *Diseases of Aquatic Organisms, 18*, pp. 143–147.

López-Romalde, S., Magariños, B., Núñez, S., Toranzo, A.E. and Romalde, J.L. (2003a). Phenotypic and genetic characterization of *Pseudomonas anguilliseptica* strains isolated from fish. *Journal of Aquatic Animal Health, 15*, pp. 39–47.

López-Romalde, S., Núñez, S., Toranzo, A.E. and Romalde, J.L. (2003b). Black spot seabream (*Pagellus bogaraveo*), a new susceptible host for *Pseudomonas anguilliseptica. Bulletin - European Association of Fish Pathologists, 23*, pp. 258–264.

López-Romalde, S., Magariños, B., Ravelo, C., Toranzo, A.E. and Romalde, J.L. (2003c). Existence of two O-serotypes in the emerging fish pathogen *Pseudomonas anguilliseptica. Veterinary Microbiology, 94*, pp. 325–333.

Lucera, A., Mastromatteo, M., Conte, A., Zambrini, A.V., Faccia, M. and Del Nobile, M.A. (2014). Effect of active coating on microbiological and sensory properties of fresh mozzarella cheese. *Food Packaging and Shelf Life, 1*, pp. 25–29.

Lund, B.M. (1983). Bacterial spoilage. In: C. Dennis, ed., *Post-Harvest Pathology of Fruits and Vegetables.* London, UK: Academic Press, pp. 219–257.

Lyczak, B., Cannon, L. and Pier, B. (2000). Establishment of Pseudomonas aeruginosa infection: lessons from a versatile opportunist. *Microbes and Infection, 2*(9), pp. 1051–1060.

Manaia, C.M., Nunes, O.C., Morais, P.V. and da Costa, M.S. (1990). Heterotrophic plate counts and the isolation of bacteria from mineral waters on selective and enrichment media. *Journal of Applied Bacteriology, 69,* pp. 871–876.

Manios, S.G., Konstantinidis, N., Gounadaki, A. and Skandamis, P.N. (2011). Single cell variability and population dynamics of *Listeria monocytogenes* and *Salmonella typhimurium* in fresh-cut salads and their sterile liquid or solidified extracts. In: E. Cummins, J.M. Frias, and V.P. Valdramidis, eds., *Seventh International Conference on Predictive Modelling in Foods e Conference Proceedings.* Dublin, Ireland: UCD, DIT, Teagasc, pp. 49–52.

Marchetti, R., Casadei, M.A. and Guerzoni, M.E. (1992). Microbial population dynamics in ready-to use vegetable salads. *Ital. J. Food Sci., 2,* pp. 97–108.

Martin, N.H. Murphy, S.C., Ralyea, R.D., Wiedmann, M. and Boor, K.J. (2011). When cheese gets the blues: *Pseudomonas fluorescens* as the causative agent of cheese spoilage. *Journal of Dairy Science, 94,* pp. 3176–3183.

Martins, M.L., Pinto, C.L., Rocha, R.B., de Araujo, E.F. and Vanetti, M.C. (2006). Genetic diversity of Gram-negative, proteolytic, psychrotrophic bacteria isolated from refrigerated raw milk. *Int. J. Food Microbiol., 111,* pp. 144–148.

McMeekin, T.A., Olley, J., Ross, T. and Ratkowsky, D.A. (1993). *Predictive microbiology: Theory and Application. Innovation in Microbiology.* Taunton, UK: Research Studies Press and John Wiley & Sons.

Mead, C. (1985). Enumeration of pseudomonads using cephaloridine-fucidin-cerimide agar (CFC). *Int. J. Food Microbiol., 2,* pp. 21–26.

Mead, C. and Adams, W. (1977). A selecive medium for the rapid isolation of pseudomonads associated with poultry meat spoilage. *Br. Poult. Sci., 18,* pp. 661–670.

Meer, R.R., Baker, J., Bodyfelt, F.W. and Griffiths, M.W. (1991). Psychrotrophic Bacillus spp. in fluid milk products: A review. *J. Food Prot., 54,* pp. 969–979.

Mena, K.D. and Gerba, C.P. (2009). Risk assessment of *Pseudomonas aeruginosa* in water. *Reviews of Environmental Contamination and Toxicology., 201,* pp. 71–115.

Mendoza-Yepes, M.J., Abellan-Lopez, O., Carrion-Ortega, J. and Marin-Inkiest, F. (1999). Inhibition of *Listeria monocytogenes* and other bacteria in Spanish soft cheese made with *Lactococcus lactis* ssp. *Diaceylactis. Journal of Food Safety, 19,* pp. 161–170.

Meyer, J.M., Geoffroy, V.A., Baida, N., Gardan, L., Izard, D., Lemanceau, P. and Palleroni, N.J. (2002). Siderophore typing, a powerful tool for the identification of fluorescent and nonfluorescent pseudomonads. *Applied and Environmental Microbiology, 68,* pp. 2745–53.

Micek, T., Lloyd, A., Ritchie, D., Reichley, R., Fraser, V. and Kollef, M. (2005). *Pseudomonas aeruginosa* bloodstream infection: importance of appropriate initial antimicrobial treatment. *Antimicro. Agents Chemother., 49,* pp. 1306–1311.

Misaghi, I.G. and Donndelinger, C.R. (1983). Limitations of tests used to detect soft-rotting fluorescent pseudomonads. *Phytopathology, 73,* pp. 1625–1628.

Molin, G., Ternstrom, A. and Ursing, J. (1986). Pseudomonas lundensis. A new bacterial species isolated from meat. *International Journal of Systematic Bacteriology, 36,* pp. 339–342.

Molina, D.N., Colon, M., Bermudez, R.H., and Ramirez-Ronda, C.H. (1991). Unusual presentation of *Pseudomonas aeruginosa* infections: A review. *Bulletin of the Puerto Rico Medical Association, 83,* pp. 160–163.

Molina, G., Pimentel, M., Pastore, G. (2013). *Pseudomonas*: a promising biocatalyst for the bioconversion of terpenes. *Applied Microbiology and Biotechnology, 97*(5), pp. 1851–1864.

Montel, M.C., Masson, F. and Talon, R. (1998). Bacterial role in flavour development. *Meat Science, 49*, pp. S111–S123.

Moore, B., Mau, M., Arascheidt, A., Bottger, C., Hutson, A., Collins, D., Van de Peer, Y., De Wachter, R. and Timmis, N. (1996). The determination and comparison of the rRNA gene sequences of species of the genus *Pseudomonas* (*sensu stricto*) and estimation of natural intrageneric relationships. *Syst. Appl. Microbiol., 19*, pp. 478–492.

Moretti, V.M., Madonia, G., Diaferia, C., Mentasti, T., Paleari, M.A., Panseri, S., Pirone, G. and Gandini, G. (2004). Chemical and microbiological parameters and sensory attributes of a typical Sicilian salami ripened in different conditions. *Meat Science, 66*, pp. 845–854.

Morn, L., Rowe M.T. and Hagan, J.A.. (1990). The effect of various heat activation treatments on fast, intermediate and slow germinating spores of Bacillus spp. *Lett. Appl. Microbiol., 10*, pp. 43–46.

Munsch-Alatossava, P. and T. Alatossava. (2006). Phenotypic characterization of raw milk-associated psychrotrophic bacteria. *Microbiol. Res., 161*, pp. 334–346.

Nakai, T. (1985). Resistance of *Pseudomonas anguilliseptica* to bactericidal action of fish serum. *Bulletin of the Japanese Society of Scientific Fisheries, 51*, pp. 1431–1436.

Nakai, T., Muroga, K. and Wakabayashi, H. (1985). First record of *Pseudomonas anguilliseptica* infection in cultured ayu, *Plecoglossus altivelis*. *Fish Pathology, 20*, pp. 481–484.

Nicas, I. and Hancock, W. (1983). *Pseudomonas aeruginosa* outer membrane permeability: isolation of a porin protein F-deficient mutante. *J. Bacteriol., 153*, pp. 281–285.

Nychas, G-J.E. and Skandamis, P. (2005). Fresh meat spoilage and modified atmosphere packaging (MAP). In: J.N. Sofos ed., *Improving the Safety of Fresh Meat*, Cambridge, UK: CRC/ Woodhead Publishing Limited, pp. 461–502.

Nychas, G., Skandamis, P., Tassou, C., and Koutsoumanis, K. (2008). Meat spoilage during distribution. *Meat Science, 78*(1–2), pp. 77–89.

Nychas, G-J. E., Skandamis, P. N., Koutsoumanis, K. P. and Baranyi, J. (2006). Effect of cell free meat extract on the growth rate of spoilage bacteria. Proceedings of the 20th international ICMFH symposium, FoodMicro 2006 – food safety and food biotechnology: Diversity and global impact (p. 122). Bologna: Italy 29/ 8–2/9 2006.

Nychas, G-J.E., Drosinos, E.H. and Board, R.G. (1998). Chemical changes in stored meat. In: R.G. Board and A. R. Davies, eds., *The Microbiology of Meat and Poultry*. London, UK: Blackie Academic and Professional, pp. 288–326.

Nychas, G-J.E. and Arkoudelos, J.S. (1990). Microbiological and physicochemical changes in minced meats under carbon dioxide, nitrogen or air at 3 °C. *International Journal of Food Science and Technology, 25*, pp. 389–398.

Nychas, G-J. E., Dillon, V.M. and Board, R.G. (1988). Glucose, the key substrate in the microbiological changes occurring in meat and certain meat products. *Biotechnology Applied Biochemistry, 10*, pp. 203–231.

Nychas, G-J.E., Marshall, D. and Sofos, J. (2007). Meat poultry and seafood. In: M.P. Doyle, L.R. Beuchat and T.J. Montville, eds., *Food Microbiology Fundamentals and Frontiers*. Washington, DC: ASM Press, chapter 6.

Palleroni, J., Kunisawa, R., Contopoulou, R. and Doudoroff, M. (1973). Nucleic acid homologies in the genus *Pseudomonas*. *Int. J. Syst. Bacteriol., 23*, pp. 333–339.

Palleroni, N.J. (2010). *Pseudomonas. Topley and Wilson's microbiology and microbial infections*. New York: John Wiley & Sons, Ltd.

Papapetropoulou, M., Iliopoulou, J. and Rodopoulou, G. (1994). Occurrence of antibiotic-resistance of *Pseudomonas* species isolated from drinking water in southern Greece. *Journal of Chemotherapy, 6*, pp. 111–116.

Parisi, S. (2003). Curve predittive per la crescita dei batteri coliformi in prodotti lattiero-caseari. *Industrie Alimentari, 421*, pp. 29–37.

Patel, J.B. (2001). 16S rRNA gene sequencing for bacterial pathogen identification in the 513 clinical laboratory. *Molecular Diagnosis, 6*, pp. 313–321.

Payment, P., Gamache, F. and Paquette, G. (1988). Microbiological and virological analysis of water from two water filtration plants and their distribution systems. *Canadian Journal of Microbiology, 34*, pp. 1304–1309.

Potvin, E., Lehoux, E., Kukavica-Ibrulj, I., Richard, L., Sanschagrin, F., Lau, W. and Levesque, C. (2003). *In vivo* functional genomics of *Pseudomonas aeruginosa* for high-throughput screening of new virulence factors and antibacterial targets. *Environ. Microbial., 5*, pp. 1294–1308.

Price, D. and Ahearn, D. (1988). Incidence and persistence of *Pseudomonas aeruginosa* in whirlpools. *Journal of Clinical Microbiology, 26*, pp. 1650–1654.

Pyle, B. and McFeters, G. (1989). Iodine sensitivity of bacteria isolated from iodinated water systems. *Canadian Journal of Microbiology, 35*, pp. 520–523.

Rahman (2002). *Handbook of Food Preservation.* New York: Marcel Dekker, Inc.

Ray, B. (2004). *Fundamental Food Microbiology.* Boca Raton, FL: CRC Press.

Ralyea, R.D., Wiedmann, M. and Boor, K.J. (1998). Bacterial tracking in a dairy production system using phenotypic and ribotyping methods. *J. Food Prot., 61*, pp. 1336–1340.

Ranieri, M.L., Huck, J.R., Sonnen, M., Barbano, D.M. and Boor, K.J. (2009). High temperature, short time pasteurization temperatures inversely affect bacterial numbers during refrigerated storage of pasteurized fluid milk. *J. Dairy Sci. 92*, pp. 4823–4832.

Reitler, R. and Seligmann, R. (1957). *Pseudomonas aeruginosa* in drinking water. *Journal of Applied Microbiology, 20*, pp. 145–150.

Relai, D. and Rosati, S. (1994). Antibiotic susceptibility and serotyping of *Pseudomonas aeruginosa* strains isolated form surface waters, thermomineral waters and clinical specimens. *Zbl Hyg., 196*, pp. 75–80.

Restaino, L., Frampton, E.W., Hemphill, J.B. and Palnikar, P. (1995). Efficacy of ozonated water against various food-related microorganisms. *Applied and Environmental Microbiology, 61*, pp. 3471–3475.

Rahme, G., et al. (2000). Plants and animals share functionally common bacterial virulence factors. *P. Natl. Acad. Sci. USA., 97*, pp. 8815–8821.

Ribas, F., Perramon, J., Terradillos, A., Frias, J. and Lucena, F. (2000). The *Pseudomonas* group as an indicator of potential regrowth in water distribution systems. *Journal of Applied Microbiology, 88*, pp. 704–710.

Ribot, E.M., Fair, M.A., Gautom, R., Cameron, D.N., Hunter, S.B., Swaminathan, B. and Barrett, T.J. (2006). Standardization of pulsedfield gel electrophoresis protocols for the subtyping of *Escherichia coli* O157:H7, *Salmonella*, and *Shigella* for PulseNet. *Foodborne Pathog. Dis., 3*, pp. 59–67.

Richards, J., et al. (1999). Nosocomial infections in pediatric intensive care units in the United States. *Pediatrics, 103*, p. 39.

Richter, R.L. and Vedamuthu, E.R. (2001). Milk and milk products. In: F.P. Downes and K. Ito, eds., *Compendium of Methods for the Microbiological Examination of Foods.* Washington, DC: American Public Health Association, pp. 483–495.

Ridgway, F. and Safarik, J. (1990). Identification and catabolic activity of well-derived gasolina-degrading bacteria from contaminated aquifer. *Appl. Environ. Microbiol. 56*, pp. 3565–3575.

Riera, E., et al. (2011). Pseudomonas aerugionosa carbapenem resistance mechanisms in Spain:impacto n the activity of imipenem meropenem and doripenem. *J. Antimicro. Chemother., 66*, pp. 2022–2027.

Riva, M., Franzetti, L. and Galli, A. (2001). Microbiological quality of shelf-life modelling of ready to eat cicorino. *J. Food Prot. 64*(2), pp. 228–234.

Rogues, A.M., Boulestreau, H., Lasheras, A., Boyer, A., Gruson, D., Merle, C., Casting, Y., Bebear, C.M. and Gachie, J.P. (2007).

Contribution of tap water to patient colonization with *Pseudomonas aeruginosa* in a medical intensive care unit. *Journal of Hospital Infection, 67*, pp. 72–78.

Romalde, J.L., López-Romalde, S., Ravelo, C., Magariños, B. and Toranzo, A.E. (2004). Development and validation of a PCR based protocol for the detection of *Pseudomonas anguilliseptica*. *Fish Pathology, 39*, pp. 33–41.

Romalde, J.L., Ravelo, C., López-Romalde, S., Avendaño-Herrera, R., Magariños, B. and Toranzo, A.E. (2005). Vaccination strategies to prevent important emerging diseases for Spanish aquaculture. In: P.J. Midtlyng, Ed., *Fish vaccinology*. Switzerland: Karger, pp. 85–95.

Ross, T., Dalgaard, P. and Tienungoon, S. (2000). Predictive modelling of the growth and survival of *Listeria* in fishery products. *International Journal of Food Microbiology, 62*, pp. 231–245.

Rusin, P.A., Rose, J.B., Haas, C.N. and Gerba, C.P. (1997). Risk assessment of opportunistic bacterial pathogens in drinking water. *Reviews of Environmental Contamination and Toxicology, 152*, pp. 57–83.

Samelis, J. (2006). Managing microbial spoilage in meat industry. In: C.W. Blackburn, ed., *Food Spoilage Microorganisms*. Cambridge, UK: Woodhead Publishing Limited, pp. 213–286.

Sands, D.C. and Hankin, L. (1975). Ecology and physiology of fluorescent pectolytic pseudomonads, *Phytopathology, 65*, pp. 921–924.

Scarpellini, M., Franzetti, L. and Galli, A. (2004). Development of PCR assay to identify *Pseudomonas fluorescens* and its biotype. FEMS Microbiology Letters, *236*, pp. 257–260.

Schillinger, J. and Du Vall Knorr, S. (2004). Drinking-water quality and issues associated with water vending machines in the city of Los Angeles. *Journal of Environmental Health, 66*, pp. 25–31.

Schokker, P. and Van Boekel, M. (1999). Kinetics of termal inactivation of the extracelular proteinase from *Pseudomonas fluorescens* 22F: Influence of pH, calcium and protein. *Journal of Agricultural Food Chemistry, 47*(4), pp. 1681–1686.

Seymour, I.J., Cole, M.B., and Coote, P.J. (1994). A substrate-mediated assay of bacterial proton efflux/influx to predict the degree of spoilage of beef mince stored at chill temperatures. *Journal of Applied Bacteriology, 76*, pp. 608–615.

Silby, M., Winstanley, C., Godfrey, S., Levy, S. and Jackson, R. (2011). *Pseudomonas* genomes: diverse and adaptable. *FEMS Microbiol Rev., 35*, pp. 652–680.

Silvestry-Rodriquez, N., Sicairos-Ruelas, A.E., Gerba, C.P. and Bright, K.R. (2007a). Silver as a disinfectant. *Reviews of Environmental Contamination and Toxicology, 191*, pp. 23–45.

Silvestry-Rodriquez, N., Bright, K.R., Uhlmann, D.R., Slack, D.C. and Gerba, C.P. (2007b). Inactivation of *Pseudomonas aeruginosa* and *Aeromonas hydrophil*a by silver in tap water. *Journal of Environmental Science and Health, 42*, pp. 1579–1584.

Simões, L., Simões, M. and Vieira, M.J. (2009). Species association increases biofilm resistance to chemical and mechanical treatments. *Water Research, 43*, pp. 229–237.

Sinigaglia, M., Bevilacqua, A., Corbo, M.R., Pati, S. and Del Nobile, M.A. (2008). Use of active compounds for prolonging the shelf life of mozzarella cheese. *International Dairy Journal, 18*, pp. 624–630.

Skandamis, P.N. and Nychas, G-J.E. (2002). Preservation of fresh meat with active and modified atmosphere packaging conditions. *International Journal of Food Microbiology, 79*, pp. 35–43.

Smith, M.A., McColloch, L.P., and Friedman, B.A. (1966). Market diseases of asparagus, onion, beans, peas, carrots, celery and related vegetables. *U.S. Dep. Agric. Agric. Handbook, 303*, p. 65.

Snouwaert, N., Brigman, K., Latour, M., Malouf, N., Boucher, C., Smithies, O. and Koller, H. (1992). An animal model for cystic fibrosis made by gene targeting. *Science, 257*, pp. 1083–1088.

Solberg, M., O'Leary, S. and Riha, E. (1972). New medium for the isolation and enumeration of pseudomonads. Applied *Microbiology, 24*(4), pp. 544–550.

Soncin, S., Chiesa, L.M., Cantoni, C. and Biondi, P.A. (2007). Preliminary study of the volatile fraction in the raw meat of pork, duck and goose. *J. Food Compos. Anal., 20*, pp. 436–439.

Sorhaug, T., and L. Stepaniak. (1997). Psychrotrophs and their enzymes in milk and dairy products: Quality aspects. *Trends Food Sci. Technol. 8*, pp. 35–41.

Stewart, D.J., Woldemariam, K., Dear, G. and Mochaba, F.M. (1983). An outbreak of "sekiten-byo" among cultured European eels, *Anguilla anguilla* L. *Scotland Journal of Fish Diseases, 6*, pp. 75–76.

Stanbridge, L.H. and Davis, A.R. (1998). The microbiology of chill-stored meat. In: R.G. Board and A.R. Davies, eds., *The microbiology of meat and poultry*. London, UK: Blackie Academic and Professional, pp. 174–219.

Stoler, J., Ahmed, H., Frimpong, L.A. and Bello, M. (2015). Presence of *Pseudomonas aeruginosa* in coliform-free sachet drinking water in Ghana. *Food Control, 55*, pp. 242–247.

Surico, G. and Lacobellis, N.S. (1978). Un Marciume batterico del Sedano (Apium graveolens) causato da Pseudomonas marginalis (Brown) Stevens. *Phytopathol. Mediterr. 17*, pp. 69–71.

Tan, W., Mahajan-Miklos, S., Ausubel, M. (1999). Killing of *Caenorhabditis elegans* by *Pseudomonas aeruginosa* used to model mammalian bacterial pathogenesis. *P Natl Acad Sci USA, 96*, pp. 715–720.

Tenover, C., Arbeit, R. and Goering, R. (1997). How to select and interpret molecular strain typing methods for epidemiological studies of bacterial infections: a review for healthcare epidemiologists. *Molecular Typing Working Group of the Society for Healthcare Epidemiology of America Infect. Control Hosp. Epidemiol., 18*, pp. 426–439.

Ternström, A., Lindberg, A.M., and Molin, G. (1993). Classification of the spoilage flora of raw and pasteurized bovine milk, with special reference to *Pseudomonas* and *Bacillus*. *Journal of Applied Bacteriology, 75*, pp. 25–34.

Toranzo, A.E. and Barja, J.L. (1993). Virulence factors of bacteria pathogenic for cold water fish. *Annual Review of Fish Diseases, 3*, pp. 5–36.

Toranzo, A.E., Magariños, B. and Romalde, J.L. (2005). A review of the main bacterial fish diseases in mariculture systems. *Aquaculture., 246*, pp. 37–61.

Tsigarida, E., Boziaris, I. S., and Nychas, G.-J.E. (2003). Bacterial synergism or antagonism in a gel cassette system. *Applied Environmental Microbiology, 69*, pp. 7204–7209.

Tsigarida, E., and Nychas, G.J.E. (2001). Ecophysiological attributes of a Lactobacillus sp. and a Pseudomonas sp. on sterile beef fillets in relation to storage temperature and film permeability. *Journal of Applied Microbiology, 90*, pp. 696–705.

van der Vossen, J.M., and Hofstra, H. (1996). DNA based typing, identification and detection systems for food spoilage microorganisms: development and implementation. *Intl. J. Food Microbiol. 33*, pp. 35–49.

Van Impe, J.F., Vercammen, D. and Van Derlinden, E. (2013). Toward a next generation of predictive models: a systems biology primer. *Food Control, 29*, pp. 336–342.

Van Tassell J.A., Martin, N.H., Murphy, S.C., Wiedmann, M., Boor, K.J. and Ivy, R.A. (2012). Evaluation of various selective media for the detection of *Pseudomonas* species in pasteurized milk. *Journal Dairy Science, 95*, pp. 1568–1574.

von Neubeck, M., Baur, C., Krewinkel, M., Stoeckel, M., Kranz, B., Stressler, T., Fischer, L., Hinrichs, J., Scherer, S. and Wenning, M. (2015). Biodiversity of refrigerated raw milk microbiota and their enzymatic spoilage potential. *International Journal of Food Microbiology, 211*, pp. 57–65.

Vasavada, C. and White, H. (1993). Symposium: developing methodology for microbial evaluation of milk and dairy products. *Journal of Dairy Science, 76*, pp. 3099–3100.

Vaz-Moreira, I., Nunes, O.C. and Manaia, C.M. (2012). Diversity and antibiotic resistance in *Pseudomonas* spp. from drinking water. *Science of the Total Environment, 426*, pp. 366–374.

Vess, R., Anderson, R. and Carr, J. (1993). The colonization of solid PVC surfaces and the acquisition of resistance to germicides by water microorganisms. *Journal of Applied Bacteriology, 74,* pp. 215–221.

Villelegas, V., et al. (2007). First identification os *Pseudomonas aeruginosa* isolates producing a KPC-type carbapenem-hydrolyzing β-lactamase. *Antimicrob. Agents. Chemother., 51,* pp. 1553–1555.

Wakabayashi, H. and Egusa, S. (1972). Characteristics of a *Pseudomonas* sp. from an epizootic of pond-cultured eels (*Anguilla japonica*). *Bulletin of the Japanese Society for the Science of Fish, 38,* pp. 577–587.

Wallner, G., Ammann, R. and Beisker, W. (1993). Optimization fluorescent *in situ* hybridization with rRNA-targeted oligonucleotide probes for flow cytometric identificaion of microorganisms. *Cytometry, 14,* pp. 136–143.

Wang, J.S. and Kelman, A. (1982). Population of pectolytic bacteria on soil in relation to depth and cropping history. (Abstr.) *Phytopathology, 72,* pp. 1140–1141.

Whiting, R.C. (1995). Microbial modelling in foods. *Critical Reviews in Food Science and Nutrition, 35,* pp. 467–494.

Warburton, D.W., Bowen, B. and Konkle, A. (1994). The survival and recovery of *Pseudomonas aeruginosa* and its effect upon *Salmonellae* in water: Methodology to test bottled water in Canada. *Canadian Journal of Microbiology, 40,* pp. 987–992.

Ward, N.R., Wolfe, R.L. and Olson, B.H. (1984). Effect of pH, application technique, and chlorine-tonitrogen ratio on disinfectant activity of inorganic chloramines with pure culture bacteria. *Applied and Environmental Microbiology., 48,* pp. 508–514.

Wiedmann, M., Weilmeier, D., Dineen, S., Ralyea, M. and Boor, J. (2000). Molecular and Phenotypic characterization of *Pseudomonas spp.* isolated from milk. *Appl. Environ. Microbiol., 66*(5), pp. 2085–2095.

Wiedmann, M., Bruce, J.L., Keating, C., Johnson, A.E., McDonough, P.L. and Batt, C.A. (1997). Ribotypes and virulence gene polymorphisms suggest three distinct Listeria monocytogenes lineages with differences in pathogenic potential. *Infect. Immun., 65,* pp. L2707–L2716.

Wiklund, T. and Bylund, G. (1990). *Pseudomonas anguilliseptica* as a pathogen of salmonid fish in Finland. *Diseases of Aquatic Organisms, 8,* pp. 13–19.

Williams, D.E., Worley, S.D., Wheatley, W.B. and Swango, L.J. (1985). Bactericidal properties of a new water disinfectant. *Applied and Environmental Microbiology, 49,* pp. 637–643.

Wolfe, R.L. (1990). Ultraviolet disinfection of potable water. *Environmental Science & Technology, 24,* pp. 768–773.

Xu, C., Song, S., Liu, L., Kong, D., Feng, M. and Wang, W. (2015). Development of an immunochromatographic strip for the rapid detection of Pseudomonas syringae pv. maculicola in broccoli and radish seeds. *Food and Agricultural Immunology., 26,* pp. 738–745.

Yahr, T.L. and Parsek, M.R. (2006). *Pseudomonas aeruginosa. The Prokaryotes, 6,* pp. 704–713.

Zagory, D. (1999). Efects of post-processing handling and packaging on microbial populations. *Post-harvest Biol. Technol., 15,* pp. 313–321.

Zhang, S., Xu, X., Wu, Q. and Zhang, J. (2013). Rapid and sensitive detection of Pseudomonas aeruginosa in bottled water by loop-mediated isothermal amplification. *European Food Research and Technology, 236,* pp. 209–215.

4

Arcobacter spp. in Food Chain—From Culture to Omics

Susana Ferreira[1], *Mónica Oleastro*[2], *and Fernanda Domingues*[1,3]

[1] *CICS-UBI - Health Sciences Research Centre, University of Beira Interior, Avenida Infante D. Henrique, 6200 - 506, Covilhã, Portugal*
[2] *National Reference Laboratory for Gastrointestinal Infections, Department of Infectious Diseases, National Institute of Health Dr. Ricardo Jorge, Avenida Padre Cruz, 1649-016, Lisboa, Portugal*
[3] *Chemistry Department, University of Beira Interior, Covilhã, Portugal*

4.1 Introduction

4.1.1 Genus *Arcobacter*

Arcobacter was first isolated in the late twentieth century by Ellis who isolated *Spirillum*-like organisms from bovine and pig foetuses (Ellis *et al.*, 1977, 1978). However, the *Arcobacter* genus was only described by Vandamme *et al.* (1991), with the reclassification of "two aerotolerant campylobacteria," at the time *Campylobacter nitrofigilis* (a nitrogen-fixing *Campylobacter* strain isolated from the roots of *Spartina alterniflora* Loisel and root-associated sediments from salt marsh, Mcclung *et al.*, 1983) and *Campylobacter cryaerophila* (animal abortions mainly, Tee *et al.*, 1988) (Table 4.1). This genus was included in the *Campylobacteriaceae* family together with *Campylobacter* and *Sulfurospirillum* genera (Vandamme and De Ley, 1991). In 1991, Kiehlbauch *et al.* (1991) isolated a group of strains from human stools and animals and named it *Campylobacter butzleri*; in 1992, the *Arcobacter* genus was extended to include the species now known as *Arcobacter butzleri* and *Arcobacter skirrowii*, isolated predominantly from preputial fluids of bulls or from bovine, porcine, and ovine aborted fetuses and diarrheic feces (Vandamme *et al.*, 1992b). Since then *Arcobacter* has grown to be a large and widely distributed genus including 23 recognized species isolated from diverse environments and hosts (Table 4.1). However, the analysis of sequences deposited in public databases suggests to the existence of potentially new species of *Arcobacter*, even though some of the sequences are from uncultured clones (Miller *et al.*, 2007). This fact, together with the description of several new *Arcobacter* species from extremely different habitats or hosts, may indicate that new species will be proposed and recognized in the future (Collado and Figueras, 2011).

Arcobacter species are gram-negative, non-spore forming, small curved rods or S-shaped or helical cells with approximately 0.2–0.9 μm wide and 0.5 to 3.6 μm long, presenting motility by means of a single polar flagellum (*A. anaerophilus* was firstly described by Sasi Jyothsna *et al.* (2013) as a non-motile species). *A. molluscorum*, *A. ellisii*, *A. suis*, and *A. anaerophilus* present few cells with filamentous form up to 7 μm long (Figueras *et al.*, 2011a, 2011b; Levican *et al.*, 2013; Sasi Jyothsna *et al.*, 2013). All genus

Foodborne Pathogens and Antibiotic Resistance, First Edition. Edited by Om V. Singh.

Table 4.1 Recognized *Arcobacter* Species.

Specie	Source	Country	Year of report	Reference
Arcobacter nitrofigilis	Roots of *Spartina alterniflora* Loisel and in root-associated sediments from salt marsh	Sapelo Island, Georgia	1983	Mcclung *et al.*, 1983
Arcobacter cryaerophilus	Bovine abortions	Several continents	1985	Neill *et al.*, 1985
Arcobacter butzleri	Humans and animals with diarrhoeal disease	Several continents	1991	Kiehlbauch *et al.*, 1991
Arcobacter skirrowii	Preputial fluids of bulls Bovine, porcine, and ovine isolates obtained from aborted fetuses and diarrhoeic faeces.	Several continents	1992	Vandamme *et al.*, 1992b
Arcobacter cibarius	Broiler carcasses	Belgium	2005	Houf *et al.*, 2005a
Arcobacter halophilus	Hypersaline lagoon	Hawaii	2005	Donachie *et al.*, 2005
Arcobacter mytili	Mussels (*Mytilus* sp) Brackish water	Spain	2009	Collado *et al.*, 2009a
Arcobacter thereius	Kidney and liver of Danish pigs abortions Cloacal content of ducks	Denmark	2009	Houf *et al.*, 2009
Arcobacter marinus	Seawater with seaweeds and starfish	Korea	2010	Kim *et al.*, 2010
Arcobacter trophiarum	Faecal samples taken rectally from fattening pigs	Belgium	2011	De Smet *et al.*, 2011a
Arcobacter defluvii	Sewage samples	Spain	2011	Collado *et al.*, 2011
Arcobacter molluscorum	Mussels (*Mytilus* sp) and oysters (2009)	Spain	2011	Figueras *et al.*, 2011a
Arcobacter ellisii	Mussels (*Mytilus* sp)	Spain	2011	Figueras *et al.*, 2011b
Arcobacter bivalviorum	Mussels (*Mytilus* sp) (collected in 2007)	Spain	2012	Levican *et al.*, 2012
Arcobacter venerupis	Clam (*Venerupis pullastra*) (collected in 2009)	Spain	2012	Levican *et al.*, 2012
Arcobacter cloacae	Mussels (*Mytilus* sp) and Sewage from the Waste Water Treatment Plant	Spain	2012	Levican *et al.*, 2013
Arcobacter suis	Pork meat (collected in 2008)	Spain	2012	Levican *et al.*, 2013
Arcobacter anaerophilus	Estuarine sediment	India	2013	Sasi Jyothsna *et al.*, 2013
Arcobacter ebronensis	Mussels	Spain	2015	Levican *et al.*, 2015
Arcobacter aquimarinus	Seawater	Spain	2015	Levican *et al.*, 2015
Arcobacter lanthieri	Pig and dairy cattle manure	Canada	2015	Whiteduck-Leveillee *et al.*, 2015
Arcobacter pacificus	Seawater	South Pacific Gyre	2015	Zhang *et al.*, 2015
Arcobacter faecis	Human waste septic tank	Canada	2016	Whiteduck-Léveillée *et al.*, 2016

members have low DNA G + C contents ranging from 24.6 to 35 mol%. *Arcobacter* spp. can be differentiated from *Campylobacter* by its ability to grow at temperatures below 30 °C and under aerobic conditions (Vandamme *et al.*, 2005), with exception of *A. anaerophilus*, which is an obligate anaerobe (Sasi Jyothsna *et al.*, 2013). All species, except *A. pacificus* (Zhang *et al.*, 2015), are oxidase positive, and catalase is absent in some species. *Arcobacter* species presents variable biochemical characteristics and some of the most important ones to differentiate between taxa within this genus are detailed in Table 4.2.

4.1.2 Human and Animal Infections Associated With the Genus *Arcobacter*

Arcobacter spp. have been associated with human and animal infections, among which *A.butzleri*, *A. cryaerophilus*, and *A. skirrowii* are known human and animal pathogens. *A. butzleri* and *A. cryaerophilus* have even been included in the list of microbes considered a serious hazard to human health by the International Commission on Microbiological Specifications for Foods (International Commission on Microbiological Specifications for Foods, 2002). In humans, *A. butzleri* and *A. cryaerophilus* infection has been associated with enteritis and occasionally with bacteremia, with *A. skirrowii* being associated with diarrheic feces (Ferreira *et al.*, 2015); however, *A. cryaerophilus* has also been found in healthy individuals (Houf and Stephan, 2007). Among *Campylobacter* spp. and related organism, *A. butzleri* has been described in several countries as the fourth most prevalent pathogen found in human diarrheic samples (Vandenberg *et al.*, 2004; Prouzet-Mauléon *et al.*, 2006; Samie *et al.*, 2007; Collado *et al.*, 2013; Ferreira *et al.*, 2014); however, the lack of surveillance averts an estimation of the real number of cases of diarrhea that may be associated with *Arcobacter* spp. infection. In a five-year survey, *A. butzleri* and *A. cryaerophilus* prevalence in fecal samples of individuals with symptoms of enteritis was of 0.7 and 0.6%, respectively. In this study, *A. thereius* was isolated for the first time in humans as the only bacterial pathogen from feces of patients with enterocolitis (Van den Abeele *et al.*, 2014). *Arcobacter* spp. has also been associated with extra-intestinal infections in humans, such as bacteremias (On *et al.*, 1995; Hsueh *et al.*, 1997; Yan *et al.*, 2000; Woo *et al.*, 2001; Lau *et al.*, 2002; Arguello *et al.*, 2015) and recently with peritonitis that occurred after repositioning of a catheter, with the patient showing fever and abdominal pain (Yap *et al.*, 2013).

Regarding fauna, *Arcobacter* spp. has been found in healthy animals, but has also been associated with animal diseases or conditions such as abortion or diarrhea (Vandamme *et al.*, 1992b; Anderson *et al.*, 1993; De Oliveira *et al.*, 1997; On *et al.*, 2002; Bath *et al.*, 2013). The isolation of aerotolerant *Campylobacter*-like organisms in bovine and porcine abortions was first described in the late 1970s (Ellis *et al.*, 1977, 1978). *A. butzleri*, *A. cryaerophilus*, and *A. skirrowii* have been isolated from aborted fetuses and placentas of bovine, porcine, and ovine sources (Vandamme *et al.*, 1992b), with a recent report suggesting *A. skirrowii* as the possible direct cause of abortion in sheep, since it was recovered from both placenta and fetuses (Bath *et al.*, 2013). The recovery of *A. butzleri*, *A. cryaerophilus*, and *A. skirrowii* from animal preputial fluid suggests venereal transmission of these bacteria as well (Ho *et al.*, 2006a). Since *Arcobacter* spp. is also detected in healthy live newborn piglets (Ho *et al.*, 2006b), it can be suggested that these bacteria may be opportunistic pathogens playing a role in the disease process as opposed to being the primary cause, although it is possible that certain strains may act as primary abortifacient pathogens, considering the high interspecies diversity (On *et al.*, 2002). Likewise, *A. butzleri*, *A. cryaerophilus*, *A. skirrowii*, and *A. thereius* have been associated with reproductive disorders, enteritis, or mastitis in domestic livestock (De Smet *et al.*, 2011c), but have also been recovered from feces of healthy animals (Van Driessche *et al.*, 2004, 2005), as well as *A. trophiarum* (De Smet *et al.*, 2011a).

Apart from reproductive disorders, *A. butzleri* has been also associated with enteritis and

Table 4.2 Differential Phenotypic Characteristics Among the 23 *Arcobacter* Species.

Characteristics	*A. nitrofigilis*	*A. cryaerophilus*	*A. butzleri*	*A. skirrowii*	*A. cibarius*	*A. halophilus*	*A. mytili*	*A. thereius*	*A. marinus*	*A. trophiarum*	*A. defluvii*	*A. molluscorum*	*A. ellisii*	*A. bivalviorum*	*A. venerupis*	*A. cloacae*	*A. suis*	*A. anaerophilus*[1]	*A. ebronensis*	*A. aquamarinus*	*A. lanthieri*	*A. pacificus*	*A. faecis*
Growth in/on																							
Air at 37°C	V(−)	V(+)	+	+	−	+	+	−	+	−	+	+	+	+	−	+	−	−	−	+	+	+	+
Microaerobic conditions at 37°C	−	V(+)	+	+	+	+	+	−	+	−	+	+	+	+	+	+	−	−	−	+	+	+	+
MacConkey Agar	−	V(−)	+	−	V	−[a]	+	V(+)	−[a]	V(+)	+	+	V(+)	−	+	+	+	ND	−	−	+	−	+
Minimal medium	−	V(−)	+	−	+	−	−	+	−[a]	−	+	−	+	−	+	V	+	ND	+	−	ND	−	−
NaCl (4%)	+	−	−	+	−	+	+	−	+	V(−)	−	+	−	+	−	−	−	+	+	−	−	+	−
Glycine (1%)	−	V(−)	V(−)	V(−)	−	+[a,b]	+	+	+	V(−)	−	−	−	−	−	−	−	+	−	−	+	ND	−
Resistance to cefoperazone (64 mg/L)	−	+	+	+	+	−[a]	−	+	−	+	V(+)	+	−	−	−	−	−	−	−	−	+	+	−
Enzyme activities																							
Catalase	+	+	(+)	+	V	−	+	+	−	+	(+)	+	+	+	+	+	+	−	−	+	+	+	+
Urease	V(+)	−	−	−	−	−	−	−	−	−	+	−	V(−)	−	+	−	−	−	+	−	−	−	−
Nitrate reduction	+	V	+	+	−	+	+	+	+	−	+	+	+	−	+	+	+	+	−	+	+	+	+
Indoxyl acetate hydrolysis	+	+	+	+	+	+	−	+	+	+	+	−	+	+	+	+	+	+	+	+	+	+	+
H_2S production	−	−	−	−	−	−	−	−	−	ND	−	−	−	−	−	−	−	+	−	−	ND	−	−

Data collected: *A. nitrofigilis* (Mcclung *et al.*, 1983; Vandamme *et al.*, 1992b; Levican *et al.*, 2013); *A. cryaerophilus, A. butzleri* and *A. skirrowii* (Vandamme *et al.*, 1992b; Levican *et al.*, 2013); *A. cibarius* (Houf *et al.*, 2005a); *A. halophilus* (Donachie *et al.*, 2005; Figueras *et al.*, 2011a); *A. mytili* (Collado *et al.*, 2009a); *A. thereius* (Houf *et al.*, 2009); *A. marinus* (Kim *et al.*, 2010; Levican *et al.*, 2013; Sasi Jyothsna *et al.*, 2013); *A. trophiarum* (De Smet *et al.*, 2011a; Figueras *et al.*, 2011a); *A. defluvii* (Collado *et al.*, 2011; Levican *et al.*, 2012); *A. molluscorum* (Figueras *et al.*, 2011a; Levican *et al.*, 2012); *A. bivalviorum* and *A. venerupis* (Levican *et al.*, 2012); *A. cloacae* (Levican *et al.*, 2013) and *A. suis* (Levican *et al.*, 2013); *A. anaerophilus* (Sasi Jyothsna *et al.*, 2013); *A. ebronensis* and *A. aquimarinus* (Levican *et al.*, 2015); *A. lanthieri* (Whiteduck-Leveillee *et al.*, 2015); *A. pacificus* (Zhang *et al.*, 2015); *A. faecis* (Whiteduck-Léveillée *et al.*, 2016).

+, ≥95% of strains positive; (+), weakly positive; −, ≤10% of strains positive; V, Variable; V(+), Variable where a majority of strains were positive; V(−), Variable where a majority of strains were negative; ND, Not Determined. All strains were Gram-stain-negative and oxidase positive, with the exception of *A. pacificus*, which was oxidase negative.[1] Obligate anaerobic growth under 100% Argon, do not grow in blood agar.

[a] Data from reference (Figueras *et al.*, 2011a), with all tested in medium supplemented with 2% NaCl; [b] Growth on glycine was previously not observed by Donachie *et al.* (2005).

diarrhea in pigs, cattle, and horses, whereas *A. skirrowii* was related with diarrhea in sheep and cattle (Ho *et al.*, 2006a). *A. butzleri* association with diarrhoea in non-human primates was also reported (Anderson *et al.*, 1993; Higgins *et al.*, 1999), with the histological evaluation of colonic specimens showing mild to moderately severe chronic active colitis (Anderson *et al.*, 1993). *Arcobacter* species were also recovered from pet cats and dogs with oral disease (Petersen *et al.*, 2007) and other health and unhealthy exotic animals, such as alpaca, rhea, rhinoceros, gorillas, or raccoons (Wesley and Miller, 2010). The observation of *Arcobacter* in both healthy and diseased animals has also been reported in other studies showing *Arcobacter* presence in swine stomach with or without gastric ulcers, and therefore, not allowing to establish a role of *Arcobacter* spp. in swine gastric ulceration (Suarez *et al.*, 1997; De Oliveira *et al.*, 2010).

In healthy animals, *Arcobacter* spp. was detected in mammals, birds, fish, and shellfish or reptiles as in healthy cattle (Van Driessche *et al.*, 2003, 2005; Öngör *et al.*, 2004; Aydin *et al.*, 2007; Shah *et al.*, 2013), pigs (Van Driessche *et al.*, 2004; De Smet *et al.*, 2011b), goats (De Smet *et al.*, 2011c; Shah *et al.*, 2013), sheep, and horses (Van Driessche *et al.*, 2003; De Smet *et al.*, 2011c), exotic animals (Wesley and Miller, 2010), domestic pets (Petersen *et al.*, 2007; Houf *et al.*, 2008), doves (Di Francesco *et al.*, 2014), domestic geese, chickens, turkeys, and ducks (Kabeya *et al.*, 2003b; Gude *et al.*, 2005; Atabay *et al.*, 2006; Andersen *et al.*, 2007; Atabay *et al.*, 2008b; Ho *et al.*, 2008; Adesiji *et al.*, 2011; Bogantes *et al.*, 2015), rainbow trout (Aydin *et al.*, 2000), edible bivalve molluscs (Collado *et al.*, 2009b, 2014; Levican *et al.*, 2014) or Lacertilia, Serpentes, and Testudines (Gilbert *et al.*, 2014).

4.1.3 Routes of Transmission of *Arcobacter* to Humans

Besides its detection on an unusually wide range of habitats and hosts, including diseased and healthy animals, human, or environmental samples, *Arcobacter* spp. are recognized food and waterborne pathogens, distributed throughout the food chain in water samples, foods (including foods of animal origin), or food processing facilities. In fact, consumption of *Arcobacter*-contaminated food or water is regarded as the most probable cause of infection by this bacterium (Wesley and Miller, 2010; Collado and Figueras, 2011). Some outbreaks have been suggested to be related with food or water, since *Arcobacter* species were found in water samples, or in patients in three outbreak reports (Rice *et al.*, 1999; Fong *et al.*, 2007; Kopilovi *et al.*, 2008). The first outbreak occurred in a Girl Scout camp in Idaho, where a breakdown in the automated chlorination system occurred at the same time as the outbreak. The well water presented fecal contamination, with *A. butzleri* being isolated from the ground water samples (Rice *et al.*, 1999). In 2004, *Arcobacter* was isolated from several groundwater samples during a groundwater-associated outbreak reported in Ohio, where a massive groundwater contamination occurred, probably with contaminants from wastewater treatment facilities and septic tanks (Fong *et al.*, 2007). *A. cryaerophilus* was found in human stool during an outbreak of acute gastroenteritis in Slovenia, that was associated with the local water distribution system, however *Arcobacter* was not confirmed as the aetiological agent (Kopilovi *et al.*, 2008). While no specific conclusions were drawn from these outbreaks, *Arcobacter* was either isolated or detected from water or stool samples from patients where disease was linked to water. In addition, an outbreak associated with roasted chicken consumption at a wedding was reported, suggesting that *A. butzleri* should be considered an agent that can cause outbreaks of foodborne illness (Lappi *et al.*, 2013). These outbreak reports strengthen the association of food and water as possible routes of transmission of *Arcobacter*. Despite this, transmission from person-to-person has also been proposed in the case of a reported outbreak of recurrent abdominal cramps associated with *A. butzleri*, where a successive timing of cases were observed (Vandamme *et al.*, 1992a), or even in a report of the isolation of *A. butzleri* from a neonate with bacteremia where the clinical data indicated a probable contraction of infection *in utero* (On *et al.*, 1995). Also, the contact with domestic pets has been suggested as a potential route for *Arcobacter* spp. human infection, due to the evidence of carriage of this microorganism by cats and dogs (Petersen *et al.*, 2007; Houf *et al.*, 2008; Fera *et al.*, 2009; Giacometti *et al.*, 2015b).

4.1.4 *Arcobacter* Species in the Food Chain

From a perspective of food safety, considering that some of the *Arcobacter* species are foodborne pathogens, its presence should be examined in all the steps of the food chain in order to better understand its distribution and possibly establish food and water consumption as a transmission route. This evaluation must be performed from primary food production to supply of food to consumer, since each element may have a potential impact on food safety.

Although there is a lack of studies focusing on *Arcobacter* detection in primary food production, some studies showed the presence of this bacterium during production, rearing or growing of primary products, mostly regarding to animal production prior to slaughter.

Throughout times the use of sewage sludge and animal manure in soils has been an agricultural practice, which may facilitate the introduction of pathogens into soil-water systems and subsequently into the food chain (Udeigwea *et al.*, 2015). *Arcobacter* spp. has been detected in animal manure and effluents (Chinivasagam *et al.*, 2007; Lévesque *et al.*, 2014; Whiteduck-Leveillee *et al.*, 2015), on manure-treated soils, also affecting groundwater (Hong *et al.*, 2013) or effluent-irrigated soils (Chinivasagam *et al.*, 2007) indicating that this practice may have a role on the introduction of *Arcobacter* on the food chain. For example, during a study on the potential use of domestic wastewater for irrigation, the authors demonstrated that this practice could lead to an increase of *Arcobacter* spp. in the phyllosphere, possibly associated with the concentration of *Arcobacter* in wastewater that could reach the same degree of magnitude than total bacteria, although this observation was not confirmed when treated wastewater was used (Yang *et al.*, 2015). Despite this data, the systematic use of these agricultural practices can represent a potential risk of food chain contamination with several pathogens, including *Arcobacter* spp.. Similarly in aquaculture, it is possible to observe the influence of animal wastes usage as fertilizer, with *Arcobacter* being present in sediment and water samples from aquaculture environment in China, with potential negative impact on human health (Xiong *et al.*, 2015). In fact, *A. cryaerophilus* infections were reported in rainbow trout farms in Turkey (Aydin *et al.*, 2000). Also, the presence of *Arcobacter* spp. on marine waters has been pointed as a potential route of contamination of shellfish (Collado *et al.*, 2008); indeed, *Arcobacter* has been detected on seafood and shellfish with high prevalence and species diversity (Collado *et al.*, 2009b; Nieva-Echevarria *et al.*, 2013; Fernandez *et al.*, 2015; Levican *et al.*, 2014, 2015; Šilha *et al.*, 2015), with several new species being for the first time isolated and characterized from these food samples (Table 4.1). Levican *et al.* (2014) reported eleven different species of *Arcobacter* in shellfish harvested from a farm of bivalve molluscs in Spain. The shellfish included mussels, oysters, clams, and bean clams showing a global *Arcobacter* spp. prevalence of 29.9%, through culturing and molecular detection methods (Levican *et al.*, 2014).

A. butzleri, *A. cryaerophilus*, and *A. skirrowii* can be detected in different types of environmental water samples such as seawater (Fera *et al.*, 2004; Collado *et al.*, 2008), estuarine waters (Fera et al., 2010), wastewater (Diergaardt et al., 2004; González *et al.*, 2007a, 2010), surface, river, canal, lakes, and spring water (Musmanno *et al.*, 1997; Diergaardt *et al.*, 2004; Morita *et al.*, 2004; Collado *et al.*, 2008, 2010; Ertas *et al.*, 2010; Lee *et al.*, 2012), but also in drinking water

(Jacob *et al.*, 1993, 1998; Ertas *et al.*, 2010), or water from farms (Gude *et al.*, 2005; De Smet *et al.*, 2011c; Shah *et al.*, 2013; Giacometti *et al.*, 2015b) and food processing plants (Gude *et al.*, 2005; Van Driessche and Houf, 2007b; Hausdorf *et al.*, 2011, 2013b; Khoshbakht *et al.*, 2014). This vast repertoire of water samples in which *Arcobacter* spp. can be detected, together with its ability to survive in these environments with scarcity of nutrients and even with low temperatures (Fera *et al.*, 2008) supports its potential as a waterborne pathogen, and therefore, a source of food contamination.

Arcobacter spp. is commonly isolated from the several steps of food animal production chain, namely from food animals on-farm, to slaughterhouses and retail. On-farm, *Arcobacter* species have been found with a wide range of prevalence: 6.2 to 40.1% in cattle (Wesley *et al.*, 2000; Van Driessche *et al.*, 2005; Shah *et al.*, 2013; Grove-White *et al.*, 2014; Giacometti *et al.*, 2015b), 0 to 10.7% in goats (De Smet *et al.*, 2011c; Shah *et al.*, 2013), 8 to 43.1% in sheep (De Smet *et al.*, 2011c; Grove-White *et al.*, 2014), 11.3 to 50% in pigs (Van Driessche *et al.*, 2004; Ho *et al.*, 2006b; De Smet *et al.*, 2011b) and 0 to 11% in live poultry (Andersen *et al.*, 2007; Amare *et al.*, 2011; Bogantes *et al.*, 2015). *A. butzleri* and *A. cryaerophilus* has been the most frequently isolated species in several animal on-farm samples (Van Driessche *et al.*, 2004; De Smet *et al.*, 2011c; Piva *et al.*, 2013; Bogantes *et al.*, 2015); however, other studies report *A. thereius* or *A. skirrowii* as the most prevalent (Ho *et al.*, 2006b; De Smet *et al.*, 2011b), with animal co-colonization seeming not to be uncommon (Van Driessche *et al.*, 2005, 2004; Ho *et al.*, 2006b; Piva *et al.*, 2013; Shah *et al.*, 2013). The prevalence figures varied between the farms, animals, sampling occasions, sampling methods, or even isolation methods, with parameters such as animal age, feed, season, and management practice being apparently associated with fecal *Arcobacter* spp. detection (Wesley *et al.*, 2000; Van Driessche *et al.*, 2005; Grove-White *et al.*, 2014).

When evaluating the prevalence of *Arcobacter* spp. in food products at retail, a high frequency of isolation is found (Table 4.3), indicating a possible contamination during processing, namely during slaughter of food animals. However, this has not yet been fully clarified and several authors have suggested different routes of contamination. The presence of *Arcobacter* spp. in the intestinal tract of health food animals with no symptoms may represent a contamination risk of carcasses fecal contamination during slaughter (Shah *et al.*, 2011a; De Smet *et al.*, 2011c), as food products of animal origin have been suggested as important sources of *Arcobacter* spp. transmission (Collado and Figueras, 2011 and references herein). Beyond food chain contamination, also the environment contamination could be a consequence of *Arcobacter* spp. presence in animal manure and faeces, namely contamination of soil and water sources (Chinivasagam *et al.*, 2007; Wilkes *et al.*, 2014). This genus has also been widely detected on farm, slaughter, or processing plants environment and equipment (Ferreira *et al.*, 2013; Rasmussen *et al.*, 2013; Serraino *et al.*, 2013a; Shah *et al.*, 2013; Serraino and Giacometti, 2014; Scarano *et al.*, 2014), which may have a potential role in contamination beyond fecal source. In fact, albeit *Arcobacter* frequent isolation from poultry carcasses and meat, it is uncommonly detected in live birds, being proposed that birds are not reservoir hosts. This suggests, therefore, post slaughter carcass contamination (Wesley and Miller 2010), associated with *Arcobacter* spp. colonization of the plant environment, instead of faecal contamination, with water likely being one possible source of contamination as well (Gude *et al.*, 2005; Andersen *et al.*, 2007; Khoshbakht *et al.*, 2014). This is supported by the rare recovery of *Arcobacter* from the intestinal contents or feces of chickens and the distribution of this genus in slaughterhouse environment; however, the contamination route was not yet clearly demonstrated (Gude *et al.*, 2005; Aydin *et al.*, 2007; Van Driessche and Houf, 2007b), with reports indicating an entrance of *Arcobacter* spp. into

Table 4.3 *Arcobacter* spp. Distribution on Food Samples at Retail or Local Markets.

Product	Source	Species detected	Prevalence (%) (positive/total)[a]	Country	References
Shellfish (mussels and clams)	Local market, with exception of mussels	*A. butzleri, A. cryaerophilus, A. skirrowii, A. mytili, A. nitrofigilis* and others arcobacters[b]	45.9% (28/61)	Spain	Collado *et al.*, 2009b
Seafood (frozen shrimp and oysters)	Local market	ND	0% (0/23)	Spain	Collado *et al.*, 2009b
Shellfish (clams and mussels)	Retailers	*A. butzleri* and *A. cryaerophilus*	73.3% (12/16)	North Spain	Nieva-Echevarria *et al.*, 2013
Shellfish (clams, mussels, oysters, razor clams, scallop and surf clams)	–	*A. butzleri, A. cryaerophilus, A. skirrowii* and *A. defluvii*	40.5% (43/106)	Chile	Collado *et al.*, 2014
Seafood (shrimp, bivalve and salmon)	Retail store	*A. butzleri, A. cryaerophilus* and *A. skirrowii*	46.2% (6/13)	Czech Republic	Šilha *et al.*, 2015
Fish meat	Supermarkets and retail shops	*A. butzleri* and *A. cryaerophilus*	34.0% (17/50)	Germany	Münch and Wochenschr, 2015
Poultry	Supermarkets	*A. butzleri, A. cryaerophilus* and *A. skirrowii*	61.7% (58/94)	Northern Ireland	Scullion *et al.*, 2006
Poultry meat	Supermarkets and retail shops	*A. butzleri*	26.8% (15/56)	Germany	Münch and Wochenschr, 2015
Fresh chicken carcasses	Retail market	*A. butzleri*	95.4% (42/44)	Turkey	Atabay *et al.*, 2003
Frozen chicken carcasses	Retail market	*A. butzleri*	22.6% (7/31)	Turkey	Atabay *et al.*, 2003
Chicken (drumsticks, edible viscera, ground or chopped breast)	Butcher shops	*A. butzleri* and *A. skirrowii*	40.0% (18/45)	Mexico	Villarruel-López *et al.*, 2003
Chicken meat	Retail shops	*A. butzleri, A. cryaerophilus* and *A. skirrowii*	23.0% (23/100)	Japan	Kabeya *et al.*, 2004
Chicken ground meat	Supermarkets	*A. butzleri* and *A. cryaerophilus*	48.8% (20/41)	Japan	Morita *et al.*, 2004

Chicken ground meat	Meat shops	*A. butzleri*	100% (10/10)	Thailand	Morita *et al.*, 2004
Chicken ground meat	Butchers and supermarkets	*A. butzleri*	72.7% (16/22)	Australia	Rivas *et al.*, 2004
Chicken carcasses	Retail shops	*A. butzleri*	68.0% (68/100)	Turkey	Aydin *et al.*, 2007
Chicken meat	Fresh market and supermarket	*A. butzleri*	38.0% (19/50)	Thailand	Vindigni et al., 2007
Chicken meat	Retail market	*A. butzleri*	64.3% (9/14)		Collado *et al.*, 2009b
Chicken (drumsticks, leg quarters and breast with skin)	Supermarkets and retail shops	*A. butzleri* and *A. cryaerophilus*	85.7% (36/42)	Spain	Pentimalli *et al.*, 2009
Chicken	Retail butchers and supermarket	*A. butzleri* and *A. cryaerophilus*	21.1% (76/360) by molecular methods; 10.6% (38/360) by cultural techniques	Republic of Korea	Lee *et al.*, 2010
Chilled chicken meat	Supermarket	*A. butzleri*	37.1% (23/62)	Malaysia	Amare *et al.*, 2011
Fresh chicken meat	Wet markets	*A. butzleri*	41% (25/61)	Malaysia	Amare *et al.*, 2011
Chicken (breast and sausage)	Retailers	*A. butzleri*	55.0% (11/20)	North Spain	Nieva-Echevarria *et al.*, 2013
Chicken viscera	Retail	*A. butzleri*, *A. cryaerophilus*, *A. skirrowii* and other arcobacters[b]	17.3% (26/150)	Costa Rica	Villalobos *et al.*, 2013
Chicken breast	Retail markets	*A. butzleri*, *A. cryaerophilus* and other arcobacters[b]	56.0% (28/50)	Costa Rica	Fallas-Padilla *et al.*, 2014
Chicken meat	Supermarkets and retail shop	*A. butzleri* and other arcobacters[b]	45.1% (23/51) by molecular methods; 41.2% (21/51) by cultural techniques	Spain	González *et al.*, 2014
Chicken meat	Retail outlet	*A. butzleri*, *A. cryaerophilus* and *A. skirrowii*[c]	31.0% (31/100) by molecular methods; 28.0% (28/100) by cultural techniques	Iran	Rahimi, 2014

(*Continued*)

Table 4.3 (Continued)

Product	Source	Species detected	Prevalence (%) (positive/total)[a]	Country	References
Chicken meat	Retail store	*A. butzleri* and *A. cryaerophilus*	80.5% (33/41)	Czech Republic	Šilha *et al.*, 2015
Chicken meat	Supermarkets	*A. butzleri* and *A. cryaerophilus*	85.7% (60/70)	Poland	Zacharow *et al.*, 2015a
Turkey carcasses	Retail shops	*A. butzleri*	4.0% (4/100)	Turkey	Aydin *et al.*, 2007
Turkey meat	Retail market	*A. butzleri* and *A. cryaerophilus*	33.3% (10/30)	Spain	Collado *et al.*, 2009b
Turkey meat	Retail outlet	*A. butzleri*	12.0% (12/100) by molecular methods; 11.0% (11/100) by cultural techniques	Iran	Rahimi, 2014
Duck meat	Retail market	*A. butzleri* and *A. cryaerophilus*	40.0% (2/5)	Spain	Collado *et al.*, 2009b
Duck meat	Retail outlet	*A. butzleri*, *A. cryaerophilus*[c] and *A. skirrowii*[c]	28.0% (14/50) by molecular methods; 11.4% (8/50) by cultural techniques	Iran	Rahimi, 2014
Duck meat	Retail store	*A. butzleri*	66.7% (8/12)	Czech Republic	Šilha *et al.*, 2015
Ostrich meat	Retail outlet	*A. butzleri*	3.3% (2/60) by molecular methods; 3.3% (2/60) by cultural techniques	Iran	Rahimi, 2014
Geese meat	Retail outlet	*A. butzleri* and *A. cryaerophilus*	10.0% (5/50) by molecular methods; 8.0% (4/50) by cultural techniques	Iran	Rahimi, 2014
Quail meat	Retail outlet	*A. butzleri*, *A. cryaerophilus* and *A. skirrowii*	12.0% (12/100) by molecular methods; 12.0% (12/100) by cultural techniques	Iran	Rahimi, 2014
Partridge meat	Retail outlet	*A. butzleri*, *A. cryaerophilus* and *A. skirrowii*[c]	11.3% (9/80) by molecular methods; 7.5% (6/80) by cultural techniques	Iran	Rahimi, 2014

Ground pork	Butcher shops	*A. butzleri*, *A. cryaerophilus* and *A. skirrowii*	51.1% (23/45)	Mexico	Villarruel-López *et al.*, 2003
Pork meat	Retail shops	*A. butzleri* and *A. cryaerophilus*	7.0% (7/100)	Japan	Kabeya *et al.*, 2004
Pork ground meat	Butchers and supermarkets	*A. butzleri*	28.6% (6/21)	Australia	Rivas *et al.*, 2004
Pork	Supermarkets	*A. butzleri*, *A. cryaerophilus*, *A. skirrowii* and other arcobacters[b]	34.6% (35/101)	Northern Ireland	Scullion *et al.*, 2006
Pork meat (minced and chop)	Bucher shops and supermarkets	*A. butzleri* and *A. cryaerophilus*	21.3% (10/47)	Belgium	Van Driessche and Houf, 2007a
Pork meat	Fresh market and supermarket	*A. butzleri*	16.0% (8/50)	Thailand	Vindigni *et al.*, 2007
Pork	Retail market	*A. butzleri*, *A. skirrowii* and others arcobacters[b]	52.9% (9/17)	Spain	Collado *et al.*, 2009b
Pork	Retail butchers and supermarket	ND	0% (0/100)	Republic of Korea	Lee *et al.*, 2010
Pork (ribs)	Retailers	*A. butzleri*	20.0% (2/10)	North Spain	Nieva-Echevarria *et al.*, 2013
Pork meat	Retail store	*A. butzleri*	3.8% (1/26)	Czech Republic	Šilha *et al.*, 2015
Pork meat	Supermarkets	*A. butzleri* and *A. cryaerophilus*	15.7% (11/70)	Poland	Zacharow *et al.*, 2015a
Mixed minced meat (pork and beef)	Supermarkets and retail shops	*A. butzleri*	2.0% (1/51)	Germany	Münch and Wochenschr, 2015
Ground beef	Butcher shops	*A. butzleri* and *A. skirrowii*	28.9% (13/45)	Mexico	Villarruel-López *et al.*, 2003
Beef meat	Retail shops	*A. butzleri*, *A. cryaerophilus* and *A. skirrowii*	2.2% (2/90)	Japan	Kabeya *et al.*, 2004
Beef ground meat	Butchers and supermarkets	*A. butzleri*	21.9% (7/32)	Australia	Rivas *et al.*, 2004
Beef	Supermarkets	*A. butzleri*, *A. cryaerophilus*, *A. skirrowii* and other arcobacters[b]	34.2% (37/108)	Northern Ireland	Scullion *et al.*, 2006

(*Continued*)

Table 4.3 (Continued)

Product	Source	Species detected	Prevalence (%) (positive/total)[a]	Country	References
Minced beef meat	Retail shops	*A. butzleri* and *A. cryaerophilus*	37.0% (10/27)	Turkey	Aydin *et al.*, 2007
Beef meat	Fresh market and supermarket	*A. butzleri*	14.0% (28/50)	Thailand	Vindigni *et al.*, 2007
Beef	Retail butchers and supermarket	ND	0% (0/106)	Republic of Korea	Lee *et al.*, 2010
Minced beef	Butcher shops and supermarkets	*A. butzleri* and *A. cryaerophilus*	9.0% (9/100)	Belgium	De Smet *et al.*, 2010
Beef meat	Stalls in wet markets	*A. butzleri*, *A. cryaerophilus* and *A. skirrowii*	30.2% (32/106)	Malaysia	Shah *et al.*, 2012a
Minced beef	Retailers	*A. butzleri*	10.0% (1/10)	North Spain	Nieva-Echevarria *et al.*, 2013
Beef meat	Retail store	ND	0.0% (0/15)	Czech Republic	Šilha *et al.*, 2015
Beef meat	Supermarkets	*A. butzleri* and *A. cryaerophilus*	20.0% (14/70)	Poland	Zacharow *et al.*, 2015a
Rabbit meat	Retail market	*A. butzleri*	10.0% (1/10)	Spain	Collado *et al.*, 2009b
Lamb ground meat	Butchers and supermarkets	*A. butzleri*	15.4% (2/13)	Australia	Rivas *et al.*, 2004
Sausage	Local market	ND	0% (0/27)	Spain	Collado *et al.*, 2009b
Eggs	Fresh market and supermarket	*A. butzleri*	2.0% (1/50)	Thailand	Vindigni *et al.*, 2007
Sheep ricotta cheese	Retail	*A. butzleri*	21.9% (7/32)	Italy	Scarano *et al.*, 2014
Fresh village cheese	Retail markets	*A. butzleri*, *A. cryaerophilus*, *A. skirrowii*, *A. nitrofigilis*, *A. cloacae*, *A. bivalviorum* and *A. cibarius*	56.0% (28/50)	Turkey	Yesilmen *et al.*, 2014
Lettuces	Retail shops	*A. butzleri*	20.0% (10/50) by molecular methods; 14.0% (7/50) by cultural techniques	Spain	González and Ferrús, 2011

[a] Prevalence of *Arcobacter* spp., if not stated, isolated by cultural techniques; [b] *Arcobacter* sp. not identified by the used methodology; [c] Specie only detected by molecular methods (PCR of enrichment); ND – not detected.

slaughterhouse through the gut content of chickens, similarly to *Campylobacter* spp. (Ho *et al.*, 2008). Cross-contamination among carcasses or through equipment and environment during slaughter processing has also been suggested as one explanation (Van Driessche and Houf, 2007a; Ho *et al.*, 2008; Ferreira *et al.*, 2013); still not being enough to fully clarify the levels of food contamination (Houf *et al.*, 2002b). In a recent longitudinal study in a dairy farm performed to identify the possible origin of *Arcobacter* spp. milk contamination, it was proposed that water may be the source of animal infection and cause of the spread in farm environment, not being fecal contamination the only or main source of milk contamination (Giacometti *et al.*, 2015b). Nevertheless, more work is necessary to demonstrate the precise route of contamination throughout the whole food chain.

At retail, an *Arcobacter* global mean prevalence of 34.4% was reported among poultry source samples, with the lowest being found for ostrich meat collected in Iran (3.3%) (Rahimi, 2014) and the highest (100%) for chicken ground meat samples from Thailand with all samples positive for *A. butzleri* (Morita *et al.*, 2004). Considering animal meat at retail, *Arcobacter* spp. was more frequently detected in poultry (3.3–100%) followed by pork (0–52.9%) and beef (0–37%); however, it has been also found in rabbit and lamb meat (Table 4.3). Furthermore, in seafood, the detection frequency was similarly high, with 34.0 to 73.3% of the samples presenting *Arcobacter* spp. contamination; however, no pathogens were detected on frozen shrimp and oysters (Collado *et al.*, 2009b).

As for vegetables, *Arcobacter* spp. was found in spinach (Hausdorf *et al.*, 2013a, 2013b) and fresh lettuces (González and Ferrús, 2011). Among the contaminating species, *A. butzleri* was detected in lettuces (González and Ferrús, 2011), and additionally to the most commonly detected species, *A. butzleri* and *A. cryaerophilus*, operational taxonomic units highly similar to *A. nitrofigilis, A. ellisii, A. suis, A. venerupis*, and unknown species were detected both in spinach and at the processing plant (Hausdorf *et al.*, 2013b). A considerable risk of vegetables contamination during the industrial washing process was previously suggested to be caused by potential pathogens suspended in the wash water probably introduced by the vegetable and dirt associated (data reported for carrots) (Hausdorf *et al.*, 2011).

Beyond the association of *Arcobacter* with raw food products, its presence was also detected in ready-to-eat food. A study addressing enteric pathogens in meals served at Bangkok restaurants recommend for tourists, showed a higher prevalence of *Arcobacter* than *Salmonella* or *Campylobacter*. The authors estimated the traveller's risk of exposure to *Arcobacter* as 13% per meal eaten, rising to 75% when 10 meals are consumed (Teague *et al.*, 2010).

The detection of *Arcobacter* in foods usually consumed raw or without subsequent heat treatment, such as vegetables, cheeses, or others, suggest that it are a potential route of transmission of this pathogen; however, the data available is still very limited and do not allows to fully perform a suitable risk evaluation.

Unlike other microorganisms, the persistence ability of *Arcobacter* spp. in food sources and plant environment is still poorly characterized. Survival of *Arcobacter* spp. in food sources has been described (Ferreira *et al.*, 2015, and references here in), with a few studies showing that some species could survive in raw ground pork (D'Sa and Harrison, 2005), on vacuum-packaged chill stored beef (Balamurugan *et al.*, 2013), in both ultrahigh-temperature or pasteurized milk and in raw cow's milk (Giacometti *et al.*, 2014). In addition, it was shown to survive during water buffalo mozzarella cheese production or storage (Serraino *et al.*, 2013b), in ricotta cheese (Giacometti *et al.*, 2015a) or in water at different temperatures usually used in food industry (Van Driessche and Houf, 2008); nevertheless, *A. butzleri* did not exhibit this ability in apple and pear purees (Lee and Choi, 2013) or in beer (Šilha *et al.*, 2013). Furthermore, several features exhibited by *Arcobacter* such the ability of survive on stainless steel, plastic, and glass surfaces (Cervenka *et al.*, 2008; Šilha *et al.*, 2014), to survive to scalding process conditions (Ho *et al.*, 2008), along with the biofilm-forming ability (Ferreira *et al.*, 2013), to tolerate biocides (Rasmussen *et al.*, 2013) or proliferate under chilled conditions (Kjeldgaard *et al.*, 2009), could be associated to its persistence in plants. In addition, *A. butzleri* is also able to survive to common sanitizing procedures (Hausdorf *et al.*, 2013b; Rasmussen *et al.*, 2013; Scarano *et al.*, 2014). The findings associated with *Arcobacter* spp. persistence and survival in food processing and food products highlight their importance as human pathogens that can be found over all the food chain, from primary production to food products directly available to consumer.

4.2 Isolation and Detection of *Arcobacter*

4.2.1 Conventional Bacteriological Culture

A diversity of methods have been exploited for the isolation of *Arcobacter* spp.; however, to date, no International Regulatory Agency presented an official recommended method for the isolation of *Arcobacter* sp. from food or other products. Traditionally, when the microbiological analysis of food products regarding pathogenic bacteria is performed, there is an isolation step through agar culture, which may be inoculated after enrichment. The reported culture methods vary from modified *Campylobacter* and *Leptospira* techniques to those involving *Arcobacter*-specific media (Merga *et al.*, 2011), with several enrichment and plating media being recommended for *Arcobacter* isolation and growth. The isolation step can be performed by the use of a selective medium or the inclusion of a filtration of the sample homogenate onto non-selective blood agar plates.

Several media and procedures for *Arcobacter*-specific isolation have been developed over the years, with diverse selective antibiotic being used in both enrichment and isolation steps. A first enrichment step has been described to be necessary in order to isolate *Arcobacter* spp. from poultry carcasses or pig abortions (Atabay and Corry, 1997; On *et al.*, 2002), with other works also reporting higher recovery rates when using an enrichment incubation (Van Driessche *et al.*, 2003; Gude *et al.*, 2005; Houf and Stephan, 2007; Van Driessche and Houf, 2007a). Nonetheless, this enrichment may be associated with a reduction in the isolation rate of *A. cryaerophilus* and *A. skirrowii* or other species probably due to antibiotic susceptibility, or even overgrown of some species over others. Houf *et al.* (2002b) reported the isolation of both *A. butzleri* and *A. cryaerophilus* species by direct plating, but when using the enrichment step, only *A. butzleri* was isolated, however, presenting a higher recovery rate than that obtained by direct plating. In a similar way, without the enrichment step, De Smet *et al.* (2011b) described a higher isolation of *A. thereius* and *A. skirrowii* through direct isolation using the selective media described by Van Driessche *et al.* (2003), however, with a lower overall recovery rate. The use of direct plating rule out the potential selection occurring during the enrichment phase, which can influence epidemiological results (Houf *et al.*, 2001a), however, diminishing the recovery of *A. butzleri*, probably due to low contamination charge (Van Driessche and Houf, 2007a).

As *Arcobacter* spp. is found in samples with a dominant accompanying flora, antimicrobial agents of various classes at diverse concentrations are usually used in both enrichment and isolation media. Some selective agents combinations have been described for application in enrichment medium (Collins *et al.*, 1996; Atabay and Corry, 1997; Johnson and Murano, 1999b; Houf *et al.*, 2001a; Van Driessche *et al.*, 2003), which are used to reduce accompanying flora, although some of the selective supplement present antibiotic concentrations that may inhibit the growth of a few *Arcobacter* species (Houf *et al.*, 2001b).

Collins *et al.* (1996) described an enrichment with the *Leptospira* medium, Ellinghausen-McCullough-Johnson-Harris Polysorbate-80 supplemented with 100 mg/L of 5-fluorouracil, followed by isolation with a modified cefsulodin-irgasan-novobiocin (CIN) agar and brain heart infusion agar supplemented with 10% of defibrinated bovine blood, cephalothin, vancomycin, and amphotericin B (CVA) agar. Both isolation media presented an efficient *Arcobacter* recovery from ground pork samples (Collins *et al.*, 1996). Also, de Boer *et al.* (1996) developed an enrichment and plating media using cefoperazone (32 mg/L), piperacillin (75 mg/L), trimethoprim (20 mg/L), and cicloheximide (100 mg/L) as selective agents for *Arcobacter* isolation from poultry, pork, and beef meats. Johnson and Murano (1999b) established a new media formulation supplemented with bile salts, cefoperazone (32 mg/L), and 5-fluorouracil (200 mg/L) as selective agents. The latter medium was compared with several broths and solid media for *Arcobacter* isolation from poultry showing to be most effective with limited contamination (Johnson and Murano, 1999a).

Through the evaluation of the susceptibility of *A. butzleri, A. cryaerophilus,* and *A. skirrowii* to antibiotics used in selective media, Houf *et al.* (2001b) showed that many of the *A. skirrowii* and *A. cryaerophilus* strains were susceptible to piperacillin at the levels used in some selective media; moreover, inclusion of cefoperazone at 32 mg/L in media could be detrimental to stressed or injured cells of *A. cryaerophilus* and *A. skirrowii.* Following this work, the same group developed a new protocol for the isolation of these three *Arcobacter* species through the incorporation of cefoperazone (16 mg/L), amphotericin B (10 mg/L), 5-fluorouracil (100 mg/L), novobiocin (32 mg/L), and trimethoprim (64 mg/L) in both enrichment and isolation media (Houf *et al.*, 2001a). These media were later modified to be used in livestock feces, by increasing novobiocin concentration (64 mg/L) and with the addition of cycloheximide (100 mg/L) for specificity improvement and fungal growth delay (Van Driessche *et al.*, 2003). Later on, it was validated for *Arcobacter* isolation from poultry intestinal content with a small increase in 5-fluorouracil concentration (125 mg/L) as well (Van Driessche and Houf, 2007b). The Houf *et al.* (2001a) selective isolation method was also validated for human stool samples and showed to ensure the growth of *A. butzleri, A. cryaerophilus,* and *A. skirrowii* and suppressed the accompanying flora without affecting the isolation of *Arcobacter* (Houf and Stephan, 2007).

Other methods using passive filtration for isolation by culture on agar plates were described for *Arcobacter*-specific isolation from biopsy samples taken from aborted porcine fetuses (On *et al.*, 2002), chicken samples (Atabay and Corry, 1997; Atabay *et al.*, 1998, 2003; Villalobos *et al.*, 2013), domestic geese (Atabay *et al.*, 2008b) or others (Aydin *et al.*, 2007; Ertas *et al.*, 2010), allowing the isolation of *A. butzleri, A. cryaerophilus,* and *A. skirrowii.*

Several studies have evaluated and compared the differences between culture methods (Atabay and Corry, 1998; Golla *et al.*, 2002; Scullion *et al.*, 2004; Hamill *et al.*, 2008; Merga *et al.*, 2011; Shah *et al.*, 2011b), however, a systematic comparison between all the recommended methods was not yet performed. Shah *et al.* (2011b) likened the methods used by De Boer *et al.* (1996) and Atabay *et al.* (1998), where the latter used an enrichment broth with cefoperazone (8 mg/L), amphotericin B (10 mg/L), and teicoplanin (4 mg/L) (CAT) supplement followed by a passive filtration step. It was found that the Atabay's method showed improved results for beef, fresh milk, and rectal swabs of cattle concerning *A. butzleri, A. cryaerophilus,* and *A. skirrowii* isolation. Oppositely, other author referred to passive filtration methods as less sensitive than the ones based on selective isolation media (Gude *et al.*, 2005). Recently, Šilha *et al.* (2015) proposed a modification of the method recommended by Atabay and Corry (1997) replacing the agar culture medium by *Campylobacter* selective agar (CASA), which improved the inhibition of competitive flora growth and facilitated the recognition of *Arcobacter* colonies (tiny brick-red colonies) for *A. butzleri*, *A. cryaerophilus*, and *A. skirrowii*. In this work, the author introduced a passive filtration step in order to overcome the use of high content of accompanying flora-supressing agents that could also suppress *Arcobacter* spp..

When comparing Collins, de Boer, and Johnson–Murano isolation methods, the latter was the most successful in *Arcobacter* detection in pork samples (Ohlendorf and Murano, 2002). Furthermore, two studies have compared the methods described by Houf *et al.* (2001a) and Johnson and Murano (1999b) in retail raw poultry and beef samples showing a similar detection of *Arcobacter* with the two methods, with an improvement in recovery through sampling of the enrichment broth at 24 h as well as at 48 h of incubation (Scullion *et al.*, 2004; Hamill *et al.*, 2008). Also, performing the incubation in a microaerobic atmosphere was associated with either a better survival and growth of *Arcobacter* spp. or better suppression of the contaminating flora (Hamill *et al.*, 2008).

Merga *et al.* (2011) performed a comparative study concerning *Arcobacter* isolation, from cattle, sheep, and badgers' fecal samples, by using five different methods corresponding to the combination of three enrichment and two isolation media. The authors tested the enrichment media reported by Houf *et al.* (2001a), Atabay and Corry (1998), and the *Campylobacter* specific enrichment broth described by Kemp *et al.* (2005) containing 5% (V/V) defibrinated horse blood, cefoperazone (20 mg/L), vancomycin (20 mg/L), trimethoprim (20 mg/L), and cycloheximide (50 mg/L). Regarding solid media, they tested the medium reported by Houf *et al.* (2001a) and a modified charcoal cefoperazone deoxycholate agar (mCCDA) with CAT supplementation (Kemp *et al.*, 2005). The most sensitive and specific method used was the enrichment broth by Houf *et al.* (2001a) combined with mCCDA with CAT supplement (Merga *et al.*, 2011). Other associations between enrichment broths and selective agars from different methods have been suggested, such as the use of the combination of Houf's enrichment selective broth (Fallas-Padilla *et al.*, 2014) or the use of De Boer selective broth followed by filtration on blood agar, with the latter broth having a better performance than the one described by Houf (Villalobos *et al.*, 2013).

Other relevant culture conditions to be taken into account are the temperature and atmosphere incubation requirements that may help to differentiate between *Campylobacter* and *Arcobacter*. Some of the *Arcobacter* species will grow optimally at 37 °C microaerophilically, whereas a temperature of 30 °C has been shown to support aerobic growth of this bacterium (Johnson and Murano, 1999b). However, lower temperatures have been used with De Boer *et al.* (1996) who applied an incubation temperature of 24 °C for *Arcobacter* isolation in order to reduce cold-sensitive microflora and favoring the selection for *Arcobacter*. Both aerobic and microaerobic conditions have been adopted for the isolation of *Arcobacter* from either clinical or environmental samples (Collado and Figueras, 2011). González *et al.* (2007a) observed a slight improvement in *Arcobacter* isolation rate when performing the enrichment step under microaerophilic conditions, without being able to draw a definitive conclusion concerning the atmospheric incubation conditions, a similar result was observed by Hamill *et al.* (2008). Nevertheless, Levican *et al.* (2014) observed an improvement in *Arcobacter* spp. recovery when using aerobic conditions in agar incubation step, suggesting however the use of both aerobic and microaerobic conditions. The authors have proposed the influence of atmospheric conditions on the diversity of species recovered, with a significant improvement in *A. butzleri* and *A. molluscorum* isolation under aerobic incubation, and the isolation of *A. thereius* and *A. defluvii* under microaerobic incubation. Beyond the increase in diversity, the use of both atmospheric conditions also influenced the higher number of genotypes recovered (Levican *et al.*, 2014).

Incubation time showed also to influence isolation of different *Arcobacter* species; for example, *A. cryaerophilus* and *A. skirrowii* required a longer enrichment and plating incubation time (Van Driessche *et al.*, 2003; Van Driessche and Houf, 2007b), with some studies reporting an increase in recovery when using both 24 and 48 h of enrichment broth incubation (Scullion *et al.*, 2004; Hamill *et al.*, 2008).

The success of bacterial cultivation is also dependent on the source of the sample, since when isolating *Arcobacter* from environmental samples, this process is highly dependent both on the distribution and on the growth state of bacteria at the time of sampling (Fera *et al.*, 2010).

The differences in recovery rates of *Arcobacter* spp. from various sources could in addition be attributed to a number of additional factors, such as geographic and seasonal divergence or other aspects associated with the hygiene at collection sites, but also to differences in the sensitivity and specificity of isolation methods used (Patyal *et al.*, 2011).

Various procedures based on *Campylobacter* isolation have been proposed, although the most frequently used isolation methods include the ones based on work published by Atabay and Corry (1997) and Houf *et al.* (2001a). In fact, the work that showed the highest diversity of *Arcobacter* species—10 recognized species, from 204 shellfish samples—used passive filtration as a subsequent step to enrichment with CAT supplement with incubation at 30 °C in aerobic atmosphere, a modified method from the one proposed by Atabay and Corry (1997) (Levican *et al.*, 2014).

In conclusion, considering all the parameters that may influence the recovery efficacy regarding both prevalence and diversity of *Arcobacter* spp., a standardized isolation method is required in order to allow a better comparison between studies, to get more reliable data and to assess the true burden of the colonization and infection with *Arcobacter*.

4.2.2 Molecular-Based Detection and Identification Methods

Phenotypical tests are frequently used for the identification of microorganisms, through the use of dichotomous keys or tables; its application for *Arcobacter* identification may, however, result in ambiguous results and even misidentification of *Arcobacter* spp.. Similarly to *Campylobacter* spp. these bacteria have fastidious growth requirements, present narrow biochemical reactivity and phenotypical similarities among species, which can jeopardize a correct identification through conventional phenotypic tests for differentiation from *Campylobacter* genus or among *Arcobacter* species (On, 1996). The use of commercial biochemical identification systems such as API CAMPY showed to lead to a misidentification of *A. butzleri* (Harrass *et al.*, 1998; Martiny *et al.*, 2011), but was able to correctly identify *A. cryaerophilus* (Martiny *et al.*, 2011). Other systems that may be used for routine laboratory identification of some species of *Campylobacter*, such as the *Neisseria–Haemophilus* identification card for Vitek 2 system, have failed to identify both *A. butzleri* and *A. cryaerophilus* (Martiny *et al.*, 2011).

Matrix-assisted laser desorption/ionization time-of-flight mass spectrometry (MALDI-TOF MS) technique has been presented as an alternative to conventional phenotypic methods and has been used to discriminate among species and to infer the existence of new species (Levican *et al.*, 2012, 2013), as well as to identify *Arcobacter* sp. (Hausdorf *et al.*, 2013b; Figueras *et al.*, 2014; Šilha *et al.*, 2015). Some recent studies analysed the use of this technique for *A. butzleri*, *A. cryaerophilus*, and *A. skirrowii* identification. Bessède *et al.* (2011) conducted a comparison study between phenotypic conventional methods (API Campy strips), molecular (Fluorescence Resonance Energy Transfer real-time PCR), and MALDI-TOF MS identification. The authors found a misidentification of all *A. butzleri* strains (n = 14) used in the study by the use of conventional methods, whereas a concordance between molecular and mass spectrometry methods was established in the identification of all isolates. Nonetheless, *A. cryaerophilus* was not able to be identified by MALDI-TOF MS (Martiny *et al.*, 2011) or gave an unconfident identification when matching with the database or even low score when compared with a type strain (Figueras *et al.*, 2014). A low identification score was also observed by Šilha *et al.* (2015) who found that among 70 *Arcobacter* spp. isolates, including *A. butzleri*, *A. cryaerophilus*, and *A. skirrowii* species, only 31.4% were identified at species-level and 68.6% of isolates were identified at genus-level only. Once the MALDI-TOF MS identification is performed by comparison with reference spectra in the database, the problems concerning with the ability of the technique to confidently identify *Arcobacter* at species-level could be explained by the lack of a proper database system with an adequate number of bacteria strains. The expansion of these databases will allow a more reliable identification and will possibly spread MALDI-TOF MS use for *Arcobacter* spp. identification. In any case, an unconfident or low score result of the strain identity should be confirmed, for instance, by molecular techniques (Figueras *et al.*, 2014).

A variety of other techniques are available for the identification of *Arcobacter* species, such as SDS-PAGE of whole cell proteins (Atabay *et al.*, 2003) or cellular phospholipid and fatty acid analysis (Jelínek *et al.*, 2006), yet molecular approaches are the most used tools (González *et al.*, 2012). DNA-based methods have been described as more suitable for the rapid detection, confirmation, and identification to a species-level of *Arcobacter* and can be more efficient and cost-effective than the classic phenotypic identification methods (González *et al.*, 2006; Figueras *et al.*, 2008; González and Ferrús, 2011; Patyal *et al.*, 2011). However, the growing number of new *Arcobacter* species has increased the doubt regarding their specificity (Douidah *et al.*, 2010), with new molecular detection assays being proposed (Table 4.4).

Genus and species PCR-based methods have been vastly used for confirmation of presumptive *Arcobacter* colonies, for *Arcobacter* detection after enrichment, or as culture independent in different sort of samples. Some of the most widely used PCR-based methods for *Arcobacter* detection include simplex and/or multiplex polymerase chain reaction (PCR), PCR-restriction fragment length polymorphism (RFLP), and real-time PCR (Table 4.4).

Molecular tools have been used extensively in the microbiological analysis of foods. Indeed, assays based on the detection of amplified products are being recommended for confirmation and identification of suspected colonies, and commercialized systems are already available for other foodborne pathogens, such as *Campylobacter* (Maurer, 2011; Gharst *et al.*, 2013).

The detection rates by PCR-based procedures has been higher than with isolation methods, probably due to a low number of cells or to the presence of nonculturable cells in samples (Fera *et al.*, 2004; González and Ferrús, 2011; Bullman *et al.*, 2012; De Boer *et al.*, 2013). One advantage of the PCR-based detection methods is the diminished time necessary for analysis even when it is necessary to include an enrichment step, which will increase the number of viable cells and so circumvent false-positive results due to the presence of dead cells and false-negative by inhibitors dilution (González and Ferrús, 2011). Nonetheless, PCR methods may fail on *Arcobacter* spp. detection even when performed after enrichment due to sample inhibitors (Ho *et al.*, 2008). Distinct parameters may influence a successful PCR detection, such as the enrichment broth used, period of enrichment incubation and the sample origin (Pejchalova *et al.*, 2006). The reaction may even be influenced by the species selection, which may occur during enrichment. Notwithstanding, PCR-based methods are increasingly used for *Arcobacter* detection in the food chain, allowing for a simultaneous detection of more than one species in the same sample.

A few studies compared different methods and applications of molecular techniques based on detection of amplified products. *Arcobacter* sp. detection rates in chicken samples and lettuces were lower when identified by conventional microbiological analysis or conventional PCR than through the use of real-time PCR, however, a previous enrichment step showed to be required for the analysis. The differences between real-time PCR and conventional PCR could be associated with different detection limits (González *et al.*, 2010; González and

Table 4.4 Molecular Methods Used for *Arcobacter* spp. Identification and Their Application in Food Chain.

Method (reference)	Gene (s) targeted	Species discriminated	Sample used	Comment	Example of applications of the method
PCR (genus-specific and species-specific PCR) (Bastyns *et al.*, 1995)	*23S rRNA*	*Arcobacter* spp., *A. butzleri*, *A. cryaerophilus* and *A. skirrowii*	Reference strains and field isolates (Isolate identification)	Primers for *A. cryaerophilus* detect both *A. cryaerophilus* and *A. skirrowii*	Raw milk and retail raw meat (isolates identification – genus-specific PCR) (Scullion *et al.*, 2006); Chicken and wastewater (detection in enrichment broth and isolates identification - genus-specific PCR) (González *et al.*, 2007a); Fresh lettuces (detection in diluted sample or after enrichment – genus-specific PCR) (González and Ferrús, 2011); Porcine samples (isolates identification – genus-specific PCR) (Scanlon *et al.*, 2013); Shellfish, meat and raw milk samples (isolates identification – genus-specific PCR) (Nieva-Echevarria *et al.*, 2013)
PCR-RFLP (Cardarelli-Leite *et al.*, 1996)	*16S rRNA*	*A. butzleri*	Animal, human and environmental strains (Isolate identification)	Equal patterns for *A. cryaerophilus*, *A. skirrowii* and *A. nitrofigilis* Restrictions endonucleases used: *Dde*I, *Hpa*II	–
PCR (Harmon and Wesley, 1996)	*16S rRNA*	*Arcobacter* spp.	Reference strains (Isolate identification)	Identify *A. butzleri*, *A. cryaerophilus* and *A. skirrowii*, but with no amplification for *A. nitrofigilis*	Water (isolates identification) (Rice *et al.*, 1999); Poultry and slaughterhouse samples (detection in enrichment broth) (Ho *et al.*, 2008); Porcine carcasses and pork products at retail (isolates identification) (Van Driessche and Houf, 2007a); Pigs on farrow-to-finish farms (isolates identification) (De Smet *et al.*, 2011b);

(*Continued*)

Table 4.4 (Continued)

Method (reference)	Gene (s) targeted	Species discriminated	Sample used	Comment	Example of applications of the method
					Human, animal and food samples (detection in enrichment broth and isolates identification) (Patyal *et al.*, 2011); Chicken viscera (isolates identification) (Villalobos *et al.*, 2013); Chicken breast meat (isolates identification) (Fallas-Padilla *et al.*, 2014)
Multiplex PCR (Harmon and Wesley, 1997)	*16S rRNA*, *23S rRNA*	*Arcobacter* spp. and *A. butzleri* simultaneously	Reference strains, poultry and porcine isolates (Isolate identification)	Genus identification of *A. butzleri*, *A. cryaerophilus* and *A. skirrowii*. Differentiates *A. butzleri* from *Arcobacter* spp.	Water (isolates identification) (Rice *et al.*, 1999); Meat and faecal samples of healthy cattle (isolates identification at genus level) (Öngör *et al.*, 2004); Meat and water (isolates identification) (Pejchalova *et al.*, 2006); Market-weight turkeys, farm and slaughter sampling (isolates identification) (Andersen *et al.*, 2007)
PCR-RFLP (Hurtado and Owen, 1997)	*23S rRNA*	*Arcobacter* genus, *Campylobacter* genus, *A. butzleri*, *A. nitrofigilis*	Field strains (Isolate identification)	Equal patterns for *A. cryaerophilus* and *A. skirrowii* Restrictions endonucleases used: *Cfo*I, *Hpa*II and *Hinf*I	Evaluation of *A. butzleri* DNA integrity (Moreno *et al.*, 2004)
PCR-RFLP (Marshall *et al.*, 1999)	*16S rRNA*	*A. butzleri*, *A. cryaerophilus* and *A. skirrowii*	Reference strains (Isolate identification)	Restrictions endonucleases used: *DdeI* and *TaqI*	Groundwater (isolates identification to genus level) (Fong *et al.*, 2007)
PCR (González *et al.*, 2000)	*16S rRNA*	*Arcobacter* spp.	Chicken meat PCR (Detection on enrichment broth)	Detection takes in account *A. butzleri*, *A. cryaerophilus* and *A. skirrowii*.	Cattle samples (isolates identification) (Merga *et al.*, 2013); Poultry and slaughterhouse samples (Detection on enrichment broth) (Khoshbakht *et al.*, 2014)

Multiplex PCR (Houf *et al.*, 2000)	*16S rRNA*, *23S rRNA*	*A. butzleri*, *A. cryaerophilus* and *A. skirrowii*	Poultry (Detection on enrichment broth and isolate identification)	Confusion between *A. cryaerophilus* and *A. skirrowii* with later reported species (Collado and Figueras, 2011), (Levican and Figueras, 2013). Detection of *Campylobacter jejuni* and *C. coli* as false positives for *A. skirrowii* (Wang *et al.*, 2014).	Chicken rearing and processing samples (Detection in enrichment broth before and after incubation, and isolates identification) (Gude *et al.*, 2005); Chicken and wastewater (detection in enrichment broth and isolates identification) (González *et al.*, 2007a); Chicken samples (Detection in enrichment broth and isolates identification) (Ho *et al.*, 2008); Meat and shellfish (isolates identification) (Collado *et al.*, 2009b); Retail meats (Detection in enrichment broth and isolates identification) (Lee *et al.*, 2010); Poultry and slaughterhouse samples (Isolates identification) (Ferreira *et al.*, 2013); Shellfish, meat and raw milk samples (isolates identification) (Nieva-Echevarria *et al.*, 2013); Fish meat, poultry meat and pork and beef minced meat (Isolates identification) (Münch and Wochenschr, 2015); Dairy farm and possible sources of contamination (Isolates identification – *A. butzleri* confirmation) (Giacometti *et al.*, 2015b)
PCR-hybridization (Rashid *et al.*, 2000)	*glyA*	*A. butzleri*	Reference strains and isolates (Isolate identification)	–	–
PCR-ELISA (Antolín *et al.*, 2001)	*16S rRNA*	*Arcobacter* spp.	Poultry (Enrichment broth)	Allows quantification from 10-10^4 cfu/g of *Arcobacter* spp. in poultry samples	–

(*Continued*)

Table 4.4 (Continued)

Method (reference)	Gene (s) targeted	Species discriminated	Sample used	Comment	Example of applications of the method
Multiplex PCR (Kabeya *et al.*, 2003a)	*23S rRNA*	*A. butzleri*, *A. cryaerophilus* 1A, *A. cryaerophilus* 1B and *A. skirrowii*	Reference strains and meat or feces isolates (Isolate identification)	DNA concentration must be strictly adjusted to 20 ng/reaction due to nonspecific amplifications with different concentrations. Produce unreliable results for all three of its targeted species (Levican and Figueras, 2013).	Cattle, swine and chicken samples (Isolates identification) (Kabeya *et al.*, 2003b); Beef, pork and chicken retail meats (Isolates identification) (Kabeya *et al.*, 2004); Broiler samples (isolates identification) (Son *et al.*, 2007)
FISH (Moreno *et al.*, 2003)	*16S rRNA*	*Arcobacter* spp.	Water and Sewage (Culture independent)	Probe from (Snaidr *et al.*, 1997) Tested to *A. butzleri*, *A. cryaerophilus*, *A. nitrofigilis*, *A. skirrowii*	Estuarine waters (Detection before and after incubation in enrichment broth) (Fera *et al.*, 2010)
PCR-RFLP (Karenlampi *et al.*, 2004)	*groEL*	*A. butzleri*	Reference strain (Isolate identification)	–	–
PCR-RFLP (González *et al.*, 2006)	*16S rRNA*, *23S rRNA*	*A. butzleri*	Reference strains and poultry and water (Isolate identification)	Used for differentiation between *Arcobacter*, *Campylobacter* and *Helicobacter* genera, with *A. butzleri* differentiation but with *A. skirrowii* and *A. cryaerophilus* sharing the same pattern Restrictions endonucleases used: *Hpa*II or *Dde*I	–
PCR (Neubauer and Hess, 2006)	*16S rRNA*	*Arcobacter* spp.	Reference strains and field isolates (Isolate identification)	Detection takes in account *A. butzleri*, *A. cryaerophilus* and *A. skirrowii*, and allows the differentiation between *Arcobacter*, *Campylobacter* and *Helicobacter* genera.	–
PCR-DGGE (Petersen *et al.* (2007)	*16S rRNA*	*A. cryaerophilus 1B*, *A. nitrofigilis*	Pets saliva (Culture independent) Reference strains and field isolates (Isolate identification)	Equal patterns for *A. butzleri*, *A. cryaerophilus* 1A and *A. skirrowii*. *A. nitrofigilis* pattern shared with *C. gracilis* and other campylobacters.	–

Taqman real-time PCR and multiplex PCR (Brightwell *et al.*, 2007)	*rpoB/C, 23S rRNA*	*A. butzleri*, *A. cryaerophilus*	Field isolates from livestock and/or meat processing (Isolate identification)	The m-PCR uses primers CRY1-CRY2 described by Houf *et al.* (2000)	–
DNA microarrays (Quiñones *et al.*, 2007)	Virulence and housekeeping genes	*A. butzleri*	Animal and human isolates and isolates from chicken package liquid (Isolate identification)	No hybridization with *A. cryaerophilus* or *A. cibarius*	–
FRET real-time PCR (Abdelbaqi *et al.*, 2007)	*gyrA*	*A. butzleri*, *A. cryaerophilus*, *A. cibarius*, *A. nitrofigilis*	Human stool (Culture independent)	Failed the identification of *A. skirrowii*	–
PCR-RFLP (Figueras *et al.*, 2008)	*16S rDNA*	*A. butzleri*, *A. cryaerophilus 1A*, *A. cryaerophilus1B*, *A. skirrowii*, *A. cibarius*, *A. nitrofigilis*, *A. halophilus*	Human, animal, food and environmental isolates (Isolate identification)	Misidentify three species (*A. trophiarum*, *A. thereius*, and some *A. cryaerophilus* strains) as *A. butzleri*. *A. suis*, and *A. defluvii* producing the same pattern; and *A. venerupis*, and *A. marinus* a very similar one (Levican and Figueras, 2013) Restriction endonuclease used: *Mse*I	Meat and shellfish (Isolates identification) (Collado *et al.*, 2009b); Fresh lettuces (Isolates identification) (González and Ferrús, 2011); Edible bivalve molluscs (Isolates identification) (Collado *et al.*, 2014); Cow milk, water buffalo milk and fresh village cheese (Isolates identification) (Yesilmen *et al.*, 2014)
PCR (Pentimalli *et al.*, 2009)	*gyrA, 16S rRNA*	*A. butzleri*, *A. cryaerophilus*, *A. skirrowii*, and *A. cibarius*	Chicken meat (Detection in enrichment broth)	Misidentification with non-targeted species being confused with *A. butzleri*, *A. cryaerophilus*, or *A. skirrowii* (Levican and Figueras, 2013)	–
MALDI-TOF MS (Alispahic *et al.*, 2010)	Proteins	*A. butzleri*, *A. cryaerophilus*, *A. skirrowii*	Reference strains and human isolates (Isolate identification)	Identification dependent of bacterial strains in database	Chicken, duck, pork and beef meats and seafood (Isolates identification) (Šilha *et al.*, 2015)

(*Continued*)

Table 4.4 (Continued)

Method (reference)	Gene (s) targeted	Species discriminated	Sample used	Comment	Example of applications of the method
Multiplex PCR (Douidah *et al.*, 2010)	*23S rRNA*	*A. butzleri*, *A. cryaerophilus*, A.skirrowii, *A. cibarius*, *A. thereius*	Reference strains (Isolate identification)	Occurs misidentification of non-targeted species as targeted ones (Levican and Figueras, 2013)	Pigs on farrow-to-finish farms (isolates identification) (De Smet *et al.*, 2011b); Chicken viscera (Isolates identification)(Villalobos *et al.*, 2013); Chicken Breast Meat (Isolates identification) (Fallas-Padilla *et al.*, 2014); Dairy farm and possible sources of contamination (Isolates identification) (Giacometti *et al.*, 2015b)
SYBR Green Real-time PCR (González *et al.*, 2010)	*23S rRNA*	*Arcobacter* spp.	Chicken and wastewater (Culture independent and detection in enrichment broth)	Uses Bastyns *et al.* (1995) primers	Fresh lettuces (detection in diluted sample or after enrichment) (González and Ferrús, 2011);
PCR (De Smet *et al.*, 2011a)	*hsp60*	*A. trophiarum*	Fattening pigs (Isolate identification)	–	–
PCR-RFLP (Figueras *et al.*, 2012)	*16S rRNA*	All recognized species, with exception of *A. anaerophilus*	Diverse origin (Isolate identification)	Update 16S RNA-RFLP previously described by (Figueras *et al.*, 2008) Restriction endonuclease used: *Mse*I, *Mnl*I and *Bfa*I	Shellfish (isolates identification) (Levican *et al.*, 2014)
Quantitative real-time PCR (SYBR® Green I and TaqMan® assay) (Hausdorf *et al.*, 2013b)	*16S rRNA*, *23S rRNA*	*Arcobacter* spp.	Spinach-processing plant samples (Culture independent)	Detects *A. bivalviorum*, *A. butzleri*, *A. cibarius*, *A. cryaerophilus*, *A. defluvii*, *A. ellisii*, *A. halophilus*, *A. marinus*, *A. molluscorum*, *A. mytili*, *A. nitrofigilis*, *A. skirrowii*, *A. thereius*, *A. trophiarum*, *and A. venerupis*. For 16S rRNA-gene assay, *A. halophilus* and *A. marinus* showed a higher Ct value and a reduced efficiency.	–

TaqMan multiplex Real-time PCR (De Boer *et al.*, 2013)	*hsp60*	*A. butzleri*	Human faeces (Culture independent)	–	–
PCR (González *et al.*, 2014)	*16S rRNA*	*Arcobacter* spp.	Chicken meat (Detection in enrichment broth)	Detects *A. butzleri*, *A. cryaerophilus*, *A. skirrowii*, *A. cibarius*, *A. trophiarum*, *A. thereius*, *A. defluvii*, *A. ellisii*, *A. mytili*, *A. molluscorum*	–
LAMP (Wang *et al.*, 2014)	*23S rRNA*	*Arcobacter* spp.	Chicken skin (Culture independent)	Detects *A. butzleri*, *A. cryaerophilus* and *A. skirrowii*. One *C. jejuni* strain showed a false-positive reaction for *Arcobacter* (Wang *et al.*, 2014).	–

RFLP, Restriction fragment length polymorphism; PCR, polymerase chain reaction; PCR-ELISA, PCR-Enzyme-linked immunosorbent assay; FISH, Fluorescent *in situ* hybridization; PCR-DGGE, PCR-denaturing gradient gel electrophoresis; FRET real-time PCR, Fluorescence resonance energy transfer real-time PCR; MALDI-TOF MS, matrix-assisted laser desorption/ionization time-of-flight mass spectrometry; LAMP, Loop-mediated isothermal amplification assay.

Ferrús, 2011). Wang *et al.* (2014) compared a loop-mediated isothermal amplification assay to conventional and multiplex PCR, with the first method showing a higher specificity, sensitivity, and a higher detection rate with shorter enrichment incubation periods than observed for PCR methods, and all molecular methods showed higher detection percentages than culture.

Levican and Figueras (2013) compared the performance of five PCR based methods with different target regions (*16S rRNA*, *23S rRNA* or *gyrA* genes): four multiplex PCR methods (Houf *et al.*, 2000; Kabeya *et al.*, 2003a; Pentimalli *et al.*, 2009; Douidah *et al.*, 2010) and one 16S rRNA-RFLP method (Figueras *et al.*, 2008) were evaluated according to its performance and ability to correctly identify the *Arcobacter* species. The authors performed the evaluation of the Douidah *et al.* (2010) method in combination with the *A. trophiarum* PCR method described by (De Smet *et al.* (2011a). They found that none of the methods evaluated were completely reliable, presenting misidentifications of both targeted and non-targeted species, thereby emphasizing the limitations of the methods. These limitations could be in part explained by the use of restricted amount of sequences to primers design and by the subsequent description of new species after PCR primers design (Levican and Figueras, 2013). From the methods described, the multiplex PCR developed by Houf *et al.* (2000) has been the more frequently used, even though being considered one of the most unreliable among the evaluated methods. Nevertheless, this method was 100% reliable for the identification of *A. butzleri*, with no misidentification with other species (Levican and Figueras, 2013). The authors have suggested that the use of an updated 16S rRNA-RFLP method (Figueras *et al.*, 2012) for the evaluation of *Arcobacter* spp. diversity in different environments. Figueras *et al.* (2012) updated the method previously described by the same group (Figueras *et al.*, 2008). They introduced two supplementary endonucleases (*Mnl*I and *Bfa*I) in addition to the previously used *Mse*I endonuclease to digest the *16S rRNA* gene, thus allowing the identification of 17 species (all recognized species at the time, except *A. anaerophilus*, *A. ebronensis*, *A. aquimarinus*, *A. lanthieri*, *A. pacificus*, and *A. faecis*). The inaccurate identification of *Arcobacter* species may influence their recognized diversity possibly with over or underestimation of some species.

For the purposes of *Arcobacter* spp. identification and prevalence evaluation, the *16S rDNA* sequencing has also been used (Woo *et al.*, 2001; Lau *et al.*, 2002; Chinivasagam *et al.*, 2007; Pejchalová *et al.*, 2008).

Fluorescence *in situ* hybridization (FISH) has been suggested as an alternative to PCR, namely for samples from aquatic environment, as it is not inactivated by sample inhibitors, does not require DNA extraction, and positive results can be directly observed in samples. Despite these advantages, the use of FISH does not improve the detection of *Arcobacter* spp., when compared with PCR, nevertheless it can provide valuable information, for example, about cell morphology (Moreno *et al.*, 2003; Fera *et al.*, 2010).

Even though high-throughput sequencing techniques have been described for their noticeable potential to be used in the epidemiologic investigation of foodborne outbreaks (Maurer, 2011), their application on *Arcobacter* spp. detection on food has not yet been reported.

Over the years, numerous methods have been suggested and developed for a more reliable *Arcobacter* spp. identification, nevertheless, the prevalence and diversity of *Arcobacter* spp. in different matrices such as water, food, and feces may be underestimated due to the limitations of the current methods to recognize all species using a stand-alone technique (Collado and Figueras, 2011).

4.2.3 Genetic Typing and Diversity

The characterization of different types of organisms beyond the species is denominated typing. Several traditional typing systems based in phenotypic differences between isolates have been used for bacteria subtyping, such as serotype, biotype, phage-type, or antimicrobial resistance typing. However, and since phenotypic typing methods are not variable enough for discriminating between closely related strains, molecular DNA-based typing methods have become widely used for bacterial strain typing due to its high resolution (Li *et al.*, 2009; Sabat *et al.*, 2013).

A number of techniques were developed for the genetic subtyping of *Arcobacter* strains, which have been used for discerning about population relatedness, distribution, and routes of transmission of this microorganism (González *et al.*, 2012). These typing methods can be organized into three main groups: DNA banding pattern-, DNA sequencing-, and DNA hybridization-based methods. The first category includes techniques that discriminate strains according to the size of fragments generated by amplification or enzymatic digestion of genomic DNA; the second group deals with the methods that assess the nucleotide sequence of target regions and the strain discrimination is centered in the study of polymorphisms of the DNA sequences; the last category mainly refers to DNA macro- and microarrays methods (Li *et al.*, 2009).

As of today, a variety of *Arcobacter* typing methods are used, including amplified fragment length polymorphism (AFLP), enterobacterial repetitive intergenic consensus-PCR (ERIC-PCR), randomly amplified polymorphic DNA-PCR (RAPD-PCR), pulsed-field gel electrophoresis (PFGE), and multi-locus sequence typing (MLST).

The AFLP method has been used to identify and explore the genetic diversity of *Arcobacter* species. AFLP profiling was used for the first time by On *et al.* (2003) for subtyping of *Arcobacter* species isolated from several sources and countries. The authors found a high degree of heterogeneity (62 profiles from 72 isolates); however, with a clonality trend within *A. butzleri*, *A. cryaerophilus*, and *A. skirrowii*. The method also allowed the identification of a potential novel *Arcobacter* taxon (On *et al.*, 2003). The same group used this methodology to evaluate the genetic diversity of *A. butzleri*, with most strains having a food-related origin (51 types from 73 isolates) (On *et al.*, 2004). Similarly, González *et al.* (2007b) reported a high genetic variation by the use of a single enzyme-AFLP based profiling, with 15 isolates from water and chicken exhibiting 11 different profile types. The discriminatory power of this technique was also used for the reassessment of *A. cryaerophilus* taxonomic status (Debruyne *et al.*, 2010). *A. cryaerophilus* is a heterogeneous species that was previously described as including two subgroups (Kiehlbauch *et al.*, 1991); however, its taxonomy was reviewed with the employment of AFLP and *heat shock protein 60* sequence analysis, and based on these methods the subgroup nomenclature was suggested to be abandoned (Debruyne *et al.*, 2010). Despite the high discriminatory power described to this AFLP-based profiling, the most used DNA-based typing method for *Arcobacter* spp. strain discrimination is ERIC-PCR.

ERIC-PCR has been widely used to evaluate the degree of heterogeneity among animal food sources (Houf *et al.*, 2002a; Van Driessche *et al.*, 2004, 2005; Aydin *et al.*, 2007; Van Driessche and Houf, 2007a; Ho *et al.*, 2008; De Smet *et al.*, 2010; Collado *et al.*, 2014; Levican *et al.*, 2014; Ramees *et al.*, 2014; Münch and Wochenschr, 2015; Zacharow *et al.*, 2015b), as well as among other sources (Collado *et al.*, 2010; Kayman *et al.*, 2012), with all the studies reporting a great level of heterogeneity. ERIC-PCR has also been used to establish the genetic relatedness of novel isolates and to help determining the taxonomic position of novel species (Houf *et al.*, 2005b; Collado *et al.*, 2009a, 2011; Houf *et al.*, 2009; Figueras *et al.*, 2011a). Beyond the diversity evaluation, this technique was also used to establish the source of contamination or possible cross-contamination during food processing. Van Driessche and Houf (2007a) reported a possible cross-contamination of pork carcasses during slaughter, and the genetic characterization of *A. butzleri* and *A. cryaerophilus* isolates indicated a contamination of raw pork during processing at retail. ERIC-PCR was also applied to evaluate the possible routes of *Arcobacter* introduction in poultry slaughterhouses (Ho *et al.*, 2008). A study by Merga *et al.* (2013) questioned the reliability of ERIC-PCR as a specific typing method due to the absence of primers binding sites in the two *A. butzleri* genome sequences studied.

Further studies on *Arcobacter* diversity were performed by employing RAPD-PCR typing to isolates from chicken carcasses (Atabay *et al.*, 2002), fresh lettuces (González and Ferrús, 2011), or domestic geese (Atabay *et al.*, 2008a). Both methods, ERIC-PCR and RAPD-PCR, were further applied for the molecular characterization of *Arcobacter* spp. isolates collected from poultry products (Houf *et al.*, 2002a), and in a poultry slaughterhouse in Belgium (Houf *et al.*, 2003). However, as the fingerprints generated with ERIC-PCR were shown to be more reproducible and complex than those obtained with RAPD-PCR (Houf *et al.*, 2002a), the latter technique was suggested to be used in the differentiation of almost identical ERIC fingerprints. RAPD-PCR typing is considered limited by its reproducibility, due to the use of a single nonspecific primer and low annealing temperatures, but is a valuable and simple technique able to discern among closely related *Arcobacter* isolates (González and Ferrús, 2011).

PFGE has been recognized as the "gold standard" for the assessment of isolates interrelationships. This method is an electrophoretic technique used for the separation of large DNA fragments by applying alternating electric fields at different angles. The purified genomic DNA sample is subjected to digestion with a rare cutting restriction endonuclease, generating large DNA fragments that are separated afterward by "pulsed-field" electrophoresis. The resulting DNA banding pattern reflects DNA polymorphisms at restriction sites, providing a macrorestriction analysis at genome level, which is used to carry out genetic comparisons among isolates (Foley *et al.*, 2009; Goering, 2010). PFGE has been successfully used for typing *Arcobacter* from foods, food animal sources, processing plants, and human patients (Rivas *et al.*, 2004; Son *et al.*, 2006; Revez *et al.*, 2013; Ferreira *et al.*, 2013; Giacometti *et al.*, 2013, 2015b; Douidah *et al.*, 2014; Scarano *et al.*, 2014), and for delineation of the transmission routes on raising of food-animals or food contamination (Hume *et al.*, 2001; Ho *et al.*, 2006b; Scarano *et al.*, 2014; Giacometti *et al.*, 2015b). This method has been used to evaluate the genotypic distribution of *Arcobacter* through fingerprinting of isolates from broiler carcasses along the processing line (Son *et al.*, 2006), in raw milk samples (Revez *et al.*, 2013), in ground chicken, pork, beef, and lamb meats from retails (Rivas *et al.*, 2004), and in samples from an artisanal dairy plant (Giacometti *et al.*, 2013), or samples from dairy cattle farms and beef samples from retail (Shah *et al.*, 2012b). The results point to a large diversity of *Arcobacter* strains, indicating the presence of multiple strains in one sample, but also the presence of common types of *Arcobacter* in different samples, suggesting possible cross-contamination or even persistence in the processing environment. A PFGE characterization conducted in a farrow-to-finish swine farm, showed great amount of genotypic variation suggesting that the animals were colonized by multiple *Arcobacter* genotypes and/or the identified isolates had undergone genomic rearrangement. In that study, the authors also found a lack of association between *Arcobacter*-positive sows and nursing piglets (Hume *et al.*, 2001). Ho *et al.* (2006b) also used PFGE for genotyping *Arcobacter* spp. to evaluate the transmission route in sows and their offspring on a breeding farm, suggesting a transplacental transmission but also a horizontal transmission from mothers to their piglets. It was proposed that the differences between the two studies were related to the serum levels of maternal antibodies in piglets (Ho *et al.*, 2006b). Genetic alteration by mutations or rearrangements in genes that are necessary under stress conditions has been suggested as a possible explanation for the high diversity of *Arcobacter* (Douidah *et al.*, 2014). PFGE was described as more discriminatory than other genotyping techniques, such as

ERIC-PCR, AFLP, or RAPD-PCR (Son *et al.*, 2006; Douidah *et al.*, 2014). Douidah *et al.* (2014) applied three typing strategies (ERIC-PCR, AFLP and PFGE) in six *Arcobacter* species (*A. butzleri, A. cryaerophilus, A. skirrowii, A. cibarius, A. thereius*, and *A. trophiarum*) showing that PFGE was the most discriminatory technique between strains. However, the authors did not found a correlation between typing pattern and the origin of the strain isolates, as previously described when using multilocus sequence typing (MLST) (Miller *et al.*, 2009).

The MLST method was first developed to overcome the lack of poor interlaboratory comparison of some of the molecular typing approaches (Sabat *et al.*, 2013). MLST is a typing method based on sequencing of fragments of housekeeping genes (Maiden *et al.*, 1998), usually six to eight *loci* are sequenced, with each unique sequence assigned as a distinct allele and each allelic combination will define a sequence type (ST) (Jolley *et al.*, 2004). The most important advantages of DNA sequencing-based genotyping over DNA banding pattern-based methods is its high reproducibility that together with the storage possibility in online databases allows genotyping comparison among laboratories (Li *et al.*, 2009). The first MLST typing scheme for five species of *Arcobacter* (*A. butzleri, A. cryaerophilus, A. skirrowii, A. cibarius*, and *A. thereius*) was introduced by Miller *et al.* (2009), using the seven *loci* previously employed for *C. jejuni* (i.e. *aspA, atpA, glnA, gltA, glyA, pgm*, and *tkt*) (Miller *et al.*, 2009). The latter study typed 374 *Arcobacter* isolates from different geographic locations and sources, identifying a large number of alleles at each *locus*, with the majority of the isolates containing a unique ST. Even though the lack of association with either host or geographical source, which may be due to infrequent observation of the same ST, common STs between samples from healthy or diarrhoeal children stool samples were observed (Miller *et al.*, 2009). Despite its advantages, so far, this scheme was only used in a few more studies. In the first one, typing of 39 *A. butzleri* isolates from animal farms was performed, and 11 different STs were detected, demonstrating a considerable level of diversity amongst *A. butzleri* isolates. Indeed, none of the STs identified were found on more than one farm or had been previously reported at the time of the study (Merga *et al.*, 2011). In another study, also performed in cattle from the United Kingdom, MLST typing displayed again a high diversity between *A. butzleri* isolates, with 43 different STs identified from 104 isolates. The observed persistent STs suggested a possible survival overtime or re-contamination by the same STs (Merga *et al.*, 2013). Typing by MLST was used by Rasmussen *et al.* (2013) to evaluate persistence and cross-contamination in a Danish broiler slaughterhouse during meat processing, finding a high strain variability, with the recurrence of two STs, indicating that persistence or cross-contamination could take place. The heterogeneity of the *A. butzleri* isolates was demonstrated by the relatively high numbers of new alleles and STs; however, it also reflects the existence of a limited number of profiles in the *Arcobacter* MLST database (Rasmussen *et al.*, 2013). The MLST scheme was also used for typing an *A. cryaerophilus* isolate from a bloody human diarrhoea sample; the isolate belonged to a new ST while sharing three known alleles with strains isolated from feces of patients with gastroenteritis (Figueras *et al.*, 2014). In Spain, Alonso *et al.* (2014) used MLST for the evaluation of heterogeneity of food isolates showing a wide allelic diversity, with a total of 34 STs being identified among 45 *A. butzleri* isolates, with no link between alleles or STs and food sources, and with only one ST having been previously reported. Recently, De Cesare *et al.* (2015) compared PFGE with MLST and showed good discriminatory capacity of both techniques regarding *A. butzleri* genetic profiles, although better for PFGE. Nevertheless, the combination between both methods showed to improve differentiation, yet increasing costs and labor. The use of these typing methods suggest the occurrence of *Arcobacter* persistence on time and cross-contamination between food and plant equipment and surfaces (De Cesare *et al.*, 2015).

Despite the results and advantages of the MLST technique, a drawback was evidenced by Merga *et al.* (2013) that obtained 410 poor quality sequences comprising *A. butzleri, A. cryaerophilus,* and *A. skirrowii* isolates, showing that some strains could not be accurately typed using MLST. This typing method demonstrated *Arcobacter* spp. heterogeneity, along with the possible cross-contamination and persistence of microorganisms, as supported by other studies using different typing approaches; however, the still limited number of isolates available in MLST database limits the assessment of additional epidemiological and genetic information.

A comparative genomic fingerprinting (CGF) using *A. butzleri* variability in accessory genome content assay was suggested for larger-scale epidemiological studies and described as less resource- and labor-intensive than MLST genotyping. This technique establishes phylogenetic clusters and discriminates between strains based on targeting representative accessory *loci* displaying variation through the genome. It is based on the presence or absence of 40 accessory genes in *A. butzleri* genome. CGF40 is presented as a rapid, high-resolution and high-throughput subtyping method of *A. butzleri*, using PCR amplification without the requirement of sequencing (Webb *et al.*, 2015).

Among the sequencing-based typing methods emerges the whole-genome sequencing (WGS), which has already been used for detection of some foodborne bacteria related outbreaks in Europe. The evolution of the high-throughput sequencing technology, bioinformatics tools, and cost reduction points toward a future use of WGS as a tool for bacterial characterization. However, some further developments in interpretation data and standard operating procedures are required to transform this potential in a more routine typing method (Sabat *et al.*, 2013).

In sum, the genus *Arcobacter* comprises several highly heterogeneous species, with the majority of them having only been recognized in recent years. This genus can be isolated from diverse environments and hosts, posing a significant threat to human health, especially through food and waterborne contamination. A significant increase in the knowledge regarding *Arcobacter* spp. prevalence, species, and genomic diversity has been observed lately, associated with a growing number of published studies dealing with *Arcobacter* distribution in either the food chain or from samples from different origins. Another significant factor that has contributed to this increase in knowledge has been the progress in the methodologies used for identification and analysis of the heterogeneity of *Arcobacter*. A future standardization of techniques will allow the successful cultivation, identification, and genotyping of a wider number of species and the collection and analysis of more reliable data. The development and standardization of these methods will allow unveiling the complex epidemiology of this pathogen and will contribute to clarify the real hazard and routes of transmission to humans.

References

Abdelbaqi, K., Buissonnière, A., Prouzet-Mauleon, V., Gresser, J., Wesley, I., Mégraud, F. and Ménard, A. (2007). Development of a real-time fluorescence resonance energy transfer PCR to detect *Arcobacter* species. *Journal of Clinical Microbiology, 45*(9), pp. 3015–3021.

Van den Abeele, A.M., Vogelaers, D., Van Hende, J. and Houf, K. (2014). Prevalence of *Arcobacter* Species among Humans, Belgium, 2008–2013. *Emerging Infectious Diseases, 20*(10), pp. 1731–1734.

Adesiji, Y.O., Coker, A.O. and Oloke, J.K. (2011). Detection of *Arcobacter* in feces of healthy chickens in Osogbo, Nigeria. *Journal of Food Protection, 74*(1), pp. 119–121.

Alispahic, M., Hummel, K., Jandreski-Cvetkovic, D., Nöbauer, K., Razzazi-Fazeli, E., Hess, M. and Hess, C. (2010). Species-specific identification and differentiation of *Arcobacter*, *Helicobacter* and *Campylobacter* by full-spectral matrix-associated laser desorption/ ionization time of flight mass spectrometry analysis. *Journal of Medical Microbiology, 59*(3), pp. 295–301.

Alonso, R., Girbau, C., Martinez-Malaxetxebarria, I. and Fernández-Astorga, A. (2014). Multilocus sequence typing reveals genetic diversity of foodborne *Arcobacter butzleri* isolates in the North of Spain. *International Journal of Food Microbiology, 191*, pp. 125–128.

Amare, L.B., Saleha, A.A., Zunita, Z., Jalila, A. and Hassan, L. (2011). Prevalence of *Arcobacter* spp. on chicken meat at retail markets and in farm chickens in Selangor, Malaysia. *Food Control, 22*(5), pp. 732–736.

Andersen, M.M.E., Wesley, I.V., Nestor, E. and Trampel, D.W. (2007). Prevalence of *Arcobacter* species in market-weight commercial turkeys. *Antonie van Leeuwenhoek, 92*(3), pp. 309–317.

Anderson, K.F., Kiehlbauch, J.A., Anderson, D.C., McClure, H.M. and Wachsmuth, I.K. (1993). *Arcobacter* (*Campylobacter*) *butzleri*-associated diarrheal illness in a nonhuman primate population. *Infection and Immunity, 61*(5), pp. 2220–2223.

Antolín, A., González, I., García, T., Hernández, P.E. and Martín, R. (2001). *Arcobacter* spp. enumeration in poultry meat using a combined PCR-ELISA assay. *Meat Science, 59*(2), pp. 169–174.

Arguello, E., Otto, C.C., Mead, P. and Babady, N.E. (2015). Bacteremia Caused by *Arcobacter butzleri* in an Immunocompromised Host. *Journal of Clinical Microbiology, 53*(4), pp. 1448–1451.

Atabay, H.I., Aydin, F., Houf, K., Sahin, M. and Vandamme, P. (2003). The prevalence of *Arcobacter* spp. on chicken carcasses sold in retail markets in Turkey, and identification of the isolates using SDS-PAGE. *International Journal of Food Microbiology, 81*(1), pp. 21–28.

Atabay, H.I., Bang, D.D., Aydin, F., Erdogan, H.M. and Madsen, M. (2002). Discrimination of *Arcobacter butzleri* isolates by polymerase chain reaction-mediated DNA fingerprinting. *Letters in Applied Microbiology, 35*(2), pp. 141–145.

Atabay, H.I. and Corry, J.E. (1997). The prevalence of campylobacters and arcobacters in broiler chickens. *Journal of Applied Microbiology, 83*(5), pp. 619–626.

Atabay, H.I. and Corry, J.E.L. (1998). Evaluation of a new arcobacter enrichment medium and comparison with two media developed for enrichment of *Campylobacter* spp. *International Journal of Food Microbiology, 41*(1), pp. 53–58.

Atabay, H.I., Corry, J.E.L. and On, S.L.W. (1998). Diversity and prevalence of *Arcobacter* spp. in broiler chickens. *Journal of Applied Microbiology, 84*(6), pp. 1007–1016.

Atabay, H.I., Unver, A., Otlu, S. and Kalaycioglu, A.T. (2008a). Genotyping of various *Arcobacter* species isolated from domestic geese by randomly amplified polymorphic DNA (RAPD) analysis. *Archiv für Lebensmittelhygiene, 59*(4), pp. 142–146.

Atabay, H.I., Unver, A., Sahin, M., Otlu, S., Elmali, M. and Yaman, H. (2008b). Isolation of various *Arcobacter* species from domestic geese (*Anser anser*). *Veterinary Microbiology, 128*(3-4), pp. 400–405.

Atabay, H.I., Wainø, M. and Madsen, M. (2006). Detection and diversity of various *Arcobacter* species in Danish poultry. *International Journal of Food Microbiology, 109*(1-2), pp. 139–145.

Aydin, F., Gümüşsoy, K.S., Atabay, H.I., Iça, T. and Abay, S. (2007). Prevalence and distribution of *Arcobacter* species in various sources in Turkey and molecular analysis of isolated strains by ERIC-PCR. *Journal of Applied Microbiology, 103*(1), pp. 27–35.

Aydin, S., Gultepe, N. and Yildiz, H. (2000). Natural and Experimental Infections of *Campylobacter cryaerophila* in Rainbow Trout: Gross Pathology, Bacteriology, Clinical Pathology and Chemotherapy. *Fish Pathology, 35*(3), pp. 117–123.

Balamurugan, S., Ahmed, R. and Chambers, J.R. (2013). Survival of *Arcobacter butzleri* on vacuum packaged chill stored beef. *Food Research International, 52*(1), pp. 503–507.

Bastyns, K., Cartuyvels, D., Chapelle, S., Vandamme, P., Goossens, H. and De Wachter, R. (1995). A Variable *23S rDNA*

Region is a Useful Discriminating Target for Genus-Specific and Species-Specific PCR Amplification in *Arcobacter* Species. *Systematic and Applied Microbiology, 18*(3), pp. 353–356.

Bath, G.F., Leask, R., Pettey, K.P. and Coetzee, D.J. (2013). Abortions in sheep associated with *Arcobacter skirrowii* infection. *Journal of the South African Veterinary Association, 84*(1), pp. 1–4.

Bessède, E., Solecki, O., Sifré, E., Labadi, L. and Mégraud, F. (2011). Identification of *Campylobacter* species and related organisms by matrix assisted laser desorption ionization-time of flight (MALDI-TOF) mass spectrometry. *Clinical Microbiology and Infection, 17*(11), pp. 1735–1739.

De Boer, E., Tilburg, J.J., Woodward, D.L., Lior, H. and Johnson, W.M. (1996). A selective medium for the isolation of *Arcobacter* from meats. *Letters in Applied Microbiology, 23*(1), pp. 64–66.

De Boer, R.F., Ott, A., Güren, P., Van Zanten, E., Van Belkum, A. and Kooistra-Smid, A.M.D. (2013). Detection of *Campylobacter* species and *Arcobacter butzleri* in stool samples by use of real-time multiplex PCR. *Journal of Clinical Microbiology, 51*(1), pp. 253–259.

Bogantes, E.V., Fallas-Padilla, K.L., Rodríguez-Rodríguez, C.E., Jaramillo, H.F. and Echandi, M.L.A. (2015). Zoonotic Species of the Genus *Arcobacter* in Poultry from Different Regions of Costa Rica. *Journal of Food Protection, 78*(4), pp. 808–811.

Brightwell, G., Mowat, E., Clemens, R., Boerema, J., Pulford, D.J. and On, S.L. (2007). Development of a multiplex and real time PCR assay for the specific detection of *Arcobacter butzleri* and *Arcobacter cryaerophilus*. *Journal of Microbiological Methods, 68*(2), pp. 318–325.

Bullman, S., O'Leary, J., Corcoran, D., Sleator, R.D. and Lucey, B. (2012). Molecular-based detection of non-culturable and emerging campylobacteria in patients presenting with gastroenteritis. *Epidemiology and Infection, 140*(04), pp. 684–688.

Cardarelli-Leite, P., Blom, K., Patton, C.M., Nicholson, M.A., Steigerwalt, A.G., Hunter, S.B., Brenner, D.J., Barrett, T.J. and Swaminathan, B. (1996). Rapid identification of *Campylobacter* species by restriction fragment length polymorphism analysis of a PCR-amplified fragment of the gene coding for 16S rRNA. *Journal of Clinical Microbiology, 34*(1), pp. 62–67.

Cervenka, L., Kristlova, J., Peskova, I., Vytrasova, J., Pejchalova, M. and Brozkova, I. (2008). Persistence of *Arcobacter butzleri* CCUG 30484 on plastic, stainless steel and glass surfaces. *Brazilian Journal of Microbiology, 39*(3), pp. 517–520.

De Cesare, A., Parisi, A., Giacometti, F., Serraino, A., Piva, S., Caruso, M., De Santis, E.P.L. and Manfreda, G. (2015). Multilocus Sequence Typing of *Arcobacter butzleri* isolates collected from dairy plants and their products, and comparison with their PFGE types. *Journal of Applied Microbiology, 120*(1), pp. 165–174.

Chinivasagam, H.N., Corney, B.G., Wright, L.L., Diallo, I.S. and Blackall, P.J. (2007). Detection of *Arcobacter* spp. in piggery effluent and effluent-irrigated soils in southeast Queensland. *Journal of Applied Microbiology, 103*(2), pp. 418–426.

Collado, L., Cleenwerck, I., Van Trappen, S., De Vos, P. and Figueras, M.J. (2009a). *Arcobacter mytili* sp. nov., an indoxyl acetate-hydrolysis-negative bacterium isolated from mussels. *International Journal of Systematic and Evolutionary Microbiology, 59*(6), pp. 1391–1396.

Collado, L. and Figueras, M.J. (2011). Taxonomy, epidemiology, and clinical relevance of the genus *Arcobacter*. *Clinical Microbiology Reviews, 24*(1), pp. 174–192.

Collado, L., Guarro, J. and Figueras, M.J. (2009b). Prevalence of *Arcobacter* in meat and shellfish. *Journal of Food Protection, 72*(5), pp. 1102–1106.

Collado, L., Gutiérrez, M., González, M. and Fernández, H. (2013). Assessment of the prevalence and diversity of emergent campylobacteria in human stool samples using a combination of traditional and molecular methods. *Diagnostic Microbiology and Infectious Disease, 75*(4), pp. 434–436.

Collado, L., Inza, I., Guarro, J. and Figueras, M.J. (2008). Presence of *Arcobacter* spp. in environmental waters correlates with high levels of fecal pollution. *Environmental Microbiology, 10*(6), pp. 1635–1640.

Collado, L., Jara, R., Vásquez, N. and Telsaint, C. (2014). Antimicrobial resistance and virulence genes of *Arcobacter* isolates recovered from edible bivalve molluscs. *Food Control, 46*, pp. 508–512.

Collado, L., Kasimir, G., Perez, U., Bosch, A., Pinto, R., Saucedo, G., Huguet, J.M. and Figueras, M.J. (2010). Occurrence and diversity of *Arcobacter* spp. along the Llobregat River catchment, at sewage effluents and in a drinking water treatment plant. *Water Research, 44*(12), pp. 3696–3702.

Collado, L., Levican, A., Perez, J. and Figueras, M.J. (2011). *Arcobacter defluvii* sp. nov., isolated from sewage samples. *International Journal of Systematic and Evolutionary Microbiology, 61*(9), pp. 2155–2161.

Collins, C.I., Wesley, I. V and Murano, E.A. (1996). Detection of *Arcobacter* spp. in ground pork by modified plating methods. *Journal of Food Protection, 59*(5), pp. 448–452.

D'Sa, E.M. and Harrison, M.A. (2005). Effect of pH, NaCl content, and temperature on growth and survival of *Arcobacter* spp. *Journal of Food Protection, 68*(1), pp. 18–25.

Debruyne, L., Houf, K., Douidah, L., De Smet, S. and Vandamme, P. (2010). Reassessment of the taxonomy of *Arcobacter cryaerophilus*. *Systematic and Applied Microbiology, 33*(1), pp. 7–14.

Diergaardt, S.M., Venter, S.N., Spreeth, A., Theron, J. and Brözel, V.S. (2004). The occurrence of campylobacters in water sources in South Africa. *Water Research, 38*, pp. 2589–2595.

Donachie, S.P., Bowman, J.P., On, S.L.W. and Alam, M. (2005). *Arcobacter halophilus* sp. nov., the first obligate halophile in the genus Arcobacter. *International Journal of Systematic and Evolutionary Microbiology, 55*(3), pp. 1271–1277.

Douidah, L., De Zutter, L., Baré, J. and Houf, K. (2014). Towards a typing strategy for *Arcobacter* species isolated from humans and animals and assessment of the *in vitro* genomic stability. *Foodborne Pathogens and Disease, 11*(4), pp. 272–80.

Douidah, L., De Zutter, L., Vandamme, P. and Houf, K. (2010). Identification of five human and mammal associated *Arcobacter* species by a novel multiplex-PCR assay. *Journal of Microbiological Methods, 80*(3), pp. 281–286.

Van Driessche, E. and Houf, K. (2007a). Characterization of the *Arcobacter* contamination on Belgian pork carcasses and raw retail pork. *International Journal of Food Microbiology, 118*(1), pp. 20–26.

Van Driessche, E. and Houf, K. (2007b). Discrepancy between the occurrence of *Arcobacter* in chickens and broiler carcass contamination. *Poultry Science, 86*(4), pp. 744–751.

Van Driessche, E. and Houf, K. (2008). Survival capacity in water of *Arcobacter* species under different temperature conditions. *Journal of Applied Microbiology, 105*(2), pp. 443–451.

Van Driessche, E., Houf, K., Van Hoof, J., De Zutter, L. and Vandamme, P. (2003). Isolation of *Arcobacter* species from animal feces. *FEMS Microbiology Letters, 229*(2), pp. 243–248.

Van Driessche, E., Houf, K., Vangroenweghe, F., Nollet, N., De Zutter, L., Vandamme, P. and Van Hoof, J. (2004). Occurrence and strain diversity of *Arcobacter* species isolated from healthy Belgian pigs. *Research in Microbiology, 155*(8), pp. 662–666.

Van Driessche, E., Houf, K., Vangroenweghe, F., De Zutter, L. and Van Hoof, J. (2005). Prevalence, enumeration and strain variation of *Arcobacter* species in the faeces of healthy cattle in Belgium. *Veterinary Microbiology, 105*(2), pp. 149–154.

Ellis, W., Neill, S., O'Brien, J., Ferguson, H. and Hanna, J. (1977). Isolation of *Spirillum/Vibrio*-like organisms from bovine fetuses. *Veterinary Record, 100*(21), pp. 451–452.

Ellis, W.A., Neill, S.D., O'Brien, J.J. and Hanna, J. (1978). Isolation of spirillum-like organisms from pig fetuses. *The Veterinary Record, 102*(5), p. 106.

Ertas, N., Dogruer, Y., Gonulalan, Z., Guner, A. and Ulger, I. (2010). Prevalence of *Arcobacter* species in drinking water, spring water, and raw milk as determined by multiplex PCR. *Journal of Food Protection, 73*(11), pp. 2099–2102.

Fallas-Padilla, K.L., Rodríguez-Rodríguez, C.E., Jaramillo, H.F. and Echandi, M.L.A. (2014). *Arcobacter*: Comparison of Isolation Methods, Diversity, and Potential Pathogenic Factors in Commercially Retailed Chicken Breast Meat from Costa Rica. *Journal of Food Protection, 77*(6), pp. 880–884.

Fera, M.T., La Camera, E., Carbone, M., Malara, D. and Pennisi, M.G. (2009). Pet cats as carriers of *Arcobacter* spp. in Southern Italy. *Journal of Applied Microbiology, 106*(5), pp. 1661–1666.

Fera, M.T., Gugliandolo, C., Lentini, V., Favaloro, A., Bonanno, D., La Camera, E. and Maugeri, T.L. (2010). Specific detection of *Arcobacter* spp. in estuarine waters of Southern Italy by PCR and fluorescent *in situ* hybridization. *Letters in Applied Microbiology, 50*(1), pp. 65–70.

Fera, M.T., Maugeri, T.L., Gugliandolo, C., Beninati, C., Camera, E., Carbone, M. and Giannone, M. (2004). Detection of *Arcobacter* spp . in the Coastal Environment of the Mediterranean Sea. *Applied and Environmental Microbiology, 70*(3), pp. 1271–1276.

Fera, M.T., Maugeri, T.L., Gugliandolo, C., La Camera, E., Lentini, V., Favaloro, A., Bonanno, D. and Carbone, M. (2008). Induction and resuscitation of viable nonculturable *Arcobacter butzleri* cells. *Applied and Environmental Microbiology, 74*(10), pp. 3266–3268.

Fernandez, H., Villanueva, M.P., Mansilla, I., Gonzalez, M. and Latif, F. (2015). *Arcobacter butzleri* and *A. cryaerophilus* in human, animals and food sources, in southern Chile. *Brazilian Journal of Microbiology, 147*, pp. 145–147.

Ferreira, S., Fraqueza, M.J., Queiroz, J.A., Domingues, F.C. and Oleastro, M. (2013). Genetic diversity, antibiotic resistance and biofilm-forming ability of *Arcobacter butzleri* isolated from poultry and environment from a Portuguese slaughterhouse. *International Journal of Food Microbiology, 162(*1), pp. 82–88.

Ferreira, S., Júlio, C., Queiroz, J.A., Domingues, F.C. and Oleastro, M. (2014). Molecular diagnosis of *Arcobacter* and *Campylobacter* in diarrhoeal samples among Portuguese patients. *Diagnostic Microbiology and Infectious Disease, 78*(3), pp. 220–225.

Ferreira, S., Queiroz, J.A., Oleastro, M. and Domingues, F.C. (2016). Insights in the pathogenesis and resistance of *Arcobacter* : A review. *Critical Reviews in Microbiology, 42*(3), pp. 364–383.

Figueras, M.J., Collado, L. and Guarro, J. (2008). A new 16S rDNA-RFLP method for the discrimination of the accepted species of *Arcobacter*. *Diagnostic Microbiology and Infectious Disease, 62*(1), pp. 11–15.

Figueras, M.J., Collado, L., Levican, A., Perez, J., Solsona, M.J. and Yustes, C. (2011a). *Arcobacter molluscorum* sp. nov., a new species isolated from shellfish. *Systematic and Applied Microbiology, 34*(2), pp. 105–109.

Figueras, M.J., Levican, A. and Collado, L. (2012). Updated 16S rRNA-RFLP method for the identification of all currently characterised *Arcobacter* spp. *BMC Microbiology, 12*, pp. 292.

Figueras, M.J., Levican, A., Collado, L., Inza, M.I. and Yustes, C. (2011b). *Arcobacter ellisii* sp. nov., isolated from mussels. *Systematic and Applied Microbiology, 34*(6), pp. 414–418.

Figueras, M.J., Levican, A., Pujol, I., Ballester, F. and Quilez, M.J.R. (2014). A severe case of persistent diarrhoea associated with *Arcobacter cryaerophilus* but attributed to *Campylobacter* sp . and a review of the clinical incidence of *Arcobacter* spp . *New Microbes and New Infections, 2*(2), pp. 31–37.

Foley, S.L., Lynne, A.M. and Nayak, R. (2009). Molecular typing methodologies for microbial source tracking and epidemiological investigations of Gram-negative bacterial foodborne pathogens. *Infection, Genetics and Evolution, 9*(4), pp. 430–440.

Fong, T.T., Mansfield, L.S., Wilson, D.L., Schwab, D.J., Molloy, S.L. and Rose, J.B. (2007). Massive microbiological groundwater contamination associated with a waterborne outbreak in Lake Erie, South Bass Island, Ohio. *Environmental Health Perspectives, 115*(6), pp. 856–864.

Di Francesco, A., Delogu, M., Giacometti, F., Stancampiano, L., Grilli, E., Guarniero, I. and Serraino, A. (2014). First detection of *Arcobacter* sp. in Eurasian collared doves (*Streptopelia decaocto*).*Veterinaria Italiana, 50*(4), pp. 313–315.

Gharst, G., Oyarzabal, O.A. and Hussain, S.K. (2013). Review of current methodologies to isolate and identify *Campylobacter* spp. from foods. *Journal of Microbiological Methods, 95*(1), pp. 84–92.

Giacometti, F., Losio, M.N. and Serraino, A. (2015a). Short communication : *Arcobacter butzleri* and *Arcobacter cryaerophilus* survival and growth in artisanal and industrial ricotta cheese. *Journal of Dairy Dcience, 98*(10), pp. 6776–6781.

Giacometti, F., Lucchi, A., Di Francesco, A., Delogu, M., Grilli, E., Guarniero, I., Stancampiano, L., Manfreda, G., Merialdi, G. and Serraino, A. (2015b). *Arcobacter butzleri, Arcobacter cryaerophilus*, and *Arcobacter skirrowii* Circulation in a Dairy Farm and Sources of Milk Contamination. *Applied and Environmental Microbiology, 81*(15), pp. 5055–5063.

Giacometti, F., Lucchi, A., Manfreda, G., Florio, D., Zanoni, R.G. and Serraino, A. (2013). Occurrence and genetic diversity of *Arcobacter butzleri* in an artisanal dairy plant in italy. *Applied and Environmental Microbiology, 79*(21), pp. 6665–6669.

Giacometti, F., Serraino, A., Pasquali, F., De Cesare, A., Bonerba, E. and Rosmini, R. (2014). Behavior of *Arcobacter butzleri* and *Arcobacter cryaerophilus* in Ultrahigh-Temperature, Pasteurized, and Raw Cow's Milk Under Different Temperature Conditions. *Foodborne Pathogens and Disease, 11*(1), pp. 15–20.

Gilbert, M.J., Kik, M., Timmerman, A.J., Severs, T.T., Kusters, J.G., Duim, B. and Wagenaar, J.A. (2014). Occurrence, diversity, and host association of intestinal *Campylobacter, Arcobacter*, and *Helicobacter* in reptiles. *PLoS ONE, 9*(7). e101599.

Goering, R. V. (2010). Pulsed field gel electrophoresis: A review of application and interpretation in the molecular epidemiology of infectious disease. *Infection, Genetics and Evolution, 10*(7), pp. 866–875.

Golla, S.C., Murano, E.A., Johnson, L.G., Tipton, N.C., Cureington, E.A., and Savell, J.W. (2002). Determination of the occurrence of *Arcobacter butzleri* in beef and dairy cattle from Texas by various isolation methods. *Journal of Food Protection, 65*(12), pp. 1849–1853.

González, A., Botella, S., Montes, R.M., Moreno, Y. and Ferrús, M.A. (2007a). Direct detection and identification of *Arcobacter* species by multiplex PCR in chicken and wastewater samples from Spain. *Journal of Food Protection, 70*(2), pp. 341–347.

González, a, Moreno, Y., González, R., Hernández, J. and Ferrús, M.A. (2006). Development of a simple and rapid method based on polymerase chain reaction-based restriction fragment length polymorphism analysis to differentiate *Helicobacter, Campylobacter*, and *Arcobacter* species. *Current Microbiology, 53*(5), pp. 416–21.

González, A., Ferrús, M.A., González, R. and Hernández, J. (2007b). Molecular fingerprinting of *Campylobacter* and *Arcobacter* isolated from chicken and water. *International Microbiology, 10*(2), pp. 85–90.

González, A. and Ferrús, M.A. (2011). Study of *Arcobacter* spp. contamination in fresh lettuces detected by different cultural and molecular methods. *International Journal of Food Microbiology, 145*(1), pp. 311–314.

González, A., Suski, J. and Ferrús, M.A. (2010). Rapid and accurate detection of *Arcobacter* contamination in commercial chicken products and wastewater samples by real-time polymerase chain reaction. *Foodborne Pathogens and Disease*, *7*(3), pp. 327–338.

González, I., Fernández-Tomé, S., García, T. and Martín, R. (2014). Genus-specific PCR assay for screening *Arcobacter* spp. in chicken meat. *Journal of the Science of Food and Agriculture*, *94*(6), pp. 1218–1224.

González, I., García, T., Antolín, A., Hernández, P.E. and Martín, R. (2000). Development of a combined PCR-culture technique for the rapid detection of *Arcobacter* spp. in chicken meat. *Letters in Applied Microbiology*, *30*(3), pp. 207–212.

González, I., García, T., Fernández, S. and Martín, R. (2012). Current Status on *Arcobacter* Research: An Update on DNA-Based Identification and Typing Methodologies. *Food Analytical Methods*, *5*(5), pp. 956–968.

Grove-White, D.H., Leatherbarrow, A.J.H., Cripps, P.J., Diggle, P.J. and French, N.P. (2014). Temporal and farm-management-associated variation in faecal pat prevalence of *Arcobacter* spp. in ruminants. *Epidemiology and Infection*, *142*(4), pp. 1196–1204.

Gude, A., Hillman, T.J., Helps, C.R., Allen, V.M. and Corry, J.E.L. (2005). Ecology of *Arcobacter* species in chicken rearing and processing. *Letters in Applied Microbiology*, *41*(1), pp. 82–87.

Hamill, S., Neill, S.D. and Madden, R.H. (2008). A comparison of media for the isolation of *Arcobacter* spp. from retail packs of beef. *Journal of Food Protection*, *71*(4), pp. 850–854.

Harmon, K.M. and Wesley, I.V. (1996). Identification of *Arcobacter* isolates by PCR. *Letters in Applied Microbiology*, *23*(4), pp. 241–244.

Harmon, K.M. and Wesley, I.V. (1997). Multiplex PCR for the identification of *Arcobacter* and differentiation of *Arcobacter butzleri* from other arcobacters. *Veterinary Microbiology*, *58*(2-4), pp. 215–227.

Harrass, B., Schwarz, S. and Wenzel, S. (1998). Identification and characterization of *Arcobacter* isolates from broilers by biochemical tests, antimicrobial resistance patterns and plasmid analysis. *Zentralblatt fur Veterinarmedizin. Reihe B. Journal of veterinary medicine. Series B*, *45*(2), pp. 87–94.

Hausdorf, L., Fröhling, A., Schlüter, O. and Klocke, M. (2011). Analysis of the bacterial community within carrot wash water. *Canadian Journal of Microbiology*, *452*, pp. 447–452.

Hausdorf, L., Mundt, K., Winzer, M., Cordes, C., Fröhling, A., Schlüter, O. and Klocke, M. (2013a). Characterization of the cultivable microbial community in a spinach-processing plant using MALDI-TOF MS. *Food Microbiology*, *34*(2), pp. 406–41.

Hausdorf, L., Neumann, M., Bergmann, I., Sobiella, K., Mundt, K., Fröhling, A., Schlüter, O. and Klocke, M. (2013b). Occurrence and genetic diversity of *Arcobacter* spp. in a spinach-processing plant and evaluation of two *Arcobacter*-specific quantitative PCR assays. *Systematic and Applied Microbiology*, *36*(4), pp. 235–243.

Higgins, R., Messier, S., Daignault, D. and Lorange, M. (1999). *Arcobacter butzleri* isolated from a diarrhoeic non-human primate. *Laboratory Animals*, *33*(1), pp. 87–90.

Ho, H.T.K., Lipman, L.J.A. and Gaastra, W. (2006a). *Arcobacter*, what is known and unknown about a potential foodborne zoonotic agent! *Veterinary Microbiology*, *115*(1-3), pp. 1–13.

Ho, H.T.K., Lipman, L.J.A. and Gaastra, W. (2008). The introduction of *Arcobacter* spp. in poultry slaughterhouses. *International Journal of Food Microbiology*, *125*(3), pp. 223–229.

Ho, T.K.H., Lipman, L.J.A., Van Der Graaf-Van Bloois, L., Van Bergen, M. and Gaastra, W. (2006b). Potential routes of acquisition of *Arcobacter* species by piglets. *Veterinary Microbiology*, *114*(1–2), pp. 123–133.

Hong, P.Y., Yannarell, A.C., Dai, Q., Ekizoglu, M. and Mackie, R.I. (2013). Monitoring the perturbation of soil and groundwater microbial communities due to pig production activities. *Applied and Environmental Microbiology, 79*(8), pp. 2620–2629.

Houf, K., Devriese, L. A., De Zutter, L., Van Hoof, J. and Vandamme, P. (2001a). Development of a new protocol for the isolation and quantification of *Arcobacter* species from poultry products. *International Journal of Food Microbiology, 71*(2-3), pp. 189–196.

Houf, K., Devriese, L.A., De Zutter, L., Van Hoof, J. and Vandamme P. (2001b). Susceptibility of *Arcobacter butzleri, Arcobacter cryaerophilus,* and *Arcobacter skirrowii* to Antimicrobial Agents Used in Selective Media. *Journal of Clinical Microbiology, 39*(4), pp. 1654–1656.

Houf, K., On, S.L.W., Coenye, T., Debruyne, L., De Smet, S. and Vandamme, P. (2009). *Arcobacter thereius* sp. nov., isolated from pigs and ducks. *International Journal of Systematic and Evolutionary Microbiology, 59*(10), pp. 2599–2604.

Houf, K., On, S.L.W., Coenye, T., Mast, J., Van Hoof, J. and Vandamme, P. (2005a). *Arcobacter cibarius* sp nov., isolated from broiler carcasses. *International Journal of Systematic and Evolutionary Microbiology, 55*, pp. 713–717.

Houf, K., On, S.L.W., Coenye, T., Mast, J., Van Hoof, J. and Vandamme, P. (2005b). *Arcobacter cibarius* sp. nov., isolated from broiler carcasses. *International Journal of Systematic and Evolutionary Microbiology, 55*(2), pp. 713–717.

Houf, K., De Smet, S., Baré, J. and Daminet, S. (2008). Dogs as carriers of the emerging pathogen *Arcobacter. Veterinary Microbiology, 130*(1–2), pp. 208–213.

Houf, K. and Stephan, R. (2007). Isolation and characterization of the emerging foodborn pathogen *Arcobacter* from human stool. *Journal of Microbiological Methods, 68*(2), pp. 408–413.

Houf, K., Tutenel, A., De Zutter, L., Van Hoof, J. and Vandamme, P. (2000). Development of a multiplex PCR assay for the simultaneous detection and identification of *Arcobacter butzleri, Arcobacter cryaerophilus* and *Arcobacter skirrowii. FEMS Microbiology Letters, 193*(1), pp. 89–94.

Houf, K., Zutter, L. De, Hoof, J. Van and Vandamme, P. (2002a). Assessment of the genetic diversity among arcobacters isolated from poultry products by using two PCR-based typing methods assessment of the genetic diversity among arcobacters isolated from poultry products by using two PCR-based typing methods. *Applied and Environmental Microbiology, 68*(5), pp. 2172–2178.

Houf, K., De Zutter, L., Van Hoof, J. and Vandamme, P. (2002b). Occurrence and distribution of *Arcobacter* species in poultry processing. *Journal of Food Protection, 65*(8), pp. 1233–1239.

Houf, K., De Zutter, L., Verbeke, B., Van Hoof, J. and Vandamme, P. (2003). Molecular characterization of *Arcobacter* isolates collected in a poultry slaughterhouse. *Journal of Food Protection, 66*(3), pp. 364–369.

Hsueh, P.R., Teng, L.J., Yang, P.C., Wang, S.K., Chang, S.C., Ho, S.W., Hsieh, W.C. and Luh, K.T. (1997). Bacteremia caused by *Arcobacter cryaerophilus* 1B. *Journal of Clinical Microbiology, 35*(2), pp. 489–491.

Hume, M.E., Harvey, R.B., Stanker, L.H., Droleskey, R.E., Poole, T.L. and Zhang, H.B. (2001). Genotypic variation among arcobacter isolates from a farrow-to-finish swine facility. *Journal of Food Protection, 64*(5), pp. 645–651.

Hurtado, A. and Owen, R.J. (1997). A molecular scheme based on 23S rRNA gene polymorphisms for rapid identification of *Campylobacter* and *Arcobacter* species. *Journal of Clinical Microbiology, 35*(9), pp. 2401–2404.

International Commission on Microbiological Specifications for Foods (ICMSF). (2002). *Microorganisms in Food 7 - Microbiological Testing in Food Safety Management.* Springer Science and Business Media.

Jacob, J., Lior, H. and Feuerpfeil, I. (1993). Isolation of *Arcobacter butzleri* from a drinking water reservoir in eastern Germany. *Zentralbl Hyg Umweltmed, 193*(6), pp. 557–562.

Jacob, J., Woodward, D., Feuerpfeil, I. and Johnson, W.M. (1998). Isolation of *Arcobacter butzleri* in raw water and drinking water treatment plants in Germany. *Zentralblatt Fur Hygiene Und Umweltmedizin, 201*(2), pp. 189–198.

Jelínek, D., Miketová, P., Khailová, L., Schram, K.H., Moore, I.M. and Vytřasová, J. (2006). Identification of *Arcobacter* species using phospholipid and total fatty acid profiles. *Folia Microbiologica, 51*(4), pp. 329–336.

Johnson, L.G. and Murano, E.A. (1999a). Comparison of three protocols for the isolation of *Arcobacter* from poultry. *Journal of Food Protection, 62*(6), pp. 610–614.

Johnson, L.G. and Murano, E.A. (1999b). Development of a new medium for the isolation of *Arcobacter* spp. *Journal of food protection, 62*(5), pp. 456–462.

Jolley, K.A., Chan, M.-S. and Maiden, M.C.J. (2004). mlstdbNet - distributed multi-locus sequence typing (MLST) databases. *BMC Bioinformatics, 5*, p. 86.

Kabeya, H., Kobayashi, Y., Maruyama, S. and Mikami, T. (2003a). One-step polymerase chain reaction-based typing of *Arcobacter* species. *International Journal of Food Microbiology, 81*(2), pp. 163–168.

Kabeya, H., Maruyama, S., Morita, Y., Kubo, M., Yamamoto, K., Arai, S., Izumi, T., Kobayashi, Y., Katsube, Y. and Mikami, T. (2003b). Distribution of *Arcobacter* species among livestock in Japan. *Veterinary Microbiology, 93*(2), pp. 153–158.

Kabeya, H., Maruyama, S., Morita, Y., Ohsuga, T., Ozawa, S., Kobayashi, Y., Abe, M., Katsube, Y. and Mikami, T. (2004). Prevalence of *Arcobacter* species in retail meats and antimicrobial susceptibility of the isolates in Japan. *International Journal of Food Microbiology, 90*(3), pp. 303–308.

Karenlampi, R.I., Tolvanen, T.P. and Hanninen, M.L. (2004). Phylogenetic Analysis and PCR-Restriction Fragment Length Polymorphism Identification of *Campylobacter* Species Based on Partial *groEL* Gene Sequences. *Journal of Clinical Microbiology, 42*, pp. 5731–5738.

Kayman, T., Abay, S., Hizlisoy, H., Ibrahim Atabay, H., Serdar Diker, K. and Aydin, F. (2012). Emerging pathogen *Arcobacter* spp. in acute gastroenteritis: Molecular identification, antibiotic susceptibilities and genotyping of the isolated arcobacters. *Journal of Medical Microbiology, 61*, pp. 1439–1444.

Kemp, R., Leatherbarrow, A.J.H., Williams, N.J., Hart, C.A., Clough, H.E., Turner, J., Wright, E.J. and French, N.P. (2005). Prevalence and Genetic Diversity of *Campylobacter* spp. in Environmental Water Samples from a 100-Square-Kilometer Predominantly Dairy Farming Area. *Applied and Environmental Microbiology, 71*, pp. 1876–1882.

Khoshbakht, R., Tabatabaei, M., Shirzad Aski, H. and Seifi, S. (2014). Occurrence of *Arcobacter* in Iranian poultry and slaughterhouse samples implicates contamination by processing equipment and procedures. *British Poultry Science, 55*(6), pp. 732–736.

Kiehlbauch, J.A., Brenner, D.J., Nicholson, M.A., Baker, C.N., Patton, C.M., Steigerwalt, A.G. and Wachsmuth, I.K. (1991). *Campylobacter butzleri* sp. nov. isolated from humans and animals with diarrheal illness. *Journal of Clinical Microbiology, 29*(2), pp. 376–385.

Kim, H.M., Hwang, C.Y. and Cho, B.C. (2010). *Arcobacter marinus* sp. nov. *International Journal of Systematic and Evolutionary Microbiology, 60*(3), pp. 531–536.

Kjeldgaard, J., Jørgensen, K. and Ingmer, H. (2009). Growth and survival at chiller temperatures of *Arcobacter butzleri*. *International Journal of Food Microbiology, 131*(2-3), pp. 256–259.

Kopilovi, B., Ucakar, V., Koren, N., Krek, M. and Kraigher, A. (2008). Waterborne outbreak of acute gastroenteritis in a costal area in Slovenia in June and July 2008. *EuroSurveillance, 13*(34). pii = 18957.

Lappi, V., Archer, J.R., Cebelinski, E., Leano, F., Besser, J.M., Klos, R.F., Medus, C., Smith, K.E., Fitzgerald, C. and Davis, J.P. (2013). An outbreak of foodborne illness among attendees of a wedding reception in Wisconsin likely caused by *Arcobacter butzleri*. *Foodborne Pathogens and Disease, 10*(3), pp. 250–255.

Lau, S.K.P., Woo, P.C.Y., Teng, J.L.L., Leung, K.W. and Yuen, K.Y. (2002). Identification by 16S ribosomal RNA gene sequencing of *Arcobacter butzleri* bacteraemia in a patient with acute gangrenous appendicitis. *Molecular pathology : MP*, *55*(3), pp. 182–185.

Lee, C., Agidi, S., Marion, J.W. and Lee, J. (2012). *Arcobacter* in Lake Erie beach waters: An emerging gastrointestinal pathogen linked with human-associated fecal contamination. *Applied and Environmental Microbiology*, *78*(16), pp. 5511–5519.

Lee, M.H., Cheon, D.-S., Choi, S., Lee, B.-H., Jung, J.-Y. and Choi, C. (2010). Prevalence of *Arcobacter* species isolated from retail meats in Korea. *Journal of Food Protection*, *73*(7), pp. 1313–1316.

Lee, M.H. and Choi, C. (2013). Survival of *Arcobacter butzleri* in Apple and Pear Purees. *Journal of Food Safety*, *33*(3), pp. 333–339.

Lévesque, C.A., Topp, E., Lapen, D.R., Talbot, G. and Khan, U.H. (2014). Draft Genome Sequences of Three *Arcobacter* Strains of Pig and Dairy Cattle Manure Origin. *GenomeAnnouncements*, *2*(3), pp. 2–3.

Levican, A., Collado, L., Aguilar, C., Yustes, C., Diéguez, A.L., Romalde, J.L. and Figueras, M.J. (2012). *Arcobacter bivalviorum* sp. nov. and *Arcobacter venerupis* sp. nov., new species isolated from shellfish. *Systematic and Applied Microbiology*, *35*(3), pp. 133–138.

Levican, A., Collado, L. and Figueras, M.J. (2013). *Arcobacter cloacae* sp. nov. and *Arcobacter suis* sp. nov., two new species isolated from food and sewage. *Systematic and Applied Microbiology*, *36*(1), pp. 22–27.

Levican, A., Collado, L., Yustes, C., Aguilar, C. and Figueras, M.J. (2014). Higher water temperature and incubation under aerobic and microaerobic conditions increase the recovery and diversity of *Arcobacter* spp. from shellfish. *Applied and Environmental Microbiology*, *80*(1), pp. 385–391.

Levican, A. and Figueras, M.J. (2013). Performance of five molecular methods for monitoring *Arcobacter* spp. *BMC Microbiology*, *13*, pp. 220.

Levican, A., Rubio-Arcos, S., Martinez-Murcia, A., Collado, L. and Figueras, M.J. (2015). *Arcobacter ebronensis* sp. nov. and *Arcobacter aquimarinus* sp. nov., two new species isolated from marine environment. *Systematic and Applied Microbiology*, *38*(1), pp. 30–35.

Li, W., Raoult, D. and Fournier, P.E. (2009). Bacterial strain typing in the genomic era. *FEMS Microbiology Reviews*, *33*(5), pp. 892–916.

Maiden, M.C., Bygraves, J.A., Feil, E., Morelli, G., Russell, J.E., Urwin, R., Zhang, Q., Zhou, J., Zurth, K., Caugant, D.A., Feavers, I.M., Achtman, M. and Spratt, B.G. (1998). Multilocus sequence typing: a portable approach to the identification of clones within populations of pathogenic microorganisms. *Proceedings of the National Academy of Sciences of the United States of America*, *95*(6), pp. 3140–3145.

Marshall, S.M., Melito, P.L., Woodward, D.L., Johnson, W.M., Rodgers, F.G. and Mulvey, M.R. (1999). Rapid identification of *Campylobacter*, *Arcobacter*, and *Helicobacter* isolates by PCR-restriction fragment length polymorphism analysis of the *16S rRNA* gene. *Journal of Clinical Microbiology*, *37*(12), pp. 4158–4160.

Martiny, D., Dediste, A., Debruyne, L., Vlaes, L., Haddou, N.B., Vandamme, P. and Vandenberg, O. (2011). Accuracy of the API Campy system, the Vitek 2 *Neisseria-Haemophilus* card and matrix-assisted laser desorption ionization time-of-flight mass spectrometry for the identification of *Campylobacter* and related organisms. *Clinical Microbiology and Infection*, *17*(7), pp. 1001–1006.

Maurer, J.J. (2011). Rapid Detection and Limitations of Molecular Techniques. *Annual Review of Food Science and Technology*, *2*(1), pp. 259–279.

Mcclung, C.R., Patriquin, D.G. and Davis, R.E. (1983). *Campylobacter nitrofigilis* Sp-Nov, a Nitrogen-Fixing Bacterium Associated with Roots of *Spartina alterniflora* Loisel. *International Journal of Systematic Bacteriology*, *33*(3), pp. 605–612.

Merga, J.Y., Leatherbarrow, A.J.H., Winstanley, C., Bennett, M., Hart, C.A., Miller, W.G. and Williams, N.J. (2011). Comparison of *Arcobacter* isolation methods, and diversity of *Arcobacter* spp. in Cheshire, United Kingdom. *Applied and Environmental Microbiology, 77*(5), pp. 1646–1650.

Merga, J.Y., Williams, N.J., Miller, W.G., Leatherbarrow, A.J.H., Bennett, M., Hall, N., Ashelford, K.E. and Winstanley, C. (2013). Exploring the Diversity of *Arcobacter butzleri* from Cattle in the UK Using MLST and Whole Genome Sequencing. *PLoS ONE, 8*(2). e55240.

Miller, W.G., Parker, C.T., Rubenfield, M., Mendz, G.L., Wösten, M.M.S.M., Ussery, D.W., Stolz, J.F., Binnewies, T.T., Hallin, P.F., Wang, G., Malek, J. a., Rogosin, A., Stanker, L.H. and Mandrell, R.E. (2007). The complete genome sequence and analysis of the epsilonproteobacterium *Arcobacter butzleri. PLoS ONE, 2*(12). e1358.

Miller, W.G., Wesley, I.V, On, S.L.W., Houf, K., Mégraud, F., Wang, G., Yee, E., Srijan, A. and Mason, C.J. (2009). First multi-locus sequence typing scheme for *Arcobacter* spp. *BMC Microbiology, 9*, pp. 196.

Moreno, Y., Alonso, J.L., Botella, S., Ferrús, M.A. and Hernández, J. (2004). Survival and injury of *Arcobacter* after artificial inoculation into drinking water. *Research in Microbiology, 155*(9), pp. 726–730.

Moreno, Y., Botella, S., Alonso, J.L., Alonso, L., Ferrús, M. A, Hernández, M. and Hernández, J. (2003). Specific Detection of *Arcobacter* and *Campylobacter* Strains in Water and Sewage by PCR and Fluorescent *In Situ* Hybridization. *Applied and Environmental Microbiology, 69*(2), pp. 1181–1186.

Morita, Y., Maruyama, S., Kabeya, H., Boonmar, S., Nimsuphan, B., Nagai, A., Kozawa, K., Nakajima, T., Mikami, T. and Kimura, H. (2004). Isolation and phylogenetic analysis of *Arcobacter* spp. in ground chicken meat and environmental water in Japan and Thailand. *Microbiology and Immunology, 48*(7), pp. 527–533.

Münch, B. and Wochenschr, T. (2015). Prevalence , virulence gene distribution and genetic diversity of *Arcobacter* in food samples in Germany. *Berliner und Münchener Tierärztliche Wochenschrift, 4*, pp. 163–168.

Musmanno, R.A., Russi, M., Lior, H. and Figura, N. (1997). *In vitro* virulence factors of *Arcobacter butzleri* strains isolated from superficial water samples. *Microbiologica, 20*(1), pp. 63–68.

Neill, S.D., Campbell, J.N., Obrien, J.J., Weatherup, S.T.C. and Ellis, W.A. (1985). Taxonomic Position of *Campylobacter-Cryaerophila* Sp-Nov. *International Journal of Systematic Bacteriology, 35*(3), pp. 342–356.

Neubauer, C. and Hess, M. (2006). Detection and identification of food-borne pathogens of the genera *Campylobacter*, *Arcobacter* and *Helicobacter* by multiplex PCR in poultry and poultry products. *Journal of Veterinary Medicine Series B, 53*, pp. pp. 376–381.

Nieva-Echevarria, B., Martinez-Malaxetxebarria, I., Girbau, C., Alonso, R. and Fernández-Astorga, A. (2013). Prevalence and genetic diversity of arcobacter in food products in the north of Spain. *Journal of Food Protection, 76*(8), pp. 1447–50.

Ohlendorf, D.S. and Murano, E.A. (2002). Prevalence of *Arcobacter* spp. in raw ground pork from several geographical regions according to various isolation methods. *Journal of Food Protection, 65*(11), pp. 1700–1705.

De Oliveira, S.J., Baetz, A.L., Wesley, I.V. and Harmon, K.M. (1997). Classification of *Arcobacter* species isolated from aborted pig fetuses and sows with reproductive problems in Brazil. *Veterinary Microbiology, 57*(4), pp. 347–354.

De Oliveira, S.J., Bernardi, R., Vogt, F., Scartezzini, M., Hepp, D. and Lunge, V. (2010). Úlceras gástricas em suínos de abate: cultivo de *Arcobacter* spp. apartir de estômagos com diferentes graus de lesões. *Acta Scientiae Veterinariae, 55*(June), pp. 351–356.

On, S.L., Stacey, A. and Smyth, J. (1995). Isolation of *Arcobacter butzleri* from a neonate with bacteraemia. *The Journal of Infection, 31*(3), pp. 225–227.

On, S.L.W. (1996). Identification methods for campylobacters, helicobacters, and related organisms. *Clinical Microbiology Reviews, 9*(3), pp. 405–422.

On, S.L.W., Atabay, H.I., Amisu, K.O., Coker, A.O. and Harrington, C.S. (2004). Genotyping and genetic diversity of *Arcobacter butzleri* by amplified fragment length polymorphism (AFLP) analysis. *Letters in Applied Microbiology, 39*(4), pp. 347–352.

On, S.L.W., Harrington, C.S. and Atabay, H.I. (2003). Differentiation of *Arcobacter* species by numerical analysis of AFLP profiles and description of a novel *Arcobacter* from pig abortions and turkey faeces. *Journal of Applied Microbiology, 95*(5), pp. 1096–1105.

On, S.L.W., Jensen, T.K., Bille-Hansen, V., Jorsal, S.E. and Vandamme, P. (2002). Prevalence and diversity of *Arcobacter* spp. isolated from the internal organs of spontaneous porcine abortions in Denmark. *Veterinary Microbiology, 85*(2), pp. 159–167.

Öngör, H., Çetinkaya, B., Açik, M.N. and Atabay, H.I. (2004). Investigation of arcobacters in meat and faecal samples of clinically healthy cattle in Turkey. *Letters in Applied Microbiology, 38*(4), pp. 339–344.

Patyal, A., Rathore, R.S., Mohan, H. V., Dhama, K. and Kumar, A. (2011). Prevalence of *Arcobacter* spp. in Humans, Animals and Foods of Animal Origin Including Sea Food from India. *Transboundary and Emerging Diseases, 58*(5), pp. 402–410.

Pejchalová, M., Dostalíková, E., Slámová, M., Brozková, I. and Vytrasová, J. (2008). Prevalence and diversity of *Arcobacter* spp. in the Czech Republic. *Journal of Food Protection, 71*(4), pp. 719–727.

Pejchalova, M., Vytrasova, J., Brozkova, I., Huskova, Z. and Cervenka, L. (2006). Occurrence of arcobacters in the Czech Republic and the influence of sample matrix on their detection using PCR. *Journal of Food and Nutrition Research, 45*(4), pp. 152–158.

Pentimalli, D., Pegels, N., García, T., Martín, R. and González, I. (2009). Specific PCR detection of *Arcobacter butzleri, Arcobacter cryaerophilus, Arcobacter skirrowii,* and *Arcobacter cibarius* in chicken meat. *Journal of Food Protection, 72*(7), pp. 1491–1495.

Petersen, R.F., Harrington, C.S., Kortegaard, H.E. and On, S.L.W. (2007). A PCR-DGGE method for detection and identification of *Campylobacter, Helicobacter, Arcobacter* and related Epsilobacteria and its application to saliva samples from humans and domestic pets. *Journal of Applied Microbiology, 103*(6), pp. 2601–2615.

Piva, S., Serraino, A., Florio, D., Giacometti, F., Pasquali, F., Manfreda, G. and Zanoni, R.G. (2013). Isolation of *Arcobacter* species in water buffaloes (*Bubalus bubalis*). *Foodborne Pathogens and Disease, 10*(5), pp. 475–477.

Prouzet-Mauléon, V., Labadi, L., Bouges, N., Ménard, A. and Mégraud, F. (2006). *Arcobacter butzleri:* Underestimated Enteropathogen. *Emerging Infectious Diseases, 12*(2), pp. 307–309.

Quiñones, B., Parker, C.T., Janda, J.M., Miller, W.G. and Mandrell, R.E. (2007). Detection and genotyping of *Arcobacter* and *Campylobacter* isolates from retail chicken samples by use of DNA oligonucleotide arrays. *Applied and Environmental Microbiology, 73*(11), pp. 3645–3655.

Rahimi, E. (2014). Prevalence and antimicrobial resistance of *Arcobacter* species isolated from poultry meat in Iran. *British Poultry Science, 55*(2), pp. 174–180.

Ramees, T.P., Rathore, R.S., Bagalkot, P.S., Sailo, B., Mohan, H.V., Kumar, A., Dhama, K. and Singh, R.K. (2014). Genotyping and genetic diversity of *Arcobacter butzleri* and *Arcobacter cryaerophilus* isolated from different sources by using ERIC-PCR from India. *Veterinary Quarterly, 34*(4), pp. 211–217.

Rashid, S.T. Al, Dakuna, I., Louie, H., Ng, D., Vandamme, P., Johnson, W. and Chan, V.L. (2000). Identification of *Campylobacter jejuni, C. coli, C. lari, C. upsaliensis, Arcobacter butzleri,* and *A. butzleri*-like species based on the *glyA* gene. *Journal of Clinical Microbiology, 38*(4), pp. 1488–1494.

Rasmussen, L.H., Kjeldgaard, J., Christensen, J.P. and Ingmer, H. (2013). Multilocus sequence typing and biocide tolerance of *Arcobacter butzleri* from Danish broiler carcasses. *BMC Research Notes*, *6*(1), pp. 322.

Revez, J., Huuskonen, M., Ruusunen, M., Lindström, M. and Hänninen, M.L. (2013). *Arcobacter* Species and Their Pulsed-Field Gel Electrophoresis Genotypes in Finnish Raw Milk during Summer 2011. *Journal of Food Protectio*, *76*(9), pp. 1630–1632.

Rice, E.W., Rodgers, M.R., Wesley, I. V., Johnson, C.H. and Tanner, S.A. (1999). Isolation of *Arcobacter butzleri* from ground water. *Letters in Applied Microbiology*, *28*(1), pp. 31–35.

Rivas, L., Fegan, N. and Vanderlinde, P. (2004). Isolation and characterisation of *Arcobacter butzleri* from meat. *International Journal of Food Microbiology*, *91*(1), pp. 31–41.

Sabat, A.J., Budimir, A., Nashev, D., Sá-Leão, R., van Dijl, J., Laurent, F., Grundmann, H. and Friedrich, A.W., on behalf of the ESCMID Study Group of Epidemiological Markers (ESGEM) (2013). Overview of molecular typing methods for outbreak detection and epidemiological surveillance. *Euro Surveillance* , *18*(4), pp. 17–30.

Samie, A., Obi, C.L., Barrett, L.J., Powell, S.M. and Guerrant, R.L. (2007). Prevalence of *Campylobacter* species, *Helicobacter pylori* and *Arcobacter* species in stool samples from the Venda region, Limpopo, South Africa: Studies using molecular diagnostic methods. *Journal of Infection*, *54*(6), pp. 558–566.

Sasi Jyothsna, T.S., Rahul, K., Ramaprasad, E.V., Sasikala, C. and Ramana, C.V. (2013). *Arcobacter anaerophilus* sp. nov., isolated from an estuarine sediment and emended description of the genus *Arcobacter*. *International Journal of Systematic and Evolutionary Microbiology*, *63*(Pt12), pp. 4619–4625.

Scanlon, K.A., Cagney, C., Walsh, D., McNulty, D., Carroll, A., McNamara, E.B., McDowell, D.A. and Duffy, G. (2013). Occurrence and characteristics of fastidious *Campylobacteraceae* species in porcine samples. *International Journal of Food Microbiology*, *163*(1), pp. 6–13.

Scarano, C., Giacometti, F., Manfreda, G., Lucchi, A., Pes, E., Spanu, C., De Santis, E.P.L. and Serraino, A. (2014). *Arcobacter butzleri* in Sheep Ricotta Cheese at Retail and Related Sources of Contamination in an Industrial Dairy Plant. *Applied and Environmental Microbiology*, *80*(22), pp. 7036–7041.

Scullion, R., Harrington, C.S. and Madden, R.H. (2004). A comparison of three methods for the isolation of *Arcobacter* spp. from retail raw poultry in Northern Ireland. *Journal of Food Protection*, *67*(4), pp. 799–804.

Scullion, R., Harrington, C.S. and Madden, R.H. (2006). Prevalence of *Arcobacter* spp. in raw milk and retail raw meats in Northern Ireland. *Journal of Food Protection*, *69*(8), pp. 1986–1990.

Serraino, A., Florio, D., Giacometti, F., Piva, S., Mion, D. and Zanoni, R.G. (2013a). Presence of *Campylobacter* and *Arcobacter* species in in-line milk filters of farms authorized to produce and sell raw milk and of a water buffalo dairy farm in Italy. *Journal of Dairy Science*, *96*(5), pp. 2801–2807.

Serraino, A. and Giacometti, F. (2014). Occurrence of *Arcobacter* species in industrial dairy plants. *Journal of Dairy Science*, *97*(4), pp. 2061–2065.

Serraino, A., Giacometti, F., Daminelli, P., Losio, M.N., Finazzi, G., Marchetti, G., Zambrini, A.V. and Rosmini, R. (2013b). Survival of *Arcobacter butzleri* during production and storage of artisan water buffalo mozzarella cheese. *Foodborne Pathogens and Disease*, *10*(9), pp. 820–824.

Shah, A.H., Saleha, A.A., Murugaiyah, M., Zunita, Z. and Memon, A.A. (2012a). Prevalence and Distribution of *Arcobacter* spp. in Raw Milk and Retail Raw Beef. *Journal of Food Protection*, *75*(8), pp. 1474–1478.

Shah, A.H., Saleha, A.A., Zunita, Z., Cheah, Y.K., Murugaiyah, M. and Korejo, N.A. (2012b). Genetic characterization of *Arcobacter* isolates from various sources. *Veterinary Microbiology*, *160*(3-4), pp. 355–361.

Shah, A.H., Saleha, A.A., Zunita, Z. and Murugaiyah, M. (2011a). *Arcobacter* - An emerging threat to animals and animal origin food products? *Trends in Food Science and Technology*, *22*(5), pp. 225–236.

Shah, A.H., Saleha, A.A., Zunita, Z., Murugaiyah, M., Aliyu, A.B. and Jafri, N. (2013). Prevalence, Distribution and Antibiotic Resistance of Emergent *Arcobacter* spp. from Clinically Healthy Cattle and Goats. *Transboundary and Emerging Diseases*, *60*(1), pp. 9–16.

Shah, A.H., Saleha, A.A., Murugaiyah, M. and Zakaria, Z. (2011b). Assessment of sensitivity, specificity and species discriminatory power of four culture-based isolation methods of *Arcobacter* spp. *African Journal of Microbiology Research*, *5*(22), pp. 3753–3759.

Šilha, D., Hrušková, L., Brožková, I., Mo, P., Jarmila, Ť.K. and Asová, V.Y.T.Ř. (2014). Survival of selected bacteria from the genus *Arcobacter* on various metallic surfaces, *53*(3), pp. 217–223.

Šilha, D., Šilhová-Hrušková, L. and Vytřasová, J. (2015). Modified isolation method of *Arcobacter* spp. from different environmental and food samples. *Folia Microbiologica*, *60*(6), pp. 515–521.

Šilha, D., Vytřasová, J., Beňová, B. and Moťková, P. (2013). Effect of selected types of beer on bacteria of the genus *Arcobacter*. *Current Microbiology*, *66*(4), pp. 368–373.

De Smet, S., Vandamme, P., De Zutter, L., On, S.L.W., Douidah, L. and Houf, K. (2011a). *Arcobacter trophiarum* sp. nov., isolated from fattening pigs. *International Journal of Systematic and Evolutionary Microbiology*, *61*(2), pp. 356–361.

De Smet, S., De Zutter, L., Debruyne, L., Vangroenweghe, F., Vandamme, P. and Houf, K. (2011b). *Arcobacter* population dynamics in pigs on farrow-to-finish farms. *Applied and Environmental Microbiology*, *77*(5), pp. 1732–1738.

De Smet, S., De Zutter, L., Van Hende, J. and Houf, K. (2010). *Arcobacter* contamination on pre- and post-chilled bovine carcasses and in minced beef at retail. *Journal of Applied Microbiology*, *108*(1), pp. 299–305.

De Smet, S., De Zutter, L. and Houf, K. (2011c). Small ruminants as carriers of the emerging foodborne pathogen *Arcobacter* on small and medium farms. *Small Ruminant Research*, *97*(1–3), pp. 124–129.

Snaidr, J., Amann, R., Huber, I., Ludwig, W. and Schleifer, K.H. (1997). Phylogenetic analysis and *in situ* identification of bacteria in activated sludge. *Applied and Environmental Microbiology*, *63*(7), pp. 2884–2896.

Son, I., Englen, M.D., Berrang, M.E., Fedorka-Cray, P.J. and Harrison, M.A. (2006). Genetic diversity of *Arcobacter* and *Campylobacter* on broiler carcasses during processing. *Journal of Food Protection*, *69*(5), pp. 1028–1033.

Son, I., Englen, M.D., Berrang, M.E., Fedorka-Cray, P.J. and Harrison, M.A. (2007). Prevalence of *Arcobacter* and *Campylobacter* broiler carcasses during processing. *International Journal of Food Microbiology*, *113*, pp. 16–22.

Suarez, D.L., Wesley, I. V. and Larson, D.J. (1997). Detection of *Arcobacter* species in gastric samples from swine. *Veterinary Microbiology*, *57*(4), pp. 325–336.

Teague, N.S., Srijan, A., Wongstitwilairoong, B., Poramathikul, K., Champathai, T., Ruksasiri, S., Pavlin, J. and Mason, C.J. (2010). Enteric pathogen sampling of tourist restaurants in Bangkok, Thailand. *Journal of Travel Medicine*, *17*(2), pp. 118–123.

Tee, W., Baird, R., Dyall-Smith, M. and Dwyer, B. (1988). *Campylobacter cryaerophila* isolated from a human. *Journal of Clinical Microbiology*, *26*(12), pp. 2469–2473.

Udeigwea, T.K., Tebohb, J.M., Ezec, P.N., Stietiyad, M.H., Kumare, V., Hendrixf, J., Mascagnif, H.J.J., Yingg, T. and Kandakjia, T. (2015). Implications of leading crop production practices on environmental quality and human health. *Journal of Environmental Management*, *151*, pp. 267–279.

Vandamme, P., Dewhirst, F.E., Paster, B.J. and On, S.L.W. (2005). Genus II. *Arcobacter*. Vandamme, Falsen, Rossau, Hoste, Segers, Tytgat and De Ley 1991a, 99VP emend. Vandamme, Vancanneyt, Pot, Mels, Hoste, Dewettinck, Vlaes, Van Den Borre, Higgins,

Hommez, Kersters, Butzler and Goossens 1992b, 355. In: D.J. Brenner, N.P. Kreig, J.T. Staley, and G.M. Garrity, eds., *Bergey's Manual of Systematic Bacteriology* New York: Springer, pp. 1161–1165.

Vandamme, P., Falsen, E., Rossau, R., Hoste, B., Segers, P., Tytgat, R. and De Ley, J. (1991). Revision of *Campylobacter*, *Helicobacter*, and *Wolinella* taxonomy: emendation of generic descriptions and proposal of *Arcobacter* gen. nov. *International Journal of Systematic Bacteriology*, *41*(1), pp. 88–103.

Vandamme, P. and De Ley, J. (1991). Proposal for a New Family, *Campylobacteraceae*. *International Journal of Systematic Bacteriology*, *41*(3), pp. 451–455.

Vandamme, P., Pugina, P., Benzi, G., Van Etterijck, R., Vlaes, L., Kersters, K., Butzler, J.P., Lior, H. and Lauwers, S. (1992a). Outbreak of recurrent abdominal cramps associated with *Arcobacter butzleri* in an Italian school. *Journal of Clinical Microbiology*, *30*(9), pp. 2335–2337.

Vandamme, P., Vancanneyt, M., Pot, B., Mels, L., Hoste, B., Dewettinck, D., Vlaes, L., van den Borre, C., Higgins, R. and Hommez, J. (1992b). Polyphasic taxonomic study of the emended genus *Arcobacter* with *Arcobacter butzleri* comb. nov. and *Arcobacter skirrowii* sp. nov., an aerotolerant bacterium isolated from veterinary specimens. *International Journal of Systematic Bacteriology*, *42*(3), pp. 344–356.

Vandenberg, O., Dediste, A., Houf, K., Ibekwem, S., Souayah, H., Cadranel, S., Douat, N., Zissis, G. and Butzler, J. (2004). *Arcobacter* Species in Humans. *Emerging Infectious Diseases*, *10*(10), pp. 1863–1867.

Villalobos, E.G., Jaramillo, H.F., Ulate, C.C. and Echandi, M.L.A. (2013). Isolation and identification of zoonotic species of genus arcobacter from chicken viscera obtained from retail distributors of the metropolitan area of San José, Costa Rica. *Journal of Food Protection*, *76*(5), pp. 879–882.

Villarruel-López, A., Márquez-González, M., Garay-Martínez, L.E., Zepeda, H., Castillo, A., Mota de la Garza, L., Murano, E.A. and Torres-Vitela, R. (2003). Isolation of *Arcobacter* spp. from retail meats and cytotoxic effects of isolates against vero cells. *Journal of Food Protection*, *66*(8), pp. 1374–1378.

Vindigni, S.M., Srijan, A., Wongstitwilairoong, B., Marcus, R., Meek, J., Riley, P.L. and Mason, C. (2007). Prevalence of foodborne microorganisms in retail foods in Thailand. *Foodborne Pathogens and Disease*, *4*(2), pp. 208–215.

Wang, X., Seo, D.J., Lee, M.H. and Choi, C. (2014). Comparison of conventional PCR, multiplex PCR, and loop-mediated isothermal amplification assays for rapid detection of *Arcobacter* species. *Journal of Clinical Microbiology*, *52*(2), pp. 557–563.

Webb, A.L., Kruczkiewicz, P., Selinger, L.B., Inglis, G.D. and Taboada, E.N. (2015). Development of a comparative genomic fingerprinting assay for rapid and high resolution genotyping of *Arcobacter butzleri*. *BMC Microbiology*, *15*(1), pp. 1–12.

Wesley, I. and Miller, W.G. (2010). *Arcobacter*: an opportunistic human food-borne pathogen? In: W.M. Scheld, M.L. Grayson, and J.M. Hughes, eds., *Emerging Infections 9* ed. Washington, DC: ASM Press, pp. 185–211.

Wesley, I.V., Wells, S.J., Harmon, K.M., Green, A., Schroeder-Tucker, L., Glover, M. and Siddique, I. (2000). Fecal shedding of *Campylobacter* and *Arcobacter* spp. in dairy cattle. *Applied and Environmental Microbiology*, *66*(5), pp. 1994–2000.

Whiteduck-Leveillee, K., Whiteduck-Leveillee, J., Cloutier, M., Tambong, J.T., Xu, R., Topp, E., Arts, M.T., Chao, J., Adam, Z., Lévesque, C.A., Lapen, D.R., Villemur, R., Talbot, G. and Khan, I.U.H. (2015). *Arcobacter lanthieri* sp. nov., isolated from pig and dairy cattle manure. *International Journal of Systematic and Evolutionary Microbiology*, *1430*(8), pp. 2709–2716.

Whiteduck-Léveillée, K., Whiteduck-Léveillée, J., Cloutier, M., Tambong, J.T., Xu, R., Topp, E., Arts, M.T., Chao, J., Adam, Z., Lévesque, C.A., Lapen, D.R., Villemur, R. and Khan, I.U.H. (2016). Identification, characterization and description of *Arcobacter faecis* sp. nov., isolated from a human waste septic tank. Systematic and Applied Microbiology, *39*(2), pp. 93–99.

Wilkes, G., Brassard, J., Edge, T.A., Gannon, V., Gottschall, N., Jokinen, C.C., Jones, T.H., Khan, I.U.H., Marti, R., Sunohara, M.D., Topp, E., Lapen, R. and Lapen, D.R. (2014). Long-term monitoring of waterborne pathogens and microbial source tracking markers in paired agricultural watersheds under controlled and conventional tile drainage management. *Applied and Environmental Microbiology, 80*(12), pp. 3708–3720.

Woo, P.C.Y., Chong, K.T.K., Leung, K.W., Que, T.L. and Yuen, K.Y. (2001). Identification of *Arcobacter cryaerophilus* isolated from a traffic accident victim with bacteremia by 16S ribosomal RNA gene sequencing. *Diagnostic Microbiology and Infectious Disease, 40*(3), pp. 125–127.

Xiong, W., Sun, Y., Zhang, T., Ding, X., Li, Y., Wang, M. and Zeng, Z. (2015). Antibiotics, Antibiotic Resistance Genes, and Bacterial Community Composition in Fresh Water Aquaculture Environment in China. *Microbial Ecology, 70*, pp. 425–432.

Yan, J., Ko, W., Huang, A. and Chen, H. (2000). *Arcobacter butzleri* bacteremia in a patient with liver cirrhosis. *Journal of the Formosan Medical Association, 99*(2), pp. 166–169.

Yang, B., Kong, X., Cui, B., Jin, D., Deng, Y., Zhuang, X., Zhuang, G. and Bai, Z. (2015). Impact of Rural Domestic Wastewater Irrigation on the Physicochemical and Microbiological Properties of Pakchoi and Soil. *Water, 7*(5), pp. 1825–1839.

Yap, D.Y., Kwan, L.P., To, K.K. and Chan, T.M. (2013). *Arcobacter* Peritonitis After Fluoroscopic Repositioning of a Tenckhoff Catheter. *Peritoneal Dialysis International, 33*(2), pp. 222–223.

Yesilmen, S., Vural, A., Erkan, M.E. and Yildirim, I.H. (2014). Prevalence and antimicrobial susceptibility of *Arcobacter* species in cow milk, water buffalo milk and fresh village cheese. *International Journal of Food Microbiology, 188*, pp. 11–14.

Zacharow, I., Bystroń, J., Wałecka-Zacharska, E., Podkowik, M. and Bania, J. (2015a). Prevalence and antimicrobial resistance of *Arcobacter butzleri* and *Arcobacter cryaerophilus* isolates from retail meat in Lower Silesia region, Poland. *Polish Journal of Veterinary Sciences, 18*(1), pp. 63–69.

Zacharow, I., N, J.B.B., B, E.W., Podkowik, M. and Bania, J. (2015b). Genetic Diversity and Incidence of Virulence-Associated Genes of *Arcobacter butzleri* and *Arcobacter cryaerophilus* Isolates from Pork , Beef , and Chicken Meat in Poland. *BioMed Research International, 2015*, p. 956507.

Zhang, Z., Yu, C., Wang, X., Yu, S. and Zhang, X.-H. (2015). *Arcobacter pacificus* sp . nov ., isolated from seawater of the South Pacific Gyre. *International Journal of Systematic and Evolutionary Microbiology, 66*, pp. 542–547.

5

Microbial Hazards and Their Implications in the Production of Table Olives

A. Valero[1], *E. Medina*[2], *and F.N. Arroyo-López*[2]

[1] *Department of Food Science and Technology, University of Cordoba, Campus de Rabanales, Edificio Darwin, 14014, Córdoba, Spain*
[2] *Food Biotechnology Department, Instituto de la Grasa (CSIC), Campus Universitario Pablo de Olavide, Building 46, Ctra, Utrera, km 1, 41013, Seville, Spain*

5.1 Table Olives: Origin, Production, and Main Types of Elaborations

Vegetables are very important for a healthy diet because they are sources of phytosterols, dietary fibers, phytochemicals, minerals, and water-soluble vitamins, among other important compounds (Gebbers, 2007). Their consumption is recommended to prevent chronic pathologies such as hypertension, coronary heart diseases, and the risk of stroke (Dauchet *et al.*, 2007; He *et al.*, 2007). The production of fermented vegetables is lost in the mists of history and the processes have been adapted over generations to favor the most robust procedures. Thereby, fermentation is responsible for many favorable effects on manufactured vegetables, removing anti-nutritional components and improving the shelf life, texture, flavor, and safety of the final products.

In particular, table olives are a traditional fermented vegetable with many centuries of history, especially in the Mediterranean basin, where this food has an enormous importance on the diet and culture of many countries. Most probably, the cultivar of olive tree (*Olea europaea*) was originated from Asia Minor. Phoenicians started disseminating this cultivar throughout the Greek mainland and Greek isles. Then, the olive spread throughout the Mediterranean countries reaching Tripoli, Tunis, Sicily, and southern Italy. Romans continued the expansion of this cultivar to the countries bordering the Mediterranean Basin. In Spain, olive growing was introduced by Phoenicians (1050 BC). Later, during the fifteenth and sixteenth centuries, Spaniards extended olive planting to America. The olive tree has recently expanded also to South Africa, Australia, and Far East (Japan). This way, table olives have currently a worldwide production and represent an important economic source for many producing countries (Arroyo-López *et al.*, 2012a).

The Trade Standard Applying to Table Olives (IOOC, 2004) defines table olives as: "*the product obtained from suitable olive cultivars, processed to remove their natural bitterness, and preserved (by natural fermentation, heat treatment or preservatives) with or without brine until consumption.*" The production of this fermented vegetable was a craft process till the middle of the twentieth century, when the mass fermentation and storage was introduced together with the pitting and pitting/stuffing automation. Then, the progressive mechanization in the industries has led to a rapid expansion of the sector.

Foodborne Pathogens and Antibiotic Resistance, First Edition. Edited by Om V. Singh.

Nowadays, the last consolidated report provided by International Olive Oil Council (IOOC), estimates that the table olive's world production reached 2,590,000 tons in the 2013/2014 season, with Spain (573,000 tons), Egypt (453,000), and Turkey (410,000 tons) as the main producers. Olive cultivars devoted for the production of table olives must meet several requirements (IOOC, 2004). Only can be used those cultivars whose volume, shape, flesh-to-stone ratio, fine flesh, taste, firmness, and ease of detachment from the stone make them particularly suitable for processing. Among the most important olive varieties in the Mediterranean basin are Manzanilla, Gordal and Hojiblanca (from Spain), Coservolea, Kalamon, Chalkidiki, Throbolea, Megaritici, and Kothreici (from Greece), Domat, Gemlik, Memeçik, Memeli, Edremit, and Sofralik (from Turkey) (Arroyo-López *et al.*, 2012a).

The olive fruit is a fleshy drupe. The range of concentration of the major components in the flesh olive fruits are, depending of the olive variety and maturity degree, moisture (650–800), fats (100–300), phenolic compounds (30–60), sugars (30–60), fiber (20–50), proteins (10–20), and ashes (10–15) (g/kg). On the contrary, olives after fermentation process are low in carbohydrates compared with other foods. Table olives should have all the benefits attributed to the olive oil, as well as other important components with proved biological activities. Thereby, olives are a good source of natural antioxidants (hydroxytyrosol, oleuropein, methyloleuropein, etc.) and triterpenic acids, all of them with unquestionable biological value. This important nutritional composition value of table olives is being recently considered (Arroyo-López *et al.*, 2016).

Because of its peculiar characteristics, presence of a bitter compound (oleuropein), olives cannot be consumed directly from the tree. For this reason, diverse processes were developed to make them palatable; although sharing the same objective, they can slightly differ between areas of production. Thereby, there are several ways to elaborate table olives, but the most prominent industrial elaborations are (Garrido-Fernández *et al.*, 1997):

i) *Spanish-style green olives.* The procedure for preparing this type of olive consists of treating the fruits with a dilute NaOH solution. The alkali is used to penetrate the olive skin and to hydrolyze the glucoside oleuropein into others not bitter compounds. Lye treatment not only has the effect of removing bitterness but also markedly increases skin permeability, which favors the release of nutrients. Then, one or two water washes are carried out to remove the excess of alkali and the fruits are immediately immersed in brines (9-12% NaCl) where lactic acid fermentation takes place.
ii) *Black ripe olives.* Fruits are previously preserved in an aqueous solution (brine, acidic water, etc.) and darkened throughout the year according to demand. Darkening consists of several treatments of dilute NaOH solutions and water washes between them. During the oxidation process, air is passed throughout the suspension of the olives in the liquid. Once the olives obtain the proper color is fixed by immersion in a lactate or gluconate iron solution.
iii) *Directly brined olives.* Fruits (green, turning color or naturally black olives) are directly brined after picking. In brine, fruits undergo a fermentation whose characteristics depend on the physicochemical conditions, cultivar, temperature, and salt content. This preparation does not include the lye treatment of fruits and thus they are characterized by a slow diffusion of olive compounds into brines. Usually, the lactic acid process is more difficult to complete due to the presence of high concentrations of polyphenols. During fermentation, the diffusion of oleuropein to brines occur, and consequently, the natural debittering of fruits progresses. In fact, the fruits are maintained in this solution until they lose their natural bitterness, at least partially.

Therefore, the preservation and preparation of table olives by fermentation is carried out by a combination of salting, natural fermentation, and acidification by lactic acid production (Arroyo-López *et al.*, 2016). Salt, mainly NaCl, has four primary roles in the preservation of olives:

i) it achieves a low water activity and consequently influences the type and extent of microbial metabolism,
ii) it increases the permeability of the fruit membranes allowing the diffusion of components into the brines,
iii) it determines the flavor of the final product, and
iv) it indirectly helps to prevent the softening of vegetable tissue. This system of processing has numerous advantages, but fundamentally fermentation offers an easy and economic way to preserve the raw material over extended periods of time by the use of microorganisms.

5.2 Importance of Microorganisms in Table Olives

The process of transforming the olive drupe into a palatable food is the result of complex biochemical reactions that are influenced by microorganisms, which determine the quality, safety, and flavor of the final product. *Enterobacteriaceae*, *Propionibacteriaceae*, lactic acid bacteria (LAB), and yeasts are among the most relevant microbial groups that can affect positively or negatively the fermentation process.

Gram-negative bacteria, mainly belonging to the *Enterobacteriaceae* family, usually appear at the onset of the process because of the high pH values reported during first days of fermentation in lye-treated olives. These microorganisms are considered as spoilage bacteria because they can produce gas pockets in the olive surface and metabolites that negatively affect the aroma of the product (Garrido-Fernández *et al.*, 1997). However, as fermentation progresses, they are quickly inhibited by the reduction in pH originated by LAB species, mainly *Lactobacillus plantarum* and *Lactobacillus pentosus*. Both species are very important because they use the sugars present in olive flesh to produce lactic acid and bacteriocins that originate the rapid and safe acidification of brines (Hurtado *et al.*, 2012). Diverse yeast species (among the most important we can foreground *Wickerhamomyces anomalus*, *Pichia membranifaciens*, *Pichia galeiformis*, *Saccharomyces cerevisiae*, *Candida boidinii*, and *Debaryomyces hansenii*) coexist with LAB practically throughout the entire process. Moreover, these microorganisms are especially relevant in directly brined green and black natural olive fermentations, where fruits are not treated with NaOH and LAB are partially inhibited due to the presence of phenolic compounds. Yeasts can have a double role, acting as desirable or undesirable microorganisms (Arroyo-López *et al.*, 2012b). These microorganisms possess many interesting technological properties such as catalase, lipolytic, β-glucosidase, and killer activities. In addition, they also consume the sugars present in brines and produce diverse types of organic acids (citric or acetic), vitamins, and aromatic compounds. However, as negative aspect, yeasts can produce spoilage of fruits during table olive storage or packaging. If these microorganisms become dominant, they can produce:

i) a product with a milder taste and a limited self preservation,
ii) a vigorous production of CO_2 that may penetrate olives and damage the fruits and swollen containers,
iii) clouding of brines, and
iv) fruit softening. Finally, during storage of fermented olives may observe the growth of species of the *Propionibacterium* genera, which increase the pH because of the production of acetic and propionic acids using the lactic acid produced during the previous phase of active fermentation.

Therefore, as in other food fermentations, it is necessary to favor the growth of desirable microorganisms and inhibit pathogen and spoilage microorganisms. Below, we detail different molecular tools to determine and know the microbial species present in table olive processing.

5.3 Molecular Methods for the Study of Microbial Populations in Table Olives

The microbial ecology of the table olives fermentation is a complex and dynamic process where microbiota suffers sequential changes along time. Study of the table olive microbiota have been performed for many researchers in different countries, types of elaboration, cultivars, and so on, in order to control the fermentation process, avoid food spoilage, select adequate starter cultures to optimize table olives elaborations and maintenance of food safety. Nowadays, the study of microbial ecology can be approached by both culture-dependent or culture-independent molecular methods.

5.3.1 Cultured-Dependent Methods

Until now, most of studies have used traditional microbiological methods to describe the predominant microbiota in table olives. Culture-dependent methods strategy consists in the isolation of microorganisms from the food matrix by classical techniques and further biochemical and/or molecular analysis. First step for cultured-dependent methods is the microbial isolation, which is determined by plating the olives brines or pulp samples onto specific solid media with selective agents for determination of the different microbial groups. Thereby, De Man, Rogosa, Sharpe (MRS) agar supplemented with sodium azide is used for the study of LAB species and YM (yeast-malt-peptone-glucose) agar with antibiotics (gentamicin sulfate, oxytetracycline, etc.) for determination of yeasts. Acetic acid bacteria is present in black ripe olive storage with acidified water, especially *Acetobacter* genera, which are detected in GYE medium (glucose-yeast extract) and confirmed in Carr broth (De Castro *et al.*, 2007). Potential spoilage microorganisms such as *Enterobacteriaceae* and butyric-acid-forming clostridia can proliferate in Spanish-style green olives and spoil the product (Fernandez Díaz, 1985) when pH is too high in early stages of brining. VRBD (Violet Red Bile Dextrose) agar is used for isolation and enumeration of *Enterobacteriacea* and anaerobic sulfite reducers Clostridia were grown on reinforced *Clostridium* medium in tubes.

As mentioned below, some foodborne pathogens can be isolated from table olives. For instance, *Listeria monocytogenes* was detected in natural green olives brines and in commercial preparation (Caggia *et al.*, 2004). Isolation was carried out with the use of an enrichment step according to USDA method (McClain and Lee, 1988). Pereira *et al.* (2008) analyzed several table olives brines of different types of elaboration in order to detect *Salmonella* according to ISO 6579 (2002) and *Staphylococcus aureus* plating on CHAPMAN (manitol-salt) agar. *Bacillus* aerobic spore-forming bacteria were detected by plating on TSA (Trypticase soy agar) (Asehraou *et al.*, 1992). Sometimes, an enrichment step is requires when target microorganism is present in low quantities in the food matrix or present a slow growth rate. The enrichment culture's environment will support the growth of a selected microorganism while inhibiting the growth of others.

Not always microbial isolates present differences in their phenotypes. Some ecological studies in table olives carried out an initial distinction by biochemical analysis (Randazzo *et al.*, 2004; Bevilacqua *et al.*, 2010). This methodology discriminates strains with same phenotypes but with different metabolic characteristics for example based on carbohydrate fermentation patterns. Despite biochemical study is time consuming and laborious, it permits the characterization of a large number of species and genera. However, molecular methods have recently started to be routinely used for the identification and discrimination of microorganisms associated with table olive processing. Partial amplification of DNA or RNA extracted directly from colonies isolated and further polymerase chain reaction (PCR) with universal primers followed by sequencing of the amplicons confers a higher accuracy in the final identification than biochemical methods.

The identification based on the amplification by PCR and partial sequencing of the 16S rRNA genes is a useful tool to classify a wide range of bacteria (Weisburg *et al.*, 1991). For yeasts, the identification through the Restriction Fragment Length Polymorphism profile (RFLP) of the 5.8S-ITS region or sequencing of the D1/D2 domains of the 26S rRNA are the most extended methods (Esteve-Zarzoso *et al.*, 1999; Kurtzman and Robnett, 1998). The resulting sequences are compared to search similarities in the relevant data banks such as GenBank from the National Center for Biotechnology Information (NCBI) using BLAST (Altschul *et al.*, 1990). Unlike traditional PCR reactions, random amplified polymorphic DNA (RAPD) and repetitive bacterial DNA elements (REP) use short primers that amplified the genomic DNA (Williams *et al.*, 1990; Gevers *et al.*, 2001). RAPD and REP create fragments with different lengths creating a specific band profile for each strain. These molecular tools are very popular for characterize and in phylogenetic studies of microbial communities. Another technique to characterize microorganisms is the study of RFLP. This method combines a PCR of rRNA genes or ITS followed to an enzymatic digestion of the amplicons. The identification is based on a comparison of the isolate band profile, unique for each microorganism; with patterns obtained from culture collection such as CECT-IATA (http://www.yeast-id.org).

An initial identification through RFLP, REP, or RAPD allows different species to be grouped while the final identification of isolates is usually confirmed by sequencing analysis. In the last years, many studies have used this molecular approach in table olives field (Hurtado *et al.*, 2008; Aponte *et al.*, 2010; Bautista-Gallego *et al.*, 2011; Franzetti *et al.*, 2011; Tófalo *et al.*, 2013; Lucena-Padrós *et al.*, 2014). Molecular techniques are rapid, easy to perform and more precise, eliminating part of the subjectivity that usually accompanies the output of the biochemical test. However, cultured-dependent methods present two main problems:

a) isolation takes days, or even weeks, to carry out and
b) only viable microorganisms are detected.

5.3.2 Culture-Independent Methods

Culture-dependent techniques do not offer a complete profile of the microbial diversity that is present in a specific ecosystem or food matrix. Only viable and culturable cells can be detected with this methodology. Hence, culture-independent methods have attracted the attention of many scientists in food science. Among culture-independent methods in food, PCR provides high sensitivity and specificity in detection of number of microorganisms, even foodborne pathogens. The sensitivity of the PCR method enables the direct detection of low concentrations of bacteria or bacterial products in food matrixes. For instance, *S. aureus* can be detected amplifying a sequence of the *nuc* genes by using two primers that targeted this gene (Brakstad *et al.*, 1992). *Listeria* species can be detected by PCR amplification and size comparison of a hypervariable internal *iap* gene fragment (Bubert *et al.*, 1999). PCR technique is ideally suited for both rapid and sensitive detection of *Salmonella* (Bailey, 1998) and specific detection of more than one serotype of *Clostridium botulinum* (Lindström *et al.*, 2001). There are numerous applications in the diagnosis of contaminated food but this method, which only detects the presence or absence of target microorganisms, is not quantitative.

Real-time polymerase chain reaction or quantitative PCR (qPCR) is a molecular technique that incorporates same principles of DNA replication used by PCR but using a fluorescent reporter molecule. Therefore, it monitors the amplification of targeted DNA during PCR reaction. The result is an elegantly simple, rapid and accurate means of tracking target cell or gene counts in samples. For instance, this technique has been used to detect and quantify the total number of yeasts species associated with table olives (Tofalo *et al.*, 2012a), and to detect biogenic amine-producing strains of LAB by directly targeting the genes responsible for biogenic amine production (Tofalo *et al.*, 2012b).

Fluorescence *in situ* hybridization (FISH) is other culture-independent method, PCR-free technique for detecting the microorganisms directly in their habitats without culture-dependent isolation of cells or culture-independent extraction of nucleic acids prior to the identification. FISH with 16S rRNA probes was used by Ercolini *et al.* (2006) for the in situ detection and identification of *L. plantarum* in table olives.

Denaturing gradient gel electrophoresis (DGGE) has received a great deal of attention for studying bacterial communities. DGGE separates PCR products along a chemical gradient in a polyacrylamide gel in relation with the content of G + C %, therefore discriminate by sequences. Recently, the studies of microbial population have been carried out in green table olives using DGGE (Abriouel *et al.*, 2011; Mucilli *et al.*, 2011; Tofalo *et al.*, 2013; Lucena-Padrós *et al.*, 2015). In spite of its widespread popularity, DGGE is a time-consuming, gradients and gels are difficult to set and run properly, and is non-quantitative, so can detect the presence of the dominant species.

Next-generation sequencing (NGS) include several technologies that achieve massively parallel sequencing of short segments amplified using universal PCR primers, mainly 16S rRNA and fungal ITS genes. NGS has many advantages compared with over first-generation profiling methods, such as DGGE and RFLP. All NGS systems sequence thousands to billions of raw reads in a single run and it has more sensitive to low-abundance OTUs, being more robust in study of diversity. The use of NGS in table olive fermentation has been useful for the study of microbial biodiversity, to observe microbial successions, impact of conditions or treatments, and so on (Cocolin *et al.*, 2013; De Angelis *et al.*, 2015).

Certainly, the best strategy to study microbial ecology is combining multiple molecular tools. Initially, a profiling tool (RAPD, REP) followed by sequencing should be used to give a broad-spectrum view of the microbial communities and biotypes present by plating, whereas DGGE and NGS may be more appropriate for in-depth studies of the microbial diversity (including non-viable or unculturable cells).

5.4 Biological Hazards in Table Olives

The origin of the microbiota in olive environment is diverse: fruits, water, pipes, pumps, vessels, equipment in general, and other devices in contact with the fruits and brine. LAB represent only a small proportion of the initial microbiota in the raw fruit and this group is responsible of the lactic fermentation of the product. Table olives elaborations have the aim to select for specific beneficial groups of microorganisms and inactivate as many of the spoilage and foodborne microorganisms as possible in order to get avoid changes during storage.

Table olives have a long history of microbial safety and fermentation occurred spontaneously in presence of desirable microorganisms. In fermented vegetable products, pH and reduced water activity (high salt content) are the main parameters that inhibit undesirable microorganisms to safeguard the health of consumers. Also, the preservative effect of many organic acids such as lactic and acetic acid produced as end products of fermentation provide an acidic environment unfavorable for the growth of many pathogenic and spoilage microorganisms. If the fermentation proceeds to completion and good manufacturing and hygiene practices are applied, sugars are completely converted into lactic acid, which drops the pH below to 4.6 units, and prevents the growth or even destroys spoilage and the pathogens of public health significance (*Enterobacteriaceae, Clostridium*, etc.) (Medina-Pradas and Arroyo-López, 2015). However, the process is not fully predictable and sometimes can lead to product spoilage or sanitary risks. Among the most relevant biological hazard in table olives we can mention the following:

5.4.1 Presence of Foodborne Pathogens

Obviously, the presence of human pathogens in table olives produces a potential health risk for consumers. In fact, several researchers have reported the presence of *L. monocytogenes* (Caggia *et al.*, 2004; RASFF portal 2012a), *S. aureus* (Asehraou *et al.*, 1992; Pereira *et al.*, 2008), and *Enterobacteriaceae* species such as *Yersinia enterocolitica* and *Escherichia coli* (Asehraou *et al.*, 1992; Franzetti *et al.*, 2011; Lucena-Padrós *et al.*, 2014) in different types of table olives and countries. However, there are not reports about illness outbreaks occasioned by these microorganisms in table olives. Particular attention should be given to the occurrence of *C. botulinum* and its toxin, since this is considered as one of the most important safety issues in fermented vegetables. Diverse outbreaks with deaths associated with table olives and recalls of suspected products have been reported (Debord *et al.*, 1920; Fenicia *et al.*, 1992; Cawthorne *et al.*, 2005; Jalava *et al.*, 2011; Pingeon *et al.*, 2011; RASFF portal, 2012b). Artisanal productions, improperly storage (pH ≥ 4.5 units) and uncorrected thermal treatment were often the origin of these outbreaks. Fortunately, table olives contain phenolic compounds with proved antimicrobial activity (Medina *et al.*, 2009), bacteriocins (Ruíz-Barba *et al.*, 1994), and other biopreservatives that make an adverse environment for the growth of undesirable microorganisms. In fact, survival of foodborne pathogens have been studied in different table olives elaborations and demonstrated that olives brines are not a favorable environment and pathogens showed a rapid decline in their populations (Grounta *et al.*, 2013; Medina *et al.*, 2013). Nevertheless, table olives can be considered microbiologically a safe food if good hygiene and manufacturing practices are correct and appropriated levels of salt (>5% p/v) and pH (<4.3) are reached in the final products.

5.4.2 Mycotoxins

Toxigenic fungal proliferation growth can occur under favorable environmental conditions and it is associated with the production of secondary metabolites that can be hazardous to humans. Many mold strains, in particular from *Aspergillus* and *Penicillium* genera, are able to grow on olives and produce various types of mycotoxins (Ochratoxin, Aflatoxin B, and Citrinin) after harvest, during drying, and storage (El Aldouni *et al.*, 2006). Contamination of table olives with fungi has been documented in cracked olives (Franzetti *et al.*, 2011), but Greek-style black olives are the most incriminated table olive elaboration (Gourama and Bullerman, 1988; Ghitakou *et al.*, 2006). Fortunately, mycotoxin levels usually found in table olives are low. Thus, efforts have to be made to prevent mold growth and mycotoxin production controlling factors as temperature, salt content, presence of aerobic conditions, and so on during olive storage and packaging.

5.4.3 Biogenic Amine Formation

The presence of these compounds in industrial fermentation brines increase throughout the brining period and is associated with food deterioration. Some microorganisms, especially those related to spoilage, have amino acid descarboxylase activity and can produce biogenic amines. The biogenic amines described in table olives with "zapatera" spoilage are putrescine, cadaverine, tiramine, and histamine (Hornero-Mendez and Garrido-Fernández, 1994). Amines concentration increase with the storage, but the water washes before pitting and packing reduces these substances. In fact, biogenic amines were only detected in Green olives (Spanish-style and natural), not in black ripe olives, but levels found in the final products are usually low and not represent a health concern (García-García *et al.*, 2001). Recently, several varieties of Greek-style table olives were subjected to study biogenic amines and producing bacteria by reverse-transciption (RT)–qPCR and low content or absence of these substances and bacteria was reported (Tofalo *et al.*, 2012b).

5.4.4 Organic Acid Degradation

Lactic acid fermentation confers organoleptic and nutritional properties to the fermented food but the most important is the acidification produced by the lactic acid production. Some spoilage microorganisms are able to metabolize the lactic acid into other organic acid with less preservative properties. For instance, *Lactobacillus buchneri* produce acetic acid from lactic acid consumption under anaerobic conditions (Johanningsmeier and McFeeters, 2013) and *Propionibacterium* species are able to convert lactic acid to propionic acid (Breidt *et al.*, 2013). "Zapatera" spoilage associated with Spanish-style green olives is characterized by the participation of species of at least two genera of bacteria, *Clostridium* and *Propionibacterium*, which can consume lactic acid when product is fermented (Montaño *et al.*, 1992). Also, oxidative yeasts can also consume the lactic and acetic acids produced during olive fermentation under aerobic conditions, but not in the absence of oxygen (Ruiz Cruz and González Cancho, 1969). These conditions raise the pH values of the brines allowing the growth of foodborne pathogens or undesirable microorganisms that cause the spoilage of table olives with the consequent loss of product quality and food safety.

5.5 Use of Starter Cultures to Reduce Biological Hazards in Table Olives

In many industries, olive fermentations still occurs spontaneously. Thus, as mentioned above, the process is not fully predictable and sometimes can lead to sanitary risks or product spoilage. To prevent these problems, and to consistently produce high-quality and safe final products, many authors have recommended strict process control of the fermentation by the use of starter cultures.

For many years, the search for starters with application in the different types of table olive elaborations has been practically focused on the technological activity of LAB. This way, the selection of LAB starters in olive fermentations was based on diverse technological criteria including homo-fermentative metabolism, high fast consumption of sugars and acidification rate, polyphenols, organic acids, high pH and salt tolerance, flavor development, wide temperature range for growth, minimum nutritional requirements, and oleuropein hydrolytic capability. A starter culture must also dominate the indigenous microbiota, which would be exerted by its fast and predominant growth under fermentation conditions. The growth of LAB reduces the pH of the medium by production of lactic acid, which added to its ability to produce antagonistic substances (bacteriocins), can inhibit growth of foodborne pathogens if an adequate pH (<4.3) is obtained (Medina-Pradas and Arroyo-López, 2015). *L. pentosus* and *L. plantarum* are among the LAB species with major applications as starter cultures in table olives (for a complete review see Corsetti *et al.* 2012).

Recently, several studies have emphasized the importance of the role played by yeasts as bio-protective agents when they are used as starter cultures during table olive processing. Apart from their technological characteristics (lipase, esterase, β-glucosidase or catalase activities, improvement of LAB growth, and the organoleptic profile of fruits), the competitive traits against undesirable microorganisms have attracted increasing attention from scientists, who proposed their application as biocontrol agents in the food industry and especially in table olives (Arroyo-López *et al.*, 2012b; Muccilli and Restuccia, 2015). These antagonistic activities rely on the production and tolerance of high concentrations of ethanol, competition for nutrients, as well as the synthesis of a large class of antimicrobial compounds (killer toxins, ethyl acetate, ethanol, etc.) that showed clearly a large spectrum of activity against foodborne pathogen and spoilage microorganisms. As mentioned above, these microorganisms could be especially effective in diverse olive preparations such as directly brined olives, where LAB growth is partially inhibited due to the presence of high concentrations of polyphenol compounds (Ruiz-Barba and Jiménez-Díaz 1993; Medina *et al.*, 2009).

Currently, the studies have focused on the exploration of microbial species with probiotic potential isolated from fermented vegetables (reviewed by Peres *et al.*, 2012 and Di Cagno *et al.*, 2013). Last study suggested that specific LAB strains collected from plant spontaneous fermentations and raw material have promising perspective of use as probiotic agents (Di Cagno *et al.*, 2013). This way, several researchers have tried to convert olives into a delivery carrier of probiotic bacteria to the human body. In table olives, the first work in this direction was carried out inoculating olive fermentations using exogenous lactobacilli strains with well-known probiotic characteristics as starters, as it was the case of the human origin strain *Lactobacillus paracasei* IMPC2.1 (Lavermicocca *et al.*, 2005; De Bellis *et al.*, 2010). This microorganism showed high adaptability to the fermentation process adhering to the surface of the olives.

Recently, diverse autochthonous LAB strains were also isolated from the fermentation of different types of table olive, and characterized according to their potentially probiotic characteristics. Among the most important probiotic characteristics, it was determined the inhibition of diverse foodborne pathogens such as *L. monocytogenes*, *S. enteritidis*, and *E. coli* with promising results (Argyri *et al.*, 2013; Bautista Gallego *et al.*, 2013; Botta *et al.*, 2014; Peres *et al.*, 2014). Studies using scanning electron microscopy techniques have proven that the ability of fruits to act as carrier was due to the formation of biofilm by microorganisms on the olive surface (Arroyo-López *et al.*, 2012c; Dominguez-Manzano *et al.*, 2012). These olive biofilms are poly-microbial communities consisting essentially of yeasts and lactobacilli, which were strongly attached to the epidermis of the fruits forming a community characterized by the excretion of a protective adhesive extracellular matrix mainly composed of exopolysaccharide. This biofilm matrix isolated from olive fermentations has shown the ability to inhibit the adhesion of the enterohemorrhagic strain *E. coli* K88 to the intestinal mucosa of pigs (Gonzalez-Ortiz *et al.*, 2013). Therefore, it could have application in both animals and humans for the treatment of food poisoning.

Thus, the last trend of the development of starters in table olives are focused on the production of mixed-multifunctional (LAB-yeasts) starters in order to improve and expand the mode of action by the use of two complementary microorganisms with different properties and with ability to inhibit foodborne pathogens.

5.6 Hazard Analysis and Critical Control Point (HACCP) System As a Useful Tool to Improve Microbial Safety and Quality of Table Olives

In table olives, the codex standard for table olives (CODEX STAN 66-1981, review 1987 and 2013) and the Trade Standard Applying to Table Olives (IOOC, 2004) recommends that the product covered by these documents must be prepared and handled in accordance with the appropriate sections of the *General Principles of Food Hygiene* (CAC/RCP 1-1969), the *Code of Hygienic Practice for Low-Acid and Acidified Low-Acid Canned Foods* (CAC/RCP 23-1979), and the *Code of Hygienic Practice for Canned Fruit and Vegetable Products* (CAC/RCP 2-1969). The product should comply with any microbiological criteria established in accordance with the *Principles for the Establishment and Application of Microbiological Criteria for Foods* (CAC/GL 21-1997). The HACCP system is a valuable tool to comply with all these hygiene criteria and good manufacturing practices (GMP) in olive processing.

The HACCP is known as a systematic and preventive procedure, internationally recognized, to tackle biological, physical, and chemical risks by means of applying prevision and prevention principles, instead of final product testing.

This system is mentioned as an annexed document within the revised text of Codex Alimentarius Commission (CAC) regarding general principles of food hygiene (CAC/RCP-1 (1069), Rev. 3 (1997 and 2003). The Codex general principles constitute a solid basis to guarantee an efficient food hygiene control, since it covers the whole production chain (from primary production to consumption).

The preventive approach is based on remarking all essential hygiene controls in each processing step. These control measures have been internationally recognized as an essential tool to ensure food safety and international trade. Subsequently, the National Advisory Committee on Microbiological Criteria for Foods (NACMCF) made the HACCP principles more concise; revised and added definitions; included sections on prerequisite programs, education, and training, and implementation and maintenance of the HACCP plan; revised and provided a more detailed explanation of the application of HACCP principles; and provided an additional decision tree for identifying critical control points (CCPs). At European level Regulation No. 852/2004 on the hygiene of foodstuffs states that food industries (among them the table olive sector) must accomplish with the requirement of elaborating, applying, and maintaining a permanent procedure based on HACCP principles.

Therefore, the HACCP concept permits a systematic approach to the identification of hazards and an assessment of the likelihood of their occurrence during the manufacture, distribution, and use of a food product, and defines measures for their control. The resulting HACCP plan can be integrated in a more general quality and safety assurance plan. In its simplest form, HACCP consists of seven principles as represented in Table 5.1. (ILSI, 2004).

The term **hazard** refers to any agent in, or condition of, food that is unacceptable because it has the potential to cause an adverse health effect. Examples of biohazards are pathogenic microorganisms and/or their toxins, or chemicals such as carcinogens and allergens and physical objects such as stones, bones, and so on that may injure the consumer.

Critical control points (CCPs) are defined as raw materials, facilities, formulations, or industrial treatments where measures can be applied to prevent or minimize the likelihood of the presence of hazards at unacceptable levels. In this sense, application of GMP—including Good Hygienic Practices (GHP)—is necessary to ensure that safe products are produced by keeping many elements of food production under control. They are sometimes referred to as "prerequisites to HACCP." For example, in the United States, the expression "prerequisite programs" is used in this context.

Critical limits are target values that indicate acceptability or non-acceptability of a given measured property (i.e., heat temperature, pH values). They indicate when acceptable (controlled) situations become unacceptable (out of control) with respect to the safety of the final product.

Monitoring is checking the conformity of the control at a CCP. It involves systematic observation, measurement, recording, and evaluation. Recording pH level during olive fermentation is an example. Corrective actions need to be taken when monitoring indicates loss of control. Corrective actions should ensure that unsafe products do not reach the consumer and should prevent, as far as possible, reoccurrence of the event.

Table 5.1 The Principles of HACCP, According to Codex Alimentarius.

Principle 1	Conduct a hazard analysis
Principle 2	Determine Critical Control Points (CCPs)
Principle 3	Establish critical limits
Principle 4	Establish monitoring systems of CCPs
Principle 5	Establish corrective actions to be taken when a given CCP is not under control
Principle 6	Establish verification procedures to confirm that the HACCP is working effectively
Principle 7	Establish documentation concerning all procedures

Verification is performed to check whether the system is correctly implemented and achieving its objectives.

Record keeping ensures that information resulting from the HACCP study and implementation (e.g., monitoring files) of the resulting HACCP plan is available for verification, review, inspection, auditing, or other purposes.

The HACCP system constitutes a scientifically sound methodology having a systematic character since it allows identifying specific hazards and needed measures for their control, in order to ensure food safety. Application of HACCP systems entails other potential benefits such as a more efficient use of resources, saving money for food industries and to give a fast response when food safety is compromised. HACCP increase the responsibility of food stakeholders. When being applied, food handlers can acquire knowledge and awareness about food safety, thus improving their handling practices and preventing food contamination. Finally, the HACCP system is definitely a useful instrument for food authorities and contributes to promote international trade and consumers' confidence.

Despite the advantages of HACCP system, food industries have to face with its implementation for a large variety of food commodities produced. Besides, for small-scale industries (<20 workers) HACCP implementation is often more difficult given the associated cost and effort. Therefore, HACCP should be adapted for being applied in different situations and adjusted to changes in equipment, processing methods, or technological achievements.

Regarding table olives sector, production chains in Mediterranean countries are sometimes based on traditional processing methods, which in most cases yield safe products as the physicochemical characteristics attained during the process (i.e., pH, acidity, salt content) can ensure the microbiological stability. However, leaving some large companies, the table olive industry remains craft, empirical, and not sufficiently controlled. As commented above, many times the processing methods still lie on the natural microorganisms present in the raw material and/or the environment. Besides, some types of table olives are not subjected to any heat treatment at the industry before consumption (pasteurization) and could thus become potential vehicles of foodborne pathogens as ready-to-eat products.

The Codex Standard for Tables Olives (CODEX-STAN 66-1981, review 2013) and the Trade Standard Applying to Table Olives (IOOC, 2004) describe the essential composition and quality factors of raw materials, intermediate and final products, as well as processing conditions and product formulations for different olive types. The standards also cover olives packed in bulk containers, which are intended for repacking into consumer size containers. Overall, several critical points must be under control during production, distribution, and retail phases in table olives. Among the most important we can foreground:

a) Hygienic quality of raw materials, which can be affected if olives are directly harvested from the soil.
b) Handling practices at primary production or during olives reception.
c) Storage facilities and conditions present at the industry.
d) Hygienic status of food-contact surfaces (i.e., washing hoppers, conveyor belts, working tables).
e) Fermentation process (acidity, salt conditions, etc.).
f) Addition of dressings (herbs, rosemary, oregano, garlic, etc.).
g) Sanitary quality of water used for olive washing and brining.
h) Chemical (authorized preservatives) or physical hazards (pieces of metal, glass) potentially present in the final product.
i) Packaging, storage, and distribution.

Hazard identification process requires consideration if this hazard should be included or not in the HACCP plan. To do this, one must use different information sources taken into account that the prerequisite programs control the environment hazards and the

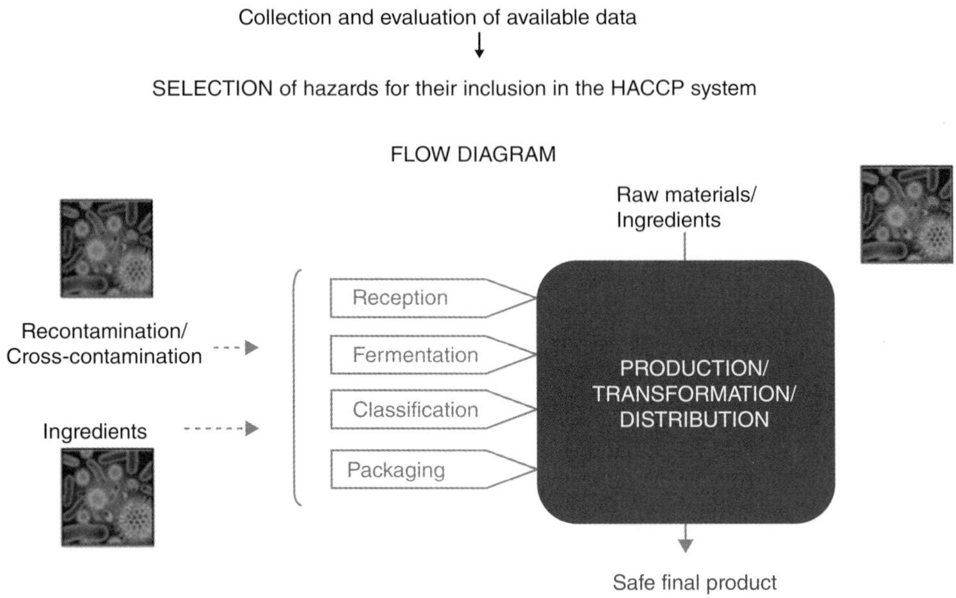

Figure 5.1 Flow-Diagram Process for Hazard Identification During Production of Table Olives. (*See insert for color representation of the figure.*)

HACCP control the process hazards. During production of table olives, critical steps should be carefully analyzed, as shown in Figure 5.1.

Afterward, a decision tree should be applied for the identification of Critical Control Points (CCPs). A general approach can be found in Figure 5.2.

For the sake of producing Spanish green table olives, different steps can be considered for their inclusion in the HACCP system. Some of them are optional, according to traditional elaboration procedures of local regions. These are namely, reception of raw material, fermentation process, sizing, washing, olives cracking, addition of olive spices (aromatic herbs, garlic, pepper, etc.) packaging and storage.

Further, critical limits must be identified by establishing minimum and maximum values for physical, chemical, and microbiological measurable parameters. This is highly useful for the determination of the acceptability of intermediate samples and final products. However, if a CCP is identified but no critical limits can be established, this point can only be considered as a CP and it has to be thoroughly controlled. For the establishment of the critical limits a revision of the legislation, internal norms of the company, literature and experimental validation must be performed. Inclusion of the critical limits, monitoring activities, corrective actions, responsibilities, documentation, and record keeping must be done when implementing an HACCP system. Table 5.2 summarizes packaging information regarding table olives. Note that for the establishment of critical limits, a certain tolerance margin (%) could also be set.

Regarding physic-chemical values, the Codex Standard for Table Olives recommends minimum limits for salt content and maximum levels for pH in the case of treated olives, natural and pasteurized olives, as well as for dehydrated and darkened olives. Overall, 5–6% of salt content, pH 4.3, and titratable acidity 0.3% are the minimal requirements for commercialization of treated and natural olives. In the case of microbiological criteria, as table olives can be considered as ready-to-eat products, they are subjected to the EU regulation No. 1441/2007 on microbiological criteria for foodstuffs. This way, according to this Regulation, *L. monocytogenes* limits are established at 100 cfu/g if food stakeholders can demonstrate that the final product

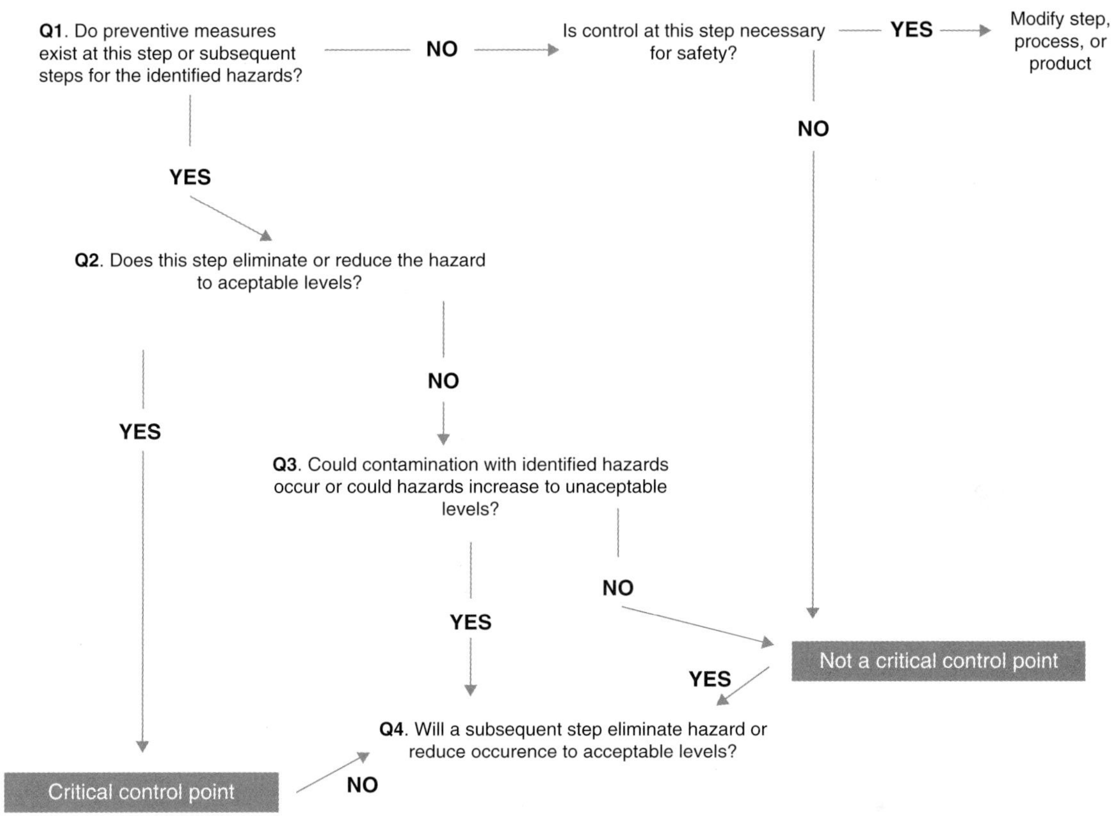

Figure 5.2 Decision Tree for Application of an HACCP System and Identify Critical Control Points.

Table 5.2 Example of Important Information to be Included About Packaging After Elaboration of Natural Table Olives Production When Performing an HACCP System.

Process step: Packaging of table olives after elaboration	
Hazard	Undesirable microbial growth due to poor storage practices
	Microbial and physic-chemical contaminants in the packaging and final product (fruit and brine)
Critical limits	pH <4.3
	Salt concentration >6%
Monitoring activity and frequency	Periodical checking of the pH and the salt concentration in final packages
	Checking to maintain the adequate level of the cover brine
Corrective action activity	Acidification and/or addition of salt
	Re-establishment of Good Manufacturing Practices
Responsibility for monitoring and corrective actions	Operator/supervisor
Documentation and record	Results of the inspections
	Analytical results
	Corrective actions

Table 5.3 Microbiological Limits Applied to Green, Fermented, Stuffed Olives and Black Olives in Brine According to the French Code (2000) for the Elaboration of Table Olives.

Microorganisms	Type of criteria	Concentration limit	Microbiological criteria
Salmonella spp.	Mandatory	Absence/25 g	M=m=0; n=5; c=0*
Coagulase positive *Staphylococcus*	Mandatory	$1{\cdot}10^2$/g	M=10 m; n=5; c=2
Anaerobes (30 °C)	Indicative	$5{\cdot}10^6$/g	M=10 m; n=5; c=2
LAB	Indicative	$1{\cdot}10^6$/g	M=10 m; n=5; c=2
E. coli	Indicative	$1{\cdot}10^2$/g	M=10 m; n=5; c=2
Faecal coliforms	Indicative	$1{\cdot}10^2$/g	M=10 m; n=5; c=2
Yeasts	Indicative	$5{\cdot}10^5$/g	M=10 m; n=5; c=2
Molds	Indicative	$5{\cdot}10^2$/g	M=10 m; n=5; c=2
Sulphite-reducing bacteria	Indicative	$1{\cdot}10^2$/g	M=10 m; n=5; c=2
C. perfringens	Indicative	$1{\cdot}10^2$/g	M=10 m; n=5; c=2

* *m* and *M* are the lower and upper microbiological limits; *n* is the number of samples to be taken from a given food lot; and *c* is the maximum allowable number of samples between m and M.

will not exceed this level at the end of shelf life. Otherwise, absence of *L. monocytogenes* in 25 g is required. In addition, the Trade Standard Applying to Table Olives (IOOC, 2004), state than when olives are tested by appropriate methods of sampling and examination, table olives shall be free from pathogenic and/or contaminant microorganisms likely to develop in the product in normal storage conditions, as well as free from substances (mycotoxins or biogenic amines) from microorganisms in amounts that may represent a hazard to health. On the contrary, fermented olives held in bulk in a covering liquid may contain microorganisms used for fermentation, notably LAB and yeasts. The number of such microorganisms in a selective culture medium may, for each one, be up to 10^9 cfu/ml of brine or per g of flesh depending on the level of fermentation. The use of these specifications is an indispensable part of the procedures based on HACCP and other measures of hygiene.

Other microbiological limits are also included in the French Code for the elaboration of table olives (Code des pratiques loyales pour les olives de table, 2000) regarding sporulated microorganisms, coagulase-positive *Staphylococcus*, coliform bacteria, anaerobes, LAB, and yeast/molds, among others. However, these criteria can vary depending on the olive type and processing methods applied. In Table 5.3, an example of these criteria is represented for fermented, stuffed olives and black olives in brine.

Despite the application of control measures and hygienic standards, several industries are set as small-scale and rely on traditional preservation and processing methods. If table olives are not pasteurized, high microbial loads can be present in the final product thus leading to a shorter shelf life. Some suggestions for corrective measures to be applied are presented in Table 5.4. Implementation of these measures should be always subject to economic viability and compatibility with the specific product.

5.7 Conclusions

Table olive is a fermented vegetable with many centuries of history in the Mediterranean basin and a long record of microbial safety. Inhibitory compounds such as polyphenols, low availability of nutrients, high salt content, low pH levels, bacteriocins, killer factors, or the addition of preservatives act as additional hurdles against

Table 5.4 Potential Problems Associated to Production of Table Olives and Putative Corrective Measures Associated.

Problem	Corrective Measure
High microbial load in environmental samples	• Installation of ozonization equipment • Ventilation and air filters • UV lamps with nocturnal activation • Design of physical barriers, mobile doors, etc., separating the different processing steps to avoid environmental contamination
Food-contact surfaces with high microbial load	• Daily cleaning with pressured hot water (70 °C) • UV lamps with nocturnal activation • Use of disinfectants and decontaminated water
Added spices with high microbial load (especially garlic and aromatic herbs)	• Scalding or sterilization of spices • Substitution and/or combination with aromatic compounds or essential oils.
Too high pH value after packaging	• Lowering pH with addition of acidifying agents
Processing steps with high probability of increasing microbial contamination to the final product (i.e., olive crashing)	• Implementation of dynamic showers with chlorinated water on conveyor belts. • Training of food handlers to increase awareness and prevent food contamination. • Use of cleaned uniforms: gloves, coats, masks, etc.
Modifications in pH levels of final products	• Measuring pH after packaging • Microbial quality improvement of raw materials and brine solutions.

grow of undesirable microorganisms, which contribute to obtaining a safe and good quality product. However, if good hygiene practices are not followed or appropriated levels of pH and salt are not obtained in the final products, growth of foodborne pathogens could occur. Because the objective of table olive producers should be to achieve zero risk in the case of illness and injury caused by microorganisms or their metabolites, industries should implement the use of starter cultures as well as good food safety management systems as HACCP for a better control of the table olive processing.

Acknowledgments

Authors have received funding from Junta de Andalucía regional Government through the PrediAlo project (AGR-7755: www.predialo.science.com.es). FNAL wishes to express thanks to the Spanish government for his Ramón y Cajal postdoctoral research contract, whereas EMP would like to thank the Junta de Andalucía for his postdoctoral research contract. European ERDF funding are greatly acknowledged for providing material and specially human resources, making possible the continuation of Risk Assessment and Management activities at national and European level by our research group AGR170 "HIBRO."

References

Abriouel, H., Benomar, N., Lucas, R. and Gálvez, A. (2011). Culture-independent study of the diversity of microbial populations in brines during fermentation of naturally fermented Aloreña green table olives. *International Journal of Food Microbiology*, *144*, pp. 487–496.

Altschull, S.F., Gish, W., Miller, W., Myers, E.W. and Lipman, D. (1990). Basic local alignment search tool. *Journal of Molecular Biology, 215*, pp. 403–410.

Aponte, M., Ventorino, V., Blaiotta, G., Volpe, G., Farina, V., Avellone, G., Lanza, C.M. and Moschetti, G. (2010). Study of green Sicilian table olive fermentations through microbiological, chemical and sensory analyses. *Food Microbiology, 27*, pp. 162–170.

Argyri, A.A., Zoumpopoulou, G., Karatzas, K.A., Tsakalidou, E., Nychas, G.J., Panagou, E.Z. and Tassou, C.C. (2013). Selection of potential probiotic lactic acid bacteria from fermented olives by in vitro tests. *Food Microbiology, 33*, pp. 282–291.

Arroyo-López, F.N., et al. (2012a). Fermentation of Olive Fruit. In Hui, Y.H. (ed.). Handbook of Plant-Based Fermented Food and Beverage Technology." Boca Ratón, FL: CRC Press, Taylor and Francis Group.

Arroyo-López, F.N., Romero-Gil, V., Bautista-Gallego, J., Rodríguez-Gómez, F., Jiménez-Díaz, R., García-García, P., Querol, A. and Garrido-Fernández, A. (2012b). Yeasts in table olive processing: desirable or spoilage microorganisms. *International Journal of Food Microbiology, 160*, pp. 42–49.

Arroyo-López F.N., García-García P., Rodríguez-Gómez F. and Garrido-Fernández A. (2016). *Olives: Types and Consumption*. In: B. Caballero, P. Finglas, and F. Toldrá, eds., *The Encyclopedia of Food and Health*, vol. 4. Oxford: Academic Press, pp. 167–170.

Asehraou, A., Faid, M. and Jana, M. (1992). Physico-chemical properties and the microflora of Moroccan black table olives. *Grasas y Aceites 43*, pp. 130–133.

Bailey, J.S. (1998). Detection of *Salmonella* cells within 24 to 26 hours in poultry samples with the Polymerase Chain Reaction BAX System. *Journal of Food Protection, 61*, pp. 792–795.

Bautista-Gallego, J., Rodriguez-Gomez, F., Barrio, E., Querol, A., Garrido- Fernandez, A. and Arroyo-López, F.N. (2011). Exploring the yeast bio diversity of green table olive industrial fermentations for technological applications. *International Journal of Food Microbiology, 147*, pp. 89–96.

Bautista Gallego, J., Arroyo-López, F.N., Rantsiou, K., Jiménez-Díaz, R., Garrido-Fernández, A. and Cocolin, L. (2013). Screening of lactic acid bacteria isolated from fermented table olives with probiotic potential. *Food Research International, 50*, pp. 135–142.

Bevilacqua, A., Perricone, M., Cannarsi, M., Corbo, M.R. and Sinigaglia, M. (2009). Technological and spoiling characteristics of the yeast microflora isolated from *Bella di Cerignola* table olives. *International Journal of Food Science and Technology, 44*, pp. 2198–2207.

Botta, C., Langerholc, T., Cencič, A. and Cocolin, L. (2014). In vitro selection and characterization of new probiotic candidates from table olive microbiota. *Plos One, 9*, p. 4.

Brakstad, O.G., Aasbakk, K. and Maeland, J. (1992). Detection of *Staphylococcus aureus* by polymerase chain reaction amplification of the *nuc* Gene. *Journal of Clinical Microbiology, 30*, pp. 1654–1660.

Breidt, F., Medina, E., Wafa, D., Pérez-Díaz, I., Franco, W., Huang, H., Johanningsmeier, S.D. and Kim, J.H. (2013). Characterization of cucumber fermentation spoilage bacteria by enrichment culture and 16S rDNA cloning. *Journal of Food Science, 78*, pp. 470–476.

Bubert, A., Hein, I., Rauch, M., Lehner, A., Yoon, B., Goebel, W. and Wagner, M. (1999). Detection and Differentiation of *Listeria* spp. by a single reaction based on multiplex PCR. *Applied and Environmental Microbiology, 64*, pp. 4688–4692.

Caggia, C., Randazzo, C.L., Salvo, M., Romeo, F. and Giudici, P. 2004. Occurrence of *Listeria monocytogenes* in green table olives. *Journal of Food Protection, 10*, pp. 2189–2194.

Cawthorne, A., Celentano, L.P., D'Ancona, F., Bella, A., Masari, M., Aniballi, F., Fenicia, L., Aureli, P., and Salmaso, S. (2005). Botulism and preserved green olives. *Emerging Infectious Diseases, 11*, pp. 781–782.

Cocolin, L., Alessandria, V., Botta, C., Gorra, R., De Filippis, F., Ercolini, D. and Rantsiou, K. (2013). NaOH-Debittering Induces Changes in Bacterial Ecology during Table Olives Fermentation. *Plos one, 8*, p. e69074.

Codex Alimentarius Commission (CAC), (1997). Principles for the establishment and application of microbiological criteria for foods. CAC/GL 21–1997.

Codex Alimentarius Commission (CAC). (1981). Codex Standards for table olives. CODEX STAN 66-1981 (Review 1987 and 2013).

Corsetti, A., Perpetuini, G., Schirone, M., Tofalo, R. and Suzzi, G. (2012). Application of starter cultures to table olive fermentation: an overview on the experimental studies. *Frontiers Microbiology, 3*, Art. 248.

Dauchet, L., Kesse-Guyot, E., Czernichow, S., Bertrais, S., Estaquio, C., Peneau, S., Vergnaud, A.C., Chat-Yung, S., Castetbon, K., Deschamps, V., Brindel, P. and Hercberg, S. (2007). Dietary patterns and blood pressure change over 5-y follow-up in the SU.VI.MAX cohort. *American Journal Clinical and Nutrition, 85*, pp. 1650–1656.

De Angelis, M., Campanella, D., Cosmai, L., Summo, C., Rizzello, C.G. and Caponio, F. (2015). Microbiota and metabolome of un-started and started Greek-type fermentation of Bella di Cerignola table olives. *Food Microbiology, 52*, pp. 18–30.

De Bellis, P., Valerio, F., Sisto, A., Lonigro, S.L. and Lavermicocca, P. (2010). Probiotic table olives: Microbial populations adhering on olive surface in fermentation sets inoculated with the probiotic strain *Lactobacillus paracasei* IMPC2.1 in an industrial plant. *International Journal of Food Microbiology, 140*, pp. 6–13.

De Castro, A., García, P., Romero, C., Brenes, M. and Garrido, A. (2007). Industrial implementation of black ripe olive storage under acid conditions. *Journal of Food Engineering, 80*, pp. 1206–1212.

Debord, G.G., Edmondson, R.B. and Thom, C. (1920). Summary of bureau of chemistry investigations of poisoning due to ripe olives. *Journal of the American Medical Association, 74*, pp. 1220–1221.

Di Cagno, R., Coda, R., De Angelis, M. and Gobbetti, M. (2013). Exploitation of vegetables and fruits through lactic acid fermentation. *Food Microbiology, 33*, pp. 1–10.

Domínguez-Manzano, J., Olmo-Ruiz, C., Bautista-Gallego, J., Arroyo-López, F.N., Garrido-Fernández, A. and Jiménez-Díaz, R. (2012). Biofilm formation on abiotic and biotic surfaces during Spanish style green table olive fermentation. *International Journal of Food Microbiology, 157*, pp. 230–238.

El Adlouni, C., Tozlovanu, M., Naman, F., Faid, M. and Pfohl-Leszkowicz, A. (2006). Preliminary data on the presence of mycotoxins (ochratoxin A, citrinin and aflatoxin B1) in black table olives "Greek style" of Moroccan origin. *Molecular Nutrition and Food Research, 50*, pp. 507–512.

Esteve-Zarzoso, B., Belloch, C., Uruburu , F. and Querol, A. (1999). Identification of yeasts by RFLP analysis of the 5.8S rRNA gene and the two ribosomal internal transcribed spacers. *International Journal of Systematic Bacteriology, 49*, pp. 329–337.

Ercolini, D., Villani, F., Aponte, M. and Mauriello, G. (2006). Fluorescence in situ hybridization detection of *Lactobacillus plantarum* group on olives to be used in natural fermentations. *International Journal of Food Microbiology, 112*, pp. 291–296.

Federation des industries condimentaires de France. (2000). Code des pratiques loyales pour les olives de table. Decembre 2000.

Fenicia, L., Ferrini, A.M., Aureli, P. and Padovan, M.T. (1992). Epidemia di botulismo da olive nere. *Industrie Alimentari, 31*, pp. 307–308.

Fernández Díaz, M.J., Castro y Ramos, R., Garrido Fernández, A., González Cancho, F., González Pellisó, F., Nosti Vega, M., Heredia Moreno, A., Mínguez Mosquera, M.I., Rejano Navarro, L., Durán Quintana, M.C., Sánchez Roldán, F., García García, P. and Castro Gómez-Millán, A. (1985). Biotecnología de la Aceituna de Mesa. Consejo Superior de Investigaciones Científicas, Gráficas Urpe, Madrid (España).

Franzetti, L., Scarpellini, M., Vecchio, A. and Planeta, D. (2011). Microbiological and safety evaluation of green table olives marketed in Italy. *Annals of Microbiology, 61*, pp. 843–851.

García-García, P., Brenes-Balbuena, M., Romero-Barranco, C. and Garrido-Fernández, A. (2001). Biogenic amines in packed table olives and pickles. *Journal of Food Protection, 64*, pp. 374–378.

Garrido Fernández, A., Fernández Díaz, M.J. and Adams, R.M. (1997). *Tables Olives: Production and Processing*. London, UK: Chapman & Hall.

Gevers, D., Huys, G. and Swings, J. (2001). Applicability of rep-PCR fingerprinting for identification of *Lactobacillus* species. *FEMS Microbiology letters, 205*, pp. 31–36.

Gebbers, J.O. (2007). Atherosclerosis, cholesterol, nutrition, and statins - a critical review. *German Medical Science, 5*, pp. 1–11.

Ghitakou, S., Koutras, K., Kanellou, E. and Markaki, P. (2006). Study of aflatoxin B1 and ochratoxin A production by natural microflora and *Aspergillus parasiticus* in black and green olives of Greek origin. *Food Microbiology, 23*, pp. 612–621.

González-Ortiz, G., Pérez, J.F., Hermes, R.G., Molist, F., Jiménez-Díaz, R. and Martín-Orúe, S. (2013). Screening the ability of natural feed ingredients to interfere with the adherence of enterotoxigenic *Escherichia coli* (ETEC) K88 to the porcine intestinal mucus. *British Journal of Nutrition, 11*, pp. 633–642.

Gourama, H. and Bullerman, L.B. (1988). Mycotoxin production by molds isolated from "Greek-style" black olives. *International Journal of Food Microbiology, 6*, pp. 81–90.

Grounta, A., Nychas, G.J.E. and Panagou, E.Z. (2013). Survival of food-borne pathogens on natural black table olives after post-processing contamination. *International Journal of Food Microbiology, 161*, pp. 197–202.

He, F.J., Nowson, C.A., Lucas, M. and MacGregor, G.A. (2007). Increased consumption of fruit and vegetables is related to a reduced risk of coronary heart disease: meta-analysis of cohort studies. *Journal Human Hypertension, 21*, pp. 717–728.

Hornero-Mendez, D. and Garrido-Fernández, A. (1994). Biogenic amines in table olives. Analysis by high-performance liquid chromatography. *Analyst, 119*, pp. 2037–2041.

Hurtado, A., Reguant, C., Esteve-Zarzoso, B., Bordons, A. and Rozès, N. (2008). Microbial population dynamics during the processing of *Arbequina* table olives. *Food Research International, 41*, pp. 738–744.

Hurtado, A., Requant, C., Bordons, A. and Rozes, N. (2012). Lactic acid bacteria from fermented olives. *Food Microbiology, 31*, pp. 1–8.

International Life Sciences Institute (ILSI). (2004). A simple guide to understanding and applying the hazard analysis critical control point concept. By M. van Schothorst, 3rd Ed. ILSI Europe Risk Analysis in Microbiology Task Force.

IOOC (International Olive Oil Council). (2004). Trade standard applying to table olives. COI/NC n°1. December 2004. Madrid, Spain.

Kurtzman, C.P. and Robnett, C.J. (1998). Identification and phylogeny of ascomycetous yeasts from analysis of nuclear large subunit (26S) ribosomal DNA partial sequences. *Antonie Van Leeuwenhoek, 73*, pp. 331–371.

Jalava, K., Selby, K., Pihlajasaari, A., Kolho, E., Dahlsten, E., Forss, N., Bäcklund, T., Korkeala, H., Honkanen-Buzalski, T., Hulkko, T., Derman, Y., Järvinen, A., Kotilainen, H., Kultanen, L., Ruutu, P., Lyytikaïnen, O. and Lindström, M. (2011). Two cases of food-borne botulism in Finland caused by conserved olives, October 2011. *Euro Surveillance, 16*(49). Available at: http://www.eurosurveillance.org/ViewArticle.aspx?ArticleId=20034 [Accessed June 08, 2016].

Johanningsmeier, S.D. and McFeeters, R.F. (2013). Metabolism of lactic acid in fermented cucumbers by *Lactobacillus buchneri* and related species, potential spoilage organisms in reduced salt fermentations. *Food Microbiology, 35*, pp. 129–135.

Lavermicocca, P., Valerio, F., Lonigro, S.L., De Angelis, M., Morelli, L., Callegai, M.L., Rizzello, C.G. and Visconti, A. (2005). Study of the adhesion and survival of *Lactobacilli* and *Bifidobacteria* on table olives with the aim of formulating a new probiotic food. *Applied and Environmental Microbiology, 71*, pp. 4233–4240.

Lindström, M., Keto, R., Markkula, A., Nevas, M., Hielm, S. and Korkeala, H. (2001). Multiplex PCR Assay for Detection and Identification of *Clostridium botulinum* Types A, B, E, and F in Food and Fecal Material. *Applied and Environmental Microbiology, 67*, pp. 5694–5699.

Lucena-Padrós, H., González, J.M., Caballero-Guerrero, B., Ruiz-Barba, J.L. and Maldonado-Barragán, A. (2014). *Propionibacterium olivae* sp. nov. and *Propionibacterium damnosum* sp. nov., isolated from spoiled packaged Spanish-style green olives. *International Journal of Systematic and Evoluationary Microbiology, 64*, pp. 2980–2985.

Lucena-Padrós, H., Jiménez, E., Maldonado-Barragán, M., Rodríguez, J.M. and Ruiz-Barba, J.L. (2015). PCR-DGGE assessment of the bacterial diversity in Spanish-style green table-olive fermentations. *International Journal of Food Microbiology, 205*, pp. 47–53.

McClain, D. and W. Lee. (1988). Development of USDA-FSIS method for isolation of *Listeria monocytogenes* from raw meat and poultry. *Journal Association of Official Analytical Chemists, 71*, pp. 660–664.

Medina, E., García, A., Romero, C., de Castro, A. and y Brenes, M. (2009). Study of the anti-lactic acid bacteria compounds in table olives. *International Journal of Food Science and Technology, 44*, pp. 1286–1291.

Medina, E., Brenes, M., Romero, C., Ramirez, E. and de Castro, A. (2013). Survival of foodborne pathogenic bacteria in table olive brines. *Food Control, 34*, pp. 719–724.

Medina-Pradas, E. and Arroyo-López, F.N. (2015). Presence of toxic microbial metabolites in table olives. *Frontiers Microbiology, 6*, Art. 873.

Montaño, A., de Castro, A., Rejano L. and Sánchez, A.H. (1992). Analysis of zapatera olives by gas and high-performance liquid chromatography. *Journal of Chromatography, 594*, pp. 259–267.

Muccilli, S., Caggia, C., Randazzo, C.L. and Restuccia, C. (2011). Yeast dynamics during the fermentation of brined green olives treated in the field with kaolin and Bordeaux mixture to control the olive fruit fly. *International Journal of Food Microbiology, 148*, pp. 15–22.

Muccilli, S. and Restuccia, C. (2015). Bioprotective role of yeasts. *Microorganisms, 3*, pp. 588–611.

Pereira, A.P., Pereira, J.A., Bento, A. and Estevinho, M.L. (2008). Microbiological characterization of table olives commercialized in Portugal in respect to safety aspects. *Food and Chemical Toxicology, 4*, pp. 2895–2902.

Peres, C.M., Peres, C., Hernández-Mendoza, A. and Xavier-Malcata, F. (2012). Review on fermented plant materials as carriers and sources of potentially probiotic lactic acid bacteria with an emphasis on table olives. *Trends Food Science and Technology, 26*, pp. 31–42.

Peres, C.M., Alves, M., Hernandez-Mendoza, A., Moreira, L., Silva, S., Bronze, M.R., Vilas-Boas, L., Peres, C. and Xavier Malcata, F. (2014). Novel isolates of lactobacilli from fermented Portuguese olive as potential probiotics. *LWT Food Science and Technology, 59*, pp. 234–246.

Pingeon, J.M., Vanbockstael, C., Popoff, M.R., King, L.A., Deschamps, B., Pradel, G., Dupont, H., Spanjaard, A., Houdard, A., Mazuet, C., Belaizi, B., Bourgeois, S., Lemgueres, S., Debbat, K., Courant, P., Quirin, R. and Malfait, P. (2011). Two outbreaks of botulism associated with consumption of green olive paste, France, September 2011. *Euro Surveillance, 16*(49). Available at: http://www.eurosurveillance.org/ViewArticle.aspx?ArticleId=20035 [Accessed June 08, 2016].

Randazzo, C.L., Restuccia, C., Romano, A.D. and Caggia, C. (2004). *Lactobacillus casei*, dominant species in naturally fermented Sicilian green olives. *International Journal of Food Microbiology, 90*, pp. 9–14.

RASFF Portal. (2012a). Reference number 2012.0703. Available at: https://webgate.ec.europa.eu/rasff-window/portal [Accessed June 08, 2016].

RASFF Portal. (2012b). Reference number 2012.1059. Available at: https://webgate.ec.europa.eu/rasff-window/portal [Accessed June 08, 2016].

Ruiz-Barba, J.L., Brenes, M., Jiménez Díaz, R., García, P. and Garrido, A. (1993). Inhibition of *Lactobacillus plantarum* by polyphenols extracted from two different kinds of olive brines. *Journal of Applied Bacteriology, 74,* pp. 15–19.

Ruíz-Barba, J.L., Cathcart, D.P., Warner, P.J. and Jimenez-Díaz, R. (1994). Use of *Lactobacillus plantarum* LPCO10, a bacteriocin producer, as a starter culture in Spanish-style green olive fermentations. *Applied and Environmental Microbiology, 60,* pp. 2059–2064.

Ruiz Cruz, J. and González Cancho, F. (1969). Methabolism of yeasts isolated from brines of Spanish-style table olives. I. The assimilation of lactic, acetic and citric acid. *Grasas y Aceites, 20,* pp. 6–11.

Tofalo, R., Schirone, M., Perpetuini, G., Suzzi, G. and Corsetti, A. (2012a). Development and application of a real-time PCR-based assay to enumerate total yeasts and *Pichia anomala, Pichia guillermondii* and *Pichia kluyveri* in fermented table olives. *Food Control, 23,* pp. 356–362.

Tofalo, R., Schirone, M., Perpetuini, G., Angelozzi, G., Suzzi, G. and Corsetti, A. (2012b). Microbiological and chemical profiles of naturally fermented table olives and brines from different Italian cultivars. *International Journal of General Molecular Microbiology, 102,* pp. 121–131.

Tofalo, R., Perpetuini, G., Schirone, M., Suzzi, G. and Corsetti. A. (2013). Yeast biota associated to naturally fermented table olives from different Italian cultivars. *International Journal of Food Microbiology, 161,* pp. 203–208.

Weisburg, W.G., Barns, S.M., Pelletier, D.A. and Genetrak, D.J.J. (1991). Ribosomal DNA Amplification for Phylogenetic Study. *Journal of Bacteriology, 173,* pp. 697–703.

Williams, J.K.G, Kubelik, A.R., Livak, K.J., Rafalski, J.A. and Tingey, S.V. (1990). DNA polymorphisms amplified by arbitrary primers are useful as genetic markers. *Nucleic Acids Research, 25,* pp. 6531–6535.

6

The Problem of Spore-Forming Bacteria in Food Preservation and Tentative Solutions

Stève Olugu Voundi[1], *Maximilienne Nyegue*[1,2], *Blaise Pascal Bougnom*[1], *and François-Xavier Etoa*[1]

[1] *University of Yaounde I, Department of Microbiology, Laboratory of Microbiology, PO Box 812, Yaounde, Cameroon*
[2] *University of Yaounde I, Department of Biochemistry, Laboratory of Phytobiochemistry and Medicinal Plant Study, PO Box 812, Yaounde, Cameroon*

6.1 Introduction

Some bacteria, under stress conditions, can temporarily and reversibly change their genetic expression, leading to a new phenotype called spore. They are termed as spore-forming bacteria. Spores or endospores are differentiated cells, highly different to vegetative cells by their structure and especially their high resistance to physico-chemical agents. This resistance allows them to survive under adverse conditions often including food sterilization process (Mckenney *et al.*, 2013).

There are about 200 species of spore-producing bacteria included in about 25 genera usually belonging to the *Bacillaceae* family (Henrique and Moran, 2007). *Bacillus* and *Clostridium* are some of the most studied (Marklandet *al.*, 2013). Due to their high representativity, pathogenic potential and high resistance, the following questions become important in the food industries: How and why does spore become resistant to sterilization conditions? What are the effective ways to destroy it?

The present chapter addresses these questions and attempts to provide answers to address these concerns. In this light, we will first describe the phenomenon by which the vegetative cell becomes a spore (sporulation), as well as the reverse process (germination). It will also be necessary to present the structure of a spore, since it is from this structure that spores derive their high resistance. We will afterward point out the problems of spore-forming bacteria in the food domain and different approaches to control them.

6.2 Sporulation

Sporulation is a differentiation process, wherebysome bacterial cells undergo significant changes in their structure and their metabolism, to adapt to harsh conditions of the environment. In simpler terms, it is a transformation of vegetative cells into completely new different forms called spores. The sporulation phenomenon is usually initiated by a significant depletion of nutrients in the bacterial environment; it lasts 7–10 hours. Under electron microscopy, sporulation occurs in several differentiation steps. Firstly, the cell divides into two unequal portions by forming a septum. Each portion contains a DNA copy of the parent cell. The small portion called pre-spore (future spore) rushes into the big one called sporangium.

Foodborne Pathogens and Antibiotic Resistance, First Edition. Edited by Om V. Singh.

At this stage, the sporulation process becomes irreversible. This phase is followed by the formation of a cortex: spore-specific peptidoglycan. Secondly, dipicolinic acid synthesis (spore-specific component), the incorporation of Ca^{2+} and acquisition of refringence. Finally, the coats are formed; the spore completes its maturation and becomes resistant to the temperature and other types of stresses. The process may end with the lysis of the sporangium, which releases the spore (Sella *et al.*, 2014). In some species, sporulation can lead to the formation of two or more spores from each vegetative cell under certain conditions (McKenney *et al.*, 2013).

In the food industry, the sporulation step is of major interest as it is the phenomenon that leads not only to the acquisition, but also to the extension of spore resistance. In fact, sporulation conditions can strongly influence the extent of spore resistance and therefore, the effectiveness of the sterilization process applied to the foods. For example, increasing temperature during sporulation enhances the heat resistance of spores (Leguerinel *et al.*, 2007). In addition, the mineral quality and quantity of the sporulation medium are major factors of resistance and subsequent spore growth in food (Setlow and Johnson, 2007). Furthermore, due to the fact that sporulation can occur both in food and the human intestinal tract, it would play an established role in food poisoning. Magnusson *et al.* (2007a, 2007b) demonstrated an example of the sporulation ability of *Bacillus cereus* in cow milk samples put for sales offer. Moreover, the works of McClane (2007) have revealed the ability of *Clostridium perfringens* to sporulate in the human small intestine.

6.3 Metabolic State of the Spore

After sporulation, the new spore goes into dormancy characterized by low metabolic activity due to extensive dehydration. Dormancy is dictated in the spore by immobilization of macromolecules and core enzymes by dipicolinate and Ca^{2+} dehydrated gel. The cortex is responsible for maintaining the dormancy state (Foster, 1994). In some bacterial species, super-dormant spores were identified, characterized by lack or very low percentage of germination of spores under favorable conditions. This property is characteristic of *Bacillus* genus and some *Clostridium* (Ghosh *et al.*, 2009). In food industries, super-dormancy generally leads to an inefficient isolation of spores during microbiological food control.

6.4 Spore Structure and Associated Mechanisms of Resistance

Under electron microscopy, bacterial spores reveal structures (Figure 6.1) that are greatly different from those of the initial vegetative cells. The high resistance of spores is linked to these structures. From the core to the tunics, all spore components contribute to preserve the spore against treatment that could damage it.

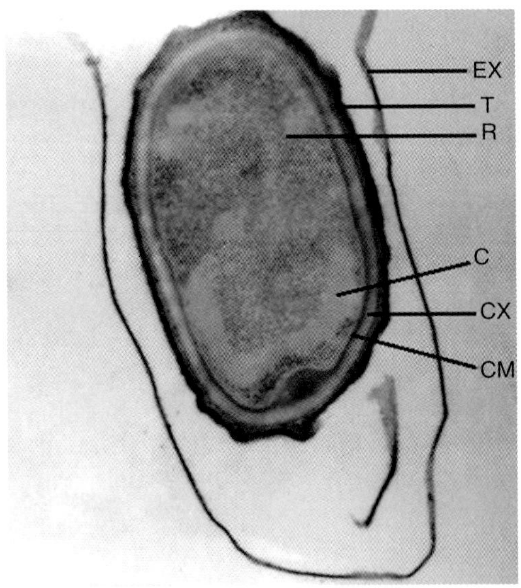

Figure 6.1 Ultrastructure of section of spore of *B. cereus* F3748 revealed at electron microscopy, magnification X 3500; EX, exosporium; T, tunic; R, ribosome; CX, cortex; C, core; CM, core membrane (Etoa, 2011). (*See insert for color representation of the figure.*)

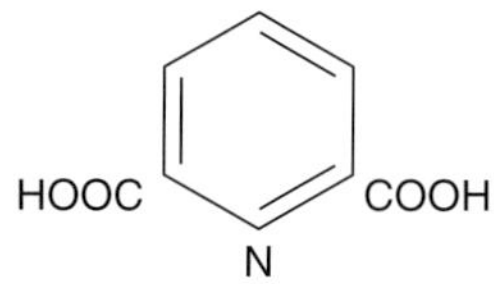

Figure 6.2 Structure of dipicolinic acid or 2,6-dicarboxylic acid pyridine. Dipicolinic acid is spore-specific component.

6.4.1 Core

The core or the cytoplasm of the spore is a gel that contains nucleic acids (RNA and DNA), ribosomes, and most spore enzymes. In contrast to vegetative cell, the core of the spore is dried (25–50% water content) and has a high content of dipicolinic acid. The dipicolinic acid or 2,6-dicarboxylic acid pyridine (Figure 6.2) alone represents 10% of the dry weight of a *Bacillus* spore (Moir, 1990). Inside the spore, dipicolinic acid is complexed to a high content of Ca^{2+}. The complex dipicolinic acid - Ca^{2+} makes the core a strong semi-crystalline structure, which, associated with a high dehydration confer a high thermal and ultraviolet (UVs) resistance of spore (Sella *et al.*, 2014). Furthermore, DNA of the spore unlike that of the vegetative cell is associated by a high quantity of small acid specific proteins (SASPs) that hold it in the alpha configuration, ensuring its resistance to UV radiations and desiccation (Sella *et al.*, 2014).

6.4.2 Membranes

The core is surrounded by two membranes. The first called inner membrane is similar to the vegetative cytoplasmic membrane. But it possesses very low lipids permeability. Germinant receptors are located on this membrane. The very low permeability of the inner membrane is an important factor in the resistance of spores to chemical agents. In fact, it acts as a restrictive barrier to carriers of molecules into the core. The more the inner membrane is impermeable, the more the spore is resistant to acids (Rose *et al.*, 2007). The second is the outer membrane, which is generally not visible under electron microscopy and its functions are not well studied. The two membranes intercalate the cell wall and the spore cortex (McKenney *et al.*, 2012).

6.4.3 Cell Wall

The spore wall is a peptidoglycan layer surrounding the inner membrane. Its structure is identical to the vegetative cell wall. Indeed, during sporogenesis, the vegetative cell wall becomes the spore wall.

6.4.4 Cortex

The cortex is a second layer of peptidoglycan located above the spore wall with structure typical to the bacterial spore. It is transparent to electrons and contains dipicolinic acid linked to Ca^{2+}. The peptidoglycan of the cortex differs from vegetative cell peptidoglycan in that 1/4 of N-acetyl muramic acid (NAM) is substituted by short peptides. Another quarter of NAM has a single residue of L-Ala. Other NAM residues are converted to D-lactam muramic (MAL), which is recognized by the cortex lytic enzyme during germination. The cortex is involved in the heat-resistance and the spore dormancy, since it is responsible for maintaining the dehydrated state of the core (Atrih and Foster, 1999).

6.4.5 Tunics

Above the cortex, are the tunics composed mainly of structural proteins and 6% of carbohydrates. Tunic proteins represent about 50–80% of total protein of a spore. About 30% of these proteins are insoluble and characterized by a high proportion of disulphide bonds. Tunics are responsible for the protection of spore against chemical agents (such as hydrogen peroxide, ozone, and sodium hypochlorite), large molecules (such as lysozymes) and bacteriophages that may cause damage of the cortex. Moreover, Riesenman and Nicholson (2000) have shown that the inner tunics were involved in the resistance of spores to UV radiation. Few works explain the involvement of tunics in the heat-resistance of the spores (Sella *et al.*, 2014).

6.4.6 Exosporium

In some spore-forming species such as *Bacillus anthrasis, Bacillus cereus, Bacillus thuringiensis, Bacillus mycoides*, and some *Clostridium*, a paracrystalline last coat called exosporium consisting of glycoproteins covers the tunics. The size and structure of this layer is specific of each strain (Henriques and Moran, 2007).

6.5 Germination of Spore

The germination of bacterial spore is a phenomenon of great relevance in food industries and in public health. Indeed, after germination, bacterial cell regains its metabolism and eventually becomes pathogenic. Germination comprises several steps consisting of spore structures degradation. It is triggered by the binding of specific compounds called germinants or activators of germination-to-germination receptors (Moir *et al.*, 2002). Activators of germination are nutrient compounds generally found in foods, belonging to the class of sugars, amino acids, and purine nucleosides. Non-nutrient compounds such as dodecylamine, dipicolinic acid, and specific peptidoglycan fragments are also able to trigger germination. The efficient effect of germinant can be enhanced by sub-lethal heat treatment of food. The process of germination of the spore takes place in two main phases: the first is an initiation phase where the spore performs a series of physiological responses due to stimulation of its receptors by germinant. The binding of the germinant induces conformational changes of receptors located on the spore inner membrane; this activates enzymes that induce the start of spore cortex's hydrolysis (Foster and Johnston, 1989). The triggered process is irreversible even if the germinant is withdrawn.

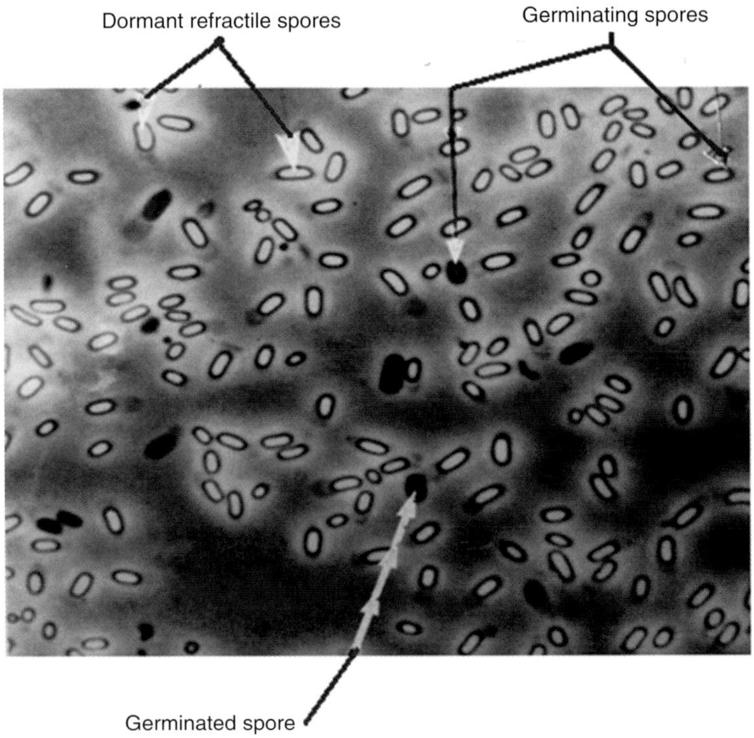

Figure 6.3 Different forms of *B. cereus* F3748 spores during germination as revealed by contrast microscopy, Magnification x 1200 (Etoa, 2011). (*See insert for color representation of the figure.*)

The second phase represents a more significant degradation. The cortex specifically undergoes complete hydrolysis and tunics are partially hydrolyzed. Likewise, SASPs are degraded by specific proteases providing a source of amino acids for protein synthesis during spore outgrowth. This results in a water intake in the core, a release of dipicolinic acid, Ca^{2+}, and a start of metabolism. Under contrast microscopy, three different forms of spores during germination could be distinguished (Figure 6.3). The refractive spores represent the spore in its dormant state, the partially dark spores represent the germinating spore and the completely dark spores pertain to germinated spores (Etoa, 2011).

The second phase is followed by the outgrowth. The germinated spore inflates and stretches 1.5-fold, as it stretches, there is a complete disintegration of the cortex and the spore wall becomes the future cell wall. Spores outgrowth involves an intense start of metabolism leading to the biosynthesis of macromolecules such as DNA, RNA, and proteins (Leyral *et al.*, 2001).

6.6 Problems of Spore-Forming Bacteria in Food Preservation

One of the fundamental problems of spore-forming bacteria is the high resistance of their spores, which limits the effectiveness of food sterilization methods. This is widely reported above and will not be discussed further. However, besides the resistance to sterilization treatments, spore-forming bacteria lead to other major concerns in public health and food preservation.

6.6.1 Total Inactivation of Spore Seem Not Possible Without Nutritional and Organoleptic Degradation of Food

Up to date, the effectiveness of any food sterilization method relies on its ability to inactivate bacterial spores (Markland *et al.*, 2013). However, to destroy spores, suitable treatments are generally violated. Heat treatments for instance are generally required at about 100 °C and are sometimes combined to pressure. Similarly, treatment by irradiation should be done at doses up to 5 kGy to ensure spore inactivation. However, it appears very difficult to apply these types of treatments without compromising the nutritional and organoleptic qualities of foods. This remains a major challenge in the food domain where currently, consumer's trends are to demand of food products with preserved natural flavor.

Furthermore, it is known that several food of high production cannot go to high heat sterilization treatments. The dairy and fruit juice industries are still using pasteurization, which, not only fail to kill spores, but sometimes rather trigger their germination. This often leads to a short shelflife of the food and requires the use of proper food preservatives. In dairy industries, to circumvent the aforementioned disadvantages, ultra-high temperature (UHT) consisting of applying high temperature up to 150 °C for few seconds was developed. UHT process ensures a well sterilization by killing both vegetative cells and spores of many species without tampering on the organoleptic and nutritional qualities of milk. However, spores of *Bacillus sporothermodurans* remain an important problem as they can survive UHT treatment and germinate in stored UHT milk reducing its shelf-life (Montanari, 2004).

6.6.2 Ubiquity and Higher Contamination Characters

Due to their high resistance to environmental stress, spores have a high ubiquitous nature and can be found in all environments (Heyndrickx, 2011). The soil is the main reservoir and may hold about 10^5-10^6 *B. cereus* spores / g (Carlin, 2011). Direct ways of food contamination by *Bacillus* and *Clostridium* spores are established from the soil. Hence, from the soil, the spore-forming bacteria will contaminate crops, as well as hay and so will be found in milk and dairy products. This could explain the large presence of *Bacillus* in dairy products and vegetables as reported by Postollec *et al.* (2010). Similarly,

Vissers (2007) found that the strongly feared contamination of *Clostridium tyrobutyricum* in cheese defects may arise from its spores present at high concentrations in the soil.

High food contamination by spore-forming bacteria was also found to have an origin in the food production chain surfaces, in materials and ingredients used in food processing contaminated with spores. Important fecal contamination is also incriminated (Reyes *et al.*, 2007).

6.6.3 Spores and Biofilms Resistance

Biofilms are complex microbial communities encapsulated in a polysaccharide matrix, colonizing several types of surfaces (solid or liquid) of the food production chain (Abee *et al.*, 2011). The presence of biofilms is a major concern for food industries, as they are difficult to eliminate and are a major source of contamination and recontamination of food. The spore-forming bacteria play a well-established role in the mechanisms of resistance and spread of biofilms. Indeed, biofilms harbor high levels of spore-forming species, and promote the activation of p-SpoA sporulation genes (Abee *et al.*, 2011). For this reason, they contain significant levels of spores, thereby limiting their destruction. This causes major health problems since biofilms can also harbor pathogenic species such as *Listeria monocytogens, Staphylococcus aureus* and *B. cereus* (Abee *et al.*, 2011; Faille *et al.*, 2014). Approximately 40% of food-borne diseases caused by bacteria in France were due to biofilm formation in food industries (Haeghebaert *et al.*, 2001).

6.6.4 Acid-Tolerant Spore Formers

Acidification of food after thermal-sterilization is a wide use method for food preservation and provides a continuous supply of food independent of the seasonal availability. Weak acids such as sorbic and benzoic acid showed the ability to inhibit the growth and germination of several microbial species and find application in acid canned foods (Van Melis *et al.*, 2012). However, this method is limited to acid tolerantspore-forming species such as *Alicyclobacillus acidoterrestris.* Its spores can withstand thermal sterilization treatment, germinate and multiply at pHs ranging from 3 to 3.5 due to the presence of saturated fatty acid ω-cyclohexane-C17:0 et ω-cyclohexane-C19:0 in their membrane. *A. acidoterrestris* is considered as one of the most involved contaminants in fruit juice spoilage (Bevilacqua *et al.*, 2008; Maldonado *et al.*, 2013). Besides, *A. acidoterrestris, Bacillus coagulans* are acidophilic and thermophylic spore-forming species able to grow in acid canned foods. They are a serious problem in the canned food industries as their spores can resist thermal treatment and are able to germinate and grow at pH between 4.1–5 (Lucasa*et al.*, 2006). During growth, *B. coagulans* cause “flat-sour” spoilage, due to the production of lactic acid without gas formation. *B. coagulans* is also able to increase the food pH to values that can allow germination of surviving *C. botulinum* spores. *B. coagulans* is the main cause of food spoilage in canned tomatoes (Lucasa *et al.*, 2006).

6.6.5 Psychrotolerant Sporeformers

Some spore-forming bacteria known as psychrotolerant sporeformers have the ability to grow at low temperatures of 3 to 5 °C, limiting the effectiveness of food preservation by refrigeration. In food industries, this is considered as an important emerging problem since in the last decades, there has-been a significant increase in the manufacture and sale of ready-to-eat or ready-to-cook foods stored under refrigerated conditions known as cooked-chilled foods or refrigerated processed foods with extended durability (REPFEDs) (Rodgers, 2003). In addition, psychrotolerant sporeformers included not only spoilage bacteria such as *Bacillus weihenstephanensis* but also some pathogenic species as *B. cereus* and *C. botilinum* Type E.

6.6.6 Pathogenic Sporeformers

Some spore-forming species are feared for their ability to cause food borne diseases (Abee *et al.*,

2011). In the *Bacillus* genus, the well-known pathogen is *B. cereus*, which is responsible for two types of infection. The first type is due the production of cereulide, an emetic heat-stable toxin remaining in the food after cooking. The cereulide is stable at temperatures above 150°C and about 8 µg/kg are sufficient to cause infections whose incubation time lasts from 30 minutes to 6 hours. The symptoms are generally nausea and vomiting. Starch-rich food such as rice and pasta are often associated to emetic infections (Markland *et al.*, 2013). In addition, *B. cereus* spores of some strains are able to withstand the acidic pH of the stomach and reach the small intestine, favorable site of hemolysin and cytotoxin production responsible for diarrheal infections. Symptoms generally observed are abdominal pain, nausea and vomiting. For this type of infection, the incubation time is between 8 to 16 hours after ingestion of protein rich food such as meat and milk products (Abee, 2010; Markland *et al.*, 2013).

Several species of *Clostridium* are also major cause of food poisoning. *C. botilinum*, classified as a category A terrorism agent (CDC, 2002; Peck, 2010) produces the most dangerous of the known neurotoxins. There are seven types of *C. botilinum* classified from A to G according to the immunological properties of the type of neurotoxin. These seven types are classified into four groups (I to IV) according to their biochemical property (proteolytic or not). Types A, B, E, and F are pathogenic to humans, whereas others infect animals, especially birds and cattle. Type E neurotoxin is produced by a psychrotrophic strain able to grow at refrigeration temperatures (3 °C) (Peck, 2010). Each year in the United States, there are approximately 55 cases, 42 hospitalizations, and 9 deaths that occur due to food-borne botulism (Scallan *et al.*, 2011). Three forms of botulism infections can be distinguished according to the mode of contamination: intoxications (common in adults) are due to the ingestion of the botulinum toxin preformed in food. Food poisoning is caused by ingesting bacteria and/or spores of *C. botilinum*. This form is more encountered in infants (0–12 months) and honey is the most offending food. There are also infections caused by inoculation of spores into wounds.

Besides the *C. botilinum*, the enterotoxigenic *Clostridium perfringens* are also frequently involved in food borne infections. Ingested at 10^5 cells/g of food, the acid pH of the stomach and intestinal flora are unable to ensure human protection against this pathogen (ANSES, 2010). During sporulation in the small intestine, *C. perfringens* releases an enterotoxin causing lysis of enterocytes and diarrhea. *C. perfringens* is also a gangrene and sepsis agent. The alpha toxin and perfringolysin are the main toxins involved in gangrene. In the year of 2007, *C. perfringens* was the third among the causes of food poisoning outbreaks identified in the framework of mandatory reporting in France (ANSES, 2010). The most offending food was meat.

Some species such as *Clostridium difficile* are listed as major cause of nosocomial diarrhea. Nearly 6000 deaths in England and Wales were attributed to *C. difficile*–associated disease in 2008 (Burns *et al.*, 2010). The infections are due to the synthesis of toxins A and B, which target the Ras superfamily of GTPases for modification via glycosylation, leading to intestinal epithelial cells degradation (Voth and Ballard, 2005). Symptoms can range from mild diarrhea to a fulminant, relapsing and potentially fatal colitis. Spores of *C. difficile* play an important role in the transmission of disease as they are resistant to chemical and physical agents and their germination can be activated by bile salt in the small intestine (Sorg and Sonenshein, 2008; Giel *et al.*, 2010).

6.6.7 Other Concerns

The existence of bacterial spores is not an industry-exclusive problem. In Cameroon and other developing countries, the quality of artisan food production is facing challenges owing to their high concentrations in spores. Feudjio (2006), for example, found spore contents above the threshold values fixed by norms in many porridge formulations for infant consumption sold in Yaoundé, Cameroon. Similar observations had been previously noticed in samples of honey, reinforcing the assertion that honey is the vector of infant botulism (Etoa and Adegoke, 1996).

6.7 Techniques of Spore Inactivation

The fight against spore-forming bacteria takes place mainly at two levels. Firstly spores can be inactivated upon food processing. Secondly during storage, spore germination and/or growth is inhibited by preservative substances.

6.7.1 Spore Inactivation Strategies

6.7.1.1 Non-Thermal Methods

Spores inactivation processes by non-thermal methods generally include the use of disinfectants, the application of high pressures or food irradiation. Disinfectants containing chlorine at 100–200 mg/L are the most used and allow for up to four decimal reductions of the initial number of spores (AFSSA, 2009). These types of treatment are mainly performed on the surfaces of the food production chain and during food washing.

High-pressure methods include treatment of food with or without packaging for 100 to 800 MPa at room or lower temperatures (50 °C) for 5 to 30 min. However, spores of some species can withstand over 1000 MPa (Merzougui *et al.*, 2013). Coupling of high pressure to heat treatments give more efficiency.

UV irradiations are one of the methods used to preserve liquid foods including orange, apple, and grape juices. UVs are capable of inactivating bacterial spores at wavelengths ranging from 200 to 280 nm (UV-C) (Hijnen *et al.*, 2006). Baysal *et al.* (2013) showed reductions of 5.5 log and 2 log CFU/ml of spores of *A. acidoterrestris* DSM 3922 respectively inoculated in apple and grape juice subjected to the 1.31 mW/cm^2 UV-C intensity. However, the limit on the application of UV remains the low penetration in food matrices with low absorption coefficients.

For over 30 years, the application of ionizing radiation is recognized by WHO as a food preservation technique without toxic effect (WHO, 1981). In the European Union, this technique is regulated by Directives n°1999/2 / EC (EU, 1999), which mentions the types of radiation (gamma radiation, X radiation, E-beam) and energy required (5 and 10 MeV respectively for the X radiation and electron-beam). The typical dose, which increases the shelf life of food, is about 5 kGy, but the effect may depend on the characteristic of each food. Ji-Hyoung *et al.* (2012) have reported that γ-radiation at 0.1; 0.2 and 0.3 KGy leads to reduction of approximately 1.3 to 1.6 log10 of *B. cereus* spores inoculated in rice. These irradiation doses completely inactivate spores when combined to NaOCl at 600 to 1000 ppm and 5 to 20 minutes of ultrasound.

6.7.1.2 Thermal Methods

Heat sterilization still remains in food industries, the usual methods or one of the most used for both vegetative cells and spores inactivation. Vegetative cells have almost the same thermal sensitivities while spores often reveal different sensitivities depending on several factors. One of the first is undoubtedly the mesophilic or thermophilic character. Indeed, the spores of mesophilic species generally show decimal reduction time $D_{121°C}$ of 0.01 to 0.1 min, usually less than those of thermophilic species between 2–5 min (Deak, 2014). Moreover, for some species, heat resistance of spores could highly dependent on environmental (pH, Aw) and physiological conditions. Four main physiological conditions are involved in the acquisition, stabilization and/or changes in the heat resistance of the spores (Etoa, 1985). They are:

- The dehydration state: the more the spore is dehydrated, the more it is heat resistant.
- The mineral composition: the heat-resistance of the spore according to the quality of mineral composition of the core is determined as

follows: Ca^{2+} > Mn^{2+}> Mg^{2+} > K^+ and Na^+ (Sella *et al.*, 2014).

- Thermal adaptation: the spores produced at a maximum temperature of growth are more heat resistant than those obtained at optimum or minimum temperature of growth.
- Heat-induced heat resistance: heat resistance of spore increases when they have previously been exposed for prolonged times to sublethal temperature.

Included among the heat treatments, the microwaves are an important alternative for sterilization, since they inactivate a wide range of bacteria including *Bacillus* and *Clostridium* vegetative cells and spores as well (Celandroni *et al.*, 2004). Microwaves offer the advantage of the possibility of processing the product in the package.

6.7.2 Spores Control by Inhibitors of Germination and Outgrowth

Some substances are currently used or strongly suggested as food preservatives with spores germination and/or outgrowth inhibitors effect. Literature usually highlights organic acids, bacteriocins, lysozymes, and essential oils.

6.7.2.1 Organic Acids

Sorbic acid at 3 mM pH 5.5 completely inhibited germination of spore of *B. cereus* ATCC 14579, whereas the less-lipophilic acetic acid, inhibited the outgrowth of spores but not germination at 15 mM (van Melis *et al.*, 2011). In addition, sodium benzoate at 500, 1000, or 2000 ppm completely inhibits the germination of spores of *A. acidoterrestris* with sporostatic effect. Furthermore, germination of *Clostridium* spores is inhibited by nitrites (Avila *et al.*, 2014). Similarly, lactate of Ca, Na, and K at concentrations ≥ 2.0% and ≥ 3.0% respectively (for the lactate of Na and K) are currently used to inhibit the germination of spores of *C. perfringens* in pork during storage at low temperature (7 °C) (Velugoti *et al.*, 2007).

Two mechanisms of action have been proposed for these effects. In their protonated forms, organic acids can easily diffuse through the spore envelopes and dissociate in the core; this leads to a strong acidification of the core, responsible for enzyme malfunctions (Bogaert and Naidu, 2000). Acids can build up in the inner membrane of the spore leadingto interactions with its function, blocking the spore outgrowth (Brul *et al.*, 2002; Chu *et al.*, 2009).

6.7.2.2 Bacteriocins

The nisin produced by *Lactococcus lactis subsp. Lactis* and reuterin produced by *Lactobacillus reuteri* are bacteriocins approved as food additives currently used as a preservative with inhibitory effect of spore outgrowth in dairy products, pasteurized, and canned products (Delves-Broughton *et al.*, 1996; Gao *et al.*, 2011). Their Minimal Inhibitory Concentrations (MIC) vary from 0.05 to 12.5 µg/ml and 0.51 to 32.5 mM respectively. These substances act by binding disulfide bonds of protein or by creating pores on the germinated spores leading to decrease in the membrane potential and oxidative metabolism (Delves-Broughton *et al.*, 1996). However, the use of these substances remain limited by the emergence of resistant strains (Mazzotta and Montville, 1997).

6.7.2.3 Lysozyme

Lysozyme inhibits the outgrowth of spores as it can hydrolyze the cell wall peptidoglycan by breaking the β-(1-4) linkages between residues of NAG and NAM. The required concentrations are generally 20 µg/mL.

6.7.2.4 Essential Oils and Other Plant Substances

The studies of Chaibi *et al.* (1997) were the first to show the inhibitory effects of germination and outgrowth of spores by essential oils. The sporostatic or sporocidal effects were observed depending on the essential oil, its concentration, and the bacterial species. Currently, essential oils are considerably studied to better understand their effectiveness, as natural preservatives with inhibitory effect of spore germination. Voundi *et al.* (2015) have recently shown the effects of essential oils *Ageratum conyzoïdes, Citrus limon, Citrus reticulata, Cymbopogon citratus, Drypetes gossweileri, Eugenia caryophyllus, Ocimum gratissimum, Thymus vulgaris, and Zingiber officinale* on the germination of *Bacillus* spores. Remarkable effects have been observed from *D. gossweileri* essential oil at 0.001 to 0.002 and from 0.004 to 0.009 mg/mL respectively on the inhibition of spore germination and cell growth.

Moreover, *C. limon* essential oil at 0.08% showed inhibition of germination and outgrowth of spores of *A. acidoterrestris* in a malt extract and lemon juice for 11 days. This activity was attributed to cinnamaldehyde one of its major compounds (Maldonado *et al.*, 2013).

Furthermore, thymol and carvarol respectively from *Oreganum* and *Thymus* essential oils at 0.6 mM inhibited the lag phase and the growth rate of *B. megaterium* spores pre-subjected to heat treatment 90°C for 25 min (Periago *et al.*, 2006).

In addition, conessine, an alkaloid derived from *Holarrhena floribunda* revealed significant inhibitory effects on the germination of *Bacillus* spores. The activity exhibited by the compound was dependent on environmental conditions (temperature and pH of treatment), and the bacterial species (Bogne, 2008; Voundi *et al.*, 2014).

6.7.2.5 Hurdle Technologies as Autcome for Efficient Control of Spores in Food Products

Hurdle technologies are employment of intelligent combination of different preservation factors or techniques to achieve multi-target, mild but reliable preservation effects. Many promising hurdles have been identified with good efficient effect on spore inactivation. The most reported are probably the combined effect of heat treatment with some antimicrobial substances as nisin and essential oil components (citral, carvacrol, and thymol) (Periago, 2006; Aouadhi *et al.*,2014; Huertas *et al.*, 2014). Moreover, weak organic acids, such as sorbic acid and acetic acid are preservatives often combined with heat treatments for more inhibition of spore germination. In addition, Ji-Hyoung *et al.* (2012) found more reduction of *B. cereus* spores inoculated in rice with combined γ-radiation and NaOCl/ultrasound treatments at small doses compared to each treatment applied alone.

References

Abee, T., et al. (2011). Biofilm formation and dispersal in Gram-positive bacteria. *Current opinion in Biotechnology, 22*, pp. 172–179.

AFSSA. (2009). Avis du 10 avril 2009 de l'Agence Française de Sécurité Sanitaire des Aliments relatif à la demande d'avis complémentaire concernant les références applicables aux denrées alimentaires en tant que critères indicateurs d'hygiène des procédés.

Agense Natonale de Sécurité Sanitaire, Alimentaire, Environnement, Travail (ANSES). (2010). *Clostridium perfringens*. Fiche de description de danger microbien transmissible par les aliments / *Clostridium perfringens*, pp. 1–3.

Aouadhi, C., et al. (2014). Inactivation of *Bacillus sporothermodurans* spores by nisin and temperature studied by design of experiments in water and milk. *Food Microbiology,38*, pp. 270–275.

Atrih, A. and Foster, S.J. (1999). The role of peptidoglycane strusture and structural dynamic during endospore dormancy and germination. *Antonie Van Leuwenhoek, 75*, pp. 299–307.

Ávila, M., et al. (2014). Inhibitory activity of reuterin, nisin, lysozyme and nitrite against vegetative cells and spores of dairy-related *Clostridium* species. *International Journal of Food Microbiology, 172*, pp. 70–75.

Baysal, A.H., Molva, C. and Unluturk, S. (2013). UV-C light inactivation and modeling kinetics of Alicyclobacillus acidoterrestris spores in white grape and apple juices. *International Journal of Food Microbiology, 166*, pp. 494–498.

Bevilacqua, S.M. and Corbo M.R. (2008). *Alicyclobacillus Acidoterrestris*: New Methods for Inhibiting Spore Germination. *International Journal of Food Microbiology, 125*(2), pp. 103–110.

Bogaert, J.C. and Naidu, A.S. (2000). Lactic acid. In:Naidu, A.S., ed.,Natural Food Antimicrobial Systems. London, UK: CRC Press, pp. 613–636.

Bogne, K. (2008). Action de *Holarrhena floribunda* et de *Acanthus montanus* sur quatre espèces de *Bacillus* endosporulantes et sur la germination de leurs spores. Thèse de Doctorat/ Ph.D, Université de Yaoundé I, pp. 48, 92–94.

Burns, D.A., Heap, J.T. and Minton N.P. (2010). *Clostridium difficile* spore germination: an update. *Research in Microbiology, 161*, pp. 730–734.

Brul, S., et al. (2002). Physiological actions of preservative agents: prospective of use of modern microbiological techniques in assessing microbial behavior in food preservation. *International Journal of Food Microbiology, 79*, pp. 55–64.

Carlin, F. (2011). Origin of bacterial spores contaminating foods. *Food Microbiology, 28*, pp. 177–182.

Centers for Disease Control and Prevention (CDC). (2003). Outbreak of a botulism type E associated with eating a beached whale— Western Alaska. *MMWR, 52*, pp. 24–26.

Celandroni F., et al. (2004). Effect of microwave radiation on *Bacillus subtilis* spores. *Journal of Applied Microbiology, 94*, pp. 1220–1227.

Chaibi, A., et al. (1997). Inhibition of germination and vegetative growth of *Bacillus cereus T andClostridium botulinum 62 A* spores by essential oils. *Food. Microbiol., 14*, pp. 161–174.

Chu, S., Hawes, J.W. and Lorigan, G.A. (2009). Solid-state NMR spectroscopic studies on the interaction of sorbic acid with phospholipid membranes at different pH levels. *Magnetic Resonance in Chemistry, 47*, pp. 654–657.

Deak, T. (2014). Food Safety Management, II. Technologies and Food Safety, Chapter 17: Thermal treatment,pp. 423–441 doi.org/10.1016/B978-0-12-381504-0.00017-2.

Delves-Broughton, J., et al. (1996). Applications of the bacteriocin, nisin. *Antonie van Leeuwenhoek, 69*, pp. 193–202.

Etoa, F.-X. (1985). Variation de la thermorésistance de la spore bactérienne (Bacillus stearothermophilus) en cours de germination. PhD dissertation, Université Pierre et Marie Curie.

Etoa, F-X. and Adegoke, G.O. (1996). Evolution of thermal resistance of spores of two *Bacillus* species in papper and honey. *Sc. Al.,16*, pp. 49–59.

Etoa, F.-X. (2011). Heat-induced resistance in Bacillus spores. Lap LAMBERT Academy publishing GmbH and Co.KG, Dudwei.

EU. 1999. Directive 1999/2/EC. On the Approximation of the Laws of the Member States Concerning Foods and Food Ingredients Treated with Ionizing Radiation (Annex III). Official Journal of the European Communities, L66, pp. 0016–0023.

Faille, C., et al. (2014). Sporulation of *Bacillus* spp. within biofilms: A potential source of contamination in food processing environments. *Food Microbiology, 40*, pp. 64–74.

Feudjio, D. (2005). Profil microbiologique de quelques aliments de patisserie artisanale vendus dans les rues de Yaoundé. Mémoire de Diplome d'Etudes Approfondies en Biochimie. Université de Yaoundé I.

Foster, S.J. (1994). The role and the regulation of cell structural dynamics during differentiation of endospore-forming bacteria. *J. Appl. Bacteriol. symposium supplement,76*, pp. 25–39.

Foster, S.J. and Johnstone, K. (1989). The trigger mechanism of bacterial spore germination. In:L. Smith, R.A. Slepecky, and P. Setlow, eds.,Regulation of Prokaryotic Development.Washington, DC: American Society for Microbiology. pp. 89–108.

Gao, Y., Qiu, W., Wu, D. and Fu, Q. (2011). Assessment of *Clostridium perfringens* spore response to high hydrostatic pressure and heat with nisin. *Appl. Biochem. Biotechnol., 164,* pp. 1083–1095.

Ghosh, S., Zhang, P., Li, Y. and Setlow, P. (2009). Superdormant spores of *Bacillus* species have elevated wet-heat resistance and temperature requirements for heat activation. *J Bacteriol., 191,* pp. 3822–3831.

Giel, J.L., et al. (2010). Metabolism of bile salts in mice influences spore germination in *Clostridium difficile. PLOS ONE, 5,* pp. 87–40.

Haeghebaert, S., et al. (2001). Les toxi-infections alimentaires collectives en France en 1998. *BEH,15,* pp. 1–12.

Henriques, O.A. and Moran, C.P. (2007). Structure and assembly of the bacterial endospore coat. *Annu Rev. Microbiol., 61,* pp. 555–588.

Heyndrickx, M. (2011). The importance of endospore-forming bacteria originating from soil for contamination of industrial food processing. *Applied and Environmental Soil Science, 2011,* pp. 1–11. doi. org/10.1155/2011/561975.

Hijnen, W.A.M., Beerendonk, E.F. and Medema, G.J. (2006). Inactivation credit of UV radiation for viruses, bacteria and protozoan (oo)cysts in water: A review. *Water Research, 40,* pp. 3–22.

Huertas, J., et al. (2014). Combined effect of natural antimicrobials and thermal treatments on *Alicyclobacillus acidoterrestris* spores. *Food Control, 35,* pp. 73–78.

Ji-Hyoung, H., Hyun-Joo, K. and Sang-Do, H. (2012). Effect of combined radiation and NaOCl/ultrasonication on reduction of *Bacillus cereus* spores in rice. *Radiation Physics and Chemistry, 81,* pp. 1177–1180.

Leguerinel, I., Couvert, O. and Mafart, P. (2007). Modelling the influence of the sporulation temperature upon the bacterial spore heat resistance, application to heating process calculation. *Int. J. Food Microbiol., 114,* pp.100–104.

Leyral, G. and Vierling, E. *(*2001*).* Hygiene et sécurité alimentaire, *3rd ed.* Paris: Doin, pp. 34, 78–84,107, 145–147.

Lucasa, R., et al.(2006). Application of the broad-spectrum bacteriocin enterocin AS-48 to inhibit *Bacillus coagulans* in canned fruit and vegetable foods. *Food and Chemical Toxicology, 44*(10), pp. 1774–1781.

Magnusson, M., Christiansson, A. and Svensson, B. (2007a). *Bacillus cereus* spores during housing of dairy cows: factors affecting contamination of raw milk. *J. Dairy Sci., 90,* pp. 2745–2754.

Magnusson, M., et al. (2007b). *Bacillus cereus* in free-stall bedding. *J. Dairy Sci., 90,* pp. 5473–5482.

Maldonado, M.C., Aban, M.P. and Navarro, A.R. (2013). Chemical and essential oil effect on *Alicyclobacillus acidoterrestris* viability. *Brazilian Journal of Microbiology,44*(4), pp. 1133–1137.

Markland, S.M., et al. (2013). Pathogenic Psychrotolerant Sporeformers: An Emerging Challenge for Low-Temperature Storage of Minimally Processed Foods. *Foodborne Pathogens and Disease, 10,* pp. 413–419.

Mazzotta, A.S. and Montville, T.J. (1997). Nisin induces changes in membrane fatty acid composition of *Listeria monocytogenes* nisin resistant strains at 10 °C and 30 °C. *J. Appl. Microbiol., 82,* pp. 23–28.

McClane, B.A. (2007). *Clostridium perfringens.* In: M.P. Doyle, L.R. Beuchat, eds., Food Microbiology.Fundamentals and Frontiers, third ed. Washington, DC: ASM Press, pp. 423–444.

McKenney, P.T., Driks, A. and Eichenberger, P. (2013). The *Bacillus subtilis* endospore: assembly and functions of the multilayered coat. *Nature rewiew/Microbiology, 11,* pp. 33–44.

Merzougui, S., Lkhider, M. and Cohen, N. (2013). *Bacillus cereus,* a real problem for food industry. *ScienceLib Edition Mersenne, 5,* pp. 2111–4706.

Moir, A. Corfe, B.M. and Behravan, J. (2002). Spore germination. *Cellular and Molecular Life Sciences, 59,* pp. 403–409.

Moir, A. (1990). The genetics of bacterial spore germination. *Annu rev. Microbiol., 44,* pp. 531–553.

Montanari, G., et al. (2004). Morphological and phonotypical characterisation of *Bacillus sporothermodurans*. *J. Appl. Microbiol.*, *97*, pp. 802–809.

Peck, M.W. (2010). *Clostridium botulinum*. In: V.K. Juneja J.N. Sofos,eds.,Pathogens and Toxins in Foods: Challenges and Interventions. Washington, DC: ASM, 2010, pp. 31–48.

Periago, P.M., et al. (2006). *Bacillus megaterium* spore germination and growth inhibition by a treatment combining heat with natural antimicrobials. *Food Technol. Biotechnol.*, *44*(1), pp. 17–23.

Postollec, F., et al. (2010). A multiparametric PCR-based tool for fast detection and identification of spore-forming bacteria in food. *International Journal of Food Microbiology*, *142*, pp. 78–88.

Riesenman, P.J. and Nicholson W.L. (2000). Role of the spore coat layers in *Bacillus subtilis* spore resistance to hydrogen peroxide, artificial UV-C, UV-B, and solar UV radiation. *Appl. Environ. Microbiol.*, *66*, pp. 620–626.

Rodgers, S. (2003). Long shelf life cook-chill technologies: food safety risks and solutions, *Food Australia*, *55*(3), pp. 80–83.

Reyes, J. E., et al. (2007). Prevalence of *Bacillus cereus* in dried milk products used by Chilean School Feeding Program. *Food Microbiology*, *24*, pp. 1–6.

Rose, R., et al.(2007). Comparison of the prop- erties of *Bacillus subtilis* spores made in liquid or on agar plates. *J Appl Microbiol.*, *103*, pp. 691–699.

Scallan, E., et al. (2011). Foodborne illness acquired in the United States—Major pathogens. *Emerg Infect. Dis.*, *17*, pp. 7–15.

Sella, S.et al.(2014). Life cycle and spore resistance of spore-forming *Bacillus atrophaeus*. *Microbiological Research*, *169*, pp. 931–939.

Setlow, P. and Johnson, E. A. (2007). Spores and their significance. In: M.P. Doyle, L.R. Beuchat, eds., Food Microbiology. Fundamentals and Frontiers. Washington DC: ASM Press, pp. 35–68.

Sorg, J.A. and Sonenshein, A.L. (2008). Bile salts and glycine as cogerminants for *Clostridium difficile* spores. *J. Bacteriol.*, *190*, pp. 2505–2512.

van Melis, C.C.J., et al.(2012). Characterizaation of germination and outgrowth of sorbic acid-stressed *Bacillus cereus* ATCC 14579 spores: phenotype and transcriptome analysis. *Food Microbiology*, *28*, pp. 275–283.

Velugotia, P.R., et al.(2007). Use of calcium, potassium, and sodium lactates to control germination and outgrowth of *Clostridium perfringens* spores during chilling of injected pork. *Food Microbiology*, *24*, pp. 687–694.

Vissers, M.M.M., et al. (2007). Concentrations of butyric acid bacteria spores in silage and relationships with aerobic deterioration. *J. Dairy Sci.*, *90*, pp. 928–936.

Voundi, O.S., et al. (2015). Effect of essential oils on germination and growth of some pathogenic and spoilage spore-forming bacteria. *Foodborne Pathogens and Disease*, *12*, pp. 551–559.

Voundi, S., et al. (2014). Activity of conessine at various temperatures and pH on inhibition of germination of *Bacillus cereus* and *Bacillus stearothermophilus* spores. *African Journal of Mirobiology*, *8*(4), pp. 320–326.

Voth, D.E. and Ballard, J.D. (2005). *Clostridium difficile* toxins: mechanism of action and role in disease. *Clin. Microbiol. Rev.*, *18*, pp. 247–263.

WHO. (1981). Wholesomeness of Irradiated Food. Joint Report FAO/IAEA/WHO Expert Committee. World Health Organization (WHO Technical Report Series, No. 659).

7

Insights into Detection and Identification of Foodborne Pathogens

Jodi Woan-Fei Law[1], *Vengadesh Letchumanan*[1,2], *Kok-Gan Chan*[2], *Bey-Hing Goh*[1,3], *and Learn-Han Lee*[1,3]

[1] *Novel Bacteria and Drug Discovery Research Group, School of Pharmacy, Monash University Malaysia, Selangor Darul Ehsan, Malaysia*
[2] *Division of Genetics and Molecular Biology, Institute of Biological Sciences, University of Malaya, Kuala Lumpur, Malaysia*
[3] *Center of Health Outcomes Research and Therapeutic Safety (Cohorts), School of Pharmaceutical Sciences, University of Phayao, Phayao, Thailand*

7.1 Introduction

Foodborne diseases have become a major public health problem worldwide due to the significantly increased incidence of foodborne diseases over the last 20 years (Oliver *et al.*, 2005). Although it is difficult to estimate the global incidence of foodborne diseases as some of the cases are under-reported especially in developing countries, but the increased incidence of foodborne diseases were reported in many parts of the world (Van de Venter, 2000). For instance, the outbreak of foodborne disease in Taiwan increased rapidly from 121 in 1995 to 177 in 1996 and since then the incidence keep rising (Chiou *et al.*, 2000). According to report from Centers for Disease Control and Prevention (CDC), approximately 48 million people in the United States get ill, 128000 people are hospitalized and 3000 people die annually due to foodborne diseases despite United States has the safest food supplies in the world (Oliver *et al.*, 2005; CDC, 2011). In addition, about a quarter of the population is at a higher risk for foodborne diseases nowadays (Oliver *et al.*, 2005).

Generally, foodborne diseases are caused by the consumption of food or water contaminated with foodborne pathogens or their toxins. Foodborne pathogens are referred to pathogens that can cause foodborne diseases in humans and they include bacteria, viruses, fungi, and parasites (Zhao *et al.*, 2014). There are 31 identified foodborne pathogens in the United States from year 2000 to 2008 and it is estimated that viruses are the primary causes of illnesses, whereas bacteria are the primary causes of hospitalizations and deaths (Scallan *et al.*, 2011). The increasing amounts of street foods and the increasing demand for minimally processed ready-to-eat products have begun to concern public health agencies on food safety assurance (Lee *et al.*, 2014). Foodborne pathogens are present in various foods such as fruits, vegetables, and ready-to-eat products that are consumed without any further treatment (Chung *et al.*, 2010; Lee *et al.*, 2014). This may lead to foodborne diseases if food safety issues are not taken into consideration. Also, foodborne diseases are often associated with the consumption of raw or undercooked foods such as seafood, meat, and poultry (Wingstrand *et al.*, 2006; Rosec *et al.*, 2012; Omurtag *et al.*, 2013). The common foodborne pathogens that are responsible for most of the foodborne disease outbreaks are *Listeria*

Foodborne Pathogens and Antibiotic Resistance, First Edition. Edited by Om V. Singh.

monocytogenes, Vibrio spp., *Escherichia coli* O157:H7, *Salmonella enterica, Bacillus cereus, Staphylococcus aureus, Campylobacter jejuni,* and *Clostridium perfringens* (Oliver *et al.*, 2005; Scallan *et al.*, 2011; Zhao *et al.*, 2014). It is essential to analyze the food for the presence of foodborne pathogens in order to ensure a safe food supply and to minimize the occurrence of foodborne diseases. Thus, microbiological food testing is important to ensure the safety of food products (Dwivedi and Jaykus, 2011; Välimaa *et al.*, 2015).

The conventional methods for detecting the foodborne bacterial pathogens present in food are based on culturing the microorganisms on agar plates followed by standard biochemical identifications (Mandal *et al.*, 2011). Conventional methods are usually inexpensive and simple but these methods can be time consuming as they depend on the ability of the microorganisms to grow in different culture media such as pre-enrichment media, selective enrichment media and selective plating media. Usually conventional methods require two to three days for preliminary identification and more than a week for confirmation of the species of the pathogens (Zhao *et al.*, 2014). Conventional methods are laborious as they require the preparation of culture media, inoculation of plates, and colony counting (Mandal *et al.*, 2011). Furthermore, conventional methods may be limited by their low sensitivity (Lee *et al.*, 2014). False-negative results may occur due to viable but non-culturable (VBNC) pathogens. The failure to detect foodborne pathogens would increase the transmission risk of pathogens.

Recently, different rapid methods with high sensitivity and specificity have been developed to overcome the limitations of conventional methods for the detection and identification of foodborne pathogens. Furthermore, researchers are still developing novel methods with improvements in terms of rapidity, sensitivity, specificity, and suitability for *in situ* analysis and distinction of the viable cell (Zhao *et al.*, 2014). Rapid detection methods are important, particularly in the food industry, as they are able to detect the presence of pathogens in raw and processed foods immediately. Rapid methods are also sensitive enough to detect pathogens that present in low numbers in the food. Sensitivity is important because a single pathogen present in food has the risk to cause infection. Rapid detection methods are more time-efficient, labor-saving, and able to reduce human errors (Mandal *et al.*, 2011; Law *et al.*, 2015a; Letchumanan *et al.*, 2014). This chapter examines these rapid detection methods and their applications in four major foodborne bacterial pathogens: *Listeria monocytogenes, Vibrio parahaemolyticus, Escherichia coli,* and *Salmonella* detection and along with their advantages and limitations.

7.1.1 *Listeria monocytogenes*

Listeria is a bacteria that was named after a British pioneer of sterile surgery, Joseph Lister (Skerman *et al.*, 1980). The only species of the genus *Listeria* (family *Listeriaceae*) that is capable of infecting both man and animals as well as being involved in known foodborne outbreaks of listeriosis is the *Listeria monocytogenes* (Monge and Arias, 1997; Cocolin *et al.*, 2002; Liu, 2006). *L. monocytogenes* is characterized as a gram-positive, rod-shaped (approximately 1–1.5 μm length and 0.5 μm width), facultative anaerobic, and non-spore-forming bacteria with low GC content (Farber and Peterkin, 1991; Liu, 2006). *L. monocytogenes* is one of the major foodborne pathogens that has been a serious threat to the food industry due to its ability to survive the most common food processing conditions such as extreme pH, high salt concentration, low water activity, and refrigeration temperatures (Liu, 2006; Zunabovic *et al.*, 2011; Jadhav *et al.*, 2012).

Generally, *L. monocytogenes* can be eliminated or reduced by pasteurization process because it cannot survive pasteurization temperatures in food processing (Jadhav *et al.*, 2012). However, post-processing or post-pasteurization contamination by *L. monocytogenes* may occur due to cross-contamination or biofilms (Jadhav *et al.*, 2012). *L. monocytogenes* can attach to food contact surfaces such as stainless steel and polystyrene during food processing and form biofilms, which is important for their survival in hostile environments (Jadhav *et al.*, 2012; Da Silva and De Martinis, 2013; Välimaa *et al.*, 2015). Biofilms may persist for a long period of time and they can tolerate high concentrations of disinfectants, sanitizers, and antimicrobials (Välimaa *et al.*, 2015). Hence, this may result in the contamination of food contact surfaces that then lead to higher risk of food contamination during and/or after processing. Food contaminated with *L. monocytogenes* has posed a great concern to the food industry as it can cause serious infection known as listeriosis when ingested. *L. monocytogenes* is recognized as an opportunistic human foodborne pathogen and is capable of causing listeriosis in all human population groups, but majority cases of human listeriosis occur in immunosuppressed individuals, pregnant women, neonates as well as the elderly (Farber and Peterkin, 1991; Liu, 2006).

Human listeriosis usually shows gastroenteritis and non-specific flu-like symptoms such as headache, chills, fatigue, muscular and joint pain in the early stages of infection. If the infection is not treated appropriately, it has the potential to develop into septicaemia, encephalitis, meningitis and may lead to death in some cases (Liu, 2006). The annual incidence of listeriosis in most of the European Union countries is estimated to be between 2 and 10 reported cases per million per year. In addition, food contaminated with *L. monocytogenes* is also one of the causes of recalls that may result in large economic losses (Jemmi and Stephan, 2006).

7.1.2 *Vibrio parahaemolyticus*

The name *Vibrio* was derived from Filippo Pacini who first isolated this microorganism from cholera patients in 1854 (Skerman *et al.*, 1980). Among the genus member is *Vibrio parahaemolyticus*, a gram-negative halophilic bacterium that is widely disseminated in estuarine, marine, and coastal surroundings (Zhang and Orth, 2013; Ceccarelli *et al.*, 2013). It belongs to the *Vibrioananceae* family (Igbinosa *et al.*, 2008). This bacterium is usually found in a free-swimming state; with its motility conferred by a single polar flagellum affixed to inert and animate surfaces including zooplankton, fish, shellfish, or any suspended matter underwater (Gode-Potratz *et al.*, 2011). While the majority of strains isolated from environmental sources are innocuous members of marine microbiota, a small number of *V. parahaemolyticus* strains are capable of causing human illness and are often associated with foodborne gastroenteritis or diarrhea (Raghunath, 2015).

V. parahaemolyticus was first recognized as a foodborne pathogen in the year 1951 in Osaka, where people frequently consume raw or uncooked seafood (Daniels *et al.*, 2000). The bacterium was isolated from victims of an outbreak of 272 infected cases and 20 deaths associated with consumption of shirasu, Japanese boiled and semi dried sardines dish (Aberoumand, 2010). Ever since then, *V. parahaemolyticus* has been commonly isolated from seafood, including shrimp, in markets in South East Asian countries (Elhadi *et al.*, 2004; Deepanjali *et al.*, 2005). *V. parahaemolyticus* has accounted for many food poisoning cases in Japan (Su and Liu, 2007; Aberoumand *et al.*, 2010; Kubota *et al.*, 2011; Hara-Kudo *et al.*, 2012), Taiwan (Yu *et al.*, 2013), China (Li *et al.*, 2014), Bangladesh (Bhuiyan *et al.*, 2002), Laos (Matsumoto *et al.*, 2000), Malaysia (Saleh *et al.*, 2011; Letchumanan *et al.*, 2015a; Letchumanan *et al.*, 2015c), Hong Kong and Indonesia (Matsumoto *et al.*, 2000). In Europe and United States, *V. parahaemolyticus* has been reported to be isolated from seafood such as shrimps, crab, oysters and clams (Lozano-Leon *et al.*, 2003; Robert-Pillot *et al.*, 2004; Martinez-Urtaza *et al.*, 2005; Newton *et al.*, 2012).

The virulent strains are discerned from avirulent strains by the presence of toxigenic genes namely, thermostable direct hemolysin (*tdh*) and/or TDH-related (*trh*) hemolysin genes (Letchumanan *et al.*, 2014). Although *V. parahaemolyticus* is commonly present in seafood, most of these isolates are regarded as non-pathogenic to human (Nishibuchi, 1995; Velazquez-Roman *et al.*, 2012; Raghunath, 2015). The strains isolated from environmental samples usually lack the pathogenic genes thermostable direct hemolysin (*tdh*) and/or TDH-related hemolysin (*trh*), which are responsible for causing diseases in human and marine animals (Deepanjali *et al.*, 2005; Canizalez-Roman *et al.*, 2011; Gutierrez West *et al.*, 2013). Previous studies reported, around 0–6% of the environmental *V. parahaemolyticus* strains carry *tdh* and/or *trh* genes (DePaola *et al.*, 2000; Vuddhakul *et al.*, 2000; Wong *et al.*, 2000; Hervio-Heath *et al.*, 2002; Alam *et al.*, 2002; Haley *et al.*, 2014). These two genes are considered major virulence factors of *V. parahaemolyticus* (Kaysner and DePaola, 2001; Zhang and Austin, 2005; Xu *et al.*, 2014).

7.1.3 *Escherichia coli*

Escherichia was named after a famous German paediatrician, Theodor Escherich, in 1885 (Skerman *et al.*, 1980). *Escherichia coli* is one of the major foodborne bacterial pathogens that has persisted during the 1990s and until the present day (Newell *et al.*, 2010). Generally, *E. coli* is a gram-negative bacillus, predominant facultative anaerobe of the human intestinal microflora that belongs to the family *Enterobacteriaceae* (Nataro and Kaper, 1998; Fukushima *et al.*, 2002). Normally, *E. coli* colonizes the gastrointestinal tract without causing any harm to the host as they both derive mutual benefit. Nevertheless, *E. coli* can cause infection in immunosuppressed host or when the host's gastrointestinal barriers are disrupted (Nataro and Kaper, 1998; Feng and Weagant, 2015). Pathogenic *E. coli* strains that causes diarrheal disease are known as diarrheagenic *E. coli* (DEC). Diarrheagenic *E. coli* strains have been divided into six categories based on specific virulence factors and phenotypic traits (Newell *et al.*, 2010; Feng and Weagant, 2015):

I. Enteropathogenic *E. coli* (EPEC),
II. Enterotoxigenic *E. coli* (ETEC),
III. Enteroinvasive *E. coli* (EIEC),
IV. Enteroaggregative *E. coli* (EAEC),
V. Diffusely adhering *E. coli* (DAEC), and
VI. Enterohemorrhagic *E. coli* (EHEC) (characterized by the production of verotoxin or Shiga toxins (Stx), also known as Vero cytotoxin-producing *E. coli* (VTEC) or Shiga toxin-producing *E. coli* (STEC))

Furthermore, *E. coli* strains are serotyped based on their somatic lipopolysaccharide O antigens, flagellar H antigens, and capsular K antigens as according to the modified Kauffman scheme (Levine, 1987; Nataro and Kaper, 1998). Among the different categories of DEC strains, VTEC/STEC is commonly associated with foodborne disease outbreaks. Serious illnesses caused by STEC include hemorrhagic colitis (bloody diarrhea) and haemolytic uremic syndrome (Newell *et al.*, 2010; Feng and Weagant, 2015). For instance, *E. coli* O157:H7 serotype is one of the STEC strains that has been a major cause of foodborne disease outbreaks in many places like the United States, Europe and Canada. In 1982, *E. coli* O157:H7 was first identified as the cause of outbreaks of hemorrhagic colitis. The early outbreaks of *E. coli* O157:H7 were often associated with contaminated ground beef as cattle are reservoirs of *E. coli* O157:H7, however, foods that are contaminated by bovine or ruminant feces such as vegetables, fruits, juices, milk, and other meat products as well as fecally contaminated drinking water can be the sources of *E. coli* O157:H7 outbreaks nowadays (Newell *et al.*, 2010). Besides, there are other STEC of non-O157 serogroups that also cause foodborne diseases worldwide such as *E. coli* serogroups O26, O45, O103, O111, O121, and O145. Due to the low infectious dose of STEC, rapid and sensitive methods are required for the detection and isolation of these pathogens from various food samples in order to prevent foodborne disease outbreaks (Newell *et al.*, 2010; Fratamico and Gehring, 2014).

7.1.4 *Salmonella*

Salmonella was named after a veterinary pathologist, Daniel Salmon, in 1900 (Skerman *et al.*, 1980). *Salmonella* infection remains a major public health concern worldwide, contributing to the economic burden for both industrialized and underdeveloped countries through the increased costs associated with surveillance, prevention and treatment of disease (Crump *et al.*, 2004). Gastroenteritis is the most common manifestation of *Salmonella* infection worldwide, followed by bacteremia and enteric fever (Majowicz *et al.*, 2010). *Salmonella* is a rod-shaped, gram-negative facultative anaerobe that belongs to the family *Enterobacteriaceae* (Barlow and Hall, 2002). Within the genus *Salmonella*, there are around 2600 serotypes that have been identified with the use of standard Kauffman–White scheme. Most of these serotypes have the ability to adapt within a variety of animal hosts, including humans (Allerberger *et al.*, 2003).

The nomenclature of *Salmonella* is controversial and still evolving. Currently, the Centers for Disease Control and Prevention (CDC) applies the nomenclatural system of *Salmonella* recommended by the World Health Organization (WHO) Collaborating Centre (Popoff *et al.*, 2003). Based on this system, the genus *Salmonella* is classified into two species: *Salmonella enterica* (type species) and *Salmonella bongori*, based on differences in their 16S rRNA sequence analysis. The type species, *Salmonella enterica*, can be further classified into six subspecies based on their genomic relatedness and biochemical properties (Reeves *et al.*, 1989). The subspecies are denoted with roman numerals:

I. *Salmonella enterica* subsp. *enterica*
II. *Salmonella enterica* subsp. *salamae*
IIIa. *Salmonella enterica* subsp. *arizonae*
IIIb. *Salmonella enterica* subsp. *diarizonae*
IV. *Salmonella enterica* subsp. *houtenae*
V. *Salmonella enterica* subsp. *indica*

Among all the subspecies of *Salmonella*, *Salmonella enterica subsp. enterica* (I) is commonly found in mammals and contributes approximately 99% of *Salmonella* infections in humans and warm-blooded animals. In contrast, the other five *Salmonella* subspecies and *Salmonella bongori* are found mainly in the environment and also in cold-blooded animals, and hence are rare in humans (Brenner *et al.*, 2000). Therefore, the development of detection method focuses more on *Salmonella enterica*.

7.2 Nucleic Acid-Based Methods

Nucleic acid-based methods operate by detecting specific DNA or RNA sequences in the target pathogen. This is done by hybridizing the target nucleic acid sequence to a synthetic oligonucleotide (probes or primers), which is complementary to the target sequence (Zhao *et al.*, 2014). There are many bacterial pathogens such as *Clostridium botulinum, Vibrio cholerae, Staphylococcus aureus*, and *Escherichia coli* O157, which produced toxins that caused foodborne diseases (Singh *et al.*, 2001; Akbulut *et al.*, 2004; Fusco *et al.*, 2011; Son *et al.*, 2014). The toxin-related genes in these pathogens can be detected by nucleic acid-based methods (Letchumanan *et al.*, 2014; Zhao *et al.*, 2014; Law *et al.*, 2015a). Furthermore, pathogens showing ambiguous phenotypic characteristics such as hippurate negative *Campylobacter jejuni* strains can be identified and confirmed by nucleic acid-based methods (Adzitey *et al.*, 2012). Nucleic-acid based methods detect the specific genes in the target pathogens, therefore preventing ambiguous or wrongly interpreted results. The nucleic acid-based methods described are simple polymerase chain reaction (PCR), multiplex polymerase chain reaction (mPCR), real-time/quantitative polymerase chain reaction (qPCR), nucleic acid sequence-based amplification (NASBA), loop-mediated isothermal amplification (LAMP), microarray technology, and next-generation sequencing (NGS).

7.2.1 Polymerase Chain Reaction (PCR)

Polymerase chain reaction (PCR) is a method with high sensitivity and specificity that is most commonly used for detection and identification of pathogenic bacteria from clinical, environmental, or seafood samples (Nelapati *et al.*, 2012; Law *et al.*, 2015a). PCR was invented about 30 years ago and it allows the detection of a single bacterial pathogen that present in food by detecting a specific target DNA sequence (Velusamy *et al.*, 2010). PCR operates by amplifying a specific target DNA sequence in a cyclic three steps process (Mandal *et al.*, 2011). Firstly, the target double-stranded DNA is denatured into single-stranded DNA at high temperature. Then, two single-stranded synthetic oligonucleotides or specific primers (forward and reverse primers) will anneal to the DNA strands. This is followed by the polymerization process, whereby the primers complementary to the single-stranded DNA are extended with the presence of deoxyribonucleotides and a thermostable DNA polymerase (Figure 7.1). The PCR amplification products are separated by agarose gel electrophoresis and visualized on the gel as bands with a DNA stain (Zhao *et al.*, 2014). PCR have been used in the detection of numerous foodborne pathogens like *Listeria monocytogenes*, *Escherichia coli* O157:H7, *Vibrio parahaemolyticus*, *Staphylococcus aureus*, *Campylobacter jejuni*, *Salmonella enterica*, and *Shigella* spp. (Cheah *et al.*, 2008; Lee *et al.*, 2008; Alves *et al.*, 2012; Chiang *et al.*, 2012; Zhou *et al.*, 2013; Letchumanan *et al.*, 2015a; Letchumanan *et al.*, 2015c).

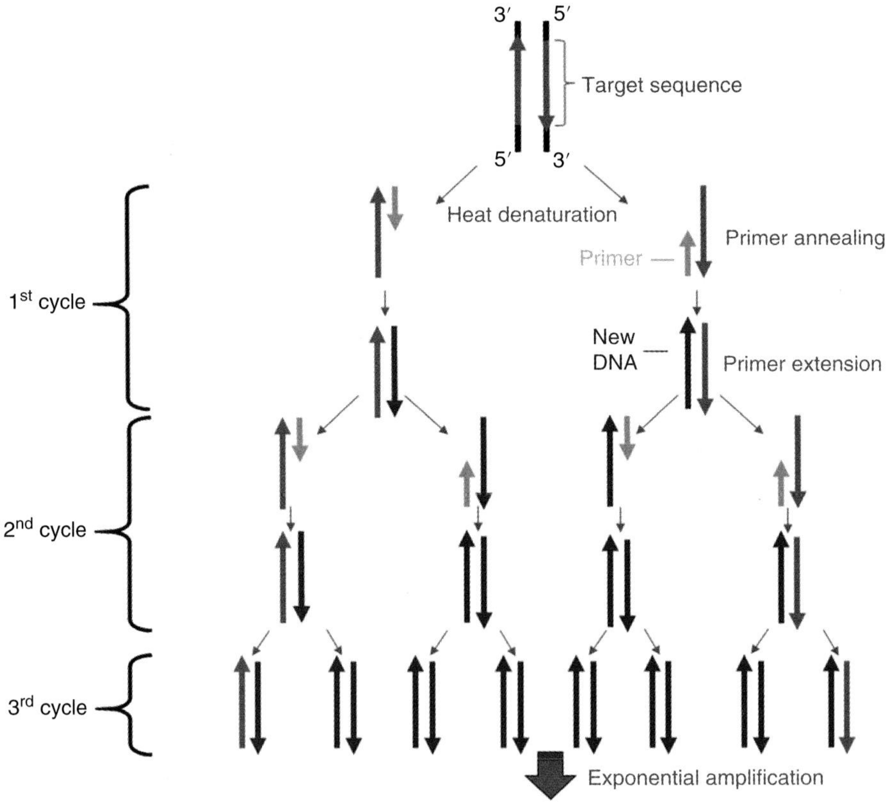

Figure 7.1 Polymerase Chain Reaction. The cyclic three steps process involves denaturation of DNA, primer annealing, and primer extension. The target double-stranded DNA is denatured under high temperature, producing two single-stranded DNA. For each single-stranded DNA, a specific primer will anneal with the target sequence in the DNA. Primer complementary to the single-stranded DNA is extended and a new strand of DNA is synthesized. These steps will be repeated in many cycles and thus producing many copies of a target DNA sequence. (*See insert for color representation of the figure.*)

7.2.1.1 *Listeria monocytogenes*

For *Listeria*, the specific detection of this genus by PCR involves PCR primers based on the highly conserved 16S rRNA sequence present in all *Listeria* spp. with a resulting 938 bp amplification product (Levin, 2003; Burbano *et al.*, 2006; Goh *et al.*, 2012; Jamali *et al.*, 2013). *L. monocytogenes* can be differentiated from other *Listeria* spp. by exploiting the molecular differences within the PCR amplified 16S rRNA gene, 23S rRNA gene and 16S-23S rRNA intergenic spacer regions (Wang *et al.*, 1992; Graham *et al.*, 1996; Sallen *et al.*, 1996; Graham *et al.*, 1997). In addition, PCR method also detects *L. monocytogenes* at the species level by targeting the virulence genes of the organism (Levin, 2003). Several virulence genes have been identified in *L. monocytogenes* and targeted for the PCR detection of the organism, for example, *hly* (*hlyA*) gene codes for listeriolysin O (LLO) (Deneer and Boychuk, 1991; Johnson *et al.*, 1992; Agersborg *et al.*, 1997; Aznar and Alarcón, 2002; Amagliani *et al.*, 2004; Burbano *et al.*, 2006), *iap* gene codes for an invasion-associated protein known as p60 (Agersborg *et al.*, 1997; Aznar and Alarcón, 2002; Swetha *et al.*, 2012), *actA* gene codes for a surface protein known as ActA, which is required for intracellular bacterial propulsion and cell to cell invasion (Moriishi *et al.*, 1998; Levin, 2003), *lma*A gene codes for *L. monocytogenes* antigen (lmaA), which also known as *Dth*-18 gene codes for delayed-type hypersensitivity protein (DTH-18 factor) (Wernars *et al.*, 1991; Johnson *et al.*, 1992; Levin, 2003), *inlA* gene codes for internalin A (Almeida and Almeida, 2000; Ingianni *et al.*, 2001; Jung *et al.*, 2003), *inlB* gene codes for internalin B (Pangallo *et al.*, 2001; Jung *et al.*, 2003), *prfA* gene codes for positive regulator factor A (PrFA) (Simon *et al.*, 1996), *pepC* codes for aminopeptidase C (Winters *et al.*, 1999), *fbp* gene codes for fibronectin-binding protein (Gilot and Content, 2002), and *plcB* Phospholipase C protein (Volokhov *et al.*, 2002). Among these targeted genes, the *hly*A gene is the most frequently chosen target gene for the PCR detection of *L. monocytogenes* (Aznar and Alarcón, 2002; Jadhav *et al.*, 2012). The *hlyA* gene codes for a protein with pore-forming activity, which is known as listeriolysin O. This protein is found to be essential for the virulence of *L. monocytogenes* as it is responsible for the lysis of phagocyte vacuole and followed by the escape of *L. monocytogenes* from the vacuole (Kathariou *et al.*, 1987; Cossart *et al.*, 1989; Levin, 2003; Liu, 2006). Besides, it has been discovered that all clinical isolates of *L. monocytogenes* have hemolytic activity due to listeriolysin O and thus the *hlyA* gene is a relevant marker for the identification of *L. monocytogenes* (Groves and Welshimer, 1977; Golsteyn-Thomas *et al.*, 1990). Additional studies that involved the use of PCR for the detection of *L. monocytogenes* in food samples are listed in Table 7.1.

7.2.1.2 *Vibrio parahaemolyticus*

For *V. parahaemolyticus*, the species detection of this pathogen by PCR involves PCR primers targeting *toxR* gene. The PCR method was developed to identify *V. parahaemolyticus* strains at the species level by targeting *toxR* gene (Vimala *et al.*, 2010; Paydar *et al.*, 2013; Suffredini *et al.*, 2014). The *toxR* gene is highly conserved in both pathogenic and non-pathogenic *V. parahaemolyticus*. The *toxR* gene stimulates the expression of thermostable direct haemolysin gene (*tdh*) and present in either pathogenic or non-pathogenic *V. parahaemolyticus* isolates (Sujeewa *et al.*, 2009). The thermolabile hemolysin (*tlh*) in *V. parahaemolyticus* is another gene that was used to develop a multiplex PCR procedure for simultaneous detection of total and virulent *V. parahaemolyticus* (Yi *et al.*, 2014). Although, the *tlh* gene is not considered a virulence factor of *V. parahaemolyticus*, the gene is reported to be a reliable marker for *V. parahaemolyticus* (Su and Liu, 2007).

There are two virulent factors have been clearly identified in association with *V. parahaemolyticus* cases: the thermostable direct hemolysin (*tdh*) and TDH-related hemolysin (*trh*) (Nishibuchi and Kaper, 1995). Thermostable direct hemolysin (*tdh*) exerts a variety of biological activities such as hemolytic activity,

Table 7.1 Examples of the Application of Nucleic Acid-Based Methods for the Detection of Various Foodborne Pathogens Present in Food Samples.

Detection Method	Foodborne Pathogens	Detection Limit	Food Matrix	Assay Time	References
Simple PCR	*L. monocytogenes*	Not stated	Naturally contaminated fish samples.	Not stated	Swetha *et al.*, 2012
	L. monocutogenes	Not stated	Naturally contaminated raw meat (chicken, beef, and fish), milk, and milk products (raw milk, cheese, and curd).	Not stated	Khan *et al.*, 2013
	L. monocytogenes	10^5 CFU/mL in pure culture. 10^1 CFU in 25 g pork. 10^1 CFU om 25 mL milk 10^1 CFU in 25 mL water.	Artificially contaminated milk, pork, and water.	18 hours including enrichment	Zhou and Jiao, 2005
Multiplex PCR	*Salmonella* spp., *Salmonella* Enteritidis	10^3 CFU/mL	Artificially and naturally contaminated chicken carcasses, minas cheese, and fresh pork sausages.	24 hours	Silva *et al.*, 2011
	STEC O26, O103, O111, O145, sorbitol fermenting O157 non-sorbitol fermenting O157.	5×10^4 CFU/mL in minced beef and sprouted seeds. 5×10^3 CFU/mL in raw-milk cheese.	Artificially contaminated minced beef, sprouted seed (soy, alfalfa, and leek), and raw-milk cheese.	24 hours	Verstraete *et al.*, 2012
	Escherichia coli O157:H7, *Listeria monocytogenes Staphylococcus aureus, Yersinia enterocolitica, Salmonella*	10^3 CFU/mL	Artificially contaminated pork.	Not stated	Guan *et al.*, 2013
	L. monocytogenes	$1.41 - 3.2 \times 10^2$ CFU/mL	Artificially contaminated milk.	Not stated	Zeng *et al.*, 2006
	L. monocytogenes	3×10^6 cells/mL without enrichment. 3×10^3 cells/mL after 24 hours enrichment.	Artificially contaminated milk.	6 hours	Rawool *et al.*, 2007
	L. monocytogenes	Not stated	Naturally contaminated deli meat samples: pork and chicken products.	Not stated	Liu *et al.*, 2015

Real-time PCR	*Salmonella enterica*	41.2 fg/PCR for *Salmonella* Typhimurium genomic DNA. 18.6 fg/PCR for *Salmonella* Enteritidis genomic DNA.	Artificially contaminated chicken, liquid egg, and peanut butter.	10 hours	Chen *et al.*, 2010
	Listeria monocytogenes, *Escherichia coli* O157, *Salmonella* spp.	<18 CFU/10 g	Artificially contaminated ground beef. Naturally contaminated beef, pork, turkey, and chicken.	24 hours	Suo *et al.*, 2010b
	Listeria monocytogenes, *Escherichia coli* O157:H7, *Salmonella* spp.	2×10^2 CFU/mL	Artificially contaminated ground pork.	24 hours	Kawasaki *et al.*, 2010
	Salmonella spp., *Listeria monocytogenes*	5 CFU/25 g	Artificially and naturally contaminated meat, fish, fruits, vegetables, dairy products, eggs, chocolate bar, omelet, lasagna, and various cooked dishes.	<30 hours	Ruiz-Rueda *et al.*, 2011
	Staphylococcus aureus, *Salmonella*, *Shigella*	9.6 CFU/g for *Staphylococcus aureus*, 2.0 CFU/g for *Salmonella* and 6.8 CFU/g for *Shigella*.	Fresh pork.	<8 hours	Ma *et al.*, 2014
	Escherichia coli O157:H7 (*stx_1* and *stx_2*)	13.35 CFU/PCR	Artificially contaminated bead slices, ground beef, salad greens, and salad dressing	Not stated.	Heller *et al.*, 2003
	L. monocytogenes (*prfA*)	5 copies of target gene per PCR	Artificially contaminated raw milk, salmon, pâté and, green-veined cheese. Naturally contaminated fish, meat, meat products, and dairy products.	Not stated	Rossmanith *et al.*, 2006
	L. monocytogenes	6 copies of *iap* gene per PCR (purified DNA)	Artificially contaminated milk.	Not stated	Hein *et al.*, 2001a
	L. monocytogenes	Not stated	Artificially contaminated pork meat	24 hours	Gattuso *et al.*, 2014

(*Continued*)

Table 7.1 (Continued)

Detection Method	Foodborne Pathogens	Detection Limit	Food Matrix	Assay Time	References
	L. monocytogenes	10^3–10^4 CFU/g or mL	Artificially contaminated soft cheese, fermented sausage, cured ham, and ready-to-eat salad. Naturally contaminated fresh meat, fresh sausages, fermented sausages, fresh cheeses, and ripened cheeses.	Not stated	Rantsiou *et al.*, 2008
	L. monocytogenes	Not stated	Leafy vegetables: collard green, cabbage, lettuce, mixed parsley, chinese cabbage, spring onion bunches, spinach, wild chicory, arugula, and watercress.	48 h	De Oliveira *et al.*, 2010
	L. monocytogenes	10 CFU/g after 6 hours enrichment. 1 CFU/g after 24 hours enrichment	Artificially contaminated chicken meat.	Not stated	Navas *et al.*, 2006
NASBA	*Escherichia coli*	40 cells/mL	Drinking water	4 hours	Min and Baeumner, 2002
	Salmonella Enteritidis	10^1 CFU/reaction	Artificially contaminated fresh meats, poultry, fish, ready-to-eat salads, and bakery products.	26 hours	D'Souza and Jaykus, 2003
	Listeria monocytogenes	400 CFU/mL	Artificially contaminated cooked ham and smoked salmon slices.	72 hours	Nadal *et al.*, 2007
	Bacillus amyloliquefaciens, Bacillus cereus, and *Bacillus circulans*	Not stated	Artificially contaminated milk.	Not stated.	Gore *et al.*, 2003
	Salmonella Enteritidis and *Salmonella* Typhimurium	<10 CFU/mL	Not stated	<90 minutes	Mollasalehi and Yazdanparast, 2013

	L. monocytogenes	<10 CFU/25 g	Artificially contaminated chicken breast meat, soft cheese, shrimps, dry sausage, minced meat (pork and beef), radish, and mushrooms.	<3 days	Uyttendaele *et al.*, 1995
LAMP	*Vibrio vulnificus*	5.4 CFU/reaction for a virulent *V. vulnificus* strain in pure culture. 2.5×10^3 CFU/g for a virulent *V. vulnificus* strain in spiked raw oyster, no enrichment. 1 CFU/g for a virulent *V. vulnificus* strain in spiked raw oyster, after 6 hours enrichment.	Artificially contaminated raw oysters.	8 hours	Han *et al.*, 2011
	Salmonella spp. and *Shigella* spp.	5 CFU/10 mL	Artificially contaminated milk.	<20 hours	Shao *et al.*, 2011
	Vibrio parahaemolyticus	10 CFU/reaction	Naturally contaminated seafood samples: fish, shrimp, and mussel.	16 hours	Wang *et al.*, 2013
	STEC O26, O45, O103, O111, O121, O145 and O157	1–20 cells/reaction in pure culture and 10^5–10^6 CFU/25 g in produce.	Artificially contaminated lettuce, spinach, and sprouts.	Not stated	Wang *et al.*, 2014
	L. monocytogenes	2.22×10^1 CFU/mL after 12 hours incubation. 2.22 CFU/mL after 24 hours incubation.	Artificially contaminated milk.	Not stated	Cho *et al.*, 2014
	L. monocytogenes	10^3 CFU/g	Artificially contaminated chicken, pork, ground beef, and milk powder.	2.5 hours	Wan *et al.*, 2012
	L. monocytogenes	8–10 cells per reaction tube	Artificially and naturally contaminated raw milk.	30 minutes	Wang *et al.*, 2011

(*Continued*)

Table 7.1 (Continued)

Detection Method	Foodborne Pathogens	Detection Limit	Food Matrix	Assay Time	References
Oligonucleotide DNA microarray	*Listeria monocytogenes*	8 logCFU/mL	Artificially contaminated milk.	Not stated	Bang *et al.*, 2013
	Escherichia coli, *Shigella* spp., *Salmonella* spp., *Proteus* sp., *Campylobacter jejuni*, *Listeria monocytogenes*, *Enterococcus faecalis*, *Yersinia enterocolitica*, *Vibrio parahaemolyticus*, *Vibrio fluvialis*, *β-hemolytic Streptococcus*, *Staphylococcus aureus*	10 CFU/mL of pure culture	Artificially and naturally contaminated pork, chicken, fish, and milk	Not stated	Huang *et al.*, 2014
NGS	*L. monocytogenes* (Whole genome)	Not stated	Deli turkey meat.	Not stated	Orsi *et al.*, 2008
		Not stated	Ready to eat meat.	Noted stated	Gilmour *et al.*, 2010
		Not stated	Quargel cheese.	Not stated	Rychli *et al.*, 2014

cytotoxicity, cardiotoxicity, and enterotoxicity in the host cell. This virulence gene is known to be pore-forming toxin due to its ability to form pores of ~2 nm in diameter on erythrocyte membrane (Matsuda *et al.*, 2010). Where else, the *trh* gene is a heat labile toxin and is reported to be immunologically similar to *tdh* gene (Honda *et al.*, 1988; Raghunath *et al.*, 2015). At present, *V. parahaemolyticus* isolates which are isolated from seafood and environmental samples are identified using PCR-based methods that amplify *toxR*, and both virulence genes; *tdh* and *trh* gene sequences (Panicker *et al.*, 2004; Yamamoto *et al.*, 2008; Paydar *et al.*, 2013; Malcolm *et al.*, 2015; Letchumanan *et al.*, 2015a; Letchumanan *et al.*, 2015c).

Recently, Li *et al.* (2016) developed a novel PCR-based approach for accurate identification of *V. parahaemolyticus* by targeting $bla_{CARB-17}$. Homologous analysis showed that $bla_{CARB-17}$-like genes are more conservative than the *tlh, toxR*, and *atpA* in identification of *V. parahaemolyticus*. In order to compare the uniqueness and specifity of the $bla_{CARB-17}$, the study used *tlh*, *atpA*, and *toxR* genes as well to detect *V. parahaemolyticus*. The results showed that PCR method based on $bla_{CARB-17}$ yeiled 100% specificity of *V. parahaemolyticus*. The PCR method based on *toxR* gene displayed similar specifity as the targeted $bla_{CARB-17}$.

7.2.1.3 *Escherichia coli*

Most of the PCR detection of *E. coli* is performed by targeting specific virulence genes responsible for their virulence factors. For instance, the *stx* genes (stx_1 and stx_2) that code for Shiga-like toxins, *eae* (*eaeA*) gene that codes for intimin, EHEC *hlyA* (*ehxA*) gene that codes for plasmid-encoded enterohemolysin or hemolysin and genes that code for heat-labile toxins (LTI, LTIIa, and LTIIb), heat-stable toxins (STI and STII), cytotoxic necrotizing factors (CNF1 and CNF2), enteroinvasive mechanisms (Einv), as well as enteroaggregative protein (Eagg) (Pollard *et al.*, 1990; Candrian *et al.*, 1991; Brian *et al.*, 1992; Gannon *et al.*, 1993; Meng *et al.*, 1996; Bastian *et al.*, 1998; Pass *et al.*, 2000; Feng *et al.*, 2001; Lorenz *et al.*, 2013). Other than the virulence genes, the *uidA* gene that codes for β-glucuronidase has been targeted for the detection of fecal coliform bacterium *E. coli* by PCR (Tsai *et al.*, 1993; Lorenz *et al.*, 2013). In an experiment conducted by Bej *et al.* (1991), a new method was developed for the detection of *E. coli* in water by PCR amplification of regions of the *uid* gene (*uidA* and *uidR*). This method was found to be sensitive and specific for *E. coli* detection. In the field of food microbiology, many studies have been conducted for the detection of diarrheagenic *E. coli* strains by PCR targeting their virulence genes because these strains are most often associated with food-borne diseases (Olsen *et al.*, 1995). Furthermore, specific detection of certain diarrheagenic *E. coli* strains such as STEC serotype O157:H7 can be done by PCR amplification of $rfbE_{O157}$ gene that codes for *E. coli* somatic antigen O157 and $fliC_{H7}$ that codes for *E. coli* structural flagella antigen H7, in addition to their virulence genes stx_1, stx_2 and *eaeA* (Wang *et al.*, 2002; Franz *et al.*, 2007).

7.2.1.4 *Salmonella*

PCR assays have improved the sensitivity of detection and identification of *Salmonella*. A PCR assay was developed by Aabo *et al.* (1995) to detect *Salmonella* in enrichment broth from minced meat and compared the results with cultural method. It was reported that the sensitivity of PCR was 92% whereas the cultural method was around 50%. There are several virulence genes of *Salmonella enterica* that express the virulence in the host. These virulence genes includes *invA*, *sef*, and *pef*, which are considered for adhesion and invasion of *Salmonella* in the host system and *spv* gene to cause systemic disease state in host cells. The *stn* virulence gene codes for enterotoxin production while both *sop* and *pip* genes are directly associated with expressions in host pathogenic processes (Murugkar *et al.*, 2003). The *invA* gene has reported to be widely utilized for the detection of *Salmonella* spp. in food samples (Chacon *et al.*, 2010). Dione *et al.* (2011) observed high prevalence (99.5%) of non-typhoidal *Salmonella* strains carrying *invA* gene. The study isolated the strains from humans, animals and food products in Gambia and Senegal. Marathe *et al.* (2012) developed a PCR to detect *hilA* gene of *Salmonella* strains. The *hilA* gene was reported to be conserved in all *Salmonella* serovars. ABD-Elghany *et al.* (2015) utilized PCR targeting three genes; *invA*, *stn*, and *spvC* genes to detect and identify *Salmonella* spp. strains. The study identified 166 isolates biochemically as *Salmonella*, and confirmed by PCR, based on the presence of *invA* and *stn* genes. However, the *spvC* gene was only detected in 42/166 of the isolates.

7.2.2 Multiplex PCR (mPCR)

Multiplex PCR offers a more rapid detection as compared to simple PCR through the simultaneous amplification of multiple gene targets. Multiplex PCR is a variant of simple PCR in which multiple gene targets are simultaneously amplified by using several sets of specific primers in a single reaction (Liu *et al.*, 2007). Primer design is very important for the development of mPCR, as the primer sets should have similar annealing temperature in order to produce a successful mPCR assay (Cheah *et al.*, 2008; Zhao *et al.*, 2014). Besides, the concentration of primers is also important in mPCR. This is because interaction may occur between the multiple primer sets in mPCR that results in primer dimers, thus, the concentration of primers may need to be adjusted to ensure the production of reliable PCR products (Zhao *et al.*, 2014). Other important factors for a successful mPCR assay include the PCR buffer concentrations, the balance between magnesium chloride and deoxynucleotide concentrations, the quantities of DNA template, cycling temperatures, and Taq DNA polymerase (Markoulatos *et al.*, 2002; Cheah *et al.*, 2008; Khoo *et al.*, 2009). Previously, mPCR was used to detect around two to three pathogens only. Now, mPCR is more advanced and it can detect up to five or more pathogens simultaneously. Multiplex PCR is employed for simultaneous detection of *L. monocytogenes*, *Salmonella* spp., *E. coli* 0157:H7, *V. parahaemolyticus*, *Staphylococcus aureus*, *Bacillus cereus*, as well as *Campylobacter jejuni* in various food samples (Lawrence and Gilmour, 1994; Gilbert *et al.*, 2003; Jofré *et al.*, 2005; Germini *et al.*, 2009; Kumar *et al.*, 2009; Yuan *et al.*, 2009; Zhang *et al.*, 2009; Zarei *et al.*, 2012; Yang *et al.*, 2013).

7.2.2.1 *Listeria monocytogenes*

A novel mPCR assay developed by Ryu *et al.* (2013) is able to detect and discriminate six *Listeria* species simultaneously in one tube with high accuracy. The *Listeria* species that were successfully identified are *L. monocytogenes*, *L. grayi*, *L. ivanovii*, *L. innocua*, *L. welshimeri*, and *L. seeligeri*. The mPCR detection limit was 7.58×10^4 copies for mixed genomic DNA. Multiplex PCR is also capable of detecting multiple virulence-associated genes of *L. monocytogenes* in a single PCR mixture. Hence, the possible failure in the detection of virulent *L. monocytogenes* can be prevented (Cooray *et al.*, 1994). In the study conducted by Cooray *et al.* (1994), *L. monocytogenes* in milk samples was successfully detected by mPCR with primers targeting three virulence-associated genes, *prfA*, *hlyA*, and *plcB*. Liu *et al.* (2007) had developed an mPCR assay targeting *inlA*, *inlC*, and *inlJ* genes for the rapid species-specific and virulence-specific determination of *L. monocytogenes*. Furthermore, the major *L. monocytogenes* serovars such as 1/2a, 1/2b, 1/2c, and 4b can be differentiated by mPCR targeting marker genes *Imo0730*, *Imo1118*, ORF2819, and ORF2110 (Doumith *et al.*, 2004; Hamdi *et al.*, 2007; Erol and Ayaz, 2011).

7.2.2.2 *Vibrio parahaemolyticus*

Rapid detection and identification of pathogenic *V. parahaemolyticus* are vital from a public health viewpoint. Therefore, many multiplex PCR assays have been developed for this purpose. Bej *et al.* (1999) reported a multiplex PCR protocol for amplification of *tlh*, *tdh*, and *trh*, which could be employed for detecting total and virulent *V. parahaemolyticus* in shellfish. The outcome of results detected *tlh* gene in all 111 strains of *V. parahaemolyticus* isolated from clinical, seafood, environmental, and oyster plants with sensitivity for detecting all three genes of at least 1–10 cells per gram of alkaline peptone water enriched sample homogenate. Many multiplex PCR protocols aiming the *toxR*, *tlh*, *tdh*, *trh*, and *fla* genes have been utilized to detect total and pathogenic *V. parahaemolyticus* from both clinical and environmental samples (Rosec *et al.*, 2009; Izumiya *et al.*, 2011; Wang *et al.*, 2011; Hossain *et al.*, 2013). The *fla* gene, which encodes polar flagellin, is an important marker to *Vibrio spp.*, such as *V. parahaemolyticus*, *V. anguillarum*, *V. cholerae*, and *V. vulnificus* (Kim and McCarter, 2000).

Kaufman *et al.* (2004) used PCR on samples of oyster mantle fluid, rather than homogenized meat, and reported that *V. parahaemolyticus* levels in the mantle fluid were highly correlated to levels in oyster tissues. Hossain *et al.* (2013) utilized multiplex PCR targeting species-specific marker *groEL* and the toxin genes *tdh* and *trh* genes in *V. parahaemolyticus* from shellfish seawater samples. In addition, multiplex PCR assays have been very popular and are utilized to differentiate *V. parahaemolyticus*, *V. cholerae*, and *V. alginolyticus* from each other (Di Pinto *et al.*, 2005; Wei *et al.*, 2014). Recently, a serogroup-*O*-specific PCR assay was utilized to detect and identify *V. parahaemolyticus* pathogens in clinical and environmental samples (Chen *et al.*, 2012b).

7.2.2.3 *Escherichia coli*

The use of mPCR for the detection of *E. coli* is very common and it often involves the amplification of more than one virulence genes in an assay. The six categories of DEC strains, which are the EPEC, ETEC, EIEC, EAEC, DAEC, and EHEC, can be detected and identified simultaneously by mPCR as demonstrated in a study carried out by Vidal *et al.* (2005). In the study, mPCR assay was designed to detect all six categories of DEC by targeting their virulence genes, which include stx_1, stx_2, and *eae* genes from STEC, *eae*, and *bfp* genes from EPEC, *lt* and *stII* genes from ETEC, *virF* and *ipaH* genes from EIEC, *daaE* gene from DAEC, and *aafII* gene from EAEC. The simultaneous detection of all categories of DEC by the mPCR assay was proven possible and highly specific.

The presence of pathogenic *E. coli* in food samples is a major public health concern. Balagué *et al.* (2006) had analyzed ready-to-eat food samples collected from supermarkets and shops in Argentina for the presence of *E. coli*. The ready-to-eat food samples included were soft cheese, cottage cheese, salads with cream or mayonnaise, food with sauces, and sandwiches (Balague *et al.*, 2006). The mPCR assay was performed by targeting stx_1, stx_2, *eaeA*, *CNF1*, *CNF2*, *Einv*, *LTI*, *STI*, and *STII* genes. The results showed that STEC strains belonged to serogroups O8, O18, O57w, O79, O44, and O128 as well as three other non-typeable strains were present in soft cheese, cottage cheese, chicken with sauce, and vegetables mayonnaise. Multiplex PCR is also capable of simultaneous detection of *E. coli* and other foodborne pathogens. Chen *et al.* (2012a) had carried out mPCR for the simultaneous detection of *E. coli* O157:H7, *Salmonella* Enteritidis, *Staphylococcus aureus*, *Shigella flexneri*, and *L. monocytogenes* using five pairs of primers targeting invasion protein (*invA*), 16S rDNA, invasion plasmid antigen H (*ipaH*), listeriolysin O (*hlyA*), and intimin (*eaeA*) gene, respectively. Lee *et al.* (2014) had carried out mPCR assay that can simultaneously detect *E. coli* O157:H7, *L. monocytogenes*, *B. cereus*, *V. parahaemolyticus*, *Salmonella* spp., and *S. aureus* in ready-to-eat food samples (e.g., sandwich, spicy chicken, kimbab, steamed pork hocks, oyster, sunsik, fresh tomato juice, sushi, and salad) and fresh lettuce. The mPCR involved the use of six primer pairs targeting the *rfbE* gene in *E. coli* O157:H7, *hly* gene in *L. monocytogenes*, *gyrB* gene in *B. cereus*, *toxR* gene in *V. parahaemolyticus*, *invA* gene in *Salmonella* spp., and *nuc* gene in *S. aureus*. The mPCR assay was able to detect a minimum of 1 CFU/mL of each pathogen after 12 hours enrichment.

7.2.2.4 *Salmonella*

Over the past years, multiplex PCR has been utilized to detect and identify Salmonella in poultry samples (Kim *et al.*, 2006; Cortez *et al.*, 2006; O'Regan *et al.*, 2008; Cheah *et al.*, 2008; Park *et al.*, 2009; Salem *et al.*, 2010; Park *et al.*, 2011). A study by Skyberg *et al.* (2006) developed a multiplex PCR to detect 17 virulence genes of *Salmonella spp.* The genes tested (*nvA, orgA, prgH, invJ, tolC, sipB, pagC, msgA, spiA, sopB, IpfC, pefA, spvB, and sifA*) were associated with *Salmonella* invasion, fimbrial production, toxin production, iron transport, and intramacrophage survival were targeted by three different PCR reactions. Eleven of the 17 genes tested (*invA, orgA, prgH, tolC, invJ, sipB, sitC, pagC, msgA, spiA*, and *iroN*) were found in all of the isolates. Another (*sopB*) was present in all isolates from sick birds and all but one isolate from healthy birds. The remaining five genes (*lpJC, cdtB, sifA, pefA*, and *spvB*) were found in 10–90% of the isolates from sick birds and 3.75–90% of the healthy birds. The study concluded that these virulence genes are widely distributed among *Salmonella* isolates of birds, regardless of whether their host origin have salmonellosis. Cortez *et al.* (2006) developed a multiplex to detect *Salmonella* spp., *Salmonella* Enteritidis, and *Salmonella* Typhimurium in chicken abattoirs targeting the genes *invA* (invasion gene of the genus *Salmonella*), *sefA* (fimbrial antigen of *Salmonella* Enteritidis), and *pefA* (plasmid-encoded fimbria of *Salmonella* Typhimurium). They concluded that the multiplex PCR is sensitive with 29 (10%) were positive for *Salmonella* spp., 16 (5.6%) for *Salmonella* Enteritidis, and 7 (2.4%) for *Salmonella* Typhimurium.

Kim *et al.* (2006) developed two five-plex PCR assays to differentiate the 30 most prevalent *Salmonella* serotypes in the US. The study used primer pairs targeting six genetic loci from *Salmonella* Typhimurium and four from *Salmonella typhi*. In a recent study, Salem *et al.* (2010) also developed two five-plex assays to detect *Salmonella* in Tunisia. Mezal *et al.* (2013) developed a multiplex PCR assay to detect 17 virulence genes (*spvB, spiA, pagC, msgA, invA, sipB, prgH, spaN, orgA, tolC, iroN, sitC, IpfC, sifA, sopB, cdtB*, and *pefA*) in 50 *Salmonella enterica* serovar Javiana isolates that were isolated from food, environmental and clinical samples. From the study, all the isolates were positive for 9 to 14 genes tested but none were positive for *pefA, spvB*, and *lpfC*, which are typically present on the *Salmonella* virulence plasmid. All isolates were positive for cytolethal distending toxin *cdtB* gene that is considered as one of the typhoid toxins of *Salmonella typhi* and is not present in the majority of *Salmonella enterica* serovars. They concluded that the detection of *cdtB* gene in several non-typhoidal *Salmonella* serovars, like *Salmonella enterica* serovar Javiana may contribute to pathogenesis of disease.

Further improvements of mPCR include the development of a novel GeXP-PCR by Zhou *et al.* (2013) for the simultaneous detection of six foodborne bacterial pathogens: *Salmonella enterica, Listeria monocytogenes, Staphylococcus aureus, Escherichia coli* O157:H7, *Shigella* spp., and *Campylobacter jejuni*. GenomeLab Gene Expression Profiler (GeXP) genetic analysis system allows high-throughput and detection of multiple pathogens in a single reaction. The analytical procedure involves primer design, PCR amplification, and capillary electrophoretic separation of PCR products instead of agarose gel electrophoresis. The GeXP multiplex PCR amplification involves the use of chimeric primers, universal primers, and capillary electrophoretic separation of PCR products instead of agarose gel electrophoresis. The chimeric primers contain a gene-specific sequence with a universal tag at the 5′ end and they are used to produce amplicons with universal tags. Then, the universal primers that contain the same sequence as the universal tags used in the chimeric primers will drive the remaining PCR reactions. The forward universal primer is covalently labeled with a fluorescent dye at the 5′ end and it is used for the detection during capillary electrophoresis (Zhou *et al.*, 2013). This method was found to be more sensitive and suitable for high-throughput analysis. The detection limit of GeXP-PCR for *Salmonella*, *Listeria monocytogenes*, *Staphylococcus aureus*, *Escherichia coli* O157:H7, *Shigella* spp., and *Campylobacter jejuni* was 420, 310, 270, 93, 85, and 66 CFU/mL respectively. Additional examples of the application of mPCR for the detection of foodborne pathogens in various food matrices are listed in Table 7.1.

7.2.3 Real-Time or Quantitative PCR (qPCR)

Real-time PCR or quantitative PCR is different from simple PCR, whereby it does not require agarose gel electrophoresis for the detection of PCR products. This method is able to monitor the PCR products formation continuously in the entire reaction by measuring the fluorescent signal produced by specific dual-labeled probes or intercalating dyes. The fluorescence intensity is proportional to the amount of PCR amplicons (Omiccioli *et al.*, 2009; Zhao *et al.*, 2014). Multiplex qPCR assay is also developed for the detection and quantification of multiple foodborne pathogens. Furthermore, commercial qPCR kits are produced due to the advantages offered by qPCR.

Several fluorescent systems have been developed for qPCR and the most commonly used fluorescent systems include SYBR green, TaqMan probes, and molecular beacons. SYBR green is a double-stranded DNA (dsDNA)-binding fluorescent dye (Hein *et al.*, 2001b). This non-sequence-specific intercalating dye emits little fluorescence and the fluorescence signal is enhanced when bound to the minor groove of the DNA double helix (Fukushima *et al.*, 2003; Levin, 2005; Singh *et al.*, 2009) (Figure 7.2).

TaqMan probes and molecular beacons are the common alternatives to SYBR green. TaqMan probes, also known as double-dye probes, are oligonucleotides that contain a fluorophore as the reporter dye at the 5′-end and the quenching dye at the 3′-end (Hein *et al.*, 2001b; Levin, 2005). The reporter dye and the quenching dye are close to each other and this prevents the emitted fluorescence of the

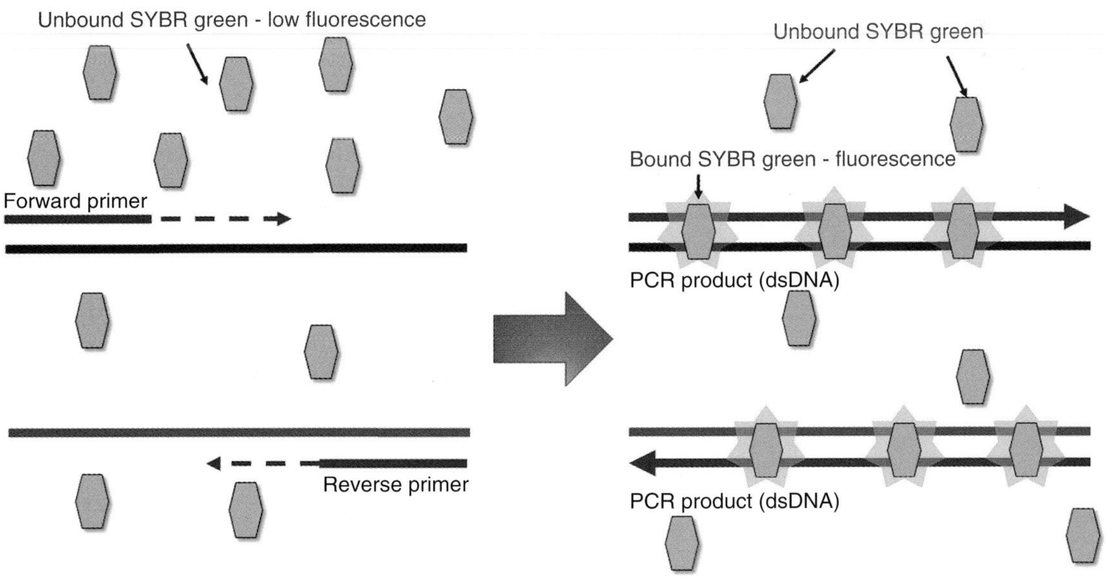

Figure 7.2 SYBR Green Fluorescent Dye Employed in qPCR. During amplification, the unbound SYBR green dye molecules emit little fluorescence. After the annealing of primers, the SYBR green dye molecules are able to bind to the double-stranded DNA and this will result in the increase of fluorescence signal. (*See insert for color representation of the figure.*)

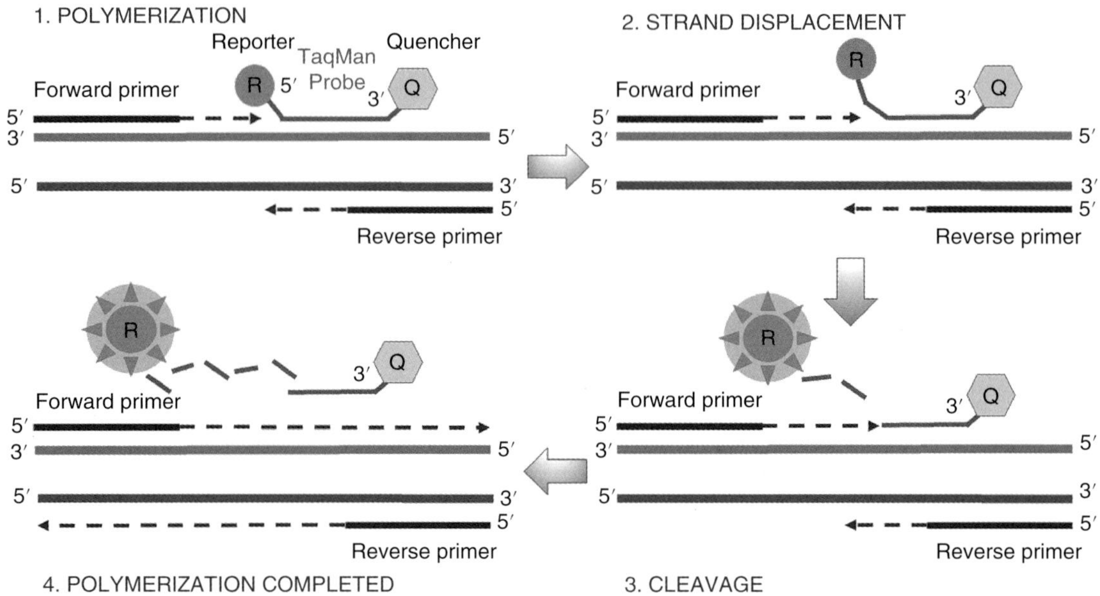

Figure 7.3 Overview of the Process Occurs to a TaqMan Probe During Extension Phase of qPCR. The TaqMan probe contains a reporter dye on the 5′end and a quencher dye on the 3′end. In this state, the fluorescence emitted by the reporter dye is prevented due to the close proximity of the quencher dye. During polymerization, the TaqMan probe binds to one of the strands of DNA between the two PCR primers. As the primer extended, the *Taq* DNA polymerase encounters the probe and cleaves the probe. The cleavage of probe separates both reporter dye and quencher dye. The separation of reporter dye from quencher dye has resulted in an increase in reporter fluorescence. The cleavage of probe also removes the probe from the target strand and this allows primer extension to continue until polymerization is completed. (*See insert for color representation of the figure.*)

reporter (Levin, 2005). TaqMan probe is complementary to a specific nucleotide sequence in one of the strands of amplicon internal to both primers and the system depends on the 5′-3′ exonuclease activity of *Taq* DNA polymerase that cleaves the probe and separates both dyes in order to generate the fluorophore signal (Patel *et al.*, 2006) (Figure 7.3).

Molecular beacons are oligonucleotide probes with hairpin/stem-and-loop configuration, in which the sequence complementary to a target sequence is present in the loop portion and the stem is produced by annealing of two complementary arm sequences. A reporter dye is attached to one end of the probe and a quencher dye is attached to the other end. Both dyes are in close proximity, which maintained by the hybrid stem, hence, no fluorescence is produced (Leone, 1998; Chen *et al.*, 2000; Levin, 2005; Patel *et al.*, 2006). Molecular beacon produces fluorescence signal upon hybridization of the probe to its complementary nucleotide sequence in the amplicon. During hybridization, the probe undergoes spontaneous conformational change that separates the two dyes and this allow fluorescence to occur (Leone, 1998; Chen *et al.*, 2000; Liming and Bhagwat, 2004; Levin, 2005) (Figure 7.4).

Among the discussed fluorescent systems for qPCR, SYBR green is simple and less costly as compared to TaqMan probes or molecular beacons (Fukushima *et al.*, 2003; Levin, 2005). Besides, the advantage of using SYBR green is that the DNA melting curve can be generated after PCR along with the calculation of the Tm value of the amplified products. This is required in order to differentiate the target products from the primer dimer formation. The primer

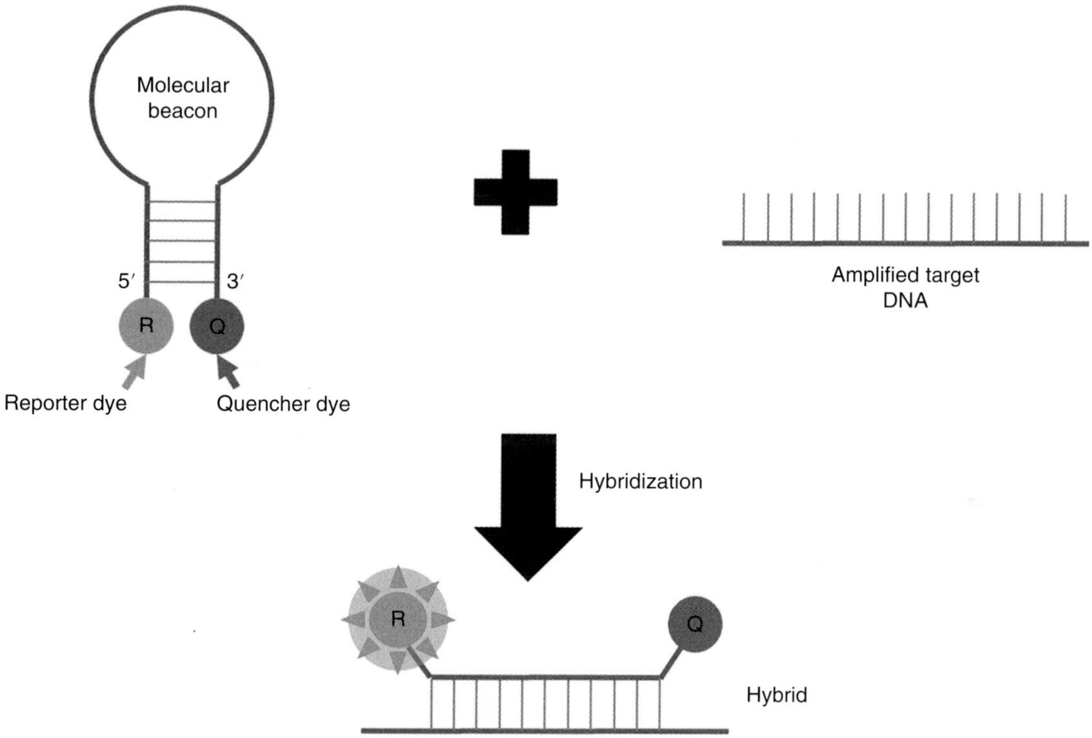

Figure 7.4 Molecular Beacon Probe with Hairpin/Stem-and-Loop Configuration. Fluorescence is not produced due to the close proximity of both reporter dye and quencher dye in the hairpin/stem-and-loop structure of molecular beacon. The loop portion of the molecular beacon will hybridize to the target sequence of DNA during the annealing step of PCR. This causes the molecular beacon to undergo a spontaneous confirmation change that separates reporter dye from the quencher dye. Hence, reporter fluorescence can be detected. (*See insert for color representation of the figure.*)

dimer products have lower Tm values as compared to the amplicons (Levin, 2005). SYBR green lacks specificity and it binds to all double-stranded DNA, thus, it can be used to detect any PCR product (Levin, 2005; Madani *et al.*, 2005). However, SYBR green dye will bind to other non-specific reaction products, which include primer dimers (Madani *et al.*, 2005). As for TaqMan probes and molecular beacons, they are sequence-specific probes and they only bind to their target sequence, hence, primer dimers will not be detected (Levin, 2005). There are some studies have shown that TaqMan-based qPCR is more sensitive compared to SYBR green or molecular beacons-based qPCR (Hein *et al.*, 2001b; Klerks *et al.*, 2004). Nevertheless, the sensitivity of PCR-based method is mainly affected by primer specificity, primer sequence and annealing temperature, rather than the choice of detection probe (Klerks *et al.*, 2004).

7.2.3.1 *Listeria monocytogenes*

Oravcová *et al.* (2005) had developed a real-time 5'-nuclease PCR targeting a sequence of the gene *actA* for the identification and quantification of *L. monocytogenes*. In this study, TaqMan probe was used for the detection and quantification of qPCR products. The successful detection of *L. monocytogenes* in fresh produce using molecular beacon-qPCR targeting the *hlyA* gene was first reported by Liming *et al.* (2004). Barbau-Piednoir *et al.* (2013) developed a combination of four qualitative SYBR Green qPCR assays for the detection and discrimination of

Listeria spp. and *L. monocytogenes* with high accuracy. In these assays, the *iap* and *prs* genes were targeted for detection of *Listeria* spp. and *hlyA* gene was targeted for detection of *L. monocytogenes.*

A novel 5′ exonuclease multiplex qPCR assay for the identification of six *Listeria* spp. including *L. monocytogenes, L. seeligeri, L. ivanovii, L. welshimeri, L. grayi,* and *L. innocua* was developed by Hage *et al.* (2014). In this study, two sets of triplex PCR were designed with one set identifying *L. seeligeri, L. welshimeri,* and *L. monocytogenes* and another set identifying *L. ivanovii, L. grayi,* and *L. innocua.* The *Listeria* species were differentiated by targeting their respective species-specific target genes and TaqMan probe was used to monitor the multiplex qPCR products.

The commercial qPCR kits available for the detection of *L. monocytogenes* are BAX® System Real-time PCR Assay *Listeria monocytogenes* (DuPont Qualicon), TaqMan® *Listeria monocytogenes* Detection Kit (Applied Biosystems), ADIAFOOD rapid pathogen detection system for *Listeria monocytogenes* (AES Chemunex), Probelia® *Listeria monocytogenes* PCR system (Biorad), LightCycler® *Listeria monocytogenes* Detection Kit (Roche/Biotecon), GeneVision® Rapid Pathogen Detection System for *Listeria monocytogenes,* CycleavePCR® *Listeria monocytogenes* (*inlA* gene) Detection Kit (TaKaRa Bio, Inc.) and iQ-Check *L. monocytogenes* kit (Bio-Rad Laboratories) (Liming *et al.*, 2004; Rodríguez-Lázaro *et al.*, 2004; Becker *et al.*, 2005; Janzten *et al.*, 2006; Maurer, 2011; Traunšek *et al.*, 2011; Liu *et al.*, 2012). The BAX® System Real-time PCR Assay *Listeria monocytogenes* is adopted by the USDA (2012) as a screening test for *L. monocytogenes.* Besides, Wan *et al.* (2003) conducted a study to compare the Probelia® *Listeria monocytogenes* PCR system and the International Organization for Standardization (ISO) method 11290-1 for the detection of *L. monocytogenes* in salmon samples. The results indicated that Probelia® *Listeria monocytogenes* PCR system is as good as the ISO method and the detection limit was approximately 20 CFU/mL broth culture for pure culture of *L. monocytogenes.*

7.2.3.2 *Vibrio parahaemolyticus*

The total and pathogenic *V. parahaemolyticus* in various seafood samples could be detected using multiplex real-time PCR along with different fluorescent probes (Ward and Bej, 2006; Nordstrom *et al.*, 2007; Tyagi *et al.*, 2009; Robert-Pillot *et al.*, 2010). Real-time PCR has the ability to process huge number of samples with speed and consistency in a single tube amplification targeting the gene (McKillip *et al.*, 2000). Ward and Bej (2006) developed a multiplex real-time PCR TaqMan assay that targeted four different genes and was capable of detecting total and pathogenic *V. parahaemolyticus,* including the pandemic O3:K6 serotype in shellfish. The gene targets included the *tdh* and *trh* genes (detection of pathogenic *V. parahaemolyticus*), ORF8 (detection of pandemic *V. parahaemolyticus* O3:K6) and *tlh* gene for the detection of total *V. parahaemolyticus.* Real-time PCR was successfully used to detect *tdh* gene (Blackstone *et al.*, 2003) and *tlh* gene (Kaufman *et al.*, 2004) in *V. parahaemolyticus* using specific primer sets and fluorogenic probes. Tyagi *et al.* (2009) developed a sensitive and highly reproducible SYBR green qPCR assay for the detection of pathogenic *tdh*-positive *V. parahaemolyticus* in tropical shellfish. The detection limit was 10^2 CFU/ml for shrimp homogenates spiked with pure culture.

A simultaneously detection of *V. parahaemolyticus, V. cholerae* and *V. vulnificus* by multiplex real-time SYBR green PCR using primers that target *vmrA, zot,* and *vuuA* genes respectively was developed by Kim *et al.* (2012). Garrido applied a multiplex real time PCR to detect pathogenic *V. parahaemolyticus* in water and food samples. The limits of detection for this method were 0.24 CFU/g for *tdh,* and 0.44 CFU/g for *trh1,* and 0.52 CFU/g for *trh2* (Garrido *et al.*, 2012). To date, many real-time

PCR assays used for the detection of pathogenic *V. parahaemolyticus* involve the use of Taqman probe based (Ward and Bej, 2006; Liu *et al.*, 2012; He *et al.*, 2014). Although there are many advantages of Taqman probe-based real-time PCR, the high cost of this method has limit the use of it in routine testing (Chen *et al.*, 2012a). He *et al.* (2014) developed an EvaGreen-based multiplex real time PCR assay for the detection of pathogenic *V. parahaemolyticus* targeting *toxR*, *tdh* and *trh* genes. The assay could unambiguously identify pathogenic *V. parahaemolyticus* in samples with a minimum detection limit of 1.4 pg genomic DNA per reaction. The study concluded that the newly developed multiplex real-time PCR assay is cost effective, practical, specific and sensitive with a high-throughput detection of pathogenic *V. parahaemolyticus* (He *et al.*, 2014).

7.2.3.3 *Escherichia coli*

A molecular beacon qPCR was employed by McKillip and Drake (2000) for the detection of *E. coli* O157:H7 in artificially contaminated skim milk, using primers that amplify region within the shiga-like toxin II (*sltII* or stx_2) gene. In addition, detection of *E. coli* O157:H7 can be carried out by multiplex qPCR with primers targeting several virulence genes. Jothikumar and Griffiths (2002) had designed a SYBR Green multiplex qPCR assay that detects *E. coli* O157:H7 by using primers that amplify the stx_1 and stx_2 genes of the organism simultaneously.

Besides, *E. coli* of different serogroups can be detected and differentiated by multiplex qPCR. For example, Fratamico *et al.* (2011) performed a study to detect Shiga toxin-producing *Escherichia coli* serogroups O26, O45, O103, O111, O121, and O145 in ground beef using multiplex qPCR with primers which target Shiga toxins (stx_1 and stx_2), intimin (*eae*), and *wzx* genes. Taqman probe with different combinations of fluorophores were used to detect the product of multiplex qPCR with detection limit around 50 CFU per PCR. Multiplex qPCR assay is also used for the detection and identification of the different categories of DEC. In the study carried out by Yang *et al.* (2007), a multiplex qPCR was developed for the detection of four categories of DEC strains (EPEC, EHEC, ETEC, and EIEC) by using primers targeting six virulence genes *lt, st, eae,* stx_1, stx_2, and *ipaH*. The detection limit was 10 CFU per assay and the overall assay time was 1.5 hours. Multiplex qPCR enables simultaneous detection of *E. coli* and other foodborne pathogens. Hu *et al.* (2014) conducted a study to detect eight foodborne bacterial pathogens which include *E. coli* O157, *Salmonella enterica* subsp. *enterica*, *Listeria monocytogenes*, *V. parahaemolyticus*, *V. vulnificus*, *Campylobacter jejuni*, *Enterobacter sakazakii*, and *Shigella* spp. by using molecular beacon based multiplex qPCR targeting *rfbE, ssaR, hlyA, toxR, vvh, gyrA, 16SrRNA*, and *ipaH* genes respectively. The detection limits of the assay ranged from $1.3 \times 10^3 - 1.6 \times 10^4$ CFU/g stool. Commercial qPCR kits can be used for the detection of *E. coli* and these include *E. coli* O157 BAX® PCR (DuPont Qualicon), and ADIAFOOD rapid pathogen detection system for *E. coli* O157 and *E. coli* O157:H7 (AES Chemunex) (Maurer, 2011).

7.2.3.4 *Salmonella*

Real-time PCR has been widely used to detect, identify and quantify *Salmonella* from poultry samples (Catarame *et al.*, 2006; Wolffs *et al.*, 2006; Kundinger *et al.*, 2008; Taha *et al.*, 2010). The detection of *Salmonella* in fresh-cut fruits and vegetables by molecular beacon qPCR targeting the invasion associated gene (*iagA*) was first reported by Liming and Bhagwat (2004). The detection limit was approximately 4 CFU/25 g of produce after enrichment. Maurischat *et al.* (2015) developed a multiplex real time PCR to detect and differentiate *Salmonella enterica* subsp. *enterica* Enteritidis, Typhimurium, and its monophasic variant. The study used Taqman probes targeting S. Enteritidis (safA), S. Typhimurium (fliA-IS200), Biphasic STM (fljB-hin), Biphasic STM (hin-ironB), and IAC (pUC18/19). The detection probability reached 100% between 3-5 CFU/25 g sample and the researchers were able to differentiate strains to *S. enterica subsp. enterica* serovars Enteritidis and Typhimurium. A multiplex melt real-time PCR assay was developed for the detection of virulent and antibiotic resistance strains of *Salmonella* using two sets of primers. The first set of primers target the virulence gene, invasion (*invA*), and tetracycline (*tetG*), streptomycin (*aadA2*), and sulphonamide (*sulI*) antibiotic resistance genes, and the second set amplifies ampicillin (*blaPSE, blaTEM*) and chloramphenicol (*floR*) resistance genes (Singh and Mustapha, 2014).

There are many examples of commercial qPCR kits for the detection of *Salmonella* which include *Salmonella* BAX™ PCR (DuPont Qualicon), AnDiaTec® *Salmonella* real-time PCR Kit (AnDia Tec), Probelia™ *Salmonella* sp. (Sanofi Diagnostics Pasteur), and TaqMan™ *Salmonella* detection kit (PE Applied Biosystems) (Maciorowski, 2005; Maurer, 2011; Park *et al.*, 2014). Kimura *et al.* (1999) evaluated the TaqMan™ *Salmonella* detection kit for the detection of *Salmonella* in shrimp and meat. The detection limit was 120 CFU/mL in pure culture. Wan *et al.* (2000) evaluated the Probelia™ *Salmonella* sp. PCR amplification and detection kits for the detection of *Salmonella* Agona in artificially contaminated milk powder and ricotta cheese. The detection limit was approximately 8–79 CFU/mL. Besides, Margot *et al.* (2013) evaluated seven commercial qPCR kits for the detection of *Salmonella*, which include BAX® system Q7 real-time PCR (DuPont Qualicon) and other commercial qPCR kit such as ADIAFOOD® *Salmonella* (AES chemunex), BIOTECON foodproof *Salmonella* Detection Kit (Biotecon Diagnostics), BioControl Assurance GDS® TM *Salmonella* (BioControl), Genedisc® Shiga Toxic *E. coli* and *Salmonella* spp. (Pall GeneDisc® Technologies), BioRad iQ-Check® *Salmonella* 2 (Biorad), and MicroSeq® *Salmonella* spp. Detection Kit (Applied Biosystems). A total of 49 *Salmonella* strains were included in this study and correctly identified by all the evaluated systems.

Conventional PCR and multiplex PCR that require agarose gel analysis for the detection of PCR products are laborious and time-consuming, thus, not suitable for high-throughput analysis and difficult to automate (Patel *et al.*, 2006; Zhou *et al.*, 2013). As for qPCR, post-PCR processing is not required. This allows low risk of cross-contamination, high-throughput analysis and automation (Fricker *et al.*, 2007; Omiccioli *et al.*, 2009). Overall, qPCR is more sensitive than conventional PCR. More examples of the detection of foodborne bacterial pathogens by qPCR are presented in Table 7.1.

There is no doubt that PCR-based detection methods are rapid, highly sensitive and specific. Nonetheless, thermocycling system is needed by these methods. Alternative methods have been developed for the amplification of nucleic acids under isothermal conditions. Two of the most commonly used isothermal nucleic acid amplification methods for the detection of foodborne pathogens are LAMP and NASBA (Law *et al.*, 2015b).

7.2.4 Nucleic Acid Sequence-Based Amplification (NASBA)

NASBA is developed by Compton (1991) in the early 1990s and it operates by amplifying nucleic acids under isothermal conditions, unlike PCR, which requires thermocycling system. NASBA is normally used for the amplification of RNA whereby the single-stranded RNA template is converted into complementary DNA (cDNA) by the reverse transcriptase during the reaction. NASBA reaction occurs at around 41 °C, involving two specific primers and three enzymes: avian myeloblastosis virus (AMV) reverse transcriptase, T7 RNA polymerase, and RNase H. The NASBA amplicons can be detected by agarose gel electrophoresis or enzyme-linked gel assay, however, these post-NASBA product detection methods are considered labor-intensive and not cost-effective (Leone, 1998; Zhao *et al.*, 2014). This leads to the development of a novel real-time NASBA which uses fluorescently labeled probes, which are molecular beacons to detect the single-stranded RNA amplicons, thus, producing a homogenous NASBA assay (Leone, 1998). Real-time NASBA has been used for the detection of various foodborne pathogens such as *Salmonella enterica*, *Vibrio cholerae*, *Staphylococcus aureus*, *Campylobacter jejuni* and *Campylobacter coli* (Simpkins *et al.*, 2000; Churruca *et al.*, 2007; Fykse *et al.*, 2007; O'Grady *et al.*, 2009).

Real-time NASBA is able to detect viable microorganisms that present in food samples through mRNA amplification and the detection of RNA targets will indicate the presence of viable cells (Simpkins *et al.*, 2000). Real-time NASBA has been used to distinguish viable from non-viable bacterial cells and RNase treatment is usually required to degrade target mRNA from dead cells before nucleic acid extraction or treating the samples with RNase-free DNase is required prior to performing the NASBA assay (Blais *et al.*, 1997; Nadal *et al.*, 2007; Dwivedi and Jaykus, 2011).

7.2.4.1 *Listeria monocytogenes*

A highly specific NASBA system was developed by Blais *et al.* (1997) for the detection of *L. monocytogenes* with primers targeting the *hlyA* mRNA sequences. The NASBA system was capable of detecting low numbers of *L. monocytogenes* (<10 CFU/g) in artificially contaminated dairy and egg products after 48 hours enrichment period. Nevertheless, false-positive results were reported and the researchers suggested that the reason for this could be due to cross-contamination of NASBA reactions with amplicons from previous amplifications performed at the same site. A molecular beacon based real-time NASBA assay for the detection of *L. monocytogenes* in cooked ham and smoked salmon slices was first described by Nadal *et al.* (2007). A sequence from the mRNA transcript of *hly* gene was used as a target for this assay and the detection limit of this assay for *L. monocytogenes* was 400 CFU/mL.

7.2.4.2 *Vibrio parahaemolyticus*

A study conducted by Xin *et al.* (2011) utilized nucleic acid sequence based amplification (NASBA) to detect and identify *V. parahaemolyticus*. The study design specific primers and probes for the *tlh* target gene of *Vibrio parahaemolyticus*. The results showed that the sensitivity of the NASBA assay was 5.1 × 102 cfu/mL, which was higher than that of PCR method without any cross reaction with other bacteria species.

7.2.4.3 *Escherichia coli*

Min and Baeumner (2002) developed a highly sensitive and specific NASBA assay for the detection of viable *E. coli* in drinking water. The detection of viable *E. coli* by NASBA was based on the amplification of target sequence from *clpB* mRNA which codes for heat shock protein. The amplicons were then quantified by electrochemiluminescence (ECL) analysis. No false-positive result, which might result due to other microorganisms or non-viable *E. coli* cells, was reported in this study. The detection limit of this NASBA assay was approximately 40 *E. coli* cells/mL and the assay time was 4 hours. The detection of viable *E. coli* in water has been further improved by using real-time NASBA. Heijnen and Medema (2009) developed a rapid molecular beacon based real-time NASBA for the detection of *E. coli* in water samples such as tap water, surface water, and treated sewage water samples. NASBA primers targeting a fragment of *clpB*-mRNA were used in this assay to detect *E. coli*. The sensitivity of the NASBA assay was approximately 1 CFU $(100\,mL)^{-1}$ and the assay time was approximately 3–4 hours.

In some cases, NASBA does not necessarily detect viable cells. According to Birch *et al.* (2001), the mRNA sequences in *E. coli* could be detected by NASBA for up to 30 hours after heat killing. The study also compared three detection techniques, PCR, Reverse Transcription PCR (RT-PCR), and NASBA for their ability to amplify mRNA or DNA targets in *E. coli* after cell death. The results suggested that NASBA was the most sensitive among the three techniques tested.

7.2.4.4 *Salmonella*

D'Souza *et al.* (2003) developed a highly sensitive and specific NASBA assay for the detection of *Salmonella enterica* from foods.

The detection of *Salmonelle* Enteritidis by NASBA was based on the amplification of target mRNA sequences of the *dnaK* gene. The amplicons were then quantified using electrochemiluminescence (ECL) analysis. The detection limit of this NASBA assay was 101 CFU of *Salmonella enterica serovar* Enteridis per amplification reaction and the assay time was 18 hours. Cook *et al.* (2002) used the similar protocol of NASBA to detect *Salmonella* Enteritidis in liquid whole egg. The finding detection limits of the inoculum levels was low as 2.8 CFU per 25 g sample after 16 h of pre-enrichment. Simpkins *et al.* (2000) developed a rapid NASBA assay to detect viable *Salmonella enterica*. The target was mRNA transcribed from the bacterial *Hsp70* homologue *dnaK* gene.

NASBA offers high-throughput analysis and it has been commercialized as kits. There are several commercial NASBA kits manufactured by Life Sciences, KIT Biomedical Research, Gen-Probe and bioMérieux (Gracias and McKillip, 2007). However, the commercial NASBA kit that is used for the detection of foodborne bacterial pathogens is mainly the Nuclisens EasyQ® Basic Kit (bioMérieux). Nuclisens EasyQ® Basic Kit can be used for the detection and identification of *L. monocytogenes*, *Salmonella enterica*, *E. coli*, and *V. cholerae* (D'Souza and Jaykus, 2003; Fykse *et al.*, 2007; Nadal *et al.*, 2007; Heijnen and Medema, 2009). More examples of the application of NASBA in food samples are shown in Table 7.1.

7.2.5 Loop-Mediated Isothermal Amplification (LAMP)

LAMP is a novel nucleic acid amplification method developed by Notomi *et al.* (2000), which provides a rapid, sensitivity, and specific detection of foodborne pathogens. LAMP is based on auto-cycling strand displacement DNA synthesis carried out by *Bst* DNA polymerase large fragment under isothermal conditions between 59°C and 65°C for 60 minutes. In LAMP, four primers comprising two inner primers and two outer primers are used to target six specific regions of target DNA. Cauliflower-like DNA structures bearing multiple loops as well as stem-loop DNAs of different sizes are the final products of LAMP. Large amount of amplicons can be produced by LAMP within 60 minutes, which is usually 10^3-fold or higher as compared to simple PCR. The LAMP amplicons can be detected by agarose gel electrophoresis or SYBR Green I dye (Wang *et al.*, 2008; Zhao *et al.*, 2014). Besides, different types of LAMP assays have been developed for the detection of foodborne pathogens. For instance, multiplex LAMP, reverse-transcription LAMP, real-time LAMP, and *in situ* LAMP (Chen *et al.*, 2008; Han and Ge, 2010; Shao *et al.*, 2011; Ye *et al.*, 2011).

LAMP has been used for the detection of various foodborne pathogens due to its rapidity and sensitivity. LAMP is proven to be more specific and sensitive as compared to conventional PCR assays for the detection of foodborne pathogens (Hara-Kudo *et al.*, 2005; Wang *et al.*, 2008; Yamazaki *et al.*, 2008). For example, Tang *et al.* (2011) conducted a sensitive and specific LAMP assay for the detection of *L. monocytogenes* with primers that target the *hlyA* gene region. In this study, the LAMP assay was evaluated against conventional PCR method for the detection of *L. monocytogenes* in food. The results indicated that LAMP assay was 100 times more sensitive than the conventional PCR assay. This is because LAMP uses four primers targeting six specific regions and it provides rapid amplification, greater yield of amplification products and lower detection limits than conventional PCR assays (Hara-Kudo *et al.*, 2005; Xu *et al.*, 2012).

To date, the availability of real-time monitoring of LAMP amplification products by the presence of turbidity or fluorescence eliminates the need for staining with ethidium bromide and gel electrophoresis. Therefore, this allows high-throughput analysis along with its high sensitivity and specificity (Yang *et al.*, 2010). Commercial LAMP kits are available for the detection of *Listeria*, *Salmonella*, *Campylobacter*, *Legionella*, and verotoxin-producing *Escherichia coli* (Mori and Notomi, 2009). The Loompamp® detection kit (Eiken Chemical Co., Ltd) is commercially available for the detection of foodborne pathogens such as *Salmonella enterica* (Ohtsuka *et al.*, 2005), *L. monocytogenes* (Wang *et al.*, 2015), *Shigella* (Song *et al.*, 2005), enteroinvasive *E. coli* (Song *et al.*, 2005), verotoxigenic *E. coli* O157 and O26 (Hara-Kudo *et al.*, 2008), and *Campylobacter* (Yamazaki *et al.*, 2009). More examples of the application of LAMP in food samples are shown in Table 7.1

7.2.5.1 *Listeria monocytogenes*

A real-time quantitative LAMP that amplifies the *hlyA* gene of *L. monocytogenes* was designed by Shan *et al.* (2012). This LAMP assay was then used to detect *L. monocytogenes* in four different types of retail food samples such as raw meat, vegetables, deli, and seafood. The study also reported that LAMP assay was more sensitive than PCR in the detection of *L. monocytogenes.* In addition, a double LAMP (dLAMP) assay was first conducted by Wu *et al.* (2014) for the detection of *L. monocytogenes* in food samples including pork, beef, chicken, mutton, shrimp, fish, and quick-frozen rice flour products. LAMP primers targeting the *hlyA* and *iap* genes of *L. monocytogenes* were used to ensure the dLAMP assay is more rapid, sensitive, and specific. The results of this study showed that dLAMP assay was more sensitive and less time consuming as compared to normal LAMP assay.

7.2.5.2 *Vibrio parahaemolyticus*

Nemoto *et al.* (2009) utilized LAMP to detect *tdh*-positive isolates of *Vibrio parahaemolyticus* targeting six regions of the *tdh* gene and compared the results to PCR for detection of *tdh* and reverse passive latex agglutination for *tdh* detection. Another LAMP assay was developed for detection of *tlh* gene and tested both with pure *V. parahaemolyticus* cultures and artificially inoculated shrimp. The assay revealed all 143 pure *Vibrio parahaemolyticus* culture were positive, whereas no LAMP product was detected from any of 33 non *V. parahaemolyticus* or 56 non-*Vibrio* isolates (Yamazaki *et al.*, 2008). Later, Yamazaki *et al.* (2010) followed up on their previous work by developing a LAMP assay to detect the *tdh* and *trh* genes in *V. parahaemolyticus* and related *Vibrio* species. LAMP assays was also successfully used targeting *rpoD* and *toxR* genes of *V. parahaemolyticus*, which resulted in positive detection of 78 *V. parahaemolyticus* strains (Nemoto *et al.*, 2011). The sensitivity of LAMP assay targeting *rpoD* and *toxR* was determined to be 3.7 and 450 CFU per test in pure culture. The rpoD-LAMP assay was combined with MPN method detection for detection of *V. parahaemolyticus* in spiked short-necked clams comparative to MPN method with a culture method using agar medium. The results showed higher sensitivity using the *rpoD*-LAMP method (Nemoto *et al.*, 2011).

A novel LAMP *in situ* method was developed for rapid detection of foodborne *V. parahaemolyticus* strains (Wang *et al.*, 2013). Recently, Zeng *et al.* (2014) developed a novel method that combines the LAMP assay with immunomagnetic separation to detect *V. parahaemolyticus* in raw oysters. The limit of detection was 0.19 CFU/g, thus providing a plat form for the comprehensive detection of pathogenic strains using a virulence-gene-specific LAMP assay (Zeng *et al.*, 2014).

7.2.5.3 *Escherichia coli*

In the field of foodborne pathogen detection, LAMP was used for the first time to detect $stxA_2$ (stx_2) gene in *Escherichia coli* O157:H7 (Maruyama *et al.*, 2003). Later, Hara-Kudo *et al.* (2007) designed a LAMP assay that can detect verotoxin-producing *E. coli* (also known as STEC) by using primers targeting *VT1* and *VT2* genes. The LAMP assay was then employed for the detection of verotoxin-producing *E. coli* in food samples such as ground beef and radish sprouts. The detection limit of the LAMP assay was found to be approximately 0.7–2.2 cells per test tube and the assay time was estimated to be within 60 minutes. In the study, it was reported that LAMP assay was more sensitive than conventional PCR assay due to faster amplification of target gene and greater yield of amplicons. A LAMP assay developed by Zhao *et al.* (2010) for the detection of *E. coli* O157 with primers targeting *rfbE*, stx_1 and stx_2 also demonstrated that LAMP assay had higher sensitivity as compared to conventional PCR assay. The detection limits of LAMP were 100 fg DNA/tube for *rfbE*, 100 fg DNA/tube for stx_1, and 10 fg DNA/tube for stx_2.

7.2.5.4 *Salmonella*

LAMP assay has been reported to be highly sensitive in detection of *Salmonella* in different food matrics, including liquid egg (Hara-Kudo *et al.*, 2005; Ohtsuka *et al.*, 2005; Ueda and Kuwabara, 2009), pork (Zhang *et al.*, 2012), meat products (Wang *et al.*, 2009), milk (Kubota *et al.*, 2013), and other poultry and vegetables. Since the *invA* gene is only present in *Salmonella spp.*, many studies have developed *Salmonella*-specific LAMP assays using *invA* target gene (Hara-Kudo *et al.*, 2005; Ueda and Kuwabara, 2009; Lu *et al.*, 2009; Ahn *et al.*, 2010; Hsieh *et al.*, 2012; Jiang *et al.*, 2012). Okamura *et al.* (2009) utilized a LAMP assay that targets fragments of O4 *Salmonella enterica*-specific gene *rfbJ*. A LAMP assay targeting *phoP* gene was designed to detect *Salmonella spp.* from minced meat of pig and raw milk. This *phoP* gene is responsible to regulate the expression genes involved in virulence and the survival of *Salmonella* from destruction by macrophage (Li *et al.*, 2009). A study developed a LAMP assay targeting *Sdf I* gene for the detection of *Salmonella enteritidis* in chicken and pork (Yang *et al.*, 2010).

Recently, Zhang *et al.* (2012) developed two sets of LAMP primers that specifically target the *Salmonella fimY* gene in strains isolated from deli meat samples. Another LAMP assay targeting *HisJ* gene was also developed by Zhang *et al.* (2012) to detect foodborne *Salmonella*. Tourlousse *et al.* (2012) applied LAMP assay targeting two genes, *invA* and *phoP*, to detect *Salmonella*.

7.2.6 Oligonucleotide DNA Microarray

The recent progress in multi-gene detection technology includes the microarray technology (Call *et al.*, 2001). Microarrays were originally used for the study of gene expression, but oligonucleotide DNA microarray has been widely used in the field of foodborne pathogen detection. Microarrays are made up of glass slides or chips coated with up to hundreds of specific oligonucleotide probes (chemically synthesized short sequences range from 25–80 bp) or PCR probes (Severgnini *et al.*, 2011). Each oligonucleotide probe is able to target a specific part of a gene sequence. In this method (Figure 7.5), the sample nucleic acid fragments (DNA, mRNA, or cDNA) are labeled with fluorescent dye, and then they are denaturated to generate single-stranded fragments. These fragments will hybridize to the array through binding to their corresponding oligonucleotide probes. The results are obtained through the visualization of the fluorescence signal produced from the probe-sample complex. The fluorescence intensity is proportional to the concentration of each labeled nucleic acid fragment (Lauri and Mariani, 2009).

An oligonucleotide DNA microarray assay that can simultaneously detect and discriminate six *Listeria* spp. including *L. monocytogenes*, *L. ivanovii*, *L. innocua*, *L. seeligeri*, *L. grayi*, and *L. welshimeri* was performed by Volokhov *et al.* (2002). The microarray assay was based on *iap*, *hly*, *inlB*, *plcA*, *plcB*, and *clpE* genes for the identification of *Listeria* species. The detection of a specific serotype can be crucial especially for *Escherichia coli*, as this pathogen has different serotypes with different level of pathogenicity that ranges from harmless strain *Escherichia coli* K-12 to deadly strain *Escherichia coli* O157:H7 (Lauri and Mariani, 2009).

Moreover, oligonucleotide DNA microarray was also used for the simultaneous detection of different foodborne bacterial pathogens which include *L. monocytogenes*, *E. coli* O157:H7, *Salmonella enterica*, and *Campylobacter jejuni* in food samples (Suo *et al.*, 2010a) (Table 7.1).

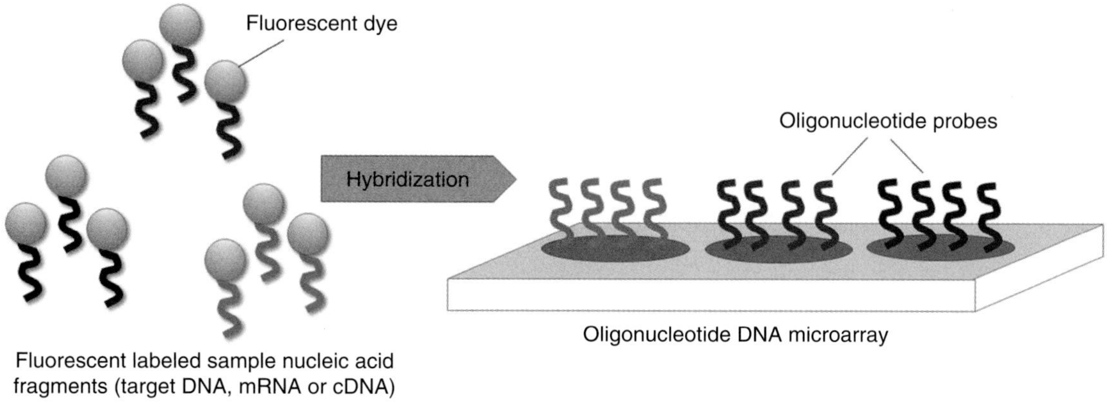

Figure 7.5 Representative Image of an Oligonucleotide DNA Microarray. The sample nucleic acid fragments (DNA, mRNA, or cDNA) that are labeled with fluorescent dye will hybridize with their corresponding oligonucleotide probes on the microarray. This generates a probe-sample complex that produces fluorescence signal. (*See insert for color representation of the figure.*)

Li *et al.* (2006) had presented the first report of detection of pathogenic *Shigella* and *Escherichia coli* serotypes by oligonucleotide DNA microarray. Wang *et al.* (2007) developed a microarray assay that allows the detection and identification of 22 foodborne pathogens. Some examples of these pathogens are *Staphylococcus aureus*, *L. monocytogenes*, *V. parahaemolyticus*, *V. cholerae*, *Campylobacter jejuni*, *Clostridium perfringens*, *Shigella* spp., *Salmonella* spp., and *Bacillus cereus*. Other studies that involved the application of oligonucleotide DNA microarray for the detection of foodborne pathogens are presented in Table 7.1.

DNA microarrays are commercially available but most of them are designed for gene expression analysis studies (Rasooly and Herold, 2008). The commercial *in situ*-synthesized arrays are high-density microarrays where short oligonucleotide probes range from 20–25 bp are synthesized directly on the surface of the microarray. In addition, multiple probes per target are included for higher sensitivity, specificity, and accuracy. These high-density microarrays require special manufacturing and they are relatively high in cost (Rasooly and Herold, 2008; Severgnini *et al.*, 2011). There are various commercially available DNA microarrays manufactured by Affymetrix, Roche NimbleGen and Agilent Technologies (Severgnini *et al.*, 2011).

Most of the commercial microarrays are not desirable for specialized application such as food microbial analysis or diagnostic laboratory because low- to medium-density array will serve as the ideal microarray platform that can provide reliable results without involving the use of complicated equipments and data management (Rasooly and Herold, 2008; Severgnini *et al.*, 2011). In this case, custom microarrays are available from the Department of Bioresources at Seibersdorf and other organizations. Custom microarrays are sensitive, specific, and less expensive than commercial microarrays (Mothershed and Whitney, 2006; Severgnini *et al.*, 2011). Nevertheless, low-density microarray is commercially available. For instance, StaphyChips® developed by Affymetrix and in collaboration with Advanced Array Technology (ATT, Eppendorf Array Technologies). StaphyChips® is able to detect a total of 15 *Staphylococcus* species, including methicillin-resistant *Staphylococcus aureus* (MRSA) (Mothershed and Whitney, 2006).

In general, DNA oligonucleotide microarray allows simultaneous identification of multiple foodborne bacterial pathogens. Therefore, it is capable of high-throughput analysis and it also has the potential to be automated (Al-Khaldi, 2002; De Boer and López, 2012; Zhou *et al.*, 2013). However, the shortcoming is that large amounts of target DNA or RNA are needed for these methods (Gasanov *et al.*, 2005).

7.2.7 Next Generation Sequencing (NGS)

The majority of bacterial genome sequences available today have been generated using the Sanger chain termination sequencing chemistries. Despite being very instrumental in the rise of the field of genomics, it is time consuming as well as resource intensive (Sanger *et al.*, 1977; Medini *et al.*, 2008). The post-Sanger era sequencing technologies, the NGS technologies, have been developed since 2005 to permit extremely rapid high-throughput whole genome sequencing (WGS) hence providing a broader application of comparative genomics (Medini *et al.*, 2008; Shendure and Ji, 2008; Letchumanan *et al.*, 2015b). Due to rapid decreasing costs for sequencers and reagents, a bacterial genome sequence can be obtained within a few days for less than US$500 (Didelot *et al.*, 2012), and more than 36,000 bacterial genome sequences are available in public databases (Reddy *et al.*, 2015). Other than serving as a detection tool, WGS is also a feasible tool for retrospective epidemiological analyses and is frequently used for the latter purpose.

Genome sequencing of several *L. monocytogenes* strains have revealed serotype- and strain-specific characteristics of *L. monocytogenes* (Orsi *et al.*, 2008; Gilmour *et al.*, 2010; Fretz *et al.*, 2010) and provided novel insights into the genomic causes underlying pathogenicity and survival in food and food processing settings (Buchrieser and Glaser, 2011). Gilmour *et al.* (2010) reported the first real-time application of WGS during an active listeriosis outbreak investigation using the hihg-throughput method to characterize two outbreak-associated isolates of *L. monocytogenes*. In 2014, Rychli *et al.* (2014) sequenced and analyzed the genomes of both outbreak strains in order to retrospectively investigate the extent of genetic diversity between the two strains. WGS analysis revealed that these two strains have distinct *in vitro* virulence potential despite originating from similar serovar (Rychli *et al.*, 2014).

The development of benchtop sequencers using NGS technology such as 454 or GS FLX™ (Roche), MiSeq (Illumina), and Ion Torrent Personal Genome Sequencer (PGM; Life Technologies) will enable bacterial WGS even in small research and clinical laboratories (Didelot *et al.*, 2012). WGS has already been actively used for the characterization of bacterial isolates in several large outbreaks in the world (Gilmour *et al.*, 2010; Reuter *et al.*, 2013) and also being used as a tool for retrospective epidemiological analyses (Orsi *et al.*, 2008; Rychli *et al.*, 2014). In the near future this technology is likely to substitute currently used typing methodologies due to its ultimate resolution and sensitivity (Sabat *et al.*, 2013). However, NGS also has its limitation. Until now, it is still too laborious and time-consuming to obtain useful data for routine surveillance (Fournier *et al.*, 2014). Library preparation and sequencing protocol requires adept and skillful technician; however, this limitations are likely to be overcome due to higher level of automation that will lead to a more streamlined process. Bioinformatics analysis and interpretation, as well as computational hardware are also another challenges to be solved especially by small laboratories (Fournier *et al.*, 2014). In addition, a fundamentally unsolved question is how the sequences should be examined for epidemiological characterization (Sabat *et al.*, 2013). There are other studies that employed NGS for the detection of foodborne pathogens, which are listed in Table 7.1.

There are many advantages of using molecular methods for the detection and identification of foodborne bacterial pathogens. For instance, the main advantages of molecular detection methods are due to their high sensitivity and specificity. Nonetheless, limitations can be found in these methods such as some molecular methods can be costly, complex, and require trained personnel. The advantages and limitations of molecular methods are summarized and listed in Table 7.2.

Table 7.2 The summary of advantages and limitations of each rapid detection methods.

Nucleic acid-based methods	Advantages	Limitations	References
Simple PCR	• High sensitivity • High specificity • Automated • Accurate and Reliable results	• Affected by PCR inhibitors • Sensitivity may be affected by non-optimized protocols • Requires DNA purification • Difficult to distinguish between viable and non-viable cells	Mandal *et al.*, 2011; Zhang, 2013; Letchumanan *et al.*, 2014; Park *et al.*, 2014; Law *et al.*, 2015a; Law *et al.*, 2015b
Multiplex PCR	• High sensitivity • High specificity • Detection of multiple foodborne pathogens • Automated • Reliable results	• Affected by PCR inhibitors • Sensitivity may be affected by non-optimized protocols • Difficult to distinguish between viable and non-viable cells • Primer design and other mPCR conditions are crucial (e.g., primer concentration, PCR buffer concentration, and quantities of DNA template)	Markoulatos *et al.*, 2002; Mandal *et al.*, 2011; Zhang, 2013; Park *et al.*, 2014; Law *et al.*, 2015a; Law *et al.*, 2015b
Real-time PCR	• High sensitivity • High specificity • Rapid cycling • Reproducible • Does not require post-amplification products processing • Real-time monitoring PCR amplification products • Assay can be multiplexed • Allows high-throughput analysis	• High cost • Difficult for multiplex real-time PCR assay • Affected by PCR inhibitors • Difficult to distinguish between viable and non-viable cells • Requires trained personnel • Cross contamination may occur	Oravcová *et al.*, 2005; Mandal *et al.*, 2011; Zhang, 2013; Letchumanan *et al.*, 2014; Park *et al.*, 2014 Law *et al.*, 2015a; Law *et al.*, 2015b

(*Continued*)

Table 7.2 (Continued)

Nucleic acid-based methods	Advantages	Limitations	References
NASBA	• Sensitive • Specific • Low cost • Does not require thermal cycling system • Able to detect viable microorganisms	• Requires viable microorganisms • Difficulties in handling RNA	Lauri and Mariani, 2009; Zhao *et al.*, 2014 Law *et al.*, 2015a; Law *et al.*, 2015b
LAMP	• High sensitivity • High specificity • Low cost • Easy to operate • Does not require thermal cycling system	• Primer design is complicated • Insufficient to detect unknown or unsequenced targets	Letchumanan *et al.*, 2014; Zhao *et al.*, 2014; Law *et al.*, 2015a; Law *et al.*, 2015b
Oligonucleotide DNA microarray	• High sensitivity • High specificity • High throughput • Enables detection of multiple foodborne pathogens • Allows detection of specific serotype • Labor-saving • Allows high-throughput analysis	• High cost • Difficult to distinguish between viable and non-viable cells • Requires trained personnel • Requires oligonucleotide probes and labeling of target genes • Requires large amount of target DNA or RNA	Lauri and Mariani, 2009; Mandal *et al.*, 2011; Park *et al.*, 2014; Law *et al.*, 2015a; Law *et al.*, 2015b
NGS	• High sensitivity and specificity • Enables simultaneous detection of multiple foodborne pathogens • Allows high-throughput analysis • Enable the analysis of whole genome of the pathogens	• Costly • Trained personnel is needed • Requires Bioinformatics skills for analysis and interpretation • Computationally intensive	Sabat *et al.*, 2013; Fournier *et al.*, 2014; Law *et al.*, 2015b

7.3 Conclusion

Conventional methods for the detection of foodborne pathogens that are based on culturing the microorganisms are selective, but they can be time-consuming and laborious. Hence, various rapid detection methods have been developed in order to overcome the limitations of conventional detection methods. Rapid methods are important for the rapid detection of foodborne pathogens in food products to prevent outbreaks of foodborne diseases and the spread of foodborne pathogens. Rapid detection methods are generally more sensitive, specific, time-efficient, labor saving, and reliable than conventional methods.

Nucleic acid-based methods such as PCR, mPCR, qPCR, DNA microarray, and NGS have high sensitivity and they are widely used for the detection of foodborne pathogens. For large number of samples, high-throughput assays such as qPCR and DNA microarray are often used for the detection of foodborne bacterial pathogens. Nevertheless, these methods require trained personnel and specialized instruments. Alternative nucleic acid-based methods such as NASBA and LAMP are available for the detection of foodborne pathogens and their toxins. NASBA and LAMP are relatively sensitive, specific, and cost efficient. They do not require thermocycling system, therefore, they are useful especially in low resource settings.

The advancements of molecular methods allow the rapid detection of various foodborne bacterial pathogens in food samples with high sensitivity and specificity while substituting the lengthy and laborious conventional detection methods. The combined use of two or more detection methods is also possible and may improve the accuracy of detecting these foodborne pathogens, but further studies on the effect of different combinations of rapid methods for foodborne pathogen detection are required. Still there are great potential for the development and application of new techniques for foodborne pathogens detection and analysis.

Acknowledgment

This work was supported by the MOSTI eScience Funds (06-02-10-SF0300 and 02-02-10-SF0215), External Industry Grant (Biotek Abadi - Vote No. GBA-808138 and GBA808813), and University of Malaya for High Impact Research Grant (UM-MOHE HIR Nature Microbiome Grant No. H-50001-A000027 and No. A000001-50001).

References

Aabo, S., Andersen, J.K. and Olsen, J.E. (1995). Research note: detection of *Salmonella* in minced meat by the polymerase chain reaction. *Letters in Applied Microbiology, 21,* pp. 180–182.

ABD-Elghany, S.M., et al. (2015). Occurrence, genetic characterization and antimicrobial resistance of *Salmonella* isolated from chicken meat and giblets. *Epidemiology Infection, 143,* pp. 997–1003.

Aberoumand, A.U. (2010). Occurrence and pathogenic *vibrios* in the marine environment of potentially pathogenic the traits of Messina, Italy. *World Journal of Fish Marine Scientific, 3,* pp. 376–382.

Adzitey, F., et al. (2012). Prevalence, antibiotic resistance and RAPD typing of *Campylobacter* species isolated from ducks, their rearing and processing environments in Penang, Malaysia. *International Journal of Food Microbiology, 154,* pp. 197–205.

Agersborg, A., Dahl, R. and Martinez, I. (1997). Sample preparation and DNA extraction procedures for polymerae chain reaction identification of *Listeria monocytogenes* in seafoods. *International Journal of Food Microbiology, 35,* pp. 275–280.

Ahn, Y.C., et al. (2010). Detection of *Salmonella* using the loop mediated isothermal amplification and real-time PCR. *Journal of Korean Chemical Society, 54,* pp. 215–221

Akbulut, D., Grant, K.A. and McLauchlin, J. (2004). Development and application of real-time PCR assays to detect fragments of the *Clostridium botulinum* types A, B, and E neurotoxin genes for investigation of human foodborne and infant botulism. *Foodborne Pathogens and Disease, 1*, pp. 247–257.

Alam, M.J., et al. (2002). Environmental investigation of potentially pathogenic *Vibrio parahaemolyticus* in the Seto-Inland Sea, Japan. *FEMS Microbiology Letters, 208*, pp. 83–87.

Al-Khaldi, S.F. (2002). DNA microarray technology used for studying foodborne pathogens and microbial habitats: minireview. *Journal of AOAC International, 85*, pp. 906–910.

Allerberger, F., et al. (2003). Occurrence of *Salmonella enterica* serovar Dublin in Austria. *Wiener medizinische Wochenschrift, 153*, pp. 148–152.

Almeida, P.F. and Almeida, R.C.C. (2000). A PCR protocol using *inl* gene as a target for specific detection of *Listeria monocytogenes. Food Control, 11*, pp. 97–101.

Alves, J., et al. (2012). Multiplex PCR for the detection of *Campylobacter* spp. and *Salmonella* spp. in chicken meat. *Journal of Food Safety, 32*, pp. 345–350.

Amagliani, G., et al. (2004). Direct detection of *Listeria monocytogenes* from milk by magnetic based DNA isolation and PCR. *Food Microbiology, 21*, pp. 597–603.

Aznar, R. and Alarcón, B. (2002). PCR detection of *Listeria monocytogenes*: a study of multiple factors affecting sensitivity. *Journal of Applied Microbiology, 95*, pp. 958–966.

Balagué, C., et al. (2006). Occurrence of non-O157 shiga toxin-producing *Escherichia coli* in ready-to-eat food from supermarkets in Argentina. *Food Microbiology, 23*, pp. 307–313.

Bang, J., et al. (2013). Development of random genomic DNA microarray for the detection and identification of *Listeria monocytogenes* in milk. *International Journal of Food Microbiology, 161*, pp. 134–141.

Barbau-Piednoir, E., et al. (2013). Development and validation of qualitative SYBR®Green real-time PCR for detection and discrimination of *Listeria* spp. and *Listeria monocytogenes. Applied Microbiology and Biotechnology, 97*, pp. 4021–4037.

Barlow, M. and Hall, B.G. (2002). Origin and evolution of the *AmpC* beta-lactamases of *Citrobacter freundii. Antimicrobial and Agents Chemotheraphy, 46*, pp. 1190–1198.

Bastian, S.N., Carle, I. and Grimont, F. (1998). Comparison of 14 PCR systems for the detection and subtyping of *stx* genes in Shiga-toxin-producing *Escherichia coli. Research in Microbiology, 149*, pp. 457–472.

Becker, B., Jordan, S. and Holzapfel, W.H. (2005). Rapid and specific detection of *Listeria monocytogenes* in smoked salmon with BAX®-PCR. *Food Control, 16*, pp. 717–721.

Bej, A.K., et al. (1991). Detection of *Escherichia coli* and *Shigella* spp. in water by using the polymerase chain reaction and gene probes for *uid. Applied and Environmental Microbiology, 57*, pp. 1013–1017.

Bej, A.K., et al. (1999). Detection of total and hemolysin-producing *Vibrio parahaemolyticus* in shell fish using multiplex PCR amplification of *tl, tdh* and *trh. Journal of Microbiology Methods, 36*, pp. 215–225.

Bhuiyan, N.A., et al. (2002). Prevalence of the pandemic genotype of *Vibrio parahaemolyticus* in Dhaka, Bangladesh, and significance of its distribution across different serotypes. *Journal of Clinical Microbiology, 40*, pp. 284–286.

Birch, L., et al. (2001). A comparison of nucleic acid amplification techniques for the assessment of bacterial viability. *Letters of Applied Microbiology, 33*, pp. 296–301.

Blackstone, G.M., et al. (2003). Detection of pathogenic *Vibrio parahaemolyticus* in oyster enrichment by real-time PCR. *Journal of Microbiology Methods, 53*, pp. 149–155.

Blais, B.W., et al. (1997). A nucleic acid sequence-based amplification system for detection of *Listeria monocytogenes hlyA* sequences. *Applied and Environmental Microbiology, 63*, pp. 310–313.

Brenner, F.W., et al. (2000). *Salmonella* nomenclature. *Journal of Clinical Microbiology, 38*, pp. 2465–2467.

Brian, M.J., et al. (1992). Polymerase chain reaction for diagnosis of enterohemorrhagic *Escherichia coli* infection and hemolytic-uremic syndrome. *Journal of Clinical Microbiology, 30*, pp. 1801–1806.

Buchrieser, C. and Glaser, P. (2011). Genomics of *Listeria monocytogenes* and other members of the genus *Listeria*. *Genomes of Foodborne and Waterborne Pathogens*. Washington DC: ASM Press, pp. 125–145.

Burbano, E., et al. (2006). Rapid DNA extraction and PCR validation for direct detection of *Listeria monocytogenes* in raw milk. *Revista MVZ Cordoba, 11*, pp. 715–724.

Call, D.R., Brockman, F.J. and Chandler, D.P. (2001). Detecting and genotyping *Escherichia coli* O157:H7 using multiplexed PCR and nucleic acid microarrays. *International Journal of Food Microbiology, 67*, pp. 71–80.

Candrian, U., et al. (1991). Detection of *Escherichia coli* and identification of enterotoxigenic strains by primer-directed enzymatic amplification of specific DNA sequences. *International Journal of Food Microbiology, 12*, pp. 339–352.

Canizalez-Roman, A., et al. (2011). Comparative evaluation of a chromogenic agar medium-PCR protocol with a conventional method for isolation of *Vibrio parahaemolyticus* strains from environmental and clinical samples. *Canadian Journal of Microbiology, 57*, pp. 136–142.

Catarame, T., et al. (2006). Comparison of a real-time polymerase chain reaction assay with a culture method for the detection of Salmonella in retail meat samples. *Journal of Food Safety, 26*, pp. 1–9.

Ceccarelli, D., et al. (2013). Distribution and dynamics of epidemic and pandemic *Vibrio parahaemolyticus* virulence factors. *Frontiers in Cellular and Infection Microbiology, 3*, pp. 97.

Centers for Disease Control and Prevention (CDC). (2011). CDC estimates of foodborne illness in the United States. http://www.cdc.gov/foodborneburden/pdfs/factsheet_a_findings_updated4-13.pdf [Accessed 1 June 2014].

Chacón, L., et al. (2010). Standardization of a PRC method for the detection of the *Salmonella spp. invA* gene in lettuce. *Revista de la Sociedad Venezolana de Microbiologia, 30*, pp. 18–23.

Cheah, Y.K., et al. (2008). Comparison of PCR fingerprinting techniques for the discrimination of *Salmonella enterica* subsp. *enterica* serovar Weltevreden isolated from indigenous vegetables in Malaysia. *World Journal of Microbiology and Biotechnology, 24*, pp. 327–335.

Chen, J., et al. (2012a). Development and evaluation of a multiplex PCR for simultaneous detection of five foodborne pathogens. *Journal of Applied Microbiology, 112*, pp. 823–30.

Chen, H.T., et al. (2008). Development of reverse transcription loop-mediated isothermal amplification for rapid detection of H9 avian influenza virus. *Journal of Virologiccal Methods, 151*, pp. 200–203.

Chen, J., et al. (2010). A real-time PCR method for the detection of *Salmonella enterica* from food using a target sequence identified by comparative genomic analysis. *International Journal of Food Microbiology, 137*, pp. 168–174.

Chen, M.B., et al. (2012b). Development of O-serogroup specific PCR assay for detection and identification of *Vibrio parahaemolyticus*. *International Journal of Food Microbiology, 159*, pp. 122–129.

Chen, W., Martinez, G. and Mulchandani, A. (2000). Molecular beacons: a real-time polymerase chain reaction assay for detecting *Salmonella*. *Analytical Biochemistry, 280*, pp. 166–172.

Chiang, Y.C., et al. (2012). Multiplex PCR and a chromogenic DNA macroarray for the detection of *Listeria monocytogenes, Staphylococcus aureus, Streptococcus agalactiae, Enterobacter sakazakii, Escherichia coli* O157:H7, *Vibrio parahaemolyticus, Salmonella* spp. and *Pseudomonas fluorescens* in milk and meat samples. *Journal of Microbiology Methods, 88*, pp. 110–116.

Chiou, C.S., et al. (2000). *Vibrio parahaemolyticus* Serovar O3:K6 as cause of unusually high incidence of food-borne disease outbreaks in Taiwan from 1996 to 1999. *Journal of Clinical Microbiology, 38,* pp. 4621–4625.

Cho, A.R., et al. (2014). Development of a loop-mediated isothermal amplification assay for detecting *Listeria monocytogenes prfA* in milk. *Food Science and Biotechnology, 23,* pp. 467–474.

Chung, M.S., Kim, C.M. and Ha, S.D. (2010). Detection and enumeration of microorganisms in ready-to-eat foods, ready-to-cook foods and fresh-cut produce in Korea. *Journal of Food Safety, 30,* pp. 480–489.

Churruca, E., et al. (2007). Detection of *Campylobacter jejuni* and *Campylobacter coli* in chicken meat samples by real-time nucleic acid sequence-based amplification with molecular beacons. *International Journal of Microbiology, 117,* pp. 85–90.

Cocolin, L., et al. (2002). Direct identification in food samples of *Listeria* spp. and *Listeria monocytogenes* by molecular methods. *Applied and Environmental Microbiology, 68,* pp. 6273–6282.

Compton, J. (1991). Nucleic acid sequence-based amplification. *Nature. 350,* pp. 91–92.

Cook, N., et al. (2002). A nucleic acid sequence-based amplification method to detect Salmonella enterica serotype Enteritidis strain PT4 in liquid whole egg. *Journal of Food Protection, 65,* pp. 1177–1178.

Cooray, K.J., et al. (1994). Detection of multiple virulence associated-genes of *Listeria monocytogenes* by PCR in artificially contaminated milk samples. *Applied and Environmental Microbiolgy, 60,* pp. 3023–3026.

Cortez, A.L.L., et al. (2006). Identification of *Salmonella* spp. isolates from chicken abattoirs by multiplex-PCR. *Research in Veterinary Medicine, 81,* pp. 340–344.

Cossart, P., et al. (1989). Listeriolysin O is essential for virulence of *Listeria monocytogenes*: direct evidence obtained by gene complementation. *Infection and Immunity, 57,* pp. 3629–3636.

Crump, J.A., Luby, S.P. and Mintz, E.D. (2004). The global burden of typhoid fever. *Bulletin of the World Health Organization, 82,* pp. 346–353.

Da Silva, E.P. and De Martinis, E.C.P. (2013). Current knowledge and perspective on biofilm formation: the case of *Listeria monocytogenes. Applied Microbiology and Biotechnology, 97,* pp. 957–968.

Daniels, N.A., et al. (2000). Emergence of a new O3:K6 *Vibrio parahaemolyticus* serotype in raw oysters. *JAMA, 284,* pp. 1541–1545.

De Boer, S.H. and López, M.M. (2012). New grower-friendly methods for plant pathogen monitoring. *Annual Review of Pytopathology, 50,* pp. 197–218.

De Oliveira, M.A., et al. (2010). Quantification of *Listeria monocytogenes* in minimally processed leafy vegetables using a combined method based on enrichment and 16S rRNA real-time PCR. *Food Microbiology, 27,* pp. 19–23.

Deepanjali, A., et al. (2005). Seasonal variation in abundance of total and pathogenic *Vibrio parahaemolyticus* bacteria in oysters along the southwest coast of India. *Applied and Environmental Microbiology, 71,* pp. 3575–3580.

Deneer, H.G. and Boychuk, I. (1991). Species-specific detection of *Listeria monocytogenes* by DNA amplification. *Applied and Environmental Microbiology, 57,* pp. 606–609.

Depaola, A., et al. (2000). Environmental investigations of *Vibrio parahaemolyticus* in oysters after outbreaks in Washington, Texas, and New York (1997 and 1998). *Applied and Environmental Microbiology, 66,* pp. 4649–4654.

Di Pinto, A., et al. (2005). A collagenase targeted multiplex PCR assay for identification of *Vibrio alginolyticus, Vibrio cholerae,* and *Vibrio parahaemolyticus. Journal of Food Protection, 68,* pp. 150–153.

Didelot, X., et al. (2012). Transforming clinical microbiology with bacterial genome sequencing. *Nature Review Genetics, 13,* pp. 601–612.

Dione, M.M., et al. (2011). Antimicrobial resistance and virulence genes of non-typhoidal Salmonella isolates in the Gambia and Senegal. *Journal of Infection in Developing Countries, 5*, pp. 765–775.

Doumith, M., et al. (2004). Differentiation of the major *Listeria monocytogenes* serovars by multiplex PCR. *Journal of Clinical Microbiology, 42*, pp. 3819–3822.

D'Souza, D.H. and Jaykus, L.A. (2003). Nucleic acid sequence based amplification for the rapid and sensitive detection of *Salmonella enterica* from foods. *Journal of Applied Mircobiology, 95*, pp. 1343–1350.

Dwivedi, H.P. and Jaykus, L.A. (2011). Detection of pathogens in foods: the current state-of-the-art and future directions. *Critical Reviews in Microbiology, 37*, pp. 40–63.

Elhadi, N., et al. (2004). Prevalence of potentially pathogenic *Vibrio* species in the seafood marketed in Malaysia. *Journal of Food Protection, 67*, pp. 1469–1475.

Erol, I. and Ayaz, N.D. (2011). Serotype distribution of *Listeria monocytogenes* isolated from Turkey meat by multiplex PCR in Turkey. *Journal of Food Protection, 31*, pp. 149–153.

Farber, J.M. and Peterkin, P.I. (1991). *Listeria monocytogenes*, a food-borne pathogen. *Microbiological Reviews, 55*, pp. 476–511.

Feng, P. and Weagant, S.D. (2015). *Bacteriological Analytical Manual (BAM) Chapter 4A: Diarrheagenic Escherichia coli*. Available at: http://www.fda.gov/Food/FoodScience Research/ LaboratoryMethods/ucm070080.htm [Accessed January 5, 2016].

Feng, P., Weagant, S.D. and Monday, S.R. (2001). Genetic analysis for virulence factors in *Escherichia coli* O104:H21 that was implicated in an outbreak of hemorrhagic colitis. *Journal of Clinical Microbiology, 39*, pp. 24–28.

Fournier, P.E., Dubourg, G. and Raoult, D. (2014). Clinical detection and characterization of bacterial pathogens in the genomics era. *Genome Medicine, 6*, pp. 114.

Franz, E., et al. (2007). Prevalence of Shiga toxin-producing *Escherichia coli stx1, stx2, eaeA and rfbE* genes and survival of *E. coli O157:H7* in manure from organic and low-input conventional dairy farms. *Applied and Environmental Microbiology, 73*, pp. 2180–2190.

Fratamico, P.M. and Gehring, A.G. (2014). *Escherichia coli* O157 and other Shiga toxin-producing *E. coli*: detection by immunemagnetic particle-based assays. In: C.A. Batt and M.L. Tortorello, eds., *Encyclopedia of Food Microbiology*. 2nd ed, New York: Elsevier, pp. 740–747.

Fratamico, P.M., et al. (2011). Detection by multiplex real-time polymerase chain reaction assays and isolation of Shiga toxin-producing *Escherichia coli* serogroups O26, O45, O103, O111, O121 and O145 in ground beef. *Foodborne Pathogens and Diseases, 8*, pp. 601–607.

Fretz, R., et al. (2010). Listeriosis outbreak caused by acid curd cheese Quargel, Austria and Germany 2009. *EuroSurveillance, 15*, pp. 1–2.

Fricker, M., et al. (2007). Diagnostic real-time PCR assays for the detection of emetic *Bacillus cereus* strains in foods and recent food-borne outbreaks. *Applied and Environmental Microbiology, 73*, pp. 1892–1898.

Fukushima, H., Tsunomori, Y. and Seki, R. (2003). Duplex real-time SYBR green PCR assays for detection of 17 species of food- or waterborne pathogens in stools. *Journal of Clinical Microbiology, 41*, pp. 5134–5146.

Fukushima, M., Kakinuma, K. and Kawaguchi, R. (2002). Phylogenetic analysis of *Salmonella, Shigella, Escherichia coli* strains on the basis of the *gyrB* gene sequence. *Journal of Clinical Microbiology, 40*, pp. 2779–2785.

Fusco, V., et al. (2011). Rapid and reliable identification of *Staphylococcus aureus* harbouring the enterotoxin gene cluster (*egc*) and quantitative detection in raw milk by real time PCR. *International Journal of Food Microbiology, 144*, pp. 528–537.

Fykse, E.M., et al. (2007). Detection of *Vibrio cholerae* by real-time nucleic acid sequence-based amplification. *Applied and Environmental Microbiology, 73*, pp. 1457–1466.

Gannon, V.P.J., et al. (1993). Detection and characterization of the eae gene of Shiga-like toxin-producing *Escherichia coli* using polymerase chain reaction. *Journal of Clinical Microbiology, 31*, pp. 1268–1274.

Garrido, A., et al. (2012). Development of a multiplex real-time PCR method for pathogenic *Vibrio parahaemolyticus* detection (*tdh*+ and *trh*+). *Food Control, 24*, pp. 128–135.

Gasanov, U., Hughes, D. and Mansbro, P.M. (2005). Methods for the isolation and identification of *Listeria* spp. and *Listeria monocytogenes*: a review. *FEMS Microbiology Reviews, 29*, pp. 851–875.

Gattuso, A., et al. (2014). Optimization of real time PCR based method for the detection of *Listeria monocytogenes* in pork meat. *International Journal of Food Microbiology, 184*, pp. 106–108.

Germini, A., et al. (2009). Simultaneous detection of *Escherichia coli* O157:H7, *Salmonella* spp., and *Listeria monocytogenes* by multiplex PCR. *Food Control, 20*, pp. 733–738.

Gilbert, C., et al. (2003). Development of a multiplex PCR assay for the specific detection of *Salmonella, Campylobacter jejuni, Escherichia coli* O157:H7, and *Listeria monocytogenes. Journal of Rapid Methods and Automation in Microbiology, 11*, pp. 61–74.

Gilmour, M.W., et al. (2010). High-throughput genome sequencing of two *Listeria monocytogenes* clinical isolates during a large foodborne outbreak. *BMC Genomics, 11*, pp. 120.

Gilot, P. and Content, J. (2002). Specific identification of *Listeria welshimeri* and *Listeria monocytogenes* by PCR assays targeting a gene encoding a fibronectin-binding protein. *Journal of Clinical Microbiology, 40*, pp. 698–703.

Gode-Potratz, C.J., et al. (2011). Surface sensing in *Vibrio parahaemolyticus* triggers a programme of gene expression that promotes colonization and virulence. *Molecular Microbiology, 79*, pp. 240–263.

Goh, S.G., et al. (2012). *Listeria monocytogenes* in retailed raw chicken meat in Malaysia. *Poultry Science, 91*, pp. 2686–2690.

Golsteyn-Thomas, et al. (1990). Sensitive and specific detection of *Listeria monocytogenes* in milk and ground beef with the polymerase chain reaction. *Applied and Environmental Microbiology, 57*, pp. 2576–2580.

Gore, H.M., et al. (2003). Real-time molecular beacon NASBA reveals *hblC* expression from *Bacillus* spp. in milk. *Biochemical and Biophysical Research Communication, 311*, pp. 386–390.

Gracias, K.S. and McKillip, J.L. (2007). Nucleic acid sequence-based amplification (NASBA) in molecular bacteriology: a procedural guide. *Journal of Rapid Methods and Automation in Microbiology, 15*, pp. 295–309.

Graham, T.A., et al. (1996). Genus- and species-specific detection of *Listeria monocytogenes* using polymerase chain reaction assays targeting the 16S/23S intergenic spacer region of the rRNA operon. *Canadian Journal of Microbiology, 42*, pp. 1155–1162.

Graham, T.A., et al. (1997). Inter- and intraspecies comparison of the 16S-23S rRNA operon intergenic spacer regions of six *Listeria* spp. *International Journal of Systematic Bacteriology, 47*, pp. 863–869.

Groves, R.D. and Welshimer, H.J. (1977). Separation of pathogenic from apathogenic *Listeria monocytogenes* by three in vitro reactions. *Journal of Clinical Microbiology, 5*, pp. 559–563.

Guan, Z.P., et al. (2013). Rapid and simultaneous analysis of five foodborne pathogenic bacteria using multiplex PCR. *European Food Research and Technology, 237*, pp. 627–637.

Gutierrez West, C.K., Klein, S.L. and Lovell, C. R. (2013). High frequency of virulence factor genes *tdh, trh, and tlh* in *Vibrio parahaemolyticus* strains isolated from a pristine estuary. *Applied and Environmental Microbiology, 79*, pp. 2247–2252.

Hage, E., et al. (2014). Identification of six *Listeria* species by real-time PCR assay. *Letters of Applied Microbiology, 58*, pp. 535–540.

Haley, B.J., et al. (2014). Molecular diversity and predictability of *Vibrio parahaemolyticus* along the Georgian coastal zone of the Black Sea. *Frontiers in Microbiology*, *5*, pp. 45.

Hamdi, T.H., et al. (2007). Identification and molecular characterization of *Listeria monocytogenes* isolated in raw milk in the region of Algiers (Algeria). *International Journal of Microbiology*, *116*, pp. 190–193.

Han, F. and Ge, B. (2010). Quantitative detection of *Vibrio vulnificus* in raw oysters by real-time-loop-mediated isothermal amplification. *International Journal of Food Microbiology*, *142*, pp. 60–66.

Han, F., Wang, F. and Ge, B. (2011). Detecting potentially virulent *Vibrio vulnificus* strains in raw oysters by quantitative loopmediated isothermal amplification. *Applied Environmental Microbiology*, *77*, pp. 2589–2595.

Hara-Kudo, Y., et al. (2005). Loop-mediated isothermal amplification for the rapid detection of *Salmonella .FEMS Microbiology Letters*, *253*, pp. 155–161.

Hara-Kudo, Y., et al. (2008). Detection of verotoxigenic *Escherichia coli* O157 and O26 in food by plating methods and LAMP method: a collaborative study. *International Journal of Microbiology*, *122*, pp. 156–161.

Hara-Kudo, Y., et al. (2007). Sensitive and rapid detection of verotoxin-producing Escherichia coli using loop-mediated isothermal amplification. *Journal of Medical Microbiology*, *56*, pp. 398–406.

Hara-Kudo, Y., et al. (2012). Characteristics of a sharp decrease in *Vibrio parahaemolyticus* infections and seafood contamination in Japan. *International Journal of Microbiology*, *157*, pp. 95–101.

He, P., et al. (2014). Multiplex real-time PCR assay for detection of pathogenic *Vibrio parahaemolyticus* strains. *Molecular and Cellular Probes*, *28*, pp. 246–250.

Heijnen, L. and Medema, G. (2009). Method for rapid detection of viable Escherichia coli in water using real-time NASBA. *Water Research*, *43*, pp. 3124–3123.

Hein, I., et al. (2001a). Detection and quantification of the *iap* gene of *Listeria monocytogenes* and *Listeria innocua* by a new real-time quantitative PCR assay. *Research in Microbiology*, *152*, pp. 37–46.

Hein, I., et al. (2001b). Comparison of different approaches to quantify *Staphylococcus aureus* cells by real-time quantitative PCR and application of this technique for examination of cheese. *Applied and Environmental Microbiolgy*, *67*, pp. 3122–3126.

Heller, L.C., et al. (2003). Comparison of methods for DNA isolation from food samples for detection of Shiga toxin-producing Escherichia coli by real-time PCR. *Applied and Environmental Microbiology*, *69*, pp. 1844–1846.

Hervio-Heath, D., et al. (2002). Occurrence of pathogenic *Vibrio* in coastal areas of France. *Journal of Applied Microbiology*, *92*, pp. 1123–1135.

Honda, T., Ni, Y.X. and Miwatani, T. (1988). Purification and characterization of a hemolysin produced by a clinical isolate of Kanagawa phenomenon-negative *Vibrio parahaemolyticus* and related to the thermostable direct hemolysin. *Infection and Immunity*, *56*, pp. 961–965.

Hossain, M.T., Kim, Y.O. and Kong, I.S. (2013). Multiplex PCR for the detection and differentiation of *Vibrio parahaemolyticus* strains using the *groEL*, *tdh* and *trh* genes. *Molecular and Cellular Probes*, *27*, pp. 171–5.

Hsieh, K., et al. (2012). Rapid, sensitive, and quantitative detection of pathogenic DNA at the point of care through microfluidic electrochemical quantitative loop mediated isothermal amplification. *Angewandte Chemie International Edition*, *51*, pp. 4896–4900.

Hu, Q., et al. (2014). A modified molecular beacons-based multiplex real-time PCR assay for simultaneous detection of eight foodborne pathogens in a single reaction and its application. *Foodborne Pathology and Diseases*, *11*, pp. 207–214.

Huang, A., et al. (2014). High-throughput detection of food-borne pathogenic bacteria using oligonucleotide microarray with quantum dots as fluorescent labels. *International Journal of Food Microbiology, 185*, pp. 27–32.

Igbinosa, E.O. and Okoh, A.I. (2008). Emerging *Vibrio* species: an unending threat to public health in developing countries. *Research in Microbiology, 159*, pp. 495–506.

Ingianni, A., et al. (2001). Rapid detection of *Listeria monocytogenes* in foods, by a combination of PCR and DNA probe. *Molecular and Cellular Probes, 15*, pp. 275–280.

Izumiya, H., et al. (2011). Multiplex PCR assay for identification of three major pathogenic *Vibrio* spp., *Vibrio cholerae, Vibrio parahaemolyticus*, and *Vibrio vulnificus. Molecular and Cellular Probes, 25*, pp. 174–176.

Jadhav, S., Bhave, M. and Palombo, E.A. (2012). Methods used for the detection and subtyping of *Listeria monocytogenes. Journal of Microbioloy Methods, 88*, pp. 327–341.

Jamali, H., Chai, L.C. and Thong, K.L. (2013). Detection and isolation of *Listeria* spp. and *Listeria monocytogenes* in ready-to-eat foods with various selective culture media. *Food Control, 32*, pp. 19–24.

Janzten, M.M., et al. (2006). Review. Specific detection of *Listeria monocytogenes* in foods using commercial methods: from chromogenic media to real-time PCR. *Spanish Journal of Agricultural Research, 4*, pp. 235–247.

Jemmi, T. and Stephan, R. (2006). *Listeria monocytogenes*: food-borne pathogen and hygiene indicator. *Scientific and Technical Review of the Office International des Epizooties, 25*, pp. 571–580.

Jiang, K., et al. (2012). A novel, sensitive, accurate multiplex loop-mediated isothermal amplification method for detection of *Salmonella spp., Shigella spp.* and *Staphylococcus aureus* in food. *Journal of Food, Agriculture and Environment, 10*, pp. 252–256.

Jofré, A., et al. (2005). Simultaneous detection of *Listeria monocytogenes* and *Salmonella* by multiplex PCR in cooked ham. *Food Microbiology, 22*, pp. 109–115.

Johnson, W.M., et al. (1992). Detection of genes coding for listeriolysin and *Listeria monocytogenes* antigen A (*lmaA*) in *Listeria* spp. by the polymerase chain reaction. *Microbial Pathogenesis, 12*, pp. 79–86.

Jothikumar, N. and Griffiths, M.W. (2002). Rapid detection of Escherichia coli O157:H7 with multiplex real-time PCR assays. *Applied and Environmental Microbiolgy, 68*, pp. 3169–3171.

Jung, Y.S., et al. (2003). Polymerase chain reaction detection of *Listeria monocytogenes* in frankfurters using oligonucleotide primers targeting the genes encoding internalin AB. *Journal of Food Protection, 66*, pp. 237–241.

Kathariou, S., et al. (1987). Tn916-induced mutations in the hemolysin determinant affecting virulence of *Listeria monocytogenes. Journal of Bacteriology, 169*, pp. 1291–1297.

Kaufman, G.E., et al. (2004). Real time PCR quantification of *Vibrio parahaemolyticus* in oysters using an alternative matrix. *Journal of Food Protection, 67*, pp. 2424–2429.

Kawasaki, S., et al. (2010). Multiplex real-time polymerase chain reaction assay for simultaneous detection and quantification of *Salmonella* species, *Listeria monocytogenes*, and *Escherichia coli* O157:H7 in ground pork samples. *Foodborne Pathogens and Disease, 7*, pp. 549–554.

Kaysner, C.A. and DePaola, A. J. (2001). *Vibrio.* In: F.P. Downes and K. Ito eds., *Compendium of Methods for the Microbiological Examination of Food.* Washington, D.C: American Public Health Association.

Khan, J.A., et al. (2013). In vitro detection of pathogenic *Listeria monocytogenes* from food sources by conventional, molecular and cell culture method. *Brazilian Journal of Microbiology, 44*, pp. 751–758.

Khoo, C.H., et al. (2009). Virulotyping of *Salmonella enterica* subsp. *enterica* isolated from indigenous vegetables and poultry meat in Malaysia using multiplex-PCR. *Antonie Van Leeuwenhoek, 96*, pp. 441–457.

Kim, H.J., et al. (2012). Simultaneous detection of pathogenic *Vibrio* species using multiplex real-time PCR. *Food Control, 23*, pp. 491–498.

Kim, S., et al. (2006). Multiplex PCR based method for identification of common clinical serotypes of *Salmonella enterica subsp. enterica*. *Journal of Clinical Microbiology, 44*, pp. 2608–3615.

Kim, Y.K. and McCarter, L.L., (2000). Analysis of the polar flagellar gene system of *Vibrio parahaemolyticus*. *Journal of Bacteriology, 182*, pp. 3693–3704.

Kimura, B., Kawasaki, S., Fujii, T., Kusunoki, J., Itoh, T., and Flood, S.J.A. (1999). Evaluation of TaqMan PCR assay for detecting *Salmonella* in raw meat and shrimp. *Journal of Food Protection, 4*, pp. 307–425.

Klerks, M.M., Zijlstra, C. and van Bruggen, A.H.C. (2004). Comparison of real-time PCR methods for detection of *Salmonella enterica* and *Escherichia coli* O157:H7, and introduction of a general internal amplification control. *Journal of Microbiology Methods, 59*, pp. 337–349.

Kubota, R., et al. (2013). Molecular diagnostics in a teacup: non-instrumented nucleic acid amplification (NINA) for rapid, low cost detection of Salmonella enterica. *Chinese Science Bulletin, 58*, pp. 1162–1168.

Kubota, K., et al. (2011). Estimating the burden of acute gastroenteritis and foodborne illness caused by *Campylobacter*, *Salmonella*, and *Vibrio parahaemolyticus* by using population-based telephone survey data, Miyagi Prefecture, Japan, 2005 to 2006. *Journal of Food Protection, 74*, pp. 1592–1598.

Kumar, T.D.K., Murali, H.S. and Batra, H.V. (2009). Simultaneous detection of pathogenic *B. cereus*, *S. aureus* and *L. monocytogenes* by multiplex PCR. *Indian Journal of Microbiology, 49*, pp. 283–289.

Kundinger, M.M., et al. (2008). Real time polymerase chain reaction quantification of *Salmonella Typhimurium hilA* expression during agitation and static incubation. *Journal of Rapid Methods and Automation Microbiolology, 16*, pp. 273–283.

Lauri, A. and Mariani, P.O. (2009). Potentials and limitations of molecular diagnostic methods in food safety. *Genes and Nutrition, 4*, pp. 1–12.

Law, J.W.F., et al. (2015a). Rapid methods for the detection of foodborne bacterial pathogens: principles, applications, advantages and limitations. *Frontiers in Microbiology, 5*, pp. 1–19.

Law, J.W.F., et al. (2015b). An insight into the isolation, enumeration, and molecular detection of *Listeria monocytogenes* in food. *Frontiers in Microbiology, 6*, pp. 1–15.

Lawrence, L.M. and Gilmour, A. (1994). Incidence of *Listeria* spp. and *Listeria monocytogenes* in a poultry processing environment and in poultry products and their rapid confirmation by multiplex PCR. *Applied and Environmental Microbiology, 60*, pp. 4600–4604.

Lee, L.H., et al. (2008). Analysis of *Salmonella* Agona and Salmonella Weltevreden in Malaysia by PCR fingerprinting and antibiotic resistance profiling. *Antonie Van Leeuwenhoek, 94*, pp. 377–387.

Lee, N., et al. (2014). A multiplex PCR assay for simultaneous detection of *Escherichia coli* O157:H7, *Bacillus cereus, Vibrio parahaemolyticus, Salmonella* spp., *Listeria monocytogenes,* and *Staphylococcus aureus* in Korea ready-to-eat food. *Foodborne Pathogens and Diseases, 11*, pp. 574–580.

Leone, G., et al. (1998). Molecular beacon probes combined with amplification by NASBA enable homogeneous, real-time detection of RNA. *Nucleic Acids Research, 26*, pp. 2150–2155.

Letchumanan, V., et al. (2015a). Prevalence and antimicrobial susceptibility of *Vibrio parahaemolyticus* isolated from retail shrimps in Malaysia. *Frontiers in Microbiology, 6*, pp. 1–11.

Letchumanan, V., Chan, K.G. and Lee, L.H. (2015b). An insight of traditional plasmid curing in *Vibrio* species. *Frontiers in Microbiology, 6*, pp. 735.

Letchumanan, V., et al. (2015c). Occurrence and Antibiotic Resistance of *Vibrio parahaemolyticus* from Shellfish in Selangor, Malaysia. *Frontiers in Microbiology, 6*, p. 1417.

Letchumanan, V., Chan, K.G. and Lee, L.H. (2014). *Vibrio parahaemolyticus*: A Review on the Pathogenesis, Prevalence and Advance Molecular Identification Techniques. *Frontiers in Microbiology, 5*, pp. 705.

Levin, R.E. (2003). Application of the polymerase chain reaction for detection of *Listeria monocytogenes* in Foods: a review of methodology. *Food Biotechnology, 17*, pp. 99–116.

Levin, R.E. (2005). The application of real-time PCR to food and agricultural systems. A review. *Food Biotechnology, 18*, pp. 97–133.

Levine, M.M. (1987). *Escherichia coli* that cause diarrhea: enterotoxigenic, enteropathogenic, enteroinvasive, enterohemorrhagic, and enteroadherent. *The Journal of Infectious Disease, 155*, pp. 377–389.

Li, X.F., et al. (2009). A loop-mediated isothermal amplification method targets the phoP gene for the detection of *Salmonella* in food samples. *International Journal of Microbiology, 13*, pp. 252–258.

Li, R., et al. (2016). A novel PCR-based approach for accurate identification of *Vibrio parahaemolyticus. Frontiers in Microbiology, 7*, pp. 1–5.

Li, Y., et al. (2006). Development of a serotype-specific DNA microarray for identification of some *Shigella* and pathogenic *Escherichia coli* strains. *Journal of Clinical Microbiology, 44*, pp. 4376–4383.

Li, Y., et al. (2014). *Vibrio parahaemolyticus*, Southern Coastal Region of China, 2007-2012. *Emerging Infectious Disease, 20*, pp. 685–688.

Liming, S.H. and Bhagwat, A.A. (2004). Application of a molecular beacon-real-time PCR technology to detect *Salmonella* species contaminating fruits and vegetables. *International Journal of Microbiology, 95*, pp. 177–187.

Liming, S.H., et al. (2004). Detection of *Listeria monocytogenes* in fresh produce using molecular beacon-real-time PCR technology. *Journal of Food Science, 69*, pp. 240–M245.

Liu, B., et al. (2012). Development of a real time PCR assay for rapid detection of *Vibrio parahaemolyticus* from seafood. *Protein and Cellular, 3*, pp. 204–212.

Liu, D. (2006). Identification, subtyping and virulence determination of *Listeria monocytogenes*, an important foodborne pathogen. *Journal of Medical Microbiology, 55*, pp. 645–659.

Liu, D., et al. (2007). A multiplex PCR for species and virulence-specific determination of *Listeria monocytogenes. Journal of Microbiology Methods, 71*, pp. 133–140.

Liu, H., et al. (2015). Rapid detection and differentiation of *Listeria monocytogenes* and *Listeria* species in deli meats by a new multiplex PCR method. *Food Control, 52*, pp. 78–84.

Liu, P., et al. (2012). A new rapid real-time PCR method for detection of *Listeria monocytogenes* targeting the *hlyA* gene. *Food Science and Technology Research, 18*, pp. 47–57.

Lorenz, S.C., et al. (2013). Prevalence of hemolysin genes and comparison of *ehxA* subtype patterns in Shiga toxin-producing *Escherichia coli* (STEC) and non-STEC strains from clinical, food, and animal sources. *Applied and Environmental Microbiology, 79*, pp. 6301–6311.

Lozano-Leon, A., et al. (2003). Identification of *tdh*-positive *Vibrio parahaemolyticus* from an outbreak associated with raw oyster consumption in Spain. *FEMS Microbiology Letters, 226*, pp. 281–284.

Lu, Y., et al. (2009). Specific detection of viable *Salmonella* cells by an ethidiummonoazide-loop mediated isothermal amplification (EMA-LAMP) method. *Journal of Health, 55*, pp. 820–824.

Ma, K., et al. (2014). Rapid and simultaneous detection of *Salmonella, Shigella*, and *Staphylococcus aureus* in fresh pork using a multiplex real-time PCR assay based on immunomagnetic separation. *Food Control, 42*, pp. 87–93.

Maciorowski, K.G. (2005). Polymerase chain reaction detection of foodborne *Salmonella* spp. in animal feeds. *Critical Reviews in Microbiology, 31*, pp. 45–53.

Madani, M., Subbotin, S.A. and Moens, M. (2005). Quantitative detection of the potato cyst nematode, *Globodera pallida*, and the beet cyst nematode, *Heterodera schachtii*, using real-time PCR with SYBR green I dye. *Molecular and Cellular Probes, 19*, pp. 81–86.

Majowicz, S.E., et al. (2010). The global burden of non-typhoidal *Salmonella* gastroenteritis. *Clinical Infectious Diseases, 50*, pp. 882–889.

Malcolm, T.T.H., et al. (2015). Detection and quantification of pathogenic *Vibrio parahaemolyticus* in shellfish by using multiplex PCR and loop-mediated isothermal amplification assay. *Food Control, 47*, pp. 664–671.

Mandal, P.K., et al. (2011). Methods for rapid detection of foodborne pathogens: an overview. *American Journal of Food Technology, 6*, pp. 87–102.

Margot, H., et al. (2013). Inclusivity, exclusivity and limit of detection of commercially available real-time PCR assays for the detection of *Salmonella. International Journal of Food Microbiology, 165*, pp. 221–226.

Markoulatos, P., Siafakas, N. and Moncany, M. (2002). Multiplex polymerase chain reaction: a practical approach. *Journal of Clinical Laboratory Analysis, 16*, pp. 47–51.

Martinez-Urtaza, J., et al. (2005). Pandemic *Vibrio parahaemolyticus* O3:K6. *European Emerging Infectious Diseases, 11*, pp. 1319–1320.

Maruyama, F., et al. (2003). Detection of bacteria carrying the stx_2 gene by *in situ* loop-mediated isothermal amplification. *Applied and Environmental Microbiology, 69*, pp. 5023–5028.

Marathe, S.A., et al. (2012). Direct detection of *Salmonella* without pre-enrichment in milk, ice-cream and fruit juice by PCR against *hilA* gene. *Food Control, 23*, pp. 559–563.

Matsuda, S., et al. (2010). Association of Vibrio parahaemolyticus thermostable direct hemolysin with lipid rafts is essential for cytotoxicity but not hemolytic activity. *Infection and Immunity, 78*, pp. 603–610.

Matsumoto, C., et al. (2000). Pandemic spread of an O3:K6 clone of *Vibrio parahaemolyticus* & emergence of related strains evidenced by Arbitrarily Primed PCR & *toxRS* sequence analysis. *Journal of Clinical Microbiology, 38*, pp. 578–585.

Maurer, J.J. (2011). Rapid detection and limitations of molecular techniques. *Annual Reviews of Food and Science Technology, 2*, pp. 259–279.

Maurischat, S., et al. (2015). Rapid detection and specific differentiation of *Salmonella enterica subsp. enterica Enteritidis, Typhimurium* and its monophasic variant 4,[5],12:i:—by real time multiplex PCR. *International Journal of Food Microbiology, 193*, pp. 8–14.

McKillip, J.L. and Drake, M. (2000). Molecular beacon polymerase chain reaction detection of *Escherichia coli 0157:H7* in milk. *Journal of Food Protection, 63*, pp. 855–859.

Medini, D., et al. (2008). Microbiology in the post-genomic era. *Natures Reviews Micrbiology, 6*, pp. 419–430.

Meng, J., et al. (1996). Polymerase chain reaction for detecting Escherichia coli O157:H7. *International Journal of Microbilogy, 32*, pp. 103–113.

Mezal, E.H., Stefanova, R. and Khan, A.A. (2013). Isolation and molecular characterization of *Salmonella enterica serovar* Javiana from food, environmental and clinical samples. *International Journal of Food Microbiology, 164*, pp. 113–118.

Min, J. and Baeumner, A.J. (2002). Highly sensitive and specific detection of viable *Escherichia coli* in drinking water. *Analytical Biochemistry, 303*, pp. 186–193.

Mollasalehi, H. and Yazdanparast, R. (2013). Development and evaluation of a novel nucleic acid sequence-based amplification method using one specific primer and one degenerate primer for simultaneous detection of *Salmonella* Enteritidis and *Salmonella* Typhimurium. *Analytica Chimica Acta, 770*, pp. 169–174.

Monge, R. and Arias, M.L. (1997). Survival of the bacterium *Listeria monocytogenes* (Listeriaceae) after addition of lemon juice or sodium hypochlorite. *Revista de Biologia Tropical, 45*, pp. 1247–1249.

Mori, Y. and Notomi, T. (2009). Loop-mediated isothermal amplification (LAMP): a rapid, accurate, and cost-effective diagnostic method for infectious diseases. *Journal of Infection and Chemotherapy, 15*, pp. 62–69.

Moriishi, K., et al. (1998). Sequence analysis of the *actA* gene of *Listeria monocytogenes* isolated from human. *Microbiology and Immunology, 42*, pp. 129–132.

Mothershed, E.A. and Whitney, A.M. (2006). Nucleic acid-based methods for the detection of bacterial pathogens: present and future considerations for clinical laboratory. *Clinica Chimica Acta, 363*, pp. 206–220.

Murugkar, H.V., Rahman, H. and Dutta, P.K. (2003). Distribution of virulence genes in Salmonella serovars isolated from man & animals. *Indian Journal of Medical Research, 117*, pp. 66–70.

Nadal, A., et al. (2007). A molecular beacon-based real time NASBA assay for detection of *Listeria monocytogenes* in food products: role of target mRNA secondary structure on NASBA design. *Journal of Microbiology Methods, 68*, pp. 623–632.

Nataro, J.P. and Kaper, J.B. (1998). Diarrheagenic *Escherichia coli. Clinical Microbilogy Reviews, 11*, pp. 142–201.

Navas, J., et al. (2006). Evaluation of effects of primary and secondary enrichment for the detection of *Listeria monocytogenes* by real-time PCR in retail ground chicken meat. *Foodborne Pathogens and Disease, 3*, pp. 347–354.

Nelapati, S., Nelapati, K. and Chinnam, B.K. (2012). *Vibrio parahaemolyticus*-An emerging foodborne pathogen-A Review. *Veterinary World, 5*, pp. 48–62.

Nemoto, J., et al. (2011). Development and evaluation of a loop mediated isothermal amplification assay for rapid and sensitive detection of *Vibrio parahaemolyticus. Journal of Food Protection, 74*, pp. 1462–1467.

Nemoto, J., et al. (2009). Rapid and specific detection of the thermostable direct hemolysin gene in *Vibrio parahaemolyticus* by loop-mediated isothermal amplification. *Journal of Food Protection, 72*, pp. 748–754.

Newell, D.G., et al. (2010). Food-borne diseases- the challenges of 20 years ago still persist while new ones continue to emerge. *International Journal of Food Microbiology, 139*, pp. S3–S15.

Newton, A., et al. (2012). Increasing rates of vibriosis in the United States, 1996-2010: review of surveillance data from 2 systems. *Clinical and Infectious Disease, 54*, pp. S391–S395.

Nishibuchi, M. and Kaper, J.B. (1995). Thermostable direct hemolysin gene of *Vibrio parahaemolyticus*: a virulence gene acquired by a marine bacterium. *Infection and Immunity, 63*, pp. 2093–2099.

Nordstrom, J.L., et al. (2007). Development of a multiplex real-time PCR assay with an internal amplification control for the detection of total and pathogenic *Vibrio parahaemolyticus* bacteria in oysters. *Applied and Environmental Microbiology, 73*, pp. 5840–5847.

Notomi, T., et al. (2000). Loop mediated isothermal amplification of DNA. *Nucleic Acids Research, 20*, pp. 63.

O'Grady, J., et al. (2009). tmRNA-a novel high-copy-number RNA diagnostic target-its application for *Staphylococcus aureus* detection using real-time NASBA. *FEMS Microbiology Letters, 301*, pp. 218–223.

O'Regan, E., et al. (2008). Development of a real-time multiplex PCR assay for the detection of multiple Salmonella serotypes in chicken samples. *BMC Microbiology, 8*, pp. 156.

Ohtsuka, K., et al. (2005). Detection of Salmonella enterica in naturally contaminated liquid eggs by loopmediated isothermal amplification, and characterization of *Salmonella* isolates. *Applied and Environmental Microbiology, 71*, pp. 6730–6735.

Okamura, M., et al. (2009). Rapid, sensitive, and specific detection of the O4 group of *Salmonella enterica* by loop-mediated isothermal amplification. *Avian Diseases, 53*, pp. 216–221.

Oliver, S.P., Jayarao, B.M. and Almeida, R.A. (2005). Foodborne pathogens in milk and the dairy farm environment: food safety and public health implications. *Foodborne Pathogens and Diseases, 4,* pp. 115–129.

Olsen, J.E., et al. (1995). Probes and polymerase chain reaction for detection of food-borne bacterial pathogens. *International Journal of Food Microbiology, 28,* pp. 1–78.

Omiccioli, E., et al. (2009). A new platform for real-time PCR detection of *Salmonella* spp., *Listeria monocytogenes* and *Escherichia coli* O157 in milk. *Food Microbiology, 26,* pp. L615–622.

Omurtag, I., et al. (2013). The risk of transfer of foodborne bacterial hazards in Turkey through the consumption of meat; risk ranking of muscle foods with the potential to transfer *Campylobacter* spp. *Food Security, 5,* pp. 117–127.

Oravcová, K., et al. (2005). Detection and quantification of *Listeria monocytogenes* by 5'-nuclease polymerase chain reaction targeting the *actA* gene. *Letters of Applied Microbiology, 42,* pp. 15–18.

Orsi, R.H., et al. (2008). Short-term genome evolution of *Listeria monocytogenes* in a non-controlled environment. *BMC Genomics, 9,* pp. 539.

Pangallo, D., et al. (2001). Detection of *Listeria monocytogenes* by polymerase chain reaction oriented to *inlB* gene. *New Microbiology, 24,* pp. 333–339.

Panicker, G., Myers, M.L. and Bej, A.K. (2004). Rapid detection of Vibrio vulnificus in shellfish and Gulf of Mexico water by real-time PCR. *Applied Environment Microbiology, 70,* pp. 498–507.

Park, S.H., et al. (2014). Current and emerging technologies for rapid detection and characterization of *Salmonella* in poultry and poultry products. *Food Microbiology, 38,* pp. 250–262.

Park, S.H., et al. (2011). Multiplex PCR assay for the detection and quantification of *Campylobacter* spp., *Escherichia coli* O157:H7, and *Salmonella* serotypes in water samples. *FEMS Microbiology Letters, 316,* pp. 7–15.

Park, S.H., et al. (2009). Identification of Salmonella enterica subspecies I, *Salmonella enterica* serovars Typhimurium, Enteritidis and Typhi using multiplex PCR. *FEMS Microbiology Letters, 301,* pp. 137–146.

Pass, M.A., Odedra, R. and Batt, R.M. (2000). Multiplex PCRs for identification of *Escherichia coli* virulence gene. *Journal of Clinical Microbiology, 38,* pp. 2001–2004.

Patel, J.R., et al. (2006). Rapid detection of *Salmonella* from hydrodynamic pressure-treated poultry using molecular beacon real-time PCR. *Food Microbiology, 23,* pp. 39–46.

Paydar, M., Teh, C.S.J. and Thong, K.L. (2013). Prevalence and characterisation of potentially virulent *Vibrio parahaemolyticus* in seafood in Malaysia using conventional methods, PCR and REP-PCR. *Food Control, 32,* pp. 13–18.

Pichler, J., et al. (2011). Lessons to be learned from an outbreak of foodborne Listeriosis, Austria 2009–2010. *Food Protection Trends, 31,* pp. 268–273.

Pollard, D.R., et al. (1990). Rapid and specific detection of verotoxin genes in *Escherichia coli* by the polymerase chain reaction. *Journal of Clinical Microbiology, 28,* pp. 540–545.

Popoff, M.Y., Bockemühl, J. and Gheesling, L.L. (2003). Supplement 2001 (no. 45) to the Kauffmann–White scheme. *Research Microbiology, 154,* pp. 173–174.

Raghunath, P. (2015). Roles of thermostable direct hemolysin (TDH) and TDH-related hemolysin (TRH) in *Vibrio parahaemolyticus. Frontiers in Microbiology, 5,* pp. 805.

Rantsiou, K., et al. (2008). Detection, quantification and vitality of *Listeria monocytogenes* in food as determined by quantitative PCR. *International Journal of Food Micrbiology, 121,* pp. 99–105.

Rasooly, A. and Herold, K.E. (2008). Food microbial pathogen detection and analysis using DNA microarray technologies. *Foodborne Pathogens and Diseases, 5,* pp. 531–550.

Rawool, D.B., et al. (2007). A multiplex PCR for detection of virulence associated genes in *Listeria monocytogenes. International Journal of Food Safety, 9,* pp. 56–62.

Reddy, T.B., et al. (2015). The Genomes OnLine Database (GOLD) v.5: a metadata management system based on a four level (meta) genome project classification. *Nucleic Acids Research, 43*, pp. D1099–106.

Reeves, M.W., et al. (1989). Clonal nature of *Salmonella typhi* and its genetic relatedness to other salmonellae as shown by multilocus enzyme electrophoresis, and proposal of *Salmonella bongori* comb. nov. *Journal of Clinical Microbiology, 27*, pp. 313–320.

Reuter, S., et al. (2013). Rapid bacterial whole-genome sequencing to enhance diagnostic and public health microbiology. *JAMA Internal Medicine, 173*, pp. 1397–1404.

Robert-Pillot, A., et al. (2010). Total and pathogenic *Vibrio parahaemolyticus* in shrimp: fast and reliable quantification by real-time PCR. *International Journal of Food Microbiology, 143*, pp. 190–197.

Robert-Pillot, A., et al. (2004). Occurrence of the *tdh* and *trh* genes in *Vibrio parahaemolyticus* isolates from waters and raw shellfish collected in two French coastal areas and from seafood imported into France. *International Journal of Food Microbiology, 91*, pp. 319–325.

Rodríguez-Lázaro, Hernández, M. and Pla, M. (2004). Simultaneous quantitative detection of *Listeria* spp. and *Listeria monocytogenes* using a duplex real-time PCR-based assay. *FEMS Microbiology Letters, 233*, pp. 257–267.

Rosec, J.P., et al. (2009). Detection of total and pathogenic *Vibrio parahaemolyticus* in shellfish: comparison of PCR protocols using pR72H or toxR targets with a culture method. *International Journal of Food Microbiology, 129*, pp. 136–145.

Rosec, J.P., et al. (2012). The international standard ISO/TS 21872-1 to study the occurrence of total and pathogenic *Vibrio parahaemolyticus* and *Vibrio cholerae* in seafood: ITS improvement by use of a chromogenic medium and PCR. *International Journal of food Micrbiology, 157*, pp. 189–194.

Rossmanith, P., et al. (2006). Detection of *Listeria monocytogenes* in food using a combined enrichment/real-time PCR method targeting the *prfA* gene. *Research Microbiology, 157*, pp. 763–771.

Ruiz-Rueda, O., et al. (2011). Multiplex real-time PCR for the simultaneous detection of *Salmonella* spp. and *Listeria monocytogenes* in food samples. *Food Analytic Method, 4*, pp. 131–138.

Rychli, K., et al. (2014). Genome sequencing of *Listeria monocytogenes* "Quargel" listeriosis outbreak strains reveals two different strains with distinct in vitro virulence potential. *PLoS One, 9*, pp. 89964.

Ryu, J., et al. (2013). Simultaneous detection of *Listeria* species isolated from meat processed foods using multiplex PCR. *Food Control, 32*, pp. 659–664.

Sabat, A.J., et al. (2013). Overview of molecular typing methods for outbreak detection and epidemiological surveillance. *European Surveillance, 18*, pp. 20380.

Saleh, Al-Othrubi, M., et al. (2011). Rapid detection and E-test antimicrobial susceptibility testing of *Vibrio parahaemolyticus* isolated from seafood and environmental sources in Malaysia. *Saudi Medical Journal, 32*, pp. 400–406.

Salem, I.B., Aouni, M. and Mzoughi, R., (2010). Two five-plex PCRs methods for identification of common *Salmonella* spp. serotypes. *Annals of Microbiology, 60*, pp. 135–141.

Sallen, B., et al. (1996). Comparative analysis of 16S and 23S rRNA sequences of *Listeria* species. *International Journal of Systematic Bacteriology, 46*, pp. 669–674.

Sanger, F., Nicklen, S. and Coulson, A.R. (1977). DNA sequencing with chain-terminating inhibitors. *Procedings of the National Academy of Sciences of the United States of America, 74*, pp. 5463–5467.

Scallan, E., et al. (2011). Foodborne illness acquired in the United States—major pathogens. *Emerging infectious Disesase, 17*, pp. 7–15.

Severgnini, M., et al. (2011). Advances in DNA microarray technology for the detection of foodborne pathogens. *Food Bioprocess Technology, 4*, pp. 936–953.

Shan, X., et al. (2012). Rapid detection of foodborne *Listeria monocytogenes* by real-time quantitative loop-mediated isothermal amplification. *Food Science and Biotechnology, 21*, pp. 101–106.

Shao, Y., et al. (2011). Development of multiplex loop-mediated isothermal amplification-RFLP (mLAMP-RFLP) to detect *Salmonella* spp. and *Shigella* spp. in milk. *International Journal of Food Microbiology, 148*, pp. 75–79.

Shendure, J. and Ji, H. (2008). Next-generation DNA sequencing. *Nature of Biotechnology, 26*, pp. 1135–1145.

Silva, D.S.P., et al. (2011). Multiplex PCR for the simultaneous detection of *Salmonella* spp. and *Salmonella* Enteritidis in food. *International Journal of Food Science and Technology, 46*, pp. 1502–1507.

Simon, M.C., Gray, D.I. and Cook, N. (1996). DNA extraction and PCR methods for the detection of *Listeria monocytogenes* in cold-smoked salmon. *Applied and Environmental Microbiology, 62*, pp. 822–824.

Simpkins, S.A., et al. (2000). An RNA transcription-based amplification technique (NASBA) for the detection of viable *Salmonella enterica*. *Letters of Applied Microbiology, 30*, pp. 75–79.

Singh, D.V., et al. (2001). Molecular analysis of *Vibrio cholerae* O1, O139, non-O1, and non-O139 strains: clonal relationships between clinical and environmental isolates. *Applied and Environmental Microbiology, 67*, pp. 910–921.

Singh, J., Batish, V.K. and Grover, S. (2009). A molecular beacon-based duplex real-time polymerase chain reaction assay for simultaneous detection of *Escherichia coli* O157:H7 and *Listeria monocytogenes* in milk and milk products. *Foodborne Pathogens and Diseases, 6*, pp. 1195–1201.

Singh, P. and Mustapha, A. (2014). Development of a real time PCR melt curve assay for simultaneous detection of virulent and antibiotic resistant *Salmonella*. *Food Microbiology, 44*, pp. 6–14.

Skerman, V.B.D., McGowan, V. and Sneath, P.H.A. (1980). Approved Lists of Bacterial Names. *International Journal of Systematic Bateriology, 30*, pp. 225–420.

Skyberg, J.A., Logue, C. M. and Nolan, L. K. (2006). Virulence genotyping of *Salmonella* spp. with multiplex PCR. *Avian Diseases, 50*, pp. 77–81.

Son, I., et al. (2014). Detection of five Shiga toxin-producing *Escherichia coli* genes with multiplex PCR. *Food Microbiology, 40*, pp. 31–40.

Song, T., et al. (2005). Sensitive and rapid detection of *Shigella* and enteroinvasive *Escherichia coli* by a loop-mediated isothermal amplification method. *FEMS Microbiology Letters, 243*, pp. 259–263.

Su, C.Y. and Liu, C. (2007). *Vibrio parahaemolyticus*: A concern of seafood safety. *Research in Microbiology, 24*, pp. 549–558.

Suffredini, E., et al. (2014). Detection and quantification of *Vibrio parahaemolyticus* in shellfish from Italian production areas. *International Journal of Microbiology, 184*, pp. 14–20.

Sujeewa, A.K.W., Norrakiah, A.S. and Laina, M. (2009). Prevalence of toxic genes of *Vibrio parahaemolyticus* in shrimps (*Penaeus monodon*) and culture environment. *International Food Research Journal, 16*, pp. 89–95.

Suo, B., et al. (2010a). Development of an oligonucleotide-based microarray to detect multiple foodborne pathogens. *Molecualr and Cellular Probes, 24*, pp. 77–86.

Suo, B., et al. (2010b). A multiplex real-time polymerase chain reaction for simultaneous detection of *Salmonella* spp., *Escherichia coli* O157, and *Listeria monocytogenes* in meat products. *Foodborne Pathogems and Diseases, 7*, pp. 619–628.

Swetha, C.S., et al. (2012). Detection of *Listeria monocytogenes* in fish samples by PCR assay. *Annals Biological Research, 3*, pp. 1880–1884.

Taha, E.G., et al. (2010). Rapid detection of *Salmonella* in chicken meat using immunomagnetic separation, CHROMagar,

ELISA and real-time polymerase chain reaction (RT-PCR). *International Journal of Poultry Science, 9*, pp. 831–835.

Tang, M.J., et al. (2011). Rapid and sensitive detection of *Listeria monocytogenes* by loop-mediated isothermal amplification. *Current Microbiology, 63*, pp. 511–516.

Tourlousse, D.M., et al. (2012). A polymer microfluidic chip for quantitative detection of multiple water- and foodborne pathogens using real-time fluorogenic loop-mediated isothermal amplification. *Biomedical Microdevices, 14*, pp. 769–778.

Traunšek, U., et al. (2011). Novel cost-efficient real-time PCR assays for detection and quantitation of *Listeria monocytogenes. Journal of Microbiology Methods, 85*, pp. 40–46.

Tsai, Y.L., Palmer, C.J. and Sangermano, L.R. (1993). Detection of *Escherichia coli* in sewage and sludge by polymerase chain reaction. *Applied and Environmental Microbiology, 59*, pp. 353–357.

Tyagi, A., et al. (2009). Detection of *Vibrio parahaemolyticus* in tropical shellfish by SYBR green real-time PCR and evaluation of three enrichment media. *International Journal of Food Microbiology, 129*, pp. 124–130.

Ueda, S. and Kuwabara, Y. (2009). The rapid detection of *Salmonella* from food samples by loop-mediated isothermal amplification (LAMP). *Biocontrol Science, 14*, pp. 73–76.

USDA. (2012). FSIS procedure for the use of a *Listeria monocytogenes* polymerase chain reaction (PCR) screening test. Effective June 2012. MLG 8A.05. Available at: http://www.fsis.usda.gov/wps/wcm/connect/1b6843c4-147f-4228-ae6a-c0993c0cf5fa/MLG-8A.pdf?MOD=AJPERES [Accessed December 3, 2015].

Uyttendaele, M., et al. (1995). Development of NASBA®, a nucleic acid amplification system, for identification of *Listeria monocytogenes* and comparison to ELISA and a modified FDA method. *International Journal of Food Microbiology, 27*, pp. 77–89.

Van de Venter, T. (2000). Emerging food-borne diseases: a global responsibility. *Food, Nutrition and Agriculture, 26*, pp. 4–13.

Velazquez-Roman, J., et al. (2012). Association of pandemic *Vibrio parahaemolyticus* O3: K6 present in the coastal environment of Northwest Mexico with cases of recurrent diarrhea between 2004 and 2010. *Applied and Environmental Microbiology, 78*, pp. 1794–1803.

Velusamy, V., et al. (2010). An overview of foodborne pathogen detection: in the perspective of biosensors. *Biotechnology Advances, 28*, pp. 232–254.

Verstraete, K., et al. (2012). Evaluation of a multiplex-PCR detection in combination with an isolation method for STEC O26, O103, O111, O145 and sorbitol fermenting O157 in food. *Food Microbiology, 29*, pp. 49–55.

Vidal, M., et al. (2005). Single multiplex PCR assay to identify simultaneously the six categories of diarrheagenic *Escherichia coli* associated with enteric infections. *Journal of Clinical Microbiology, 43*, pp. 5362–5365.

Vimala, B., Lin, T.K. and Chong, V.C. (2010). Isolation, detection and genomic differentiation of *Vibrio cholerae* and *Vibrio parahaemolyticus* in Bachok, Kelantan. *Malaysian Journal of Science, 29*, pp. 1–10.

Volokhov, D., et al. (2002). Identification of *Listeria* species by microarray-based assay. *Journal of Clinical Microbiology, 40*, pp. 4720–4728.

Vuddhakul, V., et al. (2000). Isolation of a pandemic O3:K6 clone of a *Vibrio parahaemolyticus* strain from environmental and clinical sources in Thailand. *Applied and Environmental Microbiology, 66*, pp. 2685–2689.

Välimaa, A.L., Tilsala-Timisjärvi, A. and Virtanen, E. (2015). Rapid detection and identification methods for *Listeria monocytogenes* in the food chain – a review. *Food Control, 55*, pp. 103–114.

Wan, C., et al. (2012). Development of a propidium monoazide treatment combined with loop-mediated isothermal amplification

(PMA-LAMP) assay for rapid detection of viable *Listeria monocytogenes. International Journal of Food Science and Technology, 47,* pp. 2460–2467.

Wan, J., et al. (2000). Probelia^{TM} PCR system for rapid detection of *Salmonella* in milk powder and ricotta cheese. *Letters in Applied Microbiology, 30,* pp. 267–271.

Wan, J., et al. (2003). Detection of *Listeria monocytogenes* in salmon using the Probelia polymerase chain reaction system. *Journal of Food Protection, 66,* pp. 436–440.

Wang, Y., et al. (2009). Study on loop-mediated isothermal amplification assay for detection of *Salmonella* in meat products. *Third International Conference on Bioinformatics and Biomedical Engineering,* pp. 1–6.

Wang, D., et al. (2011). Rapid detection of *Listeria monocytogenes* in raw milk with loop-mediated isothermal amplification and chemosensor. *Journal of Food Science, 76,* pp. M611–M615.

Wang, D.G., et al. (2015). Two methods for increased specificity and sensitivity in loop-mediated isothermal amplification. *Molecules, 20,* pp. 6048–6059.

Wang, F., et al. (2014). Evaluation of a loop-mediated isothermal amplification suite for the rapid, reliable, and robust detection of shiga toxin-producing *Escherichia coli* in produce. *Applied and Environmental Microbiology, 80,* pp. 2516–2525.

Wang, G., Clark, C.G. and Rodgers, F.G. (2002). Detection in *Escherichia coli* of the genes encoding the major virulence factors, the genes defining the O157:H7 serotype, and components of the Type 2 Shiga toxin family by multiplex PCR. *Journal of Clinical Microbiology, 40,* pp. 3613–3619.

Wang, L., et al. (2008). Specific and rapid detection of foodborne *Salmonella* by loop-mediated isothermal amplification method. *Food Research International, 41,* pp. 69–74.

Wang, L., et al. (2013). Detection of *Vibrio parahaemolyticus* in food samples using in situ loop-mediated isothermal amplification method. *Gene, 515,* pp. 421–425.

Wang, R.Z., et al. (2011). Detection and identification of *Vibrio parahaemolyticus* by multiplex PCR and DNA-DNA hybridization on a microarray. *Journal of Genetics and Genomics, 38,* pp. 129–135.

Wang, R., et al. (2015). The pathogenesis, detection, and prevention of *Vibrio parahaemolyticus. Frontiers in Microbiology, 6,* pp. 144.

Wang, R.F., Cao, W.W. and Johnson, M.G. (1992). 16S rRNA-based probes and polymerase chain reaction method to detect *Listeria monocytogenes* cells added to foods. *Applied and Environmtal Microbiology, 58,* pp. 2827–2831.

Wang, X.W., et al. (2007). Development and application of an oligonucleotide microarray for the detection of food-borne bacterial pathogens. *Applied Microbiology and Biotechnology, 76,* pp. 225–233.

Ward, L.N. and Bej, A.K. (2006). Detection of *Vibrio parahaemolyticus* in shellfish by use of multiplexed real-time PCR with TaqMan fluorescent probes. *Applied and Environmental Microbiology, 72,* pp. 2031–2042.

Wei, S., et al. (2014). Multiplex PCR assays for the detection of *Vibrio alginolyticus, Vibrio parahaemolyticus, Vibrio vulnificus* and *Vibrio cholerae* with an internal amplification control. *Diagnostic Microbiology and Infectious Disease, 79,* pp. 115–118.

Wernars, K., et al. (1991). Use of the polymerase chain reaction for direct detection of *Listeria monocytogenes* in soft cheese. *Journal of Appled Bacteriology, 70,* pp. 121–126.

Wingstrand, A., et al. (2006). Fresh chicken as main risk factor for campylobacteriosis, Denmark. *Emerging Infectious Diseases, 12,* pp. 280–284.

Winters, D.K., Maloney, T.P. and Johnson, M.G. (1999). Rapid detection of *Listeria monocytogenes* by a PCR assay specific for an aminopeptidase. *Molecular and Cellular Probes, 13,* pp. 127–131.

Wolffs, P., et al. (2006). Direct quantification and detection of salmonellae in biological samples without enrichment, using two-step filtration and real-time PCR. *Applied and Environmental Microbiology, 72*, pp. 3896–3900.

Wong, H.C., et al. (2000). Characterization of *Vibrio parahaemolyticus* isolates obtained from foodborne illness outbreaks during 1992 through 1995 in Taiwan. *Journal of Food Protection, 63*, pp. 900–906.

Wu, R., et al. (2014). Development of double loop-mediated isothermal amplification to detect *Listeria monocytogenes* in food. *Current Microbiology, 69*, pp. 839–845.

Xin, N., et al. (2011). Establishment of the nucleic acid sequence based amplification method of detecting *Vibrio parahaemolyticus. Chinese Journal of Preventive Veterinary Medicine, 11*, p. 14.

Xu, X., et al. (2014). Prevalence, pathogenicity and serotypes of *Vibrio parahaemolyticus* in shrimp from Chinese retail markets. *Food Control, 46*, pp. 81–85.

Xu, Z., et al. (2012). Development and application of loop-mediated isothermal amplification assays on rapid detection of various types of staphylococci strains. *Food Research International, 47*, pp. 166–173.

Yamamoto, A., et al. (2008). Quantitative modeling for risk assessment of *Vibrio parahaemolyticus* in bloody clams in southern Thailand. *International Journal of Food Microbiology, 124*, pp. 70–78.

Yamazaki, W., et al. (2008). Development of a loop mediated isothermal amplification assay for sensitive and rapid detection of *Vibrio parahaemolyticus. BMC Microbiology, 30*, pp. 163.

Yamazaki, W., et al. (2010). Development of a loop mediated isothermal amplification assay for sensitive and rapid detection of the *tdh* and *trh* genes of *Vibrio parahaemolyticus* and related *Vibrio* species. *Applied and Environmental Microbiolgy, 76*, pp. 820–828.

Yamazaki, W., et al. (2009). Comparison of loop-mediated isothermal amplification assay and conventional culture methods for detection of *Campylobacter jejuni* and *Campylobacter coli* in naturally contaminated chicken meat samples. *Applied and Evironmental Microbiology, 75*, pp. 1597–1603.

Yang, J.L., et la. (2010). Simple and rapid detection of Salmonella serovar Enteritidis under field conditions by loop-mediated isothermal amplification. *Journal of Applied Microbiology, 109*, pp. 1715–1723.

Yang, J.R., et al. (2007). Comparison between O serotyping method and multiplex real-time PCR to identify diarrheagenic *Escherichia coli* in Taiwan. *Journal of Clinical Microbiology, 45*, pp. 3620–3625.

Yang, Y., et al. (2013). Magnetic nano-beads based separation combined with propidium monoazide treatment and multiplex PCR assay for simultaneous detection of viable *Salmonella* Typhimurium, *Escherichia coli* O157:H7 and *Listeria monocytogenes* in food products. *Food Microbiology, 34*, pp. 418–424.

Ye, Y., et al. (2011). Application of *in situ* loop-mediated isothermal amplification method for detection of *Salmonella* in foods. *Food Control, 22*, pp. 438–444.

Yi, M.Y., et al. (2014). Real time loop-mediated isothermal amplification using a portable fluorescence scanner for rapid and simple detection of *Vibrio parahaemolyticus. Food Control, 41*, pp. 91–95.

Yu, W.T., et al. (2013). Prevalence of *Vibrio parahaemolyticus* in oyster and clam culturing environments in Taiwan. *International Journal of Food Microbiology, 160*, pp. 185–192.

Yuan, Y., et al. (2009). Universal primer-multiplex PCR approach for simultaneous detection of *Escherichia coli, Listeria monocytogenes*, and *Salmonella* spp. in food samples. *Journal of Food Science, 74*, pp. M446–M452.

Zarei, M., Maktabi, S. and Ghorbanpour, M. (2012). Prevalence of *Listeria monocytogenes, Vibrio parahaemolyticus, Staphylococcus aureus*, and *Salmonella* spp. in seafood products using multiplex polymerase chain reaction. *Foodborne Pathogens and Disease, 9*, pp. 108–112.

Zeng, H., et al. (2006). Multiplex PCR identification of *Listeria monocytogenes* isolates from milk and milk-processing environments. *Journal of the Science of Food and Agriculture, 86*, pp. 367–371.

Zeng, J., et al. (2014). Rapid detection of *Vibrio parahaemolyticus* in raw oysters using immunomagnetic separation combined with loop-mediated isothermal amplification. *International Journal of Food Microbiology, 174*, pp. 123–128.

Zhang, L., et al. (2012). A loop-mediated isothermal amplification method targets the *HisJ* gene for the detection of foodborne *Salmonella*. *European Food Research and Technology, 234*, pp. 1055–1062.

Zhang, D., et al. (2009). Simultaneous detection of *Listeria monocytogenes, Staphylococcus aureus, Salmonella enterica* and *Escherichia coli* O157:H7 in food samples using multiplex PCR method. *Journal of Food Safety, 29*, pp. 348–363.

Zhang, G. (2013). Foodborne pathogenic bacteria detection: an evaluation of current and developing methods. *The Meducator, 1*, pp. 15.

Zhang, L. and Orth, K. (2013). Virulence determinants for *Vibrio parahaemolyticus* infection. *Current Opinion in Microbiology, 16*, pp. 70–77.

Zhang, X.H. and Austin, B. (2005). Haemolysins in *Vibrio* species. *Journal of Applied Microbiology, 98*, pp. 1011–1019.

Zhao, X., et al. (2010). Development and application of a loop-mediated isothermal amplification method on rapid detection *Escherichia coli* O157 strains from food samples. *Molecular Biology Reports, 37*, pp. 2183–2188.

Zhao, X., et al. (2014). Advances in rapid detection methods for foodborne pathogens. *Journal of Microbiology and Biotechnology, 24*, pp. 297–312.

Zhou, B., et al. (2013). Simultaneous detection of six food-borne pathogens by multiplex PCR with GEXP analyzer. *Food Control, 32*, pp. 198–204.

Zhou, X. and Jiao, X. (2005). Polymerase chain reaction detection of *Listeria monocytogenes* using oligonucleotide primers targeting *actA* gene. *Food Control, 16*, pp. 125–130.

Zunabovic, M., Domig, K.J., and Kneifel, W. (2011). Practical relevance of methodologies for detecting and tracing of *Listeria monocytogenes* in ready-to-eat foods and manufacture environments – a review. *LWT-Food Science and Technology, 44*, pp. 351–362.

8

Rapid, Alternative Methods for *Salmonella* Detection in Food

Anna Zadernowska and Wioleta Chajęcka-Wierzchowska

University of Warmia and Mazury in Olsztyn, Poland

8.1 Introduction

Both the European Food Safety Authority (EFSA) and the Centers for Disease Control and Prevention (CDC) have identified *Salmonella* sp. as the main culprit of food poisoning around the world. The genus *Salmonella* is divided into two species, which are further divided into 6 subspecies and more than 2500 serotypes (serovars)—all of which are regarded as pathogens (Brenner *et al.*, 2000):

- *Salmonella enterica*
  - *Salmonella enterica subspecies enterica*
  - *Salmonella enterica subspecies salamae*
  - *Salmonella enterica subspecies arizonae*
  - *Salmonella enterica subspecies diarizonae*
  - *Salmonella enterica subspecies houtenae*
  - *Salmonella enterica subspecies indica*
- *Salmonella bongori*

According to the Kauffmann–White scheme, subspecies are further divided into serovars, which can be differentiated by serotyping, the traditional and standardized phenotypic method used for epidemiological typing of *Salmonella*. Serotyping is based on the antigenic variability at lipopolysaccharide moieties—O antigens, flagellar proteins—H1, H2 antigens, and capsular polysaccharides -Vi antigens (Malorny *et al.*, 2009).

The bacteria are considerably resistant to external conditions. They grow at temperatures ranging from 7 to 48 °C with an optimum growth temperature at 37 °C and at pH 4.0–9.5 with an optimal growth at pH 6.5–7.5. *Salmonella* grows optimally at a water activity of 0.995 (Ellermeier and Slauch, 2006). They die at pasteurization temperature, although they are extremely resistant to drying. They can survive for a long time in dry products in the state of anabiosis. According to study results, these rods can also survive in products with low water activity and high fat content (protective effect on cells), for example, in chocolate (Podolak *et al.*, 2010). They survive equally well in frozen food. In the opinion of the Sub-Committee on Veterinary Public Health of 15 April 2003, food that is potentially hazardous to public health includes raw meat and some products intended for eating raw, raw and undercooked poultry meat products, eggs and products containing raw eggs, as well as non-pasteurized milk and some dairy products (Commission Regulation, 2005).

8.2 Conventional Methods and Their Modifications

Determination of *Salmonella* spp. in food products always involves detecting those rods in a specific amount/volume of a product

Foodborne Pathogens and Antibiotic Resistance, First Edition. Edited by Om V. Singh.

(usually 25 g/l, very rarely 10 g/ml), but the *Salmonella* spp. count in food is not determined. The sample should reflect the image of the product from which it originates as precisely as possible. The first stage involves non-selective culturing, both in the conventional method and its modifications. This is extremely important because food production is associated with its processing, for example, heating, acidifying, which may kill or sub-lethally damage most cells. Omitting the stage of pre-culturing and inoculating the material directly on the solid medium can give false negative results. A very small number of cells or damaged cells, may result in a lack of colonies on the solid medium (Zadernowska *et. al.*, 2014b; Jasson *et al.*, 2011). When a product is stored during the period of its shelf life, bacteria can multiply to a level that is dangerous to a consumer. Although standard methods of detecting *Salmonella* spp. in food by culturing are in common use and are well-developed, they consist of several stages and the need for biochemical confirmation extends the time needed for an analysis for up to seven days (Junillon *et al.*, 2014). In the EU, *Salmonella* spp. is determined in accordance with the ISO 6579 2003 (Microbiology of food and animal feeding stuffs - Horizontal method for detection of *Salmonella* spp.) includes four stages of the detection process and depending on the need to obtain confirmations, it lasts from five to seven days:

- Pre-enrichment in non-selective liquid medium
- Selective enrichment in liquid media
- Plating on selective media
- Serological and biochemical identification of suspected colonies

There are a number of modifications to the conventional plate method, for example, using chromogenic media, in which the action of specific bacterial enzymes give colored reactions (Lauer and Martinez, 2009). Such media are offered by most media suppliers. Despite having different trade names, many of them have two chromogenic substrates, which help to identify *Salmonella* spp. rods: X-gal (5-bromo-4-chloro-3-indolyl-β-D-galactopyranoside) and magenta-caprilate (5-bromo-6-chloro-3-indolyl caprylate. X-gal, which is a substrate for ß-D-galactosidase, produced by most genera of *Enterobacteriaceae* bacteria. An enzymatic reaction of the bacteria colonies with X-gal makes them turn blue-green. Since salmonella colonies hydrolyze both X-gal and 5-bromo-6-chloro-3-indolyl caprylate, their colonies become purple (Lauer and Martinez, 2009). Using these types of media reduces the time needed to do the tests, but still they last several days (Schönenbrücher *et al.*, 2008). The long time needed for such tests makes the food trade more difficult; this applies especially to food with a short shelf life. Ready-made products cannot be released from a warehouse and placed on the market until their microbiological safety is confirmed. Apart from the long time needed to obtain results of standard tests, there are some restrictions to those that are modified by chromogenic media. The multi-stage process, readouts based on a color change, and observations of colony morphology may result in some errors, especially with a large number of microorganisms. Therefore, determination of the microorganisms in food can be done only by well-trained, experienced personnel. It is another difficulty that in food processing, microorganisms are exposed to stressful conditions that may result in their losing the ability to form colonies on solid media, although they retain their metabolic activityz—they are viable but non-culturable (VBNC)—or cause changes in their morphology, which may make it difficult or impossible to identify their growth on solid media (Jasson *et al.*, 2011; Zadernowska *et al.*, 2014b).

Apart from the culturing methods, there are very well-developed immunoenzymatic methods (ELISA) based on antigen-antibody reactions, whose result is usually read out as a change of color. There are a number of firms that offer simple tests, for example, 3 M™ TECRA™ UNIQUE SALMONELLA (NOACK), Singlepath® Salmonella (Merck Millipore).

There are also fully automatic devices used to conduct analyses by this method, for example, miniVIDAS (Biomerieux), in which 12 samples can be analyzed simultaneously. *Salmonella* spp. can be determined with VIDAS Salmonella Xpress (SLMX) immunoenzymatic tests, which operate on the principle of detecting specific antigens. The test result is read out by the fluorescent method (ELFA). The method is based on a multi-stage reaction. After its preliminary culturing, the analyzed sample is applied onto test strips, that is, a set of wells with reagents. Apart from strips, pipettes (included) with walls coated with antibodies are also put into the device. The reaction suspension in the analyzer is drawn up and pushed down by the pipettes. Antibodies bind with the antigens present in a sample and they are stained by alkaline phosphatase. During the last stage, the pipette draws in the substrate (4-methylubellipheryl phosphate). The conjugate of the enzyme catalyzes the hydrolysis of the substrate to the fluorescence product. The intensity of the fluorescence is measured at a wavelength of 450 nm. Fluorescence is measured twice. The first result shows the background fluorescence, the other is the value following incubation of the substrate with the enzyme. This is the basis for the device to calculate the test result and to interpret it as positive or negative. If a positive result is obtained, it should be confirmed by conventional methods by culturing on two plates with selective media (Chajęcka-Wierzchowska *et al.*, 2012). Although the tests are AFNOR validated, the manufacturer recommends that validation should be carried out for individual products under study (because of the "background"). A study conducted by Chajęcka-Wierzchowska *et al.* (2012) on 180 samples of contaminated meat confirmed the high sensitivity of the tests. However, studies have shown that when samples are highly contaminated with accompanying microflora, false positive results may be obtained. Such results using a miniVIDAS analyser have been obtained by other studies. This may be caused by cross-reactions with other rods of the *Enterobacteriaceae* family, which are found in the matrix (meat). The need to confirm such findings increases costs and extends the time of analyses.

8.3 Alternative Methods—Definitions, Requirements

Time is a very important factor in detecting the *Salmonella* in foods and for this reason, it is needed to use new methods that can give results in a short time. Alternative methods can be applied in food microbiological analysis provided they are validated and equivalent results are obtained in the conventional and alternative method. Alternative methods are frequently validated by respected organisations, such as AOAC, MicroVal, AFNOR, or NORDVAL. General rules and the technical protocol of validation of alternative methods in microbiological analysis of food, fodders, environmental samples, and veterinary samples are presented in the ISO 16140:2003 standard (Microbiology of food and animal feeding stuffs - Protocol for the validation of alternative methods). The standard provides the following definition of an alternative method:

Alternative method—a method of analysis for detection or estimation of the same analysis as measured by using the corresponding reference method, for a given category of products. The method can be proprietary or non-commercial and covers an entire analytical procedure, that is, from the preparation of samples to the test results, either as such, or may include references to other procedures in order to be complete. The alternative method exhibits attributes appropriate to the users' needs, for example:

- speed of analysis and/or response
- ease of execution and/or automation
- analytical properties (precision, accuracy, limit of detection, etc.)
- miniaturization
- reduction of cost

8.3.1 Molecular Methods

A considerable increase in the number of culturing methods in recent years has been accompanied by intensive development of molecular biology methods, such as PCR (polymerase chain reaction) in its all variants (Cortez *et al.*, 2006). A PCR analysis generally consists of three stages: decomposition of a double helix of DNA, the binding of starters, and the addition of a new strand catalyzed by polymerase. These are relatively simple methods, but at the same time they are sensitive and specific and they reduce the analysis time by 90%. Determination of specific fragments of nucleic acids amplified in a PCR reaction is carried out by gel electrophoresis, which can be difficult using standard tests. Moreover, some of the reagents used in the determinations are regarded as harmful to humans and the environment.

A very common modification of the PCR is the real-time measurement of the amount of PCR products, which yields qualitative results of detection of bacteria very quickly. A determination uses probes calibrated with fluorescence dyes, which bind with target-amplified fragments of nucleic acids. It is a certain limitation, especially in the food industry, that the method may also give positive results when there are dead cells in a sample under analysis. The introduction of real-time PCR in pathogen diagnostics led to various assay developments for the detection of *Salmonella* (Hein *et al.*, 2006). The majority of real-time PCR assays developed target the *inv*A, many reagents and alternative instruments for real-time PCR are available gene. Commercially enabling a broad application of this new technique in routine diagnostic laboratories (BAX® - DuPont Qualicon Wilmington, iQ-Check Salmonella II Real-Time PCR Test Kit - Bio-Rad Laboratories, TaqMan Salmonella enterica - Applied Biosystems, Foster City) (Malorny *et al.*, 2009).

In recent years, there have been several reports on the application of the LAMP technique (loop-mediated isothermal amplification). LAMP is a one-step amplification reaction that amplifies a target DNA sequence with high sensitivity and specificity under isothermal conditions. The mechanism of the LAMP reaction can be explained in three steps: an initial step, a cycling amplification step, and an elongation step (Mori and Notomi 2009). LAMP is more specific, rapid, and simple to perform than PCR. A loop-mediated isothermal amplification assay was developed and validated for the specific detection of *Salmonella* in food by Li *et al.* (2009). The authors used specifically designed primers to target within the *phoP* gene and correctly identified 66 strains of *Salmonella* and 73 non-*Salmonella* strains tested.

Molecular methods, which consist of the ability of cells to bind to fluorescence-stained probe, consisting of short oligonucleotide sequences—FISH (fluorescence *in situ* hybridization), are less developed (despite being used in medicine relatively quickly) (Bishop, 2010; Lawson *et al.*, 2011; Loonen *et al.*, 2011; Wagner *et al.*, 2003). Connection of material in a sample with a complementary probe is observed under an epifluorescence microscope by means of a flow cytometer or capillary electrophoresis. FISH produces results after 20–26 hours and is regarded as a very sensitive method (Vieira-Pinto *et al.*, 2005). Vieira-Pinto *et al.* (2007) examined samples of naturally contaminated pork by the FISH technique and by the standard method. They used the FISH technique and detected *Salmonella* spp. rods in 16 (34%) out of 47 of samples under study, whereas the bacteria were detected only in 6 samples when the standard plate method was used. The considerable differences in the number of positive samples obtained by these methods are attributed to the fact that sub-lethally damaged cells or cells in the VBNC condition can occur, which were non-culturable on solid media, but which contained the appropriate amount of rRNA necessary to bind to the probe. A study conducted by Zadernowska *et al.* (2014b) also confirmed the effectiveness of the method. Owing to the application of an oligonucleotide probe (Sal3) in the determination of *Salmonella* spp. in meat, it was possible to obtain the results in a relatively short time. Commercial tests used to determine *Salmonella* by the FISH technique are offered by Vermicon (Germany)–VIT-Salmonella kit.

8.3.2 Phage-Based Methods

Bacteriophages are viruses that infect bacteria. Their global number is estimated to be about 10^{31} of molecules with highly differentiated genotypes; they are much more numerous than bacteria A phage molecule is called a virion and it consists of a capsid (protein coat) made up of a protein or a protein and lipid, surrounding the genome. Bacteriophage genetic material consists of DNA or—less frequently—RNA in one or two strands. Apart from a capsid, most phages also have tails and filaments, which are made of proteins. Viruses do not have their own metabolism, nor can they multiply on their own; they must infiltrate a host's cell. Another important feature of phages is their high specificity toward individual bacteria, which is associated with the presence of the right receptor on the surface of a bacterial cell (Grabow, 2004; Smartt *et al.*, 2012; Serwer *et al.*, 2007). A receptor can be a lipopolysaccharide, proteins, sugar molecules, or fimbriae. Usually, one virus species can multiply only in one bacteria species or within one specific strain (Hagens and Loessner, 2007). After infecting bacteria, a bacteriophage can destroy the host's cells by lysis or it can incorporate the nucleic acid to the bacterial chromosome, thereby starting the lysogenic cycle. During the lytic cycle, a phage activates processes that lead to the creation of a daughter virus molecule, resulting in the lysis of the bacterial cell and release of daughter bacteriophages. In the lysogenic cycle, phage DNA is integrated with the bacterial chromosome. The lytic cycle is started under stress conditions. Development of genetic engineering in recent years and expansion of the knowledge of bacteriophages creates better opportunities for their practical use in detecting *Salmonella* spp. in food. Salmonella rods are detected with both phages and their purified proteins.

Application of Phages T2 phages have been used successfully to determine *Salmonella* sp., for example, by Silva de Siqueira (2003). There are also commercial tests that employ phages to determine *Salmonella* sp., for example, RapidChek SELECT *Salmonella enteritidis.* The first stage of detecting *Salmonella* spp. involves pre-culturing that aims to recreate cells that could be damaged during processing, when cells of all food-contaminating microorganisms multiply, which —with a large number of accompanying microflora during further stages of determination—can give false-positive results since the main microflora that accompany *Salmonella* spp. are other *Enterobacteriaceae* rods. Microflora growth is controlled by the addition of a mixture of bacteriophages, which attack the bacteria accompanying *Salmonella* spp., thereby enabling its intensive growth that, in turn, helps to shorten the testing time.

Application of Purified Phage Proteins Since bacteriophages can cause considerable loss in food production, for example, in production of fermented dairy beverages, using live bacteriophages in determinations may be controversial. Since some mutations may occur during the replication process, with resultant mutants attacking other bacteria species, recombinant phage proteins can be used for determinations. For example, Singh *et al.* (2010) used phage proteins to detect the presence of *Salmonella* spp. Recombinant phage proteins have also been used in a new series of tests for miniVidas (Biomerieux) - VIDAS UP Salmonella (Junnillon *et al.*, 2014). Studies conducted, for example, by Zadernowska *et al.* (2014a), have shown the tests to be highly specific even when samples are highly contaminated with accompanying microflora. Phage proteins have also been used as a biological component of biosensors for detecting *Salmonella* spp. (Vinay *et al.*, 2015).

8.3.3 Biosensors

The use of biosensors is an important step in the development of alternative methods of determination of *Salmonella* sp., as they can simultaneously determine pathogens in a large number of samples with considerable miniaturisation of equipment. Moreover, using biosensors for analyses does not require long and complicated sample preparation and operating such devices is usually simple. Biosensors are generally made up of a biological component (nucleic acids, receptors, or antibodies) that are in direct contact with an appropriate part of the equipment processing a signal generated in the biological part (McGrath *et al.*, 2012; Li *et al.*, 2012). Depending on the method of signal processing used, biosensors can be divided into four fundamental groups: optical, mass, electrochemical, and thermal. There are two categories: sensors for direct detection of a target analyte and sensors with intermediate detection. Direct detection sensors are designed such that a biospecific reaction is directly identified in real time based on the measurement of physical changes induced by the complex formation. Intermediate detection sensors are those in which a biochemical reaction occurs and its products are subsequently detected by a sensor. There is also a division into sensors operating periodically and continuously. According to IUPAC (*International Union of Pure and Applied Chemistry*), a biosensor is a self-contained integrated device that is capable of providing specific quantitative or semi-quantitative analytical information using a biological recognition element (biochemical receptor), which is retained in direct spatial contact with an electrochemical transduction element (Thevenot *et al.*, 2001) Studies of the application of biosensors in the determination of *Salmonella* spp. with antibodies as biological agents have been conducted by, for example: Guntupalli *et al.* (2007), Valadez *et al.* (2009), and Li *et al.* (2012).

8.4 Conclusions

Despite well-developed conventional methods and a number of alternative one, there is still a need for new, highly specific *Salmonella* spp. detection methods in food. Despite many methods of detection and well-developed sampling plans, this bacteria is among the prime causes of food poisoning. Constant improvement of the methods and reducing the waiting period for a result is also linked to changing lifestyles. A return to the consumption of not-highly-processed food (which, in consequence, has a relatively short shelf-life), with no preservatives, without or with minimal thermal processing, increases the risk of poisoning with the rods. Due to the intensive development of nano-biotechnology and new bioinformatic tools, the most intensive development in the nearest future is going to involve methods that employ biosensors; currently, about 85% of all commercially available biosensors are used on blood glucose assays; in a few years, biosensors used to detect pathogens in food are likely to account for a greater portion than now. The next step involves the development of "biochips," which allow for further processing of the signal received from a biosensor and "Lab-on-Chip," which are miniaturized analytical systems.

References

Bishop R. (2010). Applications of fluorescence *in situ* hybridization (FISH) in detecting genetic aberrations of medical significance. *Bioscience Horizons*, *3*, pp. 85–95.

Brenner F.W., Villar R.G., Angulo F.J., Tauxe R. and Swaminathan B. (2000). Salmonella nomenclature. *Journal of Clinical Microbiology*, *38*(7), pp. 2465–2467.

Chajęcka-Wierzchowska W., Zadernowska A., Kłębukowska L. and Łaniewska-Trokenheim Ł. (2012). *Salmonella* Detection in Poultry Meat–Validation of VIDAS Xpress Automatic Enzyme-Linked Fluorescent Immunoassay-Based Method. *Journal of Food Safety*, *32*(4), pp. 407–414.

Commission regulation (ec) no 2073/2005 of 15 November 2005 on microbiological criteria for foodstuffs.

Cortez A.L.L., Carvalho, A.C.F.B., Ikuno, A.A., Bürger K.P. and Vidal-Martins A.M.C. (2006). Identification of *Salmonella* spp. isolates from chicken abattoirs by multiplex-PCR. *Research in Veterinary Science, 81*(3), pp. 340–344.

Ellermeier G.D. and Slauch J.M. (2006). The Genus Salmonella. Chapter 3.3.7. In: Prokaryotes, 3rd ed, 6, New York: Springer, New York, pp. 123–158.

Grabow W.O.K. (2004). Bacteriophages: update on application as models for viruses in water. *Water SA, 27*(2), pp. 251–268.

Guntupalli R., Hu J., Lakshmanan R.S., Huang T.S., Barbaree, J.M. and Chin B.A. (2007). A magnetoelastic resonance biosensor immobilized with polyclonal antibody for the detection of Salmonella typhimurium. *Biosensors and Bioelectronics, 22*(7), pp. 1474–1479.

Hagens S. and Loessner, M.J. (2007). Application of bacteriophages for detection and control of foodborne pathogens. *Applied Microbiology and Biotechnology, 76*(3), p. 513–519.

Hein I., Flekna G., Krassnig M. and Wagner M. (2006). Corrigendum to Real-time PCR for the detection of Salmonella spp. in food: An alternative approach to a conventional PCR system suggested by the FOOD-PCR project. *Journal of Microbiological Methods, 66*, pp. 538–547.

ISO Microbiology of food and animal feeding stuffs. Horizontal method for the detection of *Salmonella* (ISO 6579:2003). (2003). Geneva, Switzerland: International Organization for Standardization.

ISO Microbiology of food and animal feeding stuffs - Protocol for the validation of alternative methods (ISO 16140:2003). Geneva, Switzerland: International Organization for Standardization.

Jasson V., Baert L. and Uyttendaele M. (2011). Detection of low numbers of healthy and sub-lethally injured *Salmonella* enterica in chocolate. *International Journal of Food Microbiology, 145*, pp. 488–491.

Junillon T., Mosticon D., Mallen B., Baril F., Morand L., Michel D. and Flandrois, J.P. (2014). Optimization of the reactional medium and a food impact study for a colorimetric in situ Salmonella spp. detection method. *International Journal of Food Microbiology, 181*, pp. 48–52.

Lauer W.F. and Martinez F.L. (2009). RAPID'Salmonella chromogenic medium. Performance Tested Method 050701. *Journal of AOAC International, 92*(6), pp. 1871–1875.

Lawson T.S., Connally R.E., Iredell J.R., Vemulpad S. and Piper J.A. (2011). Detection of *Staphylococcus aureus* with a fluorescence *in situ* hybridization that does not require lysostaphin. *Journal of Clinical Laboratory Analysis, 25*, pp. 142–147.

Li X., Zhang S., Zhang H., Zhang L., Tao H., Yu J., Zheng W., Liu C., Lü D., Xiang R. and Liu Y. (2009). A loop-mediated isothermal amplification method targets the *pho*P gene for the detection of *Salmonella* in food samples. *International Journal of Food Microbiology, 133*(3), pp. 252–258.

Li Q., Cheng W., Zhang D., Yu T., Yin Y., Ju H. and Ding, S. (2012). Rapid and sensitive strategy for Salmonella detection using an InvA gene-based electrochemical DNA sensor. *International Journal of Electrochemical Science, 7*(1), pp. 844–856.

Loonen A.J.M., Jansz A.R., Kreeftenberg H., Bruggeman C.A., Wolffs P.F.G. and van den Brule A.J.C. (2011). Acceleration of the direct identification of *Staphylococcus aureus* versus coagulase-negative staphylococci from blood culture material: a comparison of six bacterial DNA extraction methods. *European Journal of Clinical Microbiology & Infectious Diseases, 30*, pp. 337–342.

Malorny B., Huehn S., Dieckmann R., Krämer N. and Helmuth R. (2009). Polymerase Chain Reaction for the Rapid Detection and Serovar Identification of *Salmonella* in Food and Feeding Stuff. *Food Analytical Methods, 2*(2), pp. 81–95.

McGrath T.F., Elliott C.T. and Fodey, T.L. (2012). Biosensors for the analysis of microbiological and chemical contaminants in food. *Analytical and bioanalytical chemistry, 403*(10), pp. 75–92.

Mori Y. and Notomi T. (2009). Loop-mediated isothermal amplification (LAMP): a rapid, accurate, and cost-effective diagnostic method for infectious diseases. *Journal of Infection and Chemotherapy, 15*(2), p. 62–.

Podolak R., Enache, E., Stone W., Black D.G. and Elliott P.H. (2010). Sources and risk factors for contamination, survival, persistence, and heat resistance of Salmonella in low–moisture foods. *Journal of Food Protection, 73,* pp. 1919–1936.

Schönenbrücher V., Mallinson E.T. and Bülte M. (2008). A comparison of standard cultural methods for the detection of foodborne Salmonella species including three new chromogenic plating media. *International Journal of Food Microbiology, 123*(1–2), pp. 61–66.

Serwer P., Hayes S.J., Thomas J.A. and Hardies S.C. (2007). Propagating the missing bacteriophages: a large bacteriophage in a new class. *Virology Journal, 26,* pp. 4–21.

Silva de Siqueira R., Dodd Ch.E.R. and Rees C.E.D. (2003). Phage amplification assay as rapid method for salmonella detection. *Brazilian Journal of Microbiology, 34* (Suppl.1), pp. 118–120.

Singh A., Arya S.K., Glass N., Hanifi-Moghaddam P., Naidoo R., Szymanski C.M., Tanha J. and Evoy S. (2010). Bacteriophage tailspike proteins as molecular probes for sensitive and selective bacterial detection. *Biosensors and Bioelectronics, 26(*1), pp. 131–138.

Smartt A.E., Xu T., Jegier P., Carswell J.J., Blount S.A., Sayler G.S. and Ripp S. (2012). Pathogen detection using engineered bacteriophages. *Analytical and bioanalytical chemistry, 402*(10), pp. 3127–3146.

Thevenot D.R., Toth K., Durst R.A. and Wilson G.S. (2001). Electrochemical Biosensors: Recommended Definitions and Classification (Technical Report). *Biosensors and Bioelectronics, 16,* pp. 121–131.

Valadez A.M., Lana C.A., Tu S.I., Morgan M.T. and Bhunia, A.K. (2009). Evanescent wave fiber optic biosensor for *Salmonella* detection in food. *Sensors, 9*(7), pp. 5810–5824.

Vieira-Pinto M., Oliveira M., Bernardo F. and Martins C. (2005). Evaluation of fluorescent in situ hybridization (FISH) as a rapid screening method for detection of *Salmonella* in tonsils of slaughtered pigs for consumption: a comparison with conventional culture method. *Journal of Food Safety, 25,* pp. 109–119.

Vieira-Pinto M., Oliveira M., Bernardo F. and Martins C. (2007). Rapid detection of *Salmonella* sp. in pork samples using fluorescent in situ hybridization: a comparison with VIDAS®-SLM system and ISO 6579 cultural method. *Arquivo Brasileiro de Medicina Veterinária e Zootecnia, 59*(6), pp. 1388–1393.

Vinay M., Franche N., Grégori G., Fantino J.R., Pouillot F. and Ansaldi M. (2015). Phage-Based Fluorescent Biosensor Prototypes to Specifically Detect Enteric Bacteria Such as E. coli and Salmonella enterica Typhimurium. *PLoS One, 17*(7), p. e0131466–.

Wagner M., Horn M. and Daims H. 2003. Fluorescence *in situ* hybridization for the identification of prokaryotes. *Current Opinion in Microbiology, 6,* pp. 302–309.

Zadernowska A., Chajęcka-Wierzchowska, W. and Kłębukowska, L. (2014a). Vidas UP-Enzyme-Linked Fluorescent Immunoassay Based on Recombinant Phage Protein and Fluorescence In Situ Hybridization as Alternative Methods for Detection of Salmonella enterica Serovars in Meat. *Foodborne Pathogens and Disease, 11*(9), pp. 747–752.

Zadernowska A. and Chajecka, W. (2012). Detection of *Salmonella* spp. Presence in Food, *Salmonella* - A Dangerous Foodborne Pathogen, Dr. Barakat S M Mahmoud (Ed.), InTech, pp. 393–412.

Zadernowska A., Chajęcka-Wierzchowska W. and Kłębukowska L. (2014b). Fluorescence *in situ* hybridization as an alternative screening method for *Salmonella* sp. occurrence in chicken meat. *Żywność. Nauka. Technologia. Jakość/Food. Science Technology. Quality, 1,* pp. 92–102.

9

CRISPR-Mediated Bacterial Genome Editing in Food Safety and Industry

Michael Carroll and Xiaohui Zhou

Department of Pathobiology & Veterinary Science, University of Connecticut, Connecticut, USA

9.1 Introduction

Bacteriophages are among the most diversified and superabundant organisms in the biosphere, exceeding their hosts by a factor of 10 (Bergh *et al.*, 1989). As such, it is unsurprising that a litany of antiviral mechanisms have emerged in prokaryotes during the course of natural selection. Of these, Restriction-modification systems, considered evolutionarily primeval, are a common denominator in Monera, and have been assiduously studied and characterized (Bickle and Kruger, 1993). Restriction-modification systems employ endonucleases that transect recognized DNA motifs in phages. These target sequences are present in the host genome as well as the genome of the invading conjugative element, but methylation of the target sequence in the bacterial chromosome precludes cleavage (Bickle and Kruger, 1993), creating a self-nonself distinction mediated by cognate methyltransferase (MTase). Additionally, the secretion of polysaccharides by a bacterium limits the accessibility of a phage receptor, and the capability of a bacterium to modulate the expression of phage receptors is an important defensive property (Liu *et al.*, 2002; Hanlon *et al.*, 2001). One of the more intriguing antiviral mechanisms is abortive infection. Abortive infection occurs when the presence of a phage impinges a signal transduction in the bacterium, resulting in the activation of apoptotic mechanisms; that is, the cell nobly commits suicide, depriving the phage of the molecular machinery required for its propagation, in hopes of securing the survival of its consortia (Molineux, 1991). The ecological dynamism of phages and their hosts leads to the emergence of methods by which phages circumvent prokaryotic innate immune responses (Labrie *et al.*, 2010). Phages can produce degradative enzymes to extricate polysaccharides from phage receptors (Sutherland *et al.*, 2004), and mutative events that empower phages to enter nontraditional receptors will be naturally selected for (Doulatov *et al.*, 2007). Mutative events can also render structural differences in phage proteins, meaning a protein that would have initiated abortive infection in the host may no longer do so. If a phage's targeted motif is methylated before it is dissevered by an endonuclease, the restriction-modification system is outran, with fairly foreseeable consequences (Bickle and Kruger, 1993). A bacterial immune response that thinks and remembers as quickly as phages evolve solutions to once-competent systems would be an enviable addition to a prokaryote's armamentarium indeed.

Foodborne Pathogens and Antibiotic Resistance, First Edition. Edited by Om V. Singh.

In 1987, Japanese researchers at Osaka University studied *Escherichia coli*. They sequenced the *iap* gene, which facilitates the isozyme conversion of alkaline phosphatase. Downstream the *iap* gene, they noticed a peculiar and uncharacterized locus. "An unusual structure was found in the 3'-end flanking region of *iap*", they reported. "Five highly homologous sequences of 29 nucleotides were arranged as direct repeats with 32 nucleotides as spacing" (Ishino *et al.*, 1987). Due to the absence of a known homologous sequence in prokaryotes at the time, the researchers were victims of their place in history, unable to infer the astonishing details of their discovery, leading them to conclude that "...the biological significance of these sequences is not known" (Ishino *et al.*, 1987). This seemingly inconsequent observation was the harbinger for a biotechnology revolution that is still unfolding today, a reminder that all research is active research. It is remarkable that a marginalized paragraph of their work in 1987 can now be understood as the Chekhov's gun, anticipating the dawn of a new epoch in medicine.

Owing to the accretion of bacterial genomes in GenBank (Mojica *et al.*, 2000), loci with meaningful levels of homology to the mystery locus pondered by the Osaka researchers were identified. Scientists, being the acronym machines we are, christened this class of loci CRISPR (clustered, regularly interspaced, short palindromic repeats). Soon, it became apparent that a highly common attribute of bacteria and archaea had escaped the wry and watchful eye of science. In 2002, small, ncRNAs were studied in *Archaeoglobus fulgidus*. By isolating and sequencing those ncRNAs, the researchers ascertained them to be products of the bacteria's CRISPR locus (Tang *et al.*, 2002). In the same year, a set of proteins were determined to be related to CRISPR (Jansen *et al.*, 2002). These Cas (CRISPR-associated) proteins appeared to have an enzymatic role in nucleic acid reactions. In 2005, the spacers between DRs (direct repeats) were realized to be complementary to a host of phage and plasmid sequences (Bolotin *et al.*, 2005; Mojica *et al.*, 2005; Pourcel, Salvignol, and Vergnaud, 2005), which belied the expectation that the DRs were the functionally critical components. An inverse relationship between the number of spacers present in *Staphylococcus thermophilus* and its susceptibility to unbidden horizontal guests was established (Bolotin *et al.*, 2005).

Enough data had amassed to suggest that CRISPR-Cas systems are involved in antiphage and antiplasmid defense, mechanistically similar to RNA interference in eukaryotes. It was modeled that crRNAs (the products of spacers) complementary to invader sequences base-paired with said sequences, tagging them for destruction by Cas enzymes. Base pairing is simple and facile to design, which spurred the interest of the biotech industry in harnessing CRISPR Cas systems for programmable genome editing. Such a programmable genome editing system would presumably be child's play compared to zinc fingers and TALENS, which require the design of a completely novel endonuclease, a relatively exacting process.

The CRISPR acronym (clustered, regularly interspaced, short palindromic repeats) describes the molecular geography of the CRISPR locus. An operon containing the Cas genes required for immunization and DNA interference flanks a leading sequence, the promoter, which in turn flanks the CRISPR array. The CRISPR array is composed of DRs approximately 30–40 bp long, and the intervening spacers are approximately 20 bp.

CRISPR allows a prokaryote to remember invading conjugative dsDNA and ssDNA elements such as bacteriophages, plasmids and transposons. Three seminal studies elucidated some critical mechanistic details of CRISPR-mediated immunity, which is generally divided into three phases: acquisition, crRNA guidance, and targeting. In 2007, the function of CRISPR-Cas systems in militating phage propagation in *S. thermophilus* was evinced (Barrangou *et al.*, 2007). Barrangou and colleagues exposed *S. thermophilus* cells to phage infection and observed the consequent happenings in the CRISPR locus (Barrangou *et al.*, 2007). They found that successfully phage-resistant cells had acquired one or more new spacers that were homologous to sequences of the phage in question; it was realized as a corollary of this that CRISPR-Cas is an adaptive immune system (Barrangou *et al.*, 2007). Adaptive immune systems are a form of memory, where initial exposure to a pathogen immunizes the organism against infection by that pathogen. It does so by processing and retaining information about the pathogen. For example, dendritic cells, colloquially known as "professional" APCs (antigen-presenting cells), recognize nonself entities, gather their antigens if the number of them exceeds the threshold for their uptake, and pass the information on to B and T memory cells in humans. When the pathogen assails the cell a second time, the adaptive system utilizes the information it retained to recognize the enemy. Immunoglobulins produced by B plasma cells anneal to antigens, tagging them for phagocytosis by polymorphonuclear leukocytes such as neutrophils and macrophages. Vaccination manipulates this immunological memory by engendering the memorization of a pathogen. Contrast this to innate immune systems, which lack the capacity for immunological memory. The cardinal difference is that adaptive immune systems are able to learn about new predators and invent defenses that did not previously exist, whereas innate systems are creatures of habit, entirely nonprogressive. Prokaryotes are well endowed with innate immune systems. The discovery that CRISPR is an adaptive system is so remarkable because it is now the only such system known of in prokaryotes. During the acquisition phase of CRISPR-Cas immunity, nt (nucleotide) sequences from foreign conjugative elements (protospacers) are obtained and incorporated as spacers in the CRISPR locus. The bacterium has been inoculated against the invader from whose genome the spacer sequence was derived. The pernicious pathogen is remembered. In another important study, Brouns and colleagues discovered that host factor RNase (ribonuclease) III, contained within a ribonucleoprotein complex, was responsible for the biogenesis of crRNAs (Brouns *et al.*, 2008). The researchers determined that CRISPR-Cas immunity depends upon the presence of crRNAs (Brouns *et al.*, 2008), confirming a widely held predilection that this was the case. In a third study, Marraffini and Sontheimer demonstrated that a spacer in the CRISPR locus of *Staphylococcus epidermis* corresponded to a gene critical for the horizontal transference and propagation of staphylococcal plasmids, preventing conjugation (Marraffini and Sontheimer, 2008). Significantly, they experimentally determined that CRISPR-Cas systems, rather than being akin to eukaryotic RNA interference, provide immunity via Cas nucleases guided to the targeted DNA sequence by a crRNA transcript to which it is homologous (Marraffini and Sontheimer, 2008).

CRISPR systems all provide adaptive immunity the same way. The process consists of three phases: acquisition (Figure 9.1); crRNA targeting (Figure 9.2), and subsequently DNA interference (Figure 9.3). In spacer acquisition, a cas1:cas2 complex, bound with a target complex, finds an amenable sequence of phage or plasmid DNA

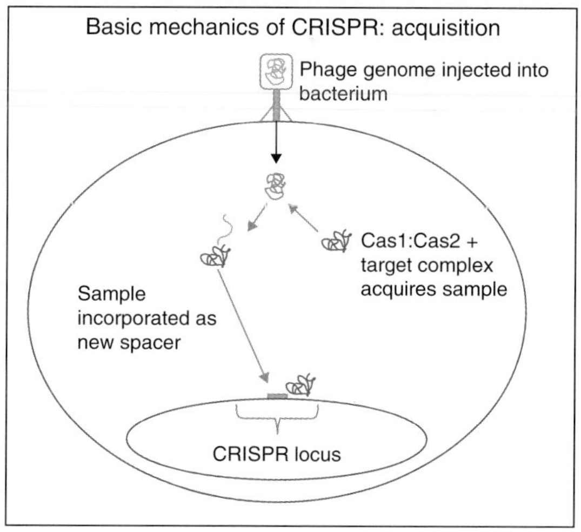

Figure 9.1 In the acquisition phase of immunity in CRISPR Types I-III, invading foreign DNA is read by a Cas1:Cas2 complex:target complex. This complex extricates an amenable sequence and incorporates it into the CRISPR locus as a new spacer, immunizing the cell from the invader. A spacer is adequate to immunize the cell if it meets the targeting requirements of the CRISPR system, particularly whether the protospacer is flanked by a PAM. (*See insert for color representation of the figure.*)

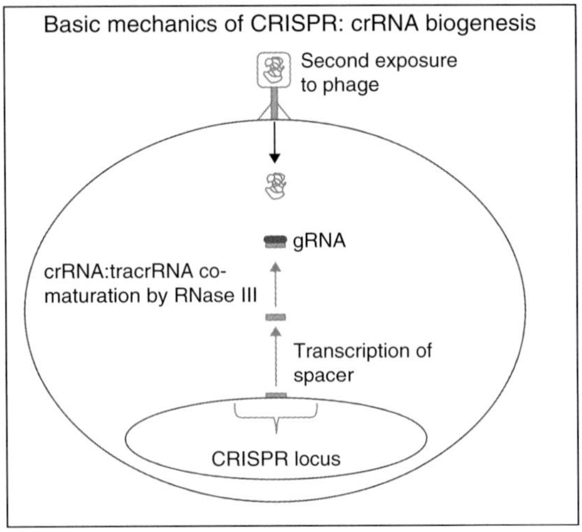

Figure 9.2 In crRNA biogenesis, the spacer is transcribed. This crRNA combines with tracrRNA, resulting in a mature guide RNA (gRNA), a process facilitated by host factor ribonuclease III. (*See insert for color representation of the figure.*)

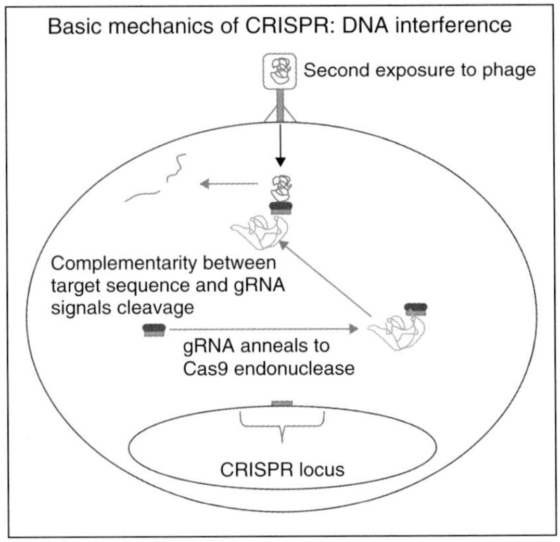

Figure 9.3 In the DNA interference phase, the gRNA anneals to Cas9, forming the functional RGE (RNA-guided endonuclease). If the target sequence is flanked by a PAM and complementary to the gRNA, they anneal, signaling a DSB by Cas9. The RuvC nuclease domain of Cas9 severs the non-target strand, and the HNH domain severs the target strand. (*See insert for color representation of the figure.*)

and excises it. Cas1 and Cas2 are common among CRISPR systems, in every case essential for acquisition. Once this sequence is pilfered, it is incorporated in the CRISPR locus as a spacer between DRs (direct repeats). If the invader enters the cell again, the crRNAs are generated from the spacer. This is mediated by host factor ribonuclease (RNase) III. During targeting, a crRNA:tracrRNA (trans-activating RNA) complex binds to a Cas endonuclease. This trifecta interrogates the DNA of an invader, and if complementarity between the spacer and a corresponding protospacer is identified, the crRNA transcript base pairs with the target. DNA interference occurs next, as the Cas nuclease induces hydrolysis of the cognate phage sequence. Following the logic of evolution, monerans lack DNA repair systems that can competently heal DSBs induced by Cas nucleases. Albeit CRISPR systems follow the same general mechanistic thrust, differences in crRNA biogenesis and targeting requisites have made a system of classification necessary. The main divisions are Types I, II, and III. In Type I systems, the Cascade complex and Cas3 exonuclease effect immunity. Cas6e, a component of the Cascade complex, is an endoribonuclease that processes precursor crRNA into mature crRNA. The Cascade complex utilizes those crRNAs to identify target sequences (Brouns *et al.*, 2008; Jore *et al.*, 2011; Wiedenheft *et al.*, 2011). A Cas8 moiety within the Cascade complex recognizes a PAM (protospacer-adjacent motif), a short sequence that must be recognized by Cas8 immediately upstream from the protospacer for the Type I system to carry out the DNA interference phase (Sashital, Wiedenheft, and Doudna, 2011). PAMs are present upstream the protospacers but not upstream from spacers in the CRISPR locus, allowing the system to distinguish self from nonself. Once the crRNA and DNA anneal, resulting in "R-loop" formation, the exonuclease Cas3 is summoned and induces ssDNA cleavage (Sinkunas *et al.*, 2013). Type I systems are the most abundant in bacteria and archaea.

Type II CRISPR systems require Cas9, RNase III, precursor RNA (pre-crRNA), and ancillary tracrRNA (trans-encoded RNA) to function (Deltcheva *et al.*, 2011). The dsRNA formed by the hybridization of complementary pre-crRNA and tracrRNA is processed by RNase III into mature crRNA guides. Cas9 forms a heteroduplex with a mature tracrRNA:crRNA complex guide, which binds to target sequences. A region of Cas9 must interact with a PAM (NGG) immediately downstream from the protospacer in order to cleave it (Anders *et al.*, 2014). An 8-12 nucleotide "seed" sequence proximal to the PAM is essential for crRNA to anneal to the protospacer. HNH-nuclease and RuvC-nuclease, domains of Cas9, endonucleolyse the target and non-target strands of the invading DNA respectively (Carte *et al.*, 2008). Type II systems are relatively rare. There are no known examples in archaea.

No PAM requirements have been observed in Type III systems. Cas6, an endoribonuclease, cleaves pre-crRNA (Carte *et al.*, 2008).The crRNAs produced become associated with either a Cas10-Csm or Cas10-Cmr complex, which is the basis for distinguishing Types III-A and III-B. The PAM domain of the Cas10 complex cleaves the non-template DNA strand. Both DNA and RNA are targeted by Type III systems, resulting in the cleavage of RNA transcripts as well. The Csm3 and Cmr4 domains of Cas10 cut RNA transcripts. In the process of the biosynthesis of crRNA by Cas6, an 8 nt-long segment of a direct repeat is included with the spacer (Carte *et al.*, 2008), resulting in a crRNA tag sequence. When the crRNA and the protospacer commence base pairing, incomplete complementarity will result due to lack of homology between the crRNA tag and the target sequence. This signals hydrolysis, whereas perfect homology would militate targeting. This protects the Type III CRISPR locus from the products it begets, precluding an autoimmune response. Type III systems are not as well characterized as Types I and II.

9.2 Application of CRISPR for Bacterial Genome Editing

CRISPR-Cas systems provide facile programmable genome editing for the experimental elucidation of biological processes. Genome editing allows researchers to identify gene products and investigate their function by inducing site-specific mutative activity and analyzing the corresponding phenotypic changes. The Type II CRISPR system, notably the endogenous system of *Streptococcus pyogenes*, has piqued the most interest as a vehicle for programmable genome editing, because only expression of Cas9, tracrRNA and crRNA is necessary for site-specific endonucleolysis. This method was advanced when researchers designed a novel chimeric sgRNA (single-guide RNA) to stand in for the tracrRNA:crRNA complex, further expediting the process (Jinek *et al.*, 2012). Type II CRISPR can be used to study the consequences of mutation. A recombinant vector containing Cas9 and a novel sgRNA, when expressed in the cell, will result in cleavage of the target sequence. If one targets a wild type gene in a bacterium, a mortal DSB is delivered by Cas9 in wild type cells. Cells that have antecedently mutated in the sequence in question will evade recognition and survive. Since genome targeting can be multiplexed in Type II CRISPR, selective pressure is now an analytic tool that can be wielded with unprecedented finesse. Genome editing using the Type II system has been frenziedly studied in eukaryotic cell lines, and in vivo in monkeys and nonviable human embryos; the attachment of a nuclear localization signal enables the vector to ingress the nuclear pore complex. CRISPR is a promising candidate for next-generation treatments of heritable human diseases. The utility of this system for bacterial genome editing and functional genomics has elicited less élan, a largely unexplored frontier thus far.

It has been observed that 22% of CRISPR loci in the *lactobacillales* clade contain spacers derived from the genomes of the cells they inhabit (Horvath *et al.*, 2009), resulting in cytotoxic autoimmunity events. These events are highly lethal, as illustrated by 3- to 5-log depletions of populations containing these unhappy accidents (Vercoe *et al.*, 2013; Jiang *et al.*, 2013). This owes in part to the ineptitude of high-fidelity prokaryotic DNA repair systems concerning Cas cleavage. A study of acquisition in *Streptococcus thermophilus* strongly connected self-targeting spacers with cell death (Oh and van Pijkeren, 2015). The existence of a self-targeting spacer promotes aberrant morphology, as CRISPR-Cas-mediated endonucleolysis of target spacers eliminates wild type cells from the population. The observation of phenotypic variants in *Pectobacterium atrosepticum* that harbor self-targeting spacers bespeaks this (Vercoe *et al.*, 2013). The potential fatal hydrolysis of a self-sequence where self-targeting CRISPR is present constitutes a powerful selection against wild types cells. The town is not big enough for the both of them. Ergo, a self-targeting CRISPR-Cas system catalyzes mutagenesis in a cell (Jiang *et al.*, 2013). This phenomena could be applied to DE (directed evolution) to rapidly generate genetic variant libraries (Vercoe *et al.*, 2013). The outcomes of specific antecedent mutations can be studied by introducing self-targeting CRISPR arrays into genetically heterogeneous populations. Such arrays could catalyze the evolution of strains in a directed manner by killing extraneous wild type cells, leaving the surviving mutants to be studied. Analytic methods centered on CRISPR could substantially deepen our knowledge of the interplay of genotype and phenotype and aid in the biochemical characterizations of gene products.

The town is big enough, on occasion. Infrequently, wild type cells are able to hinder CRISPR-Cas from acting on self-targeting spacers. Mutation in the seed sequence proximal to the PAM militates crRNA from annealing to the cognate sequence, and mutation in the PAM is a common route of CRISPR evasion (Oh and van Pijkeren, 2015; Wiedenheft *et al.*, 2011). Point mutation in or deletion of the spacer thwarts targeting as well (Vercoe *et al.*, 2013; Jiang *et al.*, 2013). Mutations that rescind crRNA biogenesis will deprive Cas nucleases of their guides, but the abrogation of the entire immune function is an exorbitant expense to the cell. In 2013, Jiang and colleagues studied *Streptococcus pneumoniae*, establishing that transfection of exogenous gDNA (genomic DNA) resulted in survival of the cell via homologous recombination (Jiang *et al.*, 2013). This beckons the ironic question of how horizontal gene transfer may be involved in the homeostasis of cells with haywire CRISPR arrays, although such events would probably be exceedingly rare.

Jiang and colleagues used Type II CRISPR-Cas9 derived from *S. pyogenes* to study the effects of self-targeting events (Jiang *et al.*, 2013). This study clarified the molecular mechanics of Type II systems and provided proof of concept for the use of Cas9 in prokaryotic genome editing. The researchers also elucidated PAM and seed requirements, and how mutations in those regions or in the protospacer influence Cas9-mediated hydrolysis. They established that double-crossover HDR (homology-directed repair) with a donor template could salvage a sequence that had been lysed by Cas9. They transformed a strain of *S. pneumoniae* containing prophage with a vector encoding a prophage-targeting Cas9:sgRNA complex, predicting that the cell would perish. Rather, HDR in the *srtA* locus salvaged the sequence. Mutations occurring in association with that process would be vertically transferred to progeny, resulting in a stable population void of prophage. The study underscored the possibility of engendering desired deletions or missense mutations in chosen loci as well (Jiang *et al.*, 2013). Another study investigated the use of Type II CRISPR-Cas9 systems for inducing mutation in user-defined loci (Semenova *et al.*, 2011). Oh and Pijkeren used a plasmid vector to express Cas9 and *recT*, a DNA-binding protein, and recombineering in *Lactobacillus reuteri* (Oh and van Pijkeren, 2015). *Lactobacillus* species are propagated industrially at high volumes to facilitate diary fermentation processes and to fortify foods with probiotic cultures, among others. Oligonucleotides containing non-targeted PAM sequences were designed to maintain the *recT* protein and obstruct Cas9 interference (Sternberg *et al.*, 2015). The researchers showed that one could be the architect of population-wide gene deletions through recombineering and selecting against the original genotype with Cas9. Oh and Pijkeren (Oh and van Pijkeren, 2015) noted that the necessity of a PAM to execute this method stymies the generalization of it to sequences without PAMs, echoing Jiang and colleagues (Jiang *et al.*, 2013). The manumission of the process from any dependence on those two obnoxious base pairings warrants research. Additionally, lactic acid bacteria are well suited to recombineering, a trait not shared by all bacteria.

Vercoe *et al.* used a Type I system endogenous to *Pectobacterium atrosepticum* recombined with self-targeting spacers (Vercoe *et al.*, 2013). Cas3 exonuclease damages DNA in a helter-skelter manner, flouting the use of Type I CRISPR to induce highly specified mutagenesis. They found that the DNA interference phase was circumvented by introducing a point mutation in the PAM. However, the transfection of a CRISPR array re-enabled cleavage. The pathogenicity island (approximately 98 kb) was wholly deleted through recombination with *attR* and *attL* sites that flank the island (Vercoe *et al.*, 2013). The researchers determined through PCR that deletion of the island occurred in wild type cells, albeit infrequently. With CRISPR as the scalpel, the genetic composition of a bacterial population can now be played with, shaped, and finely tuned with surgical precision and accuracy.

9.3 Vaccination of Industrial Microbes

Bacteriophages, which are measured in nanometers, have consequences that we measure in dollars. Bacteriophages can decimate bacterial colonies harvested for industrial purposes. Some examples of industrial microbes are *Clostridium beijerinkii*, used as a metabolic host in ethanol, butanol, and acetone production, and *Lactobacillus* spp., harvested at high volumes for diary fermentation and probiotic development (Wang *et al.*, 2015). In the metaphoric arms race between phages and their prey, the latter have evolved innate immune systems that are obsolesced by the rapidly adaptive nature of the former. Phage infections in industrial microbes are therefore culpable for quality and yield reductions, presaging rises in price for consumers and profit declines for producers. Starter cultures are homogeneous, meaning a mutant phage can singularly devastate the population (Barrangou and Selle, 2015). One method of damage control in the event of phage infection is the use of heterogeneous starter cultures, which are less likely to be wholly compromised. Starter growth rotation, the use of chelating agents in media, and steam-sterilization of industrial equipment also dissuade phage propagation (Barrangou and Selle, 2015). On the other hand, CRISPR, an adaptive immune system, can rapidly acquire new spacers, adding arrows to the quiver to confront these and other invasive MGEs (mobile genetic elements) such as resistance-conferring conjugative plasmids and transposons. The conjugation of recombinant plasmids containing CRISPR-Cas systems will immunize cells against the phages to whom the novel spacers correspond (Gasiunas *et al.*, 2012). The vaccination will be vertically and horizontally transferred until the entire population is inoculated. The acquisition and incorporation of spacers in the CRISPR locus is practically unlimited, meaning that CRISPR-mediated antiphage immunity can snowball from there. A novel spacer is wisely chosen if the corresponding protospacer represents a conserved function sequence; such a spacer would endow the microbial population with resistance to the maximal number of phages (Barrangou and Selle, 2015). Also, the targeting of sequences indispensable for the propagation of the phage would imply that any mutations that thwart DNA interference would be nonviable.

Antibiotic resistant infections are a burgeoning public health concern. Hallmark superbugs such as CRE (carbapenem-resistant Enterobacteriaceae), *Clostridium dificile, Streptococcus aeruginosa,* and MRSA (methicillin-resistant *Staphylococcus aureus*) illustrate the need for a novel approach to treatment, as the steady march toward the obsolescence of the Magic Bullet promises to retrogress modern medical capabilities, with consequences of a nature likely more dire than projected. The UK Review on Antimicrobial Resistance, published in December 2014, anticipates that the global morbidity rate associated with antibiotic-resistant infection will rise to 10 million come 2050 (HM Government, 2014), which would make it a more successful cause of death, if you will, than the cancers combined. That is mere human life, though, and I know where your concerns are more likely to lie, dear reader. Let's talk money. This much dying would mean a decrease in GDPs by 2–3.5% what they would have otherwise been in 2050, resulting in a \$100 trillion U.S. global burden (HM Government, 2014). Generations that will inherit the ramifications of supreme environmental and educational irresponsibility need not inherit this bolder-that-broke-the-camel's-back as well. This is probably a low estimate. The rendering of antibiotics inept will reap economic and social costs that are unforeseen, and these untallied costs will create a situation more imperiled than that which has been forecasted. A solution to antibiotic-resistance ought to be regarded as the Holy Grail of current research. Appropriately so, considering the very real possibility of a putative return to the Dark Ages of medicine. Due to the fact that the food industry feeds animals with as many antibiotics as we feed our children with antipsychotics, the conjugation of genetic elements that confer antibiotic resistance is naturally selected for. This compounds the problem of nosocomial infections with the problem of infection by foodborne antibiotic-resistant pathogens in contaminated foods.

By impregnating industrial microorganisms with novel exogenous CRISPR-Cas arrays, their susceptibility to injurious conjugative elements harboring antibiotic resistance genes can be greatly curtailed (Barrangou and Selle, 2015). CRISPR-Cas targeting of antibiotic resistance genes would immunize a colony from the plasmids or transposons that carry them by inducing fatal endonucleolyses. Point mutation in the target sequence is rather well tolerated by CRISPR, but mutation in the seed or PAM could enable the conjugative element to circumvent cleavage. However, mutations associated with the gene endowing antibiotic resistance would likely abrogate its ability to do so, meaning, in theory, that the conjugative genetic element is at best dead and at worst irrelevant.

In lieu of tremendous scientific alacrity regarding the industrial and clinical possibilities of CRISPR, a caveat: antibiotics elicited similar alacrity. Hopefully, duty of care is taken by the medical community to ensure that future generations need not tackle the problem of CRISPR resistance culminating in the restoration of antibiotic resistance. Let us keep our bullets magic, by avoiding an insouciant ethos for the industrial and clinical application of this system. The enthralling discoveries consistently being made about CRISPR should be accompanied by a sensible discussion about the ecological potentialities associated with the manipulation of a system we hardly knew existed five minutes ago.

9.4 Application of CRISPR in the Development of Antimicrobials

Bacterial DNA repair mechanisms are unable to cope with CRISPR-Cas mediated DNA damage. The frequency at which prokaryotes are able to salvage DNA that has been targeted by the haphazard Cas3 exonuclease of the Type I system is in the ballpark of 10^{-5}, rather low, and deletions of approximately 40 kb have been observed (Wang *et al.*, 2015). The high-fidelity HR (homologous recombination) pathway is used to repair DSBs induced by Cas9 of the Type II system, but only if a donor template is present. CRISPR-Cas targeting of coding or noncoding sequences results generally in 3- to 5-log depletions of populations (Barrangou and Selle, 2015), catching the attention of bactericidal maniacs in medical research. Gomaa and colleagues (Gomaa *et al.*, 2014) used a recombinant plasmid containing *E. coli* Type I genes and genome-targeting spacers to transform *E. coli* cells. Transformation efficiency of the self-targeting plasmid was roughly 10^5-fold lower than that of the control (Gomaa *et al.*, 2014), bespeaking the gargantuan lethality of the CRISPR-Cas system. This study also demonstrated that whether spacers corresponded to coding or noncoding loci, or conserved or dispensable genes, mass death was incurred; additionally, no indications of strand-bias were identified, and exogenous and endogenous CRISPR systems were equally bad news for bacteria (Gomaa *et al.*, 2014). Bikard and colleagues used a phagemid delivery system of S. pyogenes Type II CRISPR to selectively eliminate MRSA cells (Bikard *et al.*, 2013). A phagemid is a plasmid recombined with the replication and packing genes of a bacteriophage, in this case the staphylococcal φ-NM1 phage. The phagemid was recombined with a novel CRISPR array targeting an antibiotic resistance gene in *S. aureus* and the packing, *terL*, *rinA*, and *terS* genes of the phage. A 10^4-fold decrease in viability occurred when the recombinant phages and *S. aureus* cells were commingled (Bikard *et al.*, 2013).

The researchers assessed the antimicrobial properties of phagemid-delivered CRISPR in a mouse model, where kanamycin-resistant *S. aureus* colonies on the skin were decimated by targeting the *kanR* gene (Bikard *et al.*, 2013). In 2014, Citorik and colleagues transformed a nosocomial MRSA strain with an M13 phagemid carrying CRISPR. They designed spacers to target the pNDM-1 and pSHV-18 plasmids, which confer antimicrobial resistance to *S. aureus* (Citorik, Mimee, and Lu, 2014). Delivery of the phagemid resulted in a 10^3-fold population reduction, echoing the figure of a similar study by Bikard *et al.* (Bikard *et al.*, 2013; Citorik, Mimee, and Lu, 2014). In both studies, cells were forced to retain the plasmid due to addiction modules recombined with the plasmid, a wise application of the postsegregational killing effect. Yosef and colleagues recombined a lambda prophage with *E. coli* Type I CRISPR (Yosef *et al.*, 2015). They targeted *ndm-1* and *ctx-M-15*, genes that confer resistance to the β-lactam antibiotics, which include the penems, cephems, carbapenems and monobactams. *ndm-1* and *ctx-M-15* encode β-lactamases that disable those antibiotics by enzymatically attacking the β-ring they all contain.

Compare to *mecA*, an antibiotic resistance gene that confers β-lactam resistance to *S. aureus*. *mecA* begets PBP2a (penicillin-binding protein 2a), which facilitates the peptidoglycan cross-linkage required for cell wall biosynthesis in spite of the presence of β-lactams, which destroy bacteria by negating this pathway. Yosef and colleagues designed lytic T7 phages to kill MRSA cells, and recombined a temperate phage with Type I CRISPR-Cas and spacers targeting antibiotic resistance genes and conserved sequences of the lytic phage (Yosef *et al.*, 2015). Therefore, lysogenization with the temperate phage knocked out antibiotic resistance genes and conferred a selective advantage in the form of resistance to lytic phages. By introducing the lytic and temperate phages concomitantly, MRSA cells were either killed by lytic phages or immunized from lytic phages by the CRISPR system of the temperate phage, in which case they had also been deprived of β-lactam resistance (Yosef *et al.*, 2015). The researchers posit that such a lytic/temperate phage dynamic dyad could be used to sensitize MRSA to antibiotics. A mixture of such phages could discourage the conjugal transfer of plasmid-based antibiotic resistance genes to the avirulent bacteria of the human microflora, and could be used on hospital surfaces and medical equipment. Lysogens of the temperate phage were over four orders of magnitude more resistant to the lytic T7 phage, demonstrating that this elegant application of selective pressure potently incentivizes cells to house the temperate phage (Yosef *et al.*, 2015).

The resulting restoration of antibiotic susceptibility can be exploited accordingly. The high specificity and programmability of CRISPR-Cas systems is ideal for the design of novel therapeutics in genome editing to treat genetic disease and in the development of antimicrobials. In less than a decade of research, the versatility and power of CRISPR has captured the imaginations of scientists everywhere, and inspired methods that may presage the future of medicine.

9.5 CRISPR Delivery Systems

The endogenous DNA repair systems of eukaryotes can competently mend Cas cleavages through the non-homologous end joining (NHEJ) and HDR (homology-directed repair) pathways. If the targeted sequence were restored to wild type, it would be targeted again. Therefore, mutational activity is required, on top of being a natural consequence of those pathways. By introducing Cas9 and chimeric sgRNA to a cell, indels (insertions or deletions) are achieved in user-specified loci. Genome-editing applications of CRISPR in mammalian cells will depend on the efficacy of the delivery system whereby the active constituents are horizontally transfected.

In "Self-Assembled DNA Nanoclews for the Efficient Delivery of CRISPR-Cas9 for Genome Editing" (Sun *et al.*, 2015), the researchers modeled biology with nanotechnology to achieve efficient delivery of *Streptococcus pyogenes* Type II CRISPR, without the oncogenic collateral associated with plasmid and phage vectors incorporating their genetic material into the genome (Kotterman and Shaffer, 2014). They designed DNA nanoclews by RCA (rolling circle amplification) with "palindromic sequences encoded to drive the self-assembly of nanoparticles" (Sun *et al.*, 2015). These DNA nanoclews were devised with near complementarity to the sgRNA, which yielded the greatest efficiency of transfer, due to base-pairing between the Cas9/sgRNA complex and the nanoclew vectors.

Polyethylenimine, a cationic polymer, enveloped the nanoclews, facilitating endosomal escape (Sun *et al.*, 2015). N- and C-terminal nuclear localization signals homed the system in to the nucleus. The Cas9 was harvested in *Escherichia coli*, and treatment or control sgRNAs were used to evaluate the ability of the complex to infiltrate nuclei. The treatment sgRNA targeted EGFP (enhanced green fluorescent protein) harbored by a linearized plasmid, and the control sgRNA was extraneous, allowing simple qualification of the system. Efficient in vitro Cas9-mediated cleavage of EGFP resulted (Sun *et al.*, 2015). A U2OS cell line, which contains an EGFP gene, was used to test nanoclew-based delivery. Nuclei were imbued blue and the novel complex featured a red fluorescence tag.

CLSM (confocal laser scanning microscopy) was used to track the intracellular odysseys of the nanoclews. Within 6 hours, the nanoparticles traversed the membrane, primarily via lipid rafts and micropinocytosis, and entered the nucleus, qualified by the confluence of red and blue (Sun *et al.*, 2015). They performed a live/dead assay and identified no deleterious outcomes associated with Cas9/nanoclew concentrations from nil to 200 NM (Sun *et al.*, 2015). Frameshift caused by indels remits the function of genes. They incurred DSBs in EGFP and indels resulted from NHEJ. They established proof of concept for DNA nanoclew-based delivery, a remarkably inventive and pioneering method. It would be gainful to investigate the application of DNA nanoclews as a delivery method in prokaryotic genome editing as well. Since it has been demonstrated that no measurable decline in eukaryotic cell viability occurred in the presence of nanoclews, the method may have a high therapeutic index in the treatment of bacterial infections as well as genetic disease.

In "Temperate and lytic bacteriophages programmed to sensitize and kill antibiotic-resistant bacteria" (Yosef *et al.*, 2015), a bacteriophage vector was employed to smuggle CRISPR into eukaryotic cells. The researchers used polymerase chain reaction (PCR) to proliferate Type I CRISPR endogenous to *Escherichia coli*. The Cascade genes and Cas3 were homologously recombined with a λ prophage, whose departure from dormancy was elicited in order to lysogenize other cells, mediating the transportation of intelligently designed novel CRISPR arrays (Yosef *et al.*, 2015).

Due to cumbersome size and inherit immunogenicity, roadblocks of phage therapy, phages may not be an ideal delivery system in eukaryotes. The linearization of a plasmid vector with restriction enzymes such as EcoRI or SmaI creates blunt ends. Blunt-end ligation could be utilized to incorporate CRISPR-Cas systems into plasmids, in place of genes nonessential to conjugative transfer and replication. The recircularized plasmid could then be used as a vector. The chance integration of plasmid DNA into the host genome, which is likely to be genetically baleful, is a problem with this method. The relatively low efficiency of bacterial transformation also poses a challenge. Many bacteria are not competent to uptake plasmids, and the application of $NaCl_2$ to increase membrane permeability suffers constraints in vivo.

If these obstacles were adequately answered, the F plasmid would be an appealing subject of study. The F-pilus method of horizontal transfer, a Type IV secretion system requiring cell-to-cell contact, is rapid. When the conjugal F pilus anchors the recipient cell to the donor cell, ssDNA and accompanying proteins are injected in a syringe-like fashion through the mating channel to F- cells like wildfire. Type IV secretion systems merit investigation for the delivery of CRISPR-Cas genes in genome editing. *Helicobacter pylori* utilize Type IV secretion to impart CagA protein to gastric epithelial cells, inducing cytoskeletal alterations, and *Bordetella pertussis* use Type IV secretion to spread the pertussis toxin, causal agent of whooping cough (Christie and Vogel, 2000).

Agrobacterium tumefaciens use a Type IV system to inject oncogenic ss T-DNA into plant nuclei, resulting in crown gall disease (Christie and Vogel, 2000). The VirD2 transecterase, secreted concomitantly with ss T-DNA, features nuclear localization signals that import it into the nucleus. We are currently vague on the details of the architecture of the mating channel and molecular interactions of Type IV transfer. However, three phases of the process are putatively shared by Type IV systems within plasmids and transposons. Initially, a handful of proteins form a complex with a conserved sequence dubbed the *OriT* (origin of transfer).

Relaxase binds to the 5' end of the *OriT*. This complex, called the relaxosome, unwinds the dsDNA in the second step. Thirdly, ATPase homologues pilot the relaxosome to the Type IV secretion apparatus, and it is jettisoned into the recipient cell (Wallden, Rivera-Calzada, and Waksman, 2010). A recombinant plasmid containing *OriT* and *OriV* (origin of replication) loci could be a serviceable method of delivery of a novel CRISPR-Cas system for genome editing. Furthermore, if the genes required for the assemblage of the pilus and mating channel were incorporated in the genome of a cell, it could carry out Type IV secretion independent of a plasmid or transposon.

For instance, balsamaceous *lactobacillus* spp. could be engineered in this manner, along with a CRISPR-Cas system. Herein lies the possibility of engineering probiotics to serve as vectors for CRISPR. CRISPR arrays containing spacers, which target antibiotic resistance genes could be distributed among the human microbiome, if *lactobacillus* were empowered with a Type IV secretion system and ingested. This would militate the spread of mobile genetic elements that confer antibiotic resistance to commensal and pathogenic bacteria, a case in which eating yogurt could save your life. We are sure funding arrangements for such research could be made with Yoplait, for purely philanthropic reasons, of course.

9.6 Concluding Remarks

The high specificity and programmability of CRISPR-Cas systems is ideal for the design of novel therapeutics to treat genetic disease in Eukaryota and in the development of antimicrobials. In less than a decade of research, the versatility and power of CRISPR has captured the imaginations of scientists everywhere, and inspired methods that may presage the future of medicine. On account of its ease of design, it would be prudent to entirely supplant genome-editing methods such as zinc fingers and homing meganucleases with CRISPR. More research into the applications of nanoclew delivery systems for CRISPR is certainly warranted, and the possibility of manipulating Type IV secretion systems to deliver CRISPR needs to be explored. The vaccination of industrial microbes against harmful MGEs would benefit economies. The current body of work on CRISPR-Cas prokaryotic genome editing for functional genomics needs to be expanded. CRISPR promises to expand our antimicrobial repertoire as well. Multiple different approaches to CRISPR-Cas antimicrobials have been investigated, and should continue to be investigated, in order to create a diversified CRISPR-Cas arsenal for treatment. On a note unrelated to CRISPR, highly potent antimicrobial properties have been attributed to cannabinoids such as THC (and nonpsychoactive ones as well) for years. The search for a viable alternative to antibiotics has led us to CRISPR, but research must be done on the antimicrobial potential of the devil's lettuce, extracts from which have been demonstrated to equal or exceed vancomycin in effectiveness. Mechanical ventilation can lead to nosocomial MRSA pneumonia infections in the elderly in particular, a case where smoking weed might be the medically obvious choice for treatment, if we would only look into it in spite of cultural misgivings; the development of cannabinoid-based creams could be applied to skin infections as well, which would be easy to test in a mouse model. Laugh not—science will laugh back at you. The point of all this is not really to convey our relatively diminutive knowledge of CRISPR, or anything else; the point is to excite you, dear reader, about what we do not know, and to inspire research that answers meaningful questions about how medicine will adapt, in a rapidly changing world, in order to fulfill its commitment to the well-being of those under its charge. CRISPR-Cas systems will likely continue to surprise us as we uncover the fascinating details and possible applications of this nascent scientific and medical tool.

References

Gomaa, A.A., *et al.* (2014). Programmable Removal of Bacterial Strains by Use of Genome-Targeting CRISPR-Cas Systems. *American Society of Microbiology, 5*, p. e928–13.

Anders, C., Niewoehner, O., Duerst, A. and Jinek, M. (2014). Structural basis of PAM-dependent target DNA recognition by the Cas9 endonuclease. *Nature, 513*, pp. 569–73.

Barrangou, R., *et al.* (2007). CRISPR provides acquired resistance against viruses in prokaryotes. *Science, 325*, pp. 1709–1712.

Barrangou, R. and Selle, K. (2015). CRISPR-Based Technologies and the Future of Food Science. *Journal of Food Science, 80*, pp. 2367–2372.

Bergh, O., Borsheim, K.Y., Bratbak, G. and Heldal, M. (1989). High abundance of viruses found in aquatic environments. *Nature, 340*, pp. 467–468.

Bickle, T.A. and Kruger, D.H. (1993). Biology of DNA restriction. *Microbiology. Rev., 57*, pp. 434–350.

Bikard, D., et al. (2013). Programmable repression and activation of bacterial gene expression using an engineered CRISPR-Cas system. *Nucleic Acids Res., 41*, pp. 7429–7437.

Bolotin, A., Quinquis, B., Sorokin, A. and Ehrlich, S.D. (2005). Clustered regularly interspaced short palindromic repeats (CRISPRs) have spacers of extrachromosomal origin. *Microbiology, 151*, pp. 2551–2561.

Brouns, S.J. et al. (2008). Small CRISPR RNAs guide antiviral defense in prokaryotes. *Science, 321*, pp. 960–964.

Carte, J., Wang, R., Li, H., Terns, R.M. and Terns, M.P. (2008). Cas6 is an endoribonuclease that generates guide RNAs for invader defense in prokaryotes. *Genes Dev., 22*, pp. 3489–3496.

Christie, P.J. and Vogel, J.P. (2000). Bacterial type IV secretion: conjugation systems adapted to deliver effector molecules to host cells. *Trends in Microbiol., 8*, pp. 354–360.

Citorik, R.J., Mimee, M. and Lu, T.K. (2014). Sequence-specific antimicrobials using efficiently delivered RNA-guided nucleases. *Nat. Biotechnol., 32*, pp. 1141–1145.

Deltcheva, E., et al. (2011). CRISPR RNA maturation by trans-encoded small RNA and host factor RNase III. *Nature, 471*, pp. 602–607.

Doulatov, S., et al. (2007). Tropism switching in *Bordetella* bacteriophage defines a family of diversity-generating retroelements. *Nature, 431*, pp. 476–481.

Gasiunas, G., Barrangou, R., Horvath, P. and Siksnys, V. (2012). Cas9-crRNA ribonucleoprotein complex mediates specific DNA cleavage for adaptive immunity in bacteria. *Proc. Natl. Acad. Sci., 109*, pp. E2579–E2586.

Hanlon, G.W., Denyer, S.P., Olliff, C.J. and Ibrahim, L.J. (2001). Reduction in exopolysaccharide viscosity as an aid to bacteriophage penetration through Pseudomonas aeruginosa biofilms. *Appl. Environ. Microbiol., 67*, pp. 2746–2753.

HM Government. (2014). *Antimicrobial Resistance: Tackling a Crisis for the Health and Wealth of Nations.* Review on Antimicrobial Resistance, first publication. HM Government, Wellcome Trust, 11 Dec. 2014. Available at: http://amr-review.org/Publications/ [Accessed November 7, 2015].

Horvath, P., et al. (2009). Comparative analysis of CRISPR loci in lactic acid bacteria genomes. *Int. J. Food Microbiol., 131*, pp. 62–70.

Ishino, Y., Shinagawa, H., Makino, K., Amemura, M. and Nakata, A. (1987). Nucleotide sequence of the iap gene, responsible for alkaline phosphatase isozyme conversion in *Escherichia coli*, and identification of the gene product. *J. Bacteriol., 169*, pp. 5429–5433.

Jansen, R., Embden, J.D., Gaastra, W. and Schouls, L.M. (2002). Identification of genes that are associated with DNA repeats in prokaryotes. *Mol. Microbiol., 43*, pp. 1565–1575.

Jiang, W., et al. (2013). RNA-guided editing of bacterial genomes using CRISPR-Cas systems. *Nat. Biotechnol., 31*, pp. 233–239.

Jinek, M., et al. (2011). A programmable dual-RNA-guided DNA endonuclease in adaptive bacterial immunity. *Science, 337*, pp. 816–821.

Jore, M.M., et al. (2011). Structural basis for CRISPR RNA-guided DNA recognition by Cascade. *Nature Struct. Mol. Biol., 18*, pp. 529–536.

Kotterman, M.A., and Shaffer, D.V. (2014). Engineering adeno-associated viruses for clinical gene therapy. *Nat. Rev. Genet., 15*, pp. 445–451.

Labrie, S.J., Samson, J.E. and Moineau, S. (2010). Bacteriophage resistance mechanisms. *Nature Rev. Microbiol., 8*, pp. 317–327.

Liu, M., et al. (2002). Reverse transcriptase-mediated tropism switching in *Bordetella* bacteriophage. *Science, 295*, pp. 476–481.

Marraffini, L.A. and Sontheimer, E.J. (2008). CRISPR interference limits horizontal gene transfer in staphylococci by targeting DNA. *Science, 322*, pp. 1843–1845.

Mojica, F.J., Diez-Villasenor, C., Soria, E. and Juez, G. (2000). Biological significance of a family of regularly spaced repeats in the genomes of Archaea, Bacteria and mitochondria. *Mol. Micrbiol., 36*, pp. 244–246.

Mojica, F.J., Diez-Villasenor, C., Garcia-Martinez, J. and Soria, E. (2005). Intervening sequences of regularly spaced prokaryotic repeats derive from foreign genetic elements. *J. Mol. Evol., 60*, pp. 174–182.

Molineux, I.J. (1991). Host-parasite interactions: recent developments in the genetics of abortive phage infections. *New Biol., 3*, pp. 230–236.

Oh, J-H. and van Pijkeren, J-P. (2015). CRISPR-Cas9-assisted recombineering in *Lactobacillus reuteri. Nucleic Acids Res., 42*, p. e131.

Pourcel, C., Salvignol, G. and Vergnaud, G. (2005). CRISPR elements in *Yersinia pestis* acquire new repeats by preferential uptake of bacteriophage DNA, and provide additional tools for evolutionary studies. *Microbiology, 151*, pp. 653–663.

Sashital, D.G., Wiedenheft, B. and Doudna, J.A. (2011). Mechanism for foreign DNA selection in a bacterial adaptive immune system. *Mol. Cell, 46*, pp. 606–615.

Semenova, E., et al. (2011). Interference by clustered regularly interspaced short palindromic repeat (CRISPR) RNA is governed by a seed sequence. *Pro. Natl. Acad. Sci. USA., 108*, pp. 10098–10103.

Sinkunas, T., et al. (2013). In vitro reconstitution of Cascade-mediated CRISPR immunity in *Streptococcus thermophilus*. *EMBO J.*, *32*, pp. 385–394.

Sternberg, S.H., et al. (2013). DNA interrogation by the CRISPR RNA-guided endonuclease Cas9. *Nature*, *507*, pp. 62–67.

Sun, W., et al. (2015). Self-Assembled DNA Nanoclews for the Efficient Delivery of CRISPR-Cas9 for Genome Editing. *Angew. Chem. Int. Ed.*, *54*, pp. 12029–12033.

Sutherland, I.W., Hughes, K.A., Skillman, L.C. and Tait, K. (2004). The interaction of phage and biofilms. *FEMS Microbiol. Lett.*, *232*, pp. 1–6.

Tang, T.H., et al. (2002). Identification of 86 candidates for small non-messenger RNAs from the archaeon *Archaeoglobus fulgidus*. *Proc. Natl. Acad. Sci. USA*, *99*, pp. 7536–7541.

Vercoe, R.B., et al. (2013). Cytotoxic chromosomal targeting by CRISPR/Cas systems can reshape bacterial genomes and expel or remodel pathogenicity islands. *PLoS Genet.*, *9*, p. e1003454.

Wallden, K., Rivera-Calzada, A. and Waksman, G. (2010). Type IV secretion systems: versatility and diversity in function. *Cellular Microbiology*, *12*, pp. 1203–1212.

Wang, Y., et al. (2015). Markerless chromosomal gene deletion in *Clostridium beijerinckii* using CRISPR/Cas9 system. *J. Biotechnol.*, *200*, pp. 1–5.

Wiedenheft, B., et al. (2011). RNA-guided complex from a bacterial immune system enhances target recognition through seed sequence interactions. *Proc. Natl. Acad. Sci. USA*, *108*, pp. 10092–10097.

Yosef, I., et al. (2015). Temperate and lytic bacteriophages programmed to sensitize and kill antibiotic-resistant bacteria. *Proc. Natl. Acad. Sci.*, *112*, pp. 7267–7272.

10

Meat-borne Pathogens and Use of Natural Antimicrobials for Food Safety

Ashim Kumar Biswas[1] *and Prabhat Kumar Mandal*[2]

[1] *Division of Post-Harvest Technology, ICAR-Central Avian Research Institute, Izatnagar, Bareilly 243 122, Uttar Pradesh, India*
[2] *Dept. Livestock Products Technology, Rajiv Gandhi Institute of Veterinary Education & Research, Kurumbapet, Pondicherry, 605 009, India*

10.1 Introduction

Safety of meat and their products is a key issue in modern food trade. Consumers' expectations are healthy foods with affordable price, good shelf life, and high safety. Meat inspection requires good manufacturing practices and safety labeling that complies with the regulations. Until a few decades back, the meat industry was ruled by the unorganized sector, and the people had little knowledge to enhance productivity. In the post-WTO era, meat entrepreneurs have increased the productivity, but have little safeguard in regard to product quality, which may not meet the legislative requirements of certain standards that are setup by many developed and developing countries in the world. The use of antibiotics as growth promoters for animal food production is most discussed, since its unauthorized use gave us an inadvertent emergence of antibiotic-resistant bacteria in the food chain. Recently, the office of the United States Trade Representative has estimated that international trade has increased many folds since the signing of the General Agreement on Tariffs and Trade (GATT) in 1947. Following this agreement, the creation of the World Trade Organization (WTO) in 1995 resulted in significantly increased trade in foods of animal origin and live animals between different countries. But emergence and re-emergence of diseases due to pathogenic bacteria are the key issue of the new pattern of food trades. Some of these pathogens are so virulent that it has not been reported earlier.

According to the CDC (Center for Disease Control and Prevention, 1998) report, annual cost due to foodborne illness in the United States nearly 10 billion USD. The bacterial pathogens most frequently identified from illness associated with beef products are *Salmonella* spp., *Clostridium perfringens,* and *Staphylococcus aureus*. Interest of *Escherichia coli* O157:H7 has increased after the highly publicized outbreak of food poisoning associated with undercooked beef patties in the United States in 1993, although it was confined to North America until the mid-1990s. Likewise, the multidrug resistant *Salmonella typhimurium* DT-104 spread widely since it was first detected in the United Kingdom (JECFA, 2002).

The incidence of *Salmonella* was recorded up to 9% in red meat in India (Rao and Mahendrakar, 2003). These potentially bacterial pathogens reside on the hide or in intestinal tracts of food animals or may be originating indirectly by cross-contamination or through processing environment (Buckle *et al.*, 1989). Another

Foodborne Pathogens and Antibiotic Resistance, First Edition. Edited by Om V. Singh.

food-borne emerging diseaseis *Listeriosis*, which spread throughout the France and also in Canada, where meat and meat products were implicated as a source of *Listeria monocytogenes* (Borch and Arinder, 2002). Similarly, *Staphylococcal* food poisoning or food intoxication syndrome was first reported in 1894 but it is now a global problem in the meat industry.

10.2 Incidences of Some Important Foodborne Pathogens

10.2.1 *Salmonella*

Salmonella species are enteric bacteria and can be found in the intestinal tract of animals, including birds. Salmonellosis is one of the most common foodborne diseases worldwide and its causative agent is different variants of *Salmonella enteritic*a. Estimates suggest that tens of millions of human cases occur worldwide resulting in more than a hundred thousand deaths each year. More than 2500 different strains of Salmonella have been identified to date. In the United States, an estimated 1.028 million human cases, 19,000 hospitalizations, and 400 deaths have been reported (Scallan *et al.*, 2011). The organism is ubiquitous in nature and can survive several weeks in a dry environment and several months in water depicting hardness and survivability in unfavorable conditions.

Salmonella Typhimurium and *S. enteritidies* accounts for most non-typhoidal salmonella infection in humans in both developed and developing countries among the different serotypes of *Salmonella enterica*. They are transmitted from animals to humans in the most parts of the world, causing gastroenteritis, which can be severe in the young, the elderly, and patients with compromised immunity. As such it can frequently be isolated from raw meat and from environmental contaminants, although generally at lower numbers. Their presence in ready-to-eat foods may be a result of undercooking, poor handling practices, and cross-contamination. However, recent reports from the CDC indicate that the incidence of *Salmonella* infection in humans has shown a 9% decline in 2013 as compared to 2010–2012. In another report, an outbreak of human salmonellosis has been reported in 42 states and Puerto Rico on August 5, 2014, and a total of 300 persons infected with *Salmonella infantis*, *Salmonella newport*, or *Salmonella hadar*. The linkage of this human salmonellosis outbreak has been shown to contact with chicks, ducklings, and other live poultry. Salmonellosis caused by non-typhoidal species has shown an increasing trend in recent years with the majority of cases being caused by *S. enteritidis* and *S. typhimurium*. Symptoms in humans are characterized by acute onset of fever, abdominal pain, diarrhea, nausea, and sometimes vomiting. The onset of disease symptoms occurs 6–72 hours (usually 12–36 hours) after ingestion of *Salmonella*, and the illness lasts 2–7 days.

The incidence of *Salmonella* spp. in food animals is wide. Food-borne outbreaks of salmonellosis associated with eating of beef have been reported by Roels *et al.* (1997). Abouzzed *et al.* (2002) reported the prevalence of *Salmonella* spp. to be 4.6% in beef cattle on conducting examination of caecal contents in Canada. Patterson (1974) reported that the incidence of *Salmonella* spp was 0.34% in Northern Ireland. Similarly, Vanderlinde *et al.* (1998) reported the incidence of *Salmonella* spp. in Australian frozen bulk packed meat was 0.22%. Likewise, Scanga *et al.* (2000) did a classical study for determination of level of contaminants in raw beef trimmings and ground meat. In their study, *Salmonella* spp. was found more frequently in beef trimmings (5.2%) than culled beef cow trimmings (0%), culled dairy cow trimmings (0%), or imported trimmings. The incidences of *Salmonella* spp. in beef were also reported in the United States (Sofos *et al.*, 1999), Mexico (Chabela *et al.*, 1999), and even in India. Bachhil and Jaiswal (1988) recorded 5% of sample of fresh and frozen buffalo meat, 6.6% of minced meat, and 10% *Kabab* were positive for *Salmonella* in India.

10.2.2 *Escherichia coli*

E. Coli consists of a diverse group of bacteria. *E. coli* since its discovery by Theobald Escherich in 1885 has been receiving much greater importance due to its pathogenicity by certain strains both in man and animals. Worldwide contamination of this group of bacteria occurred in meat through soiling of the carcass and plant environment with fecal materials during the slaughter process (Johnson *et al.*, 1996). The incidence of *E. Coli* is not very variable in domestic or export beef meat regardless of the fat content in trimmings. The average *E. coli* counts (ECC) in Indian buffalo meat were 1.1$\log_{10}$ cfu/g. In another study, Harazika *et al.* (2005) screened 153 buffalo meat samples, among which 24.78% samples were positive for *E. coli*. However, enumeration data regarding various groups of *E. coli* is sparse.

Pathogenic *E. Coli* strains are categorized into pathotypes. One of the important pathotypes is Shiga toxin-producing *E. coli* (STEC), which may also be called as verocytotoxin producing *E. coli* (VTEC) or enterohemorrhagic *E. coli* (EHEC) and is the one most commonly associated with foodborne outbreaks. The other pathotypes are enterotoxigenic *E. coli* (ETEC), enteropathogenic *E. coli* (EPEC), enteroaggregative *E. coli* (EAEC), enteroinvasive *E. coli* (EIEC), and diffusely adherent *E. coli* (DAEC). The most commonly identified serotypes is *E. coli* O157. *E. coli* O157 has been reported to cause diarrhea, abdominal pain, and fever in approximately 70,000 Americans each year and may cause intense pain, high blood pressure, kidney failure, and even death.

Evidence of VTEC was found in 15 to 40% samples of ground or deboned raw beef in Canada (Acheson, 1996). Similarly, in the United Kingdom, 17% of raw beef samples contained VTEC. However, VTEC were found less frequently in continental Europe and only 1.8% of beef were recorded positive (Pierard *et al.*, 1994). Elder *et al.* (1997) noted that among 28% of cattle presented for slaughter in the Midwestern United States carried *E coli* O157:H7, only 2% of carcasses sampled were positive. Other workers also reported incidences of *E. coli* (71% serotype O157) in Sweden (Anon, 2000).

Prevalence of verotoxic *E. coli* in meat and meat products has been recorded at an alarming rate in India. In a study, Rathore (2000) reported that 89.19% of 37 *E. coli* isolates were verotoxic by vero-cell cytotoxicity assay. Smilarly, *E. coli* strains isolated from different meat and meat products revealed 15.90% isolates as verotoxigenic (Banerjee *et al.*, 2001). Hazarika *et al.* (2005) reported that 27% isolates of *E. coli* were verotoxigenic (VTEC) in vero-cell cytotoxicity assay. They further concluded that majority of VTEC isolates from meat and meat products of buffaloes were found positive for vt_2 gene (77.42%) followed by vt_1 (16.13%), whereas, both vt_1 and vt_2 were detected only in 6.45% of the VTEC isolates.

Product that test positive for *E. coli* O157:H7 cannot be sold as fresh meat as it is not safe for humans and it has to be subjected to cooking at temperature of 71°C, leading to destruction of *E. coli* O157 and is safe for human consumption. Presence of *E. coli* in poultry meat reflects fecal contamination, indicating a possible contamination by enteric pathogens. Cross-contamination during processing and human-to-food contamination via food handlers also play important role in transmission.

10.2.3 *Campylobacter*

Campylobacter can be found in the intestinal tract of wild and domesticated animals. There are 17 species and 6 sub-species belong to the genus *Campylobacter*, of which the most frequently reported in human diseases are *C. jejuni* (subspecies *jejuni*) and *C. coli* and it is the most common bacterial pathogen that cause gastroenteritis. They are most prevalent in raw meats and poultry. There has been an increase in number of outbreak of acute campylobacteriosis in humans worldwide especially in children, the elderly, and immune-suppressed patients and most cases are associated with handling raw meat, eating raw or undercooked meat, or cross-contamination to cooked foods. Recent report of CDC indicates that the *Campylobacter* infection has shown an increasing trend in the year 2013. Handling, preparation, and consumption of broiler meat may account for 20–30% of human cases of campylobacteriosis, whereas, 50–80% may be attributed to the chicken reservoir as a whole (EFSA, 2010). The intestinal tract of chicken, especially the caecum and colon, can harbour a large number of *Campylobacter* spp.; during processing, the intestinal tract may leak or rupture and the contents are transferred to the skin of the carcass. *Campylobacter* should not be present in ready-to-eat foods as consumption of food containing this pathogen may result in food-borne illness.

In poultry, the spread of *Campylobacter* in the flock is because of high levels of shedding and efficient fecal-oral transmission compounded by contaminated water and feed. Horizontal transmission from environmental sources can also occur. In broiler chickens, *C. jejuni* colonization can persist for 6-7 weeks, leading to carcass contamination at the processing plant during slaughter as a result of evisceration. Thus, the main approach in controlling *Campylobacter* contamination of poultry carcasses during processing is to minimize the spread of faecal material. Further, cross-contamination can occur during scalding, de-feathering, evisceration, washing, and chilling. In a study carried out in the Greater Washington, D.C., area from June 1999 to July 2000, majority (70.7%) of chicken samples were contaminated with *Campylobacter*.

10.2.4 *Listeria monocytogenes*

L. monocytogenes is widely distributed in the environment and can be isolated from a wide variety of foods including all meat foods and poultry. As such it is expected to be present in ready-to-eat foods that contain raw ingredients. Moreover, this organism is able to multiply slowly at 4°C. Among the six species, *L. monocytogenes* is an opportunistic pathogen in human beings as well as in various animal and poultry birds such as chickens, ducks, turkeys, geese, and so on. Chickens are thought of as reservoirs and carriers of this pathogen, whereas *L. ivanovii* primarily affects bird and animal species. However, incidences of *L. monocytogenes* in meat vary widely from 0% to 92% and the contamination mostly occurred on the surface of meat and meat products (Farber and Peterkin, 1991). The organism may also thrive in interior of muscle tissues of frozen beef. In a survey (2009) of steer/heifer carcasses in the United States it has been revealed that incidence of *L. monocytogenes* was about 4.1%. However, no *L. monocytogenes* was reported in beef carcasses from Northern Ireland (Madden *et al.*, 2001). Although, in India, *L. monocytogenes* was first isolated from sheep, later several studies suggested the presence of this organism in buffalo meat (Chaudhuri, 2001; Biswas *et al.*, 2008).

Outbreaks of listeriosis usually follow exposure to raw or uncooked poultry products but have also occurred after contaminated ready-to-eat poultry meat products were eaten. In contrast to young birds, adults may not be reservoirs of *L. monocytogenes*, but may be transiently colonized by consuming contaminated feed or water. Thus, *L. monocytogenes* may enter the packing plant at low levels in the intestine of birds, survive in biofilm, and ultimately contribute to both environmental and product contamination. The risk posed by *L. monocytogenes* is dependent on the food and how long it is stored. Foods in which all components have been cooked in the final food preparation, or have received some other listericidal treatment, should be Listeria free. Additionally, the detection of *L. monocytogenes* in foods, which have been prepared specifically for "at risk" population groups such as the elderly, immunocompromised, and infants should be considered as potentially hazardous. Foods with a long shelf life stored under refrigeration should have no *L. monocytogenes* detected in 25 g.

10.2.5 Coagulase-Positive *Staphylococci* (CPS)

Staphylococcus aureus can be routinely isolated from humans and associated environments. As such, the presence of coagulase positive staphylococci (a subgroup of *S. aureus*), is an indication of human contact. Some CPS strains produce a toxin that can cause food poisoning. Even minimal handling can result in coagulase-positive staphylococci being present in meats at low levels. This is unlikely to be a food safety concern provided that meat product is either consumed immediately or stored under temperature control. In raw or fermented meat foods, coagulase-positive staphylococci are likely to be outgrown or inhibited by the naturally occurring microorganisms. In fully cooked meat foods that support the growth of coagulase-positive staphylococci, the absence of competing microorganisms may provide the opportunity for growth and toxin production.

Coagulase-positive staphylococci appear to be of greatest concern in products such as processed meats. Unsatisfactory levels of coagulase-positive staphylococci indicate that time/temperature abuse of a food is likely to have occurred following improper handling during food preparation. A test for enterotoxin, SET, may be appropriate where levels of coagulase-positive staphylococci exceed 10^3 cfu/gram or where poor handling practices are suspected but it is likely that viable organisms may no longer be present in significant numbers. Levels of greater than or equal to 10^4 cfu/g are considered as potentially hazardous as foods with this level of contamination may result in food-borne illness if consumed.

10.2.6 *Clostridium perfringens*

C. perfringens can be isolated from a variety of environments and foods. As the spores of *C. perfringens* can survive cooking, it is of most concern in cooked foods that have been temperature abused as this provides the opportunity for spores to germinate and grow. Potential temperature abuse scenarios include holding cooked foods at warm temperatures for extended periods of time, and cooling foods too slowly. *C. perfringens* is of particular concern for cooked meats (large joints or rolled meats), gravy, and meat. *C. prefringens* type-A infection, has been reported as the most important cause of food poisoning worldwide.

The pathogen has been isolated from the intestine of the poultry birds and other large and small animals; hence, they serve as the source of contamination to varieties of meats and environment. The main symptoms in human beings include acute abdominal pain, profuse diarrhea, nausea, and rarely vomiting. *C. perfringens* causes necrotic enteritis in poultry and the mechanism of colonization of the avian small intestinal tract are the factors involved in toxin production. It is generally accepted, however, that predisposing factors are required for these bacteria to colonize and cause disease in poultry. The best known predisposing factor is mucosal damage, caused by coccidiosis. Diets with high level of indigestible, water-soluble non-starch polysaccharides, known to increase the viscosity of the intestinal contents, also predispose to necrotic enteritis.

10.2.7 Other Pathogens

Bacilus cereus can be detected in many raw foods of plant origin and in raw milk. As with *C. perfringens*, their spores will survive cooking, and poor temperature control after cooking may result in germination of the spores and subsequent growth. *B. cereus* is of greatest concern in plant- or cereal-based ready-to-eat foods and cream based sauces. Ready-to-eat foods containing raw components may contain low levels of *B. cereus*.

Vibrio parahaemolyticus, present in the marine environment and is of relevance to foods containing seafood. It is common for raw seafood to contain low levels of *V. parahaemolyticus* and this would not be an indication of poor food processing practices. Effective cooking of seafood will destroy *V. parahaemolyticus* and their presence in cooked food may be an indication of post-cooking contamination or insufficient heat treatment.

The other group of organisms that are generally considered as harmless for human health, but these being psychrotropic, are able to multiply on the product during refrigeration storage.

They develop off odour and ultimately the shelf-life of the product is shortened, though, number of spoilage bacteria as well as storage temperature history is also important in determining the shelf-life of the product. Main spoilage bacteria during aerobic storage of poultry meat under chilling conditions include *Pseudomonas* spp. as the predominant one followed by *Acinetobacter Candida* and yeasts while during modified atmospheric stoarage include *Lactobacillus, Corynaebacterium, Shewanella,* and Enterobacteriacae (psychrotropic strains).

10.3 Application of Natural Antimicrobials

10.3.1 Organic Acids

Organic acids such as acetic acid, lactic acid, propionic acid, citric acid, succinic acid, malic acid, tartaric acid, benzoic acid, sorbic acid, and their derivatives are commonly used as antimicrobial agents. As far as their spectrum of action is concerned, they are almost equally effective in ihibiting the growth of gram-positive and gram-negative bacteria. During fermentation of various foods, lactic and propionic acids are two normally produced acids. Their antimicrobial activity has been well documented. Cell walls, cell membranes, metabolic enzymes, protein synthesis systems, and genetic material are the main targets of their action against a wide range of microorganisms. The antimicrobial activity of organic acids is pH dependent and as the pH increases, their activity decreases. The reduction of intracellular pH of microorganisms changes the ability of cells to maintain the homeostasis and thus affecting the substrate transport and metabolic pathways of microorganisms. Organic acids are commonly used in the form of washes and sprays for disinfection of animal and poultry carcasses.

10.3.2 Essential Oils of Various Plant Species

Essential oils are nothing but true spices that are obtained from roots, bark, buds, seeds, or fruits of plants growing in tropical countries. The phenolic compounds that commonly occur in natural foods have been divided into three groups via simple phenols and phenolic acids (e.g., *p*- cresol, 3-ethylphenol, hydroquinone, vanillic, gallic, syringic, ellagic), hydrocinnamic acid derivatives (e.g., *p*-coumaric acid, caffeic, ferulic, sinapic), and flavonoids. Proteins, fats, salts, pH, and temperature are factors that affect the antimicrobial activity of phenolics. Their basic classification, active principles, and use in meat have been enlisted in the Table 10.1.

Besides these some other natural antimicrobial like, pepper, coriander, bay, thyme, mustard, lemon, citrus peel (lime, lemon, orange), and tobacco are a few representatives from a very long list of plant products successfully expressing antibacterial properties in various meat and meat products (Kim *et al.*, 1995). Further, Kurita *et al.* (1981) elaborated the usefulness of 11 essential oil extracts against the potential foodborne pathogens *E. coli*, *E. coli* O157:H7, *Salmonella enterica* serovar Typhimurium, *Listeria monocytogenes*, and *Vibrio vulnificus*. Hammer *et al.* (1999) demonstrated the efficacies of 52 plant oils and extracts against gram-positive and gram-negative bacteria, with some of the oils being effective against gram-positive organisms only. Dorman and Deans (2000) demonstrated antibacterial activity against 25 different genera of bacteria, including animal and plant pathogens and food poisoning and spoilage bacteria.

10.3.3 Phytoalexins

Phytoalexins are host-synthesized low molecular weight broad spectrum antimicrobial compounds whose synthesis from distant precursors is induced in plants in response to microbial infection or treatment of plant tissues with a range of naturally occurring or synthetic, artificial compounds. The molecules that signal the plants to begin the process of phytoalexin synthesis are called elicitors.

Table 10.1 Antimicrobial Activity of Different Essential Oils of Various Plant Species.

Preservative (active principles)	Use in Meat
Clove (eugenol, thymol, carvacrol)	1 and 2% clove oil as antimicrobial in frankfurters (Mytle *et al.*, 2006); 0.2 and 0.5% clove oleoresin in chicken meat marination (Carlos and Harrison, 1999); clove powder as antimicrobial in chicken nuggets (Kumar and Tanwar, 2011). Hao *et al.* (1998) applied eugenol to beef slices or cooked chicken and it was proved that eugenol inhibited the growth of *A. hydrophila* and *L. monocytogenes*.
Ginger (6-gingerol and its derivatives)	2.5% ginger extract as tenderizing, antimicrobial in smoked spent hen meat (Naveena and Mendiratta, 2001).
Garlic (allicin, diallyl sulfide, allyl sulfide, diallyl trisulfide, and propyl sulfide)	Garlic powder and garlic oil as antimicrobial in raw chicken sausage (Sallam *et al.*, 2004); garlic extract as growth inhibitor of facultative aerobic, mesophilic, and fecal coliforms on the surface of poultry carcasses (Oliveira *et al.*, 2005).
Nutmeg (macelignan, d-camphene)	Nutmeg showed bactericidal activity against *E. coli* O157:H7 in barbecued chicken (Shekarforoush *et al.*, 2007).
Oregano (rosmarinic acid, carvacrol, thymol, and glycosides)	Oregano oil in meat reduced the microbial growth during refrigerated storage (Burt 2004; Chouliara *et al.*, 2007).
Sage (carnosol, rosmanol, rosemadiol, epirosmanol, isorosmanol, galdosol, methyl carnosate, and carnosic acid)	Sage extract alone or in combination with sodium isoascorbate resulted in reduced mesophilic bacteria and coliforms counts in raw vacuum-packaged turkey meatballs (Karpinska-Tymoszczyk, 2007).
Cinnamon (cinnamaldehyde, eugenol, cinnamicaldehyde)	Cinnamon powder @ 2% and 4% (w/w) level extended the shelf life of minced chicken meat at room temperature and refrigeration temperature (Yadav *et al.*, 2002).
Pomegranate (tannins, antho-cyanins, and flavonoids)	0.02% phenolics solution as antimicrobial in chicken meat (Vaithiyanathan *et al.*, 2011).
Rosemary (carnosol, carnosic acid, rosmanol, isorosmanol, rosmariquinone, rosmaridiphenol, and rosmary-diphenol)	Rosemary extract for antimicrobial activity (4% w/w) in chicken frankfurters (Riznar *et al.*, 2006).

10.3.4 Lactoferrin and Activated Lactoferrin

Lactoferrin (LF) is an 80-kDa iron chelating bacteriostatic protein and it is a component of phagosome. Lactoferrin consists of about 600–800 contiguous amino acid residues depending on its source. It exhibits defense and various biological activities including antibacterial activities (Dionysius*et al.*, 1993). Its antimicrobial activity seems to be due to sequester iron with high affinity and bacterial growth is inhibited by the iron deficiency (Arnold *et al.*, 1977). After elucidation of globular structure of LF, it is predicted that LF has additional antibacterial mechanism for bacteria. Highly cationic parts in LF interacts with porins and lipid-A component of lipopolysaccharide layer of gramnegatives, hence, LF-mediated LPS release resulted in disruption of bacterial cell wall integrity and permeability (Naidu, 2003). However, exact action mechanism of lactoferrin yet not been fully elucidated. It is thought that LF binds to positively charged molecules on bacterial surface of gram positives such as lipoteichoic acid (Gonzalez-Chavez *et al.*, 2008). Lactoferricin inhibited the growth of *Klebsiella pneumoniae, Pseudomonas aeruginosa, Listeria monocytogenes,* and *Staphylococcus aureus.* Bovine-lactoferrin (bLF) is currently used by the food industry as a supplement in yoghurt, skim milk, infant formula, and drinks. Recombinant human lactoferrin has been produced by using organisms such as *Saccharomyces* sp, *Aspergillus* sp (Lonnerdal and Iyer, 1995), transgenic rice (Bethell and Huang, 2004), and transgenic mice (Nuijens *et al.*, 1997).

10.3.5 Lactoperoxidase

Lactoperoxidase system consists of thiocyanate and hydrogen peroxide and shows a strong antimicrobial action. Hydrogen peroxide may be added either directly or indirectly into the system. The hypothiocyanate ion is the major intermediate oxidation product of thiocyanate and the principal antimicrobial agent generated by lactoperoxidase in milk.

10.3.6 Lysozyme

Lysozyme was discovered by Alexander Flemming in 1922. It is β-1, 4-N-acetyl-muramidase. It is found in many mammals, birds, fish, insects, plants, and viruses. It constitutes 3-4% of albumen proteins. The concentration of lysozyme in hen, duck, and goose egg is 3000, 1000, and 1200 μg/g. Hen egg white lysozyme is classified as Generally Recognised as Safe (GRAS) in the United states and is approved for use in some foods in the European Union, Japan, and Australia (Losso *et al.*, 2000). Lysozyme from the hen egg is extracted commercially by using cation exchange chromatography. Mostly it causes the cell death of gram-positive bacteria. The enzyme makes contact with the peptidoglycan in the cell wall, the polymer gets cleaved, which in turn leads to osmotic uptake of water and eventual rupturing of cytoplasmic membrane. Since its main cite of action is peptidoglycan in the cell wall, gram-negative bacteria are resistant to it due to the absence of peptidoglycan. It has been documented that in combination with a cation-binding agent the antimicrobial activity of lysozyme gets increased. The use of lysozyme as a food preservative requires recombinant DNA technology and protein engineering so as to reduce the production costs and improve its activity. Arima *et al.* (1997) has revealed that up to 500 mg/L of recombinant lysozyme can be produced but this is still very low amount for its commercial use at industry level. Lysozyme activity is affected by various conditions like type of food, processing, and storage conditions. Lysozyme has also been evaluated for its use as an antimicrobial packaging component.

Pre-surface application of nisin and lysozyme reduced the growth of *Listeria monocytogenes* during refrigerated storage of low-fat turkey bologna (Mangalassary *et al.*, 2008). Ntzimani *et al.* (2010) used as natural antimicrobial in combination with rosemary, oregano oil, and EDTA in semi-cooked chicken meat fillets).The effect of lysozyme action on microbiological stability of confectioned raw poultry meat was investigated (Kijowski *et al.*, 2002); treatment of chicken skin with 2–100 Mm TSP and lysozyme (10 μg/ml) reduced viable numbers of attached *Campylobacter jejuni, E. coli, Pseudomonas fluorescens,* and *S. enteritidis* by up to 99.9% depending upon target organism and temperature (Carneiro de Melo *et al.*, 1998). Samuelson *et al.* (1985) also found that the combination of lysozyme and EDTA inhibited serovar typhimurium on poultry.

10.3.7 Avidin

Avidin protein constitutes about 0.05% of the total proteins in the albumen of hen's eggs. Avidin is a glycoprotein whose one molecule binds to four molecules of vitamin biotin. Hence, it inhibits the growth of some bacteria and yeasts that have a requirement for biotin, with the primary mechanism being the nutrient deprivation. According to another scientific study, it has been suggested that it can also bind the porin proteins of *E. coli.* Thus it can also inhibit microbes in vitro by interfering with transport through porins.

10.3.8 Chitosan and Chitinases

Chitosan is a natural polymer obtained from exoskeletons of crustaceans and also from cell walls of fungi and insects, but commercially it is produced from deacetylation of chitin. Chitin, chitosan, and their modified derivatives have been applied in medicine, agriculture, food, and non-food industries. In food preservation, chitosan inhibits the growth of food-borne bacteria, yeasts, and molds. The minimum inhibitory concentration (MIC) for bacteria and yeasts vary widely depending upon the molecular weight of the polymer, degree of acetylation, pH, temperature, and presence of interfering substances such as fats and proteins. Usually in combination with the other antimicrobials like benzoates, the MIC of chitosan can be reduced.

Chitinases and other hydrolytic enzymes are also having the antimicrobial activity. These enzymes are produced by plants, insects, protozoa, and microorganisms. Plants produce endo-acting chitinases, whereas bacteria produce exo-acting chitinases. The antifungal activity of chitinases and glucanases is well known.

10.3.9 Bacteriocins

The term "Bacteriocins" was first coined by Jacob *et al.* (1953) to describe certain protein substances made by microorganisms to inhibit similar or closely related microbes. These are usually high molecular weight proteins, entered into the cell using specific cell-surface receptors, thereby killing bacterial cells by forming ion channels and inhibiting protein synthesis in the target cell (Françoise, 2010). Bacteriocins are produced by strains of all genera of lactic acid bacteria of relevance to meats and act against bacteria closely related to the producer organisms. Bacteriocinogenic lactic acid bacteria and/or their isolated bacteriocins are considered safe additives (GRAS), useful to control the frequent development of spoilage and pathogenic microbes.

Among all bacteriocins, nisin probably is the most common bacteriocin used as a food preservative in the United States. It has been reported that addition of nisin to the scalding water to treat any *L. monocytogenes* present in the water or on the surface of the turkey skin (32), a 1-log reduction of *Listeria* in the nisin-treated samples occurred and was followed by further reduction after refrigeration. Gram-negative bacteria such as *Salmonella* are less sensitive to nisin than are gram-positive bacteria. However, when used with EDTA or Tween 80, the activity of nisin against the pathogen can be increased. The source of bacteriocins has been given in Table 10.2.

The bacteriocins can be classified under various broad categories:

- **Class Ia:** These are small peptides (<5 kDa) called as lantibiotics, with some unusual amino acids like lanthionine, methyl-lanthionine, dehydrobutyrine, and dehydroalanine, for example, nisin (Shefet *et al.* 1995). Lantibiotics are also produced by the genera other than LAB.
- **Class Ib:** These are small globular peptides with no net charge or net negative charge, for example, mersacidin.
- **Class IIa:** These are small heat-stable peptides (non -antibiotics), that are synthesized in a form of precursor, which is processed after two glycine residues and are mainly active against *Listeria monocytogens*, for example, pediocin PA-1, sakacins A and P, leucocin A.
- **Class IIb:** Two-component system: two different peptides required to form an active poration complex, for example, some other bacteriocins like lactococcins G and F, lactacin F, plantaricin FE & JK.
- **Class III:** These are large heat-sensitive peptide molecules, for example, helveticins J and V-1829, lactacins A and B, and acidophilucin A.
- **Class IV:** These are the peptide molecules having lipid and carbohydrate moieties, for example, leuconocin, lactacin 27, and pediocin SJ-I.

The action of a bacteriocin in a food system depends upon its chemical composition and the physical condition. The action of bacteriocins

Table 10.2 Antimicrobials of Microbial Origin Use in Poultry Meat.

Bacteriocin	Producing organisms
Lactocidin, Acidophilin, Lactacin B	*Lactobacillus acidophilus*
Acidocin B, Acidolin J	*Lactobacillus acidophilus*
Plantacin B and N, Plantaricin C	*Lactobacillus plantarum*
Plantaricin SA6, Plantaricin LP84	*Lactobacillus plantarum*
Nisin, Lacticin, Lactococin G	*Lactococcus lactic* subsp. *lactis*
Diplococcin, Lactococcin	*Lactococcus lactic* subsp. *cremoris*
Lactostrepcin 5	*Lactococcus lactic* subsp. *cremoris*
Sakacin, Sakacin P	*Lactobacillus sake*
Bulgaricin	*Lactobacillus delbruceckii* subsp. *bulgaricus*
Curvacin	*Lactobacillus curvatus*
Helveticin J	*Lactobacillus helveticus*
Leucocin	*Leuconostocgelidum*
Lactacin F	*Lactobacillus johnsonii*
Bifidin	*Bifidobacteriumbifidum*
Leuconocin S	*Leuconostocparamesenteroides*
Caseicin A	*Lactobacillus caseii*
PediocinAcH, Pediocin PA-1	*Pediococcusacidilactici*
Propionicin, Propionicin PLG-1	*Propionibacteriumthoenii*

may be reduced by the binding of bacteriocin molecules to food components, by the destabilizing action of proteases and other enzymes, and by an uneven distribution in the food matrix (Schillinger *et al.*, 1996). Use of bacteriocins in meat and meat products is somewhat difficult than the liquid foods because of different reasons, such as uneven distribution or because bacteriocins can partition to the fat phase or may bind to meat proteins. Since lactic acid bacteria (LAB) are commonly found fauna in meat, hence, the antimicrobial roles of bacteriocins have been well documented. All bacteriocins act well in meat systems, but the role of nisin is well studied (Cleveland *et al.*, 2001).

Nisin was first discovered in England in 1928 as a result of difficulties experienced during cheese making. Nisin may be ineffective in meat due to high pH (Rayman *et al.*, 1983) or improper distribution and interference by meat components such as phospholipids (de Vuyst and Vandamme, 1994). Nisin in combination with chelating agents showed potent bactericidal activity against gram-negative bacteria including *Salmonella* and *E. coli*, and has led to the investigation whether this combination can be applied to meat and poultry carcasses (Blackburn *et al.*, 1989; Delves-Broughton, 1993; Cutter and Siragussa, 1995). It may also inhibit *Brochothrix thermosphacta* in cold meat-binding system (Cutter and Siragusa 1998). The use of nisin in spray form for decontamination of meat contact surfaces along the processing lines has also been documented. The nisin concentrate "Nisaplin" is used commercially as a preservative in dairy products, canned foods, cured meats, and other segments of fermentation industry. While working with turkey carcasses, Mahadeo and Tatini (1994) documented that nisin at 2.5 μg/L provided synergistic effect to kill *L. monocytogenes* by heat (52°C for 3 mins) in the scalding water. Shefet *et al.* (1995) successfully used a combination mixture of nisin (100 μg/ml) with chelating agents (EDTA, citric acid) and a surfactant to reduce *S. typhimurium* contamination on chicken drumsticks and to extend the shelflife of refrigerated chicken drumsticks. Bell and DeLacy (1986) documented that degree of nisin recovery was poor from minced meat and meat emulsions and it was affected by fat content in it. The use of nisin in raw meat systems is quite limited due to their limited activity against gram-negative bacteria. The use of nisin is better understood in case of sausages and pate meats where homogenization of meat allows its even distribution in the meat system. But nisin is more effective in lean meat and the type of phosphate used in meat affected the activity of nisin against lactic acid bacteria. Nisin at the level of 6.25–25 μg/g effectively inhibited the growth of a mixed lactic acid bacteria and extended the shelflife of vacuum packaged sausages.

Other bacteriocins such as leucocin A, enterocins, sakacins, and the carnobactericins A and B also prolong the shelf life of fresh meat. Pediocins are more effective than nisin in inhibiting bacteria in meat and are even more effective in dairy products. Dipping meat in 5,000 A.U. of crude Pediocin PA-1 per ml decreases the viability of *L. monocytogenes* 100- to 1000-fold. Pediocin AcH at 1350 A.U/ml reduces listeria populations in ground beef and sausages by 1 to 7 logs. Pediocin AcH is more effective at 4°C than at 25 °C against *L. monocytogenes* in wiener exudates. Pediocin AcH (PA-1) successfully controlled the growth of *L. monocytogenes* in raw chicken (Goff *et al.*, 1996). Pediocin may be used in chicken before cooking for maximum effectiveness. Pediocin AcH inhibits *L. monocytogenes* in chicken summer sausages. Besides this, Pediocin PA-1 has also been expressed in the yeast *Saccharomyces cerevisiae* to improve preservation of wine, bread and other food products where yeast is used.

10.3.10 Bacteriophages

Bacteriophages are the viruses that infect the bacteria. Application of phages has been investigated extensively against antibiotic resistant bacteria (Yosef *et al.*, 2014). It has been reported that lytic phage (virulent phage) when it infects a host bacterium replicates much faster than the host cell and dies, simultaneously liberating a large number of progeny phages, which are each then ready to start another cycle by infecting new neighboring bacterial cells. The whole cycle can be completed in 30–40 min. So, lytic phages seem to be considered as natural antimicrobial controlling agents. The lysogenic phages also called "temperate" or dormant phages, which may take the form of a "prophage" by integrating with the viral DNA in the host chromosome. They become a part of the host cell and replicate along with the host chromosome for many generations, coexisting as opposed to lysing the host cell (Jassim and Limoges, 2013). By this way dormant phage provides additional benefits of immunity against infection by further phage particles of the same type, ensuring that there is only one copy of phage per bacterial cell.

Since the 1960s, numerous studies were carried out for the potential application to treat bacterial infections, and it was reported that bacteriophages can destroy their host cells recovered from fish, meats, and skim milk. Phages have been extremely effective at treating a number of bacterial infections in controlled animal studies, especially as a biocontrol agent in the prevention of foodborne illnesses, due to its target specificity, rapid bacterial killing, and ability to self-replicate (Biswas *et al.*, 2002; Hawkins *et al.*, 2010). Phage therapy is like other methods of biological control with some comfort in the reduction of the use of chemical agents against pathogens (Fujiwara *et al.*, 2011). Goode *et al.* (2003) documented that application of specific phages on excised chicken skin can reduce *S. enteritidis* and *Campylobacter jejuni*. Atterbury *et al.* (2007) demonstrated the efficacy of host specific phages to reduce the numbers of *C. jejuni* on artificially contaminated chicken skin. In another study, it has been found that aerosol spraying and intramuscular injection have given the best results over using oral delivery of phages via direct administration or addition to drinking water and/or feed (Sillankorva *et al.*, 2012). This may be due to gastric pH levels preventing the proliferation of phages. Wagenaar *et al.* (2005) reported that phage therapy (biocontrol) reduces *Campylobacter jejuni* colonization of broiler chickens. Several studies have also addressed the use of phages to decrease *Campylobacter* and *Salmonella* concentrations on poultry (Goode *et al.*, 2003; Atterbury *et al.*, 2007). Some important limitation of this technique are a limited host range, the requirement for threshold numbers of the bacterial targets, phage-resistant mutants, and the potential for the transduction of undesirable characteristics from one bacterial strain to another. Phages have been shown to reduce the numbers of *E. coli* O157:H7 and *Salmonellae* on fresh poultry. The apparent reduction of numbers of viable *E. coli* from fresh poultry is also possible using coliphages. As a whole, phages can be bactericidal, they can increase in number responding to the incidence of pathogens over the course of treatment, and tend to only minimally disrupt normal flora. They are equally effective against antibiotic-resistant bacteria, are often easily discovered, seem to be capable of disrupting bacterial biofilms, and can have low inherent toxicities. The exploitation of phages as a realistic approach in the control of pathogens has attracted considerable interest in recent years (Jassim *et al.*, 2012), because of the emergence of antibiotic-resistant bacteria.

10.3.11 Ozone (O_3)

The application of ozone to control pathogens on animal carcasses has been tried. Ozone was first discovered by Schonbe in in 1840 (USA), but use of gaseous ozone as a food preservative was started in 1909 in Germany for the storage of meats. Due to multiple benefits of ozone, the U.S. FDA has also cleared long pending approval of ozone for direct use on foods. Application of gaseous ozone are now in need to better utilize the long known potential of ozone as an antimicrobial agent. It has been reported that ozone is a gas at ambient pressure and temperature that dissolves readily in water at refrigeration temperatures and acidic pH values (Gordon, 1995). As the pH or temperature is elevated, ozone decomposes into a variety of free radicals. As a result, ozone is a potent oxidizing agent and effective in microbial inactivation. However, mechanism of action for bacterial cell death during ozone treatment is attributed to changes in cellular permeability followed by cell lysis. The cell membrane is believed to be the primary point of attack at the level of glucoproteins, glycolipids, or proteins containing sulfhydryl groups (Greene *et al.*, 1993). The targeted amino acids in order of susceptibility include cysteine, methionine, tryptophan, tyrosine, histidine, cystine, and phenylalanine (Menzel, 1971). Other amino acids in proteins were unaffected.

However, ozone action is influenced by properties of treated medium. On a stainless steel surface, the bactericidal properties of gaseous ozone were greater than in the presence of meat-based broth (Moore *et al.*, 2000). It was reported that gram-negative organisms were most sensitive to gaseous ozone than gram-positive bacteria (Moore *et al.*, 2000). However, it was also reported that there were no differences in resistance to ozone that correlated with the differences in the cell wall structure of the bacteria (Broadwater *et al.*, 1973). It appears that ozone indiscriminately and electrophilicaly attacks all organic molecules. Bovine serum albumin adsorbed alumina (Al_2O_3) surfaces could be removed more easily following partial decomposition by ozone (Urano and Fukuzaki, 2001).

The effectiveness of ozone depends on the residual ozone remaining to react with the microorganisms after demands have been met. But ozone has been shown to be most effective against weakly attached bacteria, but these contaminants also are easy to rinse away. Microbial contaminants on meat present challenges. When beef cuts were challenged with *Escherichia coli* O157:H7, *Listeria monocytogenes*, and *Clostridium perfringens*, and rinsed with 3 ppm ozone in water for 5 min, populations of these pathogens decreased by 0.96, 0.98, and 1.2 $\log_{10}$/g, respectively. Meat has an uneven cut surface of polysaccharides, collagen, and muscle tissue with bacteria embedded in gaps and crevices.

Ozone is potentially useful in disinfecting poultry carcasses or recycling of chilled water. It has been observed that an ozonated water spray produced the largest reduction (1.23 $\log_{10}$) in *Salmonella* Enteritidies, regardless of the attachment times, on skinless chicken breasts (Dave, 1999). A study evaluating the effects of ozone on poultry meat organisms in water reported 1- to 5-$\log_{10}$ reductions, depending on the presence and amount of soluble organic material (Yang and Chen, 1979). Kaess and Weidemann (1968) reported that count of *Pseudomonas* spp. and *Candida scotti* on contaminated beef decreased significantly at >2 µg/L gaseous ozone and the lag phase of *Thamnidium* spp. and *Penicillium* spp. increased but their growth rate did not change. The color of the muscle surface treated with <0.6 µg/L ozone did not differ from that of the control treatment. Ozone has been tested in the process of tenderizing meats to control surface microflora (*Pseudomonas* spp., spores, *Salmonellae*, *Staphylococcus* spp.). Further study on spraying beef brisket fat with hydrogen peroxide (50 g/L) solution and ozonated water (5 g/L) was effective in reducing bacterial contamination, when compared to treatments with trisodium phosphate (120 g/L), acetic acid (20 g/L), and commercial sanitizer (3 g/L) (Gorman *et al.*, 1995). Reagan *et al.* (1996) conducted a study to compare procedures and inventions for eliminating physical and bacterial contamination from beef carcasses. Rinsing with ozonated water (0.3 to 2.3 ml/L) reduced aerobic plate counts by 1.3 log cfu/cm^2, which was approximately equivalent to conventional washing in reducing bacterial populations on beef.

Rusch and Kramer (1989) used ozone for the treatment of airborne microorganisms on the surface of meat stored at 2.5 to 6°C and 92 to 95% RH. The treatment halted growth of several *Enterobacteriaceae* but not that of *Pseudomonas* spp. Dondo *et al.* (1992) evaluated ozone usage for beef kept in a refrigerator. Ozone stopped the growth of surface contaminants during several days of storage, improved the sensory quality, and decreased the formation of total volatile N compounds. Horvath *et al.* (1985) indicated that in the presence of ozone, growth of microflora on meat surfaces decreased at refrigeration temperatures; however, no inhibitory effect was observed if the meat was heavily contaminated. But it has been reported that some inorganic colloidal particles protect ozone from decomposition and enhance its potency against microbial contaminants. Food-grade silica gel, for example, was very effective in binding and stabilizing ozone. Silica-bound ozone was a much more potent antimicrobial agent than was aqueous form of the sanitizer, particularly in the presence of ozone-demanding organic loads. Compared to aqueous or gaseous ozone, the silica-stabilized agent exerts its antimicrobial action over longer contact times. It is likely, therefore, that stabilized ozone may be useful in meat applications.

10.3.12 Hydrogen Peroxide

Hydrogen peroxide (H_2O_2) is GRAS for use in food products as a bleaching, oxidizing, reducing, and antimicrobial agent. The effectiveness of H_2O_2 to destroy the microorganisms depends on several factors including pH, temperature, and environmental conditions. Under normal conditions, H_2O_2 itself gets decomposed into oxygen and water thereby leaving no toxic residues in it. It has bactericidal or bacteriostatic effect, which is based on the formation of radicals that damage nucleic acids, proteins, and lipids. Hydrogen peroxide @ 0.5% w/v in water acts as poultry carcass decontaminant (Bolder, 1997). A combination spray of hydrogen peroxide and sodium bicarbonate resulted in increased shelflife of poultry (Fletcher *et al.*, 1993). The application of hydrogen peroxide seems to be an effective and safe method for carcass decontamination to control the spread of pathogens.

10.3.13 Electrolyzed Water

As an alternative to antibiotic, studies on efficacy of electrolyzed water (EW) on foodborne pathogens were conducted. Electrolyzed (EO) water is generated by electrolysis of a dilute salt (NaCl) solution in an electrolysis chamber where anode and cathode electrodes are separated by a membrane. It is presumed that on the anode side, acidic EO water is generated and has strong bactericidal effect on most known pathogenic bacteria, due to its low pH, high oxidation–reduction potential (ORP) (about 1100 mV), and the presence of hypochlorous acid (Kim *et al.*, 2000). Muriana (2005) reported that EW is lethal to exposed cells of various foodborne pathogens, resulting in significant reductions when placed in contact only for 2 minutes. Additional tests were also examined with shorter contact times on strong biofilm-forming strains of *Listeria monocytogenes*. In the same study, the researcher also proposed that the EW can give greater than a 6-log reduction of *E. coli* O157, *Listeria monocytogenes*, and *Salmonella* Enteritidis suspended in the peptone water solution than the corresponding non-EW within 2 min. In another study, Park *et al.* (2002) reported that EO water was very effective not only in reducing the population of *C. jejuni* on chicken, but also could prevent cross-contamination of processing environments due to no survival of *C. jejuni* in the EO water after washing treatment. The advantages of using EO water in the poultry processing are:

1) it is a non-thermal treatment for microbial inactivation;
2) no other chemicals except very dilute NaCl solution are required to generate the EO water;
3) it has strong antimicrobial effect to prevent cross-contamination of processing environments;
4) it is produced on site and on demand with the concentration for direct usage and no dilution from concentrated chemicals is needed;
5) it is a less potential health hazard to the worker due to the lack of need to handle concentrated chemicals for microbial inactivation.

10.4 Regulatory Aspects of Natural Antimicrobials

According to FDA regulations, there should not be residual H_2O_2 present in the foods under any circumstances. LAB are GRAS for the production of fermented foods. GRAS status was conferred by U.S. Food and Drug Administration (FDA) and is desirable because it allows a compound to be used in a specific application without additional regulatory approval. Nisin has been approved for use in all the foods internationally. The Joint Food and Agriculture/World Health Organization accepted nisin as a food additive in 1969 and set the maximum intake level as 33,000 I.U per kg body weight. That is why many countries allow nisin in a variety of products, sometimes with no restrictions as to maximum level. In the recent years, the U.S. FDA has given approval of Listeria-specific phage preparations for food additives. It is known that phages only infect and lyse bacterial cells and are harmless to mammalians (USFDA, 2006). This has eventually led to the development of a phage-related product, which received regulatory approval from the U.S. FDA in 2011, as a natural antimicrobial for use in agro-food industry as GRAS and by US-FSIS as safe for use in animals (Klumpp and Loessner, 2013). In general, although the safety of phages has been strongly suggested by human phage therapy, it should be noted that some phages, notably when in the form of lysogens (prophages), have been recognized as important contributors to bacterial virulence, or as vectors in horizontal gene transfer through transduction (Verheust *et al.*, 2010). Recently the U.S. FDA has given the approval for the use of O_3 for foodborne pathogens reduction and other bacterial decontamination techniques.

10.5 Health Benefits of Natural Antimicrobials

Natural antimicrobials, although primarily used for killing or inhibiting number of bacterial species, have several beneficial effects. The use of LAB as dietary adjunct to ameliorate lactose intolerance is well documented. LAB in fermented foods in addition to their antimicrobial activity, produce desirable changes in texture, flavor, and taste of the foods. Since these are commonly present in various fermented foods it is assumed that these bacteria do not pose any health risk to man and are designated as GRAS, (Jeevaratnam *et al.*, 2005). Curry leaves commonly used as flavoring agent in many traditional dishes are gaining attention because of its medicinal values such as anticarcinogenic (Chevallier, 1996), antiplatelets, and vasorelaxing effects (Wu *et al.*, 1998). Prochaska and Talalay (1992) demonstrated that phenolics and their derivatives present in green and black tea have protective effects against liver cirrhosis, emphysema, and in cancer. Nonetheless, addition of plant phenols susceptible secondary amines and amides to form potent carcinogenic nitrosamines and nitrosamides, invivo as well as invitro during different processing aids of cured and smoked meat products. Similarly hypertensive effects of oleuropein in olive leaves and fruits were also reported. Addition of bacteriocins to meat and milk products has been permitted by the U.S. FDA and WHO. Moreover, nisin, pediocin, and lactococcin are more active and stable at low pH and high temperature, respectively. So they have been given GRAS status and are naturally present in many food products. Ginger is anti-inflammatory and prevents against ovarian and colorectal cancer. Garlic reduces the risk of cancer, stroke, and aging. Bioactive milk peptides may act as opioid agonists, opioid antagonists, antihypertensive agents, immunostimulants, antithrombotic, mineral carriers, and so on (Shah, 2000). Chitosan acts as a fat binder and thereby helps to lose body weight. It is also shown to have beneficial effect on cholesterol, ulcers, osteoporosis, and blood pressure.

10.6 Summary

Each natural antimicrobial has its inherent advantages and disadvantages. The most acceptable method is one that should not induce alterations in chemical and nutritional constituents, as well as maintain all the sensory attributes of meat and meat products. Plant-derived essential oils, lactoferrin and activated lactoferrin, lysozymes, prebiotics, bacteriocins, bacteriophases, ozone, and electrolyte water are important tool in preserving the foods from the hazardous and detrimental effects of microorganisms. Among various natural preservatives, only few bacteriocins are used. This may be partially due to the fact that newly discovered broad-spectrum bacteriocins have yet to be fully characterized and officially approved. Hence, it can be concluded from the above discussion that natural preservatives are not only effective, but are also safe for use in the food supply. Moreover, it appears that their pertinent exploitation will produce biopreservation systems that are optimum for each food. In future, natural preservatives are going to play an important role in food preservation. However, it is quite doubtful that any newly discovered synthetic antimicrobials will be approved for use in foods by worldwide regulatory agencies. Hence, the future of natural antioxidants and antimicrobials lies in novel uses.

References

Abouzeed, Y.M., Hariharan, H., Poppe, C. and Kibenge, F.S.B. (2002). Characterization of *Salmonella* isolates from beef cattle, broiler chicken and human sources on Prince Edward Island. *Journal of Comparative Microbiology, Immunology and Infectious Diseases, 23*, pp. 253–266.

Acheson, D.W.K. (1996). Detection of shiga-like toxin producing *Escherichia coli* in ground beef and milk by commercial enzyme linked immunosorbent assay. *Journal of Food Protection, 59,* pp. 344–349.

Anon. (2000). Annual epidemiological report. Swedish Institute of Infectious Disease Control.

Arima, H., Ibrahim, H.R., Kinoshita, T., and Kato, A. (1997). Bactericidal action of lysozymes attached with various sizes of hydrophobic peptides to the C-terminal using genetic modification. *FEBS Letter, 415,* pp. 114–118.

Arnold, R.R., Cole, M.F. and McGhee, J.R. (1977). A bactericidal effect for human lactoferrin. *Science, 197,* pp. 263–265.

Atterbury, R.J., et al. (2007). Bacteriophage therapy to reduce Salmonella colonisation of broiler chickens. *Applied Environmental Microbiology, 73,* pp. 4543–4549.

Bachhil, V.N. and Jaiswal, T.N. (1988). Occurrence of *Salmonella* in meats. *Journal of Food Science and Technology, 25,* pp. 310–312.

Banerjee, R., Kapoor, K.N., Agarwal, R.K. and Ghatak, S. (2001). Verotoxin producing *Escherichia coli* (VETO) in foods of animal origin. *Journal of Food Science and Technology, 38,* pp. 82–84.

Bell, R.G. and DeLacy, K.M. (1986). Factors influencing the determination of nisin in meat products. *Journal of Food Technology, 21,* pp. 1–7.

Bethell, D.R. and Huang, J. (2004). Recombinant human lactoferrin treatment for global health issues: iron deficiency and acute diarrhea. *BioMetals, 17,* pp. 337–342.

Biswas, B., et al. (2002). Bacteriophage therapy rescues mice bacteremic from a clinical isolate of vancomycin-resistant Enterococcus faecium. *Infection and Immunology, 70,* pp. 204–210.

Biswas, A.K., et al. (2008).Microbial profiles of frozen trimmings and silver sides prepared at Indian buffalo meatpacking plants. *Meat science, 80,* pp. 418–422.

Blackburn, P., Polak, J., Gusik, S. and Rubino, S.D. (1989). Nisin compositions for use as enhanced, broad range bacteriocins. International application No. PCT/US89/02625; Applied Microbiology Inc., New York, USA.

Bolder, N.M. (1997). Decontamination of meat and poultry carcasses. *Trends in Food Science and Technology, 8,* pp. 221–227.

Borch, E. and Arinder, P. (2002). Bacteriological safety issues in red meat and ready-to-eat meat products, as well as control measures. *Meat Science, 62,* pp. 381–390.

Broadwater, W.T., Hoehn, R.C. and King, P.H. (1973). Sensitivity of three selected bacterial species to ozone. *Applied Microbiology, 26,* pp. 391–393.

Buckle K.A., et al. (1989). *Foodborne microorganisms of public health significance, 4th edn.* Australian Institute of Food Science and Technology, Food Microbiology Group, Australia.

Burt, S. (2004). Essential oils: Their antibacterial properties and potential applications in foods-A review. *International Journal of Food Microbiology, 94,* pp. 223–253.

Carlos, A.M.A. and Harrison, M.A. (1999). Inhibition of selected microorganisms in marinated chicken by pimento leaf oil and clove oleoresin. *Journal of Applied Poultry Reseseach, 8,* pp. 100–109.

CDC (Centers for Disease Control and Prevention). (1998). Preventing emerging infectious diseases; a strategy from 21st century. Overview of update CDC plan, *Morbidity Mortality Weekly Report, 47* (No. RR-15).

Chabela, M.L.P., Serrano, G.M.R., Calderon, P.I. and Guerrer, I. (1999). Microbial spoilage of meats offered for retail sale in Mexico City. *Meat Science, 51,* pp. 279–282.

Chaudhari, S.P. (2001). Immunodiagnostic potential of PI-PLC virulence marker of *L. monocytogenes* in listeric infection in buffaloes. Ph.D. thesis submitted to the Deemed University, IVRI, Izatnagar, India.

Chevallier, A. (1996). The encyclopedia of medicinal plants. Darling Kindersley, London, UK.

Chouliara, E., et al. (2007). Combined effect of oregano essential oil and modified atmosphere packaging on shelf-life extension of fresh chicken breast meat stored at 4°C. *Food Microbiology, 24,* pp. 607–617.

Cleveland, J., Montville, T.J., Nes, I.F. and Chikindas, M.L. (2001). Bacteriocins: safe, natural antimicrobials for food preservation. *International Journal of Food Microbiology, 71*, pp. 1–20.

Cutter, C.N. and Siragusa, G.R. (1995). Population reductions of gram negative pathogens following treatments with nisin and chelators under various conditions. *Journal of Food Protection, 58*, pp. 977–983.

Cutter, C.N. and Siragusa, G.R. (1998). Incorporation of nisin into a meat binding system to inhibit bacteria on beef surfaces. *Letter of Applied Microbiology, 27*, pp. 19–23.

Dave, S. (1999). Inactivation of *Salmonella enteritidis* on the surface boneless, skinless chicken breasts by aqueous ozone. M.S. Thesis. The Ohio State University, pp. 58–80.

Delves-Broughton, J. (1993). The use of EDTA to enhance the efficacy of nisin towards Gram negative bacteria. *International Journal of Biodeterioration and Biodegradation, 32*, pp. 87–97.

de Vuyst, L. and Vandamme, E. (1994). Nisin, a lantibiotic produced by *Lactococcus lactis* subsp. lactis: properties, biosynthesis and applications. In: L. de Vuyst and E. Vandamme, Editors, *Bacteriocins of Lactic Acid Bacteria Microbiology, Genetics and Applications*, Blackie Academic and Professional, London, pp. 151–221.

Dionysius, D.A., Grieve, P.A.and Milne, J.M. (1993). Forms of lactoferrin: Their antibacterial effect on Enterotoxigenic *Escherichia coli*. *Journal of DairyScience, 76*, pp. 2597–2606.

Dondo, A., Nachtman, C., Doglione, L., Rosso, A. and Genetti, A. (1992). Foods: their preservation by combined use of refrigeration and ozone. *Ingegneria Alimentare Conserve Animali, 8*, pp. 16–25.

Dorman, H.J.D. and Deans, S.G. (2000). Antimicrobial agentsfrom plants: Antibacterial activity of plant volatile oils. *Journal of AppliedMicrobiology, 88*, pp. 308–316.

EFSA (European Food Safety Authority). (2010). Analysis of the baseline survey on the prevalence of Campylobacter in broiler batches and of Campylobacter and Salmonella on broiler carcasses, in the EU, 2008. *The EFSA Journal, 8*(8), p. 1522–.

Elder, R.O., et al. (1997). Correlation of enterohaemorrhagic *E coli* O157 in feaces, hide and carcasses of beef cattle during processing. *Proceeding of the National Academy of Sciences of the USA.*, pp. 2999–3003.

Farber, J.M. and Peterkin, P.I. (1991). *Listeria monocytogenes,* a foodborne pathogen. *Microbial Review, 55*, pp. 476–511.

Fletcher, D.L., Russell, S.M., Walker, J.M. and Baily, J.S. (1993). An evaluation of a rinse procedure using sodium bicarbonate and hydrogen peroxide on the recovery of bacteria from broiler carcasses. *Poultry Science, 72*, pp. 2152–2156.

Françoise, L. (2010).Occurrence and role of lactic acid bacteria in seafood products. *Food Microbiology, 27*(6), pp. 698–709.

Fujiwara, A., et al. (2011). Biocontrol of Ralstonia solanacearum by treatment with lytic bacteriophages. *Applied Environmental Microbiology, 77*(12), pp. 4155–4162.

Goff, J.H., Bhunia, A.K. and Johnson, M.G. (1996). Complete inhibition of low levels of *Listeria monocytogenes* on refrigerated chicken meat with Pediocin AcH bound to heat-killed *Pediococcus acidilactici* cells. *Journal of Food Protection, 59*, pp. 1187–1192.

Gonzalez-Chavez, S.A., Arévalo-Gallegos, S. and Rascón-Cruz, Q. (2008). Lactoferrin: structure, function and applications. *Journal of Antimicrobial Agents.* doi:10.1016/j.ijantimicag.2008.07.020.

Goode, D., Allen, V.M. and Barrow, P.A. (2003). Reduction of experimental Salmonella and Campylobacter contamination of chicken skin by application of lytic bacteriophages. *Applied Environmental Microbiology, 69*(8), pp. 5032–5036.

Gordon, G. (1995). The chemistry and reactions of ozone in our environment. *Progress in Nuclear Energy, 29*, pp. 89–96.

Gorman, B.M., et al. (1995). Evaluation of hand trimmings, various sanitizing agents, and hot water spray washing as decontamination interventions for beef brisket adipose tissue. *Journal of Food Protection, 58*, pp. 899–907.

Greene, A.K., Few, B.K. and Serafini, J.C. (1993). A comparison of ozonation and chlorination for disinfection of stainless steel surfaces. *Journal of Dairy Science, 76,* pp. 3617–3620.

Hammer, K.A., Carson, C.F. and Riley, T.V. (1999). Antimicrobialactivity of essential oils and their plant extracts. *Journal of Applied Microbiology, 86,* pp. 985–990.

Hao, Y.Y., Brackett, R.E. and Doyle, M.P. (1998). Efficacy of plant extracts in inhibiting *Aeromonas hydrophila* and *Listeria monoctyogenes* in refrigerated, cooked poultry. *Food Microbiology, 15,* pp. 367–378.

Hawkins, C., Harper, D., Burch, D., Anggard, E., Soothill, J. (2010). Topical treatment of Pseudomonas aeruginosa otitis of dogs with a bacteriophage mixture: a before/after clinical trial. *Veterinary Microbiology, 146*(3–4), pp. 309–313.

Hazarika, R.A., et al. (2005). Detection and characterization of verotoxin producing *E. coli* (VTEC) isolated from buffalo meat. *Journal of Food* Safety, *24*(4), pp. 281–290.

Horvath, M., Bilitzky, L. and Huttner, J. (1985). Field of utilization of ozone, 257–316. In: R.J.H. Clack, ed., *Ozone*. New York: Elsevier Science Publishing Co.

Jacob, F., Lwoff, A., Siminowitch, A. and Wollman, E. (1953). Definition de quelques termes relatives a la lysogenie. *Annales des Institute Pasteur Paris, 84,* pp. 222–224.

Jassim, S.A.A., Abdulamir, A.S. and Abu Bakar, F. (2012). Novel phage-based bio-processing of pathogenic Escherichia coli and its biofilms. *World Journal of Microbiology and Biotechnology, 28,* pp. 47–60.

Jassim, S.A.A. and Limoges, R.G. (2013). Impact of external forces on cyanophage–host interactions in aquatic ecosystems. *World Journal of Microbiology and Biotechnology, 29*(10), pp. 1751–1762.

Jeevaratnam, K., Jamuna, M. and Bawa, A.S. (2005). Biological preservation of foods-Bacteriocins of Lactic acid bacteria. *Indian Journal Biotechnology, 4,* pp. 446–454.

JECFA. (2002). Future trends in veterinary public health. *Fifty-eight report of the Joint FAO/WHO Expert Committee on Food Additives.* Report of a WHO study group. WHO technical Report Series. No. 907.

Johnson, R.P., et al. (1996). Growing concerns and recent outbreaks involving non-O157: H7 serotypes of verotoxigenic *Escherichia coli. Journal of Food Protection, 59,* pp. 1112–1122.

Kaess, G. and Weidemamm, J.F. (1968). Ozone treatment of chilled beef. Effects of low concentrations of ozone on microbial spoilage and surface colour of beef. *Journal of Food Technology, 3,* pp. 325–334.

Karpińska-Tymoszczyk, M. (2007). Effects of sage extract (Salvia officinalis l.) and a mixture of sage extract and sodium isoascorbate on the quality and shelf life of vacuum-packed turkey meatballs. *Journal of Muscle Foods, 18,* pp. 420–434.

Kijowski, J., Marciszewska, C. and Cegielska-Radziejewska, R. (2002). Quality and microbiological stability of chilled chicken breast muscles treated with lysozyme solution. *Pollish Journal of Food and Nutrition Science, 11,* pp. 47–54.

Kim, J., Marshall, M.R. and Wei, C. (1995). Antibacterial activity ofsome essential oils components against five foodborne pathogens. *Journal of Agriculture Food Chemistry, 43,* pp. 2839–2845.

Kim, C., Hung, Y.C. and Brackett, R.E. (2000). Roles of oxidation– reduction potential in electrolyzed oxidizing and chemically modified water for the inactivation of food-related pathogens. *Journal of Food Protection, 63,* pp. 19–24.

Klumpp, J. and Loessner, M.J. (2013). Listeria phages genomes, evolution, and application. *Bacteriophage, 3*(3), p. e26861–. Available at: https://www.landesbioscience.com/journals/bacteriophage/2013BACTERIOPHA GE0033R.pdf [Accessed June 8, 2016].

Kumar, D. and Tanwar, V.K. (2011). Utilization of clove powder as phytopreservative for chicken nuggets preparation. *Journal of Stored Products & Postharvest Research, 2,* pp. 11–14.

Kurita, N., Miyaji,M., Kurane, R. and Takahara, Y. (1981). Antifungalactivity of compounds of essential oils. *Agricultural and Biological Chemistry, 45,* pp. 945–952.

Lonnerdal, B. and Iyer, S. (1995). Lactoferrin: molecular structure and biological function. *Annual Review of Nutrition, 15,* pp. 93–110.

Losso, J.N., Nakai, S. and Charter E.A. (2000). Lysozyme. In: *Natural Food Antimicrobial Systems* (ed. A.S. Naidu), CRC Press, Boca Raton, pp. 185–210.

Madden, R.H., et al. (2001). Occurrence of *Escherichia coli* O157: H7, *Listeria monocytogenes, Salmonella* and *Campylobacter* spp. on beef carcasses in Northern Ireland. *Meat Science, 58,* pp. 343–346.

Mahadeo, M. and Tatini, S.R. (1994). The potential use of nisin to control *Listeria monocytogenes* in poultry. *Letters in Appled Microbiolgy, 18,* pp. 323–326.

Mangalassary, S., et al. (2008). Effect of combining nisin and/or lysozyme with in-package pasteurization for control of *Listeria monocytogenes* in ready-to-eat turkey bologna during refrigerated storage. *Food Microbiology, 25,* pp. 866–870.

Menzel, D.B. (1971). Oxidation of biologically active reducing substances by ozone. *Arch. Environ. Health., 23,* pp. 149–153.

Moore, G., Griffith, C. and Peters, A. (2000). Bactericidal properties of ozone and its potential application as terminal disinfectant. *Journal of Food Protection, 63,* pp. 1100–1106.

Muriana, P.M. (2005). Efficacy of electrolyzed water on foodborne pathogens of concern tothe meat and poultry processing industry. A report prepared for SanAquel LLC502 Industrial Road Bristow, OK 74010.

Mytle, N., et al. (2006). Antimicrobial activity of clove (*Syzgium aromaticum*) oil in inhibiting *Listeria monocytogenes* on chicken frankfurters. *Food Control, 17,* pp. 102–107.

Naidu, A.S., et al. (2003). "Activated Lactoferrin: Natural antimicrobial for food safety." *AGROFoodindustry hi-tech,* May/June.

Naveena, B.M. and Mendiratta, S.K. (2001). Tenderisation of spent hen meat using ginger extract. *British Poultry Science, 42,* pp. 344–49.

Nuijens, J.H., et al. (1997). Characterization of recombinant human lactoferrin secreted in milk of transgenic mice. *The Journal of BiologicalChemistry, 272,* pp. 8802–8807.

Oliveira, K.A.M., et al. (2005). Aqueous garlic extract and microbiological quality of refrigerated poultry meat. *Journal of Food Processing and Preservation, 29,* pp. 98–08.

Park,H., Hung, Y-C and Brackett, R.E. (2002). Antimicrobial effect of electrolyzed water for inactivating Campylobacter jejuni during poultry washing. *International Journal of Food Microbiology, 72,* pp. 77–83.

Patterson, J.T. (1974). Microbiology of frozen meat in 60 lb boxes. Agricultural Research Council, MRI Symposium No. 3, Meat Freezing Why & How? Langford, Bristol.

Pierard, D., Van Damme, Z., Stevans, D., Moriau, L. and Lauwers, S. (1994). Detection of verocytotoxin-producing *Escherichia coli* in meat in Belgium. In: Kormali, M.A. and Gaglio, A.G., eds., *Recent Advances in Verocytotoxin Producing Escherichia coli infection.* Elsevier Science. Amsterdam.

Prochaska, H.J. and Talalay, P. (1992). Phenolic antioxidants as inducers of anticarcinogenic enzymes. In Phenolic compounds in Food and Their Effect on Health I, Analysis, Occurance, and Chemistry (eds Chi-Tang Ho, Chang Y. Lee, Mou-Tuan Huang). ACS symposium series, Wsahington, DC.

Rao, D.N. and Mahendrakar, N.S. (2003). Challenges of the quality control of Indian meat and meat products in view of global market. In: *First convention of Indian Meat Science Association and National Symposium on Impact of Globalization on Indian meat industry.* 11th–12th December, 2003, pp. 18–26.

Rathore, R.S. (2000). Studies on enterohaemorrhagic *Escherichia coli* with special reference to detection of virulence genes. Ph.D. thesis submitted to Deemed University, IVRI, Izatnagar, India.

Rayman, K., Malik, N. and Hurst, A. (1983). Failure of nisin to inhibit outgrowth of *Clostridium botulinum* in a model cured meat system. *Appl Environ Microbiol, 46,* pp. 1450–1452.

Reagan, J.O., et al. (1996). Trimming and washing of beef carcasses as a method of improving microbiological quality of meat. *Journal of Food Protection, 59,* pp. 751–756.

Riznar, K., et al. (2006). Antioxidant and antimicrobial activity of rosemary extract in chicken frankfurters. *Journal of Food Science, 71,* pp. C425–C429.

Roels, T.H., et al. (1997). Incomplete sanitation of a meat grinder and ingestion of new ground beef: contributing factors to a large outbreak of *S. typhimurium* infection. *Epidemiological Infection, 119,* pp. 127–134.

Rusch, A. and Kraemer, J. (1989). Influence of instruments for elimination of microorganisms on surface bacterial contamination of fresh meat and on airborne microorganisms in cold stores with increased RH. *Archiv. Lebensmittelhygiene., 40,* pp. 61–65.

Sallam, K.I., Ishioroshi, M. and Samejima, K. (2004). Antioxidant and antimicrobial effects of garlic in chicken sausage. *LWT- Food Science & Technology, 37,* pp. 849–855.

Scallan, E., et al. (2011). Food-borne illness acquired in the United States--major pathogens. *Emerging Infectious Diseases,17*(1), pp. 7–15.

Samuelson, K.J., Rupnow, J.H. and Froning, G.W. (1985). The effect of lysozyme and EDTA on Salmonella on broiler parts. *Poult Science, 64,* pp. 1488–1490.

Scanga, J.A., et al. (2000). Microbiological contamination of raw beef trimmings and ground beef. *Meat Science, 56,* pp. 145–152.

Schillinger, U., Geisen, R. and Holzapfel, W.H. (1996). Potential of antagonistic microorganisms and bacteriocins for the biological preservation of foods. *Trends in Food Science and Technology, 7,* pp. 158–164.

Shah, N.P. (2000). Effects of milk-derived bioactives: an overview. *British Journal of Nutrition, 84,* pp. S3–S10.

Shefet, S.M., Sheldon, B.W. and Klaenhammer, T.R. (1995). Efficacy of optimized nisin based treatments to inhibit Salmonella typhimurium and extend shelf-life of broiler carcasses. *Journal of Food Protection, 58,* pp. 1077–1082.

Shekarforoush, S.S., et al. (2007). Effects of storage temperatures and essential oils of oregano and nutmeg on the growth and survival of Escherichia coli O157:H7 in barbecued chicken used in Iran. *Food Control, 18,* pp. 1428–1433.

Sillankorva, S.M., Oliveira, H. and Azeredo, J. (2012). Bacteriophages and their role in food safety. *International Journal of Microbiology,* doi:10.1155/2012/863945.

Sofos, J.N., et al. (1999). Sources and extent of microbiological contamination of beef carcass in seven United States slaughtering plants. *Journal of Food Protection, 62,* pp. 140–145.

Urano H. and Fukuzaki S. (2001). Facilitation of cleaning of alumina surfaces fouled with heat treated bovine serum albumin by ozone treatment. *Journal of Food Protection, 64,* pp. 108–112.

USFDA. (2006). Food additives permitted for direct addition to food for human consumption; bacteriophage preparation. FDA, Washington, DC. Available at: http://www.fda.gov/OHRMS/DOCKETS/98fr/cf0559.pdf [Accessed 3 August 2006].

Vaithiyanathan, S., et al. (2011). Effect of dipping in pomegranate (*Punica granatum*) fruit juice phenolic solution on the shelf life of chicken meat under refrigerated storage (4°C). *Meat Science, 88,* pp. 409–414.

Vanderlinde, P.B., Shay, B. and Murray, J. (1998). Microbial quality of Australian beef carcass meat and frozen bulk packed beef. *Journal of Food Protection, 61*(4), pp. 437–443.

Verheust, C., et al. (2010). Contained use of bacteriophages: risk assessment and biosafety recommendations. *Applied Biosafety, 15*(1), pp. 32–44.

Wagenaar, J.A., et al. (2005). Phage therapy reduces *Campylobacter jejuni* colonization in broilers. *Veterinary Microbiology, 109*(3–4), pp. 275–283.

Wu, T.S., et al. (1998). Platelet aggregation inhibitor from *Murraya keonigii*. *Phytotherapy Research, 12*, pp. 380–382.

Yadav, A.S., et al. (2002). Effect of garlic extract and cinnamon powder on microbial profile and shelf-life of minced chicken meat. *Indian Journal of Poultry Science, 37*, pp. 72–77.

Yang, P.P.W. and Chen, T.C. (1979). Stability of ozone and its germicidal properties on poultry meat microorganisms in liquid phase. *Journal of Food Science, 44*, pp. 501–504.

Yosef, I., et al. (2014). Different approaches for using bacteriophages against antibiotic-resistant bacteria. *Bacteriophage, 4*, p. e28491.

11

Foodborne Pathogens and Their Apparent Linkage with Antibiotic Resistance

Mariah L. Cole and Om V. Singh

Division of Biological and Health Sciences, University of Pittsburgh, Bradford, Pennsylvania, USA

11.1 Introduction

Food provides proteins, carbohydrates, fats, and other elements necessary for human or animal survival. Food is defined under the Food, Drug, and Cosmetic Act (FD&C) as articles used for food or drink for man or other animals, chewing gum, or articles used as a component of any other such article (FDA, 2012). Food can become spoiled or adulterated due to microbial contamination (i.e., bacteria, fungi) and other physical or chemical alterations. The FD&C act defines food adulteration as "if food bears or contains, an unsafe food additive; an unsafe new animal drug; an unsafe color additive; it consists, in whole or in part, of any filthy, putrid, or decomposed substance" (FDA, 2012). The occurrence of microorganisms in adulterated food causes serious illnesses including gastroenteritis in humans. The overuse of antibiotics and related compounds has produced microbes that can sustain life in the extreme environment of antibiotics (i.e., antibiotophiles) (Gabani *et al.*, 2012; Woappi *et al.*, 2013, 2016). These extremophiles are emerging as pathogens in the food chain.

The spoilage of food is directed by the biochemical activity of microorganisms in the food, some of which may be able to adapt to extreme environments (e.g., freezing, drying, and pressure). Food spoilage can be confirmed through taste, smell, touch, and sight; signs include off-flavors and odors, slime formation, visible mold growth, and color deterioration. Microorganisms' alterations of food products can be positive or negative. The positive aspects are seen in dairy products such as cheeses and yogurt, whereas the negative aspects are seen in food spoilage (e.g., putrefaction, fermentation, or rancidity) (Ummeayman, 2009). Most food spoilage microorganisms release their own enzymes (proteins) into their surroundings, allowing them to absorb the products that are being released due to biochemical reactions of food exposure (Ummeayman, 2009). Different preservation and treatment strategies have been designed to limit the amount of microbial presence in food items for safer human consumption.

To limit the amount of pathogenic microorganisms in ready-to-go food, bagged food, canned food, and raw food items, a variety of methods have been developed. Heat and cold preservation, chemical preservation, biopreservation, nonthermal and thermal ultra-high pressurization, and irradiation methods allow a variety of food products to be consumed without the presence of pathogens. However, natural food preservation methods that use fewer chemicals and additives are among the most

Foodborne Pathogens and Antibiotic Resistance, First Edition. Edited by Om V. Singh.

favorable, as they do not cause harm to the consumer and they preserve the natural flavors and nutrients from food products (Reis *et al.*, 2012).

The U.S. Food and Drug Administration (FDA) recommends a variety of techniques to detect the presence of pathogenic microorganisms. Bacteriological analytical methods involve sampling specific amounts of food, based on three factors: consumer group, contamination during manufacturing, and the food history. In addition to microbiological testing, the "omics"-based approaches have been recommended for further analysis. The tools of systems biology are genomics, proteomics, metabolomics, and bioinformatics; genomics and proteomics are useful to identify how a microorganism causes food spoilage, including the understanding of metabolic networks as well as cellular responses (protein and gene expression) in food-spoilage microorganisms. Genomics represents the study of all genes in a genome, and has better capabilities than other technologies to differentiate between organisms. Currently, it is being used as a way to identify foodborne pathogens that cause illnesses in efforts to greatly reduce the spread of these foodborne pathogens. Proteomics is the study of all proteins in a cell; it provides discrete information about the protein function and stress processes during food preservation, allowing us to understand how certain proteins may respond under stress (Capozzi *et al.*, 2009; Hörmann *et al.*, 2006).

The use of antibiotics has been important for treating infectious diseases in humans and animals. Antibiotics are created by specific bacteria or fungi, chemically synthesized by chemists, or modified chemically to increase their strength (semi-synthetic) (Salyers and Whitt, 2005). Each individual antibiotic works in a different way, but with the same end result: the inactivation or killing of pathogenic bacteria (Bhunia, 2008; Franco *et al.*, 2009; Allen *et al.*, 2014; Sharma *et al.*, 2013). Consequently, overuse of antibiotics can cause bacteria to develop a means of protection against them, that is, resistance genes. The overuse and misuse of antibiotics has become prevalent in the past few decades and has led to the emerging evolution of antibiotic-resistant bacteria (Bhunia, 2008; Gabani *et al.*, 2012; Woappi *et al.*, 2013, 2016; Berman and Riley, 2013; Wellington *et al.*, 2013). This resistance has emerged from the force of selective pressure, evolving over repeated periods of exposure to antibiotics.

Antibiotic residues in water and foods of animal and plant origin represent a major concern, as consistent exposure to antibiotics is a serious hazard for both humans and the environment. Food acts as a transport vessel for antibiotic-resistant bacteria and resistance genes to humans through agricultural products (beef, chicken, pork, and plant products), agricultural runoff into irrigation water (used on vegetation and soil), and the food-packaging process (Verraes *et al.*, 2013; Durso *et al.*, 2014). There are three major types of microorganisms associated with food: beneficial microorganisms (i.e., *Lactobacillus acidophilus* and *Lactococcus lactis*, used in dairy products), spoilage microorgansims (i.e., *Pseudomonas* spp and *Bacillus*), and disease-causing microorganisms (i.e., *Escherichia coli* and *Salmonella enterica*) (Bhunia, 2008). Foodborne pathogenic microorganisms are one of the main reasons for recalls of food products. In the current article, we discuss the biochemistry of food spoilage, preservation of food items, specific foodborne pathogens, the systems biology of food spoilage, antibiotics and antibiotic resistance mechanisms, antibiotic resistance among foodborne organisms, and alternatives to antibiotics, as well as new approaches for fighting foodborne pathogens before human consumption.

11.2 Food Spoilage

11.2.1 Biochemistry of Food Spoilage

Food spoilage occurs when the original nutritional value, texture, and flavor of a food are damaged due to intrinsic and extrinsic factors that make it harmful to the consumer. Food can become spoiled biologically, enzymatically

Table 11.1 Foodborne Pathogens Affecting Essential and Non-Essential Amino Acids in Food Spoilage.

Microorganisms	Enzymes	Amino acids	Volatile products
Enterobacteriaceae	Cystine desulfhydrase	Cysteine	H_2S
Lactococci, Lactobacilli, Brevibacteria *Pseudomonas ovalis*	Methionine Lyase	Methionine	Methyl mercaptans
E. coli	Tryptophanase	Tryptophan	Indole Cadaverine
Bacillus sp., *Listeria* sp.,	Decarboxylase	Lysine	Putrescine
Lactic acid bacteria Filamentous fungi		Arginine Histidine	Histamine

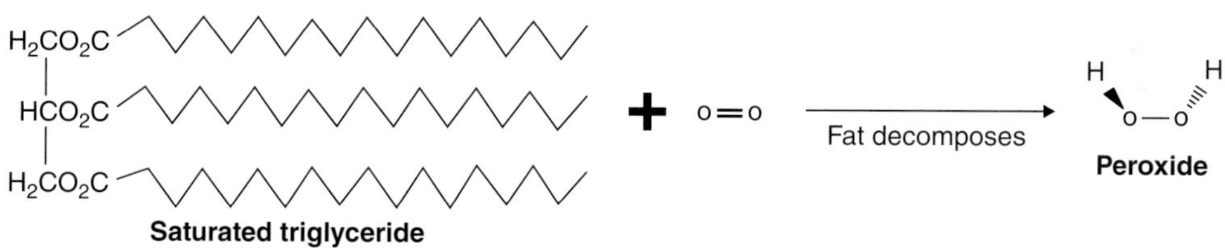

Figure 11.1 Oxidation of fats: rancidity. A saturated fat reacts with an oxygen molecule, which decomposes the fat and produces peroxidase, which causes the rancidity of spoiled food items.

(autolysis), oxidatively, or by freezing, heating, drying, or pressure changes; spoiled food can be detected by sight, smell, taste, texture, and special laboratory techniques (Food Spoilage & Preservation, 2009). Protein degradation, an enzymatic reaction, involves proteases, which cleave protein chains into amino acids; an important protease involved in food spoilage is protease plasmin. Protease plasmin has properties that allow it to survive in pasteurization temperature conditions, causing dairy protein degradation, coagulation, and gelatinization in meat products. Additional enzymes that cause food spoilage are ascorbic acid oxidase, lipoxygenase, polyphenoloxidase in vegetables, pectic enzymes, polyphenoloxidase, peroxidases in fruits, and lipase and thiaminase in milk and meats respectively. Another form of protein degradation that causes spoilage in food products is putrefaction, or anaerobic microorganisms decomposing animal proteins. Certain microbes have the ability to simplify amino acids by removing attached groups, causing deleterious effects in food. Table 11.1 summarizes how foodborne pathogens affect amino acids.

Fruits and vegetables that are noticeably brown and soft have been affected by phenol oxidase enzymatic activity from within, and those that are not noticeably browning are affected by a reaction involving certain amino acids (lysine) and reducing sugars (Tomás-Barberán and Espín, 2001; Ummeayman, 2009). Fruits and vegetables also spoil due to the breakdown of carbohydrates (degradation of plant cell wall), which make up the majority of these products. Lipid deterioration occurs in food products containing fats and oils; the rate of oxidation of unsaturated fats is affected by the number of double bonds in the fatty acids in addition to triglycerides' effectiveness due to light and heat (Ummeayman, 2009). The generic reaction of fatty acid decomposition by oxygen is summarized in Figure 11.1. Consumption of contaminated food products without proper care can cause adverse health issues in both humans and animals, including food poisoning.

11.2.2 Microorganisms and Food Spoilage

A large variety of food products are already exposed to bacteria (mostly gram-negative) that cause food spoilage in their natural habitats, especially meat and vegetable products (Chaillou *et al.*, 2014; Widsten *et al.*, 2014). One newly arising pathogenic vehicle for humans is fruit juice. Fruit juices have shown the pathogenic presence of enterohaemorrhagic *Escherichia coli* O157, *Salmonella*, *Cryptosporidium*, spoilage microorganisms such as *Alicyclobacillus acidoterrestris*, fungi growth, *Propionibacterium cyclohexanicum*, *Byssochlamys fulva*, *B. nivea*, *Neosartory fischeria*, and *Talaromyces* species. Meat and seafood show dominant bacterial species of *Lactobacillales*, *Bacillales*, *Enterobacterales*, *Pseudomonadales*, *Shewanella putrefaciens*, and *Vibronales* (Aneja *et al.*, 2014, Chaillou *et al.*, 2014; Widsten *et al.*, 2014). Some microorganisms in fermented food products have been identified as containing antibiotic resistance determinant reservoirs (lactic acid bacteria) as well as antibiotic resistance phenotypes (*S. thermophiles* and *L. helveticus*) (Devirgiliis *et al.*, 2014).

Staphylococcus aureus is a major foodborne (meat and poultry) pathogen that causes approximately 241,000 illnesses in the United States annually, and is related to MRSA; even though half of the population is infected with *S. aureus*, according to the Centers for Disease Control and Prevention (CDC), only 1.5% of the population carries MRSA (Doyle *et al.*, 2012). MRSA cells have been found in meat, poultry, milk, salads, and ready-to-eat foods and cheeses, and can cause staphylococcal enterocolitis in people whose normal flora has been destroyed due to antibiotic usage; however, this can be prevented by proper hygiene, decontamination, cooking, and storage of food items.

Listeria monocytogenes is a foodborne gram-positive bacterium found primarily in dairy, meat, precooked meals, ready-to-eat foods, fish, and vegetables (Barbosa *et al.*, 2013). *L. monocytogenes* causes listeriosis. In many case studies, a vast number of *L. monocytogenes* strains have acquired resistance to tetracycline, ciprofloxacin, nitrofurantoin, and erythromycin (Barbosa *et al.*, 2013). A third common foodborne infection is *E. coli*. Shiga toxin-producing *E. coli* (STEC O157A), which causes diarrhea and abdominal cramping, infects approximately 63,000 people annually in the United States (Marder, 2014). Figure 11.2 shows a Venn diagram of the most popular foodborne pathogens associated with food products. All three types of food products show the presence of *Salmonella*, *L. monocytogenes*, *Pseudomonas* spp., and *Shigella* spp. (Bhunia, 2008).

Marder *et al.* (2014) studied outbreaks of Shiga toxin-producing *E. coli* (STEC) O157 associated with leafy green vegetables (spinach and lettuce) and salads, and revealed that the microbiota of retail spinach contained DNA sequences for antibiotic resistance (Berman *et al.*, 2013). Antibiotic resistance in foodborne pathogens can horizontally transfer from food products containing non-pathogenic bacteria that are antibiotic resistant (Devirgiliis *et al.*, 2014). *Salmonella enterica* also causes food poisoning, and along with other enterococcal strains, has been found in ready-to-eat salad vegetables and shown high antibiotic resistance (Taban *et al.*, 2013; Macovei *et al.*, 2007; Pesavento *et al.*, 2014). Van *et al.* (2007) studied 180 raw food samples and tested for *Salmonella* and *E. coli*; 60.8% of meat samples and 18.0% of shellfish samples contained *Salmonella spp.*, and more than 90% of the food samples tested positive for *E. coli* contamination. The study showed 50.5% of *E. coli* isolates being resistant to antibiotics (ampicillin, tetracycline, and kanamycin), in addition to 13% of *Salmonella spp.* Asgharpour *et al.* (2014) showed that at least 18 different strains of *Salmonella spp.* carried integron class 1, allowing the strains to transfer antibiotic resistance to other bacterial species. Rasheed *et al.* (2014) found drug-resistant *E. coli* in raw chicken, vegetable salad, raw meat, raw egg, and unpasteurized milk; the researchers also discovered ESBL producers in vegetable

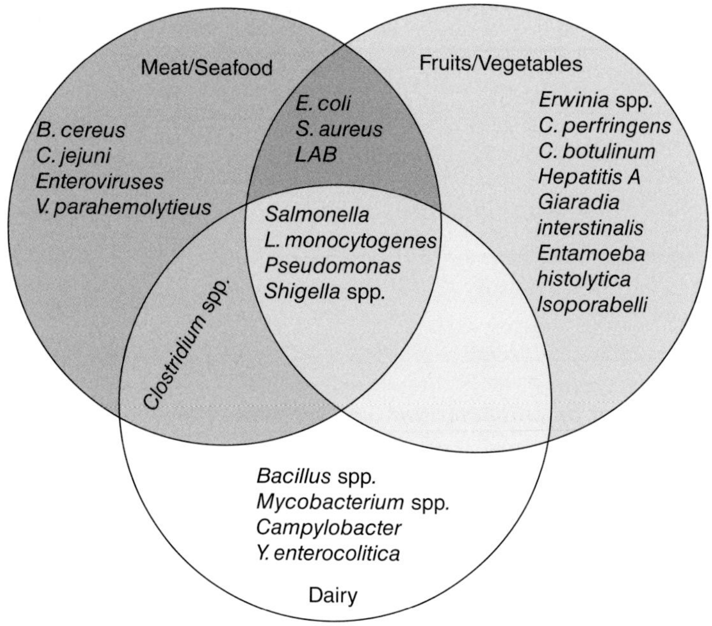

Figure 11.2 Venn diagram of the most common foodborne pathogens associated with meat, seafood, fruits, vegetables, and dairy products. (*See insert for color representation of the figure.*)

salads and raw chicken. Antimicrobial-resistant *Pantoea (Enterobacter) agglomerans*, a gram-negative bacillus, has even been observed in powdered infant milk (Mardaneh and Dallal, 2013).

Major outbreaks of foodborne pathogens in the United States in the last decade are summarized in Tables 11.2 (meat and fish products) and 11.3 (fruit, vegetables, dairy, and nut products).

11.2.3 Factors Affecting and Controlling Food Spoilage

Multiple intrinsic and extrinsic factors, such as nutrient type, pH, moisture content, oxidation, reduction potential, inhibitors, temperature, proteins, carbohydrates, and fats, play important roles in the breakdown of food matrix that influences microorganisms to grow (Ummeayman, 2009; Food Spoilage & Preservation, 2009; Dave *et al.*, 2011). pH is a critical factor in the deterioration of meat products. When animals are slaughtered, the glycogen content is reduced and the pH alters in respect to the amount of lactic acid in muscles. Higher-pH meat will become darker, dryer, and firmer, and vice versa for lower pH. Since favorable growth of most bacteria occurs below normal pH range due to protein degradation, meat products are more susceptible to spoilage at lower pH (Dave *et al.*, 2011). Some microbes are able to grow in high acid concentrations but not at high heat (such as yeasts and molds), excluding thermophiles (Ummeayman, 2009). Therefore, the storage temperature greatly affects microbial growth. Cerveny *et al.* (2009) showed that *Pseudomonas* spp., among gram-negative bacteria, and Enterobacteriaceae were generally found in meat items that had been refrigerated, whereas Enterococci and yeasts were found in raw products due to curing salt resistance (Dave *et al.*, 2011).

Food spoilage remains a health risk, and innovative preservation techniques for controlling food spoilage are still being developed. Highly perishable food products (meat, fruit, milk, vegetables) and semi-perishable foods (potatoes, nuts) are the main focus for innovative preservation (Ananou *et al.*, 2006).

Table 11.2 Major Outbreaks of Microbes in the U.S. Food Chain (Meat and Fish Products) in the Last Decade.*

Year	Food Type	Major Pathogen
Meat		
2015	Costco Rotisserie Chicken	*Escherichia coli* O157:H7
2015	Pork	*Salmonella I 4,[5],12:i:- and Salmonella infantis*
2015	Raw, Frozen, Stuffed, Chicken entrees	*Salmonella enteritis*
2014	Ground Beef	*Escherichia coli* O157:H7
2014/2013	Chicken	*Salmonella* Heidelberg
2014/2013/2012	Live Poultry	*Salmonella* Infantis, *Salmonella* Newport, and *Salmonella* Hadar
2013	Ground Beef	*Salmonella* Typhimurium
2012	Ground beef	*Salmonella* Enteritidis
2011	Ground beef	*Salmonella* Typhimurium
2011	Kosher Broiled Chicken Livers	*Salmonella* Heidelberg
2011	Ground Turkey	*Salmonella* Heidelberg
2011	Turkey burgers	*Salmonella* Hadar
2011	Lebanon bologna	*Escherichia coli* O157:H7
2010/2009/2008/2007	Beef	*Escherichia coli* O157:H7
Fish		
2015	Frozen Raw Tuna	*Salmonella Paratyphi B variant L(+) tartrate(+) and Salmonella Weltevreden*
2012	Raw Tuna	*Salmonella* Bareilly and *Salmonella* Nchanga
2014	Smoked Nova Salmon	*Listeria monocytogenes*
2013	Raw Shellfish	*Vibrio parahaemolyticus*

* Information obtained from Foodborne illness outbreak database available at: http://www.outbreakdatabase.com/and Foodborne outbreak online database (FOOD Tool) available at http://wwwn.cdc.gov/foodborneoutbreaks/ [Accessed June 16, 2016].

Preservation methods include heat methods (pasteurization, boiling, canning), cold preservation (refrigeration at 32°–38°F or freezing at < 32°F/0°C), biopreservatives (fermentation), reduction of moisture availability (drying, smoking, osmotic pressure, freeze-drying), chemical preservation (smoking, sodium benzoate, nitrates, sulfites), oxygen exclusion (replacing oxygen with nitrogen), ultra-high pressure (111,000 psi), and irradiation (which is limited to chicken, fruits, vegetables, and ground beef products) (Ananou *et al.*, 2006; Food Spoilage & Preservation, 2009; Aneja *et al.*, 2014). Biopreservation has been considered a valuable tool because it preserves food without chemicals or additives, using natural or controlled microbiota and/or their antimicrobial compounds (Ananou *et al.*, 2007; Montville and Chikindas, 2007; Swarnadyuti *et al.*, 2014).

Thermal processing is a technique used to make enzymes inactive in fruit juices, but at the cost of flavor loss, which reduces sales. Thus, other techniques must be considered to maintain consumers' interest. Nonthermal preservative

Table 11.3 Major Outbreaks of Microbes in the U.S. Food Chain (Fruit, Vegetables, Dairy, and Nut Products) in the Last Decade.*

Year	Food Type	Major Pathogen
Fruit		
2013	Fresh Produce	*Cyclospora*
2012	Mangos	*Salmonella*
2012	Cantaloupes	*Salmonella* Typhimurium and *Salmonella* Newport
2011	Whole, fresh, imported Papaya	*Salmonella* Agona
2011	Del monte cantaloupe	*Salmonella* Panama
2009	Cantaloupe	*Salmonella* Litchfield
Vegetables		
2015	Cucumbers	*Salmonella* Poona
2014	Mung Bean Sprouts	*Salmonella* enteritidis
2014	Cilantro	*Cyclospora*
2014/2012	Sprouts	*Listeria monocytogenes* and *Salmonella* Cubana
2014/2012	Raw Clover Sprouts	*E. coli* O121/*E. coli* O26
2013	Lettuce (Ready-to-eat salads)	*E. coli* O157:H7
2013	Ready-to-eat salads	*Escherichia coli* O157:H7
2013	Cucumbers	*Salmonella* Saintpaul
2012/2006	Fresh Spinach/Spring mix	*E. coli* O157:H7
2011	Romaine Lettuce	*Escherichia coli* O157:H7
2010	Alfalfa sprouts	*Salmonella* Newport
2009	Raw produce	*Salmonella* Saintpaul
2006	Tomatoes	*Salmonella* Typhimurium
Dairy/Nuts		
2015	Blue Bell ice cream	*Listeria monocytogenes*
2015	Soft Cheeses	*Listeria monocytogenes*
2014	Hispanic-style cheese	*Listeria*
2014	Nut Butter	*Salmonella* Braenderup
2014	Cashew Cheese	*Salmonella*
2013/2009	Pistachios	*Salmonella* (multiple types)
2013/2012	Cheeses	*Listeria monocytogenes*
2012	Peanut butter	*Salmonella* Bredeney
2011	Pine nuts	*Salmonella*

* Information obtained from Foodborne illness outbreak database available at: http://www.outbreakdatabase.com/and Foodborne outbreak online database (FOOD Tool) available at http://wwwn.cdc.gov/foodborneoutbreaks/ [Accessed June 16, 2016].

methods have been approved by the FDA, have positive effects on consumer sales rates, and are proven to be safe as well. The nonthermal methods include high-intensity pulsed electric fields, high hydrostatic pressure, high-pressure homogenization, ultraviolet, ultrasound, and irradiation (Aneja *et al.*, 2014). However, some of the nonthermal methods (high hydrostatic pressure in juices) can alter protein structures and polysaccharides, changing the food's functions and causing harmful effects in humans (Aneja *et al.*, 2014).

The chemical preservatives sodium benzoate and potassium sorbate are widely used in food (Aneja *et al.*, 2014). Other chemical preservatives are natural and use multiple mechanisms to inactivate specific microorganisms, such as lysozyme attack, disintegration of peptidoglycan cell wall in gram-positive bacteria, or a lactoperoxidase system that creates both hypothiocyanite and hypothiocyanous acid, which damage multiple components of the microbial cell. Chitosan creates a polycationic that interacts with anionic components and disfigures the outer cell membrane, which allows intracellular components to leak out into the extracellular area. Nisin inhibits gram-positive bacteria by forming a pore in the cytoplasmic membrane that results in the loss of essential ions, amino acids, and ATP (Aneja *et al.*, 2014; Suganthi *et al.*, 2012; Huang *et al.*, 2012).

Foods lose nutrients if they are sterilized by heat or preserved by drying or adding salt or sugar. Refrigeration decreases the microbial growth rate, but does not sterilize food; similarly, antibiotics retard and/or inhibit microbial growth. Both are effective in controlling food spoilage. While antibiotics are yet to be accepted universally as preservatives, they were used commercially in the early years (Vaughn and Stewart, 1960). Antibiotics produced from bacteria are of keen interest to the food industry for their potential as bio-preservatives (Mehta *et al.*, 2013; Subramanian and Smith, 2015). Bacteriocins produced by lactic acid bacteria are among many potential bio-preservatives. As an antimicrobial peptide, pediocins produced by *Pediococcus acidilactici* revealed strong activity against food spoilage and pathogenic bacteria (Mehta *et al.*, 2013). Among natural preservatives, liquid smoke has shown tremendous antimicrobial properties. Most common foodborne pathogens such as *Salmonella, Staphylococcus, Listeria monocytogenes*, and pathogenic *Escherichia coli* have been reported to be sensitive to liquid smoke in vitro and in food systems (Lingbeck *et al.*, 2014).

Plant-based extracts (bioactive phenols, flavonoids, phenolic acids) have been proposed as a possible innovative way of controlling and suppressing spoilage bacteria, which can effectively help with the transport of food materials from overseas (Widsten *et al.*, 2014). Widsten *et al.* (2014) reported minimal inhibitory concentration (MIC) findings showing that mango seed extract (MaSd) and tannic acid (TaAc) were able to successfully inhibit the majority of both fish and lamb spoilage bacteria as well as other pathogenic bacteria (*S. aureus, C. estertheticum, B. thermosphacta*, and *L. monocytogenes*).

11.3 Food Processing and Microbial Contamination

As described by the CDC, food processing is "changing plants or animals into what we recognize and buy as food" (CDC, 2014). Food processing procedures can cause the introduction of pathogenic microorganisms into large quantities of food products. One contaminated product can contaminate surrounding products and lead to a pathogenic outbreak (CDC, 2014). Cross-contamination and improper temperature control are the two main factors involved in foodborne sicknesses. These microorganisms are undetectable to human eyes and may grow to millions in one hour of exposure to the right conditions (CDC, 2014). Meat, vegetables, fruits, and dairy products are packaged and processed in different ways; however, contamination happens similarly in each food type.

11.3.1 Meat

Many infections by foodborne pathogens are caused by meat processing. The public believes that the main route of food contamination is during slaughter when feces spill out from the animal; however, most contamination comes from the animal's hide, which has been in contact with the surrounding environment (Durso *et al.*, 2014). Enteroviruses and rotaviruses, found in both cattle and pigs, are fecal-orally transmitted (Koopmans *et al.*, 2008). Meat, dairy, and vegetable products are processed in large-scale industries all around the world.

It is important to consider food contamination during the process of slaughtering (Chambers *et al.*, 2001). In the United Kingdom, prior to slaughter, the animal has to be cleared of any diseases or health problems, and rested overnight to reduce stress, referred to as a holding period (Chambers *et al.*, 2001). At the time of slaughter, the animals are taken out into a confined area, with little stress and noise as possible, and restrained (stunning box) before they are stunned. There are three types of stunning methods that can be performed: percussion stunning (percussive shock to the brain), electrical stunning (for pigs and poultry; an induced electroplectic shock to the brain), and carbon dioxide gas stunning (Chambers *et al.*, 2001). A sharp knife is then used to sever the main neck blood vessels, in order to drain the blood (Chambers *et al.*, 2001). During this process, contamination could occur if the knife is not thoroughly cleaned. Afterward, microbes can also enter the meat during packaging.

11.3.2 Fruits, Vegetables, and Dairy Products

Processing for fruits, vegetables, and dairy is much less extensive than meat processing. Raw fruits and vegetables are mainly washed, cleaned, and sorted. In addition, some produce may also be trimmed, sliced, shredded, or bagged (CDC, 2014). Fruits and vegetables can become contaminated through a variety of ways; these include the usage of contaminated water for irrigation or packaging, distribution contamination through prolonged exposure to sunlight while waiting for shipping, transport in vehicles that were not properly sanitized, food workers failing to wash hands after using a restroom, and improper washing of kitchen utensils (CDC, 2014).

Milk products are pasteurized, or exposed to high temperatures for a certain amount of time, to reduce human illnesses due to pathogenic contamination passed from an infected cow (CDC, 2014; International Dairy Foods Association, 2014). Statistical data from the CDC between 1998 and 2011 showed that 79% of exposure to *E. coli*, *Salmonella*, and *Listeria* in dairy was due to the consumption of raw milk (CDC, 2014).

11.4 Foodborne Pathogens and Antibiotic Resistance

11.4.1 Antibiotic Usage

An antibiotic is an agent or chemical compound that kills or inhibits the growth of bacteria without having harmful effects on humans (Salyers and Whitt, 2005). Antibiotics have been taken for granted since their discovery, and consequently bacteria are currently developing antibiotic resistance. The ability to resist the effect of an antibiotic (i.e., growth inhibition) and continue to grow beyond the normal susceptibility of the bacterial species can be confirmed via the identification of an antibiotic resistance gene and through a variety of laboratory procedures (Verraes *et al.*, 2013). There are three major types of resistance in bacterial species: epidemiological, pharmacological, and clinical resistance (Verraes *et al.*, 2013).

Nearly 24.6 million pounds of antibiotics are used annually on animals, and 3 million pounds in humans. The World Health Organization (WHO) composed a list of important human antimicrobial medicinal drugs consisting of 32 drug classes containing approximately 260 individual drugs, whereas WHO only allows the use of 39 individual drugs (aminoglycosides, macrolides, and penicillin) for agricultural cattle, swine, and poultry (Fair *et al.*, 2014; Durso *et al.*, 2014). Resistance has been seen to common antibiotics used on a daily basis, such as sulphonamides, tetracycline, amoxicillin, ampicillin, penicillin, and trimethoprim, among others (Chowdhury *et al.*, 2014). An example of extreme antibiotic resistance is found in the once defeated tuberculosis; there are at least 450,000 new cases of tuberculosis strains becoming multi-drug-resistant annually (Hede, 2014). MRSA, which was grounded by most European countries, has started spreading to other countries and developing more resistance. In 2002 MRSA levels in Norway were very low, at 0.4%; the levels increased to 225% in 2012. A similar trend was observed in Slovakia, with a low percentage in 2002 that dramatically climbed to 165% resistance in 2012 (Hede, 2014). The development of new antibiotics greatly increases the resistance of bacterial pathogens in many countries, but they continue to be used daily in most.

Antibiotics have been added to farm animals' daily feed and water as growth promoters and infection fighters for many decades now. In the early 1980s, researchers discovered that overusage of these antibiotics in farm animals increased antibiotic resistance genes in both pathogenic and non-pathogenic bacteria found in agricultural animal intestines as well as their surrounding environments (Devirgilii *et al.*, 2014; Cheng *et al.*, 2014). Research has shown that methicillin-resistant *Staphylococcus aureus* (MRSA) was prevalent in 12% of food products in the Netherlands, including beef, veal, lamb, and pork, as well as dairy products purchased in Italy; additionally, farmed fish and market shrimp have shown human pathogens carrying antibiotic resistance genes (Marshall *et al.*, 2011).

The European Union recognized the potential consequences of antibiotic overuse and banned growth-promoting antimicrobials in farm animals in 2006. The United States FDA proposed an action plan in 2013 to reduce the usage of antibiotics in farm animals (Hede, 2014; Fair *et al.*, 2014; Chowdhury *et al.*, 2014; Petternel *et al.*, 2014).

11.4.2 Modes of Antibiotic Resistance

A bacterial species that has acquired antibiotic resistance will be unaffected by antibiotic compounds. Such new strains may develop as a result of a mutation or through the most common means of genetic transfer: horizontal gene transfer (HGT) (Marshall and Levy, 2011). Some pathogens become resistant to more than one drug due to transposons having the ability to carry more than one resistant gene, with the largest carrier of multi-antibiotic resistance genes being the mercury resistance transposons (Chowdhury *et al.*, 2014).

Antibiotic resistance can operate via a variety of mechanisms. The main three categories are as follows: inactivation of the antibiotic, changes in the permeability of the bacterial cell membrane, and modification of the bacterial cell wall (Frye *et al.*, 2013; Verraes *et al.*, 2013). However, the most common mode is the enzymatic degradation of antibiotics via microbial enzymes, for example, the use of acetyltransferases, nucleotidyltransferases, and phosphotransferases to inactivate the antibiotic (Verraes *et al.*, 2013).

HGT is an important part of bacteria's evolutionary changes and can occur through soil, water, human and animal digestive tracts, and food sources. HGT has the ability to contaminate intestinal bacteria and alter strains that were deemed safe into harmful strains. It can be detected through two approaches: phylogenetic and compositional (Rossi *et al.*, 2014). Microorganisms used for fermenting food products and probiotics have an increased chance of HGT (Rossi *et al.*, 2014). Bacteria have a greater chance of becoming multi-drug-resistant by exchanging genes through mobile genetic structures such as plasmids, integrons, and transposons; additionally, recent research has shown transposons mediating between species (Verraes *et al.*, 2013; Devirgiliis *et al.*, 2013).

Among antibiotics, penicillin inhibits the growth of gram-positive bacteria by inhibiting specific enzymes in the cell wall (peptidoglycan). However, some gram-positive bacteria are able to produce betalactamase, or penicillinase (an enzyme used to tear apart penicillin) as a fighting tactic; these bacteria are named *Enterococcus*.

Streptomycin is an effective antibiotic used to treat tuberculosis and functions by blocking ribosomal protein synthesis, binding next to the smaller ribosomal subunit. The smaller subunit then misreads the sequence, producing random proteins, which eventually leads to bacterial cell death (Mgrdichian *et al.*, 2013). Even though streptomycin was the first antibiotic to fight tuberculosis, studies have shown that some strains of the pathogenic *Mycobacterium tuberculosis* carry a mutation in their rRNA genes (*rpsL* and *rrs*) that is found to be related to streptomycin resistance.

Ampicillin uses the same mechanism as penicillin, but has an amino group attached. Due to the presence of the amino group, it has the ability to penetrate both gram-negative and gram-positive bacteria, inhibiting transpeptidase, which bacteria require to synthesize their cellular walls, causing the cell to lyse and be destroyed (Sharma *et al.*, 2013). *Haemophilus influenza* causes meningitis (type b, capsulated type) and is commonly treated with ampicillin. It has been shown not to have beta-lactamase present, allowing for resistance to antibiotics that use beta-lactamase as a mechanism of attack, including ampicillin (King, 2012; Hasegawa *et al.*, 2006). Also, specific strains of *E. coli* (KL99 Self) and sensitive *S. enterica* serovar Typhimurium strains are able to grow in the presence of ampicillin (Perlin *et al.*, 2009).

Chloramphenicol is an antibiotic that is less commonly used due to its toxicity and carcinogenetic effects in humans; however, it is now being used more often due to arising resistance to commonly used antibiotics. Chloramphenicol works by reversibly binding to the 50S subunit of the 70S ribosome (which inhibits peptidyl transferase), causing damage to the mitochondria (Li, 2009). Studies revealed that *Pseudomonas putida* KT2440 (an infectious bacterium that is multi-resistant) acquired resistance to chloramphenicol at 25ug/mL (Fernandez *et al.*, 2012). Table 11.4 summarizes popular antibiotics and their modes of action against varying pathogens.

11.4.3 Antibiotic Resistance Among Foodborne Pathogens

A variety of pathogenic organisms have acquired antibiotic resistance and become multi-drug-resistant over the years. *Salmonella enterica* causes over 2 million human infections

Table 11.4 Popular Antibiotics, Their Mechanism of Attack, and Common Resistant Pathogens.

Antibiotic	Mechanism of attack	Resistant pathogens
Penicillin	Inhibits enzyme in cell wall (peptidoglycan)	*Enterococcus*
Streptomycin	Blocks ribosomal protein synthesis	*Mycobacterium tuberculosis*
Ampicillin	Inhibits enzyme in cell wall (peptidoglycan)	*Haemophillus influenza*, *E. coli* strains, *S. enterica*
Chloramphenicol	Inhibits peptidyl transferase, damaging mitochondria	*Pseudomonas putida* KT2240
Tetracycline	Inhibit protein synthesis by blocking attachment of aminoacyl-tRNA	*Salmonella*
Methicillin	Inhibition of bacterial cell wall synthesis by binding penicillin-binding proteins.	*Staphlococcus aureus*

in the United States each year and is generally transmitted through contaminated food; it affects multiple host organisms (humans and other animals, including food-producing animals), causing mild to severe diseases (i.e., gastroenteritis, typhoid fever, and systemic infections) (Frye and Jackson, 2013; Bhunia, 2008; Igbinosa, 2014). *Salmonella* has also been shown to have the ability to acquire resistance to multiple antibiotics (i.e., tetracycline, ampicillin, ceftriaxone, and carbenicillin), especially beta-lactams. Multidrug resistance in *Salmonella* strains could occur from exposure to biocides and overusage of antibiotics. The last resort in antibiotic treatment could be aminoglycosides or folic acid pathway inhibitors (Frye and Jackson, 2013; Marshall and Levy, 2011; Verraes *et al.*, 2013). *Salmonella* contains a genomic island (i.e., Serovars Typhimurium, Paratyphi, and agona [SGI-1]) that encodes genes for resistance against multiple antibiotics (ampicillin, chloramphenicol, streptomycin, sulfonamides, and tetracycline) (Bhunia, 2008).

E. coli strains inhabit the intestinal tracts of humans as commensal bacteria and are nonpathogenic; however, many pathogenic strains (*E. coli* O157:H7 and *E. coli* O157: NM) have been found in contaminated water, soil, meat, fruits, and vegetable products, which humans consume, causing gastroenteritis, dysentery, urinary tract infections, pneumonia, and other problems (Bhunia, 2008). *E. coli* is often found in the intestinal tracts of pigs, and has acquired the ability to resist tetracyclines, aminoglycosides, sulphonamide-trimethoprim, and ampicillin (Barton, 2014).

11.4.4 Selective Pressure of Antibiotics in Agricultural Bacteria

The overuse of antibiotics in agriculture exerts selective pressure on bacteria to take on antibiotic resistance genes, which may be horizontally transferred from one bacterium to another in soil, intestines, meat, vegetables, or into humans as meal (Durso *et al.*, 2014). Tetracycline and erythromycin are common agricultural antibiotics used to identify antibiotic resistance in foodborne lactic acid bacteria; these resistance genes correspond with transposable elements, allowing the resistance to be transferred to other bacteria (Devirgilii *et al.*, 2014.) Fluoroquinolone-resistant pathogens (*E. coli* and *Salmonella*) and vancomycin-resistant (*van*A) *Enterococci* have been shown to arise due to the overusage of these antibiotics in food-producing animals. In addition, ampicillin and tetracycline resistance are generally higher in countries using fluoroquinolones in agricultural animals. However, countries that have banned fluoroquinolones, that is, Australia, show a reduced resistance rate in human samples (Fair *et al.*, 2014; Barton, 2014). *van*A enterococci were identified in humans after consumption of meat from pigs raised using avoparcin as a growth promoter. The usage of tylosin as a growth promoter in pigs leads to *Campylobacter coli* resistance to macrolids. Isolates of an *E. coli* strain were recognized to carry multi-drug-resistant plasmids for both fluoroquinolone (*qnrS1*) and CTX-M beta-lactamase ($bla_{\text{CTX-M-14}}$) (Barton, 2014). The CTX-M extended spectrum beta lactamases (ESBLs) were first discovered in cattle using PCR. Notably, the genes have been found to be associated with mobile genetic elements and other resistance genes; additionally, *Kluyvera ascorbata*, an enteric bacterium, is the original source for the encoded gene CTX-M-5-ESBLs (Allen, 2014). The CTX-M-15 gene was found on a plasmid in *E. coli* genomes, along with a stx_2 Shiga toxic gene and an antibiotic resistance gene cluster (Chowdhury *et al.*, 2014).

Metagenomic studies show a plethora of antibiotic resistance genes in agricultural fecal matter (Allen, 2014). Colistin, an old antibiotic used against gram-negative bacteria in poultry and pigs, is also used as a last resort; however, its nephrotoxicity remains a concern. Studies have revealed resistance to colistin in sick animals: in Belgium, 34 *E. coli* strains from calves showed a 14.7% resistance; in the UK, 2480 *E. coli* strains from pigs showed a 34.1% resistance; in Japan, 118 *E. coli* strains showed a 35.6% resistance; and lastly, in China, 60 *E. coli* strains from pigs showed a 33.3% resistance (Kempf *et al.*, 2013).

Notably, fecal *E. coli* samples from humans and agricultural animals in China showed that 84.1% and 75.5% respectively carried the *aaa*C2 gene for resistance to gentamicin (Kempf *et al.*, 2013). A study in Japan revealed that *E. coli* in farm animals showed multi-drug resistance to penicillin, tetracyclines, and streptomycin (Harada *et al.*, 2010). Another study in Austria found multi-drug-resistant bacteria in minced meat from livestock animals that shared human genetic traits and a large diversity of ESBL genes (Petternel *et al.*, 2014).

Unpublished data from our lab exploring microbial occurrence in baby food, canned food, and bagged food revealed a total of 112 isolates from all three products; 21 of these were identified through 16S rRNA sequencing, revealing *Bacillus* sp., *Staphylococcus* sp., *Delftia* sp., *Staphylococcus* sp., *Serratia* sp., *Arthrobacter* sp., *Acinetobacter* sp., *Pseudomonas* sp., *Sphingobacterium* sp., *Exiquobacterium* sp., *Paenibacillus* sp., *Micrococcus* sp., and *Methylobacterium* sp. (Figure 11.3A–S). A diverse number of these isolates showed antibiotic resistance to ampicillin, streptomycin, chloramphenicol, and kanamycin at concentrations of 100 ug/mL, 500 ug/mL, or 1000 ug/mL. Ampicillin revealed the most microbial resistance, with a total resistance of 57.14% from the total isolates, followed by kanamycin, with a microbial resistance of 26.79%.

11.4.5 Antibiotic Resistance Genes Prominent in Animal Environments

Environment plays a huge role in the emergence of antibiotic resistance. Bacterial isolates have been identified to contain antibiotic resistance in pig farm environments, which is a problem not just for the pigs, but also for humans and other animals going into that environment (Barton, 2014). Livestock farms are filled with animal waste, including human (feces and urine), which is recycled in an inexpensive way compared to purchasing mineral fertilizers. Land that has been spread with animal manure serves as a reservoir for antibiotic-resistant bacteria, which can transfer their resistance

(A)

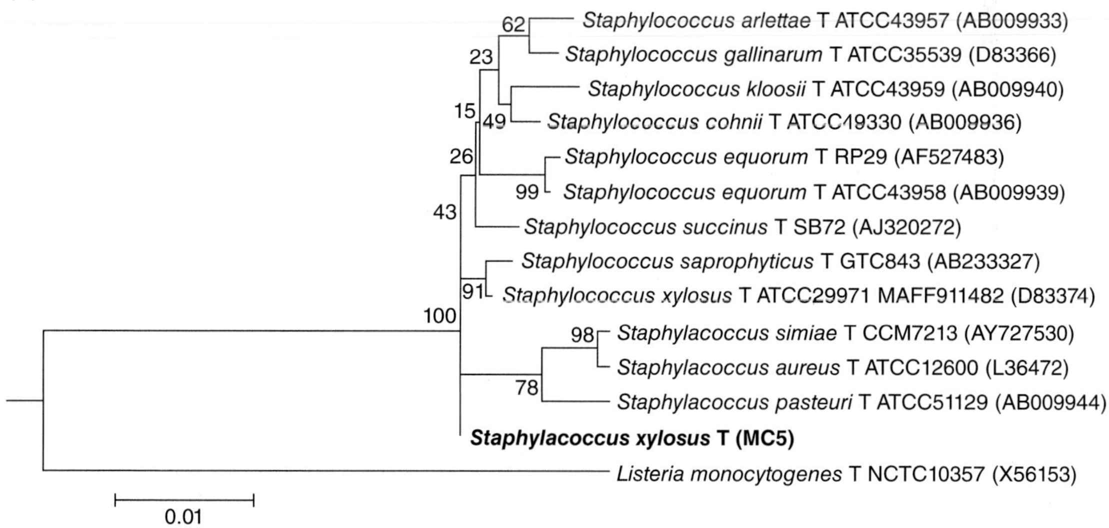

Figure 11.3A–S 16S rRNA gene sequence based neighbor-joining phylogenetic tree showing relationship between food-borne pathogen isolates (A–S) showing close proximity with respective homologues. Different outgroups were used for each isolates. Numbers at the nodes indicate levels of bootstrap support based on neighbor-joining analysis of 1000 resampled datasets. GenBank accession numbers are given in parenthesis.

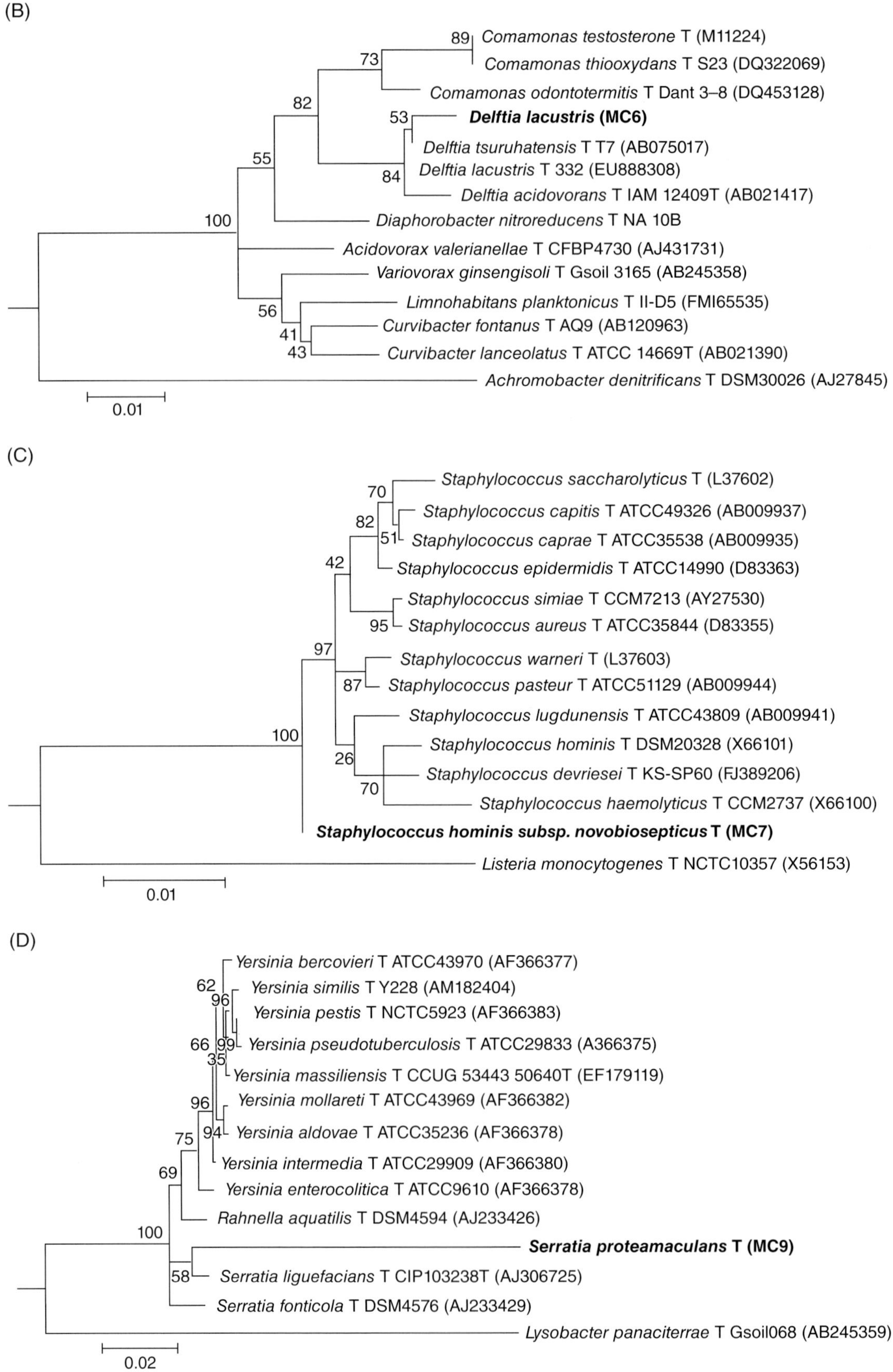

Figure 11.3 (Continued)

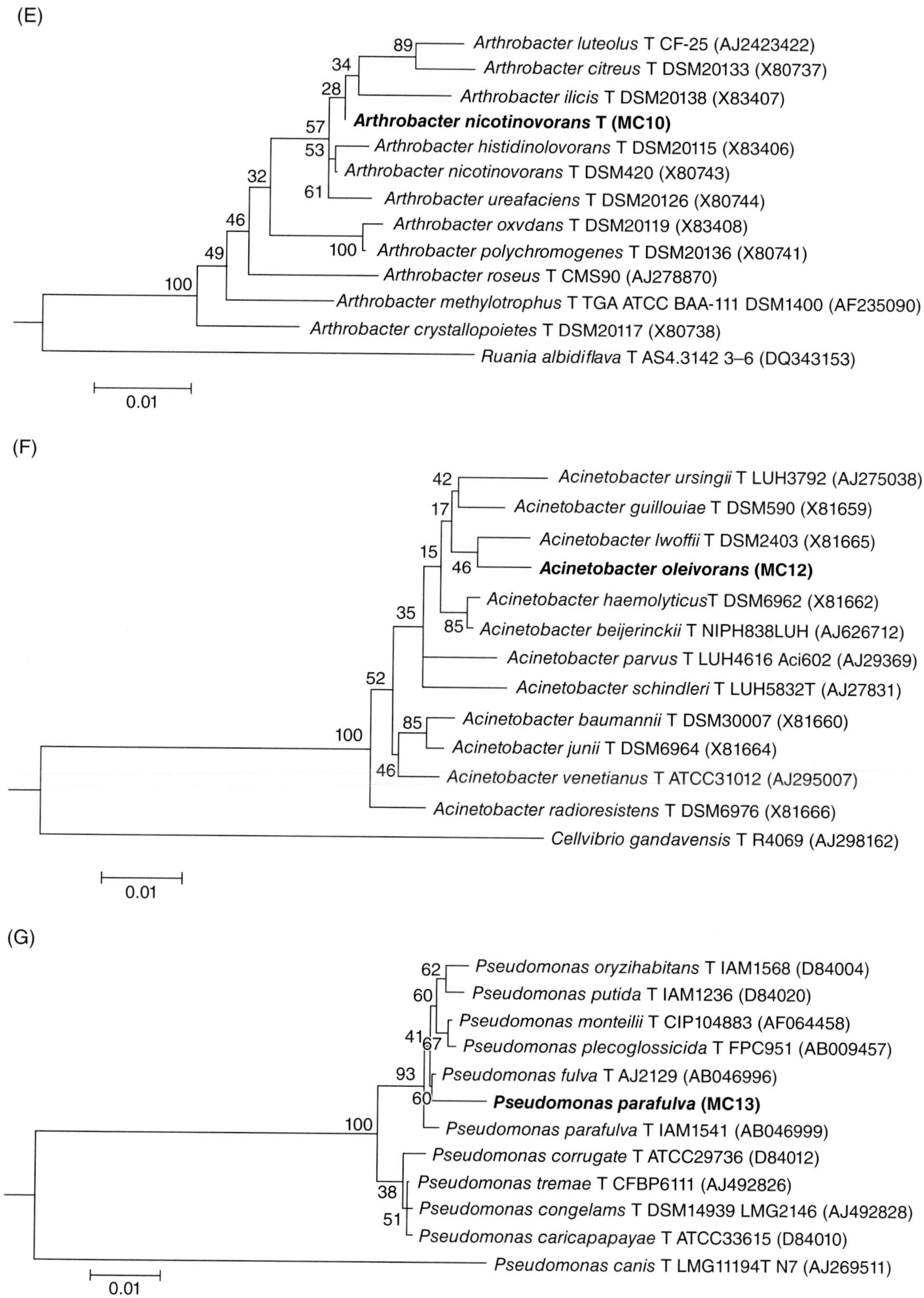

Figure 11.3 (Continued)

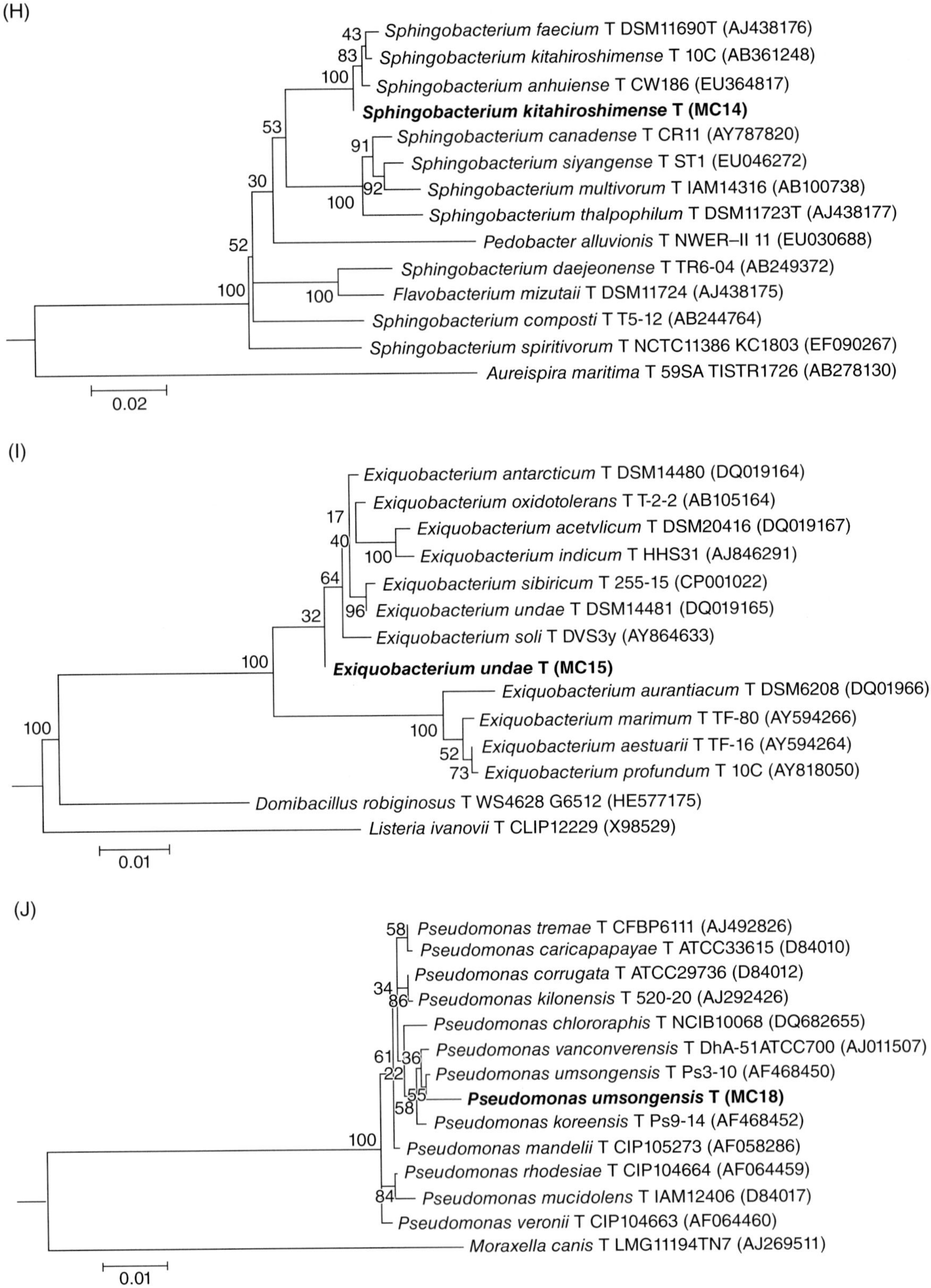

Figure 11.3 (Continued)

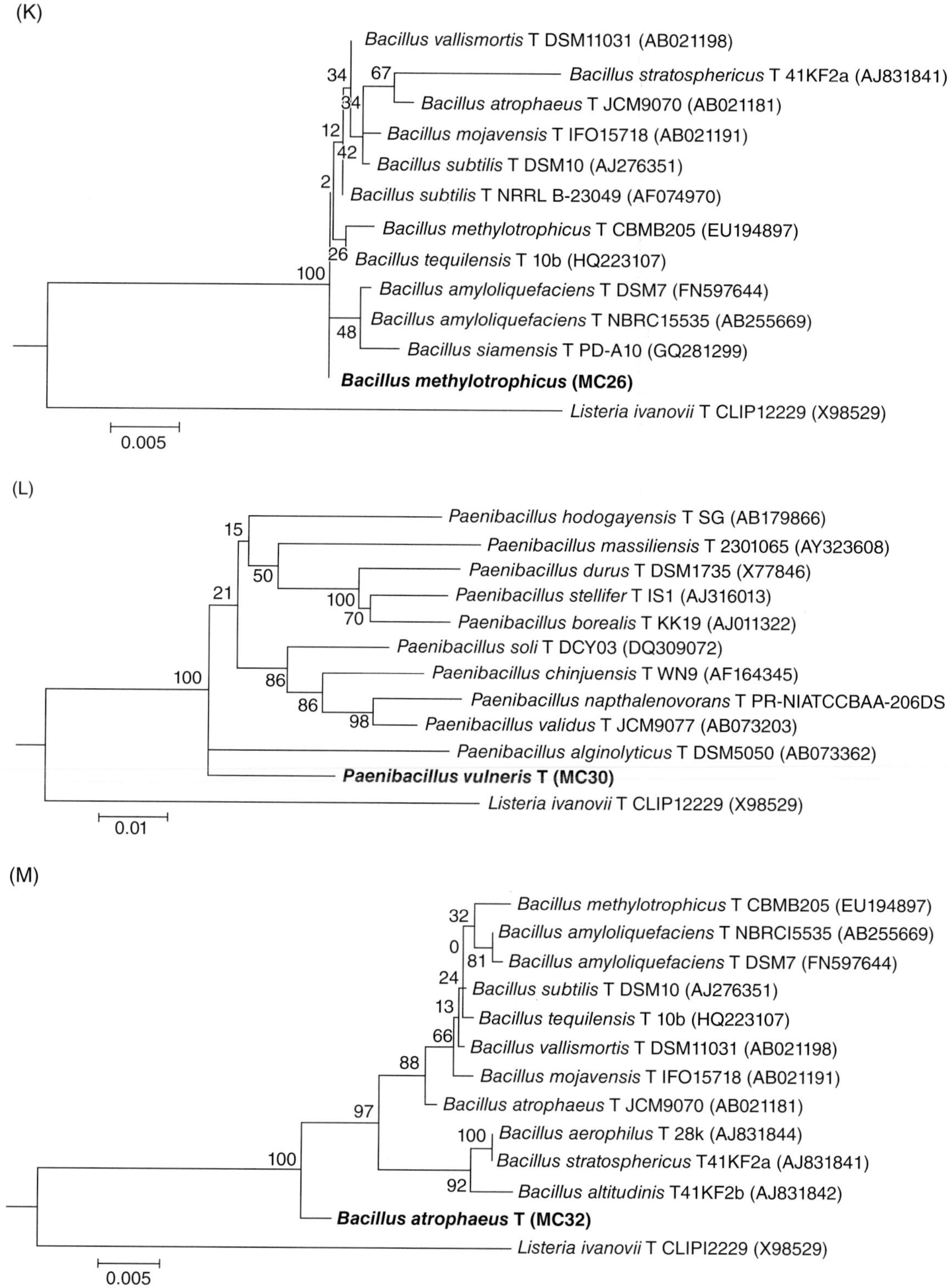

Figure 11.3 (Continued)

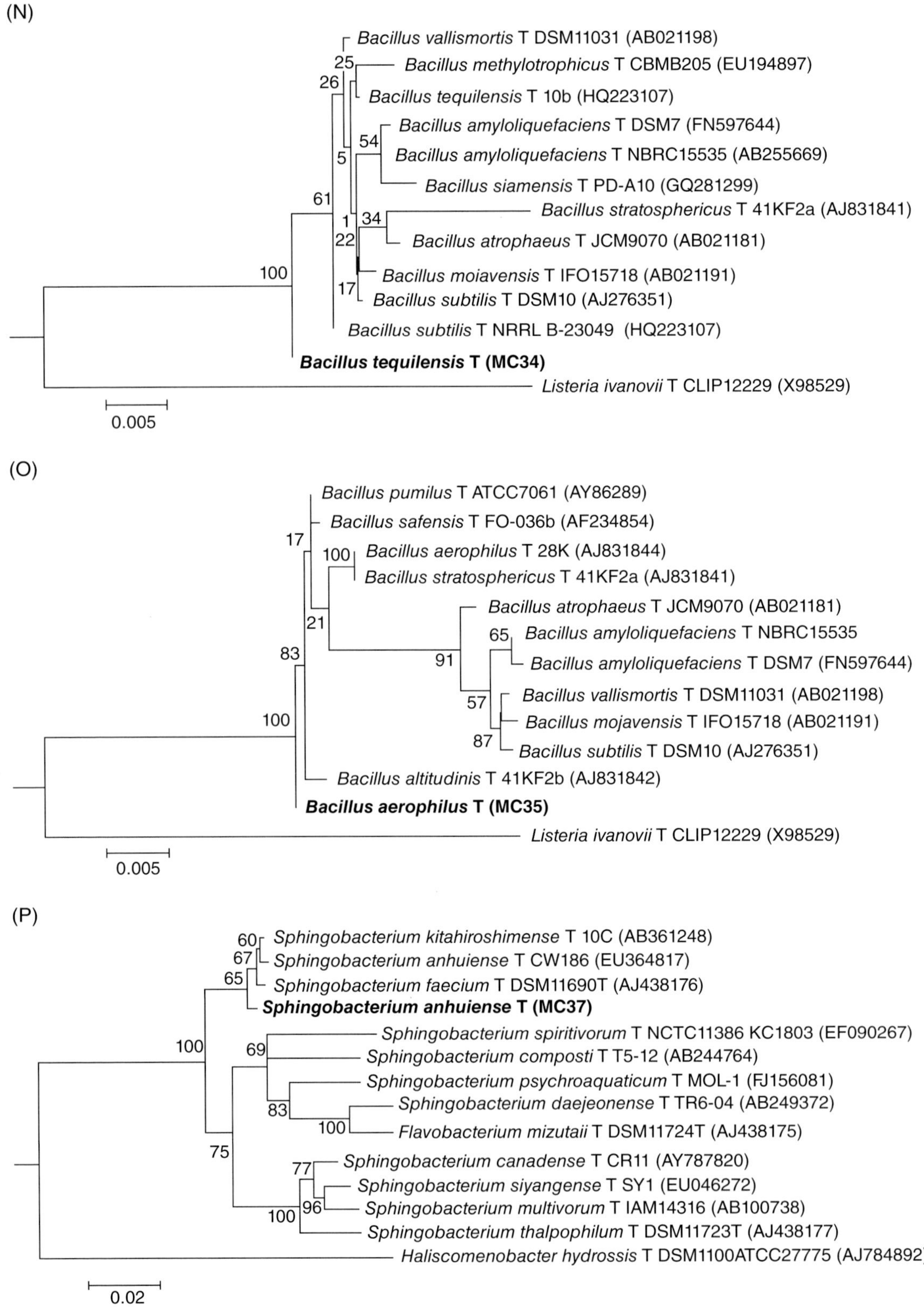

Figure 11.3 (Continued)

(Q)

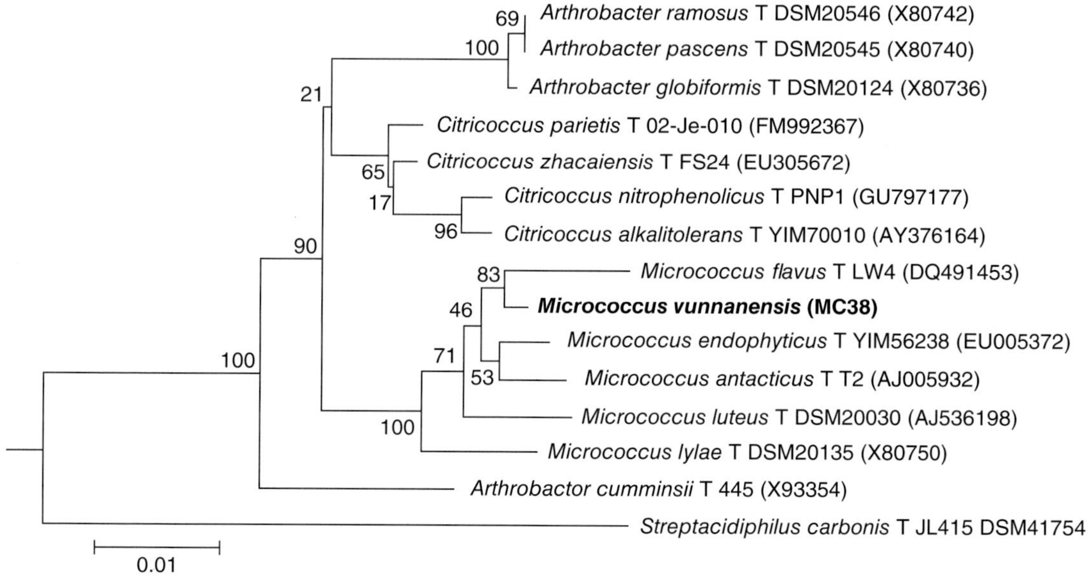

(R)

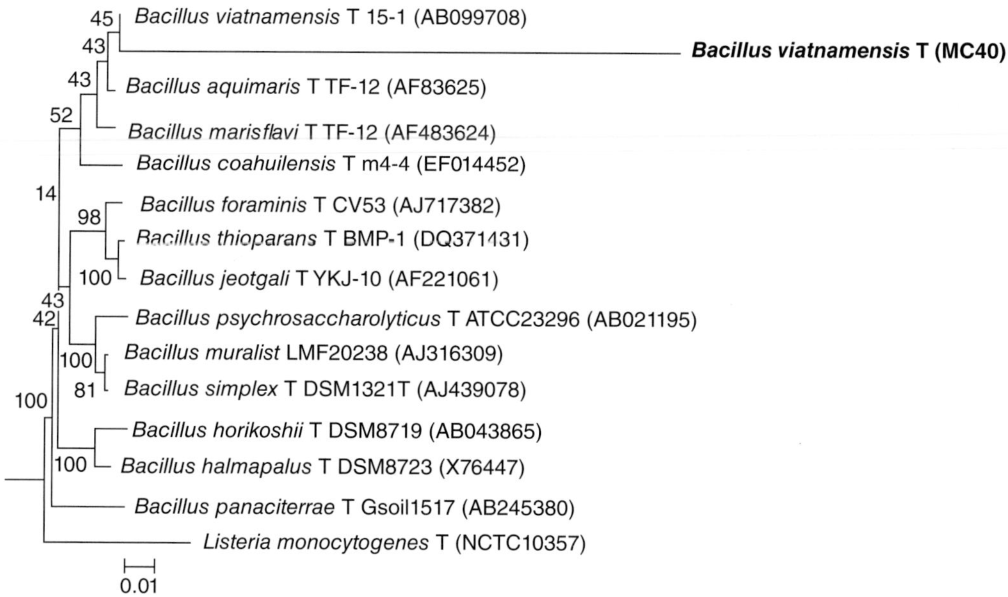

Figure 11.3 (Continued)

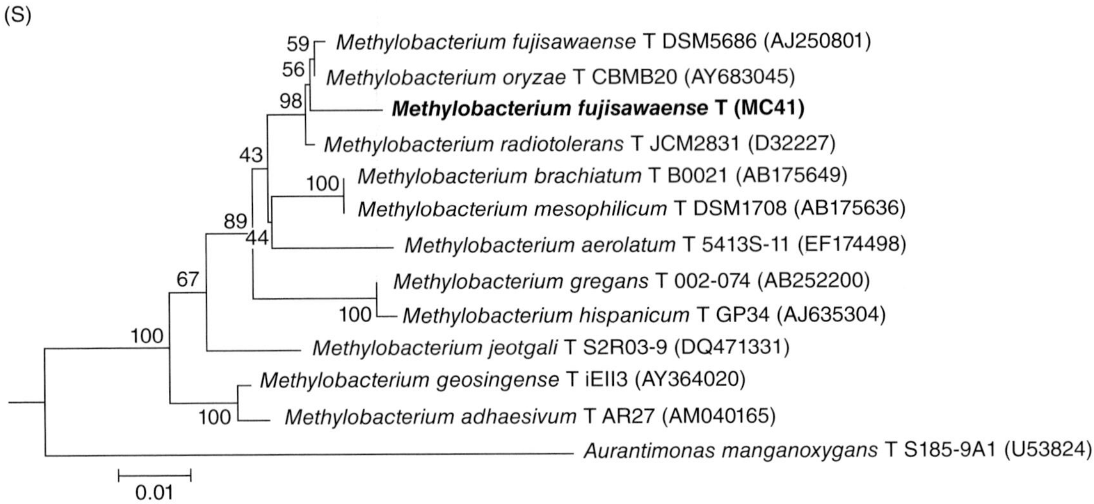

Figure 11.3 (Continued)

genes to other bacteria inhabiting that soil area (Marti *et al.*, 2014). Vegetable products, soil, water, and leaves on or near farm areas have shown *Enterococcus* spp., antibiotic resistance (Micallef *et al.*, 2013).

Fields in Germany and the United Kingdom showed an increase of the antibiotic resistance genes *sul*1, *sul*2, and *int*1, respectively, due to the usage of swine manure and sewage sludge, which contained these genes and transferred them to the surrounding soils (Marti *et al.*, 2014). Wastewater for irrigation also serves as a reservoir of pathogenic bacteria, of which some are antimicrobial-resistant if not multi-drug-resistant. The antibiotic-resistant bacteria, some of which are pathogenic to humans, are disseminated throughout the environment from the wastewater (Broszat *et al.*, 2014). Figure 11.4 shows the pathway of antibiotic-resistant pathogens via both human and agricultural transports to the public and environment. The first path reveals use of antibiotics by humans in hospitals and other care facilities, which led to infection of the residential flora in the intestinal tract and eventually to drug resistance via horizontal gene transfer. Humans who are routinely exposed to farms and slaughterhouses, such as veterinarians, farm workers, and anyone in contact with those workers, are at risk of being infected with antibiotic-resistant bacteria, providing a vehicle for antibiotic resistance genes (Marshall *et al.*, 2011).

11.5 Antibiotics and Alternatives

A variety of foodborne pathogens are evolving resistance to traditional antibiotics. Several innovative techniques are being tested to overcome this issue, including antibiotics found naturally in the environment. Among these techniques are direct-fed microbials (probiotics), live microbial feed supplements, prebiotics, vaccination, cationic peptides, cytokines, bacteriocins, antimicrobial peptides (bacteriocins), bacteriophages, organic acids, plant extracts, and using dietary control to limit the antibiotic load (Amalaradjou and Bhunia, 2013; Diarra and Malouin, 2014; Cheng *et al.*, 2014). There are also different approaches including natural products (bioactive molecules) such as berries that can act as an antiviral or immunostimulant, positively associate with the microbiota located in the intestines, or contain

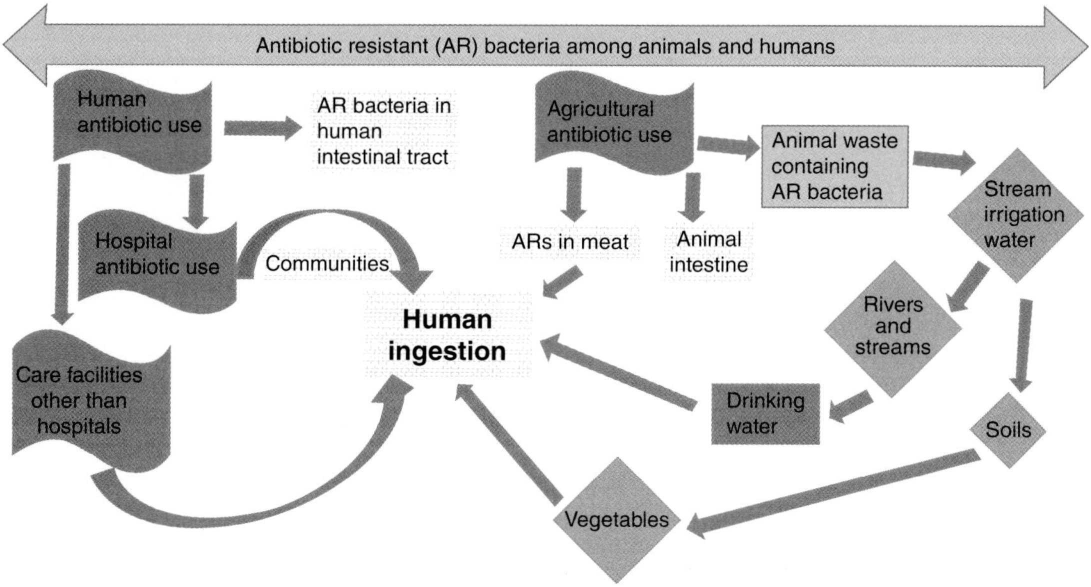

Figure 11.4 Routes of antibiotic resistant bacteria toward human consumption from varying conditions including food. (*See insert for color representation of the figure.*)

flavonoids (inhibit adhesion of specific pathogens and prevent low-density lipid oxidation). However, bioactive material could not be easily extracted from the berries, and that is why they were only used in addition to antibiotics (Diarra and Malouin, 2014).

Another alternative is the bacterial intracellular second messenger molecule, cyclic diguanosine monophosphate (C-DI-GMP). C-DI-GMP has the ability to control the pathogenic infection process by controlling microbial movement, cell adhesion, and biofilm production in pathogens. C-DI-GMP is an extremely potent immunostimulatory agent that has been proven useful as a mucosal adjuvant (Diarra and Malouin, 2014). Phage therapy is another promising practice that has been used in treatment of foodborne pathogens in both animals and plants (Allen *et al.*, 2014). Phage therapy involves the usage of a virus to infect the bacterial host. However, although bacterial pathogens have low resistance to phages, the phages can alter the commensal bacteria in the host's system (Allen *et al.*, 2014). Antimicrobial peptides (AMPs) may prove to be another viable alternative; however, their toxicity limits their therapeutic implications. A subgroup of AMPs, bacteriocins, are less toxic, select for less resistance than antibiotics among bacteria, and can be used as a food preservative; they are effective against human multidrug-resistant pathogens including a vancomycin-resistant *Enterococcus* strain (Allen *et al.*, 2014).

11.5.1 Drawbacks of Antibiotic Alternatives

In studies of some antibiotic alternatives, such as zinc (Zn) and copper (Cu), used for nutritional value in farm animals and for diarrheal disease prevention, enteral bacteria (commensal and pathogenic) have been found to develop resistance to these elements (Yazdankhah *et al.*, 2014). These studies showed that many opportunistic bacterial isolates from animals revealed higher MIC values to Zn and Cu compared to background isolates (Table 11.5) (Yazdankhah *et al.*, 2014). In another study, the removal of growth

Table 11.5 Layout of the Disadvantages of a Few Alternatives to Antibiotics.

Alternative	Disadvantage	Reference
Zn and Cu metal ions	Potentiality for resistance	Yazdankhah, *et al.*, 2014
Bacterial Vaccinations	May develop to full virulence	Cheng, G., *et al.*, 2014
Bacteriophages	Host strain specific, leads to new microbe strains and resistance, temperature dependent, only useful in early stage of disease	Cheng, G., *et al.*, 2014
Antimicrobial peptides	Highly costly, genetic engineering failure, toxicity, bacterial resistance	Cheng, G., *et al.*, 2014
Probiotics	Inactivity during transportation, pH specific, unsafe bacterial species	Cheng, G., *et al.*, 2014
Prebiotics	Cannot kill or inhibit bacteria, side effects, expensive	Cheng, G., *et al.*, 2014

promoters and introduction of Zn showed Zn-resistant methicillin-resistant *Staphylococcus aureus* strains (Singer *et al.*, 2014). The drawbacks of *Salmonella* vaccinations, another possible alternative, include the fact that the *Salmonella* strain stays in the host for an extended time period and may develop to full virulence.

The use of bacteriophages has its own drawbacks: as the phages are specific to the host strain, the individual's immune system may recognize the phage as an intruder and destroy or inactivate the phage. In addition, the bacteriophages may transfer their genomic DNA into the host and could leave new microbes to develop resistance to the phages due to mutations and other environmental factors (Table 11.5) (Cheng *et al.*, 2014).

AMPs may seem like a better alternative to antibiotics due to their bactericidal effects; however, they are expensive compared to antibiotics. Therefore, genetic engineering approaches have been tried on AMPs, but the structures of the newly synthesized AMPs cannot be guaranteed to present the same bactericidal effects as the natural AMPs. In addition, a variety of AMPs are hemolytic and toxic to eukaryotic cells with a narrow antibacterial spectrum, and the fear remains that bacteria will acquire resistance to AMPs by limiting the receptors on AMPs (Table 11.5) (Cheng *et al.*, 2014).

Probiotics and prebiotics have also drawn attention as possible alternatives to antibiotics. Probiotics contain a variety of unsafe bacterial species. Once food is transported and processed, probiotics are inactivated. Also, they require a specific pH; most probiotics cannot survive at low pH levels. Prebiotics cannot work the same way as antibiotics, as they lack the ability to kill bacteria or inhibit their growth, and have side effects such as bloating and diarrhea (Table 11.5) (Cheng *et al.*, 2014).

11.6 Genomics and Proteomics of Foodborne Pathogens and Antibiotic Resistance

The genome (i.e., the entire gene set in a cell) and proteome (i.e., the entire protein set in a cell) offer a great amount of information that can be used to identify food spoilage organisms and antibiotic resistance genes. These systems-biology-based approaches can identify food-spoilage organisms rapidly and effectively, and help us understand antibiotic resistance at a molecular level by identifying genes and proteins in relation to resistance.

11.6.1 Genomics

The genome consists of all the genetic information of a living organism. The organism transfers that information through a process called transcription and produces the transcriptome (mRNA population). Whole gene transcript analysis via microarray technology is now an approachable technique to understand disease and biological processes (e.g., food spoilage and biofilm production).

The field of foodborne pathogen genomics is rapidly growing and has already shown evidence of promise. Studies involving microarray analysis on *E. coli* O157:H7 have revealed stress-responsive genes being induced as well as translational and attachment genes being repressed during *E. coli* infection (Puttamreddy *et al.*, 2008). Comparative microarray studies conducted on regulators of gene expression of *S. typhimurium* in the gastrointestinal tract have shown that genes located within the SPI-1 and SPI-2 pathogenicity islands are upregulated in mutant strains of *S. typhimurium*, causing the infection to continue (Puttamreddy *et al.*, 2008). A transcriptional analysis of *C. jejuni* heat shock protein HspR showed heat shock chaperones being upregulated, leaving the bacteria unable to reproduce after specific temperatures; additionally, studies also revealed *hspR* mutants' association with flagella formation of *C. jejuni*, showing a downregulation, an indication of the inability to protrude through epithelial INT-407 cells (Puttamreddy *et al.*, 2008).

Antimicrobial genes can be determined by analyzing the genome sequence of bacterial species. Moreno *et al.* (2014) studied the genomes of *Bacillus* and *Paenibacillus* food spoilage bacteria and found 20 resistance genes, including resistance to beta-lactams, aminoglycosides, copper, chloramphenicol, and vancomycin. Whole-genome sequencing has enhanced identifications of microbial strains by allowing the opportunity to compare microbial genomes with different strains.

Comparative genomics confirmed both diploid and triploid genomes in isolates of the spoilage yeast *Brettanomyces bruxellensis*, as well as structural genome variations (duplications and deletions) among beneficial food strains and spoilage food strains (Crauwels *et al.*, 2014). Crauwels *et al.* (2014) identified phylogenetic relationships based on the nuclear large subunit (LSU) rRNA gene sequences of *Brettanomyces*, using DNA fingerprinting assays (random amplified polymorphic DNA-PCR, arbitrarily primed PCR, and PCR). Mols *et al.* (2007) used a phenotypic microarray to identify stress levels and resistance to chemicals and antibiotics in strains of the foodborne pathogen *B. cereus*, showing how the metabolic capabilities of each strain were important. The comparative analysis of the two *B. cereus* strains' genomes using the INPARANOID database showed ATCC 14579 containing 5,234 open reading frames and 6,035 domains, whereas ATCC 10987 contained 5,603 open reading frames and 4,315 domains, proving comparative genomics is an innovative technique for identification of multiple genome variations (Mols *et al.*, 2007). Fu *et al.* (2014) used qRT-PCR, a genomic approach, to confirm 10 genes (*pfaC, pyrB, alr2, ribH, plA2, pheS, rpoD, groES, dnaK,* and *groEL*) responsible for stress responses to high-pressure food preservation. Genotyping using microarray-based comparative genome hybridization (M-CGH) is another method to detect food spoilage organisms and compare almost identical bacterial species (Caspers *et al.*, 2011).

11.6.2 Proteomics

Tools and techniques used for proteomic approaches include 2D gel electrophoresis, SDS-PAGE, mass spectrometry, and database searches (Gabani *et al.*, 2012). Rapid foodborne pathogen identification is an important factor in the fight to reduce food-related illnesses. Matrix-assisted laser desorption ionization–Time of flight (MALDI-TOF) is efficient in that it takes less time and requires only a small sample to provide effective results; it can be used for the identification of bacterial species as well as creating phyloproteomic relationships to compare different strains (Qunitela-Baluja *et al.*, 2012). Qunitela-Baluja *et al.* (2012) studied the food-spoilage pathogen *Enterococcus* and determined a genus-specific biomarker with a *m/z* 4426 +/− 1 using MALDI-TOF. Along with the MALDI-TOF, a protein cluster program, SPECLUST, was used in another study for the identification of five mass peaks *m/z* ranging from 2200-6000 for the foodborne pathogen *Streptococcus paruberis*. The distinct spectral fingerprints allow for the comparison to known foodborne bacterial profiles (Fernandez-No *et al.*, 2012).

Fu *et al.* (2014) used a MALDI-TOF proteomic approach to reveal the specific molecular and biological processes of *Vibrio parahaemolyticus* when subjected to high hydrostatic pressure (HHP). This study identified 26 differentially expressed proteins, including membrane structure, function, biosynthesis, cellular, transcription, chaperon, and heat-shock proteins as well as metabolism-associated proteins, of which 8 were increased and 15 were decreased under HHP conditions. Abhyankar *et al.* (2011) used a gel-free proteomic approach with LC-MS/MS to identify 55 bacterial coat proteins from the spore-forming bacterium *Bacillus subtilis*. Lima *et al.* (2013) reviewed proteomics-based studies of most bacterial species, and discussed alterations occur in energy and nitrogen metabolism regulation, glucan biosynthesis, amino acid, protein, and nucleotide synthesis is being affected; and various proteins show a stress response after exposing the microorganisms to the antibiotics. In the past, many proteomics based tools have assisted to provide realistic image of the bacterial resistance against antibiotics.

11.7 Conclusion

With over 76 million illnesses a year in the United States (FDA, 2014), food-derived illnesses in humans are a major concern for public health. They can be caused through the contamination and spoilage of food products by foodborne bacterial pathogens. This contamination originates from farming (irrigation water), manufacturing (slaughtering), packaging (cross-contamination), distribution (dirty vehicles), and preparation (unsanitary cooking utensils). Food can be spoiled biochemically, enzymatically, oxidatively, or by freezing, heating, drying, or pressure changes, and spoilage can be detected through human senses such as sight (browning), smell (putrid), taste (sour), and touch (soft/hard). In order to limit contamination, diverse food preservation methods have been developed.

Bacterial infections are traditionally treated with antibiotics (such as penicillin, ampicillin, chloramphenicol, kanomycin, and streptomycin). However, over the years, bacteria have developed ways to resist the effects of antibiotics and transferred them horizontally (to unrelated bacterial species) and vertically (to related bacterial species).

Systems biology tools, that is, genomics and proteomics, offer novel ways to identify foodborne pathogens as well as their related resistance genes and proteins. In order to effectively and efficiently detect foodborne pathogens, new industrial tools and procedures are being evaluated. The impact of antibiotic resistance can be lessened by the development of alternative strategies to fight bacterial pathogens, such as prebiotics, vaccinations, cytokines, antimicrobial peptides, bacteriophages, and plant extracts, among others.

Acknowledgments

The authors are thankful for the institutional assistance.

References

Abhyankuar, W., et al. (2011). Gel-free proteomic identification of the *Bacillus subtilis* insoluble spore coat protein fraction. *Proteomics, 11*, pp. 4541–4550.

Allen, K.H., et al. (2014). Finding alternatives to antibiotics. *Annal NY Acad Sci., 1323*, pp. 91–100.

Allen, K.H. (2014). Antibiotic resistance gene discovery in food-producing animals. *Curr Opin Microbiol., 19*, pp. 25–29.

Amalaradjou, M.A.R. and Bhunia, A. (2013). Bioengineered probiotics, a strategic approach to control enteric infections. *Bioengineered., 4*, pp. 379–387.

Ananou, S., et al. (2007). Biopreservation, an ecological to improve the safety and shelf-life of foods. *Communicating Current Education and Educational Topics and Trends in Microbiology. FORMATEX*, pp. 475–486.

Aneja, K.R., et al. (2014). Emerging preservation techniques for controlling spoilage and pathogenic microorganisms in fruit juices. *Int J Microbiol., 2014*, Article ID 758942–.

Asgharpour, F., et al. (2014). Investigation of class I integron in *Salmonella infantis* and its association with drug resistance. *Jundishapur J Microbiol., 7*, p. e10019–.

Barbosa, J., et al. (2013). Evaluation of antibiotic resistance patterns of food and clinical *Listeria monocytogenes* isolates in Portugal. *Foodborne Pathog Dis, 10*, pp. 861–866.

Barton, D.M. (2014). Impact of antibiotic use in the swine industry. *Curr Opin Microbiol., 19*, pp. 9–15.

Berman, H.F. and Riley, W.L. (2013). Identification of novel antimicrobial resistance genes from microbiota on retail spinach. *BMC Microbiol., 13*, p. 272–.

Bhunia, A.K. (2008). *Foodborne Microbial Pathogens*. New York, NY: SpringerVerlag.

Broszat, M., et al. (2014). Wasterwater irrigation increases the abundance of potentially harmful gammaproteobacteria in soils in Mezquital Valley, Mexico. *Appl Environ Microbiol., 80*, pp. 5282–5291.

Caspers, M.P., et al. 2011. A mixed-species microarray for identification of food spoilage bacilli. *Food Microbiol., 28*, pp. 245–251.

Centers for Disease Control and Prevention. (2014). *Haemophilus influenzae* Disease (Including Hib). Available at: http://www.cdc.gov/hi-disease/[Accessed June 16, 2016].

Cerveny, J., Meyer, J.D. and Hall, P.A. (2009). Microbiological spoilage of meat and poultry products. In: W.H. Sperber, M.P. Doyle, eds., *Compendium of the microbiological spoilage, of foods and beaverages. Food Microbiology and Food Safety*, New York: Springer Science and Business Media, pp. 69–686.

Chaillou, S., et al. (2014). Origin and ecological selection of core and food-specific bacterial communities associated with meat and seafood spoilage. *The ISME Journal., 9*, pp. 1105–1118.

Chambers, P.G. and Grandin, T. (2001). Food and Agriculture Organization of the United Nations Regional Office for Asia and the Pacific. *Human Society International*. Available at: http://www.fao.org/docrep/003/x6909e/x6909e00.htm#Contents [Access date November 17, 2015].

Cheng, G., et al. (2014). Antibiotic alternatives: the substitution of antibiotics in animal husbandry. *Front Microbiol., 5*, p. 217–.

Chowdhury, P.R., et al. (2014). Genomic interplay in bacterial communities: implications for growth promoting practices in animal husbandry. *Front Microbiol., 5*, p. 394–.

Crauwels, S., et al. (2014). Assessing genetic diversity among Brettanomyces Yeasts by DNA fingerprinting and whole-genome sequencing. *Appl Environ Microbiol., 80*, pp. 4398–4413.

Dave, D. and Ghaly, A.E. (2011). Meat spoilage mechanisms and preservation techniques: a critical review. *Am J Agr Biol Sci., 6*, pp. 486–510.

Devirgiliis, C., Zinno, P. and Perozzi G. (2013). Update on antibiotic resistance in foodborne *Lactobacillus* and *Lactococcus* species. *Front Microbiol., 4*, p. 301–.

Devirgiliis, C., et al. (2014). Functional screening of antibiotic resistance genes from a representative metagenomic library of food fermenting microbiota. *BioMed Res Int., 2014*, Article ID 290967.

Diarra, S.M. and Malouin, F. (2014). Antibiotics in Canadian poultry productions and anticipated alternatives. *Front Microbiol., 5*, p. 282–.

Doyle, E.M., Hartmaan, A.F. and Wong, L.C.A. (2012). Methicillin-resistant staphylococci: implications for our food supply? *Anim Health Res Rev., 13,* pp. 157–180.

Durso, M.L. and Cook, L.K. (2014). Impacts of antibiotic use in agriculture: what are the benefits and risks? *Curr Opin Microbiol., 19,* pp. 37–44.

Fair, R.J. and Tor, Y. (2014). Antibiotics and bacterial resistance in the 21st century. *Perspect Medicin Chemi, 6,* pp. 25–64.

Fernández, M., et al. (2012). Mechanisms of resistance to Chloramphenicol in *Pseudomonas putida* KT2440. *Antimicrob Agents Chemother., 56,* pp. 1001–1009.

Fernandez-No, I.C., et al. (2012). Isolation and characterization of *Streptococcus parauberis* from vacuum-packaging refrigerated seafood products. *Food Microbiol., 30,* pp. 91–97.

Food and Drug Administration (FDA). U.S. (2012). Regulatory information. Sec 201. [21 U.S.C.321]. Available at: http://www.fda.gov/RegulatoryInformation/Legislation/FederalFoodDrugandCosmeticActFDCAct/FDCActChaptersIandIIShortTitleandDefinitions/ucm086297.htm [Accessed June 16, 2016].

Food spoilage and preservation. (2009). EHVH 441: Food Protection.

Franco, B.E., et al. (2009). The determinants of the antibiotic resistance process. *Infect Drug Resist., 2,* pp. 1–11.

Frye, J.G. and Jackson, C.R. (2013). Genetic mechanisms of antimicrobial resistance identified in *Salmonella enterica, Escherichia coli,* and *Enteroccocus* spp. Isolated from U.S. food animals. *Front Microbiol., 4,* pp. 135–.

Fu, L.L., et al. (2014). Proteomic identification of responsive proteins of *Vibrio parahaemolyticus* under high hydrostatic pressure. *J Sci Food Agric., 94,* pp. 2630–2638.

Gabani, P., Prakash, D. and Singh, O.V. (2012). Emergence of antibiotic-resistant extremophiles (AREs). *Extremophiles., 16,* pp. 697–713.

Harada, K, and Asai, T. (2010). Role of antimicrobial selective pressure and secondary factors on antimicrobial resistance prevalence in *Escherichia coli* from food-producing animals in Japan. *J Biomed Biotechnol.,2010,* Article ID 180682.

Hasegawa, K., et al. (2006). High prevalence of type b beta-lactamase-non-producing ampicillin-resistant *Haemophilus influenzae* in meningitis: the situation in Japan where Hib vaccine has not been introduced. *J Antimicrob Chemother., 57,* pp. 1077–1082.

Hede, K. (2014). Antibiotic resistance: An infectious arms race. *Nature, 509,* pp. S2–S3.

Hörmann, S., et al. (2006). Comparative proteome approache to characterize the high-pressure stress response of *Lactobacillus sanfranciscensis* DSM 20451. *Proteomics., 6,* pp. 1878–1885.

Huang, J., et al. (2012). Chitosan-based edible coatings for quality preservation of postharvest whiteleg shrimp (*Litopenaeus vannamei*). *J Food Sci., 77,* pp. 491–496.

Igbinosa, I.H. (2015). Prevalence and detection of antibiotic-resistant determinant in *Salmonella* isolated from food-producing animals. *Trop Anim Health Prod., 47,* pp. 37–43.

International Dairy Foods Association. (2014). Pasteurization. *Making a difference for dairy.* Available at: http://www.idfa.org/news-views/media-kits/milk/pasteurization [Accessed June16, 2016].

Kempf, I., et al. (2013). What do we know about resistance to colistin in Enterobacteriaceae in avian and pig production in Europe? *Int J Antimicrob Agents., 42,* pp. 379–383.

King, P. (2012). *Haemophilus influenzae* and the lung (*Haemophilus influenzae* and the lung). *Clin Transl Med., 1,* p. 10–.

Koopmans, M.P.G., Cliver, D.O. and Bosch, A. (2008). *Food-borne viruses: progress and challenges.* Washington, DC: ASM Press.

Li, C., et al. (2010). Chloramphenicol causes mitochondrial stress, decreases ATP biosynthesis, induces matrix metalloproteinase-13 expression, and soild-tumor cell invasion. *Toxicol Sci., 116,* pp. 140–150.

Lima, T.B., et al. (2013). Bacterial resistance mechanism: what proteomics can elucidate. *FASEB J., 27,* pp. 1291–1303.

Lingbeck, J.M., et al. (2014). Functionality of liquid smoke as an all-natural antimicrobial in food preservation. *Meat Sci., 97,* 197–206.

Macovei, L. and Zurek, L. (2007). Influx of enterococci and associated antibiotic resistance and virulence genes from ready-to-eat food to the human digestive tract. *Appl Environ Microbiol., 73,* pp. 6740–6747.

Mardaneh, J. and Dallal, M.M. (2013). Isolation, identification and antimicrobial susceptibility of Pantoea (Enterobacter) agglomerans isolated from consumed powdered infant formula milk (PIF) in NICU ward: First report from Iran. *Iran J Microbiol., 3,* pp. 263–267.

Marder, P.E., et al. (2014). Multistate outbreak of *Escherichia coli* O157:H7 associated with bagged salad. *Foodborne Pathog Dis., 11,* pp. 593–595.

Marshall, B.M. and Levy, S.B. (2011). Food animals and antimicrobials: impacts on human health. *Clin Microbiol Rev., 24,* pp. 718–733.

Marti, R., et al. (2014). Safely coupling livestock and crop production systems: how rapidly do antibiotic resistance genes dissipate in soil following a commercial application of swine or dairy manure. *Appl Environ Microbiol., 80,* pp. 3258–3265.

Mehta, R., et al. (2013). Bio-preservative and therapeutic potential of pediocin: recent trends and future perspectives. *Recent Pat Biotechnol., 7,* pp. 172–178.

Mgrdichian, L. (2013). Molecular structure reveals how the antibiotic streptomycin works. U.S. Department of Energy. Brookhaven National Laboratory. Available at: https://www.bnl.gov/newsroom/news.php?a=24251 [Accessed October 31, 2014].

Micallef, S.A., et al. (2013). Diversity, distribution and antibiotic resistance of *Enterococcus* spp. Recovered from tomatoes, leaves, water and soil on U.S. Mid-Atlantic farms. *Food Microbiol., 36,* pp. 465–474.

Mols, M., et al. (2007). Metabolic capacity of *Bacillus cereus* strains ATCC 14579 and ATCC 10987 interlinked with comparative genomics. *Environ Microbiol., 9,* pp. 2933–2944.

Montville, T.J. and Chikindas, M.L. (2007). Biopreservation of Foods. In: M.P. Doyle, L.R. Beuchat, eds., *Food Microbiology: Fundamentals and Frontiers, 3rd Edition.* ASM Press, pp. 747–764.

Moreno, S.A., et al. (2014). Genomic comparison of sporeforming bacilli isolated from milk. *BMC Genomics., 15,* p. 26–.

Perlin, M.H., et al. (2009). Protection of *Salmonella* by ampicillin-resistant *Escherichia coli* in the presence of other wise lethal drug concentrations. *Proc Biol Sci., 276,* pp. 3759–3768.

Pesavento, G., et al. (2014). Prevalence and antibiotic resistance of *Enterococcus* spp. Isolated from retail cheese, ready-to-eat salads, ham, and raw meat. *Food Microbiol., 41,* pp. 1–7.

Petternel, C., et al. (2014). Isolation and characterization of multi-drug resistant bacteria from minced meat in Austria. *Food Microbiol., 44,* pp. 41–46.

Puttamreddy, S., et al. (2008). Transcriptome analysis of organisms with food safety relevance. *Foodborne Pathogs Dis., 5,* pp. 517–529.

Quintela-Baluja, M., et al. (2013). Characterization of different food-isolated *Enterococcus* strains by MALDI-TOF mass fingerprinting. *Electrophoresis, 34,* pp. 2240–2250.

Rasheed, U.M., Thajuddin, N. and Jamil, K. (2014). Antimicrobial resistance in strains of *Escherichia coli* isolated from food sources. *Rev Inst Med Trop Sao Paulo., 56,* pp. 341–346.

Reis, J.A., et al. (2012). Lactic Acid bacteria antimicrobial compounds: characteristics and applications. *Food Eng Rev., 4,* pp. 124–140.

Rossi, F., et al. (2014). Review: Horizontal gene transfer among micoorganisms in food: current knowledge and future perspectives. *Food Microbiol., 42,* pp. 232–243.

Salyers, A.A. and Whitt, D.D. (2005). *Revenge of the Microbes.* Washington, D.C: ASM Press.

Sharma, S.K., Singh, L. and Singh, S. (2013). Comparative study between penicillin and ampicillin. *Sch J App Med Sci., 1*, pp. 291–294.

Singer, S.R. and Williams-Nguyen, J. (2014). Human health impacts of antibiotic use in agriculture: A push for improved causal inference. *Curr Opin Microbio., 19*, pp. 1–8.

Subramanian, S. and Smith, D.L. (2015). Bacteriocins from the rhizosphere microbiome—from an agriculture perspective. *Front Plant Sci., 6*, Article 909.

Suganthi, V., et al. (2012). Lantibiotic Nisin: Natural preservative from *Lactococcus Lactis*. *Int Res J Pharm., 3*, pp. 13–19.

Swarnadyuti, N., et al. (2014). Role of biopreservation in improving food safety and storage. *J Engineer Research Appl., 4*, pp. 26–32.

Taban, B.M., et al. (2013). Characterization of antibiotic resistance in *Salmonella enterica* isolates determined from ready-to-eat (RTE) salad vegetables. *Braz J Micro., 44*, pp. 385–391.

Tomás-Barberá, F.A. and Espin, J.C. (2001). Phenolic compounds and related enzymes as determinants of quality in fruits and vegetables. *J Sci Food Agricul., 81*, pp. 853–876.

Ummeayman, R. (2009). Biochemistry of Food Spoilage. Protein Foods and Nutrition Development Association of India. Available at: http://www.pfndai.com/food_safety.htm [Accessed December 1, 2015].

Van, T.T.H., et al. (2007). Antibiotic resistance in food-borne bacterial contaminants in Vietnam. *Appl Environ Microbiol., 73*, pp. 7906–7911.

Vaughn, R.H. and Stewart, G.F. (1960). Antibiotics as food preservatives. *JAMA, 174*, pp. 1308–1310.

Verraes, C., et al. (2013). Antimicrobial resistance in the food chain: A review. *Int J Environ. Res Public Health., 10*, pp. 2643–2669.

Wellington, E.M.H., et al. (2013). The role of the natural environment in the emergence of antibiotic resistance in Gram-negative bacteria. *The Lancet: Inf Dis., 13*, pp. 155–65.

Widsten, P., et al. (2014). Tannins and extracts of fruit by-products: antibacterial activity against foodborne bacteria and antioxidant capacity. *J Agric Food Chem., 62*, pp. 11146–11156.

Yazdankhah, S., Rudi, K. and Bernhoft, A. (2014). Zinc and copper in animal feed development of resistance and co-resistance to antimicrobial agents in bacteria of animal origin. *Microb Ecol Health Dis., 25*, 25862–. Available at http://www.ncbi.nlm.nih.gov/pmc/articles/PMC4179321/pdf/MEHD-25-25862.pdf (Accessed June 16, 2016).

Woappi, Y., Gabani, P., Singh, A. and Singh, O.V. (2016). Antibiotrophs: The complexity of antibiotic-subsisting and antibiotic-resistant microorganisms. *Crit Rev Microbiol., 42*, pp. 1549–7828.

Woappi, Y., Gabani, P. and Singh, O.V. (2013). Emergence of antibiotic-producing microorganisms in residential versus recreational microenvironments. *Br Microbiol Res J., 3*, 280–294.

12

Antimicrobial Food Additives and Disinfectants: Mode of Action and Microbial Resistance Mechanisms

*Meera Surendran Nair, Indu Upadhyaya, Mary Anne Roshni Amalaradjou, and Kumar Venkitanarayanan**

Department of Animal Science, University of Connecticut, Connecticut, USA

12.1 Introduction

Beyond the need for microbial control in human and animal medicine, there exists a significant requirement for pathogen exclusion in the food industry. In this regard, the food industry has been using a variety of non-antibiotic based antimicrobials, including additives and disinfectants for controlling foodborne spoilage and pathogenic microorganisms. In fact, some of these antimicrobial additives such as salt, nitrites and sulfites have been in use for 50–100 years. Despite the long history of usage of non-antibiotic antimicrobials in foods, there is little information on the development of microbial resistance to these antimicrobials (Davidson and Harrison, 2002). However, increasing incidence of antibiotic resistance among pathogens and the reliance on non-antibiotic based antimicrobials as major hurdles to microorganisms in foods raise concerns over the potential for development of microbial resistance to these non-antibiotics.

Resistance mechanisms exhibited by foodborne pathogens enable them to adapt and evolve with the changing environmental conditions. Hence emergence of resistance to food preservatives and disinfectants are of particular concern to the food industry. Furthermore, for these non-antibiotics to be effective control strategies against foodborne microorganisms, it is critical to understand the potential for resistance development and the underlying mechanisms that contribute to resistance. Therefore, this chapter will focus on the commonly used antimicrobial food preservatives and disinfectants, their mode of action, microbial resistance to these non-antibiotics, and the underlying resistance mechanisms. Further, this chapter will briefly discuss the potential application of these antimicrobials as alternatives to control antibiotic resistant foodborne pathogens.

12.2 Food Additives

The federal Food, Drug, and Cosmetic Act (FDCA) defines food additives as "any substance the intended use of which results or may reasonably be expected to result, directly or indirectly, in its becoming a component or otherwise affecting the characteristic of any food (including any substance intended for use in producing, manufacturing, packing, processing, preparing, treating, packaging, transporting, or holding food; and including any source of

* Address all correspondence to: kumar.venkitanarayanan@uconn.edu

Foodborne Pathogens and Antibiotic Resistance, First Edition. Edited by Om V. Singh.

Table 12.1 Indirect Antimicrobial Preservatives.

Indirect antimicrobial preservatives	Major functions in food	Mode of action	Application	References
Sodium chloride (Salt)	Flavorant, fruit, vegetable and meat preservative	Lower the water activity in foods, Potentiated the antimicrobial action of antimicrobial preservatives like potassium sorbate	Meat products, cured meat, fruit and vegetable products, seafood	Davidson *et al.*, 2005; Naidu, 2000; Taylor and Henny, 2010
Polyphosphates	Antioxidant in meat, emulsifier in dairy products, mineral supplementation in cereals	Chelate cations essential for microbial growth, leakage of essential minerals from cells, loss of osmoregulation, delayed botulin toxin production with sorbate and nitrite	Dairy Products, cereals, meat products	Davidson *et al.*, 2005; Naidu, 2000; Wagner and Busta, 1983
Alcohol	Solvent in flavoring extracts, fortification of wine	Denature proteins in the bacterial cytoplasm at high concentration	Wine, Packaged foods	Davidson *et al.*, 2005; Russell, 2003
Nitrates and nitrites	Flavoring agent	Inhibit glucose degradation enzymes and attach to the microbial dehydrogenase system, inhibitory to *C. botulinum*	Salami, Bacon, Beef jerky	Leuck, 1996; Woods *et al.*, 1981; Quastel and Woolridge, 1927

radiation intended for any such use); if such substance is not GRAS (generally recognized as safe) or sanctioned prior to 1958 or otherwise excluded from the definition of food additives" (FFDCA Section 201 s, 1958). In brief, food additives are substances added to food to enhance safety, maintain nutritional quality and improve texture, flavor, and appearance of food. It includes preservatives, sweeteners, color additives, flavors and spices, flavor enhancers, nutrient supplements, emulsifiers, texturing agents, acidulants, and enzymes (FDA, 2013). Of the above-mentioned additives, food preservatives, enzymes, and acidulants have been demonstrated to possess antimicrobial properties. This section of the chapter will briefly discuss antimicrobial additives and preservatives. Table 12.1 and 12.2 provide a comprehensive list of antimicrobial additives, their mode of action and application.

12.2.1 Antimicrobial Preservatives and Additives

One of the main challenges to food production is the maintenance of food quality and stability during a product's whole shelf life (Brul and Coote, 1999). In this regard, preservatives are primarily incorporated in foods to retard spoilage, enhance shelf stability, and preserve flavor. Besides their inhibitory action on spoilage microorganisms, many preservatives also exert antimicrobial action against foodborne pathogens, thereby preventing food poisoning and illnesses. Antimicrobial preservatives make the environment unfavorable for microbes to survive by bringing down the pH, increasing the salt content, and reducing the water activity. Utilization of preservatives like salt, sugar, vinegar and alcohol, or smoking, drying, freezing, and pickling are among the most common practices in food preservation (Inetianbor *et al.*, 2015). In addition, newer alternatives, including natural antimicrobials derived from plants, animals, and microbes have been explored for potential application in foods.

Antimicrobial preservatives can be placed under two categories, namely, direct and indirect preservatives. Direct antimicrobial preservatives include GRAS compounds that have a long history of use in foods, like salt, sorbates, and benzoates. The indirect antimicrobial preservatives comprise additives that are primarily added to foods to preserve flavor and color, but not primarily for microbial growth inhibition. However, these indirect preservatives also possess some antimicrobial action (Jay, 1995). Moreover, some additives potentiate the inhibitory effects of other preservatives rather than producing antimicrobial effects by themselves. These indirect antimicrobials can also hold GRAS status and mainly include antioxidants, flavor enhancers, emulsifiers, and chelating agents (Davidson *et al.*, 2005). Table 12.1 lists the common indirect antimicrobial preservatives and their conventional functions in the food as well as their mechanisms of indirect antimicrobial action. Similarly Table 12.2 lists the various direct antimicrobial preservatives and their mechanisms of action.

12.3 Mode of Action and Resistance to Antimicrobial Food Preservatives

In general, there are physical (sterilization, pasteurization, freezing, drying, and irradiation), chemical (direct and indirect antimicrobial preservatives) and biological (bacteriocins, protective cultures) methods of food preservation (Leuck *et al.*, 1980). In the context of this chapter, the mode of action and resistance mechanisms against chemical and biological preservatives along with examples of reported resistance development are described. Antimicrobial additives, including preservatives are found to act on DNA, protein synthesis, enzyme activity, cell wall, and cell membrane and nutrient transport mechanism of microorganisms. Occasionally, cumulative effects of these mechanisms could result in complete eradication of

Table 12.2 Direct Antimicrobial Preservatives.

Direct antimicrobial preservatives	Compounds	Mechanism of action	Application	References
Weak organic acids	Acetic acid, lactic acid, benzoic acid, sorbic acid	Membrane disruption, inhibition of essential metabolic reactions stress on intracellular pH homeostasis, accumulation of toxic anions	Meat products, pickles, sauces, ketchup, fruit sugar preserves, jams, jellies	Freese *et al.*, 1973; Krebs *et al.*, 1983; Salmond *et al.*, 1984; Eklund, 1985; Russell and Gould, 2003
Peroxides	Hydrogen peroxide	Generation of superoxide radicals, sporicidal	Milk, syrups	Toledo, 1975; Ananthaswamy and Eisenstark, 1977; Juven and Pierson, 1996
Plant-derived biomolecules	Oregano, thyme, clove, basil, cumin, caraway, isothiocynates of onion and garlic, carvacrol, thymol and trans-cinnamaldehyde	Inactivation of extracellular enzymes through oxidative cleavage of disulphide bonds, membrane disruption, reduction of the intracellular ATP content	Meat products, fruit and vegetable products, sausages	Delaquis and Mazza, 1995; Helander *et al.*, 1998; Hirasa and Takemasa, 1998; Koga *et al.*, 1999; Rai and Bai, 2014
Bacteriocins	Nisin, lacticin, pediocin, natamycin	Increase in the lag phase, impede cell wall biosynthesis	Meat products, dairy products, vegetable and fish products	Szabo and Cahill, 1998; Cleveland *et al.*, 2001; Ananou *et al.*, 2007
Enzymes	Lysozymes	Acts on the peptidoglycan layer of the cell membrane and causes cell lysis	Fresh fruits and vegetables, seafoods, meats and sausages, potato salad, tofu bean	Proctor *et al.*, 1988; Masschalck and Michiels, 2003

spoilage and pathogenic organisms in foods. In addition to the direct killing action, certain additives impede microbial growth by reducing water activity, altering pH of the system, or by preventing oxygen access to foods. Nevertheless, the efficiency of preservatives is influenced solely by dose-effect interaction. Therefore, adequate amounts of preservatives need to be applied in a practical food preservation system (Leuck *et al.*, 1980). It is also important to note that all preservatives are not effective against all spoilage and pathogenic organisms. Further the antimicrobial efficacy of a preservative is also influenced by several factors in the food, including pH, water activity, solubility, redox potential of the substrate, and food ingredients (Leuck *et al.*, 1980).

12.3.1 Chemical Preservatives

12.3.1.1 Weak Organic Acids

Mode of Action Organic acids such as acetic acid, lactic acid, benzoic acid, and sorbic acid are the most common preservatives used in the food industry. As depicted in Table 12.1, organic acids help preserve foods by inhibiting bacterial and fungal spoilage (Brul and Coote, 1999). The mechanism of action of these acids is primarily pH dependent. Weak organic acids at low pH exist in an undissociated state that enables permeation through the microbial plasma membrane. Upon entry into the cell, they encounter a higher pH, which leads to an intracellular accumulation of anions and protons (Booth and Kroll, 1989). This in turn results in membrane disruption (Stratford and Anslow, 1996; Bracey *et al.*, 1998), inhibition of essential metabolic reactions (Krebs *et al.*, 1983), potential stress on intracellular pH homeostasis (Cole and Keenan, 1987; Bracey *et al.*, 1998), and the accumulation of toxic anions (Eklund, 1985). Food acidulants such as the above-mentioned organic acids are found to be more effective against gram-positive than gram-negative bacteria, primarily due to the differences in the cell wall structure (Brul and Coote, 1999). In addition to their antimicrobial action against vegetative cells, organic acids have been reported to inhibit the germination and outgrowth of bacterial spores (Sofos and Busta, 1981; Blocher and Busta, 1985; Brul and Coote, 1999).

Resistance Mechanisms Although susceptible to organic acids, several foodborne pathogens demonstrate intrinsic and adaptive tolerance mechanisms to acid stress encountered during food processing. For example, in the case of enterohemorrhagic *Escherichia coli* (EHEC), a major acid tolerant foodborne pathogen, the tolerance traits are critical for its survival in various foods and in the bovine intestinal tract. Three distinct systems have been identified in the low pH–induced acid environment (Foster, 2004; Lin *et al.*, 1996). The sigma factor S (σS) and the global regulatory protein called cAMP receptor protein are required for acid tolerance at a pH of 2.5 in the oxidatively metabolizing bacterial cells. However, cells that have been grown in minimal media containing glucose are unable to survive when exposed to pH 2.5 unless the medium is supplemented either with glutamate (glutamate-dependent acid resistance) or arginine (arginine-dependent acid resistance). The last two systems are controlled by an array of amino acid decarboxylase and antiporters (Foster, 2004).

Similarly, *Salmonella* Typhimurium is known to possess an adaptive acid tolerance mechanism comprising σS-dependent acid shock proteins and pH homeostasis–dependent lysine decarboxylase and antiporters (Park *et al.*, 1996). Similar tolerance responses to acid challenge in *Listeria monocytogenes* have also been reported (O'Driscoll *et al.*, 1996). These researchers also found that the acid-adapted *Listeria* showed enhanced tolerance to thermal and osmotic stress. The increased tolerance to acid stress in *L. monocytogenes* is mediated by the glutamate decarboxylase (GAD) system (Cotter *et al.*, 2001; Gandhi and Chikindas, 2007) and a two-component regulatory system, LisR-LisK (Arenas *et al.*, 2013). The glutamate decarboxylase system facilitates intracellular pH homeostasis by neutralizing protons in a decarboxylation reaction that generates gamma-aminobutyrate from glutamate (Feehily and Karatzas, 2013).

Additionally, innate resistance to sorbic acid salts has been observed in bacteria such as *Sporolactobacillus*, certain *Pseudomonas* sp., molds, including *Aspergillus, Fusarium, Geotrichum, Mucor, Penicillium,* and yeasts like *Candida, Saccharomyces,* and *Torulopsis* (Davidson *et al.*, 2002). Bacteria, including *Bacillus, Corynebacterium, Micrococcus,* and mold such as *Aspergillus,* can degrade benzoates and benzoic acid to succinic acid and acetyl Coenzyme A (beta ketoadipate pathway). Moreover, enzymatic degradation of esters of p-hydroxybenzoic acid by *Pseudomonas aeruginosa* and organic acids such as sorbic acid by several foodborne fungi have been documented as effective acid-resistance mechanisms (Hugo and Foster, 1964; Samson *et al.*, 1995). Likewise, the acid resistance of some of the spoilage yeasts has been found to be dependent on specific H^+ -pumping ATPases (Holyoak *et al.*, 1996). In line with this, Piper *et al.* (1998) found that the activation of an ATP-binding cassette transporter, Pdr12 in *Saccharomyces cerevisiae* enabled the yeast to extrude the anions produced by the dissociated acids.

12.3.1.2 Peroxides

Mode of Action Hydrogen peroxide is one of the most common preservatives added to raw milk, whey, corn starch, dried eggs, fish marinades, and in aseptic food packaging (Russel, 1991; Brul and Coote, 1999). Besides its use as a preservative, peroxides are also employed as disinfectants in the food industry. Hydrogen peroxide (H_2O_2) is a known bactericidal agent co-dependent on the pH and temperature (Cords and Dychdala, 1993) of the food surface. These parameters play an important role especially in determining the sporicidal efficacy of hydrogen peroxide. Whereas H_2O_2 was found to be weakly sporicidal at room temperatures, it exerted a potent effect at higher temperatures (Toledo, 1975; Block, 2001). Higher concentrations of H_2O_2 (10 to 30%) and longer contact times are required for sporicidal activity (Russell, 1991), although this effect is significantly increased in the gaseous phase.

Whereas the sporicidal mechanism by which hydrogen peroxide kills spores is yet to be studied extensively, killing of vegetative bacteria (Anathaswamy and Eisenstark, 1977; Imlay and Linn, 1988) and fungi (Frankenberg *et al.*, 1993) is known to involve DNA damage. H_2O_2 acts as an oxidant by producing hydroxyl free radicals (•OH), which attack essential cell components, including lipids, proteins, and DNA. Moreover, it has been shown that exposed sulfhydryl groups and double bonds are particularly targeted (Block, 2001). H_2O_2 is considered environmentally friendly, because it can rapidly degrade into water and oxygen. Although pure solutions are generally stable, most contain stabilizers to prevent decomposition.

Additionally, the lactoperoxidase system in milk, requiring hydrogen peroxide and thiocyanate, has profound antimicrobial effects against bacteria and fungi (Russell, 1991; de Wit and van Hooydonk, 1996; Brul and Coote, 1999). When conditions are suitable, the lactoperoxidase reaction generates a short-lived singlet oxygen species, resulting in biocidal effects (reviewed in Tatsouzawa *et al.*, 1998). Moreover, incomplete reduction of molecular oxygen leads to generation of superoxide radicals that are antimicrobial. Furthermore, both gram-negative (Borch *et al.*, 1989; Wray and McLaren, 1987) and gram-positive bacteria (Oram and reiter, 1966; Siragusa and Johnson, 1989) are effectively inhibited by the lactoperoxidase system (Brul and Coote, 1999).

Resistance Mechanisms Most catalase-positive bacteria utilize the enzyme catalase to degrade hydrogen peroxide, thereby protecting themselves against the killing action of peroxide (Ma and Eaton, 1992). However, in the case of bacterial spores, small acid-soluble peptides produced during spore formation protect them from peroxide-mediated inactivation (Setlow, 1992). However, this protective effect varies between bacterial species. For example, the spore coats of *Bacillus* sp. are not found to be as resistant or efficient enough to survive peroxide stress as that of some *Clostridium* sp. (Bayliss and Waites, 1976). Conversely, yeast cells have several adaptive enzymatic and non-enzymatic antioxidant defense systems that are activated in response to prior exposure to peroxide and help overcome subsequent lethal stress. It involves dismutation of superoxide anions, decomposition of hydrogen peroxide, and inhibition of the Fenton reaction (Moradas-Ferreira *et al.*, 1996).

12.3.1.3 Plant-Derived Antimicrobials

Mode of Action Plant-derived compounds represent an important source of safe, effective, and natural antimicrobials. Plant extracts have traditionally been used as food preservatives, flavor enhancers, and dietary supplements to prevent food spoilage and improve health. The antimicrobial activity of several plant-derived compounds has been documented and their antimicrobial components have been characterized (Cowan, 1999; Savoia, 2012; Burt, 2004, Upadhyay *et al.*, 2014). At bactericidal concentrations, plant compounds disintegrate bacterial cell wall and cell membrane, causing leakage of cellular contents, disruption of proton motive force and enzyme dysfunction eventually causing cell death (Burt, 2004; Lewis and Ausubel, 2006).

The antimicrobial activity of plant-derived antimicrobials is primarily mediated by their hydrophobic components, which disrupt cell membrane proteins and associated structures, making the cells more permeable (Sikkema 1994; Knobloch, 1989). In addition, some plant compounds may stimulate the growth of pseudo-mycelia affecting the synthesis of bacterial structural components (Conner and Beuchat, 1984). More recently, research has revealed that plant-derived antimicrobials at low or sub-lethal concentrations affect microbial virulence by modulating the expression of critical genes and proteins (Goh *et al.*, 2002; Tsui *et al.*, 2004; Upadhyay *et al.*, 2012; Azizkhani *et al.*, 2013; Qiu *et al.*, 2010a,b; Amalaradjou *et al.*, 2014; Ananda Baskaran and Venkitanarayanan, 2014).

Resistance Mechanisms Although information on the resistance mechanisms against plant-derived antimicrobials is very limited, there are a few studies on the adaptation of bacteria to antimicrobials like carvacrol (Ultee *et al.*, 2000). When grown in the presence of non-lethal concentrations of carvacrol, *B. cereus* showed a decreased sensitivity to the antimicrobial. They found that adapted bacterial cells had increased cell membrane fluidity and fatty acid composition. They also noted a rapid increase in the phase transition temperature (T_m) while the cells were getting adapted to carvacrol. However, the increased addition of the antimicrobial to adapted cells gradually decreased the T_m to the normal range observed with non-adapted cells in the absence of carvacrol. Similarly, certain strains of *Vibrio parahaemolyticus* are found to be resistant to essential oils from basil and sage (Koga *et al.*, 1999). *V. parahaemolyticus* at the stationary growth phase was found to display a greater resistance to these essential oils compared to the cells at the exponential phase. Based on the background information, Koga *et al.* (1999) hypothesized that expression of several stationary phase survival genes and/or the non-growing conditions might have influenced the tolerance attribute. Likewise, *Staphylococcus aureus* was found to be resistant to berberine and palmitine due to its ability to actively extrude amphipathic cations (Hsieh *et al.*, 1998).

12.3.1.4 Sodium Chloride

Mode of Action Although table salt has been used as a food preservative for centuries, it still maintains its prominence among preservatives. In comparison to other preservatives, hardly any legal restrictions exist in the use of common salt due to its status as a vital dietary constituent (Leuck, 1980). However, owing to the unacceptable taste restrictions, it is practically unachievable to protect foods entirely by using salt alone from unwanted microbial growth.

Therefore, salt is used in combination with other physical and chemical preservative methods to achieve the desirable effect (Kushner, 1971). The principal mode of action of salt is associated with its ability to lower the water activity of foods. The resulting reduction in water activity renders the food system unfavorable for microbial life. Since this action is similar to drying, using salt in foods is often called as "chemical drying" (Leuck, 1980). Sinskey (1980) reported lethal accumulation of amino acids in bacteria when the water activity was lowered. Furthermore, studies showed that salt reduced the solubility of oxygen in water, thereby inhibiting the survival of aerobic organisms (Leuck, 1980).

Resistance Mechanisms Based on their inherent salt tolerance level, microorganisms are categorized as halophilic (up to 1–5%), moderately halophilic (5–20%), and extremely halophilic (up to 30%). Common examples of halotolerant species are *Torula* yeasts and *Staphylococcus* sp. Ganjian *et al.* (2012) and Haleh (2012) found that the antimicrobial resistance of *S. aureus* was increased when exposed to sublethal concentrations of salt. The response of most microorganisms to osmotic stress involves both physiological and genetic changes, called osmo-adaptation (Sleator and Hill, 2002). *Listeria* sp. are known to survive in a salt concentration of approximately 30%, whereas *Salmonella* is found to be highly sensitive to high salt environments. Sigma factor B (σ B) is one of the important general stress response element of *L. monocytogenes*, which also plays a significant role in salt adaptation (Gandhi and Chikindas, 2007). Bergholz *et al.* (2012) reported that short-term adaptation to salt stress increased the expression of carboxypeptidase, a peptidoglycan-modifying enzyme in *Listeria*. Additionally, salt shock proteins (Ssp) are found to be induced and overexpressed in *L. monocytogenes* during the adaptation period. Studies have also demonstrated that some stress acclimation proteins (Sap) continue to get expressed even after conditions return to normal.

12.3.1.5 Other Antimicrobial Additives

Polyphosphates Polyphosphates are mainly used in meat and dairy products to stabilize meat emulsions and to disperse proteins, respectively (Davidson *et al.*, 2005). Their antimicrobial action is attributed to their direct interaction with bacterial cell membrane (Wood *et al.*, 1988). It has also been shown that polyphosphates exert an antimicrobial action by chelating essential cations. Wagner and Busta (1985) noted inhibition of a toxin activating protease and reported an altered morphology of vegetative cells of *Clostridium botulinum* in the presence of polyphosphates. The antimicrobial action of polyphosphates depends on several contributing factors such as concentration, pH, nature of the substrate, combination with physical methods, and the length of polyphosphate chain. It was found that longer chain phosphates were more effective than shorter ones. Gram-positive bacteria and molds are found to be effectively inhibited, whereas gram-negative organisms displayed resistance to an extent (Lorencová *et al.*, 2012; Naidu, 2000). This higher susceptibility of gram-positive bacteria is attributed to their increased requirement for cations, which also serve as targets for polyphosphate binding in the cell membrane (Naidu, 2000).

Alcohol Several alcohols such as ethyl alcohol (ethanol, alcohol), isopropyl alcohol (isopropanol, propan-2ol), and n-propanol are widely used as food preservatives and flavoring extracts. Alcohols possess a broad spectrum of antimicrobial activity against vegetative bacteria, viruses, and fungi. However, their ability to inhibit sporulation and spore germination is found to be reversible (Trujillo and Laible, 1970; Mcdonnell and Russell, 1999). Their antimicrobial activity is mainly mediated by cell membrane damage, rapid protein denaturation, and cell lysis (Dagely *et al.*, 1950;

Sykes, 1939). The most effective antimicrobial alcohol concentration is found to be 60 to 70% and at this higher concentration, ethanol denatures the protoplasmic proteins of microorganisms (Russell, 2003; Davidson *et al.*, 2005). However, at a concentration of 10%, ethanol inhibits the growth of most microorganisms.

Due to the non-specific denaturing of protoplasmic proteins, alcohol is found to be effective against all types of microorganisms. In general bacteria are found to be more sensitive to alcohol than yeasts. Ethanol is also found to enhance the antimicrobial property of sorbic acid. For instance, ethanol at a concentration of 5% was able to sensitize bacteria to low pH and osmotic stress (Jay, 2012). On the other hand, *Streptococcus faecalis*, lactic acid bacteria and certain yeasts were found to be able to tolerate up to 20% alcohol. Although the information on resistance mechanisms against alcohol is limited, there are few reports on the development of phenotypic adaptation to alcohols (Chopra, 1987; Ingram, 1976). Ingram (1976) reported adaptive membrane alterations in *E. coli* K12 on exposure to alcohols. The fatty acid composition of the membrane was significantly altered when the bacteria were grown in the presence of alcohols (Ingram, 1976).

Nitrates and Nitrites Nitrates and nitrites are primarily added to meat products to enhance flavor, maintain characteristic pink color of cured meat, and prevent spoilage (Leuck, 1980). Furthermore, nitrite is also known to exert an inhibitory effect on *C. botulinum* and botulin toxin. The production of nitrous acid and oxides of nitrogen are responsible for this antimicrobial action. Additionally, nitrites are known to inhibit glucose degradation enzymes and attach to the microbial dehydrogenase system (Woods *et al.*, 1981; Quastel and Woolridge, 1927). However, since nitrite concentrations used in the meat industry (80–160 mg/kg) are not inhibitory to bacterial contaminants, common salt is added along with nitrite to lower the microbial load in foods.

12.3.2 Biological Preservatives

12.3.2.1 Bacteriocins

Bacteriocins are ribosomal synthesized antibacterial proteins secreted by bacteria that inhibit the growth of other bacteria. The wide array of known bacteriocins are mainly produced by lactic acid bacteria, and are found in several fermented and non-fermented food products (Cleveland *et al.*, 2001). Bacteriocins have a broad spectrum of action against gram-positive bacteria, including foodborne pathogens such as *L. monocytogenes* and *S. aureus*, and food spoilage microorganisms such as *Clostridium tyrobutyricum* (Deegan *et al.*, 2006; Ryan *et al.*, 1996; Galvin *et al.*, 1999). Their general mode of antimicrobial action involves pore formation in the membrane of targeted bacteria, depletion of transmembrane potential and the pH gradient eventually leading to the leakage of cellular components and cell death (Cleveland *et al.*, 2001).

12.3.2.2 Nisin

Mode of Action Nisin is an antimicrobial polypeptide produced by *Lactococcus lactis* subsp. Lactis that exhibits a narrow spectrum of inhibitory activity against gram-positive bacterial cells and spores. It is approved globally for use in foods. The susceptible bacterial population includes *Bacillus, Clostridium, Listeria*, and *Streptococcus* (Cleveland *et al.*, 2001; Davidson *et al.*, 2002). Nisin binds to phospholipids in the lipid bilayer and leads to pore formation in the bacterial cell membrane. This results in membrane disruption and extrusion of intracellular components, leading to depletion of the proton motive force and cell death (Cleveland *et al.*, 2001).

Resistance Mechanisms The resistance mechanism against nisin is mainly associated with the enzymatic degradation of the polypeptide or alteration in bacterial membrane susceptibility (Montville *et al.*, 2001). Hurst and Hoover (1993) reported the production of the enzyme, "nisinase" in *Streptococcus thermophilus, Lactobacillus plantarum* and certain *Bacillus* species that neutralizes the antimicrobial activity of nisin. Additionally, the development of nisin resistant *Liste*ria mutants was observed by Crandall and Montville (1998). These resistant strains had an altered cell membrane composition with a decrease in membrane anionic phospholipids (cardiolipin and phosphatidyl glycerol) and an increase in phosphatidylethanolamine content. This change in phospholipid content resulted in a decreased net negative charge, which hindered nisin binding. There are also reports on increased long chain fatty acid composition in cell membrane of nisin mutants, which in turn alter the cell fluidity and stabilization (Ming and Daeschel, 1993; Mazzotta and Montville, 1997).

12.3.2.3 Enzymes

Mode of Action The naturally occurring enzyme, lysozyme, is globally approved for use as a preservative in foods. Lysozyme is known to be inhibitory to Gram positive bacteria (Cagri *et al.*, 2004). Lysozyme acts mainly on the peptidoglycan layer of the bacterial cell membrane. Specifically, the antimicrobial activity is due to its N-acetylmuramoylhydrolase enzyme activity which results in the hydrolysis of peptidoglycan and lysis of cells (Masschalck and Michiels, 2003). However, the lipid bilayer of the Gram negative bacteria prevents the action of the enzyme (Doyle *et al.*, 2006).

Resistance Mechanism In order to survive the lysosomal activity, certain gram-positive bacterial species like *Neisseria gonorrhoeae, Proteus mirabilis,* and *S. aureus* alter the components of peptidoglycan post translationally (Clarke and Dupont, 1992). In *S. aureus*, resistance to lysozyme is attributed to the O-acetylation of the peptidoglycan layer, modifications on teichoic acids in the cell wall and higher level of peptide cross-linking (Bera *et al.*, 2007). Additionally, in other gram positives, the surface attachment polymers like capsular polysaccharides and the incorporation of D-Aspartic acid in the bacterial peptidoglycan cross bridge (Eg: *Lactococcus lactis*) impede lysozyme action. N-deacetylation of the acetamido group of the hexosamine residues of peptidoglycan and production of protein-inhibitors specific to lysozyme are the other possible mechanisms that could induce resistance to lysozyme (Benkerroum, 2008; Bera *et al.*, 2007). In addition, certain bacterial species such as *E. coli*, produce a lysozyme-binding protein which can inactivate the enzyme (Deckers *et al.*, 2004).

12.4 Disinfectants

Disinfection of food-processing premises and equipment is an integral part of the Good Hygienic Practice (GHP) (FDA, 2014). The primary aim of disinfection is to effectively eliminate microorganisms on food-contact surfaces, thereby preventing contamination of food products with pathogens and spoilage organisms (FDA, 2014). Disinfection eliminates many or all pathogenic microorganisms, except bacterial spores on inanimate objects, hence rendering it non-sporicidal (CDC, 2015). Based on the time of exposure and concentration of the disinfectant, a desired effect can be reasonably achieved. At higher concentrations but with shorter exposure periods, disinfectants will kill all microorganisms except large numbers of bacterial spores; they are called *high-level disinfectants* (CDC, 2008a). *Low-level disinfectants* can eliminate most vegetative bacteria, selected fungi, and few viruses in a practical period of time (≤10 minutes). *Intermediate-level disinfectants* might be cidal for mycobacteria, vegetative bacteria, most viruses, and most fungi, but do not necessarily kill bacterial spores (CDC, 2008a).

There are six major classes of disinfectants, namely quaternary ammonium compounds (QAC), halogens (iodophors, chlorine), peroxygens, aldehydes, phenols, and biguanides. Several factors can affect the potency of disinfectants commonly used in the food industry, which include

1) prior cleaning of the object
2) organic and inorganic load present
3) type and level of microbial contamination
4) concentration of and exposure time to the disinfectant
5) physical nature of the object (e.g., crevices, hinges, and lumens)
6) presence of biofilms
7) temperature and pH of the disinfection process and also
8) relative humidity of the sterilization process (e.g., ethylene oxide).

Disinfectants in general act through multiple, concentration dependent targets and result in cell wall and membrane destruction (Russell, 2003). Among the different factors that influence disinfectant efficacy, concentration plays a critical role. Biocides with higher concentration component (n or η : measures the effect of concentration or dilution on biocide activity); for example, phenols and alcohol lose their activity on dilution, whereas those with lower values like quaternary ammonium compounds (QAC), chlorhexidine, and glutaraldehyde can sustain their activity even at higher dilutions (Russell, 2003).

Based on the above factors and the mechanism of action, there exists a wide application of disinfectants in the food industry. Among the different disinfectant classes mentioned above, QACs, chlorine-releasing compounds (halogens), and peroxygens are most commonly used in the food industry. Table 12.3 summarizes the mode of action and application of the most common disinfectants used in the food industry.

12.5 Mode of Action and Resistance to Disinfectants

According to the Environmental Protection Agency (EPA, 2004), disinfectants are chemical agents used in hospitals, food-processing facilities, and restaurants to destroy or inactivate undesirable vegetative bacteria and fungi. Disinfectants have a multi-targeted mode of action. They can penetrate the gram-negative bacterial membrane by displacement of outer membrane cations and enter the gram-positive cells by passive diffusion. They are also known to inhibit protein synthesis, inactivate specific enzyme activity, and damage DNA (Russell, 2003). Due to this multi-pronged rapid action, mutations resulting in acquired resistance to biocides are less likely to happen (Doyle, 2006). However, exposure to biocides activates intrinsic resistance mechanisms that result in reduced uptake of biocides by altering the permeability of physical barriers (spores, mycobacteria) and induction of efflux pumps (Russell, 2003). The following section discusses the mode of action and microbial resistance mechanisms against different categories of disinfectants.

12.5.1 Quaternary Ammonium Compounds (QAC)

Mode of Action QACs or "Quats" are commonly used for surface disinfection in households, food plants, and as deodorizers. QACs are primarily surface-active agents (surfactants) having two regions in their molecular structures: a water-repellent (hydrophobic) hydrocarbon group and a water-attracting (hydrophilic or polar) group (McDonnell and Russell, 1999). They are non-corrosive and are frequently chosen as the disinfectant of choice at food plants (Holah *et al.*, 2002). They can be further classified based on the charge or absence of ionization of the hydrophilic group, as cationic, anionic, nonionic, and ampholytic (amphoteric) compounds. The most effective QAC compounds are the cationic agents, which are often referred to as exemplary quaternary ammonium compounds (QACs), and are the most useful antiseptics and disinfectants (Frier, 1971).

QACs are known to exhibit antimicrobial activity as membrane active agents (Hugo and Frier, 1969), thereby targeting predominantly the cytoplasmic (inner) membrane in bacteria

Table 12.3 Commonly Used Disinfectants in Food Industry.

Disinfectant	Mechanism of action	Major target groups/ organisms	Application	References
Quaternary ammonium salts (Quats)	Positively charged, attach to negatively charged bacterial membrane, leading to membrane leakage, initiation of autolysis.	Methanogens, *Staphylococcus aureus*, *Listeria monocytogenes*, *Escherichia coli*, *Salmonella*	Disinfectant, surfactant in food plants/ waste water plants	Hotchkiss, 1944; Ioannou *et al.*, 2007
Halogens Chlorine	Oxidation of thiol groups to disulfides, sulfoxides or disulfoxides Inhibition of DNA synthesis Inhibition of DNA synthesis	*Staphylococcus aureus*, *Listeria monocytogenes*, *Escherichia coli.*	Hard surface disinfection of food surfaces	McDonnell and Russell, 1999
Sodium hypochlorite (Hypochlorous acid)	Production of hypochlorite ion, disrupts oxidative phosphorylation and membrane associated activity	*Salmonella*, *Bacillus subtilis*	Hard surface disinfection	McDonnell and Russell, 1999
Peroxygens Hydrogen peroxide	Formation of free hydroxyl radicals (OH), which oxidize thiol groups in proteins and enzymes	Gram-positive bacteria (*L. monocytogenes*)	Disinfection, sterilization, antisepsis	McDonnell and Russell, 1999
Peracetic acid	Denaturation of proteins and enzymes and Increase cell wall permeability by disrupting sulfydryl and sulfur bonds.		Low temperature, liquid sterilization	Baldry and Fraser, 1988

or the plasma membrane in yeasts (Maillard, 2002, Ikeda *et al.*, 1984). Moreover, Salton (1968) proposed a sequence of events to elaborate the effect of QACs on microorganisms, which included: (a) adsorption and penetration of the agent into the cell wall; (b) reaction with the cytoplasmic membrane (lipid or protein) followed by membrane disorganization; (c) leakage of intracellular low-molecular-weight material; (d) degradation of proteins and nucleic acids; and (e) wall lysis caused by autolytic enzymes (Salton, 1968; Mcdonnell and Russell, 1999). These events lead to loss of structural organization and integrity of the bacterial cytoplasmic membrane, thereby causing severe damage to bacteria (Denyer, 1995). QACs have also been studied as sporostatic agents that inhibit the development of a vegetative cell from a germinated spore but not the actual germination processes itself (development from dormancy to a metabolically active state (Russell, 1990).

Resistance Mechanisms Bacterial resistance mechanisms elaborated against QACs are well known (Chapman, 2003). Research studies indicate that the majority of QAC-resistant isolates have membrane-bound, proton-motive efflux pumps that contribute to the resistance to cationic surfactants. These efflux pumps are found in both gram-positive and gram-negative bacteria. These pumps are found to actively extrude QACs, which leads to resistance (Chapman, 2003). In a study conducted by Aase *et al.* (2000), a proton motive force–dependent efflux pump was expressed in benzalkonium chloride resistant *Listeria* strains. Heir *et al.* (1995, 1999) reported resistance development to QAC in *Staphylococci*, which was also mediated through a multi-drug efflux pump.

In addition to the proton motive pumps, presence of integrons bearing QAC-resistant genes in different bacterial populations have been implicated in resistance development. These integrons are found to be transferred among bacterial populations by horizontal transfer (Lucey *et al.*, 2000). Furthermore, Lemaitre *et al.* (1998) isolated strains of *L. monocytogenes* from poultry, which were resistant to benzalkonium chloride. It was identified that the plasmid-borne resistance in these strains was transferred to other *Listeria sp.* and *S. aureus*. Plasmid-mediated antiseptic and disinfectant resistance against QACs is also reported in several other bacterial species, including *E. coli*, *P. aeruginosa*, *S. aureus*, and *B. cepacia* (Rossouw and Rowbury, 1984; Sutton and Jacoby, 1978; Heir *et al.*, 1995; Gillespie *et al.*, 1989; Shields *et al.*, 1995). Gram-negative and gram-positive bacteria, mycobacteria, and bacterial spores are also reported to show some intrinsic resistance to QACs (Mcdonnell and Russell, 1999). The outer membrane of gram-negative bacteria, waxy cell wall of mycobacteria, spore coat, and cortex of bacterial spores act as barriers that limit the intracellular entry of QACs (McDonnell and Russell, 1999).

12.5.2 Chlorine and Chlorine-Releasing Compounds (Halogens)

Mode of Action Chlorine and chlorine-releasing compounds are the most commonly employed disinfectants in the food industry. Many studies and reviews have highlighted the chemical, physical, and microbiological properties of chlorine and chlorine releasing agents (Bloomfield, 1996; Dychdala, 1991; McDonnell and Russell, 1999). Some of the important types of chlorine-releasing agents include sodium hypochlorite, chlorine dioxide, and chloramine-T (N-chloro tosylamide).

Sodium hypochlorite is the household bleach that is most commonly used for hard surface disinfection (McDonnell and Russell, 1999). Sodium hypochlorite ionizes to produce Na^+ and hypochlorite ion, OCl^- and establishes an equilibrium with the hypochlorous acid (HOCl) in aqueous solvents. Hypochlorous acid is considered as the active moiety responsible for bacterial inactivation by chlorine compounds (Bloomfield, 1996; Dychdala, 1991). Therefore, chlorine-releasing compounds act as potent oxidizing agents that denature cellular proteins and impede metabolic activity (Bloomfield, 1996). By forming chlorinated derivatives of nucleotide bases, chlorine compounds inactivate the bacterial DNA as well (Dennis *et al.*, 1979, Dukan and Touati, 1996). Additionally, at higher concentrations, chlorine reactive agents are found to be sporicidal and virucidal. As a result of the biocidal treatment, spores lose their refractivity, the spore coat gets separated from the cortex, and eventually lysis occurs (Kulikovsky *et al.*, 1975).

Between pH 4 and 7, chlorine exists predominantly as HClO, the active moiety, whereas above pH 9, OCl_2 predominates. Moreover, chlorine-releasing agents are potent oxidizing agents destroying the cellular activity of proteins (Bloomfield, 1996). When the percentage of undissolved HOCl is the highest, there is maximum antimicrobial activity (Dychdala, 1991). Moreover, hypochlorous acid has also been found to disrupt oxidative phosphorylation (Barette Jr. *et al.*, 1989) and other membrane-associated activity (Campers and MCfetters, 1979). However, chlorine dioxide is predominately associated with the inhibition of bacterial protein synthesis (Benarde *et al.*, 1967). Chlorine-based/releasing agents are one of the few disinfectants, which at higher concentrations are sporicidal (Bloomfield and Arthur, 1992).

Resistance Mechanisms Under varying circumstances, bacteria, fungi, virus, and protozoa were found to be resistant to hypochlorite (McDonnell and Russell, 1999). Generally vegetative cells are less resistant than bacterial spores. Studies have revealed the development of varying degrees of resistance associated with *coli-aerogenes* when disinfected with chlorine (Heathman *et al.*, 1936). In case of bacterial spores, the altered molecular configuration of spore coat proteins protected the sulfhydryl groups from the action of the biocide (Odlaug, 1981). Stewart and Olson (1992) and Lisle *et al.* (1998) showed that when starved for carbon, gram-negative bacteria like *E. coli* and *Klebsiella* are more tolerant to chlorine. Likewise, during the different stages of poultry processing, Mokgatla *et al.* (1998) isolated *Salmonella* strains resistant to chlorine. According to these authors, this increased resistance to chlorine was due to the enhanced production of catalase enzyme, reduced activity of membrane-bound dehydrogenases, and decreased DNA damage.

12.5.3 Peroxygens

Mode of Action

Hydrogen Peroxide Hydrogen peroxide (H_2O_2) is a commonly used biocide for disinfection, sterilization, and antisepsis. It is commercially available in a variety of concentrations ranging from 3 to 90%. H_2O_2 demonstrates a broad-spectrum efficacy against viruses, bacteria, yeasts, and bacterial spores (Block, 2001). Additional details on the mechanism of action and resistance mechanisms of peroxide agents have been discussed under the preservatives section.

Peracetic Acid Peracetic acid (PAA) (CH_3COOOH) is another potent disinfectant more effective than hydrogen peroxide. It is known as a sporicidal, bactericidal, virucidal, and fungicidal agent at low concentrations (0.3%) (Block, 2001). PAA decomposes to safe byproducts (acetic acid and oxygen), but has the added advantages of being free from decomposition by peroxidases, unlike H_2O_2, and remaining active in the presence of organic loads (Malchesky, 1993; Srey *et al.*, 2013). Similar to H_2O_2, PAA presumably denatures proteins and enzymes and increases cell wall permeability by disrupting sulfhydryl and sulfur bonds (Baldry and Fraser, 1988). Several studies have shown the effectiveness of PAA against various microorganisms (Marques *et al.*, 2007; Salvia *et al.*, 2011), including biofilms of *L. monocytogenes*, *S. aureus* and *P. aeruginosa* (Cabeca *et al.*, 2008; Tote *et al.*, 2010).

Resistance Mechanisms Since the antibacterial action of peroxygens is mediated by the generation of oxygen radicals, the resistance mechanism observed in surviving microbes is attributed to the expression of their cellular defense systems (Greenberg, 1991; Dukan and Touati, 1996). For instance, in *E. coli*, in response to free radicals, a multigene system involving *soxRS* and *oxyR* is activated resulting in the synthesis of catalase, superoxide dismutase, and alkyl hydro-peroxidases. These enzymes degrade free radicals and help repair radical-induced cellular damage (Demple, 1996). In addition, phenotypic tolerance endowed by certain bacterial populations in growing biofilms help the cells to survive radical induced damage when exposed to peroxides (Chapman, 2003).

12.6 Plant-Derived Antimicrobials as Alternatives

Since the early discovery of antibiotics, there has been a rising concern of resistance development in microorganisms. The excessive and inappropriate use of antibiotics in medicine and agriculture has led to the emergence of antimicrobial resistant bacteria. The situation is further worsened by the scientific challenges, time, and cost associated with the development of new and effective antimicrobials. Furthermore, the food chain has been considered as the main route of transmission of antibiotic-resistant bacteria between animal and human population (Witte, 1997; Mathur and Singh, 2005). Therefore there is a crucial need for the development of safe and effective alternatives to antibiotics for use in animal agriculture and food production. Among the different non-antibiotics discussed in this chapter, plant-derived antimicrobials could serve as potential alternatives to antibiotics.

- *Food animals:*
  Plant-derived antimicrobials have been evaluated as alternatives to antibiotics for both growth promotion and disease prevention in food animals. Several studies have demonstrated the efficacy of in-feed supplementation of plant compounds such as trans-cinnamaldehyde, eugenol, carvacrol, and thymol in reducing the colonization of foodborne pathogens, including *Salmonella* Enteritidis and *Campylobacter* sp. in layer and broiler chicken (Upadhyaya *et al.*, 2015; Kollanoor-Johny *et al.*, 2012a, b; Arsi *et al.*, 2014; Wagle *et al.*, 2014). Similar results have been obtained with other foodborne pathogens. For example, purple prairie clover (Dalea purpurea Vent.) was effective in reducing the fecal shedding of *E. coli* O157:H7 in lambs (Huang *et al.*, 2015). Similarly, essential oils thymol, carvacrol, trans-cinnamaldehyde, and capsaicin were found to be effective in decreasing the colonization of *Clostridium perfringens* and pathogenic *E. coli* in broilers (Jamroz *et al.*, 2003, 2005; Mitsch *et al.*, 2004).

  Salmonella enterica serovar Heidelberg (SH) is one of the most commonly detected *Salmonella enterica* serovars in swine and poultry (Foley *et al.*, 2008; CDC, 2008b). SH strains resistant to multiple antibiotics, including streptomycin, kanamycin, ceftriaxone, ciprofloxacin, and other beta-lactam antibiotics have been isolated from poultry meat. Plant compounds have been found effective in reducing the colonization of drug-resistant strains of SH in poultry. Administration of blends of essential oil (carvacrol, thymol, eucalyptol, and lemon) in drinking water at (0.05%) was found to reduce the cecal colonization of the pathogen in broilers (Alali *et al.*, 2013). Moreover, the essential oil blend also increased weight gain and improved feed conversion ratio in the birds. Similar results were obtained with a blend of thymol and cinnamaldehyde (Amerah *et al.*, 2012). Interestingly, these researchers also observed a decrease in the horizontal transfer of the pathogen between birds, which plays a crucial role in the spread, and survival of pathogens in farms.

  In addition to their efficacy in reducing pathogens colonization and/or survival in food animals, plant compounds such as beta-resorcylic acid, caprylic acid, sodium caprylate, and *trans*-cinnamaldehyde have been effective in reducing pathogens on cattle hide (Baskaran *et al.*, 2013), in cattle drinking water (Amalaradjou *et al.*, 2006; Charles *et al.*, 2008), and on beef (Oussalah *et al.*, 2006). In another study, Wells and colleagues (2005) reported that addition of 0.5% *trans*-cinnamic acid and 0.5% *para*-coumaric acid reduced the survival of *E. coli* O157:H7 in bovine feces. These studies suggest that plant compounds can be used as alternatives to antibiotics to control pathogens in food animals and the farm environment.

- *Environment:*
  Antimicrobial resistance in human pathogens is linked to resistance reservoirs present in farms animals and also in the wider environment. Human microbial pathogens represent only a small fraction of the extensive diversity of bacteria that are exposed to antibiotics (Woolhouse and Gaunt, 2007). The vast majority include microbiota of farm animals as well as the large numbers present in the environment such as soil (Hollowell, 2015), activated sludge (Guo and Zhang, 2012), wastewater effluent plants (Iweriebor *et al.*, 2015), and lakes (Czekalski, 2013). While significant research is required to test the application of plant compounds in these niches, evidence exists on the efficacy of plant-derived antimicrobials against bacterial biofilms in the environment. Bacterial biofilms provide the microbes with better survival opportunities through enhanced resistance toward environmental stresses and antimicrobials (Penesyan *et al.*, 2015). Therefore, this mode of microbial life has to be given due consideration in the process of developing novel strategies to combat drug resistant microbes.

  Plant compounds have been shown to exhibit significant anti-biofilm efficacy against major nosocomial pathogens including *S. aureus* (Al-Bakri *et al.*, 2010; Nostro *et al.*, 2007; Walencka *et al.*, 2007) *P. aeruginosa* (Wu *et al.*, 2011), and foodborne pathogens such as *Cronobacter sakazakii* (Amalaradjou *et al.*, 2011a, b), *E. coli* O157:H7 (Lee *et al.*, 2011, 2013, 2014), and *L. monocytogenes* (Upadhyay *et al.*, 2013; Desai *et al.*, 2012). Thus, plant compounds could be used as an eco-friendly biosanitizer for controlling pathogens in environments such as farms or food-processing areas, thereby thwarting infection risks and antibiotic resistance development.

12.6.1 Role of Plant-Derived Antimicrobials in Reducing Antibiotic Resistance

In addition to their ability to control antibiotic resistant pathogens, plant compounds have been recently identified to increase antibiotic sensitivity in resistant microorganisms. The efficacy of plant compounds in increasing the sensitivity of microbes to antibiotics could be due to their efficacy in modulating the drug-resistance strategies of microbes. Inhibition of efflux pumps (Nikaido, 1994), modification of antibiotic target sites (Spratt, 1994), and enzymatic degradation of antimicrobials (Davies, 1994) are the major strategies employed by microbes to resist the action of antibiotics. Efflux pumps reduce cytoplasmic drug accumulation by pumping out compounds out of the inner membrane to the periplasmic space or directly to the external environment (Horiyama *et al.*, 2010). In gram-positive bacteria, the main efflux pumps belong to the multidrug and toxic compound extrusion (MATE) family (Alvarez-Ortega *et al.*, 2013). In contrast, in gram-negative microbes, the resistance-nodulation-division (RND) family that forms an efficient channel between inner membrane, periplasmic space and outer membrane, along with AcrAb-TolC pumps, helps in extruding antimicrobial compounds (Horiyama *et al.*, 2010; Alvarez-Ortega *et al.*, 2013).

Several studies have highlighted the efficacy of plant-derived antimicrobials in reducing efflux pump activity in gram-positive and gram-negative bacteria (Cherigo *et al.*, 2008; Kumar *et al.*, 2008; Holler *et al.*, 2012; Piddock, 2006). In a recent study, five plant-derived compounds, namely, trans-cinnamaldehyde, carvacrol, thymol, eugenol, and Beta-resorcylic acid were found to increase the sensitivity of *Salmonella* Typhimurium DT104 to five antibiotics (Johny *et al.*, 2010). Since the mechanism of antimicrobial resistance in *S.* Typhimurium DT104 is primarily mediated by interaction between specific antibiotics and AcrAB-TolC efflux pump, the aforementioned plant compounds could be acting through modulation of these efflux pumps to increase bacterial sensitivity to antibiotics (Quinn *et al.*, 2006).

Beta-lactamase is an enzyme produced by drug-resistant microbes that imparts resistance against -lactam antibiotics (Frere, 1995). Several plant compounds that exert an inhibitory effect on beta-lactamases have been identified (Jiminez-Valera *et al.*, 1987). *Garcinia lucida* and *Bridelia micrantha* are medical plants found in Cameroon. Extracts from these plants have been shown to exert significant anti beta-lactamase activity (Gangoué-Piébojiand *et al.*, 2009). Similarly, epigallocatechin gallate was found to increase the sensitivity of *S. aureus* to penicillin and augment the antimicrobial efficacy of ampicillin and sulbactam against methicillin resistant *S. aureus* (Zhao *et al.*, 2003).

12.7 Conclusion

Although most discussions on resistance and resistance mechanisms are targeted towards antibiotics, it is also critical to examine microbial resistance to other antimicrobials. With particular reference to the food industry, a significant amount of antimicrobials in use are non-antibiotics. These antimicrobials serve as primary hurdles to control pathogens in foods. Therefore, it is crucial that the food industry is aware of the potential for development of resistance to these antimicrobials and the underlying mechanisms. This knowledge is imperative to the development and application of effective control strategies to limit the potential survival and growth of foodborne pathogens. In this regard, more research is required to determine (a) the occurrence and mechanism of antimicrobial resistance in food systems, (b) development of potential alternatives to current intervention strategies, and (c) development and application of strategies aimed at reducing resistance development (Davidson and Harrison, 2002). The availability of next-generation sequencing tools coupled with advancements in microbial genomics, transcriptomics, and metabolomics would enable scientists to better understand the resistance mechanisms and devise potential novel compounds as antimicrobials for sustained application in the food industry.

References

Aase, B., *et al.* (2000). Occurrence of and a Possible Mechanism for Resistance to a Quaternary Ammonium Compound in *Listeria Monocytogenes. International Journal of Food Microbiology*, *62*(1), pp. 57–63.

Alali, W.Q., *et al.* (2013). Effect of essential oil compound on shedding and colonization of *Salmonella enterica* serovar Heidelberg in broilers. *Poultry Science*, *92*(3), pp. 836–841.

Al-Bakri, A.G., Othman, G. and Afifi, F.U. (2010). Determination of the antibiofilm, antiadhesive, and anti-MRSA activities of seven Salvia species. *Pharmacognosy Magazine*, *6*(24), pp. 264.

Alvarez-Ortega, C., Olivares, J. and Martínez, J.L. (2013). RND multidrug efflux pumps: what are they good for? *Frontiers in Microbiology*, (*4*), pp. 1–11

Amerah, A.M., Mathis, G. and Hofacre, C.L. (2012). Effect of xylanase and a blend of essential oils on performance and *Salmonella* colonization of broiler chickens challenged with *Salmonella Heidelberg. Poultry Science*, *91*(4), pp. 943–947.

Amalaradjou, M.A.R., *et al.* (2006). Inactivation of *Escherichia coli* O157: H7 in cattle drinking water by sodium caprylate. *Journal of Food Protection*, *69*(9), pp. 2248–2252.

Amalaradjou, M.A.R., Narayanan, A. and Venkitanarayanan, K. (2011a).

Trans-cinnamaldehyde decreases attachment and invasion of uropathogenic *Escherichia coli* in urinary tract epithelial cells by modulating virulence gene expression. *The Journal of Urology, 185*(4), pp. 1526–1531.

Amalaradjou, M.A.R. and Venkitanarayanan, K. (2011b). Effect of trans-cinnamaldehyde on inhibition and inactivation of *Cronobacter sakazakii* biofilm on abiotic surfaces. *Journal of Food Protection, 74*(2), pp. 200–208.

Amalaradjou, M.A.R., Kim, K.S. and Venkitanarayanan, K. (2014). Sub-inhibitory concentrations of trans-cinnamaldehyde attenuate virulence in *Cronobacter sakazakii* in vitro. *International Journal of Molecular Sciences, 15*(5), pp. 8639–8655.

Ananda Baskaran, S. and Venkitanarayanan, K. (2014). Plant-derived antimicrobials reduce *E. coli* O157: H7 virulence factors critical for colonization in cattle gastrointestinal tract in vitro. *BioMed Research International, 2014*, pp. 1–10.

Ananthaswamy, H.N. and Eisenstark, A. (1977). Repair of hydrogen peroxide-induced single-strand breaks in *Escherichia coli* deoxyribonucleic acid. *Journal of Bacteriology, 130*(1), pp. 187–191.

Ananou, S., et al. (2007). Biopreservation, an ecological approach to improve the safety and shelf-life of foods. *Communicating Current Research and Educational Topics and Trends in Applied Microbiology, 1*, pp. 475–486.

Arenas, J., et al. (2013). Involvement of three meningococcal surface-exposed proteins, the heparin-binding protein NhbA, the α-peptide of IgA protease and the autotransporter protease NalP, in initiation of biofilm formation. *Molecular Microbiology, 87*(2), pp. 254–268.

Arsi, K., et al. (2014). The efficacy of the natural plant extracts, thymol and carvacrol against *Campylobacter* colonization in broiler chickens. *Journal of Food Safety, 34*(4), pp. 321–325.

Azizkhani, M., et al. (2013). Effects of Zataria multiflora Boiss. essential oil on growth and gene expression of enterotoxins A, C and E in *Staphylococcus aureus* ATCC 29213. *International Journal of Food Microbiology, 163*(2), pp. 159–165.

Barrette Jr, W.C., et al. (1989). General mechanism for the bacterial toxicity of hypochlorous acid: abolition of ATP production. *Biochemistry, 28*(23), pp. 9172–9178.

Baskaran, S.A., et al. (2013). Inactivation of *Escherichia coli* O157: H7 on Cattle Hides by Caprylic Acid and β-Resorcylic Acid. *Journal of Food Protection, 76*(2), pp. 318–322.

Bayliss, C.E. and Waites, W.M. (1976). The Effect of Hydrogen Peroxide on Spores of *Clostridium bifermentans. Journal of General Microbiology, 96*(2), pp. 401–407.

Benarde, M.A., Snow, W.B. and Olivieri, V.P. (1967). Chlorine dioxide disinfection temperature effects. *Journal of Applied Bacteriology, 30*(1), pp. 159–167.

Benkerroum, N. (2008). Antimicrobial activity of lysozyme with special relevance to milk. *African Journal of Biotechnology, 7*(25), pp. 4856–4867.

Bera, A., et al. (2007). Influence of wall teichoic acid on lysozyme resistance in *Staphylococcus aureus. Journal of Bacteriology, 189*(1), pp. 280–283.

Bergholz, T.M., et al. (2012). *Listeria monocytogenes* shows temperature-dependent and-independent responses to salt stress, including responses that induce cross-protection against other stresses. *Applied and Environmental Microbiology, 78*(8), pp. 2602–2612.

Blocher, J.C. and Busta, F.F. (1985). Multiple modes of inhibition of spore germination and outgrowth by reduced pH and sorbate. *Journal of Applied Bacteriology, 59*(5), pp. 469–478.

Block, S.S. (2001). Peroxygen compounds. *Disinfection, sterilization, and preservation*. Philadelphia: Lippincott Williams & Wilkins, pp. 185–204.

Bloomfield, S.F. (1996). Chlorine and iodine formulations. *Handbook of disinfectants and antiseptics*. New York: Marcel Dekker, pp. 133–58.

Bloomfield, S.F. and Arthur, M. (1992). Interaction of Bacillus subtilis spores with sodium hypochlorite, sodium dichloroisocyanurate and chloramine-T. *Journal of Applied Bacteriology, 72*(2), pp. 166–172.

Booth, I.R., Kroll, R.G. and Gould, G.W. (1989). The Preservation of Foods by Low pH. *Mechanisms of Action of Food Preservation Procedures*. London: Elsevier Applied Science Publishers Ltd, pp. 119–160.

Borch, E., et al. (1989). Antibacterial effect of the lactoperoxidase/thiocyanate/hydrogen peroxide system against strains of *Campylobacter* isolated from poultry. *Journal of Food Protection, 52*(9), pp. 638–641.

Bracey, D., Holyoak, C.D. and Coote, P.J. (1998). Comparison of the inhibitory effect of sorbic acid and amphotericin B on *Saccharomyces cerevisiae*: is growth inhibition dependent on reduced intracellular pH. *Journal of Applied Microbiology, 85*(6), pp. 1056–1066.

Baldry, M.G.C. and Fraser, J.A.L. (1988). Disinfection with peroxygens. *Crit. Rep. Appl. Chem., 22*, pp. 91–116.

Brul, S. and Coote, P. (1999). Preservative Agents in Foods: Mode of Action and Microbial Resistance Mechanisms. *International Journal of Food Microbiology, 50*(1), pp. 1–17.

Burt, S. (2004). Essential oils: their antibacterial properties and potential applications in foods—a review. *International Journal of Food Microbiology, 94*(3), pp. 223–253.

Cabeça, T.K., Pizzolitto, A.C. and Pizzolitto, E.L. (2008). Assessment of action of disinfectants against *Listeria monocytogenes* biofilms. *Alimentos e Nutrição Araraquara, 17*(2), pp. 121–125.

Cagri, A., Ustunol, Z. and Ryser, E.T. (2004). Antimicrobial Edible Films and Coatings. *Journal of Food Protection, 67*(4), pp. 833–848.

Camper, A.K. and Mcfeters, G.A. (1979). Chlorine injury and the enumeration of waterborne coliform bacteria. *Applied and Environmental Microbiology, 37*(3), pp. 633–641.

CDC. (2008a). Centers for Disease Control and Prevention. Guideline for disinfection and sterilization. Available at: http://www.cdc.gov/hicpac/pdf/guidelines/Disinfection_Nov_2008.pdf [Accessed November 28, 2015].

CDC. (2008b). Centers for Disease Control and Prevention 2008. Salmonella surveillance: annual summary, 2006. Atlanta, GA: U.S. Department of Health and Human Services.

CDC. (2015). Centers for Disease Control and Prevention - Disinfection of Medical Devices - HAI. Available at: http://www.cdc.gov/HAI/prevent/sd_medicalDevices.html [Accessed November 28, 2015].

Chapman, J.S. (2003). Disinfectant resistance mechanisms, cross-resistance, and co-resistance. *International biodeterioration & Biodegradation, 51*(4), pp. 271–276.

Charles, A.S., et al. (2008). Reduction of *Escherichia coli* O157: H7 in cattle drinking-water by trans-cinnamaldehyde. *Foodborne Pathogens and Disease, 5*(6), pp. 763–771.

Chérigo, L., et al. (2008). Inhibitors of Bacterial Multidrug Efflux Pumps from the Resin Glycosides of Ipomoea murucoides. *Journal of Natural Products, 71*(6), pp. 1037–1045.

Chopra, I. (1987). Microbial resistance to veterinary disinfectants and antiseptics. Disinfection in veterinary and farm animal practice. Oxford: Blackwell Scientific Publications Ltd., pp. 43–65.

Clarke, A.J. and Dupont, C. (1992). O-Acetylated Peptidoglycan: Its Occurrence, Pathobiological Significance, and Biosynthesis. *Canadian Journal of Microbiology, 38*(2), pp. 85–91.

Cleveland, J., et al. (2001). Bacteriocins: safe, natural antimicrobials for food preservation. *International Journal of Food Microbiology, 71*(1), pp. 1–20.

Conner, D., and Beuchat, L.R. (1984). Effects of essential oils from plants on growth of food spoilage yeasts. *Journal of Food Science, 49*(2), pp. 429–434.

Cole, M.B. and Keenan, M.H.J. (1987). Effects of weak acids and external pH on the intracellular pH of *Zygosaccharomyces bailii*, and its implications in weak-acid resistance. *Yeast, 3*(1), pp. 23–32.

Cords, B.R. and Dychdala, G.R. (1993). Sanitizers: halogens, surface-active agents, and peroxides. *Food Science and Technology*. New York: Marcel Dekker, pp. 469–469.

Cotter, P.D., GM Gahan, C.G.M. and Hill, C. (2001). A glutamate decarboxylase system protects *Listeria monocytogenes* in gastric fluid. *Molecular Microbiology, 40*(2), pp. 465–475.

Cowan, M.M. (1999). Plant products as antimicrobial agents. *Clinical Microbiology Reviews, 12*(4), pp. 564–582.

Crandall, A.D. and Montville, T.J. (1998). Nisin resistance in *Listeria monocytogenes* ATCC 700302 is a complex phenotype. *Applied and Environmental Microbiology, 64*(1), pp. 231–237.

Czekalski, N. (2013). Sources, Spreading and Fate of Antibiotic Resistance Genes and Resistant Bacteria in Vidy Bay, Lake Geneva, Switzerland. Ph.D Thesis. Swiss Federal Institute of Technology.

Dagley, S., Dawes, E.A. and Morrison, G.A. (1950). Inhibition of growth of *Aerobacter aerogenes*: the mode of action of phenols, alcohols, acetone, and ethyl acetate. Journal of Bacteriology, *60*(4), p. 369–.

Davidson, P. M., and Harrison, M.A. (2002). Resistance and adaptation to food antimicrobials, sanitizers, and other process controls. *Food Technology, 56*(11), pp. 69–78.

Davidson, P.M., Juneja, V.K. and Branen, J.K. (2002). Antimicrobial agents. *Food Science and Technology, 2002*, pp. 563–620.

Davidson, P.M, Sofos, J.N., and Branen, A.L., eds. (2005). *Antimicrobials in Food*. New York: CRC Press.

Davies, J. (1994). Inactivation of antibiotics and the dissemination of resistance genes. *Science, 264*(5157), pp. 375–382.

Deckers, D., et al. (2004). Periplasmic Lysozyme Inhibitor Contributes to Lysozyme Resistance in *Escherichia Coli. Cellular and Molecular Life Sciences CMLS, 61*(10), pp. 1229–1237.

Deegan, L.H., Cotter, P.D., Hill, C., and Ross, P. (2006). Bacteriocins: biological tools for bio-preservation and shelf-life extension. *International Dairy Journal, 16*(9), pp. 1058–1071.

Delaquis, P.J. and Mazza, G. (1995). Antimicrobial Properties of Isothiocyanates in Food Preservation. *Food Technology, 49*(11), pp. 73–84.

Demple, B. (1996). Redox signaling and gene control in the *Escherichia coli* soxRS oxidative stress regulon—a review. *Gene, 179*(1), pp. 53–57.

Dennis, W.H., Olivieri, V.P. and Kruse, C.W. (1979). The reaction of nucleotides with aqueous hypochlorous acid. *Water Research, 13*(4), pp. 357–362.

Denyer, S.P. (1995). Mechanisms of action of antibacterial biocides. *International Biodeterioration & Biodegradation, 36*(3), pp. 227–245.

De Wit, J.N. and Van Hooydonk, A.C.M. (1996). Structure, functions and applications of lactoperoxidase in natural antimicrobial systems. *Nederlands melk en Zuiveltijdschrift, 50*(2), pp. 227–244.

Desai, M.A., et al. (2012). Reduction of Listeria monocytogenes biofilms on stainless steel and polystyrene surfaces by essential oils. *Journal of Food Protection, 75*(7), pp. 1332–1337.

Doyle, M.P., et al. (2006). Antimicrobial Resistance: Implications for the Food System. *Compr. Rev. Food Sci. Food Saf., 5*, pp. 71–137.

Dukan, S. and Touati, D. (1996). Hypochlorous acid stress in *Escherichia coli*: resistance, DNA damage, and comparison with hydrogen peroxide stress. *Journal of Bacteriology, 178*(21), pp. 6145–6150.

Dychdala, G.R. (1991). Chlorine and chlorine compounds, pp. 131–151. In: S.S. Block, ed., *Disinfection, sterilization, and preservation, 4th ed.* Philadelphia, PA: Lea & Febiger.

Eklund, T. (1989). Organic acids and esters. *Mechanisms of Action of Food Preservation Procedures*. New York: Elsevier Applied Science Publishers Ltd. pp. 161–200.

EPA. (2004). U.S. Environmental Protection Agency. Pesticides: Regulating

Pesticides – Registering Pesticides. Available at: www.epa.gov/cgi bin/epaprint only.cgi [Accessed December 8, 2004].

FDA. (2013). Food and Drug Administration, U.S. Overview of food ingredients, additives & colors. Available at: http://www.fda.gov/Food/IngredientsPackagingLabeling/FoodAdditivesIngredients/ucm094211.htm [Accessed March 21, 2013].

FDA. (2014). Food and Drug Administration, U.S. Guidance For Industry: Guide To Minimize Microbial Food Safety Hazards Of Fresh-Cut Fruits And Vegetables. Available at: http://www.fda.gov/Food/GuidanceRegulation/GuidanceDocumentsRegulatoryInformation/ProducePlantProducts/ucm064458.htm#ch7 [Accessed November 28, 2015].

FFDCA. (1958). Federal Food, Drug, and Cosmetic Act. Section 201(s). Washington, D.C.: U.S. Government Printing Office.

Feehily, C., O'Byrne, C.P. and Karatzas, K.A.G. (2013). Functional γ-aminobutyrate shunt in *Listeria monocytogenes*: role in acid tolerance and succinate biosynthesis. *Applied and Environmental Microbiology, 79*(1), pp. 74–80.

Foley, S.L., Lynne, A.M., and Nayak, R. (2008). Challenges: Prevalence in swine and poultry and potential pathogenicity of such isolates. *Journal of Animal Science, 86*(14), pp. E149–E162.

Foster, J.W. (2004). *Escherichia Coli* Acid Resistance: Tales of an Amateur Acidophile. *Nature Reviews Microbiology, 2*(11), pp. 898–907.

Frankenberg, D., Frankenberg-Schwager, M. and Harbich, R. (1993). Mechanisms of oxygen radiosensitization in irradiated yeast: i. dna double-strand breakage. *International Journal of Radiation Biology, 64*(5), pp. 511–521.

Frere, J.M. (1995). Beta-lactamases and bacterial resistance to antibiotics. *Molecular Microbiology, 16*(3), pp. 385–395.

Frier, M. (1971). Derivatives of 4-amino-quinaldinium and 8-hydroxyquinoline. *Inhibition and Destruction of the Microbial Cell*. London: Academic Press Ltd. pp. 107–120.

Freese, E., Sheu, C.W. and Galliers, E. (1973). Function of lipophilic acids as antimicrobial food additives. *Nature, 241*, pp. 321–325.

Galvin, M., Hill, C. and Ross, R.P. (1999). Lacticin 3147 displays activity in buffer against Gram-positive bacterial pathogens which appear insensitive in standard plate assays. Letters in Applied Microbiology, *28*(5), pp. 355–358.

Gandhi, M. and Chikindas, M.L. (2007). *Listeria:* a foodborne pathogen that knows how to survive. *International Journal of Food Microbiology, 113*, pp. 1–15.

Gangoue-Pieboji, J., et al. (2009). The in-vitro antimicrobial activity of some medicinal plants against β-lactam-resistant bacteria. *The Journal of Infection in Developing Countries, 3*(9), pp. 671–680.

Ganjian, H., et al. (2012). Effects of salt stress on the antimicrobial drug resistance and protein profile of Staphylococcus aureus. *Jundishapur Journal of Microbiology, 5*(1), pp. 328–331

Goh, E.-B., et al. (2002). Transcriptional modulation of bacterial gene expression by subinhibitory concentrations of antibiotics. *Proceedings of the National Academy of Sciences, 99*(26), pp. 17025–17030.

Gillespie, M.T, Lyon, B.R. and Skurray, R.A. (1989). Gentamicin and Antiseptic Resistance in Epidemic Methicillin-Resistant *Staphylococcus Aureus. The Lancet, 333*(8636), p. 503–.

Greenberg, J.T., et al. (1991). Activation of oxidative stress genes by mutations at the soxQ/cfxB/marA locus of *Escherichia coli. Journal of Bacteriology, 173*(14), pp. 4433–4439.

Guo, F. and Zhang, T. (2012). Profiling bulking and foaming bacteria in activated sludge by high throughput sequencing. *Water Research, 46*(8), pp. 2772–2782.

Haleh, G., et al. (2012). Effects of Salt Stress on the Antimicrobial Drug Resistance and Protein Pattern of *Staphylococcus aureus. Jundishapur Journal of Microbiology, 1*, pp. 328–331.

Heathman, L.S., Pierce, G.O., and Kabler, P. (1936). Resistance of various strains of *E. typhi* and coli aerogenes to chlorine and chloramine.

Public Health Reports (1896–1970), *51*(40), pp. 1367–1387.

Heir, E., Sundheim, G. and Holck, A.L. (1995). Resistance to quaternary ammonium compounds in *Staphylococcus* spp. Isolated from the food industry and nucleotide sequence of the resistance plasmid pst827. *Journal of Applied Bacteriology, 79*(2), pp. 149–156.

Heir, E., Sundheim, G. and Holck, A.L. (1999). Identification and Characterization of Quaternary Ammonium Compound Resistant *Staphylococci* from the Food Industry. *International Journal of Food Microbiology, 48*(3), pp. 211–219.

Helander, I.M., et al. (1998). Characterization of the action of selected essential oil components on gram-negative bacteria. *Journal of Agricultural and Food Chemistry, 46*(9), pp. 3590–3595.

Hirasa, K. and Takemasa, M. (1998). Antimicrobial and antioxidant properties of spices. *Spice Science and Technology, 80*(7) pp. 163–200.

Holah, J.T., et al. (2002). Biocide use in the food industry and the disinfectant resistance of persistent strains of *Listeria monocytogenes* and *Escherichia coli*. *Journal of Applied Microbiology, 92*(s1), pp. 111S–120S.

Holler, J.G., et al. (2012). Novel inhibitory activity of the Staphylococcus aureus NorA efflux pump by a kaempferol rhamnoside isolated from Persea lingue Nees. *Journal of Antimicrobial Chemotherapy, 67*(5), pp. 1138–1144.

Hollowell, A.C., et al. (2015). Native California soils are selective reservoirs for multidrug-resistant bacteria. *Environmental Microbiology Reports, 7*(3), pp. 442–449.

Holyoak, C.D., et al. (1996). Activity of the plasma membrane h (+)-atpase and optimal glycolytic flux are required for rapid adaptation and growth of saccharomyces cerevisiae in the presence of the weak-acid preservative sorbic acid. *Applied and Environmental Microbiology, 62*(9), pp. 3158–3164.

Horiyama, T., Yamaguchi, A. and Nishino, K. (2010). TolC dependency of multidrug efflux systems in *Salmonella* enterica serovar Typhimurium. *Journal of Antimicrobial Chemotherapy, 65*(7), pp. 1372–1376.

Hotchkiss, R.D. (1944). Gramicidin, tyrocidine, and tyrothricin. *Adv. Enzymol. Relat. Areas Mol. Biol., 4*, pp. 153–199.

Hsieh, P.-C., et al. (1998). Bacteria lacking a multidrug pump: a sensitive tool for drug discovery. *Proceedings of the National Academy of Sciences, 95*(12), pp. 6602–6606.

Huang, Q.Q., et al. (2015). Effects of purple prairie clover (Dalea purpurea Vent.) on feed intake, nutrient digestibility and faecal shedding of *Escherichia coli* O157: H7 in lambs. *Animal Feed Science and Technology, 207*, pp. 51–61.

Hugo, W.B. and Foster, J.H.S. (1964). Growth of *Pseudomonas Aeruginosa* in Solutions of Esters of p-hydroxybenzoic Acid. *Journal of Pharmacy and Pharmacology, 16*(3), p. 209–.

Hugo, W.B. and Frier, M. (1969). Mode of action of the antibacterial compound dequalinium acetate. *Applied Microbiology, 17*(1), pp. 118–127.

Hurst, A. and Hoover, D.G. (1993). Nisin. *In Antimicrobials in Foods*. New York: Marcel Dekker, pp. 369–369.

Ikeda, T.O.M.I.K.I., et al. (1984). Interaction of a polymeric biguanide biocide with phospholipid membranes. *Biochimica et Biophysica Acta (BBA)-Biomembranes, 769*(1), pp. 57–66.

Imlay, J.A. and Linn, S. (1988). DNA damage and oxygen radical toxicity. *Science, 240*, pp. 1302–1309.

Inetianbor, J.E., Yakubu, J.M. and Ezeonu, S.C. (2015). Effects of food additives and preservatives on man—a review. *Asian Journal of Science and Technology, 6*(2), pp. 1118–1135.

Ingram, L.O. (1976). Adaptation of membrane lipids to alcohols. *Journal of Bacteriology, 125*(2), pp. 670–678.

Ioannou, C.J., Hanlon, G.W. and Stephen P. Denyer, S.P. (2007). Action of disinfectant quaternary ammonium compounds against

Staphylococcus aureus. *Antimicrobial Agents and Chemotherapy, 51*(1), pp. 296–306.

Iweriebor, B.C., et al. (2015). Antibiotic Susceptibilities of *Enterococcus* Species Isolated from Hospital and Domestic Wastewater Effluents in Alice, Eastern Cape Province of South Africa. *International Journal of Environmental Research and Public Health, 12*(4), pp. 4231–4246.

Jamroz, D., et al. (2003). The influence of phytogenic extracts on performance, nutrient digestibility, carcass characteristics, and gut microbial status in broiler chickens. *Journal of Animal and Feed sciences, 12*(3), pp. 583–596.

Jamroz, D., et al. (2005). Use of active substances of plant origin in chicken diets based on maize and locally grown cereals. *British Poultry Science, 46*(4), pp. 485–493.

Jay, J.M. (1995). Antimicrobial food preservatives. *In Handbook of Biocide and Preservative Use,* Netherlands: Springer, pp. 334–348.

Jay, J.M. (2012). Intrinsic and extrinsic parameters of foods that affect microbial growth. In Modern Food Microbiology. New York: Springer Publications, pp. 38–66.

Jimenez-Valera, M., Ruiz-Bravo, A. and Ramos-Cormenzana, A. (1987). Inhibition of β-lactamases from *Yersinia enterocolitica* by plant extracts. *Journal of Antimicrobial Chemotherapy, 19*(1), pp. 31–37.

Johny, A.K., Hoagland, T. and Venkitanarayanan, K. (2010). Effect of subinhibitory concentrations of plant-derived molecules in increasing the sensitivity of multidrug-resistant *Salmonella* enterica serovar Typhimurium DT104 to antibiotics. *Foodborne Pathogens and Disease, 7*(10), pp. 1165–1170.

Juven, B.J. and Pierson, M.D. (1996). Antibacterial effects of hydrogen peroxide and methods for its detection and quantitation. *Journal of Food Protection, 59*(11), pp. 1233–1241.

Knobloch, K., et al. (1986). Progress in essential oil research. Berlin, Walter de Gruyther, p. 429–.

Koga, T., Hirota, N. and Takumi, K. (1999). Bactericidal activities of essential oils of basil and sage against a range of bacteria and the effect of these essential oils on *Vibrio parahaemolyticus. Microbiological Research, 154*(3), pp. 267–273.

Kollanoor-Johny, A., et al. (2012a). Reduction of *Salmonella enterica* serovar Enteritidis colonization in 20-day-old broiler chickens by the plant-derived compounds trans-cinnamaldehyde and eugenol. *Applied and Environmental Microbiology, 78*(8), pp. 2981–2987.

Kollanoor-Johny, A., et al. (2012b). Caprylic acid reduces *Salmonella* Enteritidis populations in various segments of digestive tract and internal organs of 3-and 6-week-old broiler chickens, therapeutically. *Poultry Science, 91*(7), pp. 1686–1694.

Kulikovsky, A., Pankratz, H.S. and Sadoff, H.L. (1975). Ultrastructural and chemical changes in spores of Bacillus cereus after action of disinfectants. *Journal of Applied Bacteriology, 38*(1), pp. 39–46.

Kumar, A., et al. (2008). Novel structural analogues of piperine as inhibitors of the NorA efflux pump of *Staphylococcus aureus. Journal of Antimicrobial Chemotherapy, 61(*6), pp. 1270–1276.

Kushner, D.J. (1971). Life in extreme environments (Microorganism life in extreme high temperature, pH and solute concentration environments, noting salt effect on enzyme activity). *Chemical Evolution and the Origin of Life.* New York. Springer Publications, pp. 485–491.

Krebs, H.A., et al. (1983). Studies on the mechanism of the antifungal action of benzoate. *Journal of Biochemistry, 214,* pp. 657–663.

Lee, J.-H., Cho, M.H, and Lee, J. (2011). 3-Indolylacetonitrile Decreases Escherichia coli O157: H7 Biofilm Formation and *Pseudomonas aeruginosa* Virulence. *Environmental Microbiology, 13*(1), pp. 62–73.

Lee, J.-H., et al. (2013). Diverse plant extracts and trans-resveratrol inhibit biofilm formation and swarming of *Escherichia coli* O157: H7. *Biofouling, 29*(10), pp. 1189–1203.

Lee, J.-H., et al. (2014). Coumarins reduce biofilm formation and the virulence of *Escherichia coli* O157: H7. *Phytomedicine, 21*(8), pp. 1037–1042.

Lemaitre, J.-P., et la. (1998). Plasmid-mediated resistance to antimicrobial agents among listeriae. *Journal of Food Protection, 61*(11), pp. 1459–1464.

Lewis, K. and Ausubel, F.M. (2006). Prospects for plant-derived antibacterials. *Nature Biotechnology, 24*(12), pp. 1504–1507.

Leuck, E. (1980). *Antimicrobial food additives. Characteristics. Uses. Effects.* Berlin: Springer, pp. 175–182.

Lin, J., et al. (1996). Mechanisms of Acid Resistance in Enterohemorrhagic *Escherichia Coli. Applied and Environmental Microbiology, 62*(9), pp. 3094–3100.

Lisle, J.T. (1998). Effects of Starvation on Physiological Activity and Chlorine Disinfection Resistance in *Escherichia coli* O157: H7. *Applied and Environmental Microbiology, 64*(12), pp. 4658–4662.

Lorencová, E., et al. (2012). Antibacterial effect of phosphates and polyphosphates with different chain length. *Journal of Environmental Science and Health, Part A, 47*(14), pp. 2241–2245.

Lucey, B.D. et al. (2000). Integron like structures in *Campylobacter* spp. of human and animal origin. Emerging Infectious Diseases, 6(1), p. 50–.

Ma, M. and Eaton, J.W. (1992). Multicellular Oxidant Defense in Unicellular Organisms. *Proceedings of the National Academy of Sciences of the United States of America, 89*(17), pp. 7924–7928.

Maillard, J-Y. (2002). Bacterial target sites for biocide action. *Journal of Applied Microbiology, 92*(s1), pp. 16S–27S.

Malchesky, P.S. (1993). Peracetic acid and its application to medical instrument sterilization. *Artificial Organs, 17*(3), pp. 147–152.

Marques, S.C., et al. (2007). Formation of biofilms by *Staphylococcus aureus* on stainless steel and glass surfaces and its resistance to some selected chemical sanitizers. *Brazilian Journal of Microbiology, 38*(3), pp. 538–543.

Masschalck, B. and Michiels, C.W. (2003). Antimicrobial properties of lysozyme in relation to foodborne vegetative bacteria. *Critical Reviews in Microbiology, 29*(3), pp. 191–214.

Mathur, S. and Singh, R. (2005). Antibiotic resistance in food lactic acid bacteria—a review. *International Journal of Food Microbiology, 105*(3), pp. 281–295.

Mazzotta, A.S. and Montville, T.J. (1997). Nisin induces changes in membrane fatty acid composition of *Listeria monocytogenes* nisin-resistant strains at 10°c and 30°c. *Journal of Applied Microbiology, 82*(1), pp. 32–38.

McDonnell, G. and Russell, A.D. (1999). Antiseptics and disinfectants: activity, action, and resistance. *Clinical Microbiology Reviews, 12*(1), pp. 147–179.

Ming, X. and Daeschel, M.A. (1993). Nisin resistance of foodborne bacteria and the specific resistance responses of *Listeria Monocytogenes* Scott A. *Journal of Food Protection, 56*(11), pp. 944–948.

Mitsch, P., et al. (2004). The effect of two different blends of essential oil components on the proliferation of Clostridium perfringens in the intestines of broiler chickens. *Poultry Science, 83*(4), pp. 669–675.

Mokgatla, R.M., Brözel, V.S. and Gouws, P.A. (1998). Isolation of *Salmonella* resistant to hypochlorous acid from a poultry abattoir. *Letters in Applied Microbiology, 27*(6), pp. 379–382.

Montville, T.J., Winkowski, K. and Chikindas, M.L. (2001). Biologically based preservation systems. In: M.P. Doyle, L.R. Beuchat, and T.J. Montville, eds., *Food Microbiology: Fundamentals and Frontiers, 2nd ed.* Washington, DC: ASM Press, pp. 629–647.

Moradas-Ferreira, P., et al. (1996). The molecular defences against reactive oxygen species in yeast. *Molecular Microbiology, 19*(4), pp. 651–658.

Naidu, A.S., ed. (2000). *Natural Food Antimicrobial Systems.* New York: CRC press.

Nikaido, H. (1994). Prevention of drug access to bacterial targets: permeability barriers and active efflux. *Science, 264*(5157), pp. 382–388.

Nostro, A., et al. (2007). Effects of oregano, carvacrol and thymol on *Staphylococcus aureus* and *Staphylococcus epidermidis* biofilms. *Journal of Medical Microbiology, 56*(4), pp. 519–523.

O'Driscoll, B., Gahan, C.G. and Hill, C. (1996). Adaptive acid tolerance response in *Listeria monocytogenes*: isolation of an acid-tolerant mutant which demonstrates increased virulence. *Applied and Environmental Microbiology, 62*(5), pp. 1693–1698.

Odlaug, T.E. (1981). Antimicrobial activity of halogens. *Journal of Food Protection, 44*(8), pp. 608–613.

Oram, J.D. and Reiter, B. (1966). The inhibition of *Streptococci* by lactoperoxidase, thiocyanate and hydrogen peroxide. The oxidation of thiocyanate and the nature of the inhibitory compound. *Biochem. J., 100,* pp. 382–388.

Oussalah, M., et al. (2006). Antimicrobial effects of alginate-based film containing essential oils for the preservation of whole beef muscle. *Journal of Food Protection, 69*(10), pp. 2364–2369.

Park, Y.-K., et al. (1996). Internal ph crisis, lysine decarboxylase and the acid tolerance response of *salmonella* typhimurium. *Molecular Microbiology, 20*(3), pp. 605–611.

Penesyan, A., Gillings, M. and Paulsen, I.T. (2015). Antibiotic Discovery: Combatting Bacterial Resistance in Cells and in Biofilm Communities. *Molecules, 20*(4), pp. 5286–5298.

Piddock, L.J.V. (2006). Multidrug-resistance efflux pumps? not just for resistance. *Nature Reviews Microbiology, 4*(8), pp. 629–636.

Piper, P., et al. (1998). The pdr12 ABC transporter is required for the development of weak organic acid resistance in yeast. *The EMBO Journal, 17*(15), pp. 4257–4265.

Proctor, V.A., Cunningham, F.E. and Fung, D.Y.C. (1988). The chemistry of lysozyme and its use as a food preservative and a pharmaceutical. *Critical Reviews in Food Science & Nutrition, 26*(4), pp. 359–395.

Qiu, J., et al. (2010a). Eugenol reduces the expression of virulence-related exoproteins in *Staphylococcus aureus. Applied and Environmental Microbiology, 76*(17), pp. 5846–5851.

Qiu, J., et al. (2010b). Subinhibitory concentrations of thymol reduce enterotoxins A and B and alpha-hemolysin production in *Staphylococcus aureus* isolates. *PLoS One, 5*(3), pp. e9736.

Quastel, J.H. and Wooldridge, W.R. (1927). The effects of chemical and physical changes in environment on resting bacteria. *Biochemical Journal, 21*(1), p. 148.

Quinn, T., et al. (2006). Multi-drug resistance in *Salmonella* enterica: efflux mechanisms and their relationships with the development of chromosomal resistance gene clusters. *Current Drug Targets, 7*(7), pp. 849–860.

Rai, V.R. and Bai, J.A. (2014). *Microbial Food Safety and Preservation Techniques.* New York: CRC Press.

Rossouw, F.T. and Rowbury, R.J. (1984). Effects of the resistance plasmid R124 on the level of the ompf outer membrane protein and on the response of *Escherichia Coli* to environmental agents. *Journal of Applied Bacteriology, 56*(1), pp. 63–79.

Russell, A.D. (1990). Bacterial spores and chemical sporicidal agents. *Clinical Microbiology Reviews, 3*(2), pp. 99–119.

Russell, A.D. (1991). Mechanisms of Bacterial Resistance to non-antibiotics: Food Additives and Food and Pharmaceutical Preservatives. *Journal of Applied Bacteriology, 71*(3), pp. 191–201.

Russell, A.D. (2003). Biocide use and antibiotic resistance: the relevance of laboratory findings to clinical and environmental situations. *The Lancet Infectious Diseases, 3*(12), pp. 794–803.

Russell, N.J. and Gould, G.W. (2003). *Food Preservatives.* New York: Springer Science & Business Media.

Ryan, M.P., et al. (1996). An application in cheddar cheese manufacture for a strain of *Lactococcus lactis* producing a novel broad-spectrum bacteriocin, lacticin 3147.

Applied and Environmental Microbiology, 62(2), pp. 612–619.

Salmond, C.V., Kroll, R.G., and Booth, I.R. (1984). The Effect of Food Preservatives on pH Homeostasis in Escherichia Coli. *Journal of General Microbiology, 130*(11), pp. 2845–2850.

Salton, M.R.J. (1968). Lytic agents, cell permeability, and monolayer penetrability. *The Journal of General Physiology, 52*(1), pp. 227–252.

Salvia, A.C.R.D., et al. (2011). Effectiveness of 2% peracetic acid for the disinfection of gutta-percha cones. *Brazilian Oral Research, 25*(1), pp. 23–27.

Samson, R.A., et al. (1995). *Introduction to Food-Borne Fungi*. Wageningen (Netherlands): CBS.

Savoia, D. (2012). Plant-derived antimicrobial compounds: alternatives to antibiotics. *Future Microbiology, I*(8), pp. 979–990.

Setlow, P. (1992). I Will Survive: Protecting and Repairing Spore DNA. *Journal of Bacteriology, 174*(9), pp. 2737–2741.

Shields, M.S., et al. (1995). TOM, a New Aromatic Degradative Plasmid from *Burkholderia (Pseudomonas) cepacia* G4. *Applied and Environmental Microbiology, 61*(4), pp. 1352–1356.

Sikkema, J., De Bont, J.A., and Poolman, B. (1995). Mechanisms of membrane toxicity of hydrocarbons. *Microbiological Reviews, 59*(2), pp. 201–222.

Sinskey, A.J. (1980). *Mode of action and effective application. Developments in food preservatives*. London: Applied Science Publishers Ltd, pp. 127–129.

Siragusa, G.R. and Johnson, M.G. (1989). Inhibition of *Listeria monocytogenes* growth by the lactoperoxidase-thiocyanate-H2O2 antimicrobial system. *Applied and Environmental Microbiology, 55*(11), pp. 2802–2805.

Sofos, J.N., and Busta, F.F. (1981). Antimicrobial activity of sorbate. *Journal of Food Protection., 44*(8), pp. 614–622.

Sleator, R.D. and Hill, C. (2002). Bacterial osmoadaptation: the role of osmolytes in bacterial stress and virulence. *FEMS Microbiology Reviews, 26*(1), pp. 49–71.

Spratt, B.G. (1994). Resistance to antibiotics mediated by target alterations. *Science, 264*(5157), pp. 388–393.

Srey, S., Jahid, I.K. and Ha, S.-D. (2013). Biofilm formation in food industries: a food safety concern. *Food Control, 31*(2), pp. 572–585.

Stewart, M.H. and Olson, B.H. (1992). Impact of growth conditions on resistance of *Klebsiella pneumoniae* to chloramines. *Applied and Environmental Microbiology, 58*(8), pp. 2649–2653.

Stratford, M. and Anslow, P.A. (1996). Comparison of the inhibitory action of weak-acid preservatives, uncouplers, and medium chain fatty acids. *FEMS Microb Lett., 142*, pp. 53–58.

Sutton, L. and Jacoby, G.A. (1978). Plasmid-Determined Resistance to Hexachlorophene in *Pseudomonas Aeruginosa*. *Antimicrobial Agents and Chemotherapy, 13*(4), p. 634.

Sykes, G. (1939). The influence of germicides on the dehydrogenases of Bact. Coli: Part I. The succinic acid dehydrogenase of Bact. Coli. *Journal of Hygiene, 39*(4), pp. 463–469.

Szabo, E.A., and Cahill, M.E. (1998). The combined effects of modified atmosphere, temperature, nisin and ALTA™ 2341 on the growth of *Listeria monocytogenes*. *International Journal of Food Microbiology, 43*(1), pp. 21–31.

Tatsuzawa, H., et al. (1998). Inactivation of Bacterial Respiratory Chain Enzymes by Singlet Oxygen. *FEBS Letters, 439*(3), pp. 329–333.

Taylor, C.L. and Henry, J.E. (2010). *Strategies to Reduce Sodium Intake in the United States*. Washington, DC: National Academies Press.

Toledo, R.T. (1975). Chemical sterilants for aseptic-packaging. *Food Technology, 29*(5), p. 102.

Tote, K., et al. (2010). Inhibitory effect of biocides on the viable masses and matrices of *Staphylococcus aureus* and *Pseudomonas aeruginosa* biofilms. *Applied and Environmental Microbiology, 76*(10), pp. 3135–3142.

Trujillo, R., and Laible, N. (1970). Reversible inhibition of spore germination by alcohols. *Applied Microbiology, 20*(4), pp. 620–623.

Tsui, W.H.W., et al. (2004). Dual effects of MLS antibiotics: transcriptional modulation and interactions on the ribosome. *Chemistry & Biology, 11*(9), pp. 1307–1316.

Ultee, A., et al. (2000). Adaptation of the Food-Borne Pathogen *Bacillus Cereus* to Carvacrol. *Archives of Microbiology, 174*(4), pp. 233–238.

Upadhyay, A., et al. (2012). Plant-derived antimicrobials reduce *Listeria monocytogenes* virulence factors in vitro, and down-regulate expression of virulence genes. *International Journal of Food Microbiology, 157*(1), pp. 88–94.

Upadhyay, A., et al. (2013). Antibiofilm effect of plant derived antimicrobials on *Listeria monocytogenes. Food Microbiology, 36*(1), pp. 79–89.

Upadhyay, A., et al. (2014). Combating pathogenic microorganisms using plant-derived antimicrobials: A Minireview of the mechanistic basis. *Biomed Research International, 2014*, pp. 1–18.

Upadhyaya, I., et al. (2015). In-feed supplementation of trans-cinnamaldehyde reduces layer-chicken egg-borne transmission of *Salmonella* enterica Serovar Enteritidis. *Applied and Environmental Microbiology, 81*(9), pp. 2985–2994.

Wagner, M.K. and Busta, F.F. (1985). Inhibition of *Clostridium botulinum* 52A toxicity and protease activity by sodium acid pyrophosphate in media systems. *Applied and Environmental Microbiology, 50*(1), pp. 16–20.

Wagle B.R., et al. (2014). Efficacy of the natural compound, β-resorcylic acid, against *Campylobacte*r colonization in poultry. *Poultry Science, 93*(E-Suppl. 1).

Walencka, E. (2007). Salvipisone and aethiopinone from *Salvia sclarea* hairy roots modulate staphylococcal antibiotic resistance and express anti-biofilm activity. *Planta Medica, 73*(6), pp. 545–551.

Wells, J.E., Berry, E.D. and Varel, V.H. (2005). Effects of common forage phenolic acids on *Escherichia coli* O157: H7 viability in bovine feces. *Applied and Environmental Microbiology, 71*(12), pp. 7974–7979.

Witte, W. (1997). Impact of antibiotic use in animal feeding on resistance of bacterial pathogens in humans. *Antibiotic Resistance: Origins, Evolution, Selection and Spread.* West Sussex: John Wiley and Sons, Ltd, pp. 61–75.

Wood, H.G. and Clark, J.E. (1988). Biological aspects of inorganic polyphosphates. *Annual Review of Biochemistry, 57*(1), pp. 235–260.

Woods, L.F.J., Wood, J.M. and Gibbs, P.A. (1981). The involvement of nitric oxide in the inhibition of the phosphoroclastic system in *Clostridium sporogenes* by sodium nitrite. *Journal of General Microbiology, 125*(2), pp. 399–406.

Woolhouse, M. and Gaunt, E. (2007). Ecological origins of novel human pathogens. *Critical Reviews in Microbiology*, 33(4), pp. 231–242.

Wray, C. and McLaren, I. (1987). A note on the effect of the lactoperoxidase systems on salmonellas *in vitro* and *in vivo. Journal of Applied Bacteriology* 62, (2), pp. 115–118.

Wu, H., et al. (2011). Effects of ginseng on *Pseudomonas aeruginosa* motility and biofilm formation. *FEMS Immunology & Medical Microbiology, 62*(1), pp. 49–56.

Zhao, W.-H., et al. (2003). Restoration of antibacterial activity of β-lactams by epigallocatechin gallate against β-lactamase-producing species depending on location of β-lactamase. *Journal of Pharmacy and Pharmacology, 55*(6), pp. 735–740.

13

Molecular Biology of Multidrug Resistance Efflux Pumps of the Major Facilitator Superfamily from Bacterial Food Pathogens

Ranjana K.C.[1], Ugina Shrestha[1], Sanath Kumar[2], Indrika Ranaweera[1], Prathusha Kakarla[1], Mun Mun Mukherjee[1], Sharla R. Barr[1], Alberto J. Hernandez[1], T. Mark Willmon[1], Bailey C. Benham[1], and Manuel F. Varela[1]

[1] *Department of Biology, Eastern New Mexico University, Portales, New Mexico, USA*
[2] *QC Laboratory, Harvest and Post Harvest Technology Division, Central Institute of Fisheries Education (CIFE), Mumbai, India*

13.1 Foodborne Bacterial Pathogens

Foodborne pathogens remain a major human health concern worldwide. Although food production systems have undergone enormous transformation from traditional ways to scientifically sound and technologically robust systems that produce safe, healthy, and nutritionally balanced foods, foodborne pathogens still pose a great challenge to human health in the modern world. In the United States, it is estimated that foodborne bacterial agents are responsible for 3.6 million infections annually, which accounts for approximately 39% of all foodborne infections (Scallan *et al.*, 2011).

The morbidity and mortality associated with foodborne bacterial infections are much higher in comparison to viruses and parasites. Nearly 60% of the hospitalized cases of foodborne illnesses are of bacterial origin (Mead *et al.*, 1999). The manifestations of foodborne infections can be diverse depending on the pathogen and the physiology of the host. Certain bacterial infections are mild and self-limiting, whereas others could be severe and life threatening. Some of the most serious consequences of foodborne bacterial infections are manifested as hemolytic uremic syndrome, Guillain-Barré Syndrome, reactive arthritis, meningitis, chronic inflammatory bowel disease, renal failure, and neurological conditions (Thomson *et al.*, 1995; Rocourt *et al.*, 2003; Tam *et al.*, 2003).

The bacterial agents of infection live in association with food animals and birds as commensals or parasites and as a consequence, are found in meat (Knechtges, 2011). Infections occur when contaminated foods are either consumed raw or insufficiently cooked to destroy the bacterial pathogens. Foodborne bacteria are opportunistic pathogens, and several factors, such as the virulence of the pathogen, its quantity in food, and antimicrobial resistance, can all determine the nature and severity of the illnesses. Additionally, the age and health status of the human host can affect the severity of an illness.

Although a large number of bacterial pathogens of animal origin are capable of causing human infections and intoxications, only a few of them are responsible for a majority of these gastrointestinal disturbances. These bacteria include *Salmonella enterica*, *Campylobacter* spp., Shiga toxin-producing *Escherichia coli* (STEC), *Shigella*

Foodborne Pathogens and Antibiotic Resistance, First Edition. Edited by Om V. Singh.

spp., *Listeria monocytogenes*, *Staphylococcus aureus*, *Vibrio cholerae*, and *Yersinia* spp. (Swartz, 2002).

Many of these pathogens are associated with poultry, cattle, and swine, except the vibrios, which are exclusively found in marine fish and shellfish. *Salmonella*, for example, is associated with poultry and could be found in eggs and chicken (Bryan and Doyle, 1995). Similarly, poultry is a major source for *Campylobacter* infections (Sahin *et al.*, 2002). Others such as STEC and *Yersinia* are associated with cattle and swine (Swartz, 2002). While infections through contaminated meat and meat products are directly related to the presence of bacterial pathogens in the food animal, several other foods such as fruits, vegetables, milk, and milk products, such as cheese and ice cream, have also been implicated in numerous outbreaks due to secondary contamination during various stages of food manufacturing, transportation, and even during preparation of food by the consumer.

The use of animal and poultry manures in agriculture results in the contamination of agricultural products with bacterial pathogens. Reduction of pathogen load in food animals, personal hygiene, proper handling, low temperature food preservation, and adequately cooking food to kill pathogens are among several strategies that can mitigate incidences of foodborne infections.

The problem of foodborne infections is further confounded by several pathogens becoming multidrug-resistant (MDR). Antibiotics are used in animal husbandry and aquaculture as prophylactics against outbreak of bacterial infections, growth enhancement, and better food production (McEwen and Fedorka-Cray, 2002). However, the use of antibiotics for purposes other than treatment of specific bacterial infections has resulted in the development and spread of multidrug resistance in animal-borne human pathogens (van den Bogaard and Stobberingh, 2000). The use of antibiotics in farms has been responsible for the development of resistance in *Salmonella* spp., *C. jejuni*, and several ESBL-producing gram-negative bacteria (van den Bogaard and Stobberingh, 2000; Spika *et al.*, 1987; Smith *et al.*, 1999a; Kluytmans *et al.*, 2013). Farm animals serving as reservoirs of antibiotic resistant bacteria are a serious concern for transmission to humans through contaminated food (Silbergeld *et al.*, 2008).

13.1.1 *Escherichia coli*

Variants of *Escherichia coli* are amongst the most versatile of bacterial pathogens (Kaper *et al.*, 2004). This microorganism typically exists as part of the gut microbiome but the possession of virulence factors enables infection within the intestinal and urinary tracts of hosts (Kaper *et al.*, 2004; Gill *et al.*, 2006). The adaptability of these bacteria has led to a number of nomenclature designations based on pathogenicity (Kaper *et al.*, 2004). Enteropathogenic *E. coli* (EPEC) causes diarrhea through the physical attachment of bacterial cells to microvilli in the intestinal lining (Chen and Frankel, 2005). Enterotoxigenic *E. coli* (ETEC) produces plasmid-associated enterotoxins some of which can cause symptoms that mimic those of cholera (Kaper *et al.*, 2004; Qadri *et al.*, 2005). Another pathogen, enterohemorrhagic *E. coli* (EHEC) produces verotoxins that are fundamentally similar to *Shigella* toxin and are responsible for producing bloody diarrhea (Elder *et al.*, 2000; Hussein and Sakuma, 2005). Hemorrhagic colitis and hemolytic-uremic syndrome are both conditions that may result from infection with EHEC strains. The latter condition can present with kidney failure, anemia, and thrombocytopenia (Vallance *et al.*, 2002; Kaper *et al.*, 2004). These pathogenic strains present with diarrhea, vomiting, abdominal cramping, and sometimes fever (Kaper *et al.*, 2004). Typically, hosts who have succumbed to illness by *E. coli* are not treated with antibiotics, but rather with rehydration therapy and with dialysis in the case of disrupted renal functions, to aid in recovery (Qadri *et al.*, 2005). On the other hand, the presence of pathogenic bacteria is cause for concern with respect to the development of antimicrobial resistance, since drug-resistant bacterial variants are known to cause illness in farm animals (Nagy and Fekete, 1999), particularly during weaning, and such hosts are likely to be exposed to antimicrobial agents (Silbergeld *et al.*, 2008).

13.1.2 *Staphylococcus aureus*

S. aureus is a gram-positive extracellularly-growing bacterium that causes severe disease in human and veterinary medicine (Lowy, 1998; Safdar and Bradley, 2008). *S. aureus* produces virulence factors that facilitate attachment, colonization, cell-cell interactions, immune evasion, and tissue damage (Mulligan *et al.*, 1993; Murray *et al.*, 2012). The bacterial pathogen colonizes the skin and mucosa and subsequently invades multiple organs within the body. Therefore, *S. aureus* may be the causative agent for a large variety of infections, ranging from mild infections to life-threatening diseases, including abscesses of various organs, pneumonia, osteomyelitis, endocarditis, arthritis, and sepsis (Gordon and Lowy, 2008). In livestock, *S. aureus* primarily causes mastitis, skin and soft tissue infections, and less frequently locomotor system infections (Fitzgerald, 2012).

Healthcare-associated infections (i.e., nosocomial infections) due to *S. aureus* are reported more frequently in small, non-human, domesticated animals and horses. Furthermore, opportunistic *S. aureus* may cause severe human illnesses due to the production of virulence factors, such as exotoxins and enterotoxins, which are associated with intoxications, including food poisoning, scalded skin syndrome, and toxic shock syndrome (Projan and Novick, 1997).

Treatment of invasive and systemic infections largely relies on antimicrobial agents, such as penicillinase-stable β-lactams (e.g., isoxazolyl penicillins, cephalosporins) for infections in human clinical settings. Bacterial resistance to these anti-bacterial agents has drastically limited the availability of therapeutic options, leading to an emerging problem on a global scale (Chambers and DeLeo, 2009). Presently, the majority of *S. aureus* clinical isolates are resistant to methicillin, and some variants have further acquired resistance to more than 20 distinctive antimicrobial agents (Paulsen *et al.*, 1997).

The antimicrobial agent vancomycin, a glycopeptide antibiotic, had in some cases become the sole remaining effective therapeutic agent against certain multidrug-resistant *S. aureus* strains, including methicillin resistant *S. aureus* (MRSA) and methicillin-resistant *S. epidermidis* (MRSE) (Walsh, 1999). However, due to the recent emergences of *S. aureus* variants with intermediate level (VISA) and high level resistances to vancomycin (VRSA), the effectiveness of vancomycin has been compromised (Appelbaum, 2006; Smith *et al.*, 1999b).

Therefore, before considering antimicrobial treatment of MRSA infections in livestock or humans, the risk of development of further resistance should be considered, and the risk/benefit assessment for antimicrobial treatment needs to be compared with alternatives (Catry *et al.*, 2010). The recent emergence of novel hypervirulent community-acquired *S. aureus* variants has emphasized the critical need for new methods of control and treatment (Stevens, 2003; Vandenesch *et al.*, 2003).

13.1.3 *Vibrio cholerae*

The causative agent of the disease cholera in food animals and humans is the comma-shaped gram-negative toxigenic bacterium *V. cholerae*, which secretes a potent enterotoxin known as the cholera toxin (CT) (Harris *et al.*, 2012; Rabbani and Greenough, 1999). The virulent strains of *V. cholerae*, categorized as O1 and O139, cause diarrhea by affecting the epithelial cells of the gastrointestinal tract, thereby leading to the loss of body fluids such as water and electrolytes.

The presence of the genes encoding CT, the *Vibrio* pathogenicity island (VPI), and the toxin co-regulated pilus (TCP) within the O1 serogroups of *V. cholerae* make these strains more virulent and potentially pandemic than their non-O1 counterparts (Robins and Mekalanos,

2014; Ritchie and Waldor, 2009). In the past 200 years, virulent strains of O1 *V. cholerae* have led to millions of fatalities in Asia and Africa, with the spread of endemic and pandemic disease cholera evolving as a major public health concern (Mandal *et al.*, 2011).

Improper sanitary conditions and contaminated water sources promote the spread of cholera disease, and lack of treatment could lead to mortality due to excessive vomiting and diarrhea (Harris *et al.*, 2012). Along with oral rehydration, the affected individual could also be prescribed with antibiotics, such as cotrimoxazole, erythromycin, tetracycline, chloramphenicol, furazolidone, and norfloxacin as chemotherapy for the disease. The widespread use and misuse of antibacterial agents, however, has enabled bacterial variants to develop bacterial resistance against multiple antimicrobial agents, potentially confounding effective therapy (Ghosh and Ramamurthy, 2011; Andersen *et al.*, 2015).

Antimicrobial efflux pumps actively transport antibacterial agents out of the bacterial cell enabling the bacterial pathogens to gain resistance against clinically important antibacterial agents (Andersen *et al.*, 2015; Smith *et al.*, 2009). These multidrug resistant bacterial variants potentially compromise the chemotherapeutic efficacy against cholera (Kitaoka *et al.*, 2011).

13.1.4 *Salmonella enterica*

S. enterica is a facultative anaerobic gram-negative bacterium belonging to the Enterobacteriaceae family and the Proteobacteria phylum. *S. enterica* is rod-shaped, motile, and non-endospore forming, with size dimensions of 0.7–1.5 by 2.0–5.0 μm (Su and Chiu, 2007; Murray and Jorgensen, 2007). *S. enterica* resides as a normal resident flora within the gut microbiome of carriers and vertebrates. These pathogens can infect other individuals causing multiple salmonelloses (Callaway *et al.*, 2008). Salmonellosis is a foodborne disease that leads to significant hospitalizations and mortality (Threlfall *et al.*, 1996). *S. enterica* has become the second most common bacterial cause of foodborne gastroenteritis cases worldwide, where non-typhoidal *Salmonella* spp., are estimated annually to cause 1,000,000 cases, 19,000 hospitalizations, and 380 deaths; *S. enterica* serotype Typhi is estimated to cause 1,800 illnesses, 200 hospitalizations, and no deaths each year (Sanchez-Vargas *et al.*, 2011). According to the Centers for Disease Control and Prevention (CDC), approximately 42,000 salmonellosis cases are reported every year in the United States (Swaminathan *et al.*, 2006). *Salmonella* infection causes 1.4 million estimated human cases in the United States and is a causative agent in 23,000 hospitalizations and 450 deaths from non-typhoidal *Salmonella* foodborne disease cases annually (Gunn *et al.*, 2014; Voetsch *et al.*, 2004).

Chemotherapy is not the first choice of treatment in salmonellosis as the infection is often self-limiting (Nishino *et al.*, 2009). Replacement of electrolytes and fluid is considered to be the best treatment during gastric infections (Waddington *et al.*, 2014). Currently, ceftriaxone is the agent of choice for chemotherapy in severe cases (Zhao *et al.*, 2003). Antimicrobial chemotherapies using ampicillin, chloramphenicol, and trimethoprim sulfamethoxazole (cotrimoxazole) are indicated for severe foodborne and waterborne enteric fevers and salmonellosis if *S. enterica* is the causative agent (Smith *et al.*, 2010).

Antibiotic treatments (ciprofloxacin, cotrimoxazole, ampicillin, or cephalosporins) are used to treat bacteremia, typhoid fever, and gastroenteritis in neonates, children, the elderly, and immunocompromised individuals (Harish *et al.*, 2011). The increasing incidences of antibiotic resistant clinical isolates of *S. enterica* enhance the risk of therapeutic failure in cases of life-threatening salmonellosis (Andersen *et al.*, 2015; Nishino *et al.*, 2009). As most of the bacterial strains are resistant to these drugs, fluoroquinolones, and cephalosporins are thus prescribed (Threlfall, 2002). Recently, resistances to fluoroquinolones and cephalosporins have been reported, indicating the current need to reestablish clinical chemotherapeutic efficacies (Threlfall, 2002; Smith *et al.*, 2010; Kumar *et al.*, 2008).

13.2 Major Classes of Clinically Important Antibacterial Agents

There are several major classes of antibacterial agents; the most common antibacterial activity is inhibition of cell wall synthesis (Murray *et al.*, 2012; Walsh, 2003). Most of the antibiotics included in this group are β-lactam antibiotics such as penicillins, cephalosporins, cephamycins, carbapenems, monobactams, and they are often combined with β-lactamase inhibitors. These antibacterials are so named because they have a common β-lactam ring structure. Other antibiotics that inhibit cell wall synthesis are vancomycin, daptomycin, and bacitracin. β-lactam antibiotics bind to specific penicillin binding proteins (PBPs) in the bacterial cell wall, inhibiting the assembly of peptidoglycan chains. Autolysins are, in turn, activated, resulting in the degradation of the cell wall and leading to bacterial cell death. Therefore, β-lactam antibiotics act as bactericidal agents.

The second largest group of antimicrobial agents includes antibiotics that inhibit protein synthesis (Walsh, 2003; Murray *et al.*, 2012). The aminoglycosides isolated from *Streptomyces* species include streptomycin, neomycin, and kanamycin. Aminoglycosides isolated from *Micromonospora* species include gentamycin and sisomicin, and amikacin and netilmicin are synthetic derivatives of kanamycin and sisomicin, respectively. These antimicrobial agents act by passing through the bacterial outer membrane of the cell wall and the cytoplasmic membrane into the cytoplasm, where they inhibit bacterial protein synthesis by irreversibly binding to the 30S ribosomal proteins. Tetracyclines (tetracycline, doxycycline, minocycline), broad-spectrum bacteriostatic antibiotics, also inhibit protein synthesis by reversibly binding to 30S ribosomal subunits. Chloramphenicols act as bacteriostatic antibiotics by binding reversibly to the peptidyl transferase component of the 50S ribosomal subunit, thus inhibiting protein synthesis (translation) by blocking peptide elongation. Macrolides (erythromycin, azithromycin, clarithromycin) act by reversibly binding to the 23S ribosomal RNA of the 50S ribosomal subunit, which blocks polypeptide elongation. Clindamycin blocks protein elongation by binding to the 50S ribosome.

Other major antibiotic groups consist of nucleic acid synthesis inhibitors (quinolones, rifampicin, and rifabutin) and antimetabolites (sulfonamides, trimethoprim, dapsone, *p*-aminosalicylic acid) (Walsh, 2003; Murray *et al.*, 2012). Quinolones inhibit bacterial DNA gyrases, which are required for DNA replication, recombination, and repair. Rifampicin binds to DNA-dependent RNA polymerase and inhibits the initiation of RNA synthesis. Sulfonamides compete with *p*-aminobenzoic acid preventing the synthesis of folic acid. Trimethoprim interferes with folic acid metabolism by inhibiting dihydrofolate reductase thereby preventing the conversion of dihydrofolate to tetrahydrofolate. Dapsone and *p*-amino salicylic acid also act as antifolates.

13.3 Antimicrobial Agents Used in Food Animals for Treatment of Infections

In the United States, the Food and Drug Administration (FDA) recognizes the necessity of therapeutic usage of antimicrobials in food-producing animals. Therapeutic use of antimicrobials is defined by the U.S. FDA as the usage of antibiotics for treatment, control, or prevention of animal diseases (Landers *et al.*, 2012). Table 13.1 summarizes the antimicrobial agents used therapeutically in the United States (Rogers and Haines, 2005). In 2009, the U.S. FDA reported that 13.1 million kilograms of antibiotics were sold for food animal use compared to 3.3 million kilograms sold for human medical use (FDA, 2013).

Since an initial recommendation by the Swann Committee in 1969, several groups including the U.S. FDA and World Health Organization (WHO) have suggested that veterinarians avoid giving antimicrobial agents that are used in human therapy or those which

Table 13.1 Antimicrobial Agents Approved for Therapeutic Use in the United States (Rogers and Haines, 2005).

Antimicrobial (Class and Drug)	Cattle (Beef)	Cattle (Dairy)	Fish	Fowl	Honey Bees	Poultry	Sheep	Swine
Aminoglycosides								
Gentamicin	✓	✓				✓		
Neomycin	✓	✓				✓	✓	✓
Spectinomycin	✓					✓		✓
Streptomycin	✓	✓				✓		✓
Aminopenicillins								
Ampicillin	✓	✓						✓
Amoxicillin	✓	✓						✓
Cephalosporins (3rd gen)								
Ceftiofur	✓	✓				✓	✓	✓
Fluoroquinolones								
Enrofloxacin	✓					✓		
Lincosamides								
Lincomycin Hydrochloride						✓		✓
Macrolides								
Erythromycin	✓	✓				✓		✓
Tylosin	✓	✓				✓		✓
Tilmicosin	✓	✓					✓	✓
Penicillins								
Cloxacillin sodium		✓						
Penicillin G procaine	✓	✓				✓	✓	✓
Penicillin G benzathine	✓							
Peptides								
Bacitracin	✓	✓				✓	✓	✓
Sulfonamides								
Sulfadimethoxine	✓	✓		✓		✓	✓	✓
Sulfamethazine	✓	✓				✓		✓
Sulfanitran						✓		
Sulfaquinoxaline Sulfathiazole	✓	✓				✓		✓
Streptogramins								
Virginiamycin	✓					✓		✓
Tetracyclines								
Chlortetracycline	✓	✓				✓	✓	✓
Oxytetracycline	✓	✓	✓		✓	✓	✓	✓
Tetracycline hydrochloride	✓	✓				✓	✓	✓

promote cross-resistance to food-animals (Kemper, 2008). The United States amended the Food, Drug and Cosmetic Act in 1996 to allow for a new class of Veterinary Feed Directive (VFD) antibiotics that can only be administered by a licensed veterinarian with a VFD (Tollefson *et al.*, 1997). The U.S. FDA occasionally amends its animal drug regulations regarding VFD drugs to better protect human and animal health.

13.4 Antimicrobial Agents Used in Food Animals for Prophylaxis

Antimicrobial agents may be used clinically to treat bacterial infections in food animals, but are more extensively used in agriculture to provide prophylaxis as well as to enhance the growth of healthy food animals (Aarestrup, 1999; Silbergeld *et al.*, 2008). Prevention or prophylaxis refers to the administration of antimicrobial agents to healthy individuals who are considered at risk for infection but in which antimicrobials are administered prior to infectious disease onset as well as prior to the isolation or identification of the causative agents (Marshall and Levy, 2011).

In agriculture, antimicrobial agents are administered to food animals for treatment, prevention, or control of bacterial infectious disease, as well as in growth promotion of food animals (Marshall and Levy, 2011). Sub-therapeutic levels of antibiotics, such as penicillin or tetracycline, are sometimes provided to food animals in their feed in order to enhance their feed/weight efficiency ratios (Marshall and Levy, 2011). From an evolutionary standpoint, antimicrobial use in human clinical medicine, veterinary medicine, agriculture, and in water and soil environments, has resulted in the selection of bacterial variants that are not only single-drug resistant but are multiple drug resistant, an observation that has been extensively documented (Levy and Marshall, 2004; Marshall and Levy, 2011; Silbergeld *et al.*, 2008; Garcia-Migura *et al.*, 2014).

The presence and emergence of genetic determinants encoding antimicrobial resistance factors have been widely studied in these settings (Summers, 2006; Silbergeld *et al.*, 2008; Burgos *et al.*, 2005; Jones *et al.*, 2011; Peng *et al.*, 2008). In farms that have routinely used antibiotics, it has been observed that drug resistant bacteria reside in the gut microbiomes of farm workers and food animals more frequently than those in individuals and food animals from farms that do not use antibiotics (Marshall and Levy, 2011; Silbergeld *et al.*, 2008).

13.5 Antimicrobial Agents Used in Food Animals for Growth Enhancement

Since the early 1950s, antimicrobial agents have been utilized both in poultry and livestock (Anderson *et al.*, 2003). Sub-therapeutic concentrations of antimicrobials have been used for growth promotion and disease prevention, whereas therapeutic concentrations of antimicrobials have been used to treat animals infected with bacterial pathogens (Laxminarayan *et al.*, 2015). These low–dose antibiotics that are used in the food industry for the purpose of performance enhancement of food animals are usually called antibiotic growth promoters (AGPs) (Lin, 2011; Modi *et al.*, 2011). Growth enhancement with sulfonamide antibiotics was first observed in the chicken in 1946 (Moore *et al.*, 1946). The significant growth in chickens, cattle, and pigs with aureomycin (chlortetracycline) was the foundation of antibiotic growth promotion in animals for more than 60 years (Jukes, 1985).

In the United States, at least 17 classes of antimicrobial agents including tetracycline, penicillins, macrolides, lincomycin, and virginiamycin are approved for the improvement of growth and feed efficiency (Anderson *et al.*, 2003). There are four hypotheses that have been proposed to explain the mechanism of action for these AGPs (Feighner and Dashkevicz, 1987):

(i) nutrients are protected by reduction of competing microorganisms;

(ii) nutrients are more efficiently absorbed because of a thinning of the small intestinal barrier;
(iii) the antibiotics decrease the production of growth-depressing toxins or metabolites by intestinal bacteria; and
(iv) there is a reduction or elimination in the incidence of sub-clinical intestinal infections.

Presently, it is unclear which of these postulates, if any, are responsible for the observed performance and growth enhancements observed in food animals.

Bambermycin (flavomycin) is a growth-promoting antibacterial agent used in animal feeds. Bambermycin is primarily active against gram-positive microorganisms and functions by inhibiting peptidoglycan synthesis (Huber and Nesemann, 1968). The streptogramins, a two-component class of antimicrobial agents, consist of component A and component B, which act synergistically. Bacterial cell death by the streptogramins occurs due to the inhibition of protein synthesis (Cocito *et al.*, 1997). Several of these streptogramins, that is, virginiamycin, pristinamycin, and quinupristin/dalfopristin have been marketed either as clinical therapeutics or as growth-promoting agents (Butaye *et al.*, 2003).

Avilamycin, on the other hand, is used only for the promotion of growth in food animals. Avilamycin binds to the 30S subunit of the ribosome and is effective mainly against gram-positive microorganisms (Wolf, 1973). Bacitracin is used both as a therapeutic and growth-promoting agent (Butaye *et al.*, 2003), and it inhibits bacterial cell wall formation, being active mostly against gram-positive bacteria (Siewert and Strominger, 1967). Ionophore antibiotics are divided into the neutral, the carboxylic, and the channel-forming quasi-ionophores. The carboxylic ionophores such as monensin (in bovine) and salinomycin (in porcine) are incorporated into animal feed for growth promotion. These ionophores interfere with ion transport systems of both prokaryotic and eukaryotic cells (Butaye *et al.*, 2003).

The quinoxalines, such as carbadox and olaquidox, are active against gram-negative bacteria and function by inhibiting DNA synthesis (English and Dunegan, 1970; Suter *et al.*, 1978). Although these quinoxalines are considered to be growth promoters in their own right, they are used mostly as therapeutic agents for the prevention of swine dysentery (Butaye *et al.*, 2003).

Efrotomycin is used only as a growth promoter. Its use, however, is limited, being inactive against gram-negative microorganisms and staphylococci (Landini *et al.*, 1993). The mode of action of efrotomycin involves the formation of a non-dissociable complex between the ribosome and elongation factor Tu (EF-Tu), thus inhibiting bacterial growth (Landini *et al.*, 1992).

There are two positive influences of antimicrobial growth promoters on farm productivity. The growth and feed efficiency of animals are enhanced with the increased synthesis of vitamins and other growth factors (Dibner and Richards, 2005; Modi *et al.*, 2011) along with increases in labor and capital productivities due to the use of antimicrobials for hygiene management practices in animal housing or transportation (MacDonald and Wang, 2011; Key and McBride, 2014). Emergence of antimicrobial resistance determinants, however, has been largely attributed to widespread use of antimicrobial agents in food animals when used prophylactically or for growth promotion, and dissemination of antimicrobial resistant bacteria through populations has been facilitated by genetic transfer mechanisms (Silbergeld *et al.*, 2008; Summers, 2006; Aarestrup, 1999). These areas of investigation and agriculture practices continue to be important and intensively studied fields.

13.6 Mechanisms of Bacterial Resistance to Antimicrobial Agents

The occurrence of antibiotic resistant bacteria in food producing animals is, by far, due to excessive use and misuse of antibiotics in animals for growth enhancement and disease prevention purposes. Antimicrobial agents inhibit susceptible bacteria while selecting variants that are not only resistant to the antimicrobials used but such variants are also often multiple drug resistant (Marshall and Levy, 2011). There are various cell molecular mechanisms to which bacteria can acquire antibiotic resistance, such as (a) drug target modification (b) enzymatic degradation of antimicrobial agents (c) reduced drug permeability and (d) active efflux of drugs from bacterial pathogens (Kumar and Varela, 2013; Floyd *et al.*, 2013; Blair *et al.*, 2015, He, Thorpe, *et al.*, 2011; He, Zhang, *et al.*, 2011).

13.6.1 Drug Target Modification

Many antibiotic resistance genes have been found in food producing animals and their DNA sequences determined. The sulfonamide resistance gene *sul1*II has been isolated from R plasmids of *E. coli* from pigs (Sunde *et al.*, 1998). It appears that this gene has spread between various ecological niches within a short time as there is no sign of a point mutation which would otherwise indicate genetic drift (Sunde *et al.*, 1998).

Likewise, aminoglycoside resistance is due to the modification of aminoglycoside molecules by enzymes that are encoded by a variety of gene cassettes in class 1 integrons. These genes encode acetyltransferases and phosphotransferases which then modify the aminoglycoside molecules. For example, the *add*A1 cassette can be found in *E. coli* from domesticated birds (Bass *et al.*, 1999), in *V. cholerae*, and the fish pathogen *Aeromonas salmonicida* (L'Abee-Lund and Sorum, 2001). Similarly, trimethoprim resistance genes from gram-negative bacteria are located as gene cassettes (family 1 of *dfr* genes) on integrons (Huovinen *et al.*, 1995). These genes are also found on plasmids in many gram-negative and in gram-positive bacteria (Huovinen *et al.*, 1995).

13.6.2 Reduced Drug Permeability

In gram-negative bacteria, resistance to aminoglycosides and tetracyclines are due to decreased cellular entry of the antimicrobials into bacterial cells, often by reducing expression of solute transporter entry systems (Chopra and Roberts, 2001; Nelson and Levy, 2011; Wright, 2005). On the other hand, aminoglycosides may be enzymatically destroyed in resistant strains, and in tetracycline-resistant gram-positive bacteria, ribosomal protection proteins with homology to the elongation factors EF-Tu and EF-G are produced, in addition to the integral membrane drug efflux pumps for these antimicrobials (Chopra and Roberts, 2001; Wright, 2005).

13.6.3 Enzymatic Inactivation of Drugs

Bacteria have evolved powerful mechanisms of neutralizing antibiotics by means of enzymatic hydrolysis of drugs, group transfer, and redox process (Davies, 1994). The β-lactam group of antibiotics such as penicillins and cephalosporins are inactivated by β-lactamases, which hydrolyze the β-lactam ring of these antibiotics rendering them ineffective. TEM-1 was the first β-lactamase enzyme discovered in *E. coli* from a patient named Temoniera in Greece (Datta and Kontomichalou, 1965), followed by SHV-1 β-lactamase (sulfhydryl variable active site) (Livermore, 1995). TEM-1 and SHV-1 share 68% amino acid sequence similarity. The TEM-2 enzyme differs from TEM-1 by only one amino acid (Paterson and Bonomo, 2005). TEM-1, TEM-2, and SHV-1 are narrow-spectrum β-lactamases highly active against ampicillin and first generation cephalosporins, but not against oxyimino cephalosporins.

Subsequently, several variants of β-lactamases with broader substrate range were identified in gram-negative bacteria, and all of these have evolved from either TEM or SHV enzymes due to point mutations in the genes encoding them under antibiotic pressure (Philippon *et al.*, 1989). TEM and SHV β-lactamases were thought to be the most prevalent β-lactamases until the discovery of a third type, a cefotaxime-hydrolyzing enzyme (CTX-M), in *Klebsiella pneumoniae* (Bonnet, 2004). CTX-M preferentially hydrolyzes cefotaxime and has only 40% similarity with TEM and SHV enzymes (Tzouvelekis *et al.*, 2000).

Currently, more than 150 amino acid variants of CTX-M enzymes belonging to five major phylogenetic groups—CTX-M-1, CTX-M-2, CTX-M-8, CTX-M-9, CTX-M-25—have been identified, with CTX-M-15 being the most prevalent worldwide (Pitout and Laupland, 2008; Rossolini *et al.*, 2008). CTX-M enzymes have replaced TEM and SHV as the most prevalent ESBL enzymes in the Enterobacteriaceae family of bacteria. These enzymes are collectively called extended spectrum β-lactamases (ESBLs) owing to their ability to hydrolyze penicillins and broad-spectrum β-lactams such as cefotaxime, enzymes, and monobactams such as aztreonam (Philippon *et al.*, 1989), but not cephamycins and carbapenems (Jacoby and Carreras, 1990). β-lactamase inhibitors such as clavulanate, sulbactam, and tazobactam inhibit ESBLs.

β-lactamases are classified into four groups (A-D) in Ambler's scheme of classification based on their amino acid homology (Ambler, 1980). Subsequently a classification scheme was proposed (Bush *et al.*, 1995) based on functional homology and susceptibility to inhibitors. The Bush-Jacoby-Medeiros classification scheme consists of four major groups—Group I (Ambler Class C) β-lactamases (also known as AmpC enzymes), Group 2 (Ambler Class A) enzymes, Group 3 (Ambler Class B) enzymes, and Group 4 β-lactamases, which are further divided into several subgroups (a–f) (Paterson and Bonomo, 2005). Enzymes belonging to classes A and C occur more frequently than others. ESBLs belong to the Ambler class A group and subgroup 2be of the Bush-Jacoby-Medeiros classification scheme.

Members of the family Enterobacteriaceae such as *E. coli* and *Klebsiella pneumoniae* are the major producers of ESBL among Gram-negative bacteria (Nathisuwan *et al.*, 2001). ESBL-producing non-Enterobacteriaceae such as *Acinetobacter* and *Pseudomonas aeruginosa* are also increasingly being reported in hospitals (Nordmann and Guibert, 1998). Many of these also produce AmpC cephalosporinases, which are resistant to β-lactamase inhibitors (Minami *et al.*, 1980). Unlike ESBLs, AmpC enzymes can hydrolyze cephamycins and are resistant to most β-lactams except carbapenems and fourth-generation cephalosporins (Sanders *et al.*, 1996). AmpC lactamases belong to Ambler's class C and to group 1 of the Bush-Jacoby-Medeiros classification (Jacoby and Carreras, 1990).

Chromosomal- or plasmid-encoded AmpC enzymes are either constitutive or inducible. Constitutively expressed, low levels of AmpC enzymes do not confer significant antibiotic resistance. However, mutations in AmpC-encoding genes can result in constitutive, high-level expression of the enzyme. Alternatively, when an inducible AmpC enzyme is present, the presence of β-lactam antibiotics can trigger the production of enzyme to a level necessary to confer antibiotic resistance (Jacoby and Carreras, 1990). Several members of the Enterobacteriaceae family produce inducible AmpC enzymes resulting in high levels of resistance to penicillins and cephalosporins (Bauernfeind *et al.*, 1998). AmpC enzymes are susceptible to inhibition by oxacillin, cloxacillin, and aztreonam.

The ability of ESBL-producing bacteria such as *E. coli* and *K. pneumoniae* to resist multiple cephalosporins and expand their resistance spectrum quickly has tremendously constricted the chemotherapeutic options against them (Talbot *et al.*, 2006). Apart from fluoroquinolones, carbapenems, tigecycline, and colistin are advocated for treating infections with ESBL-producing enterobacteria. However, ESBL-producing bacteria resistant to multiple fluoroquinolones and carbapenems can potentially lead to treatment failures.

The other important enzymatic mechanisms of antibiotic resistance involve group transfer and redox processes. Transfer of functional groups such as acyl, ribosyl, phosphoryl, or thiol groups to the antibiotics aminoglycosides, fosfomycin, macrolides, lincomycin, and chloramphenicol result in irreversible structural alteration of these antibiotics and loss of antimicrobial activities (Davies, 1994; Wright, 2005).

Some examples of such enzymes include chloramphenicol acetyltransferases (CATs) (Schwarz *et al.*, 2004), aminoglycoside acetyltransferases (AACs), streptogramin acetyl transferases (VATs), aminoglycoside phosphotransferases (APHs), macrolide kinases (MPHs), ADP-ribosyltransferases of rifampin (ARRs), nucleotidyl transferases (ANTs and Lin) of aminoglycosides, plus lincomycin and macrolide glycosyltransferases (Wright, 2005).

The redox mechanism is relatively less prevalent compared other bacterial mechanisms of enzyme-based antibiotic resistance. The redox process alters the $Mg^{+}2$ sequestering ability of tetracycline by transfer of a single hydroxyl group at position 11a by a flavin-dependent monoxygenase enzyme TetX. The modified tetracycline is less able to bind to its bacterial targets resulting in loss of its activity (Park and Levy, 1988; Yang *et al.*, 2004). Resistance to tigecycline by a transposon-encoded TetX has also been reported (Moore *et al.*, 2005).

13.6.4 Active Drug Efflux

An important resistance mechanism against antimicrobial agents and other toxic compounds utilized by bacteria is active drug efflux (Levy, 1992b). Drug efflux pumps, known as integral membrane solute transporters, prevent the accumulation of drugs inside bacterial cells, thus reducing their intracellular concentrations to sub- or non-inhibitory levels (Kumar and Varela, 2013). The slow phase of the antibiotic efflux provides sufficient time for bacteria to adapt to the antibiotics, resulting in drug resistance due to mutations or alteration of antibiotic targets (Li and Nikaido, 2004). Furthermore, the resistance to antimicrobials can be enhanced by the over-expression of drug efflux pumps (Okusu *et al.*, 1996).

These efflux pumps may interact with a wide range of completely unrelated pharmacological classes, and, therefore, have broad substrate specificities (Van Bambeke *et al.*, 2000). Bacterial efflux pumps are divided into several categories, based on the physiological mechanisms by which drugs are transported across the biological membrane - often against concentration gradients. In the first category, primary active transporters export drugs across the membrane by using the hydrolysis of ATP as the driving energy (Higgins, 2007). The secondary active transporters export the drug across the membrane by using the energy stored in ion gradients that were in turn generated by respiration (Henderson, 1991; Henderson *et al.*, 1993; Krämer, 1994; Poolman and Konings, 1993). The phosphotransferase system (PTS) is the third category that transports drugs with an associated phosphorylation of the drug (Kumar *et al.*, 2011; Parr and Saier, 1992).

In addition to single-drug efflux pumps (Levy, 2002; Alekshun and Levy, 2007; Varela and Griffith, 1993; Varela *et al.*, 1995), bacteria may also express multi-drug efflux pumps to extrude structurally-different antimicrobial agents (Kumar and Varela, 2012; Alekshun and Levy, 2007; Bhardwaj and Mohanty, 2012; George, 1996). Multi-drug efflux pumps actively extrude antibiotics from bacterial cells, thus becoming multi-drug resistant (Saidijam *et al.*, 2011; Webber and Piddock, 2003). These multi-drug efflux pumps represent a serious health concern due to a diminishment in the efficacy of chemotherapy for infection caused by bacteria that harbor these pumps (Gootz, 2010; Kumar and Schweizer, 2005). Efflux-mediated resistance also provides baseline resistance for the emergence of further resistance mechanisms such as drug inactivation or drug target modification (Ian-Zhi and Hiroshi, 2009).

The bacterial drug efflux transporters are divided into five different superfamilies based on the source of energy, amino acid sequence, predicted secondary protein structures and phylogenetic relationships: (i) the major facilitator superfamily (MFS), (ii) the small multi-drug resistance superfamily (SMR), (iii) the multiple antimicrobial extrusion protein superfamily (MATE), (iv) the ATP-binding cassette superfamily (ABC), and (v) the resistance-nodulation-cell division superfamily (RND) (Kumar and Varela, 2012; Kumar, Mukherjee, and Varela, 2013; Floyd *et al.*, 2013).

13.7 The Major Facilitator Superfamily of Solute Transporters

The major facilitator superfamily contains thousands of solute transporters that are related in terms of their primary sequences and protein structures but differ with respect to the nature of their solutes, mode of energization, and direction of solute transport across the membrane (Marger and Saier, 1993). The superfamily was discovered by Henderson and coworkers when they observed that sugar transporters from different organisms were related to each other and to other antimicrobial transporters (Henderson, 1990; Maiden *et al.*, 1987).

As more of these solute transporters were included within the growing superfamily, it became clear that these seemingly distinct solute transporters shared highly conserved amino acid sequence motifs (Henderson, 1990), some motifs of which are critically important in the solute transport cycle (Jiang *et al.*, 2013) or in hinge conformational changes inherent in substrate translocation across the membrane (Yaffe *et al.*, 2013). Several closely studied and key transporters of the major facilitator superfamily show that the overall protein structures are similar to each other (Yin *et al.*, 2006; Jiang *et al.*, 2013; Deng *et al.*, 2014), suggesting that these various solute transporters operate by a similar mechanism to catalyze translocation across the biological membrane (Griffith *et al.*, 1992). Presently, the transporters of this superfamily are categorized in a very useful Transporter Classification Database (TCD) http://www.tcdb.org (Saier *et al.*, 2014).

13.8 Key Bacterial Multidrug Efflux Pump Systems of the Major Facilitator Superfamily

13.8.1 *Escherichia coli* MFS Efflux Pumps

Table 13.2 shows characterized MFS efflux pumps from the bacterium *E. coli*. One of the best understood multidrug efflux pumps of the major superfamily from *E. coli* is EmrD, which was originally discovered as a determinant that conferred adaptation to a condition of low energy shock, and thus, resistance to uncouplers of oxidative phosphorylation (Naroditskaya *et al.*, 1993), suggesting that EmrD catalyzed the transport of *m*-chloro carbonyl cyanide phenylhydrazone (CCCP).

Subsequently, evidence was further provided to indicate that EmrD conferred two-fold resistances against benzalkonium chloride, a quaternary ammonium compound, and sodium dodecyl sulfate, a detergent (Nishino and Yamaguchi, 2001). In 2006, a high-resolution crystal structure for EmrD was determined (Yin *et al.*, 2006). The general structural features of EmrD include a set of 12 transmembrane α-helices where the N- and C-termini of

Table 13.2 MFS Multidrug Efflux Pumps of *Escherichia coli.*

Pump	No. Amino Acids	No. TMS	Substrates	References
EmrD	394	12	CCCP, tetrachlorosalicylanilide, sodium dodecyl sulfate (SDS), benzalkonium, and quaternary ammonium compound (QAC)	Naroditskaya *et al.*, 1993; Yin *et al.*, 2006
EmrB	513	12	carbonyl cyanide m-chlorophenylhydrazone (CCCP), nalidixic acid	Lomovskaya and Lewis, 1992
MdfA	410	12	benzalkonium, chloramphenicol, ciprofloxacin, daunomycin, erythromycin, ethidium, isopropyl-1-thio-β-D-galactopyranoside (IPTG), kanamycin, neomycin, norfloxacin, puromycin, rhodamine, rifampin, tetraphenylphosphonium, thiamphenicol, tetracycline, dequalinium	Edgar and Bibi, 1997

the protein reside within the cytoplasmic side of the inner-membrane of *E. coli* (Yin *et al.*, 2006). The EmrD crystal structure further shows two relatively asymmetric N- and C-terminal halves or bundles (but with a two-fold molecular axis of symmetry) in which the channel is closed to the aqueous periplasmic side of the membrane, presumably in a closed intermediate state between the various conformations that occur during the transport cycle (Yin *et al.*, 2006).

Another well studied multidrug efflux pump of the major facilitator superfamily from *E. coli* is known as MdfA, which confers resistance in host cells to an extremely diverse array of structurally distinct antimicrobial substrates, such as benzalkonium, chloramphenicol, ciprofloxacin, daunomycin, erythromycin, ethidium, kanamycin, neomycin, norfloxacin, puromycin, tetracycline, rifampin, rhodamine, and tetraphenylphosphonium (Edgar and Bibi, 1997; Mine *et al.*, 1998). MdfA has also been referred to as Cmr (Nilsen *et al.*, 1996) and shown to catalyze proton-dependent antiport of chloramphenicol across the membrane (Mine *et al.*, 1998).

In light of the ability to transport such a broad spectrum of structurally different drug substrates, it was shown that MdfA could simultaneously bind two or more of these substrates at different binding sites in the transporter (Lewinson and Bibi, 2001). MdfA was later shown to mediate proton-dependent transport of chloramphenicol and thiamphenicol accompanied by a net charge movement across the membrane (i.e., electrogenic drug/proton antiport) as well as a proton-dependent transport of monovalent substrates, such as Hoechst 33342, accompanied by no net charge movement, that is, electroneutral drug/proton antiport (Lewinson *et al.*, 2003). In a closer examination of the conformational changes that occur during multidrug transport, it was shown that structurally distinct drugs, whether substrates of MdfA, could trigger the changes in transporter conformation, suggesting that occupation of the drug binding pocket with drug molecules confers substrate promiscuity and drives the catalytic mechanism (Fluman *et al.*, 2009).

Future work will be necessary to determine whether this promiscuous conformation switching discovered in MdfA also occurs in other multidrug transporters of the major superfamily. More recently, it was shown that by either random mutagenesis or by rational design, the drug/proton stoichiometry of MdfA could be altered to become a divalent cation–mediated antiporter, indicating that multidrug efflux pumps of the major facilitator superfamily could be promiscuous in terms of the proton motive force, in addition to their antimicrobial drug substrates (Tirosh *et al.*, 2012).

Clearly, these molecular aspects pertaining to multidrug efflux in *E. coli* and the homology shared between transporters of the major facilitator superfamily may thus be relevant to many other solute transporters, either in clinical medicine to circumvent drug resistance or in biotechnology to allow transport of economically desirable substrates.

13.8.2 *Staphylococcus aureus* MFS Efflux Pumps

Recently reviewed by Costa *et al.* (2013), eight physiologically studied multidrug resistant efflux pumps have so far been established to be members of the major facilitator superfamily in the gram-positive microorganism *S. aureus*, a serious pathogen (see Table 13.3). Of these, two key multidrug efflux pumps are NorA (Piddock, 1999) and QacA (Brown and Skurray, 2001). NorA from *S. aureus* was discovered in the late 1980s (Ubukata *et al.*, 1989; Yoshida *et al.*, 1990) and shown to confer resistance to multiple quinolones and fluoroquinolones (Yoshida *et al.*, 1990). NorA was also demonstrated to be homologous not only to another multidrug efflux pump, Bmr, from *Bacillus subtilis* (Neyfakh, 1992) but also to the sub-family of single-drug antiporters for tetracycline, TetA (Levy, 1992a; Nelson and Levy, 2011). The plasmid-encoded determinant for QacA of *S. aureus* confers resistance to acriflavine, ethidium, quaternary ammonium compounds, propamidine isethionate, and diamidinodiphenylamine (Tennent *et al.*, 1989). Interestingly, QacA was

Table 13.3 MFS Multidrug Efflux Pump of *Staphylococcus aureus*.

Pump	No. Amino Acids	No. TMS	Substrates	References
QacA	514	14	quaternary ammonium compounds: benzalkonium and cetrimide; damidines: propamidine isethionate, diamidinophenylamine dihydrochloride	Tennent *et al.*, 1989
NorA	388	12	norfloxacin, ciprofloxacin, rhodamine, ethidium and quaternary ammonium compounds (e.g., benzalkonium chloride and tetraphenylphosphonium)	Ubukata, Itoh-Yamashita, and Konno, 1989; Ohshita, Hiramatsu, and Yokota, 1990; Yoshida *et al.*, 1990
MdeA	479	14	tetraphenylphosphonium chloride, norfloxacin, rhodamine 6G, doxorubicin, daunorubicin	Huang *et al.*, 2004
LmrS	480	14	linezolid, tetraphenylphosphonium chloride, trimethoprim, florfenicol, chloramphenicol, erythromycin, streptomycin, fusidic acid, kanamycin, lincomycin, gatifloxacin, oxytetracycline	Floyd *et al.*, 2010

shown to be related to sugar transporters of the major facilitator superfamily (Rouch *et al.*, 1990).

Furthermore, QacA represents an excellent first example of a major facilitator superfamily multidrug efflux pump for which physiological evidence exists in which fusions of QacA with enzymatic reporters alkaline phosphatase and β-galactosidase show that QacA possesses 14 transmembrane domains (Paulsen *et al.*, 1996). Structure-function analyses have shown the importance of charged amino acids (Hassan *et al.*, 2007; Xu *et al.*, 2006) and tyrosine residues (Wu *et al.*, 2008) in mediating drug transport in QacA. This transporter continues to be a clinically important factor in infections caused by multidrug resistant *S. aureus* pathogens (Wassenaar *et al.*, 2015).

More recently, the multidrug efflux pump, LmrS, from a clinical isolate of methicillin resistant *S. aureus* was discovered (Floyd *et al.*, 2010) and demonstrated to confer resistance to multiple structurally distinct antimicrobial substrates, such as linezolid, ethidium, trimethoprim, florfenicol, streptomycin, fusidic acid, kanamycin, lincomycin, gatifloxacin, oxytetracycline, rifampin, cetrimide, and sulfadimethoxine. It was further demonstrated that LmrS from *S. aureus* was able to catalyze the active transport of ethidium from host cells (Floyd *et al.*, 2010). LmrS contains 14 predicted transmembrane segments and appears to be related to LmrB of *B. subtilis*, EmrB of *E. coli*, and VceB of *V. cholerae* (Floyd *et al.*, 2010). Because of the serious nature of clinical infections caused by *S. aureus* and its drug resistant variants, these and other efflux pumps of the major facilitator superfamily will become increasingly important as targets for modulation (Andersen *et al.*, 2015; Grkovic *et al.*, 2002; Schindler *et al.*, 2013; Stavri *et al.*, 2007; Kumar *et al.*, 2013).

13.8.3 *Vibrio cholerae* MFS Efflux Pumps

Thus far, only two multidrug efflux pumps belonging to the major facilitator superfamily from the important pathogen *V. cholerae* have been physiologically characterized (Table 13.4). The first such transporter VceB is well studied (Colmer *et al.*, 1998; Woolley *et al.*, 2005). VceB, encoded within the *vceCAB* operon, is predicted to harbor 14 transmembrane domains and confer resistance to CCCP, deoxycholate,

Table 13.4 MFS Multidrug Efflux Pump of *Vibrio cholerae*.

Pump	No. Amino Acids	No. TMS	Substrates	References
EmrD-3	379	12	linezolid, tetraphenylphosphonium chloride, rifampin, ethidium bromide, minocycline, erythromycin, trimethoprim, chloramphenicol, rhodamine 6G, oxytetracycline, tetracycline, nalidixic acid, florfenicol	Smith, Kumar, and Varela, 2009
VceB	117	14	chloramphenicol, nalidixic acid, CCCP	Colmer, Fralick, and Hamood, 1998

nalidixic acid, and chloramphenicol (Colmer *et al.*, 1998). In the same study, it was found that the inner membrane bound VceB works with TolC in the outer membrane and VceA in the periplasm to form a tripartite complex to expel its substrates from the cell (Colmer *et al.*, 1998; Woolley *et al.*, 2005), similar to the case with EmrAB (Lomovskaya *et al.*, 1995).

The more recently discovered multidrug efflux pump, EmrD-3 from *V. cholerae*, belongs to the major facilitator superfamily and confers resistance to many structurally diverse antimicrobial agents such as linezolid, rifampin, ethidium, minocycline, erythromycin, trimethoprim, chloramphenicol, rhodamine, oxytetracycline, tetracycline, nalidixic acid, and florfenicol (Smith *et al.*, 2009). This newer determinant was predicted to harbor 12 transmembrane domains and shown to mediate efflux of ethidium (Smith *et al.*, 2009). Both VceB and EmrD-3 would make suitable targets for modulation in order to eventually restore the efficacy of antimicrobial agents to which *V. cholerae* are presently resistant (Varela *et al.*, 2013).

13.8.4 *Salmonella enterica* MFS Efflux Pumps

Of the drug efflux pumps of the major facilitator superfamily thought to reside in *S. enterica* (Andersen *et al.*, 2015), only a relatively few have been characterized physiologically (see Table 13.5). One of these particular *Salmonella* transport systems, EmrAB, has been characterized at the molecular level (Nishino *et al.*, 2006). The determinant encoding EmrB from a fluoroquinolone-resistant mutant of *S. enterica* serovar Typhimurium was observed to be overexpressed (Chen *et al.*, 2007), confer resistance to deoxycholate, nalidixic acid, novobiocin, and triclosan (Rensch *et al.*, 2014), and require TolC on the outer membrane and EmrA in the periplasm to form a tripartite complex with EmrB to produce drug resistance activity (Horiyama *et al.*, 2010). This *S. enterica* EmrAB-TolC multidrug efflux system is reminiscent to that known in *E. coli* (Lomovskaya and Lewis, 1992). On the other hand, a related multidrug efflux pump MdfA from *S. enterica* does not appear to require TolC (Horiyama *et al.*, 2010), even though MdfA is a member of the major facilitator superfamily and presumably resides in the inner membrane (Nishino *et al.*, 2006). Cells harboring the MdfA efflux pump were demonstrated to confer resistance to chloramphenicol, doxorubicin, norfloxacin, and tetracycline (Nishino *et al.*, 2006).

Recently, additional multidrug efflux pumps belonging to the major facilitator superfamily were studied physiologically for their potential to confer the multidrug resistance phenotype. For example, SmvA confers resistance to acriflavine, ethidium, malachite green, and pyronin B, and requires TolC in order to do so (Hongo *et al.*, 1994; Villagra *et al.*, 2008). FloR confers resistance to florfenicol, chloramphenicol, and thiamphenicol (Braibant *et al.*, 2005). In a separate study, evidence showed that STY0901 transports benzalkonium chloride, and STY2458 confers resistance to doxycycline, tetracycline, and kanamycin (Shaheen *et al.*, 2015). Interestingly, in preliminary phylogenetic

Table 13.5 MFS Multidrug Efflux Pump of *Salmonella enterica*.

Pump	No. Amino Acids	No. TMS	Substrates	References
EmrB	513	12	fluoroquinolones	Chen *et al.*, 2007; Balleste-Delpierre *et al.*, 2014
SmvA	126	14	acriflavine, ethidium bromide, malachite green, pyronin B	Hongo *et al.*, 1994; Villagra *et al.*, 2008
FloR	404	12	chloramphenicol, florfenicol, thiamphenicol	Braibant *et al.*, 2005
MdfA	302	12	tetracycline, chloramphenicol, norfloxacin, doxorubicin	Nishino, Latifi, and Groisman, 2006
STY0901	403	12	benzalkonium chloride	Shaheen *et al.*, 2015
STY2458	396	12	tetracycline, doxycycline, kanamycin	Shaheen *et al.*, 2015

studies of multidrug efflux pumps from the major facilitator superfamily (Andersen *et al.*, 2015; Shrestha *et al.*, 2015) that are encoded on the *S. enterica* genome (McClelland *et al.*, 2001; Parkhill *et al.*, 2001), approximately 46 belong to the major facilitator superfamily, most of which have not been characterized. In conclusion, therefore, much molecular physiological work remains to be completed in order to fully understand the scope of drug and multidrug resistances conferred by solute transporters of this important superfamily.

13.9 Future Directions

As effective chemotherapeutic efforts are compromised by drug resistant and especially multidrug resistant bacterial food pathogens, infections caused by such antimicrobial resistant bacteria will continue to remain serious public health concerns (Marshall and Levy, 2011; Levy, 2001). In addition to improving widespread education with respect to personal hygiene, food handling, food preparation, and proper hand washing, it will also become necessary to invoke proper antimicrobial stewardship in clinical medicine (MacDougall and Polk, 2005), in veterinary medicine (Guardabassi and Prescott, 2015; Prescott, 2014) and in agriculture (Woolhouse *et al.*, 2015). A new push to develop antimicrobial agents will be necessary and efficacious (Ling *et al.*, 2015) especially antimicrobials with novel modes of action. Comparative genomic analysis between pathogenic bacteria and their non-virulent counterparts may allow the discovery of new cellular and molecular targets (Mukherjee *et al.*, 2014).

Due to the presence of solute transporter systems in all living organisms, modulation of multidrug drug efflux pumps, either through regulation of gene expression or through efflux pump inhibition are promising avenues (Kumar, Mukherjee, and Varela, 2013; Lewis, 2001). In particular, proteins of the major facilitator superfamily are found in living organisms across all known taxa (Saier *et al.*, 1999; Pao, Paulsen, and Saier, 1998; Fluman and Bibi, 2009). Therefore, these bacterial multidrug efflux pumps require further study.

Towards molecular physiological studies of major facilitator superfamily proteins, several key areas are of particular interest. First, study of the structure-function relationships of these transporters will enhance our understanding of their mechanisms for substrate translocation across the membrane, and such molecular structures will make good targets for development of new inhibitors. Mutational analysis and elucidation of additional crystal structures of these multidrug efflux pumps may enhance our insight into this important question.

Second, it will be important to have a good understanding of the molecular mechanisms responsible for determining the modes of driving energy in a transporter, whether passive or active; and if secondary active transport is operational in a drug transporter, it will be important to understand how the energy stored in the ion gradients generated by respiration are coupled to the transport and accumulation of drug substrate on one or other side of the membrane.

Third, it will be necessary to understand the transporter machinery involved in dictating the direction of solute transport, that is, symport versus antiport. Fourth, despite the immense effort directed at comprehending how the efflux pumps have such diverse and seemingly extreme substrate promiscuity, it remains a mystery, nonetheless, how the multidrug efflux pumps determine substrate number (single drug versus multiple drugs).

Fifth, it is further unclear how each of these transporters accomplish their multiple drug specificities. Sixth, it is poorly understood how the efflux pumps translocate multiple substrates through a pump while keeping unwanted substrates out. Seventh, we do not know, yet, what determines the ion specificity (proton versus sodium) in the antimicrobial efflux pumps of the major facilitator superfamily. Eighth, it will continue to be important to closely study the evolution of the multidrug efflux pumps as they move through living populations and influence foodborne infectious disease. Ninth, on a higher plane, it is still unknown to what extent a common molecular mechanism exists for drug transport catalysis, especially considering that these efflux pumps collectively harbor such a diverse array of substrates. In short, there remains plenty of work for present and future investigators.

Acknowledgments

This publication was supported in part by a grant from the National Institute of General Medical Sciences (P20GM103451) of the National Institutes of Health and by an internal research grant, ENMU.

References

Aarestrup, F.M. (1999). Association between the consumption of antimicrobial agents in animal husbandry and the occurrence of resistant bacteria among food animals. *Int J Antimicrob Agents, 12*(4), pp.279–85.

Alekshun, M.N. and Levy, S.B. (2007). Molecular mechanisms of antibacterial multidrug resistance. *Cell, 128*(6), pp.1037–1050.

Ambler, R.P. (1980). The structure of β-lactamases. *Philos Trans R Soc Lond B Biol Sci., 289*(1036), pp. 321–31.

Andersen, J.L., et al. (2015). Multidrug efflux pumps from Enterobacteriaceae, *Vibrio cholerae* and *Staphylococcus aureus* bacterial food pathogens. *Int J Environ Res Public Health, 12*(2), pp. 1487–547.

Anderson, A.D., et al. (2003). Public health consequences of use of antimicrobial agents in food animals in the United States. *Microbial Drug Resistance, 9*(4), pp. 373–379.

Appelbaum, P.C. (2006). The emergence of vancomycin-intermediate and vancomycin-resistant *Staphylococcus aureus. Clinical Microbiology and Infection, 12*(s1), pp. 16–23.

Balleste-Delpierre, C., et al. (2014). Molecular study of quinolone resistance mechanisms and clonal relationship of *Salmonella enterica* clinical isolates. *Int J Antimicrob Agents, 43*(2), pp. 121–125.

Bass, L., et al. (1999). Incidence and characterization of integrons, genetic elements mediating multiple-drug resistance, in avian *Escherichia coli. Antimicrob Agents Chemother., 43*(12), pp. 2925–2929.

Bauernfeind, A., Chong, Y. and Lee, K. (1998). Plasmid-encoded AmpC β-lactamases: how far have we gone 10 years after the discovery? *Yonsei Med J.*, *39*(6), pp. 520–525.

Bhardwaj, A. and Mohanty, P. (2012). Bacterial efflux pumps involved in multidrug resistance and their inhibitors: rejuvinating the antimicrobial chemotherapy. *Recent Patents on Anti-Infective Drug Discovery*, *7*(1), pp. 73–89.

Blair, J., et al. (2015). Molecular mechanisms of antibiotic resistance. *Nature Reviews Microbiology*, *13*(1), pp. 42–51.

Bonnet, R. (2004). Growing group of extended-spectrum β-lactamases: the CTX-M enzymes. *Antimicrob Agents Chemother*, *48*(1), pp. 1–14.

Braibant, M., et al. (2005). Structural and functional study of the phenicol-specific efflux pump FloR belonging to the major facilitator superfamily. *Antimicrob Agents Chemother*, *49*(7), pp. 2965–2971.

Brown, M.H. and Skurray, R.A. (2001). Staphylococcal multidrug efflux protein QacA. *J Mol Microbiol Biotechnol*, *3*(2), pp. 163–170.

Bryan, F.L. and Doyle, M.P. (1995). Health risks and consequences of *Salmonella* and *Campylobacter jejuni* in raw poultry. *Journal of Food Protection*, *58*(3), pp. 326–344.

Burgos, J.M., Ellington, B.A. and Varela, M.F. (2005). Presence of multidrug-resistant enteric bacteria in dairy farm topsoil. *J Dairy Sci*, *88*(4), pp. 1391–1398.

Bush, K., Jacoby, G.A. and Medeiros, A.A. (1995). A functional classification scheme for β-lactamases and its correlation with molecular structure. *Antimicrob Agents Chemother.*, *39*(6), pp. 1211–1233.

Butaye, P., Devriese, L.A. and Haesebrouck, F. (2003). Antimicrobial growth promoters used in animal feed: effects of less well known antibiotics on gram-positive bacteria. *Clinical Microbiology Reviews*, *16*(2), pp. 175–188.

Callaway, TR, et al. (2008). Gastrointestinal microbial ecology and the safety of our food supply as related to *Salmonella*. *Journal of Animal Science*, *86*(14 suppl), pp. E163–E172.

Catry, B., et al. (2010). Reflection paper on MRSA in food-producing and companion animals: epidemiology and control options for human and animal health. *Epidemiology and Infection*, *138*(05), pp. 626–644.

Chambers, H.F. and DeLeo, F.R. (2009). Waves of resistance: *Staphylococcus aureus* in the antibiotic era. *Nature Reviews Microbiology*, *7*(9), pp. 629–641.

Chen, H.D. and Frankel, G. (2005). Enteropathogenic *Escherichia coli*: unravelling pathogenesis. *FEMS Microbiol Rev.*, *29*(1), pp. 83–98.

Chen, S., et al. (2007). Contribution of target gene mutations and efflux to decreased susceptibility of *Salmonella enterica* serovar typhimurium to fluoroquinolones and other antimicrobials. *Antimicrob Agents Chemother.*, *51*(2), pp. 535–542.

Chopra, I. and Roberts, M. (2001). Tetracycline antibiotics: mode of action, applications, molecular biology, and epidemiology of bacterial resistance. *Microbiol Mol Biol Rev.*, *65*(2), pp. 232–260.

Cocito, C., et al. (1997). Inhibition of protein synthesis by streptogramins and related antibiotics. *Journal of Antimicrobial Chemotherapy*, *39*(suppl 1), pp. 7–13.

Colmer, J. A., Fralick, J.A. and Hamood, A.N. (1998). Isolation and characterization of a putative multidrug resistance pump from *Vibrio cholerae*. *Mol Microbiol.*, *27*(1), pp. 63–72.

Costa, S.S., et al. (2013). Multidrug Efflux Pumps in *Staphylococcus aureus*: an Update. *Open Microbiol J.*, *7*, pp. 59–71.

Datta, N. and Kontomichalou, P. (1965). Penicillinase synthesis controlled by infectious R factors in Enterobacteriaceae. *Nature*, *208*(5007), pp. 239–241.

Davies, J. (1994). Inactivation of antibiotics and the dissemination of resistance genes. *Science*, *264*(5157), pp. 375–382.

Deng, D., et al. (2014). Crystal structure of the human glucose transporter GLUT1. *Nature*, *510*(7503), pp. 121–125.

Dibner, J.J. and Richards, J.D. (2005). Antibiotic growth promoters in agriculture: history and mode of action. *Poultry Science*, *84*(4), pp. 634–643.

Edgar, R. and Bibi, E. (1997). MdfA, an *Escherichia coli* multidrug resistance protein

with an extraordinarily broad spectrum of drug recognition. *J Bacteriol., 179*(7), pp. 2274–2280.
Elder, R.O. et al. (2000). Correlation of enterohemorrhagic *Escherichia coli* O157 prevalence in feces, hides, and carcasses of beef cattle during processing. *Proceedings of the National Academy of Sciences, 97*(7), pp. 2999–3003.
English, A.R. and Dunegan, C.M. (1970). Quinoxaline-1, 4-di-N Oxides I. Inhibition of Deoxyribonucleic Acid Synthesis in *Escherichia coli* by 2, 3-Dihydroxymethyl-quinoxaline-1, 4-di-N-oxide. *Experimental Biology and Medicine, 133*(2), pp. 398–400.
Feighner, S.D. and Dashkevicz, M.P. (1987). Subtherapeutic levels of antibiotics in poultry feeds and their effects on weight gain, feed efficiency, and bacterial cholyltaurine hydrolase activity. *Applied and Environmental Microbiology, 53*(2), pp. 331–336.
Fitzgerald, J.R. (2012). Livestock-associated *Staphylococcus aureus*: origin, evolution and public health threat. *Trends in Microbiology, 20*(4), pp. 192–198.
Floyd, J.L., et al. (2010). LmrS is a multidrug efflux pump of the major facilitator superfamily from *Staphylococcus aureus*. *Antimicrob Agents Chemother., 54*(12), pp. 5406–5412.
Floyd, J.T., et al. (2013). A review of the molecular mechanisms of drug efflux in pathogenic bacteria: A structure-function perspective. *Recent Research Developments in Membrane Biology, 3*, pp. 15–66.
Fluman, N. and Bibi, E. (2009). Bacterial multidrug transport through the lens of the major facilitator superfamily. *Biochim Biophys Acta, 1794*(5), pp. 738–747.
Fluman, N., et al. (2009). A promiscuous conformational switch in the secondary multidrug transporter MdfA. *J Biol Chem., 284*(47), pp. 32296–32304.
Food and Drug Admistration. (2013). Summary report on antimicrobials sold or distributed for use in food-producing animals. US Food and Drug Administration, Center for Veterinary Medicine, Washington, DC, pp. 1–27.
Garcia-Migura, L., et al. (2014). Antimicrobial resistance of zoonotic and commensal bacteria in Europe: the missing link between consumption and resistance in veterinary medicine. *Vet Microbiol., 170*(1–2), pp. 1–9.
George, A.M. (1996). Multidrug resistance in enteric and other Gram-negative bacteria. *FEMS Microbiology Letters, 139*(1), pp. 1–10.
Ghosh, A. and Ramamurthy, T. (2011). Antimicrobials & cholera: are we stranded? *Indian J Med Res., 133*, pp. 225–231.
Gill, S.R., et al. (2006). Metagenomic analysis of the human distal gut microbiome. *Science, 312*(5778), pp. 1355–1359.
Gootz, T.D. (2010). The global problem of antibiotic resistance. *Critical Reviews in Immunology, 30*(1), pp. 79–93.
Gordon, R.J. and Lowy, F.D. (2008). Pathogenesis of methicillin-resistant *Staphylococcus aureus* infection. *Clinical Infectious Diseases, 46*(Suppl 5), pp. S350–S359.
Griffith, J.K., et al. (1992). Membrane transport proteins: implications of sequence comparisons. *Curr Opin Cell Biol., 4*(4), pp. 684–695.
Grkovic, S., Brown, M.H. and Skurray, R.A. (2002). Regulation of bacterial drug export systems. *Microbiol Mol Biol Rev., 66*(4), pp. 671–701, table of contents.
Guardabassi, L. and Prescott, J.F. (2015). Antimicrobial stewardship in small animal veterinary practice: from theory to practice. *Vet Clin North Am Small Anim Pract., 45*(2), pp. 361–376, vii.
Gunn, J.S., et al. (2014). *Salmonella* chronic carriage: epidemiology, diagnosis, and gallbladder persistence. *Trends Microbiol., 22*(11), pp. 648–655.
Harish, B.N. and Menezes, G.A. (2011). Antimicrobial resistance in typhoidal salmonellae. *Indian J Med Microbiol., 29*(3), pp. 223–229.
Harris, J.B., et al. (2012). Cholera. *Lancet, 379*(9835), pp. 2466–2476.
Hassan, K.A., Skurray, R.A. and Brown, M.H. (2007). Transmembrane helix 12 of the *Staphylococcus aureus* multidrug transporter QacA lines the bivalent cationic drug binding pocket. *J Bacteriol., 189*(24), pp. 9131–9134.

He, G. X., et al. (2011). EmmdR, a new member of the MATE family of multidrug transporters, extrudes quinolones from *Enterobacter cloacae. Arch Microbiol., 193*(10), pp. 759–765.

He, G.X., et al. (2011). SugE, a new member of the SMR family of transporters, contributes to antimicrobial resistance in *Enterobacter cloacae. Antimicrob Agents Chemother., 55*(8), pp. 3954–3957.

Henderson, P. (1991). Studies of translocation catalysis. *Bioscience Reports, 11*, pp. 477–538.

Henderson, P. J. (1990). Proton-linked sugar transport systems in bacteria. *J Bioenerg Biomembr., 22*(4), pp. 525–569.

Henderson, P.J., et al. (1993). Homologous sugar-transport proteins in microbes and man. *Biochem Soc Trans., 21*(4), pp. 1002–1006.

Higgins, C.F. (2007). Multiple molecular mechanisms for multidrug resistance transporters. *Nature, 446*(7137), pp. 749–757.

Hongo, E., et al. (1994). The methyl viologen-resistance-encoding gene *smvA* of *Salmonella typhimurium. Gene, 148*(1), pp. 173–174.

Horiyama, T., Yamaguchi, A. and Nishino, K. (2010). TolC dependency of multidrug efflux systems in *Salmonella enterica* serovar Typhimurium. *J Antimicrob Chemother., 65*(7), pp. 1372–1376.

Huang, J., et al. (2004). Novel chromosomally encoded multidrug efflux transporter MdeA in *Staphylococcus aureus. Antimicrob Agents Chemother., 48*(3), pp. 909–917.

Huber, G. and Nesemann, G. (1968). Moenomycin, an inhibitor of cell wall synthesis. *Biochemical and Biophysical Research Communications, 30*(1), pp. 7–13.

Huovinen, P., et al. (1995). Trimethoprim and sulfonamide resistance. *Antimicrob Agents Chemother., 39*(2), pp. 279–289.

Hussein, H.S. and Sakuma, T. (2005). Invited review: prevalence of Shiga toxin-producing *Escherichia coli* in dairy cattle and their products. *Journal of Dairy Science, 88*(2), pp. 450–465.

Ian-Zhi, L. and Hiroshi, N. (2009). Efflux-mediated drug resistance in bacteria: an update. *Drugs, 69*(12), pp. 1555–1623.

Jacoby, G.A. and Carreras, I. (1990). Activities of β-lactam antibiotics against *Escherichia coli* strains producing extended-spectrum β-lactamases. *Antimicrob Agents Chemother., 34*(5), pp. 858–862.

Jiang, D., et al. (2013). Structure of the YajR transporter suggests a transport mechanism based on the conserved motif A. *Proc Natl Acad Sci USA, 110*(36), pp. 14664–14669.

Jones, S.E., et al. (2011). Dairy farm age and resistance to antimicrobial agents in *Escherichia coli* isolated from dairy topsoil. *Curr Microbiol., 62*(4), pp. 1139–1146.

Jukes, T.H. (1985). Some historical notes on chlortetracycline. *Reviews of Infectious Diseases, 7*(5), pp. 702–707.

Kaper, J.B., Nataro, J.P. and Mobley, H.L. (2004). Pathogenic *Escherichia coli. Nat Rev Microbiol., 2*(2), pp. 123–140.

Kemper, N. (2008). Veterinary antibiotics in the aquatic and terrestrial environment. *Ecological Indicators, 8*(1), pp. 1–13.

Key, N. and McBride, W.D. (2014). Sub-therapeutic antibiotics and the efficiency of US hog farms. *American Journal of Agricultural Economics, 96*(3), pp. 831–850.

Kitaoka, M., et al. (2011). Antibiotic resistance mechanisms of *Vibrio cholerae. J Med Microbiol., 60*(Pt 4), pp. 397–407.

Kluytmans, J.A., et al. (2013). Extended-spectrum β-lactamase-producing *Escherichia coli* from retail chicken meat and humans: comparison of strains, plasmids, resistance genes, and virulence factors. *Clin Infect Dis., 56*(4), pp. 478–487.

Knechtges, P.L. (2011). Food borne infectious and microbial agents. In: *Food Safety: Theory and Practice.* Jones & Bartlett Publishers.

Krämer, R. (1994). Functional principles of solute transport systems: concepts and perspectives. *Biochimica et Biophysica Acta (BBA)-Bioenergetics, 1185*(1), pp. 1–34.

Kumar, A. and Schweizer, H.P. (2005). Bacterial resistance to antibiotics: active efflux and reduced uptake. *Advanced Drug Delivery Reviews, 57*(10), pp. 1486–1513.

Kumar, S. and Varela, M.F. (2013). Molecular mechanisms of bacterial resistance to antimicrobial agents. In: A. Méndez-Vilas, ed., *Microbial Pathogens and Strategies for Combating Them: Science, Technology and Education*. Formatex Research Center, Inc., pp. 522–534.

Kumar, S., Mukherjee, M.M. and Varela, M.F. (2013). Modulation of bacterial multidrug resistance efflux pumps of the major facilitator superfamily. *Int J Bacteriol., 2013*, pp. 1–15.

Kumar, S., Rizvi, M. and Berry, N. (2008). Rising prevalence of enteric fever due to multidrug-resistant *Salmonella*: an epidemiological study. *J Med Microbiol., 57*(Pt 10), pp. 1247–1250.

Kumar, S., et al. (2011). Cloning and molecular analysis of a mannitol operon of phosphoenolpyruvate-dependent phosphotransferase (PTS) type from *Vibrio cholerae* O395. *Archives of Microbiology, 193*(3), pp. 201–208.

Kumar, S. and Varela, M.F. (2012). Biochemistry of bacterial multidrug efflux pumps. *International Journal of Molecular Sciences, 13*(4), pp. 4484–4495.

L'Abee-Lund, T.M. and Sorum, H. (2001). Class 1 integrons mediate antibiotic resistance in the fish pathogen *Aeromonas salmonicida* worldwide. *Microb Drug Resist., 7*(3), pp. 263–272.

Landers, T.F., et al. (2012). A review of antibiotic use in food animals: perspective, policy, and potential. *Public Health Rep., 127*(1), pp. 4–22.

Landini, P., et al. (1992). Inhibition of bacterial protein synthesis by elongation-factor-Tu-binding antibiotics MDL 62,879 and efrotomycin. *Biochem. J., 283*, pp. 649–652.

Landini, P., et al. (1993). Sensitivity of elongation factor Tu (EF-Tu) from different bacterial species to the antibiotics efrotomycin, pulvomycin and MDL 62879. *Journal of General Microbiology, 139*(4), pp. 769–774.

Laxminarayan, R., Van Boeckel, T. and A. Teillant. (2015). The Economic Costs of Withdrawing Antimicrobial Growth Promoters from the Livestock Sector. *OECD Food, Agriculture and Fisheries Papers, 78*, pp. 1–42.

Levy, S.B. (1992a). Active efflux mechanisms for antimicrobial resistance. *Antimicrob Agents Chemother 36* (4), pp. 695–703.

Levy, S.B. (2001). Antibiotic resistance: consequences of inaction. *Clin Infect Dis., 33*(Suppl 3), pp. S124–S129.

Levy, S.B. and Marshall, B. (2004). Antibacterial resistance worldwide: causes, challenges and responses. *Nat Med., 10*(12 Suppl), pp. S122–S129.

Levy, S.B. (2002). Active efflux, a common mechanism for biocide and antibiotic resistance. *Journal of Applied Microbiology, 92*(s1), pp. 65S–71S.

Levy, S.B. (1992b). Active efflux mechanisms for antimicrobial resistance. *Antimicrobial Agents and Chemotherapy, 36*(4), pp. 695.

Lewinson, O., et al. (2003). The *Escherichia coli* multidrug transporter MdfA catalyzes both electrogenic and electroneutral transport reactions. *Proc Natl Acad Sci USA, 100*(4), pp. 1667–1672.

Lewinson, O. and Bibi, E. (2001). Evidence for simultaneous binding of dissimilar substrates by the *Escherichia coli* multidrug transporter MdfA. *Biochemistry, 40*(42), pp. 12612–12618.

Lewis, K. (2001). In search of natural substrates and inhibitors of MDR pumps. *J Mol Microbiol Biotechnol., 3*(2), pp. 247–254.

Li, X.Z., and Nikaido, H. (2004). Efflux-mediated drug resistance in bacteria. *Drugs, 64*(2), pp. 159–204.

Lin, J. (2011). Effect of antibiotic growth promoters on intestinal microbiota in food animals: a novel model for studying the relationship between gut microbiota and human obesity? *Frontiers in Microbiology, 2*, pp. 1–3.

Ling, L.L., et al. (2015). A new antibiotic kills pathogens without detectable resistance. *Nature, 517*(7535), pp. 455–459.

Livermore, D.M. (1995). β-Lactamases in laboratory and clinical resistance. *Clin Microbiol Rev., 8*(4), pp. 557–584.

Lomovskaya, O. and K. Lewis, K. (1992). Emr, an *Escherichia coli* locus for multidrug resistance. *Proc Natl Acad Sci USA., 89*(19), pp. 8938–8942.

Lomovskaya, O., Lewis, K. and Matin, A. (1995). EmrR is a negative regulator of the *Escherichia coli* multidrug resistance pump EmrAB. *J Bacteriol., 177*(9), pp. 2328–2334.

Lowy, F.D. (1998). *Staphylococcus aureus* infections. *N Engl J Med., 339*(8), pp. 520–532.

MacDonald, J.M. and Wang, S.L. (2011). Foregoing sub-therapeutic antibiotics: the impact on broiler grow-out operations. *Applied Economic Perspectives and Policy, 33*(1), pp. 79–98.

MacDougall, C. and Polk, R.E. (2005). Antimicrobial stewardship programs in health care systems. *Clin Microbiol Rev., 18*(4), pp. 638–656.

Maiden, M.C., et al. (1987). Mammalian and bacterial sugar transport proteins are homologous. *Nature, 325*(6105), pp. 641–643.

Mandal, S., Mandal, M.D. and Pal, N.K. (2011). Cholera: a great global concern. *Asian Pac J Trop Med., 4*(7), pp. 573–580.

Marger, M.D. and Saier, M.H., Jr. (1993). A major superfamily of transmembrane facilitators that catalyse uniport, symport and antiport. *Trends Biochem Sci., 18*(1), pp. 13–20.

Marshall, B.M. and Levy, S.B. (2011). Food animals and antimicrobials: impacts on human health. *Clin Microbiol Rev., 24*(4), pp. 718–733.

McClelland, M., et al. (2001). Complete genome sequence of *Salmonella enterica* serovar Typhimurium LT2. *Nature, 413*(6858), pp. 852–856.

McEwen, S.A. and Fedorka-Cray, P.J. (2002). Antimicrobial use and resistance in animals. *Clin Infect Dis., 34*(Suppl 3), pp. S93–S106.

Mead, P. S., et al. (1999). Food-related illness and death in the United States. *Emerg Infect Dis., 5*(5), pp. 607–625.

Minami, S., et al. (1980). Induction of β-lactamase by various β-lactam antibiotics in *Enterobacter cloacae. Antimicrob Agents Chemother., 18*(3), pp. 382–385.

Mine, T., et al. (1998). Evidence for chloramphenicol/H^+ antiport in Cmr (MdfA) system of *Escherichia coli* and properties of the antiporter. *J Biochem., 124*(1), pp. 187–193.

Modi, C.M., et al. (2011). Growth promoting use of antimicrobial agents in animals. *Journal Applied Pharmaceutical Science, 1*(8), pp. 33–36.

Moore, I.F., Hughes, D.W. and Wright, G.D. (2005). Tigecycline is modified by the flavin-dependent monooxygenase TetX. *Biochemistry, 44*(35), pp. 11829–11835.

Moore, P.R., et al. (1946). Use of sulfasuxidine, streptothricin, and streptomycin in nutritional studies with the chick. *J. Biol. Chem., 165*(2), pp. 437–441.

Mukherjee, M., et al. (2014). Comparative genome analysis of non-toxigenic non-O1 versus toxigenic O1 *Vibrio cholerae. Genome Discov., 2*(1), pp. 1–15.

Mulligan, M.E., et al. (1993). Methicillin-resistant *Staphylococcus aureus*: a consensus review of the microbiology, pathogenesis, and epidemiology with implications for prevention and management. *The American Journal of Medicine, 94*(3), pp. 313–328.

Murray, B. and Jorgensen, L. (eds.). (2007). *Manual of Clinical Microbiology*, 9th ed. Washington, DC: American Society for Microbiology.

Murray, P.R., Rosenthal, K.S. and Pfaller, M.A. (2012). *Medical Microbiology*. New York: Elsevier Health Sciences.

Nagy, B. and Fekete, P.Z. (1999). Enterotoxigenic *Escherichia coli* (ETEC) in farm animals. *Vet Res., 30*(2-3), pp. 259–284.

Naroditskaya, V., et al. (1993). An *E. coli* gene *emrD* is involved in adaptation to low energy shock. *Biochem Biophys Res Commun., 196*(2), pp. 803–809.

Nathisuwan, S., Burgess, D.S. and Lewis, J.S. 2nd. (2001). Extended-spectrum β-lactamases: epidemiology, detection, and treatment. *Pharmacotherapy, 21*(8), pp. 920–928.

Nelson, M.L., and Levy, S.B. (2011). The history of the tetracyclines. *Ann N Y Acad Sci., 1241*, pp. 17–32.

Neyfakh, A.A. (1992). The multidrug efflux transporter of *Bacillus subtilis* is a structural and functional homolog of the *Staphylococcus* NorA protein. *Antimicrob Agents Chemother., 36*(2), pp. 484–485.

Nilsen, I.W., et al. (1996). Isolation of *cmr*, a novel *Escherichia coli* chloramphenicol resistance gene encoding a putative efflux pump. *J Bacteriol., 178*(11), pp. 3188–3193.

Nishino, K., Latifi, T. and Groisman, E.A. (2006). Virulence and drug resistance roles of multidrug efflux systems of *Salmonella enterica* serovar Typhimurium. *Mol Microbiol., 59*(1), pp. 126–141.

Nishino, K. Nikaido, E. and Yamaguchi, A. (2009). Regulation and physiological function of multidrug efflux pumps in *Escherichia coli* and *Salmonella*. *Biochim Biophys Acta., 1794*(5), pp. 834–843.

Nishino, K. and Yamaguchi, A. (2001). Analysis of a complete library of putative drug transporter genes in *Escherichia coli*. *J Bacteriol., 183*(20), pp. 5803–5812.

Nordmann, P. and Guibert, M. (1998). Extended-spectrum β-lactamases in *Pseudomonas aeruginosa*. *J Antimicrob Chemother., 42*(2), pp. 128–131.

Ohshita, Y., Hiramatsu, K. and Yokota, T. (1990). A point mutation in *norA* gene is responsible for quinolone resistance in *Staphylococcus aureus*. *Biochem Biophys Res Commun., 172*(3), pp. 1028–1034.

Okusu, H., Ma, D. and Nikaido, H. (1996). AcrAB efflux pump plays a major role in the antibiotic resistance phenotype of *Escherichia coli* multiple-antibiotic-resistance (Mar) mutants. *Journal of Bacteriology, 178*(1), pp. 306–308.

Pao, S.S., Paulsen, I.T. and Saier, M.H., Jr. (1998). Major facilitator superfamily. *Microbiol Mol Biol Rev., 62*(1), pp. 1–34.

Park, B.H. and Levy, S.B. (1988). The cryptic tetracycline resistance determinant on Tn4400 mediates tetracycline degradation as well as tetracycline efflux. *Antimicrob Agents Chemother, 32*(12), pp. 1797–1800.

Parkhill, J., et al. (2001). Complete genome sequence of a multiple drug resistant *Salmonella enterica* serovar Typhi CT18. *Nature, 413*(6858), pp. 848–852.

Parr, T.R. and Saier, M.H. (1992). The bacterial phosphotransferase system as a potential vehicle for the entry of novel antibiotics. *Research in Microbiology, 143*(5), pp. 443–447.

Paterson, D.L. and Bonomo, R.A. (2005). Extended-spectrum β-lactamases: a clinical update. *Clin Microbiol Rev., 18*(4), pp. 657–686.

Paulsen, I.T., et al. (1996). Multidrug resistance proteins QacA and QacB from *Staphylococcus aureus*: membrane topology and identification of residues involved in substrate specificity. *Proc Natl Acad Sci USA, 93*(8), pp. 3630–3635.

Paulsen, I.T., Firth, N. and Skurray, R.A. (1997). Resistance to antimicrobial agents other than β-lactams. In: *The Staphylococci in Human Disease*, pp. 175–212.

Peng, Y., et al. (2008). Pasteurized whole milk confers reduced susceptibilities to the antimicrobial agents trimethoprim, gatifloxacin, cefotaxime and tetracycline via the *marRAB* locus in *Escherichia coli*. *J Dairy Res., 75*(4), pp. 491–496.

Philippon, A., Labia, R. and Jacoby. G. (1989). Extended-spectrum β-lactamases. *Antimicrob Agents Chemother 33*(8), pp. 1131–6.

Piddock, L.J. (1999). Mechanisms of fluoroquinolone resistance: an update 1994-1998. *Drugs, 58*(Suppl 2), pp. 11–18.

Pitout, J.D. and Laupland, K.B. (2008). Extended-spectrum β-lactamase-producing Enterobacteriaceae: an emerging public-health concern. *Lancet Infect Dis., 8*(3), pp. 159–166.

Poolman, B. and Konings, W.N. (1993). Secondary solute transport in bacteria. *Biochimica et Biophysica Acta (BBA)-Bioenergetics, 1183*(1), pp. 5–39.

Prescott, J.F. (2014). The resistance tsunami, antimicrobial stewardship, and the golden age of microbiology. *Vet Microbiol., 171*(3–4), pp. 273–278.

Projan, S.J. and Novick, R.P. (1997). The molecular basis of pathogenicity. *The staphylococci in human disease*. New York, NY: Churchill Livingstone, pp. 55–81.

Qadri, F., et al. (2005). Enterotoxigenic *Escherichia coli* in developing countries: epidemiology, microbiology, clinical features, treatment, and prevention. *Clin Microbiol Rev.*, *18*(3), pp. 465–483.

Rabbani, G.H. and Greenough, W.B., 3rd. (1999). Food as a vehicle of transmission of cholera. *J Diarrhoeal Dis Res.*, *17*(1), pp. 1–9.

Rensch, U., et al. (2014). *Salmonella enterica* serovar Typhimurium multidrug efflux pumps EmrAB and AcrEF support the major efflux system AcrAB in decreased susceptibility to triclosan. *Int J Antimicrob Agents.*, *44*(2), pp. 179–180.

Ritchie, J.M. and Waldor, M.K. (2009). *Vibrio cholerae* interactions with the gastrointestinal tract: lessons from animal studies. *Curr Top Microbiol Immunol.*, *337*, pp. 37–59.

Robins, W.P. and Mekalanos, J.J. (2014). Genomic science in understanding cholera outbreaks and evolution of *Vibrio cholerae* as a human pathogen. *Curr Top Microbiol Immunol.*, *379*, pp. 211–29.

Rocourt, J., et al. (2003). *The Present State of Foodborne Disease in OECD Countries*. Geneva: WHO, p. 1.

Rogers, S. and Haines, J. (2005). Detecting and mitigating the environmental impact of fecal pathogens originating from confined animal feeding operations: review. EPA/600/R-06/021, pp. 1–185.

Rossolini, G.M., D'Andrea, M.M. and Mugnaioli, C. (2008). The spread of CTX-M-type extended-spectrum β-lactamases. *Clin Microbiol Infect.*, *14*(Suppl 1), pp. 33–41.

Rouch, D.A., et al. (1990). Efflux-mediated antiseptic resistance gene *qacA* from *Staphylococcus aureus*: common ancestry with tetracycline- and sugar-transport proteins. *Mol Microbiol.*, *4*(12), pp. 2051–2062.

Safdar, N. and E. A. Bradley, E.A. (2008). The risk of infection after nasal colonization with *Staphylococcus aureus*. *Am J Med.*, *121*(4), pp. 310–315.

Sahin, O., Morishita, T.Y. and Zhang, Q. (2002). *Campylobacter* colonization in poultry: sources of infection and modes of transmission. *Anim Health Res Rev.*, *3*(2), pp. 95–105.

Saidijam, M., et al. (2011). The MFS Efflux Proteins of Gram-Positive and Gram-Negative Bacteria. *Advances in Enzymology and Related Areas of Molecular Biology*, *77*, pp. 147–166.

Saier, M.H., Jr., et al. (1999). The major facilitator superfamily. *J Mol Microbiol Biotechnol.*, *1*(2), pp. 257–279.

Saier, M.H., Jr., et al. (2014). The transporter classification database. *Nucleic Acids Res.*, *42*(Database issue), pp. D251–D258.

Sanchez-Vargas, F.M., Abu-El-Haija, M.A. and Gomez-Duarte, O.G. (2011). *Salmonella* infections: an update on epidemiology, management, and prevention. *Travel Med Infect Dis.*, *9*(6), pp. 263–277.

Sanders, W.E., Jr., Tenney, J.H. and Kessler, R.E. (1996). Efficacy of cefepime in the treatment of infections due to multiply resistant *Enterobacter* species. *Clin Infect Dis.*, *23*(3), pp. 454–461.

Scallan, E., et al. (2011). Foodborne illness acquired in the United States--major pathogens. *Emerg Infect Dis.*, *17*(1), pp. 7–15.

Schindler, B.D., Jacinto, P. and Kaatz, G.W. (2013). Inhibition of drug efflux pumps in *Staphylococcus aureus*: current status of potentiating existing antibiotics. *Future Microbiol.*, *8*(4), pp. 491–507.

Schwarz, S., et al. (2004). Molecular basis of bacterial resistance to chloramphenicol and florfenicol. *FEMS Microbiol Rev.*, *28*(5), pp. 519–542.

Shaheen, A., et al. (2015). Characterization of putative multidrug resistance transporters of the major facilitator-superfamily expressed in *Salmonella* Typhi. *J Infect Chemother.*, *21*(5), pp. 357–362.

Shrestha, U., et al. (2015). Multidrug resistance efflux pumps of *Salmonella enterica*,. In: C.B. Beckett, ed., *Salmonella: Prevalence, Risk*

Factors and Treatment Options. New York: Nova Science Publishers, Inc., pp. 1–30.

Siewert, G. and Strominger, J.L. (1967). Bacitracin: an inhibitor of the dephosphorylation of lipid pyrophosphate, an intermediate in the biosynthesis of the peptidoglycan of bacterial cell walls. *Proceedings of the National Academy of Sciences of the United States of America*, *57*(3), pp. 767.

Silbergeld, E.K., Graham, J. and Price, L.B. (2008). Industrial food animal production, antimicrobial resistance, and human health. *Annu Rev Public Health.*, *29*, pp. 151–69.

Smith, K.E., et al. (1999a). Quinolone-resistant *Campylobacter jejuni* infections in Minnesota, 1992-1998. Investigation Team. *N Engl J Med.*, *340*(20), pp. 1525–1532.

Smith, K.P., et al. (2010). Elucidation of antimicrobial susceptibility profiles and genotyping of *Salmonella enterica* isolates from clinical cases of Salmonellosis in New Mexico in 2008. *World J Microbiol Biotechnol.*, *26*(6), pp. 1025–1031.

Smith, K.P., Kumar, S. and Varela, M.F. (2009). Identification, cloning, and functional characterization of EmrD-3, a putative multidrug efflux pump of the major facilitator superfamily from *Vibrio cholerae* O395. *Arch Microbiol.*, *191*(12), pp. 903–911.

Smith, T.L., et al. (1999b). Emergence of vancomycin resistance in *Staphylococcus aureus*. *New England Journal of Medicine*, *340*(7), pp. 493–501.

Spika, J.S., et al. (1987). Chloramphenicol-resistant *Salmonella* newport traced through hamburger to dairy farms. A major persisting source of human salmonellosis in California. *N Engl J Med.*, *316*(10), pp. 565–570.

Stavri, M., Piddock, L.J. and Gibbons, S. (2007). Bacterial efflux pump inhibitors from natural sources. *J Antimicrob Chemother.*, *59*(6), pp. 1247–1260.

Stevens, D.L. (2003). Community-acquired *Staphylococcus aureus* infections: increasing virulence and emerging methicillin resistance in the new millennium. *Current Opinion in Infectious Diseases*, *16*(3), pp. 189–191.

Su, L. and Chiu, C. (2007). *Salmonella*: clinical importance and evolution of nomenclature. *Chang Gung Medical Journal*, *30*(3), p. 210.

Summers, A.O. (2006). Genetic linkage and horizontal gene transfer, the roots of the antibiotic multi-resistance problem. *Anim Biotechnol.*, *17*(2), pp. 125–135.

Sunde, M., et al. (1998). Antibiotic resistance in *Escherichia coli* of the normal intestinal flora of swine. *Microb Drug Resist.*, *4*(4), pp. 289–299.

Suter, W., Rosselet, A. and Knüsel, F. (1978). Mode of action of quindoxin and substituted quinoxaline-di-N-oxides on *Escherichia coli*. *Antimicrobial Agents and Chemotherapy*, *13*(5), pp. 770–783.

Swaminathan, B., Barrett, T.J. and Fields, P. (2006). Surveillance for human *Salmonella* infections in the United States. *J AOAC Int.*, *89*(2), pp. 553–559.

Swartz, M.N. (2002). Human diseases caused by foodborne pathogens of animal origin. *Clin Infect Dis.*, *34*(Suppl 3), pp. S111–S122.

Talbot, G.H., et al. (2006). Bad bugs need drugs: an update on the development pipeline from the Antimicrobial Availability Task Force of the Infectious Diseases Society of America. *Clin Infect Dis.*, *42*(5), pp. 657–668.

Tam, C.C., Rodrigues, L.C. and O'Brien, S.J. (2003). Guillain-Barré syndrome associated with *Campylobacter jejuni* infection in England, 2000–2001. *Clin Infect Dis.*, *37*(2), pp. 307–310.

Tennent, J.M., et al. (1989). Physical and biochemical characterization of the *qacA* gene encoding antiseptic and disinfectant resistance in *Staphylococcus aureus*. *J Gen Microbiol.*, *135*(1), pp. 1–10.

Thomson, G.T., et al. (1995). Post-*Salmonella* reactive arthritis: late clinical sequelae in a point source cohort. *Am J Med.*, *98*(1), pp. 13–21.

Threlfall, E.J. (2002). Antimicrobial drug resistance in *Salmonella*: problems and perspectives in food- and water-borne infections. *FEMS Microbiol Rev*, *26*(2), pp. 141–148.

Threlfall, E.J., et al. (1996). Increasing spectrum of resistance in multiresistant *Salmonella typhimurium*. *The Lancet*, *347*(9007), pp. 1053–1054.

Tirosh, O., et al. (2012). Manipulating the drug/proton antiport stoichiometry of the secondary multidrug transporter MdfA. *Proc Natl Acad Sci USA.*, *109*(31), pp. 12473–12478.

Tollefson, L., Altekruse, S.F. and Potter, M.E. (1997). Therapeutic antibiotics in animal feeds and antibiotic resistance. *Rev Sci Tech.*, *16*(2), pp. 709–715.

Tzouvelekis, L.S., et al. (2000). CTX-M-type β-lactamases: an emerging group of extended-spectrum enzymes. *Int J Antimicrob Agents.*, *14*(2), pp. 137–142.

Ubukata, K., Itoh-Yamashita, N. and Konno, M. (1989). Cloning and expression of the *norA* gene for fluoroquinolone resistance in *Staphylococcus aureus*. *Antimicrob Agents Chemother.*, *33*(9), pp. 1535–1539.

Vallance, B.A., et al. (2002). Enteropathogenic and enterohemorrhagic *Escherichia coli* infections: emerging themes in pathogenesis and prevention. *Can J Gastroenterol.*, *16*(11), pp. 771–778.

Van Bambeke, F., Balzi, E. and Tulkens, P.M. (2000). Antibiotic efflux pumps. *Biochemical Pharmacology*, *60*(4), pp. 457–470.

van den Bogaard, A.E., and Stobberingh. E.E. (2000). Epidemiology of resistance to antibiotics. Links between animals and humans. *Int J Antimicrob Agents*, *14*(4), pp. 327–35.

Vandenesch, F., et al. (2003). Community-acquired methicillin-resistant *Staphylococcus aureus* carrying Panton-Valentine leukocidin genes: worldwide emergence. *Emerging Infectious Diseases*, *9*(8), pp. 978–984.

Varela, M.F., Kumar, S. and He, G. (2013). Potential for inhibition of bacterial efflux pumps in multidrug-resistant *Vibrio cholerae*. *Indian J. Med. Res.*, *138*, pp. 285–287.

Varela, M.F., Sansom, C.E. and Griffith, J.K. (1995). Mutational analysis and molecular modelling of an amino acid sequence motif conserved in antiporters but not symporters in a transporter superfamily. *Molecular Membrane Biology*, *12*(4), pp. 313–319.

Varela, M.F. and Griffith, J.K. (1993). Nucleotide and deduced protein sequences of the class D tetracycline resistance determinant: relationship to other antimicrobial transport proteins. *Antimicrobial Agents and Chemotherapy*, *37*(6), pp. 1253–1258.

Villagra, N.A., et al. (2008). SmvA, and not AcrB, is the major efflux pump for acriflavine and related compounds in *Salmonella enterica* serovar Typhimurium. *J Antimicrob Chemother.*, *62*(6), pp. 1273–1276.

Voetsch, A.C., et al. (2004). FoodNet estimate of the burden of illness caused by nontyphoidal *Salmonella* infections in the United States. *Clin Infect Dis.*, *38*(Suppl 3), pp. S127–S134.

Waddington, C.S., Darton, T.C. and Pollard, A.J. (2014). The challenge of enteric fever. *J Infect*, *68*(Suppl 1), pp. S38–S50.

Walsh, C. (2003). *Antibiotics: Actions, Origins, Resistance*. Washington, DC: American Society for Microbiology Press.

Walsh, C. (1999). Deconstructing vancomycin. *Science*, *284*(5413), pp. 442–443.

Wassenaar, T.M., et al. (2015). Review and phylogenetic analysis of *qac* genes that reduce susceptibility to quaternary ammonium compounds in Staphylococcus species. *Eur J Microbiol Immunol (Bp).*, *5*(1), pp. 44–61.

Webber, M.A. and Piddock, L.V.J. (2003). The importance of efflux pumps in bacterial antibiotic resistance. *Journal of Antimicrobial Chemotherapy*, *51*(1), pp. 9–11.

Wolf, H. (1973). Avilamycin, an inhibitor of the 30 S ribosomal subunits function. *FEBS Letters*, *36*(2), pp. 181–186.

Woolhouse, M., et al. (2015). Antimicrobial resistance in humans, livestock and the wider environment. *Philos Trans R Soc Lond B Biol Sci.*, *370*(1670), pp. 1–7.

Woolley, R.C., et al. (2005). Characterization of the *Vibrio cholerae vceCAB* multiple-drug resistance efflux operon in *Escherichia coli*. *J Bacteriol.*, *187*(15), pp. 5500–5503.

Wright, G.D. (2005). Bacterial resistance to antibiotics: enzymatic degradation and modification. *Adv Drug Deliv Rev.*, *57*(10), pp. 1451–1470.

Wu, J., et al. (2008). Functional analyses reveal an important role for tyrosine residues in the staphylococcal multidrug efflux protein QacA. *BMC Microbiol*, *8*, pp. 1–7.

Xu, Z., et al. (2006). Role of transmembrane segment 10 in efflux mediated by the staphylococcal multidrug transport protein QacA. *J. Biol. Chem.*, *281*(2), pp. 792–799

Yaffe, D., et al. (2013). Identification of molecular hinge points mediating alternating access in the vesicular monoamine transporter VMAT2. *Proc Natl Acad Sci USA*, *110*(15), pp. E1332–E1341.

Yang, W., I. et al. (2004). TetX is a flavin-dependent monooxygenase conferring resistance to tetracycline antibiotics. *J Biol Chem*, *279*(50), pp. 52346–52352.

Yin, Y., et al. (2006). Structure of the multidrug transporter EmrD from *Escherichia coli*. *Science*, *312*(5774), pp. 741–744.

Yoshida, H., et al. (1990). Nucleotide sequence and characterization of the *Staphylococcus aureus norA* gene, which confers resistance to quinolones. *J Bacteriol*, *172*(12), pp. 6942–6949.

Zhao, S., et al. (2003). Characterization of *Salmonella enterica* Serotype Newport Isolated from Humans and Food Animals. *Journal of Clinical Microbiology*, *41*(12), pp. 5366–5371.

14

Prevalence, Evolution, and Dissemination of Antibiotic Resistance in *Salmonella*

Brian W. Brunelle[1], Bradley L. Bearson[2], and Heather K. Allen[1]*

[1] *Food Safety and Enteric Pathogens Research Unit, National Animal Disease Center, ARS, USDA, Ames, IA, USA*
[2] *Agroecosystems Management Research Unit, National Laboratory for Agriculture and the Environment, ARS, USDA, Ames, IA, USA*

14.1 Introduction

Salmonella is a ubiquitous bacterial pathogen that causes mild to severe intestinal disease in humans and animals. Over 2500 serotypes are recognized by host immune systems as unique antagonists, and subspecies of *Salmonella* are designated based on this serotyping. At the highest classification level, *Salmonella* are grouped into serotypes that are restricted to humans (Typhi and Paratyphi) and those that have reservoirs in animals (nontyphoidal) (Goldberg and Rubin, 1988). Although typhoid fever caused by *S.* Typhi can be a life-threatening systemic illness, it is estimated that less than 6000 cases occur in the United States each year. Vaccines are available for the prevention of typhoid fever, and the majority of U.S. cases (~75%) are associated with foreign travel. Nontyphoidal *Salmonella* frequently colonize the gastrointestinal tracts of food-producing animals and poultry in the United States with no effective interventions for prevention of disease in humans. Because of their significance to food safety and to antibiotic resistance gene dissemination, nontyphoidal serotypes of *Salmonella* are the focus of this chapter.

Nontyphoidal *Salmonella* is a common source of bacterial gastrointestinal disease in humans, with an estimated annual worldwide toll of 94 million cases and 155,000 deaths (Majowicz *et al.*, 2010). Within the United States, nontyphoidal *Salmonella* is the leading cause of hospitalizations and death due to foodborne disease (CDC, 2013b), and is estimated to cause over 1 million illnesses and 350 deaths each year (Economic Research Service, 2014). A significant contributing factor to its morbidity and mortality is the widespread prevalence of antibiotic resistance in nontyphoidal *Salmonella*. Antibiotic-resistant *Salmonella* are associated with higher rates of disease severity compared to sensitive isolates, although the underlying mechanisms for this clinical observation have not been fully elucidated (Molbak, 2005; Varma *et al.*, 2005a; Varma *et al.*, 2005b).

The majority of recent outbreaks of antibiotic resistant *Salmonella* have occurred from contaminated retail meat (Table 14.1), but other sources include pet rodents, companion animals, and farm environments (Fey *et al.*, 2000; Wright *et al.*, 2005; Swanson *et al.*, 2007). Because of the multiple paths to outbreaks and the ensuing severity of disease, the CDC has classified antibiotic-resistant *Salmonella* as a Serious Threat Level pathogen (CDC, 2013a). The World Health Organization (WHO, 2011) and other public health entities have called for research on antibiotic resistance and the development of alternative treatment strategies.

Foodborne Pathogens and Antibiotic Resistance, First Edition. Edited by Om V. Singh.

Table 14.1 CDC List of Recent MDR *Salmonella* Outbreaks in the United States.

Year	Serotype	Source	Antibiotic Resistance
2015	I 4,[5],12:i:-	pork	ampicillin, streptomycin, sulfisoxazole, and tetracycline
2014	Heidelberg	chicken	ceftriaxone plus a combination of ampicillin, cefoxitin, ceftiofur, amoxicillin/clavulanic acid, sulfisoxazole, tetracycline and trimethoprim/sulfamethoxazole
2013	Heidelberg	chicken	combinations of the following: ampicillin, chloramphenicol, gentamicin, kanamycin, streptomycin, sulfisoxazole, and tetracycline
2013	Heidelberg	chicken	amoxicillin/clavulanic acid, ampicillin, and ceftriaxone
2011	Typhimurium	ground beef	amoxicillin/clavulanic acid, ampicillin, ceftriaxone, cefoxitin, kanamycin, streptomycin, sulfisoxazole, and tetracycline
2011	Heidelberg	ground turkey	ampicillin, streptomycin, tetracycline, and gentamicin
2011	Hadar	turkey burgers	ampicillin, amoxicillin/clavulanate, cephalothin, and tetracycline

14.2 Antibiotic Resistance Prevalence Among *Salmonella* Serotypes

Antibiotic-resistant *Salmonella* has been a public health concern for over 40 years (Gill and Hook, 1966; Baldwin, 1970; Pocurull *et al.*, 1971; Bissett *et al.*, 1974; Lawrence and Hoeprich, 1975), and the prevalence of resistance continues to persist in nontyphoidal *Salmonella* according to data collected by the National Antimicrobial Resistance Monitoring System (NARMS) (CDC, 2015). NARMS was established in 1996 by several U.S. agencies (CDC, USDA, FDA) to monitor antibiotic resistance in zoonotic pathogens in humans, food producing animals, and retail meats. Comparing the average prevalence of resistance in nontyphoidal *Salmonella* from the most recent human NARMS data (2013) to the previous five years (2008–2012), resistance to one or more classes of antibiotics significantly increased to 19.2% from 15.7% (CDC, 2015). Interestingly, data from a 1967 survey of 400 *Salmonella* isolates show a similar pattern to the contemporary data, with 22% of isolates resistant to one or more antibiotics (Schroeder *et al.*, 1968).

The various nontyphoidal *Salmonella* serotypes differ in their prevalence for the number of antibiotic classes to which they are resistant. For 2013, 9.8% of 2,178 nontyphoidal serotypes isolated from humans with clinical disease were considered multidrug-resistant (MDR) as defined by resistance to three or more antibiotic classes (CDC, 2015). Resistance to four antibiotic classes was the most frequent MDR combination for the *Salmonella* serotypes. Typhimurium and Newport are two of the most frequently isolated serotypes from humans, and the 2013 NARMS data shows 16.9% and 5.7% of these isolates, respectively, to be multidrug-resistant (Table 14.2).

The serotypes with the highest percentage of multidrug-resistance include Choleraesuis (100%), Dublin (92%), I 4,[5],12:i:- (51%), and Heidelberg (33.3%) (Table 14.2). Fortunately, serotypes Choleraesuis and Dublin are not commonly isolated from patients with clinical disease. An interesting trend in the nontyphoidal *Salmonella* serotype data for antibiotic resistance is the recent increases in multidrug-resistance for serotypes I 4,[5],12:i:- and Heidelberg. In 2004, only 8.3% of I 4,[5],12:i:- and 13.0% of Heidelberg NARMS isolates

Table 14.2 *Salmonella* Serotypes with Frequent Resistance to Multiple Antibiotic Classes from NARMS 2013[a] (CDC, 2015).

Serotype	Total # Isolates[b]	# MDR[c]	% MDR[d]	Number of Antibiotic Classes[e]					
				3	4	5	6	7	8
Typhimurium	325	55	16.9%	7 (14.9%)	8 (10%)	32 (69.6%)	1 (16.7%)	7 (21.2%)	
Newport	209	12	5.7%	2 (4.2%)				10 (30.3%)	
I 4,[5],12:i:-	127	65	51.2%	3 (6.4%)	59 (73.7%)	1 (2.2%)	1 (16.7%)	1 (3.0%)	
Heidelberg	60	20	33.3%	15 (31.9%)	1 (1.2%)	3 (6.5%)		1 (3.0%)	
Agona	28	2	7.1%			1 (2.2%)			1 (50%)
Dublin	12	11	91.7%					10 (30.3%)	1 (50%)
Choleraesuis	2	2	100%				2 (33.3%)		

[a] Only the most frequent serotypes with resistance to 3–8 antibiotic classes are shown.
[b] Total # of human isolates of a given serotype analyzed by NARMS in 2013 of 2,178 nontyphoidal isolates.
[c] Total # of isolates for a given serotype that were MDR of 214 nontyphoidal isolates.
[d] % of isolates for a given serotype that were MDR.
[e] The # of isolates of a given serotype are shown along with the % prevalence of a serotype for the class combination shown compared to all other nontyphoidal serotypes.

were MDR. However, since 2004 the incidence of MDR within serotypes Heidelberg and I 4,[5],12:i:- has dramatically increased 2.6- and 6.2-fold, respectively. Due to these increases in MDR, serotypes Typhimurium, I 4,[5],12:i:-, and Heidelberg made up 65.4% of all nontyphoidal MDR *Salmonella* NARMS isolates in 2013 (Table 14.2).

Multidrug-resistant *S.* Typhimurium is of particular importance to health and food safety because of its high prevalence in both humans and livestock. Over a 10-year period (2004–2013) (CDC, 2015), *S.* Typhimurium accounted for 16.9% of all nontyphoidal *Salmonella* in humans, with 28.1% of these isolates being MDR (Table 14.3). This is in contrast to a much lower prevalence of antibiotic resistance in nontyphoidal *Salmonella*, with data from the same ten-year period demonstrating that 10.2% of the total nontyphoidal isolates tested were MDR (Table 14.3); excluding the *S.* Typhimurium isolates, the amount of MDR nontyphoidal isolates drops to 6.5%. *S.* Typhimurium has the highest prevalence of resistant nontyphoidal isolates in humans over this time, and in fact, was responsible for nearly half of all MDR nontyphoidal isolates (CDC, 2015).

The increased frequency of antibiotic resistance observed in *S.* Typhimurium compared to nontyphoidal *Salmonella* is generally true regardless of the type of sample monitored by NARMS (Tables 14.4 and 14.5) (FDA, 2013a). For instance, the frequency of human nontyphoidal *Salmonella* isolates resistant to ampicillin, chloramphenicol, streptomycin, and tetracycline ranged between 7–14%. In contrast, the frequency of resistance to the same four antibiotics in *S.* Typhimurium ranged from

Table 14.3 Percent of Total Nontyphoidal *Salmonella*, *S.* Typhimurium, *S.* I 4,[5],12:i:-, and *S.* Heidelberg Isolates from Humans that were Identified as Antibiotic Resistant over a 10-year Period Calculated from the NARMS Human Isolate Report (CDC, 2015).

Isolates	% Nontyphoidal	% Typhimurium	% I 4,[5],12:i:-	% Heidelberg
Susceptible	82.6	64.8	68.9	58.7
Resistant	17.4	35.2	31.1	41.3
≥2 classes	13.1	32.3	24.3	33.0
≥3 classes	10.2	28.1	21.4	21.8
≥4 classes	7.6	24.6	17.8	7.2
≥5 classes	5.8	21.4	2.6	5.2

Table 14.4 Percent of Total Nontyphoidal *Salmonella* Isolates Identified as Antibiotic Resistant to the Following Antibiotics Over a 12-year Period Calculated from the Most Recent National Antimicrobial Resistance Monitoring System Executive Report (FDA, 2013a).

	Human	Chicken	Turkey	Cattle	Swine	Retail Chicken	Ground Turkey	Ground Beef	Pork Chops
Amoxicillin/ Clavulanic acid	3.6	10.5	6.2	15.2	3.0	24.9	8.3	5.5	2.1
Ampicillin	11.4	13.4	24.1	21.5	14.8	32.6	40.5	18.1	24.3
Cefoxitin	3.3	9.8	5.4	14.0	2.6	22.3	7.3	4.7	3.6
Ceftiofur	3.4	10.5	5.7	15.0	2.6	20.2	5.9	3.9	2.9
Ceftriaxone	3.3	10.4	5.6	15.1	2.5	17.2	4.1	2.4	2.9
Chloramphenicol	7.2	2.2	4.2	19.1	10.4	1.0	1.6	7.9	9.3
Ciprofloxacin	0.2	0.0	0.1	0.0	0.0	0.0	0.0	0.0	0.0
Gentamicin	1.7	6.3	18.7	2.3	1.4	5.5	23.9	5.5	10.0
Kanamycin	3.0	2.9	18.0	9.1	6.3	9.8	17.4	7.1	7.1
Nalidixic Acid	2.1	0.3	3.1	0.8	0.1	0.1	0.1	0.0	0.0
Streptomycin	11.5	24.0	36.9	22.9	33.9	28.5	42.9	21.3	36.4
Sulfisoxazole	12.1	11.1	30.0	22.4	32.2	23.8	5.7	2.4	19.3
Tetracycline	13.7	29.1	57.8	29.9	54.5	37.8	37.9	10.2	30.0
Trimethoprim/ Sulfamethoxazole	1.6	0.3	1.6	2.8	1.5	0.2	0.4	0.0	1.4

24–34% in humans. Additionally noteworthy is that resistance to these four antibiotics among *S.* Typhimurium isolates is frequently higher in isolates from food-producing animals compared to isolates from humans (Figure 14.1). This shows that agricultural animals are a reservoir for MDR *S.* Typhimurium, and factors in the agricultural ecosystem (Stokes and Gillings, 2011; Allen and Stanton, 2014) and in *Salmonella* itself have selected for this reservoir.

Table 14.5 Percent of Total *Salmonella* Typhimurium Isolates Identified as Antibiotic Resistant to the Following Antibiotics Over a 12-Year Period Calculated from the Most Recent National Antimicrobial Resistance Monitoring System Executive Report (FDA, 2013a).

	Human	Chicken	Turkey	Cattle	Swine	Retail Chicken	Ground Turkey	Ground Beef	Pork Chops
Amoxicillin/ Clavulanic acid	5.4	27.1	26.4	19.1	3.4	50.8	37.5	0.0	0.0
Ampicillin	31.6	35.9	59.3	63.1	69.9	62.9	66.7	22.2	39.3
Cefoxitin	4.3	25.4	24.2	16.0	1.9	43.1	33.3	0.0	0.0
Ceftiofur	4.6	27.0	24.2	18.2	1.5	42.7	29.2	0.0	0.0
Ceftriaxone	4.5	27.0	24.2	18.2	1.2	38.2	33.3	0.0	0.0
Chloramphenicol	24.4	7.5	52.7	45.6	56.3	1.4	12.5	22.2	35.7
Ciprofloxacin	0.1	0.1	0.0	0.0	0.0	0.0	0.0	0.0	0.0
Gentamicin	2.0	6.6	40.7	1.2	2.9	2.4	8.3	0.0	3.6
Kanamycin	6.3	7.8	34.1	23.7	9.5	23.7	8.3	0.0	14.3
Nalidixic Acid	1.1	0.5	26.4	0.6	0.7	0.0	0.0	0.0	0.0
Streptomycin	31.1	17.3	70.3	57.7	72.1	17.3	37.5	22.2	67.9
Sulfisoxazole	34.4	42.8	74.7	59.5	78.2	64.7	37.5	0.0	39.3
Tetracycline	33.0	41.1	78.0	59.1	87.6	71.0	54.2	0.0	46.4
Trimethoprim/ Sulfamethoxazole	2.5	0.4	2.2	2.9	0.0	0.2	0.0	0.0	3.6

14.3 Antibiotic Treatment of *Salmonella*

The acquisition of multiple antibiotic resistance genes by individual nontyphoidal serotypes has led to the development of MDR strains and a reduction in infection treatment options. The CDC considers the development of antibiotic resistance to fluoroquinolones and third-generation cephalosporins in nontyphoidal *Salmonella* a serious concern because fluoroquinolones (e.g., ciprofloxacin) and third-generation cephalosporins (e.g., ceftriaxone) are the primary course of treatment for severe nontyphoidal *Salmonella* infections (CDC, 2013a).

In addition to direct selection for resistance against fluoroquinolones, nalidixic acid resistance can potentiate resistance to fluoroquinolones, as resistance to both antibiotics is selected through changes to the same target (Heisig and Tschorny, 1994; Cebrian *et al.*, 2003). Thus far, the prevalence of resistance to nalidixic acid and ceftriaxone for nontyphoidal *Salmonella* isolates in the United States is low, with 3% of isolates resistant to each antibiotic, respectively (CDC, 2015). The resistance of *Salmonella* serotypes to both of these antibiotic classes has increased since 1996 when the resistance prevalence was 0.0% for ceftriaxone and 0.4% for nalidixic acid (CDC, 2008). Due to the importance of fluoroquinolones and ceftriaxone for treating *Salmonella* infections, the CDC advises that continual action is essential to prevent further increases in the dissemination of MDR *Salmonella* and the genes that confer multidrug-resistance.

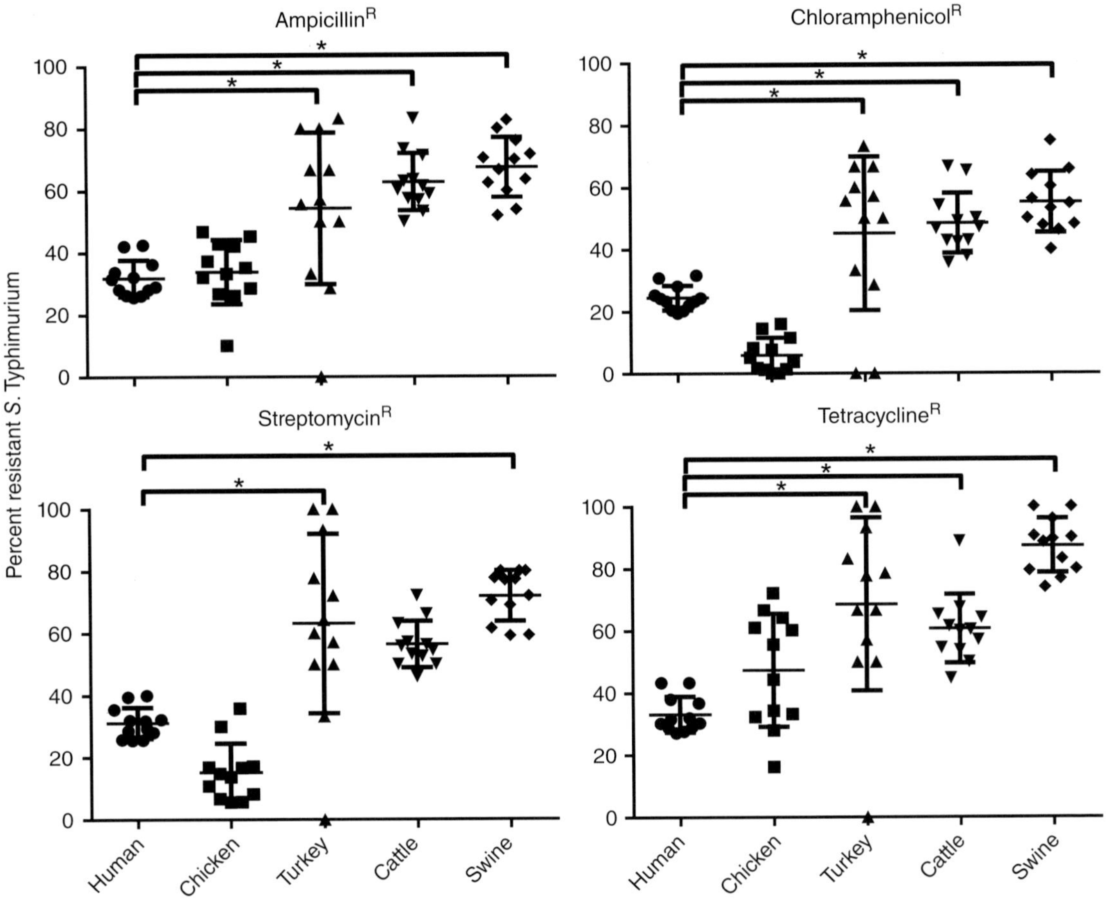

Figure 14.1 Dot Plot of Percent Resistant *Salmonella* Typhimurium in Each of 12 Years of NARMS Data in Humans, Livestock, and Poultry. Vertical lines within each plot show the standard deviation around the mean for each set of dots. Statistically significant differences ($P < 0.05$) between humans and agricultural animals were determined by the Kruskal-Wallis test with Dunn's multiple comparison correction and are shown with an asterisk.

14.4 Antibiotics and Resistance Mechanisms

NARMS tests *Salmonella* for resistance to antibiotics of several different classes, including aminoglycosides, β-lactams, folate pathway inhibitors, phenicols, quinolones, and tetracyclines. Each of these classes of antibiotics target a different bacterial function in the cell, and bacteria have developed multiple overlapping strategies to resist these antibiotics. Mechanisms employed by *Salmonella* to mitigate the effects of antibiotics are: restricting entry into the cell, inactivation, prevention from accessing the target, or exportation (Table 14.6).

14.4.1 Prevention of Antibiotic Entry

So-called innate resistance mechanisms can prevent the antibiotic from affecting its target. One innate resistance mechanism is failure of the antibiotic to enter the bacterial cell, which can be due to the size of the antibiotic or to the down-regulation of porin activity that controls influx of certain compounds. The major porins

Table 14.6 General Mechanisms of Antibiotic Resistance Utilized by *Salmonella*.

Mechanism	Antibiotic class	Commonly associated genes	References
Prevent entry (decreased porin activity)	β-lactams	*ompC, ompF*	Medeiros *et al.*, 1987; *Bellido et al.*, 1989; Armand-Lefevre *et al.*, 2003; Su *et al.*, 2012
Enzymatic inactivation	Aminoglycosides	*strAB, aac, aad, ant, aph*	Shaw *et al.*, 1993; Ramirez and Tolmasky, 2010; Frye and Jackson, 2013
	β-lactams	*bla*	Kalman and Barriere, 1990; Garau *et al.*, 1997; Li *et al.*, 2007
	Phenicols	*cat*	Schwarz *et al.*, 2004
Target site modification			
a. Direct (target site)	Quinolones	*gyrAB, parCE*	Chen *et al.*, 2007; Redgrave *et al.*, 2014Skold, 2001
	Folate pathway inhibitors	*sul1-3, dhr, dhfr*	
b. Indirect (access to target site)	Quinolones	*qnr*	Tran *et al.*, 2005; Jacoby *et al.*, 2014
Efflux pumps			
a. Enhanced innate multidrug pumps	Quinolones	*marAB*	Alekshun and Levy, 1999; Chen *et al.*, 2007; Redgrave *et al.*, 2014; Blair *et al.*, 2015
b. Acquired single-antibiotic pump	Phenicols	*floR*	Schwarz *et al.*, 2004; Braibant *et al.*, 2005; Glenn *et al.*, 2011
	Tetracyclines	*tetA-D, G*	Chopra and Roberts, 2001; Frye and Jackson, 2013

in *Salmonella* are encoded by *ompC* and *ompF*, and clinical isolates have been identified with mutations in these genes that increased resistance to β-lactams by decreasing uptake of the antibiotic (Medeiros *et al.*, 1987; Bellido *et al.*, 1989; Armand-Lefevre *et al.*, 2003; Su *et al.*, 2012).

14.4.2 Inactivation

One mechanism *Salmonella* exploits to counter antibiotic exposure is to enzymatically inactivate the antibiotic. *Salmonella* encodes three different methods of enzymatic inactivation for one class of antibiotics, the aminoglycosides, through the utilization of acetyltransferases (which include the *aac* family of genes), nucleotidyltrasferases (*aad, ant*), and phosphotransferases (*aph, str*) (Shaw *et al.*, 1993; Ramirez and Tolmasky, 2010; Frye and Jackson, 2013). Each enzyme acts by catalyzing the modification of an amine or hydroxyl group on an aminoglycoside, thereby rendering the antibiotic inactive. *Salmonella* can also modify and inactivate a member of another antibiotic class, chloramphenicol, through the enzyme encoded by the chloramphenicol acetyltransferase (*cat*) gene; however, this enzyme is ineffective on the analogous antibiotic florfenicol (Schwarz *et al.*, 2004).

Further mechanisms of antibiotic deactivation encoded by *Salmonella* are β-lactamase enzymes, which act on β-lactam antibiotics such as amoxicillin and penicillin. The common chemical structure of β-lactam antibiotics is a four-membered ring that contains an amide, and chemical differences among antibiotics in this class are based on attached moieties (Livermore, 1995). The ring can be cleaved, thus inactivating the antibiotic, by β-lactamases encoded by members of the *bla* gene family. New β-lactams with different attached moieties have been identified and created to evade enzymatic inactivation, including four generations of β-lactams in the cephalosporin family (Kalman and Barriere, 1990; Garau *et al.*, 1997). However, the selective pressure applied by new antibiotics containing structural modifications eventually selects for modified *bla* resistance genes. Mutations in *bla* that improve fitness are maintained in the *Salmonella* population, resulting in a vast diversity of *bla* gene derivatives.

There are primarily six different categories of *bla* genes that have been observed in *Salmonella* (bla_{CMY}, $bla_{CTX\text{-}M}$, bla_{OXA}, bla_{PSE}, bla_{SHV}, and bla_{TEM}), and each of these have numerous variants based on sequence polymorphisms (Li *et al.*, 2007). An increasing number of *Salmonella* isolates from humans and livestock have the bla_{CMY2} gene, which mediates resistance to ceftriaxone, the third-generation cephalosporin of primary importance for treating *Salmonella* infections (Frye *et al.*, 2008; Glenn *et al.*, 2011; Glenn *et al.*, 2013). Therefore, *Salmonella* contributes to β-lactamase evolution due to their exposure to antibiotic selective pressure in both human medicine and animal agriculture, making them an important reservoir for β-lactamases of clinical relevance.

14.4.3 Target Site Modification

After an antibiotic enters a cell, another method to avoid impairment is to alter or restrict accessibility of the compound to its target site. One way this can occur is by directly changing the target site, thereby neutralizing the specific effect of the antibiotic. This strategy has been utilized in *Salmonella* to resist exposure to sulfonamides that inhibit the folate pathway. Sulfonamides, such as sulfisoxazole, bind to dihydropteroate and prevent synthesis of the folic acid required by bacteria to survive (Wormser *et al.*, 1982). The resistance-based *sul* gene encodes an analog of dihydropteroate that lacks the binding region utilized by sulfonamides, but can still function properly in the folate synthesis pathway (Skold, 2001), thereby enabling *Salmonella* to flourish even in the presence of this antibiotic.

Salmonella also evades another class of antibiotics, the quinolones, by encoding modifications that alter the antibiotic binding site in the target. Quinolones function by targeting and binding to DNA gyrase and topoisomerase (Drlica and Zhao, 1997). However, the accumulation of point mutations in the genes that encode gyrase (*gyrA* and *gyrB*) and topoisomerase (*parC* and *parE*) alters amino acid residues in the respective proteins to inhibit the binding of quinolones and fluoroquinolones (Chen *et al.*, 2007; Redgrave *et al.*, 2014). An alternative mechanism of resistance to quinolones in *Salmonella* is impeding access to the active site from the antibiotic.

The quinolone resistance gene (*qnr*) produces a peptide that can bind to DNA gyrase and topoisomerase, and by doing so, protects the quinolone-binding site from the antibiotic (Tran *et al.*, 2005; Jacoby *et al.*, 2014). The peptide produced by the *qnr* gene mediates a low-level of quinolone resistance compared to the mutations that alter the binding sites on DNA gyrase and topoisomerase. However, the presence of *qnr* in isolates that lack such mutations enables the bacteria to survive fluoroquinolone exposure while simultaneously providing selective pressure for nucleotide changes that could facilitate increased resistance (Hooper and Jacoby, 2015).

14.4.4 Efflux Pumps

The third mechanism of antibiotic resistance encoded by genes in *Salmonella* is efflux, which is the active export of the antibiotic from the cell. In general, two different efflux pump systems function in bacteria: one is non-specific and transports a multitude of different compounds, and the other is specific for a particular chemical structure, including those of certain antibiotics. All *Salmonella* encode the multidrug AcrAB-TolC efflux pump system, which confers resistance to a low concentration of a wide variety of antimicrobials, including antibiotics, heavy metals, cleaning compounds, and bile (Li and Nikaido, 2004; Schuldiner, 2006; Baucheron *et al.*, 2014). This system is also thought to play a role in *Salmonella* virulence (Virlogeux-Payant *et al.*, 2008; Webber *et al.*, 2009), colonization (Buckley *et al.*, 2006), and biofilm formation (Baugh *et al.*, 2012). The AcrAB-TolC efflux pumps can be augmented to provide enhanced resistance to quinolones and fluoroquinolones (Alcaine *et al.*, 2007; Nikaido and Pages, 2012). Mutations in transcriptional regulators of efflux pump gene expression (such as *marRAB*) can lead to the overexpression of *acrAB* and *tolC*, which results in increased efflux activity and a corresponding increase in resistance to higher concentrations of quinolone compounds (Alekshun and Levy, 1999; Chen *et al.*, 2007; Redgrave *et al.*, 2014; Blair *et al.*, 2015).

Antibiotic-specific efflux pumps are not present in every *Salmonella* isolate, but can be acquired exogenously via several different means. Examples of specialized pumps in *Salmonella* include the *cml* and *floR* genes that encode efflux pumps for both chloramphenicol and florfenicol, and the *tetA-D* and *G genes*, each of which encode a different antiporter pump for tetracycline (Chopra and Roberts, 2001; Gebreyes *et al.*, 2004; Schwarz *et al.*, 2004; Braibant *et al.*, 2005; Glenn *et al.*, 2011; Frye and Jackson, 2013). A single *Salmonella* isolate may harbor one or more tetracycline resistance genes, demonstrating the functional redundancy and rich reservoir for resistance in *Salmonella*. In addition, specialized pumps are important because they typically confer resistance to much higher concentrations of antibiotics than the general efflux pump systems.

An *S.* Typhimurium isolate expressing the specific transporter TetA was much more resistant to tetracycline (minimum inhibitory concentration [MIC] of 256 µg/ml) than a different isolate expressing TetG (MIC of 64 µg/ml), whereas an isolate lacking a dedicated Tet efflux pump had an MIC of ~2 µg/ml due to the generalized efflux system (Brunelle *et al.*, 2015). However, efflux pumps are not necessarily as effective as inactivation enzymes, as was observed in a *Salmonella* isolate exposed to chloramphenicol that encoded *floR* (MIC of 128 µg/ml) compared to another isolate that encoded *cat* (MIC >512 µg/ml) (Brunelle *et al.*, 2015). The methods of resistance and the specific genes present in any particular isolate or population is important when considering sub-MIC antibiotic levels and defining therapeutic doses.

14.5 Evolution and Transfer of Antibiotic Resistance

Multiple factors contribute to the selection and evolution of antibiotic resistance genes in *Salmonella*, impeding the creation of efficacious treatment strategies. The selection for diverse antibiotic resistance genes as harbored by *Salmonella* is the most notorious effect of antibiotics because it directly limits the utility of antibiotic therapy (Hughes and Andersson, 2012). However, antibiotics have many other effects on bacterial communities (Looft and Allen, 2012; Stecher *et al.*, 2013; Allen and Stanton, 2014; Brunelle *et al.*, 2015). Recent research has shown that antibiotics additionally promote resistance in *Salmonella* by induction of multiple mechanisms of evolution: gene duplication and amplification (GDA), conjugation, and transduction.

14.5.1 Gene Duplication and Amplification

GDA leads to increased copies of a gene and occurs during normal bacterial replication via either *recA*-dependent or *recA*-independent methods (Andersson and Hughes, 2009; Sandegren and Andersson, 2009; Reams *et al.*, 2012). The increase in the copy number of a gene is only observed during selection conditions, such as nutrient limitation or antibiotic exposure. It has been shown that inhibitory levels of antibiotics can select for *Salmonella* isolates containing increased copies of a resistance gene, which confers a higher MIC (Nielsen *et al.*, 2005; Sun *et al.*, 2009; Lee *et al.*, 2015).

Furthermore, duplication of a resistance gene within a genome can lead to the creation of new alleles through point mutations (Pettersson *et al.*, 2009), which is one explanation for the many iterations of each *bla* gene that exists in *Salmonella* (Andersson and Hughes, 2009; Sandegren and Andersson, 2009). New *bla* genes could occur when a high concentration of a β-lactam antibiotic selects for increased copies of the *bla* gene after a GDA event. The extra copies would confer a fitness advantage that enables the *Salmonella* isolate to survive, as well as provide the opportunity for point mutations to occur in one or more copies of the *bla* gene. Those mutations that conferred increased resistance could then provide a selective advantage and remain in the population.

Changes due to a GDA event, however, are difficult to identify because the extra gene copies are an acute effect of selective pressure. Duplicates are not maintained following the removal of selective pressure, thereby making it difficult to determine if GDA was the mechanism by which resistance gene sequence divergence occurred in a clinical isolate. The temporary nature of gene duplications also makes it difficult to determine if GDA-enhanced resistance influenced a clinical isolate in a host, which has health implications as GDA events in resistance genes have been identified in patients treated with antibiotics (Musher *et al.*, 2002; Brochet *et al.*, 2008; Duvernay *et al.*, 2011).

14.5.2 Conjugation

Conjugation is the direct transfer of foreign DNA from a donor bacterium to a recipient bacterium using a pilus, which functions as a conduit between the two adjacent bacterial cells (Kelly *et al.*, 2009a, b). Conjugation requires specific transfer proteins for pilus formation and mobilization of DNA from the donor bacterium. Therefore, only bacteria that contain DNA encoding transfer and mobilization genes are proficient for conjugation. Many *Salmonella* plasmids encode the *tra* family of transfer genes and can conjugate both in the laboratory and *in vivo* (Jarolmen and Kemp, 1969; Archambaud *et al.*, 1991; Ferguson *et al.*, 2002; Daniels *et al.*, 2009).

Lateral gene transfer of antimicrobial resistance determinants between the *Enterobacteriaceae*, including *Salmonella*, was demonstrated by multiple investigators in the 1960s. For example, Gill and Hook found that 71% of antibiotic resistant clinical isolates, including MDR *Salmonella* serotypes Typhimurium, Heidelberg, and Newport, were able to simultaneously transfer multiple antibiotic resistance factors to *Escherichia coli* K12 (Gill and Hook, 1966). These and other experiments demonstrating antibiotic resistance transfer among *Salmonella* and the *Enterobacteriaceae* highlighted the public health hazard of antibiotic resistant pathogens in the 1970s (Baldwin, 1970; Pocurull *et al.*, 1971; Lawrence and Hoeprich, 1975).

In addition to its direct role in resistance gene transfer, conjugation is additionally relevant to the dissemination of resistance genes because it can be induced by sub-MIC antibiotic concentrations, as shown in bacteria that are closely related to *Salmonella*. One such example is the observation that sub-MIC levels of fluoroquinolones stimulated conjugation in *E. coli* (López *et al.*, 2007). Additionally, the conjugative transfer of single-stranded plasmid DNA itself induced the SOS response, integrase expression, and increased frequency of recombination in *Vibrio cholerae* (Baharoglu *et al.*, 2010). Further research is needed to determine the role of sub-MIC antibiotic induction of conjugation in the dissemination of resistance genes among *Salmonella*.

14.5.3 Transduction

Transduction is the injection of foreign DNA by a bacteriophage (phage) into a recipient cell. For transduction to occur, the donor bacterium must be infected with a bacteriophage to facilitate packaging of DNA into phage heads. Once the phages are assembled in the donor bacterium, phage-mediated lysis kills the donor bacterial host and releases the infectious virions.

Two lifestyles of phages are lytic phages, which are always seeking a bacterial host to prey upon, and temperate phages, which have a dormancy phase (lysogeny) within a bacterial host prior to a lytic cycle. Bacteriophage that are integrated into the bacterial host DNA during lysogeny are termed prophage. The genomes of *Salmonella* spp. often contain four to six prophages (McClelland *et al.*, 2001; Brussow *et al.*, 2004; Cooke *et al.*, 2008). Certain prophages, including P22-like prophage, randomly package non-phage DNA in a size similar to the genome of the bacteriophage performing the event; the transfer of random, non-phage DNA to a recipient bacterium is termed generalized transduction (Weinstock, 2002). For P22-like prophage, the length of DNA that can be transferred is ~43 kb (Casjens and Hayden, 1988), thus allowing a small plasmid or ~1% of the *Salmonella* genome to be mobilized in an individual phage head.

Some *Salmonella* Typhimurium strains, such as DT104 and DT120, contain a P22-like prophage integrated in their genomes that can be activated and perform generalized transduction (Schmieger and Schicklmaier, 1999; Cooke *et al.*, 2008; Bearson *et al.*, 2014). *Salmonella* isolates within multiple serotypes may contain generalized transducing prophage. *Salmonella* isolates in the SARA reference collection (Beltran *et al.*, 1991) shown to perform generalized transduction include Typhimurium, Heidelberg, Muenchen, Paratyphi B, and Saintpaul (Schicklmaier and Schmieger, 1995). In addition, genome scanning previously indicated that generalized transducing prophages are present in a variety of *Salmonella* serotypes, and their presence may enhance bacterial strain evolution due to gene transfer (Bearson *et al.*, 2014). To provide additional support for this hypothesis, prophage genomes with generalized transduction potential were identified by searching for genes homologous to the integrase gene (*int*) of prophage 1, a generalized transducing prophage in Typhimurium DT104 with similarity to bacteriophage P22. Five candidate prophage genomes were aligned and compared to the DT104 prophage 1 reference (Figure 14.2). The prophages in serotypes Heidelberg and Pullorum are similar to the P22-like prophage of DT104, as shown in Figure 14.2 by the presence

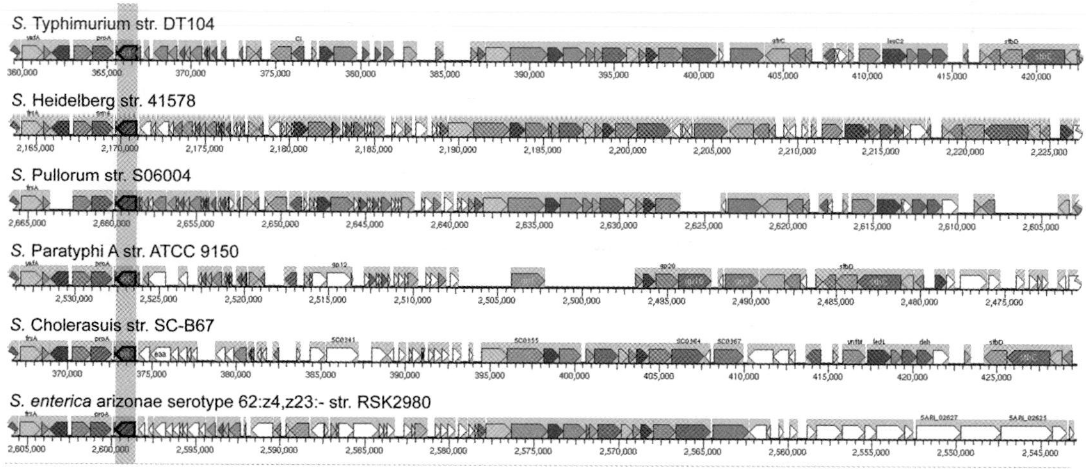

Figure 14.2 Open Reading Frame Maps of Potential Generalized Transducing Prophage Compared to Prophage 1 of *Salmonella* Typhimurium DT104. Alignments were made to the gene encoding the prophage 1 integrase (*int*, DT104_03231), highlighted with a blue box across all alignments. Prophage 1 (P22-like), a generalized transducing prophage in DT104 (NC_022569), is integrated into *thrW* (DT104_t00005) at genome coordinates 365, 588–406, 646 (Cooke *et al.*, 2008). Gene color indicates orthologous groups. The alignment was generated using Pathway Tools version 19.5 by SRI International (Menlo Park, CA) (Karp *et al.*, 2002). (*See insert for color representation of the figure.*)

of many orthologous genes and a similar gene arrangement. Although a core set of orthologous genes is present in all six serotypes, the prophages in Paratyphi A, Choleraesuis, and *S. enterica* arizonae serotype 62: z4,z23:- appear to be more distantly related since a greater number of genes are different. These data suggest that diverse *Salmonella* harbor prophages capable of generalized transduction, and further experimentation is required to test this finding.

Factors that cause lysogenic phages, or prophages, to enter a lytic cycle include stress and DNA damage. Sub-inhibitory concentrations of antibiotics that cause DNA damage have been shown to induce prophages in an SOS-dependent manner, including ciprofloxacin (a fluoroquinolone) in *Pseuodomonas aeruginosa* (Cirz *et al.*, 2006); fluoroquinolones, trimethoprim, and furazolidone in Shiga-toxin *Escherichia coli* (STEC) (Kimmitt *et al.*, 2000); and ciprofloxacin and trimethoprim in *Staphylococcus aureus* (Goerke *et al.*, 2006; Cirz *et al.*, 2007). In addition to DNA-damaging antibiotics, sub-inhibitory concentrations of β-lactam antibiotics induced prophages in *S. aureus* in an SOS-dependent manner, and virulence factors were shown to be transduced (Maiques *et al.*, 2006). In *Salmonella*, fluoroquinolones (ciprofloxacin, enrofloxacin, and danofloxacin) stimulated phage-mediated transduction of a plasmid-borne kanamycin resistance gene in two *S.* Typhimurium strains (DT104 and DT120), and the associated increase in *recA* transcription suggests this induction was dependent on the SOS response (Bearson and Brunelle, 2015). Additionally, the antibiotic carbadox was shown to induce prophages in *S.* Typhimurium, including generalized transducing phages that package portions of the bacterial chromosome or plasmids that carry antibiotic resistance genes (Bearson *et al.*, 2014).

Carbadox is an in-feed agricultural antibiotic important for the control of bacterial diseases in swine. It is a member of the quinoxaline 1,4-dioxide compounds, which like the quinolones exert mutagenicity via induction of the SOS response by causing RecA-mediated cleavage of LexA and subsequent error-prone DNA repair in *E. coli* and *Salmonella* (Nunoshiba and Nishioka, 1989; Little, 1993).

Antibiotics that have no analog in human medicine, such as cardadox, can still be approved for use in livestock as they do not directly select for resistance genes of clinical importance (FDA, 2013b). However, in addition to an antibiotic's direct role in selecting for antibiotic resistance genes, the data showing that carbadox induces prophages that transfer antibiotic resistance genes to susceptible bacterial hosts (Bearson *et al.*, 2014) suggest that resistance gene selection is not the only effect worthy of regulatory consideration. An unintended consequence of bacterial exposure to certain antimicrobials like carbadox and fluoroquinolones is the promotion of phage transduction and resistance gene transmission that may enhance bacterial strain evolution.

14.6 Co-localization of Resistance Genes

The genes that encode antibiotic-specific pumps, inactivation enzymes, resistance analogs, and certain protective peptides are often located on integrons in transposons, plasmids, or both. This allows a resistance gene to move either via the whole plasmid, or just a sub-section, into neighboring bacteria by conjugation or distant cells by transduction. In addition, some *Salmonella* plasmids that encode resistance genes have been found to also contain genes for conjugation (Carattoli, 2003). Movement of resistance genes within a single bacterium from one chromosomal location to another has also occurred, and together with horizontal gene transfer has led to the creation of clusters of resistance genes that are referred to as either cassettes or resistance islands (Miriagou *et al.*, 2006).

Co-localization of fitness genes creates a clinical problem because selection for one antibiotic resistance gene will lead to selection of all genes within that cluster. This has been observed in *S. Typhimurium* DT104 isolates that encode resistance genes to five antibiotics within a ~12 kb region that is part of a transposon located in the chromosome (Boyd *et al.*, 2001). Evidence also suggests that this transposon, which is part of the 43 kb *Salmonella* Genomic Island (SGI-1), has been horizontally acquired by other serotypes (Doublet *et al.*, 2005). Additionally, a resistance gene cluster has been identified that has co-integrated with the *Salmonella* virulence plasmid (Fluit, 2005). This is problematic as even in the absence of antibiotic pressure, the resistance genes in such a plasmid can be co-selected and maintained by its virulence attributes.

14.7 Conclusions

The existence of MDR *Salmonella* has been recognized since the 1960s (Gill and Hook, 1966). The potential for transfer of resistance genes into *Salmonella* has also been appreciated for decades; the transfer of antibiotic resistance genes into *Salmonella* from *E. coli* was shown in 1969, both in vitro and in vivo, in the absence of antibiotic selective pressure (Smith, 1970). Historical data such as these indicate that MDR *Salmonella* is a persistent problem rather than an emerging concern. Contemporary research has demonstrated that certain antibiotics enhance the evolution and transfer of resistance genes, illuminating one aspect of the challenge associated with combating MDR *Salmonella*. Epidemiological data showing the ubiquity of MDR *Salmonella* and its transmission between the agricultural reservoir and humans further explains the challenge in curtailing its prevalence. The derivation of efficacious intervention strategies will require the consideration of these properties of MDR *Salmonella* to minimize further antibiotic resistance development, transmission, and enhancement.

Several intervention strategies have been investigated to reduce the incidence of MDR *Salmonella* infections in humans. Research into developing safe and effective probiotics and vaccines to prevent infection, as well as novel therapeutics, will help enable the continued use of antibiotics in patients with severe bacterial illness. In addition, the Food and Drug Administration has issued a Guidance for Industry that calls for the voluntary phasing out of medically important antibiotics in animal food and water (FDA, 2012; 2013b). It will be interesting to observe if and how the ensuing decrease in the use of these types of antibiotics affects the prevalence of MDR *Salmonella* in food-producing animals.

Acknowledgments

We would like to thank Dr. Shawn Bearson for her critical review of the manuscript. USDA is an equal opportunity provider and employer.

References

Alcaine, S.D., Warnick, L.D. and Wiedmann, M. (2007). Antimicrobial resistance in nontyphoidal *Salmonella*. *J Food Prot.*, *70*, pp. 780–790.

Alekshun, M.N. and Levy, S.B. (1999). The mar regulon: multiple resistance to antibiotics and other toxic chemicals. *Trends Microbiol.*, *7*, pp. 410–413.

Allen, H.K. and Stanton, T.B. (2014). Altered egos: antibiotic effects on food animal microbiomes. *Annual Reviews in Microbiology*, *68*, pp. 297–315.

Andersson, D.I. and Hughes, D. (2009). Gene amplification and adaptive evolution in bacteria. *Annu Rev Genet.*, *43*, pp. 167–195.

Archambaud, M., et al. (1991). Possible in-vivo transfer of beta-lactamase TEM-3 from *Klebsiella pneumoniae* to *Salmonella kedougou*. *J Antimicrob Chemother.*, *27*, pp. 427–436.

Armand-Lefevre, L., *et al.* (2003). Imipenem resistance in *Salmonella enterica* serovar Wien related to porin loss and CMY-4 beta-lactamase production. *Antimicrob Agents Chemother.*, *47*, pp. 1165–1168.

Baharoglu, Z., Bikard, D. and Mazel, D. (2010). Conjugative DNA transfer induces the bacterial SOS response and promotes antibiotic resistance development through integron activation. *PLoS Genet.*, *6*, pp. e1001165–e1001165.

Baldwin, R.A. (1970). The development of transferable drug resistance in *Salmonella* and its public health implications. *J Am Vet Med Assoc.*, *157*, pp. 1841–1853.

Baucheron, S., *et al.* (2014). Bile-mediated activation of the *acrAB* and *tolC* multidrug efflux genes occurs mainly through transcriptional derepression of *ramA* in *Salmonella enterica* serovar Typhimurium. *J Antimicrob Chemother.*, *69*, pp. 2400–2406.

Baugh, S., *et al.* (2012). Loss of or inhibition of all multidrug resistance efflux pumps of *Salmonella enterica* serovar Typhimurium results in impaired ability to form a biofilm. *J Antimicrob Chemother*, *67*, pp. 2409–2417.

Bearson, B.L., *et al.* (2014). The agricultural antibiotic carbadox induces phage-mediated gene transfer in *Salmonella*. *Front Microbiol*, *5*, article 52.

Bearson, B.L. and Brunelle, B.W. (2015). Fluoroquinolone induction of phage-mediated gene transfer in multidrug-resistant *Salmonella*. *International Journal of Antimicrobial Agents*, *46*, pp. 201–204.

Bellido, F., *et al.* (1989). Permeability and penicillin-binding protein alterations in *Salmonella muenchen*: stepwise resistance acquired during beta-lactam therapy. *Antimicrob Agents Chemother.*, *33*, pp. 1113–1115.

Beltran, P., *et al.* (1991). Reference collection of strains of the *Salmonella typhimurium* complex from natural populations. *J Gen Microbiol.*, *137*, pp. 601–606.

Bissett, M.L., Abbott, S.L. and Wood, R.M. (1974). Antimicrobial resistance and R factors in *Salmonella* isolated in California (1971-1972). *Antimicrob Agents Chemother.*, *5*, pp. 161–168.

Blair, J.M., *et al.* (2015). Molecular mechanisms of antibiotic resistance. *Nat Rev Microbiol.*, *13*, pp. 42–51.

Boyd, D., *et al.* (2001). Complete nucleotide sequence of a 43-kilobase genomic island associated with the multidrug resistance region of *Salmonella enterica* serovar Typhimurium DT104 and its identification in phage type DT120 and serovar Agona. *J Bacteriol.*, *183*, pp. 5725–5732.

Braibant, M., *et al.* (2005). Structural and functional study of the phenicol-specific efflux pump FloR belonging to the major facilitator superfamily. *Antimicrob Agents Chemother.*, *49*, pp. 2965–2971.

Brochet, M., *et al.* (2008). A naturally occurring gene amplification leading to sulfonamide and trimethoprim resistance in *Streptococcus agalactiae*. *J Bacteriol.*, *190*, pp. 672–680.

Brunelle, B.W., Bearson, B.L. and Bearson, S.M. (2015). Chloramphenicol and tetracycline decrease motility and increase invasion and attachment gene expression in specific isolates of multidrug-resistant *Salmonella enterica* serovar Typhimurium. *Front Microbiol*, *5*, p. 801.

Brussow, H., Canchaya, C. and Hardt, W.D. (2004). Phages and the evolution of bacterial pathogens: from genomic rearrangements to lysogenic conversion. *Microbiol Mol Biol Rev*, *68*, pp. 560–602, table of contents.

Buckley, A.M., *et al.* (2006). The AcrAB-TolC efflux system of *Salmonella enterica* serovar Typhimurium plays a role in pathogenesis. *Cell Microbiol.*, *8*, pp. 847–856.

Carattoli, A. (2003). Plasmid-mediated antimicrobial resistance in *Salmonella enterica*. *Curr Issues Mol Biol.*, *5*, pp. 113–122.

Casjens, S. and Hayden, M. (1988). Analysis in vivo of the bacteriophage P22 headful nuclease. *J Mol Biol.*, *199*, pp. 467–474.

CDC. (2008). National Antimicrobial Resistance Monitoring System for Enteric Bacteria (NARMS): Human Isolates Final Report, 2005. Atlanta, GA: U.S. Department of Health and Human Services, Centers for Disease Control and Prevention.

CDC. (2013a). Antibiotic Resistance Threats in the United States, 2013. Atlanta, GA: U.S. Department of Health and Human Services, Centers for Disease Control and Prevention.

CDC. (2013b). FoodNet 2011 Surveillance Report. Atlanta, GA: U.S. Department of Health and Human Services, Centers for Disease Control and Prevention.

CDC. (2015). National Antimicrobial Resistance Monitoring System for Enteric Bacteria (NARMS): Human Isolates Final Report, 2013. Atlanta, GA: U.S. Department of Health and Human Services, Centers for Disease Control and Prevention.

Cebrian, L., *et al.* (2003). Characterisation of *Salmonella* spp. mutants produced by exposure to various fluoroquinolones. *Int J Antimicrob Agents, 22,* pp. 134–139.

Chen, S., *et al.* (2007). Contribution of target gene mutations and efflux to decreased susceptibility of *Salmonella enterica* serovar Typhimurium to fluoroquinolones and other antimicrobials. *Antimicrob Agents Chemother., 51,* pp. 535–542.

Chopra, I. and Roberts, M. (2001). Tetracycline antibiotics: mode of action, applications, molecular biology, and epidemiology of bacterial resistance. *Microbiol Mol Biol Rev., 65,* pp. 232–260; second page, table of contents.

Cirz, R.T., *et al.* (2007). Complete and SOS-mediated response of *Staphylococcus aureus* to the antibiotic ciprofloxacin. *Journal of Bacteriology, 189,* pp. 531–539.

Cirz, R.T., *et al.* (2006). Defining the *Pseudomonas aeruginosa* SOS response and its role in the global response to the antibiotic ciprofloxacin. *Journal of Bacteriology, 188,* pp. 7101–7110.

Cooke, F.J., et al. (2008). Characterization of the genomes of a diverse collection of *Salmonella enterica* serovar Typhimurium definitive phage type 104. *J Bacteriol, 190,* pp. 8155–8162.

Daniels, J.B., *et al.* (2009). Role of ceftiofur in selection and dissemination of blaCMY-2-mediated cephalosporin resistance in *Salmonella enterica* and commensal *Escherichia coli* isolates from cattle. *Appl Environ Microbiol., 75,* pp. 3648–3655.

Doublet, B., *et al.* (2005). The *Salmonella* genomic island 1 is an integrative mobilizable element. *Mol Microbiol., 55,* pp. 1911–1924.

Drlica, K. and Zhao, X. (1997). DNA gyrase, topoisomerase IV, and the 4-quinolones. *Microbiol Mol Biol Rev., 61,* pp. 377–392.

Duvernay, C., et al. (2011). Duplication of the chromosomal *bla*SHV-11 gene in a clinical hypermutable strain of *Klebsiella pneumoniae. Microbiology, 157,* pp. 496–503.

Economic Research Service. (2014). Foodborne Illness Cost Calculator: *Salmonella.* Washington, D.C.: United States Department of Agriculture.

FDA. (2012). The Judicious use of medically important antimicrobial drugs in food-producing animals: Guidance for Industry #209. Available at: http://www.fda.gov/downloads/AnimalVeterinary/GuidanceComplianceEnforcement/GuidanceforIndustry/UCM216936.pdf [Accessed 15 October 2015].

FDA. (2013a). National Antimicrobial Resistance Monitoring System – Enteric Bacteria (NARMS): 2011 Executive Report. Rockville, MD: U.S. Department of Health and Human Services, Food and Drug Administration.

FDA. (2013b). New Animal Drugs and New Animal Drug Combination Products Administered in or on Medicated Feed or Drinking Water of Food-Producing Animals: Recommendations for Drug Sponsors for Voluntarily Aligning Product Use Conditions with GFI #209: Guidance for Industry #213. Available at: http://www.fda.gov/downloads/AnimalVeterinary/GuidanceComplianceEnforcement/GuidanceforIndustry/UCM299624.pdf [Accessed 15 October 2015].

Ferguson, G.C., Heinemann, J.A. and Kennedy, M.A. (2002). Gene transfer between *Salmonella enterica* serovar Typhimurium inside epithelial cells. *J Bacteriol., 184,* pp. 2235–2242.

Fey, P.D., et al. (2000). Ceftriaxone-resistant *Salmonella* infection acquired by a child from cattle. *N Engl J Med., 342,* pp. 1242–1249.

Fluit, A.C. (2005). Towards more virulent and antibiotic-resistant *Salmonella*? *FEMS Immunology and Medical Microbiology, 43,* pp. 1–11.

Frye, J.G., et al. (2008). Analysis of *Salmonella enterica* with reduced susceptibility to the third-generation cephalosporin ceftriaxone isolated from U.S. cattle during 2000–2004. *Microb Drug Resist.*, *14*, pp. 251–258.

Frye, J.G. and Jackson, C.R. (2013). Genetic mechanisms of antimicrobial resistance identified in *Salmonella enterica*, *Escherichia coli*, and *Enteroccocus* spp. isolated from U.S. food animals. *Front Microbiol.*, *4*, p. 135–.

Garau, J., et al. (1997). Fourth-generation cephalosporins: a review of in vitro activity, pharmacokinetics, pharmacodynamics and clinical utility. *Clinical Microbiology and Infection*, *3*, Supplement 1, pp. S87–S101.

Gebreyes, W.A., et al. (2004). Characterization of antimicrobial-resistant phenotypes and genotypes among *Salmonella enterica* recovered from pigs on farms, from transport trucks, and from pigs after slaughter. *J Food Prot.*, *67*, pp. 698–705.

Gill, F.A. and Hook, E.W. (1966). *Salmonella* strains with transferable antimicrobial resistance. *JAMA*, *198*, pp. 1267–1269.

Glenn, L.M., et al. (2013). Antimicrobial resistance genes in multidrug-resistant *Salmonella enterica* isolated from animals, retail meats, and humans in the United States and Canada. *Microb Drug Resist.*, *19*, pp. 175–184.

Glenn, L.M., et al. (2011). Analysis of antimicrobial resistance genes detected in multidrug-resistant *Salmonella enterica* serovar Typhimurium isolated from food animals. *Microb Drug Resist.*, *17*, pp. 407–418.

Goerke, C., Köller, J. and Wolz, C. (2006). Ciprofloxacin and trimethoprim cause phage induction and virulence modulation in *Staphylococcus aureus*. *Antimicrobial Agents and Chemotherapy*, *50*, pp. 171–177.

Goldberg, M.B. and Rubin, R.H. (1988). The spectrum of *Salmonella* infection. *Infect Dis Clin North Am.*, *2*, pp. 571–598.

Heisig, P. and Tschorny, R. (1994). Characterization of fluoroquinolone-resistant mutants of *Escherichia coli* selected in vitro. *Antimicrob Agents Chemother.*, *38*, pp. 1284–1291.

Hooper, D.C. and Jacoby, G.A. (2015). Mechanisms of drug resistance: quinolone resistance. *Ann N Y Acad Sci.*, *1354*, pp. 12–31.

Hughes, D. and Andersson, D.I. (2012). Selection of resistance at lethal and non-lethal antibiotic concentrations. *Current Opinion in Microbiology*, *15*, pp. 555–60.

Jacoby, G.A., Strahilevitz, J. and Hooper, D.C. (2014). Plasmid-mediated quinolone resistance. *Microbiol Spectr.*, *2*(5).

Jarolmen, H. and Kemp, G. (1969). R factor transmission in vivo. *J Bacteriol.*, *99*, pp. 487–490.

Kalman, D. and Barriere, S.L. (1990). Review of the pharmacology, pharmacokinetics, and clinical use of cephalosporins. *Tex Heart Inst J.*, *17*, pp. 203–215.

Karp, P.D., Paley, S. and Romero, P. 2002. The Pathway Tools software. *Bioinformatics*, *18*(Suppl 1), pp. S225–S232.

Kelly, B.G., Vespermann, A. and Bolton, D.J. (2009a). Horizontal gene transfer of virulence determinants in selected bacterial foodborne pathogens. *Food Chem Toxicol.*, *47*(5), pp. 969–977.

Kelly, B.G., Vespermann, A. and Bolton, D.J. (2009b). The role of horizontal gene transfer in the evolution of selected foodborne bacterial pathogens. *Food Chem Toxicol.*, *47*, pp. 951–968.

Kimmitt, P.T., Harwood, C.R. and Barer, M.R. (2000). Toxin gene expression by shiga toxin-producing *Escherichia coli*: the role of antibiotics and the bacterial SOS response. *Emerging infectious diseases*, *6*, p. 458.

Lawrence, R.M. and Hoeprich, P.D. (1975). Microbial development of drug resistance: mechanisms and clinical significance. *CRC Crit Rev Clin Lab Sci.*, *5*, pp. 365–386.

Lee, K., et al. (2015). Extensive amplification of GI-VII-6, a multidrug resistance genomic island of *Salmonella enterica* serovar Typhimurium, increases resistance to extended-spectrum cephalosporins. *Front Microbiol.*, *6*, p. 78.

Li, X.Z., et al. (2007). beta-Lactam resistance and beta-lactamases in bacteria of animal origin. *Vet Microbiol 121*, pp. 197–214.

Li, X.Z. and Nikaido, H. (2004). Efflux-mediated drug resistance in bacteria. *Drugs, 64*, pp. 159–204.

Little, J.W. (1993). LexA cleavage and other self-processing reactions. *J Bacteriol., 175*, pp. 4943–4950.

Livermore, D.M. (1995). beta-Lactamases in laboratory and clinical resistance. *Clin Microbiol Rev., 8*, pp. 557–584.

Looft, T. and Allen, H.K. (2012). Collateral effects of antibiotics on mammalian gut microbiomes. *Gut Microbes, 3*, pp. 463–467.

López, E., et al. (2007). Antibiotic-mediated recombination: ciprofloxacin stimulates SOS-independent recombination of divergent sequences in *Escherichia coli*. *Molecular Microbiology, 64*, pp. 83–93.

Maiques, E., et al. (2006). β-Lactam antibiotics induce the SOS response and horizontal transfer of virulence factors in *Staphylococcus aureus*. *Journal of Bacteriology, 188*, pp. 2726–2729.

Majowicz, S.E., et al. (2010). The global burden of nontyphoidal *Salmonella* gastroenteritis. *Clin Infect Dis., 50*, pp. 882–889.

McClelland, M., et al. (2001). Complete genome sequence of *Salmonella enterica* serovar Typhimurium LT2. *Nature, 413*, pp. 852–856.

Medeiros, A.A., et al. (1987). Loss of OmpC porin in a strain of *Salmonella typhimurium* causes increased resistance to cephalosporins during therapy. *J Infect Dis., 156*, pp. 751–757.

Miriagou, V., Carattoli, A. and Fanning, S. (2006). Antimicrobial resistance islands: resistance gene clusters in *Salmonella* chromosome and plasmids. *Microbes Infect., 8*, pp. 1923–1930.

Molbak, K. (2005). Human health consequences of antimicrobial drug-resistant *Salmonella* and other foodborne pathogens. *Clin Infect Dis., 41*, pp. 1613–1620.

Musher, D.M., et al. (2002). Emergence of macrolide resistance during treatment of pneumococcal pneumonia. *N Engl J Med., 346*, pp. 630–631.

Nielsen, N.M., et al. (2005). Characterisation of rotavirus strains among hospitalised and non-hospitalised children in Guinea-Bissau, 2002 A high frequency of mixed infections with serotype G8. *J Clin Virol., 34*, pp. 13–21.

Nikaido, H. and Pages, J.M. (2012). Broad-specificity efflux pumps and their role in multidrug resistance of Gram-negative bacteria. *FEMS Microbiol Rev., 36*, pp. 340–363.

Nunoshiba, T. and Nishioka, H. (1989). Genotoxicity of quinoxaline 1, 4-dioxide derivatives in *Escherichia coli* and *Salmonella* typhimurium. *Mutation Research/DNA Repair, 217*, pp. 203–209.

Pettersson, M.E., et al. (2009). Evolution of new gene functions: simulation and analysis of the amplification model. *Genetica, 135*, pp. 309–324.

Pocurull, D.W., Gaines, S.A. and Mercer, H.D. (1971). Survey of infectious multiple drug resistance among *Salmonella* isolated from animals in the United States. *Appl Microbiol., 21*, pp. 358–362.

Ramirez, M.S. and Tolmasky, M.E. (2010). Aminoglycoside modifying enzymes. *Drug Resist Updat., 13*, pp. 151–171.

Reams, A.B., et al. (2012). Multiple pathways of duplication formation with and without recombination (RecA) in *Salmonella enterica*. *Genetics, 192*, pp. 397–415.

Redgrave, L.S., et al. (2014). Fluoroquinolone resistance: mechanisms, impact on bacteria, and role in evolutionary success. *Trends Microbiol., 22*, pp. 438–445.

Sandegren, L. and Andersson, D.I. (2009). Bacterial gene amplification: implications for the evolution of antibiotic resistance. *Nat Rev Microbiol., 7*, pp. 578–588.

Schicklmaier, P. and Schmieger, H. (1995). Frequency of generalized transducing phages in natural isolates of the *Salmonella typhimurium* complex. *Appl Environ Microbiol., 61*, pp. 1637–1640.

Schmieger, H. and Schicklmaier, P. (1999). Transduction of multiple drug resistance of *Salmonella enterica* serovar typhimurium

DT104. *FEMS Microbiol Lett., 170,* pp. 251–256.

Schroeder, S.A., Terry, P.M. and Bennett, J.V. (1968). Antibiotic resistance and transfer factor in *Salmonella,* United States 1967. *Journal of the American Medical Association., 205,* pp. 903–906.

Schuldiner, S. (2006). Structural biology: the ins and outs of drug transport. *Nature, 443,* pp. 156–157.

Schwarz, S., et al. (2004). Molecular basis of bacterial resistance to chloramphenicol and florfenicol. *FEMS Microbiol Rev., 28,* pp. 519–542.

Shaw, K.J., et al. (1993). Molecular genetics of aminoglycoside resistance genes and familial relationships of the aminoglycoside-modifying enzymes. *Microbiol Rev., 57,* pp. 138–163.

Skold, O. (2001). Resistance to trimethoprim and sulfonamides. *Vet Res., 32,* pp. 261–273.

Smith, H.W. (1970). The transfer of antibiotic resistance between strains of enterobacteria in chicken, calves and pigs. *Journal of Medical Microbiology, 3,* pp. 165–180.

Stecher, B., Maier, L. and Hardt, W.D. (2013). 'Blooming' in the gut: how dysbiosis might contribute to pathogen evolution. *Nat Rev Microbiol., 11,* pp. 277–284.

Stokes, H.W. and Gillings, M.R. (2011). Gene flow, mobile genetic elements and the recruitment of antibiotic resistance genes into Gram-negative pathogens. *FEMS Microbiology Reviews, 35,* pp. 790–819.

Su, L.H., Wu, T.L. and Chiu, C.H. (2012). Development of carbapenem resistance during therapy for non-typhoid *Salmonella* infection. *Clin Microbiol Infect., 18,* pp. E91–E94.

Sun, S., et al. (2009). Contribution of gene amplification to evolution of increased antibiotic resistance in *Salmonella typhimurium. Genetics, 182,* pp. 1183–1195.

Swanson, S.J., et al. (2007). Multidrug-resistant *Salmonella enterica* serotype Typhimurium associated with pet rodents. *N Engl J Med., 356,* pp. 21–28.

Tran, J.H., Jacoby, G.A. and Hooper, D.C. (2005). Interaction of the plasmid-encoded quinolone resistance protein QnrA with *Escherichia coli* topoisomerase IV. *Antimicrob Agents Chemother., 49,* pp. 3050–3052.

Varma, J.K., et al. (2005a). Hospitalization and antimicrobial resistance in *Salmonella* outbreaks, 1984-2002. *Emerg Infect Dis., 11,* pp. 943–946.

Varma, J.K., et al. (2005b). Antimicrobial-resistant nontyphoidal *Salmonella* is associated with excess bloodstream infections and hospitalizations. *J Infect Dis., 191,* pp. 554–561.

Virlogeux-Payant, I., et al. (2008). TolC, but not AcrB, is involved in the invasiveness of multidrug-resistant *Salmonella enterica* serovar Typhimurium by increasing type III secretion system-1 expression. *Int J Med Microbiol., 298,* pp. 561–569.

Webber, M.A., et al. (2009). The global consequence of disruption of the AcrAB-TolC efflux pump in *Salmonella enterica* includes reduced expression of SPI-1 and other attributes required to infect the host. *J Bacteriol., 191,* pp. 4276–4285.

Weinstock, G.M. (2002). "Transduction in Gram-Negative Bacteria," in *Modern Microbial Genetics.* New York: John Wiley & Sons, Inc., pp. 561–579.

WHO. (2011). Critically important antimicrobials for human medicine, 3rd revision. *WHO Advisory Group on Integrated Surveillance of Antimicrobial Resistance,* pp. 1–31.

Wormser, G.P., Keusch, G.T. and Heel, R.C. (1982). Co-trimoxazole (trimethoprim-sulfamethoxazole): an updated review of its antibacterial activity and clinical efficacy. *Drugs, 24,* pp. 459–518.

Wright, J.G., et al. (2005). Multidrug-resistant *Salmonella* Typhimurium in four animal facilities. *Emerg Infect Dis., 11,* pp. 1235–1241.

15

Antibiotic Resistance of Coagulase-Positive and Coagulase-Negative Staphylococci Isolated From Food

Wioleta Chajęcka-Wierzchowska and Anna Zadernowska

University of Warmia and Mazury in Olsztyn, Poland

15.1 Characteristics of the Genus *Staphylococcus*

The genus *Staphylococcus* includes spherical or slightly oval cocci with a diameter of around 0.5–1.5 µm. On slides, they are usually visible as irregular characteristic grape-like clusters, but they can also be arranged singly or in pairs. Staphylococci are gram-positive, non-motile and non-spore-forming. They are considered relative anaerobes, which grow faster under aerobic conditions and only a few strains produce capsules (Götz *et al.*, 2006). Their distinguishing feature is high tolerance to different physical and chemical factors. They show growth across a wide pH range (from 4 to 9), but they develop better in alkaline medium (Harris *et al.*, 2002). The optimum temperature is 35–41°C and the temperature range at which their growth is possible is 6–48°C (Yang *et al.*, 2001). At a temperature of 60°C, they die only after half an hour or an hour of heating. The ability of staphylococci to grow at a low water activity ($a_w = 0.83$) is of high importance for food safety (Qi and Miller, 2000). Staphylococci also tolerate high salt concentrations (up to 25% NaCl), though the preferred concentration is 7–10% NaCl (Bang *et al.*, 2008). Moreover, staphylococci grow in the presence of 40% bile salt. They also withstand a 1% phenol concentration for 15 minutes and die at 2% phenol and in the presence of chlorine compounds.

Genus *Staphylococcus* is a broad group of microorganisms, among which 47 species and 24 subspecies have been described. The basic division of these bacteria distinguishes two groups: coagulase-positive staphylococci (CPS) and coagulase-negative staphylococci (CNS). This division results from the ability of *S. aureus* to clot plasma observed many years ago and the lack of this property in other species known at the time.

15.2 Coagulase-Positive Staphylococci

Coagulase production characterizes such *Staphylococcus* species as *S. aureus, S. intermedius, S. pseudintermedius, S. delphini, S. lutrae, S. schleiferi* subsp. *Coagulans*, and also some strains of *S. hyicus* (Chrobak *et al.*, 2009). The best-known and characterized species is *S. aureus*. It is one of the leading etiologic agents of hospital infections. It can cause purulent skin and soft tissue infections, including post-operative and traumatic wound infections, bone marrow and bone inflammations, pneumonias, myocardium and endocardium inflammations,

Foodborne Pathogens and Antibiotic Resistance, First Edition. Edited by Om V. Singh.

cerebrospinal meningitis, urinary system infections, and bacteremias (Ryan and Ray, 2004). Toxins produced by *S. aureus* are one of the most frequent causes of (often massive) bacterial food poisonings (Hennekinne *et al.*, 2012). The pathogenic nature of *S. aureus* is related to the high genotypic and phenotypic heterogeneity of its strains. This results mainly from the extraordinary abilities of staphylococci to exchange genetic material through mobile genetic elements. Their virulence factors, such as heat-resistant nuclease, catalase, coagulase, hemolysins, protein A, lipase, leukocidin, staphylokinase, toxic shock toxin, and exfoliative toxins A and B are of great importance in the pathogenicity of staphylococci (Cunha *et al.*, 2006).

15.3 Coagulase-Negative Staphylococci

Coagulase-negative staphylococci are among the most widespread microorganisms in nature. They occur most often on the skin and mucous membranes of humans and animals, they also populate the alimentary tract and the respiratory system. Contrary to *S. aureus*, coagulase-negative strains were regarded as harmless to humans for many years (Queck and Otto, 2008). They were classified as microflora unable to cause disease. A consistent increase in the role of coagulase-negative staphylococci in causing hospital infections has been observed in recent years (Rogers *et al.*, 2009). This concerns mainly the species *S. epidermidis, S. haemolyticus*, and *S. saprophyticus*. Improving research techniques based on genotypic analyses have contributed to revealing the pathogenic nature of this group of staphylococci. A number of virulence factors originally regarded as characteristic of coagulase-positive staphylococci have also been identified in strains not producing coagulase (Akinjogunla and Enabulele, 2010). The most often found factors include haemolysins, leukocidins and exfoliative toxins A and B.

Contrary to coagulase-positive staphylococci, in which the presence of factors related to toxicity is considered dominant, coagulase-negative staphylococci have more developed adhesion ability (Queck and Otto, 2008). The ability to produce biofilm on the surface of damaged tissues and on abiotic surfaces is regarded as the main cause of virulence for this group of bacteria (Simojoki *et al.*, 2012). It turns out that coagulase-negative staphylococci can be the etiologic agent of many infections, mainly in people with reduced immunity. Moreover, they are believed to be a reservoir of genes conditioning antibiotic resistance (Chajęcka-Wierzchowska *et al.*, 2015).

15.4 Genetic Mechanisms Conditioning Antibiotic Resistance of Staphylococci

Resistance to an antibiotic is the ability of a microorganism to oppose its bacteriostatic or bactericidal action. Bacteria display natural (innate) and acquired resistance. Natural resistance is most often the result of ineffective antibiotic penetration through the cell wall and membrane structures due to the presence of elements inhibiting the possibilities of its reaching the destination. Such resistance is usually chromosomally encoded and therefore non-transmissible. While innate resistance is a characteristic feature of a given species, related to the information encoded in the bacterial chromosome, acquired resistance appears as a result of changes in the genome. Such changes can occur as a consequence of point (random) mutations, which are most often errors in the nucleotide sequence of genetic material or can be the result of acquiring a gene or gene sets determining antibiotic resistance from other resistant bacteria. This latter phenomenon occurs much more often, and they both consequently lead to permanent resistance inheritance and also its spread by horizontal transfer (Savage *et al.*, 2013). As a result of horizontal gene transfer (HGT), new resistance genes appear in cells, carried on mobile genetic elements.

HGT allows bacteria to quickly acquire a complex of new properties during one recombination process and is the key element in their evolution. It often leads to the formation of so-called multi-drug resistant strains (MDR), that is, showing insensitivity to at least one antibiotic from three or more classes, active toward a given microorganism species. Strains with acquired resistance have a selective advantage under strong antibiotic pressure conditions. The appearance of multi-drug resistant strains is favored by simultaneous encoding of several different resistance genes on the same genetic element.

The phenomenon of the so-called "cross-resistance" is often observed in staphylococci. This term most often defines insensitivity to all or some antibiotics belonging to the same chemical group (e.g., aminoglycosides, macrolides, β-lactams). Cross-resistance can also occur toward antibiotics belonging to different groups (e.g. macrolides and lincosamines) if the binding sites of these antibiotics are close to each other. Genetic linkage within one genetic element—a plasmid or transposon—also plays an important role. An example can be the presence on plasmids in *S. aureus* of genes encoding resistance to quaternary ammonium compounds (QAC) and at the same time gentamicin or penicillin resistance genes. This is a very unfavorable phenomenon because the widening use of disinfectants in the food industry can lead to a parallel increase in antibiotic resistance.

The direct cause of bacterial resistance to different kinds of compounds is the presence and expression of relevant genes in the cell. Two resistance types are distinguished: constitutive and inducible. Constitutive resistance occurs if the genes encoding resistance to an antibiotic show continuous expression, regardless of the appearance of the inductor or its absence. Inducible resistance requires the presence of an inducing substance, in this case, the antibiotic, but it can occur in a subinhibitory concentration. Genes of resistance to different antibacterial compounds can be located on the chromosome and also on extrachromosomal elements, such as plasmids or transposons (Firth and Skurray, 1998). Plasmids are most often circular, self-replicating DNA molecules, inherited independently of the chromosome. Besides genes responsible for independent replication and regulatory genes, plasmids can contain a number of genes responsible for different phenotypic properties of bacteria, for example, resistance to antibacterial compounds and the ability to conjugate or produce toxins (McCarthy and Lindsay, 2012). Plasmids often carry resistance genes, single or organized in integrons, which are part of transposons.

Integrons are defined as natural localized recombination systems, permitting to include in their structure genes in the form of gene cassettes and their expression. They are not mobile (as gene cassettes) but, because they are often encountered on conjugative plasmids or transposons, they are regarded as structures participating in horizontal gene transfer. Insertion sequences (IS) and transposons are mobile genetic elements which, because of their ability to often change their location within the chromosome, are described as "jumping genes." The aim of their movement is most probably to place these elements in other, mobile genetic units, such as plasmids or conjugative transposons, which are capable of independent movement between bacteria (Kozitskaya *et al.*, 2004).

Staphylococcal plasmids were divided into three basic classes and a provisional fourth class which includes plasmids unclassifiable in any of the first three ones (Novick, 1990; Paulsen *et al.*, 1997). Class I includes small plasmids (1.3–4.6 kbp; 1 kbp means a thousand base pairs), occurring in 10–50 copies per cell, they contain most often one (and rarely two) resistance determinants. They include plasmids conditioning resistance to chloramphenicol, tetracyclines, aminoglycosides and MLS_B (macrolides, lincosamides, streptogramins B). Class II plasmids are larger (15–46 kbp) and occur only in 4–6 copies in a cell. They include, among others, many plasmids containing genes responsible for the production of β-lactamases, plasmids conditioning resistance to aminoglycosides and trimethoprim. Class III includes large conjugative plasmids (30–60 kbp) often containing determinants of resistance to aminoglycosides, mupirocin, but also to disinfectants.

15.4.1 β-Lactam Antibiotic Resistance

β-lactam antibiotics have a bactericidal effect, inhibiting the synthesis of the main bacterial cell wall component—peptidoglycan. Peptidoglycan consists of polysaccharide chains and peptide chains forming cross-links between each other. Cross-link formation is catalyzed by enzymes located in bacterial cell membrane, which are called penicillin-binding proteins (PBP) because of their ability to bind to β-lactam antibiotics (among others, penicillin). Antibiotic binding to a PBP protein causes inhibition of its enzymatic function, and thus, peptidoglycan synthesis inhibition. Bacteria deprived of cross-linked peptidoglycan become sensitive to environmental conditions and die, destroyed by their own autolytic enzymes.

Staphylococci have developed several resistance mechanisms to β-lactam antibiotics. One of them is their production of β-lactamases (penicillinases). These enzymes hydrolyse the antibiotic β-lactam ring, which, as a result, does not inhibit enzymatic functions of penicillin-binding proteins (PBP). Four penicillinase groups have been distinguished: A, B, C, and D. Genes encoding these enzymes are rarely on mobile genetic elements such as plasmids or transposons, they are most often located on the bacterial chromosome (Fuda *et al.*, 2005). These enzymes are encoded by the gene *blaZ* and two regulatory genes *blaR1* and *blaI*.

Most isolated *Staphylococcus aureus* strains (80–90%) have the ability to produce induced beta-lactamase of plasmid origin, which leads to the appearance of penicillin/ampicillin resistance. For this reason, semisynthetic penicillins resistant to staphylococcal penicillinase and first- and second-generation cephalosporins are used in non-hospital infections (Hiramatsu, 2001).

Since β-lactamase production in staphylococci is most often encoded by genes located on plasmids and in transposons, less often in the chromosome, it is a property easily transferred between strains. Besides the genes responsible for β-lactamase production, penicillinase plasmids can have genes determining resistance to heavy metal salts, quaternary ammonium compounds, antibiotics (erythromycin, kanamycin) and ethidium bromide.

The second mechanism of staphylococcal resistance to β-lactam antibiotics is the synthesis of a penicillin-binding protein with an altered, weak affinity for antibiotics from this group. The altered protein, PBP2a (PBP2'), is encoded by the gene complex *mecI*/*mecR*/*mecA*. These genes are located within the SCCmec cassette (staphylococcocal cassette chromosome *mec*) on the bacterial chromosome. This cassette is a so-called "genomic island" (GI). Five types of these cassettes have been described so far, each differing in the configuration of genes contained therein and in size. From a clinical point of view, methicillin-resistant staphylococci currently present a very significant therapeutic issue (Huber *et al.*, 2011). According to the literature data, there is a risk that coagulase-negative staphylococci are a source of SCCmec (Leonard and Markey, 2008). This hypothesis is supported by the complete homology of SCCmec cassettes in the investigated coagulase-positive and coagulase-negative staphylococci. Studies have indicated that a type IV of the SCCmec cassette, initially detected only in *S. epidermidis*, has also started to occur, over time, in *S. aureus* (Chrobak *et al.*, 2009; Zhang, 2009).

Resistance to β-lactam antibiotics may also be related to mutations in the genes encoding PBP2, PBP3, and PBP4 proteins. This mutation is triggered by a selective pressure imposed by an antibiotic. The results are a penicillin-binding protein showing a lower affinity to β-lactam antibiotics. This type of resistance in *Staphylococcus aureus* is called MODSA (modified oxacillin-resistant *S. aureus*).

Overproduction or synthesis of an enzyme with a modified substrate spectrum is another resistance mechanism detected mainly in *S. aureus. S. aureus* strains, called BORSA, display resistance to methicillin and other β-lactams, which is known as borderline resistance. These strains do not have a *mec*A gene (Maalej *et al.*, 2012). Similarly to MODSA, BORSA strains show a lower resistance to β-lactams than PBP2a-protein producing *S. aureus*.

15.4.2 Resistance to Glycopeptides

Increasing bacterial resistance to β-lactam antibiotics has prompted a need to find other effective medicines. Much hope was given to vancomycin (introduced in 1956) and teicoplanine marketed in 1978. Similarly to β-lactam, a bactericidal action of glycopeptide antibiotics is associated with inhibitin peptidoglycan synthesis. This process takes place via attaching an antibiotic to the lateral peptide chains, which prevents cross-links from being structured. Glycopeptides also disrupt the RNA synthesis and the cell membrane permeability (Yim *et al.*, 2014).

Resistance is related to Van ligase production and these ligases catalyse the synthesis of D-alanine-D-serine. This dipeptide is then built into peptidoglycane precursors that have a lower affinity to vancomycine than precursors whose chain ends with D-alanyle-D-alanine (Gardete and Tomasz, 2014).

Vancomycin-resistant staphylococcal strains appeared in 2002 and, most probably, they acquired the resistance genes from enterococci. A Tn1546 transposable element (TE, or transposon) has been detected therein and it encodes the *van*A operone genes. The MI values for vancomycin of the given strain were 1024 mg/L and for teicoplanine—32 mg/L. In enterococci, the *van*A system is located on the Tnn1546 transposon, of 10,851 bp, and included in the Tn3-like transposons. Tn1546 may be incorporated into both the chromosome or a large conjugative plasmide. The vanA operon genes in *S. aureus*, similar to *Enterococcus* sp., undergo regulation and the expressed phenotype of VanA resistance to glycopeptides is an inductive system (Levine, 2006).

15.4.3 Resistance to Aminoglycosides

The acquired resistance to aminoglycosides may be related to a mutation in the ribosomes, although most often it is an effect of blocking the binding between an antibiotic and a target site, namely, the 30S subunit of the bacterial ribosome as a result of an antibiotic modification by attaching additional chemical residues (acyl, nucleotide, or phosphoryl) to its particle (Woodford, 2005). Aminoglycoside-modifying enzymes (AME) may present an acetylase (AAC), phosphotransferase (APH) or nucleotidyltransferase (ANT) activity. The genes of AME enzymes are transferred onto mobile genetic elements, namely, plasmids and transposons, which favors a horizontal spread of resistance. Most aminoglycoside-resistant enterococci have a bifunctional enzyme with an acetylase-phosphotransferase/ phosphotransferase (Aac(6')-Aph(2")) activity, which modifies all aminoglycosides except for streptomycine (Rice, 1998). The *aph(2")-Ia-aac(6')*Ie gene is located both on the chromosome and on plasmids. Less often, the resistance is determined by two genes encoding the insensitivity to aminoglycosides other than gentamycin: *aph(3")-IIIa and ant(4')-Ia*. The *ant(4')-Ia* gene encodes adenosine aminoglycoside transferase, which results in resistance to low concentrations of tobramycin, kanamycin, neomycin, and amikacin (Lascols *et al.*, 2011). It may be located on both small (e.g., pUB110) and large plasmids. The *aph(3')-IIIa* gene encodes aminoglycoside phosphotransferase (Aph(3')-IIIa), which phenotypically presents as the resistance to kanamycin. This gene is most often situated on small plasmids.

15.4.4 Resistance to Tetracyclines

Tetracyclines are bacteriostatic antibiotics that inhibit the synthesis of proteins in the bacterial cells. Their mode of action consists in an antibiotic binding with the 30S ribosome and in blocking aminoacyl-tRNA on the m-RNA acceptor sites. This prevents translation and disturbs energy processes in the cell. Tetracycline is a relatively inexpensive antibiotic with a wide spectrum of activity, and as such, it is often used as a preventive measure in veterinary medicine (Martinez, 2009).

The tetracycline-inhibiting mechanisms, developed by *Staphylococcus*, consist in attaching the *tet*M or *tet*O resistance genes to bacterial ribosomes, which blocks the binding of an antibiotic to the ribosome. The removal of an antibiotic from the cell immediately after its penetration is another defensive mechanism. The *tet*L and *tet*K genes, which encode the Tet(K) and Tet(L) protein respectively, determine the active transport, the so-called "efflux mechanism," of an antibiotic out of the cell (Roberts, 2005). These proteins belong to the MFS group (major facilitator superfamily) and act as proton pumps, exchanging the antibiotic-Mg2+ complex for a proton. This results in a reduction of tetracycline concentration inside the cell, which hinders reaching the target site (the ribosome) by antibiotics.

In staphylococci, the *tet*M gene is most often found and it is located on the bacterial chromosome. The *tet*M gene may be transferred on conjugative transposons. Its presence is often conjugated with the *erm*B that encodes the resistance to macrolide antibiotics. Such location of the genes explains a phenomenon consisting in the common occurrence of the concomitant resistance to macrolides and/or chloramphenicol in tetracycline-resistant isolates. The *tet*K gene most often occurs on small plasmids and may be integrated into the chromosome. It is responsible for encoding the membrane protein whose action is to actively remove all tetracyclines, except for minocycline, from the cell. In recent years, Tet(S) and Tet(U) proteins have been also described and detected in vancomycin-resistant staphylococci (VRSA) (Młynarczyk, 2008).

15.4.5 Resistance to Macrolides, Lincosamides, and Streptogramins B (MLS$_B$)

Macrolides, lincosamides, and streptogramins are the antibiotics used both in human and veterinary medicine. These antimicrobials are also widely administered in production animal treatments (Collignon *et al.*, 2009).

For macrolides such as erythromycin, the mode of action consists in inhibiting the synthesis of bacterial proteins by reversible binding to the 50S subunit of the bacterial ribosome, which disrupts peptidyl-tRNA translocation (Portillo *et al.*, 2000). This action is mainly of a bacteriostatic nature. The presented mechanisms of resistance to MLSB antibiotics include the following: active removal of antibiotics from the cell, enzymatic inactivation, or modification by a target site mutation (23S mRNA protein). The latter dictates the phenotype of resistance to all macrolides, lincosamides, and streptogramins (MLSB). Due to the overlapping binding sites, attachment of lincosamides and streptogramins B is inhibited, which leads to the cross-resistance to all MLSB. The MLSB mechanism is related to the occurrence of ermA, ermB, or ermC genes encoding N-6-methyltransferase that methylates adenine in 23S rRNA, which is the site of action for an antibiotic (Zou *et al.*, 2011). The ermA gene is inductive and found on transposons located on the chromosome. The ermB gene, in turn, is constitutive and usually occurs on the transposons incorporated into plasmids, whereas the ermC gene may be inductive or constitutive and is located on small plasmids (Thumu and Halami, 2012). Induction is possible even at sub-inhibitive concentrations of erythromycin (0.01–0.1 μg/ml) (Matsuoka *et al.*, 2002). In methicillin-resistant staphylococci, the ermA gene is predominant, whereas in methicillin-sensitive staphs, the presence of ermC is most often detected (Schmitz *et al.*, 2000).

An enzymatic modification of an antibiotic is most often related to the activity of nucleotidyltransferase encoded by the *lnu*(A) and *lnu*(B) genes. Such strains demonstrate a phenotypic resistance to lincosamides. However, this type of resistance is extremely rare, both in coagulase-positive and coagulase-negative staphylococci.

The membrane ABC proteins (ATP-binding cassette) are responsible for the active removal of an antibiotic from the staphylococcal cells. These proteins are encoded by the *msr*A gene, located on plasmids and determine the resistance to macrolides and streptogramins (Portillo *et al.*, 2000). This is a result of the active pumping of macrolides with the 14- and 15-unit lactone ring and of streptogramins B out of the cell. Macrolides with the 16-unit lactone ring and streptogramins A remain active.

15.4.6 Resistance to Oxazolidinones

The mode of action of linezolid is believed to be unique since it inhibits the earliest stage of bacterial protein synthesis instead of the stage of chain extension as typical of other antibiotics. By binding to the 50S ribosomal subunit (23SrRNA) in a peptidyltransferase area, linezolid prevent the translation-inducing complex from being formed.

Even though linezolid was marketed recently, MRSA and VRE strains have been already observed to be resistant to this drug. Most resistance cases have been caused by alterations to the binding site for linezolid at the 50S bacterial ribosome subunit (Lawrence *et al.*, 2008). A replacement of a 2576 guanine with thymine (a G2576U mutation in the closed loop of the V 23S domain of ribosomal RNA) is the most common mutation among linezolid-resistant *S. aureus* (Halle *et al.*, 2005; Bersos *et al.*, 2004). A mutation involving the exchange of a 2500 thymine for adenine has been also reported (Meka *et al.*, 2004). The degree of resistance depends on the number of mutated allels, yet a mutation in at least one allel is sufficient for the induction of resistance (Jankowski *et al.*, 2006).

15.5 Food as a Source of Antibiotic-Resistant Staphylococci

Antibiotics are used for prophylactic purposes in swine, cattle, rabbit, and poultry production as well as in vegetable and fruit cultivation and beekeeping (Ding and He, 2010) and, thus, the meat of farm animals, fruit, vegetables, and water may be a source of resistant strains. The presence of antibiotic-resistant strains among food-derived microorganisms suggests that it may play a much more important role in transferring the antibiotic resistance encoding genes than previously thought. From a food safety point of view, the presence of antibiotic-resistant staphylococci should be considered on a number of levels.

There are studies suggesting that *S. aureus* strains found in food have the capacity to penetrate the bloodstream and, as such, present a risk of systemic infections and, although this may mainly concern people with impaired immunity, such risk still exists. The food chain may be also a key site for resistance transmission between the environment and humans (Gazzola *et al.*, 2012; Marshall *et al.*, 2009; Wang, 2012; Wang *et al.*, 2006). Most studies on this issue have focused on the pathogens presenting a direct risk to human health. However, bacteria entering the gastrointestinal tract with ingested food may also transfer the antibiotic resistance genes, which presents a risk for their transfer to other microorganisms, including pathogenic ones. This mainly applies to food consumed without processing (e.g., of a ready-to-eat type) that, if it contains antibiotic-resistant staphylococci, may serve as a source of transferable resistance genes to the strains that colonize the gastrointestinal tract in humans. Moreover, the skin surface and mucous membranes may be colonized during food ingestion or preparation. If the food (such as raw mat) containing antibiotic-resistant staphylococci is correctly heat-treated, a potential risk related to its consumption virtually does not exist. However, any direct contact with meat during its preparation for heat treating may result in colonization of the human body (Kluytmans, 2010).

15.5.1 Prevalence of Resistant *S. aureus* in Food

The frequency at which MRSA is isolated from food varies a lot, ranging from few to several dozen percent. These microorganisms have been most often isolated from milk and dairy products (Kwon *et al.*, 2005; Normanno *et al.*, 2007) and raw meat products (Kitaj *et al.*, 2005; Pu and Wang, 2011). It is estimated that the presence of staphylococci in such products mainly results from their capacity to survive and multiply in a wide pH range and NaCl concentrations (Le Loir, 2003).

The occurrence of MRSA strains in food presents the highest risk, although it should be noted that methicillin-sensitive staphylococci may also serve as a reservoir of resistance genes to the antibiotics that are therapeutically important in human medicine. In studies carried out in Italy, it has been found that staphylococci isolated from food were resistant to fluoroquinolones, tetracycline, erythromycin, bacitracin, oxacillin, penicillin, and ampicillin (Gundogan 2005). In Japan, *S. aureus* strains isolated from raw fish and sashimi demonstrated resistance to macrolides, aminoglycosides, and penicillins. The presence of genes encoding resistance to aminoglycosides, tetracyclines, β-lactams, macrolides, lincosamides, and streptogramin B has also been confirmed (Hammad *et al.*, 2012).

In Turkey, 413 milk, dairy product, meat, and meat product samples were examined. The presence of *S. aureus* was detected in 33.4% of the samples. Of 138 investigated isolates, all demonstrated resistance to at least a number of 9 tested antibiotics. Up to 92.7% of the strains showed resistance to penicillin. Most isolates were also resistant to sulfamethoxasole-trimethoprim (71%), clindamycin (60.8%), erythromycin (58.6%), fusidic acid (52.2%), and teicoplanin (44.2%). One of the strains (isolated from feta cheese) demonstrated resistance to all investigated antibiotics and was the only one resistant to mupirocin (Guven *et al.*, 2010).

However, it has been shown that antibiotic-resistant *S. aureus* is found not only in food of animal origin. In studies conducted South Korea, resistance to at least one tested antibiotic was found in all *S. aureus* isolated from leafy vegetables. Furthermore, 35.85% of the isolates demonstrated multi-drug resistance (MDR). Among *S. aureus* isolates, 28.30% were phenotypically resistant to oxacillin, of which the *mec*A gene was detected in 73.3%. The studies on strains isolated from vegetables were compared with clinical isolates and it was found that *S. aureus* derived from plant materials showed resistance to quinupristun/dalfopristin and linezolid, which was not detected in human isolates (Hong *et al.*, 2015).

MRSA presence in ready-to-eat food is a significant issue, as this type of food is not heat-treated before consumption. Based on the author's own research carried out on 858 food samples (cheese, cold cuts, sausages, smoked fish, salads), MRSA was detected in 35 tested samples. All staphylococci showing phenotypic resistance had the *mec*A gene (Chajęcka-Wierzchowska *et al.*, 2014). A slightly lower proportion of MRSA was determined in 1,634 milk and cheese samples by researchers from Italy (Normanno *et al.*, 2007). In Japan, Hammad *et al.* (2012) tested 120 sashimi samples, of which they isolated 10 MRSA strains.

In wide-scale studies conducted in Italy, a minor percentage of methycillin-resistant *S. aureus* was identified in 2,162 food samples. The presence of *S. aureus* was 17.1%, of which only two strains isolated from milk destined for dairy products showed resistance to methicillin. These strains were simultaneously resistant to tetracycline, had a *spa* t899 type and did not possess the gene encoding Panton-Valentine leukocidin (Traversa *et al.*, 2015).

S. aureus strains were the staphylococci most commonly isolated from milk and cheese samples investigated by Al-Khafaji and Flayyih (2015). They constituted 48% of all isolated staphylococci. As many as 80 *S. aureus* isolates

(82.47%) were resistant to methicillin, 8 strains (8.24%)—to vancomycin, and 18 strains (19%)—to tetracycline. In MRSA strains, the presence of *femA* gene (96 isolates; 98.96%) and *mecA* gene (91 isolates; 93.81%) was detected. In studies carried out by Carfora *et al.* (2015) on 554 samples of raw milk and dairy products from cow, goat and sheep milk, three MRSA strains were isolated. These isolates originated from products made of sheep milk. All isolates represented a spa t127 type, sequential type 1, clonal complex 1, and SCCmec type IVa (Carfora *et al.*, 2015).

15.5.2 Coagulase-Negative Staphylococci

Studies on the antibiotic resistance in coagulase-negative staphylococci (CNS) are reported in a much smaller number of publications. Nevertheless, in the existing reports it has been shown that coagulase-negative staphylococci are also a reservoir of resistance genes. This phenomenon is explained by the common presence of CNS on the skin. During antibiotic treatments administered to animals, the concentration of an antibiotic on the skin is often subinhibitive, which may induce the resistance. Bacteria found on the skin in animals (*S. xylosus, S. capitis,* and *S. equorum*) may then contaminate milk or the meat of production animals. Antibiotic resistance genes may be transferred from coagulase-negative staphylococci to pathogenic microorganisms of the same or another genus. According to the literature, it is probable that coagulase-negative staphylococci are a source of SCCmec cassettes (Leonard and Markey, 2008). This hypothesis is, for instance, supported by the complete homology of SCCmec cassettes in the investigated coagulase-positive and coagulase-negative staphylococci. Studies have indicated that a type IV of the SCCmec cassette that was initially determined only in *S. epidermidis*, also started in time to occur in *S. aureus* (Chrobak *et al.*, 2009; Zhang, 2009). Studies on coagulase-negative staphylococci isolated from dairy products have demonstrated that the same genes that were detected in pathogenic *S. aureus* were also found in the genomes of coagulase-negative staphylococcal isolates. This applied to the *cat* gene (encoding chloramphenicol transferase) and the genes encoding the resistance to tetracyclines and responsible for the efflux mechanism (*tetK*) identified in *S. xylosus* and *S. caprae* (Perreten *et al.*, 1998; Irlingen, 2008).

Among CNS, *S. carnosus, S. xylosus S. condimenti, S. equorum, S. piscifermentans,* and *S. succinus* play an important role in food production. *S. carnosus* and *S. xylosus* strains are used as starter cultures in fermented meat products (meat, salami). However, it is found that staphylococci of these species show antibiotic resistance. The resistance is usually more commonly detected in *S. xylosus* strains (*S. xylosus, S. equorum, S. succinus*) as compared to the *S. carnosus* group (*S. carnosus, S. condimenti and S. piscifermentans*) (Resch *et al.*, 2008). The increased prevalence of resistant strains in the *S. xylosus* group may be explained by their closer relationship to an opportunistic pathogen such as *S. saprophyticus* (Shah *et al.*, 2007).

Resch *et al.* (2008), while investigating 330 coagulase-negative staphylococcal strains used as starter cultures, found that over 90% of strains belonging to *S. succinus* and *S. xylosus* demonstrated multi-drug resistance. Resistance to lincomycin, penicillin, fusidic acid, oxacillin, ampicillin, and tetracycline was most often identified. The highest percentage of resistant strains originated from cheese (87%) and sausages (83%) and meat fermentation cultures (93%).

Antibiotic-resistant coagulase-negative staphylococci have also been detected in water for human consumption that met all microbiological standards (Faria *et al.*, 2009). Staphylococci with a lower resistance to β-lactam antibiotics, tetracycline, clindamycin, and erythromycin were identified. Resistance to erythromycin was most often reported, which was associated with the efflux pump encoded by the *msrA* gene.

Al-Khafaji and Flayyih (2015) isolated 200 staphylococci from 300 investigated cheese and milk samples, of which 103 strains belonged to coagulase-negative staphylococci of the following species: *S. chromogenes* (82 isolates; 41%) and *S. epidermidis* (21 isolates; 11%). Resistance to methicillin was detected in 64 strains (62.13%), to tetracycline—in 28 isolates (27.18%), whereas 5 CNS (4.85%) were resistant to vancomycin. The presence of *mec*A gene was confirmed in 67% of CNS strains. Methicillin resistance in *S. epidermidis* is believed to be an important marker distinguishing the commercial and pathogenic strains, apart from the ability to produce biofilm and the presence of *ica*ADBC operon (Prasad *et al.*, 2012).

Marthy *et al.* (2012) analyzed samples of traditionally fermented Swiss sausages (23 samples), meat products obtained with spontaneous fermentation (21 samples), and 16 commercial fermentation starters. Of 132 isolated CNS, 49% showed resistance to at least one antibiotic, most often to tetracycline and penicillin. The resistance of coagulase-negative staphylococci isolated from food to β-lactam antibiotics is one of the most common types (Moon *et al.*, 2007; Marino *et al.*, 2011). In particular, this applies to penicillin G and ampicillin. Such frequent resistance to penicillin and tetracycline among isolates derived from food of animal origin may be explained by the fact that penicillins are still the most common antibiotics in animal production (Riesen and Perreten, 2009). β-lactam antibiotics are widely administered in mastitis cases (Sawant *et al.*, 2009). The resistance to penicillins is encoded on plasmids, which may explain its rapid spread among strains. Since penicillin resistance is plasmidic, it spreads out very quickly to several other strains, whereas methicillin resistance is chromosomic and its diffusion is therefore slower (Pesavento *et al.*, 2007).

In studies by Marthy *et al.* (2012), the highest proportion of resistant strains was identified in in *S. warneri* (78%), *S. capitis* (75%), and *S. epidermidis* (67%) strains. Among strains used as fermentation starters, resistance was more often shown by the isolates from *S. xylosus* species (57%) than *S. carnosus* (19%). The resistance to penicillin, tetracycline, erythromycin, and fusidic acid was identified only in *S. xylosus*. In *S. carnosus* isolates, only streptomycin and trimethoprim resistance was detected. In the same studies, the presence of *tet*(K), *bla*Z, *mph*(C), *lnu*(A), and *str* genes was confirmed in 17 selected *S. xylosus* strains and 8 *S. equorum* subsp. *equorum* DSM 20674 T strains.

Based on the author's own research, it was demonstrated that ready-to-eat food that does not require heat-treatment before consumption may be a significant source of antibiotic-resistant staphylococci. These bacteria were identified in 888 ready-to-eat food samples. Out of 113 isolates, *S. xylosus, S. saprophyticus*, and *S. epidermidis* were found to be the second-most predominate species (after *S. aureus*). Over half of the isolates (54.9%) demonstrated resistance to at least one antibiotic, of which some showed multi-drug resistance. Most isolates were resistant to cefoxitin (49.6%), followed by clindamycin (39.3%), tigecycline (27.4%), quinupristun/dalfopristin (22.2%), rifampicin (20.5%), tetracycline (17.9%), and erythromycin (8.5%). All methicillin-resistant strains had the *mec*A gene. The tet(M), tet(L), and tet(K) genes were the most common ones that encoded tetracycline resistance (Chajęcka-Wierzchowska *et al.*, 2014).

Based on the author's own research conducted on 146 samples of animal-derived food, 58 CNS strains were isolated and classified as *S. xylosus* (29 strains), *S. epidermidis* (16 strains), *S. lentus* (7 strains), *S. saprophyticus* (4 strains), *S. hyicus* (1 strain), and *S. simulans* (1 strain). The highest proportion of isolates were resistant to cefoxitin (43.1%), followed by clindamycin (36.2%), tigecyclin (24.1%), rifampicin (17.2%), and erythromycin (13.8%). All methicillin-resistant staphylococci possessed the *mec*A gene. Multi-drug resistance (MDR) was demonstrated by 32.2% of the strains. The tetracycline-resistant isolates most often had the tet(M) gene and simultaneously the Tn916-Tn1545 transposon. Among the isolates resistant to erythromycin, the presence of *erm*(C) or *msr*(A/B) genes was found (Chajęcka-Wierzchowska *et al.*, 2015). Although coagulase-negative staphylococci are not believed to be foodborne pathogenic microorganisms, their presence may have an impact on public health due to the potential for spreading antibiotic resistance. Our studies have demonstrated that ready-to-eat food may present an important source of antibiotic-resistant strains that additionally possess the genes encoding resistance to essential antibiotics and that such genes may be easily transmitted onto mobile genetic elements.

15.6 Summary

The presented data support the conclusion that food may present an important fomite for transmission of antibiotic-resistant staphylococci. This paper emphasizes the growing antibiotic resistance in coagulase-negative staphylococci, which are usually overlooked in food research and are less often a target for scientists. It was demonstrated that both groups of staphylococci may serve as a reservoir of the resistance of genes to different classes of antibiotics, most often including β-lactams, tetracyclines, penicillins, and macrolides. Because of the expanded and robust mechanism of horizontal gene transfer, staphylococci may spread them easily onto other strains.

The above-mentioned data should also be a prerequisite for more accurate selection of strains for fermentation starters used to manufacture fermented meat products (such as meat or salami). It has been demonstrated that *S. carnosus* and *S. xylosus* strains also show resistance to a number of different groups of antibiotics.

References

Akinjogunla O.J. and Enabulele I.O. (2010). Virulence Factors, Plasmid Profiling and Curing analysis of Multidrug Resistant Staphylococcus aureus and Coagulase negative Staphylococcus spp. isolated from Patients with Acute Otitis Media. *Journal of American Science, 6*(11), pp. 1022–1033.

Al-Khafaji M.H. and Flayyih M.T. (2015). Relationship between methicillin resistance and the presence of *femA* and *mecA* genes in coagulase positive and negative staphylococci isolated from milk and cheese. *World Journal of experimental Experimental Biosciences, 3,* pp. 50–56.

Bang W., Hanson D.J. and Drake M.A. (2008). Effect of salt and sodium nutrite on growth and enterotoxin production of Staphylococcus aureus during the production of airdried fresh pork sausage. *Journal of Food Protection, 71,* pp. 191–195.

Bersos Z., Maniati M., Kontos F., Petinaki E., Maniatis A., et al. (2004). First report of a linezolid–resistant vancomycin–resistant Enterococcus faecium strain in Greece. *Journal Antimicrobial Chemotherapy, 53,* pp. 685–686.

Carfora V., Caprioli A., Marri N., Sagrafoli D., Boselli C., Giacinti G., Giangolini G., Sorbara L., Dottarelli S., Battisi A., Amatiste S., et al. (2015). Enterotoxin genes, enterotoxin production, and methicillin resistance in Staphylococcus aureus isolated from milk and dairy products in Central Italy. *International Dairy Journal, 42,* pp. 12–15.

Chajęcka-Wierzchowska W., Zadernowska A., Nalepa B., Sierpińska M., Łaniewska-Trokenheim Ł., et al. (2014). Retail Ready-to-Eat Food as a Potential Vehicle for Staphylococcus spp. Harboring Antibiotic Resistance Genes. *Journal of Food Protection, 77*(6), pp. 993–998.

Chajęcka-Wierzchowska W., Zadernowska A., Nalepa B., Sierpińska M., Łaniewska-Trokenheim Ł., et al. (2015). Coagulase-negative staphylococci (CoNS) isolated from ready-to-eat food of animal origin-phenotypic and genotypic antibiotic resistance. *Food Microbiology, 46,* pp. 222–226.

Chrobak D., Kizerwetter-Swida M., Rzewuska M., Binek M., et al. (2009). Pet-aquired methicillin-resistant Staphylococcus intermedius (PA-MRSI) as a potential reservoir for mecA gene. *Advanced in Microbiology, 48,* pp. 235–242.

Collignon P., Powers J.H., Chiller T.M., Aidara-Kane A., Aarestrup F.M., et al. (2009). World Health Organization ranking of antimicrobials according to their importance in human medicine: A critical step for developing risk management strategies for the use of antimicrobials in food production animals. *Clinical Infection Diseases, 49*, pp. 132–141.

Cunha Mde L., Rugolo L.M., and Lopes C.A. (2006). Study of virulence factors in coagulase-negative staphylococci isolated from newborns. *Memórias do Instituto Oswaldo Cruz, 101*(6), pp. 661–668.

Ding C. and He J. (2010). Effect of antibiotics in the environment on microbial populations. *Applied Microbiology and Biotechnology, 87*, pp. 925–941.

Faria C., Vaz-Moreira I., Serapicos E., Nunes O.C., Manaia C., et al. (2009). Antibiotic resistance in coagulase negative staphylococci isolated from wastewater and drinking water. *Science of the Total Environment, 407*, pp. 3876–3882.

Firth N. and Skurray R.A. (1998). Mobile elements in the evolution and spread of multiple-drug resistance in staphylococci. *Drug Resistance Updates, 1*, pp. 49–58.

Fuda C.C., Fisher I.F. and Mobashery S. (2005). Beta-lactam resistance in Staphylococcus aureus: the adaptive resistance of a plastic genome. *Cellural and Molecular Life Science, 62*(22), pp. 2617–2633.

Gardete S. and Tomasz A. (2014). Mechanisms of vancomycin resistance in Staphylococcus aureus. *Journal of Clinical Investigation, 124*(7), pp. 2836–2840.

Gazzola S., Fontana C., Bassi D., Cocconcelli P.S., et al. (2012). Assessment of tetracycline and erythromycin resistance transfer during sausage fermentation by culture-dependent and -independent methods. *Food Microbiology, 30*, pp. 348–354.

Götz F., Bannerman T. and Schleifer K.H. (2006). The Genera Staphylococcus and Macrococcus. *Prokaryotes, 4*, pp. 5–75.

Gundogan N., Citak S. and Turan E. (2006). Slime production, DNasse avtivity and antibiotic resistance of Staphylococcus aureus isolated from raw milk, pasteurized milk and ice cream samples. *Food Control, 17*, pp. 389–392.

Guven K., Mutlu M.B., Gulbandilar A., Cakir P., et al. (2010). Occurrence and characterization of *Staphylococcus aureus* isolated from meat and dairy products consumed in Turkey. *Journal of Food Safety, 30*, pp. 196–212.

Halle E., Padberg J., Rosseau S., Klare I., Werner G., Witte W., et al. (2005). Linezolid–resistant Enterococcus faecium and Enterococcus faecalis isolated from a septic patent: report of first isolates in Germany. *Infection, 32*, pp. 182–183.

Hammad A.M., Hammad T., Fujii T., Watanabe W., Shimamoto T., et al. 2012. Occurrence and characteristics of methicillin resistant and -susceptible Staphylococcus aureus and methicillin resistant coagulase-negative staphylococci from Japanese retail ready-to-eat raw fish. *International Journal of Food Microbiology, 156*, pp. 286–289.

Harris L.G., Foster S.J. and Richards R.G. (2002). An introduction to Staphylococcus aureus and techniques for identyfying and quantifying S. aureus adhesion in relation to adhesion to biomaterials: review. *European Cell and Material, 4*, pp. 39–60.

Hennekinne J.A., de Buyser M.L. and Dragacci S. (2012). Staphylococcus aureus and its food poisoning toxins: characterization and outbreak investigation. *FEMS Microbiology Reviews, 36*, pp. 815–836.

Hiramatsu K. (2001). Vancomycin-resistant Staphylococcus aureus: a new model of antibiotic resistance. *The Lancet Infectious Diseases, 1*(3), pp. 147–155.

Hong J., Kim Y., Kim J., Heu S., Kim S.R., Kim K.P., Roh., E., et al. (2015). Genetic Diversity and antibiotic resistance patterns of Staphylococcus aureus isolated from leaf vegetables in Korea. *Journal of Food Science, 80*(7), pp. M1526–M1531.

Huber H., Ziegler D., Pfluger V., Vogel G., Zweifel C., Stephan R., et al. (2011). Prevalence and characteristics of methicillin-resistant coagulase-negative staphylococci from livestock, chicken carcasses, bulk tank milk, minced meat, and contact persons. *BMC Veterinary Research, (7)*, p. 6.

Irlingen F. (2008). Safety assessment of dairy microorganisms: Coagulase-negative staphylococci. *International Journal of Food Microbiology, 126,* pp. 302–310.

Jankowski A. and Stefanik W. 2006. Linezolid – nowy antybiotyk w leczeniu zakażeń wywołanych bakteriami Gram-dodatnimi. *Wiadomości Lekarskie, 59,* pp. 727–731.

Kluytmans J.A. (2010). Methicillin-resistant Staphylococcus aureus in food products: cause for concern or case for complacency?. *Clinical Microbiology and Infection, 16*(1), pp. 11–15.

Kozitskaya S., Cho S.H, Dietrich K., Marre R., Naber K., Ziebuhr W., et al. (2004). The bacterial insertion sequence element IS256 occurs preferentially in nosocomial Staphylococcus epidermidis isolates: association with biofilm formation and resistance to aminoglycosides. *Infection and Immunity, 72*(2), pp. 1210–1215.

Kwon N.H., Park K.T., Moon J.S., Jung W.K., Kim S.H., Kim J.M., Hong S.K., Koo H.C., Joo Y.S., Park Y.H., et al. (2005). Staphylococcal cassette chromosome *mec* (SCC*mec*) characterization and molecular analysis for methicillin-resistant *Staphylococcus aureus* and novel SCC*mec* subtype IVa isolated from bovine milk in Korea. *Journal of Antimicrobial Chemotherapy, 56,* pp. 624–632.

Lascols C., Legrand P., Mérens A., Leclercq R., Muller-Serieys C., Drugeon H.B., Kitzis M.D., Reverdy M.E., Roussel-Delvallez M., Moubareck C., Brémont S., Miara A., Gjoklaj M., Soussy C.J., et al. (2011). In vitro antibacterial activity of ceftobiprole against clinical isolates from French teaching hospitals: proposition of zone diameter breakpoints. *International Journal of Antimicrobial Agents, 37,* pp. 235–239.

Lawrence L., Danese P., DeVito F., Franceschi F., Sutciffe J., et al. (2008). In Vitro vitro Activities activities of the Rx-01 Oxazolidinones oxazolidinones against Hospital hospital and Community community pPathogens. *Antimicrobial Agents Chemotheraphy, 52,* pp. 1653–1662.

Le Loir Y., Baron F. and Gautier M. (2003). Staphylococcus aureus and Food Poisoning. *Genetic Molecular research Research, 2,* pp. 63–76.

Leonard F.C. and Markey B.K. (2008). Methicillin-resistant Staphylococcus aureus in animals: A review. *The Veterinary Journal, 175,* pp. 27–36.

Levine D.P. (2006). Vancomycin: a history. *Clinical Infection Diseases, 42*(suppl 1), pp. S5–S12.

Maalej S.M, Rhimi F.M, Fines M., Mnif B., Leclercq R., Hammami A., et al. (2012). Analysis of borderline oxacillin-resistant Staphylococcus aureus (BORSA) strains isolated in Tunisia. *Journal of Clinical Microbiology, 50*(10), pp. 3345–3348.

Marino M., Frigo F., Bartolomeoli I., Maifreni M., et al. (2011). Safety-related properties of staphylococci isolated from food and food environments. *Journal of Applied Microbiology, 110,* pp. 550–561.

Marshall B.M.O., Dorothy J. and Levy S.B. (2009). Commensals: underappreciated reservoir of antibiotic resistance. *Microbe, 4,* pp. 231–238.

Marthy E., Buchs J., Eugster-Meier E., Lacroix C., Meile L., et al. (2012). Identification of staphylococci and dominant lactic acid bacteria in spontaneously fermented Swiss meat products using PCR-RFLP. *Food Microbiology, 29,* pp. 157–166.

Martinez J.L. (2009). Environmental pollution by antibiotics and by antibiotic resistance determinants. *Environmental Pollution, 157,* pp. 2893–2902.

Matsuoka M., Inoue M., Nakajima Y., Endo Y., et al. (2002). New *erm* gene in *Staphylococcus aureus* clinical isolates. *Antimicrobial Agents and Chemotherapy, 46,* pp. 211–215.

McCarthy A.J. and Lindsay J.A. (2012). The distribution of plasmids that carry virulence and resistance genes in Staphylococcus aureus is lineage associated. *BMC Microbiology, 12,* p. 104.

Meka V.G., Pillai S.K., Sakoulas G., Wennersten C., Venkataraman L., DeGirolami P.C., Eliopoulos G.M., Moellering R.C., Gold H.S., et al. (2004). Linezolid resistance in sequential Staphylococcus aureus isolates associated with a T2500A mutation in the 23S rRNA gene and loss of a single copy of rRNA. *The Journal of Infectious Diseases, 190,* pp. 311–317.

Młynarczyk A. and Młynarczyk G. (2008). Molecular mechanisms of resistance to antibacterial drugs in Staphylococcus aureus. *Advances in Microbiology, 47*(3), pp. 423–431.

Moon J.S., Lee A.R,, Jaw S.H., Kang H, M., Joo Y. S., Park Y. H., Kim M.N., Koo H.C., et al. (2007). Comparison of antibiogram, staphylococcal enterotoxin productivity, and coagulase genotypes among *Staphylococcus aureus* isolated from animal and vegetable sources in Korea. *Journal of Food Protection, 70,* pp. 2541–2548.

Normanno G., Corrente M., La Salandra G., Dambrosio A., Quaglia N.C., Parisi A., Greco G., Bellacicco A.l., Virgilio S., Celano G.V., et al. (2007). Methicillin-resistant Staphylococcus aureus (MRSA) in foods of animal origin product in Italy. *International Journal of Food Microbiology, 117,* pp. 219–222.

Normanno G., Corrente M., La Sandra G., Dambrosio A., Quaglia N.C., Parisi A., Greco G., Bellacicco A. L., Virgilio S., Celano G.V., et al. (2007). Methicillin-resistant *Staphylococcus aureus* in foods of animal origin product in Italy. *International Journal of Food Microbiology, 117,* pp. 219–222.

Novick R.P. (1990). The Staphylococcus as a molecular genetic system. In: P. Red. P.R. Novick, R. and A. Skurray., eds., *W: Molecular biology of the staphylococci.* New York: VCH Publishers, Inc., pp. 1–37.

Paulsen I.T., Firth N. and Skurray R.A. (1997). Resistance to antimicrobial agents other than b -lactams. In: K. Red, B.B. Crossley, and G.L. Archer, eds., *The Staphylococci in human disease.* New York: Churchill Livingston. pp. 175–212.

Perreten V., Giampa N., Schuler-Schmid U., Teuber M., et al. (1998). Antibiotic resistance genes in coagulase-negative staphylococci isolated from food. *Systematic and Applied Microbiology, 21*(1), pp. 113–120.

Pesavento G., Ducci B., Comodo N., Lo Nostro A., et al. (2007). Antimicrobial resistance profile Staphylococcus aureus isolated from raw meat: A research 136 for methicillin resistant Staphylococcus aureus (MRSA). *Food Control, 18,* pp. 196–200.

Portillo A., Ruiz-Larrea F., Zarazaga M., Alonso A., Martinez J. L. Torres C., et al. (2000). Macrolide resistance genes in Enterococcus spp. *Antimicrobial Agents and Chemotherapy, 44,* pp. 967–971.

Prasad S., Nayak N., Satpathy G., Nag H.L., Venkatesh P., Ramakrishnan S., Ghose S., Nag T.C., et al. (2012). Molecular & phenotypic characterization of *Staphylococcus epidermidis* in implant related infections. *The Indian Journal of Medical Research, 136*(3), pp. 483–490.

Pu S., Wang F. and Ge B. (2011). Characterization of toxin genes and antimicrobial susceptibility of Staphylococcus aureus isolates from Louisiana retail meats. *Foodborne Pathogens and Disease, 8,* pp. 8299–8306.

Qi Y. and Miller K.J. (2000). Effect of low water activity on staphylococcal enterotoxin A and B biosynthesis. *Journal of Food Protection, 63*(4), pp. 473–478.

Queck S.Y. and Otto M. (2008). *Staphylococcus epidermidis* and other coagulase-negative staphylococci. In: J.A. Lindsay, ed, *Staphylococcus Molecular Genetics.* Norfolk, UK: Caister Academic Press.

Resch S., Nagel F. and Christian B. (2008). Antibiotic resistance of coagulase-negative staphylococci associated with food and used in starter cultures. *International Journal of Food Microbiology, 127,* pp. 99–104.

Rice L.B. (1998). Tn916 family conjugative transposons and dissemination of antimicrobial resistance determinants. *Antimicrobial Agents and Chemotherapy, 42,* pp. 1871–1877.

Riesen A. and Perreten V. (2009). Antibiotic resistance and genetic diversity in Staphylococcus aureus from slaughter pigs in Switzerland. *Schweizer Archiv fuer Tierheilkunde, 151,* pp. 425–431.

Roberts M.C. (2005). Update on acquired tetracycline resistance genes. *FEMS Microbiology Letters, 245,* pp. 195–203.

Rogers, K.L, Fey, P.D. and Rupp, M.E. (2009). Coagulase-negative staphylococcal infections. *Infectious Disease Clinics of North America, 23*(1), pp. 73–98.

Ryan K.J. and Ray C.G. (2004). Sherris Medical Microbiology: An Introduction to Infectious Diseases, 4th Ed. New York: McGraw McGraw-Hill Publisher.

Savage V.J., Chopra, I. and O'Neil A.J. (2013). Staphylococcus aureus Biofilms biofilms Promote promote Horizontal horizontal Transfer transfer of Antibiotic antibiotic Resistanceresistance. *Antimicrob. Agents Chemotherapy, 57*(4), pp. 1968–1970.

Sawant, A.A, Gillespie, B.E. and Oliver, S.P. (2009). Antimicrobial susceptibility of coagulase-negative Staphylococcus species isolated from bovine milk. *Veterinary Microbiology, 134,* pp. 73–81.

Schmitz, F.J, Sadurski R., Kray A., Boosa M., Geisela R., Köhrera K., Verhoef J., Fluit A.C., et al. (2000). Prevalence of macrolide-resistance genes in Staphylococcus aureus and Enterococcus faecium isolates from 24 European university hospitals. *Journal of Antimicrobial Chemotherapy, 45,* pp. 891–894.

Shah M.M., Iihara H., Noda M., Song S.X., Nhung P.H, Ohkusu K., Kawamura Y., Ezaki T., et al. (2007). Dna gene sequence-based assay for species identification and phylogenetic grouping in the genus Staphylococcus. *International Journal of Systematic and Evolutionary Microbiology, 57,* pp. 25–30.

Simojoki H., Hyvönen P., Plumed Ferrer C., Taponen S., Pyörälä S., et al. (2012). Is the biofilm formation and slime producing ability of coagulase-negative staphylococci associated with the persistence and severity of intramammary infection? *Veterinary Microbiology, 158*(3–4), pp. 344–352.

Thumu S.C.R. and Halami P.M. (2012). Acquired resistance to macrolide–lincosamide–streptogramin antibiotics in lactic acid bacteria of food origin. *Indian Journal of Microbiology, 52,* pp. 530–537.

Traversa A., Gariano G.R., Gallina S., Bianchi D.M., Orusa R., Domenis L., Cavallerio P., Fossati L., Serra R., Decastelli L., et al. (2015). Methicillin resistance in Staphylococcus aureus strains isolated from food and wild animal carcasses in Italy. *Food Microbiology, 52,* pp. 154–158.

Wang H. (2012). The transfer of antibiotic resistance from food to humans: facts, implications and future directions. *Scientific and Technical Review, 31,* pp. 249–260.

Wang H., Manuzon H., Lehman M., Wan M., Luo K., Wittum H., Yousef A., Bakaletz L.O., et al. (2006). Food commensal microbes as a potentially important avenue in transmitting antibiotic resistance genes. *FEMS Microbiology Letters, 254,* pp. 226–231.

Woodford N. (2005). Biological counterstrike: antibiotic resistance mechanisms of Gram-positive cocci. *Clinical Microbiology and Infection, 11*(3), pp. 2–21.

Yang S.E., Yu R.C. and Chou C.C. (2001). Infuence of holding temperature on the growth and survival of Salmonella and Staphylococcus aureus and the production of staphylococcal eneroxin in egg products. *Internationnal Journal of Food Microbiology, 63,* pp. 99–107.

Yim G., Thaker M.N., Koteva K., Wright G., et al. (2014). Glycopeptide antibiotic biosynthesis. *The Journal of Antibiotics, 67,* pp. 31–41.

Zhang K., McClure J.A., Elsayed S. Conly J.M., et al. (2009). Novel Staphylococcal staphylococcal Cassette cassette Chromosome chromosome mec Typetype, Tentatively tentatively Designated designated Type type VIII, Harboring harboring Class class A mec and tType 4 ccr Gene gene Complexes complexes in a Canadian Epidemic epidemic Strain strain of Methicillinmethicillin-Resistant resistant Staphylococcus aureus. *Antimicrobial Agents Chemotherapy, 53*(2), pp. 531–540.

Zou L.K., Wang H.N., Zeng B., Li J.N., Li X.T., Zhang A.Y., Zhou Y.S., Yang X., Xu C.W., Xia Q.Q., et al. (2011). Erythromycin resistance and virulence genes in Enterococcus faecalis from swine in China. *New Microbiologica, 34,* pp. 73–80.

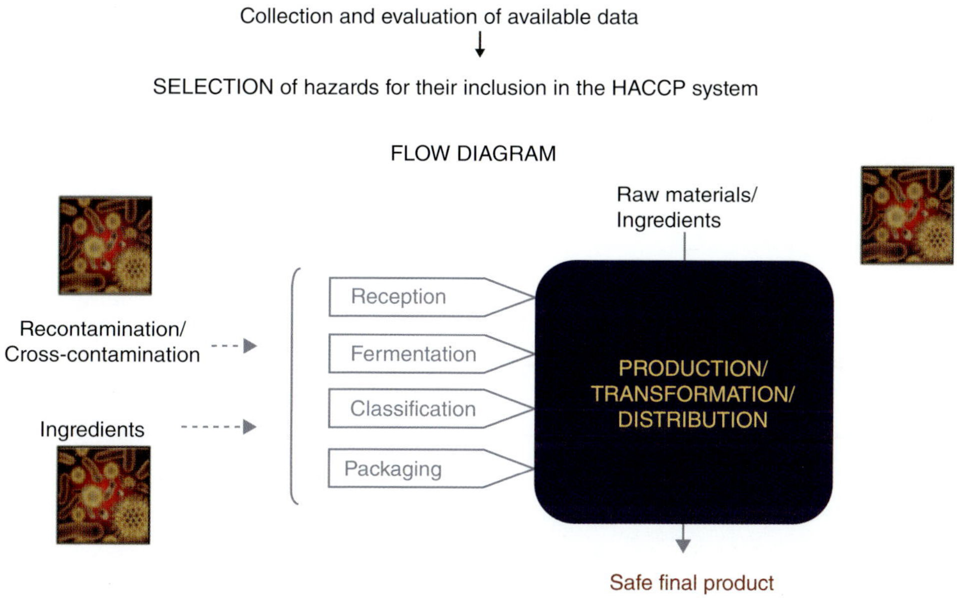

Figure 5.1 Flow-Diagram Process for Hazard Identification During Production of Table Olives.

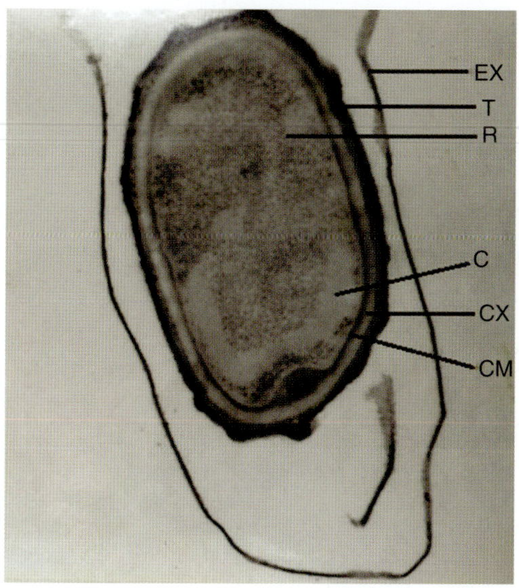

Figure 6.1 Ultrastructure of section of spore of *B. cereus* F3748 revealed at electron microscopy, magnification X 3500; EX, exosporium; T, tunic; R, ribosome; CX, cortex; C, core; CM, core membrane (Etoa, 2011).

Foodborne Pathogens and Antibiotic Resistance, First Edition. Edited by Om V. Singh.

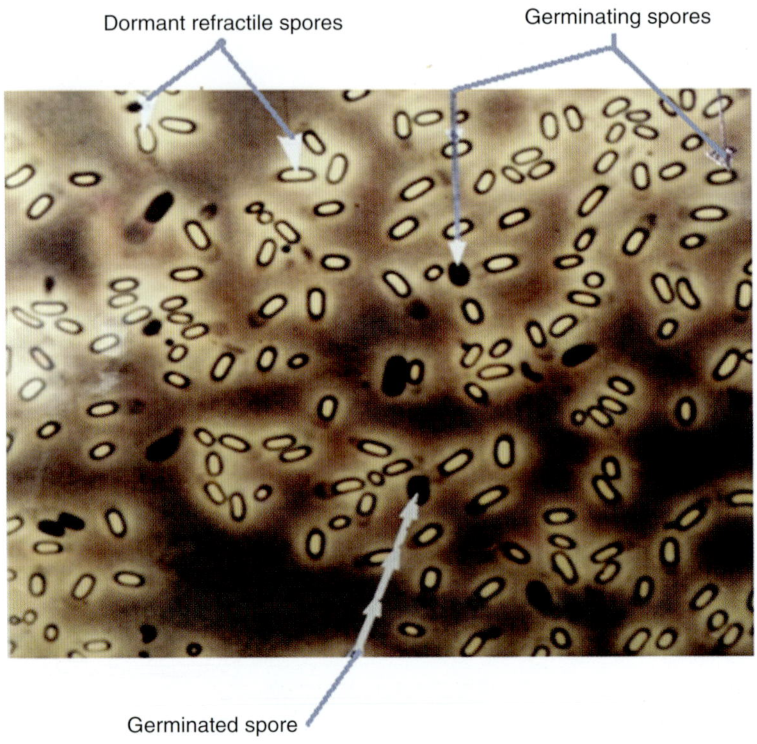

Figure 6.3 Different forms of *B. cereus* F3748 spores during germination as revealed by contrast microscopy, Magnification x 1200 (Etoa, 2011).

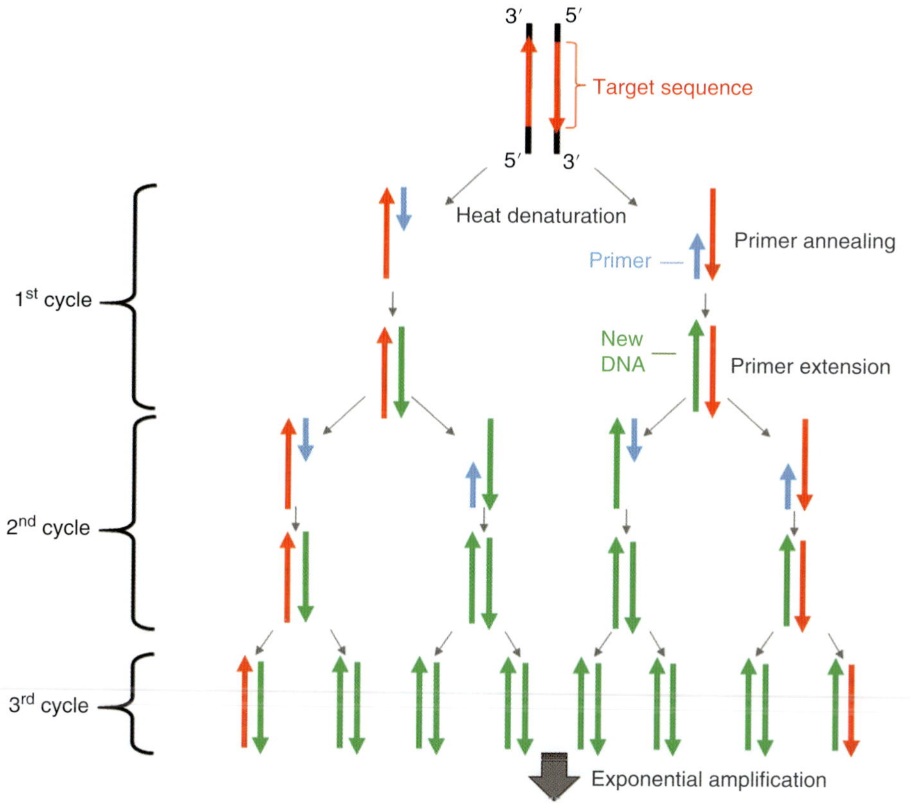

Figure 7.1 Polymerase Chain Reaction.The cyclic three steps process involves denaturation of DNA, primer annealing, and primer extension. The target double-stranded DNA is denatured under high temperature, producing two single-stranded DNA. For each single-stranded DNA, a specific primer will anneal with the target sequence in the DNA. Primer complementary to the single-stranded DNA is extended and a new strand of DNA is synthesized. These steps will be repeated in many cycles and thus producing many copies of a target DNA sequence.

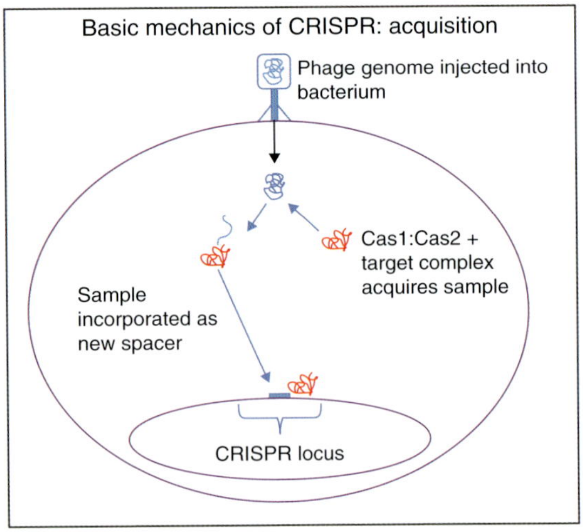

Figure 9.1 In the acquisition phase of immunity in CRISPR Types I-III, invading foreign DNA is read by a Cas1:Cas2 complex:target complex. This complex extricates an amenable sequence and incorporates it into the CRISPR locus as a new spacer, immunizing the cell from the invader. A spacer is adequate to immunize the cell if it meets the targeting requirements of the CRISPR system, particularly whether the protospacer is flanked by a PAM.

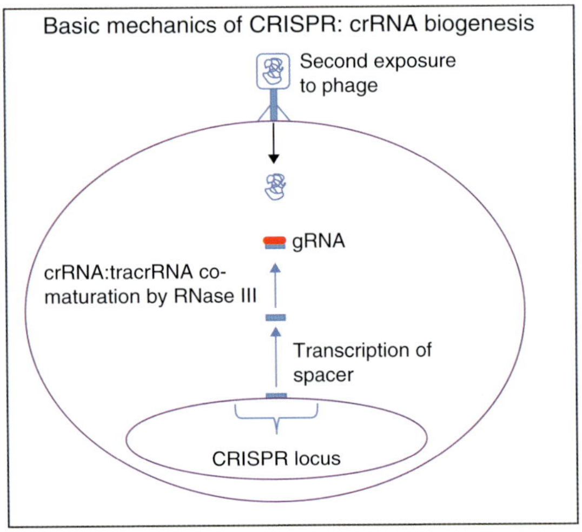

Figure 9.2 In crRNA biogenesis, the spacer is transcribed. This crRNA combines with tracrRNA, resulting in a mature guide RNA (gRNA), a process facilitated by host factor ribonuclease III.

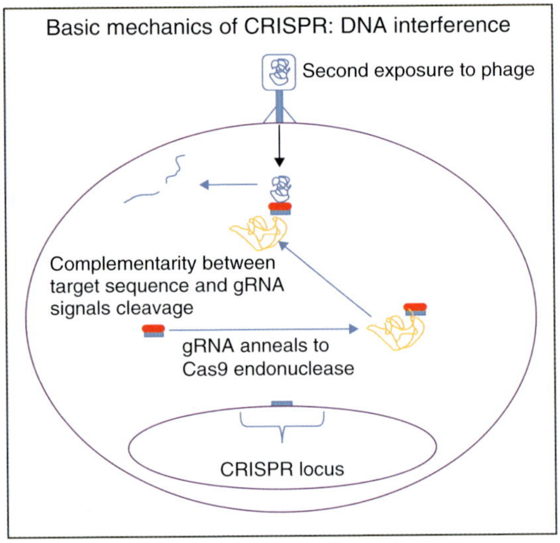

Figure 9.3 In the DNA interference phase, the gRNA anneals to Cas9, forming the functional RGE (RNA-guided endonuclease). If the target sequence is flanked by a PAM and complementary to the gRNA, they anneal, signaling a DSB by Cas9. The RuvC nuclease domain of Cas9 severs the non-target strand, and the HNH domain severs the target strand.

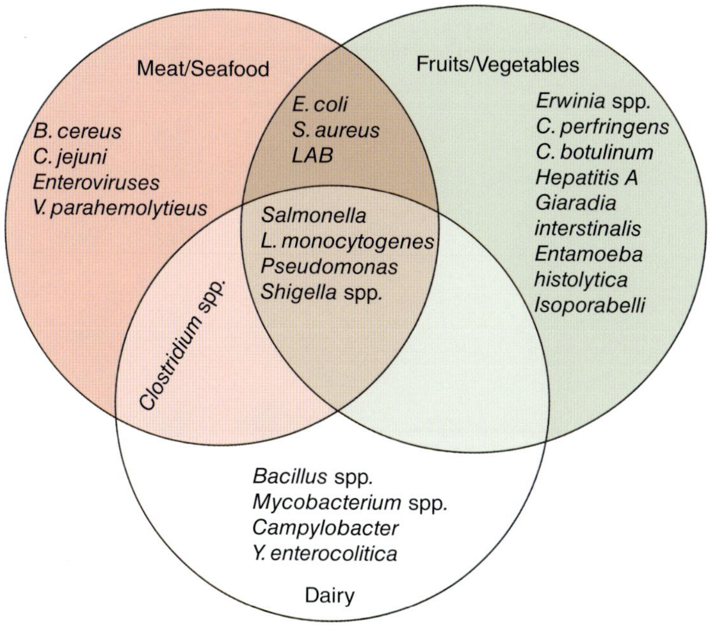

Figure 11.2 Venn diagram of the most common foodborne pathogens associated with meat, seafood, fruits, vegetables, and dairy products.

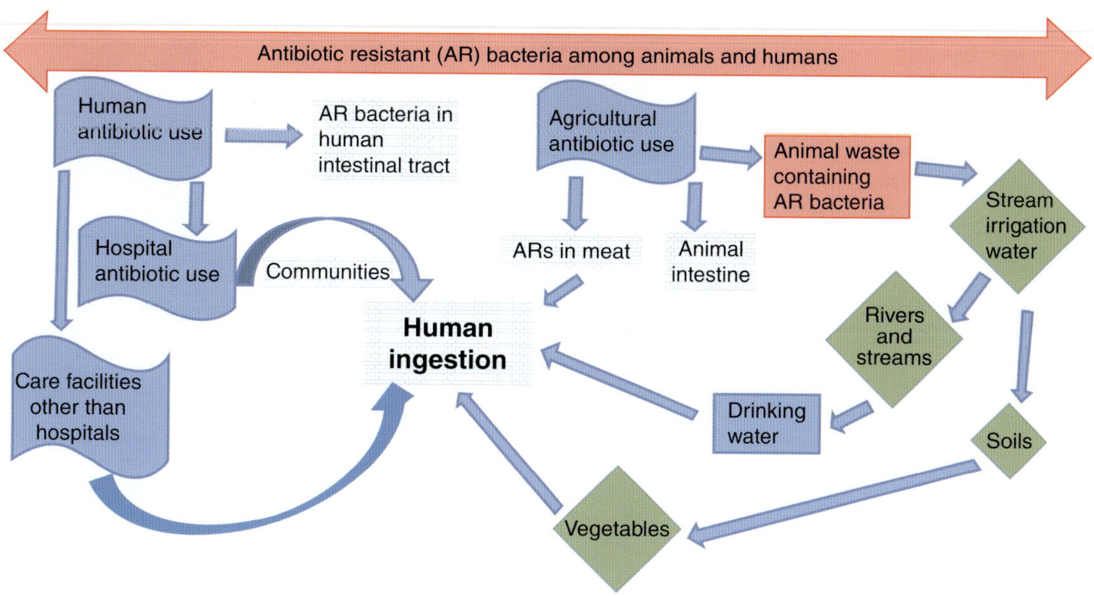

Figure 11.4 Routes of antibiotic resistant bacteria toward human consumption from varying conditions including food.

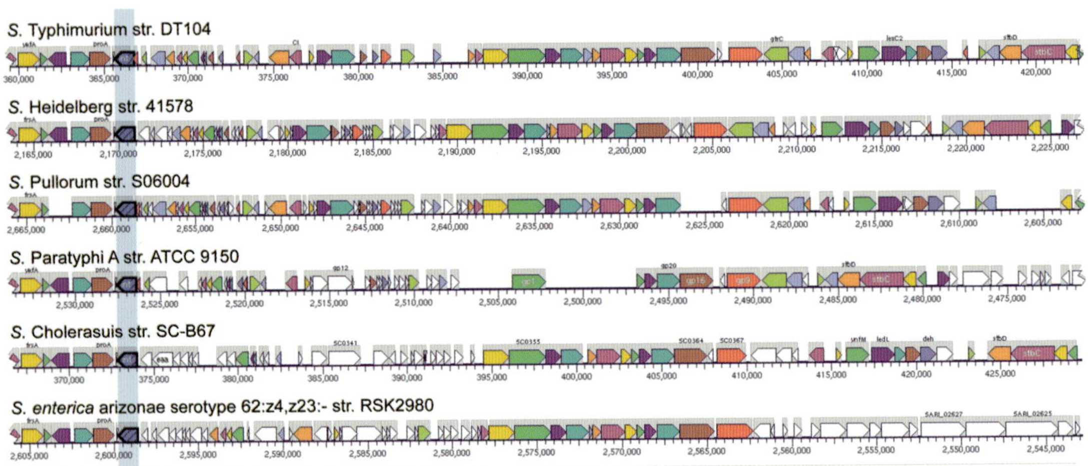

Figure 14.2 Open Reading Frame Maps of Potential Generalized Transducing Prophage Compared to Prophage 1 of *Salmonella* Typhimurium DT104. Alignments were made to the gene encoding the prophage 1 integrase (*int*, DT104_03231), highlighted with a blue box across all alignments. Prophage 1 (P22-like), a generalized transducing prophage in DT104 (NC_022569), is integrated into *thrW* (DT104_t00005) at genome coordinates 365,588–406,646 (Cooke *et al*., 2008). Gene color indicates orthologous groups. The alignment was generated using Pathway Tools version 19.5 by SRI International (Menlo Park, CA) (Karp *et al*., 2002).

16

Antibiotic Resistance in *Enterococcus* spp. Friend or Foe?

Vangelis Economou[1], Hercules Sakkas[2], Georgios Delis[3], and Panagiota Gousia[2]

[1] *Laboratory of Animal Food Products Hygiene – Veterinary Public Health, School of Veterinary Medicine, Faculty of Health Sciences, Aristotle University of Thessaloniki, Thessaloniki, 54124, Greece*
[2] *Food-Water Microbiology Group, Department of Microbiology, School of Medicine, Faculty of Health Sciences, University of Ioannina, Ioannina, 45110, Greece*
[3] *Laboratory of Pharmacology, School of Veterinary Medicine, Faculty of Health Sciences, Aristotle University of Thessaloniki, Thessaloniki, 54124, Greece*

16.1 Introduction

Enterococcal species have been used for decades for food fermentation and preservation. Still, they have emerged as nosocomial– and community-acquired pathogens (Werner, 2012; Guerrero-Ramos *et al.*, 2016). Much of the attention they have received is due to the fact that they exhibit resistance to various antibiotics, with the resistance being both intrinsic and acquired. The transfer of antibiotic resistance is performed via mobile genetic elements such as plasmids or transposons, or through chromosomal exchange or mutations (Guerrero-Ramos *et al.*, 2016). The exchange of resistance traits of enterococci has not been limited to its own species, but involved bacteria of other species, namely *Staphylococcus* spp. and *Listeria* spp. The exchange rate is expected to be in higher in close relation habitats like the intestinal tract. Considering these traits, *Enterococcus* spp. are regarded as reservoirs of antimicrobial resistance genes (Aarestrup *et al.*, 2000a), and considered as indicators of antibiotic resistance (Capita and Alonso-Calleja, 2013).

The *Enterococcus* species *E. faecium* and *E. faecalis* are used as indicators of antimicrobial resistance in Gram-positive bacteria, since they are present in animal feces and are of importance to human medicine, especially the vancomycin-resistant *E. faecium* (EFSA, 2015). The occurrence of *E. faecalis* and *E. faecium* in animal gastrointestinal tract can provide a reservoir for human contamination, directly by contact with fecal material or indirectly by animal products. Furthermore the trends of antimicrobial resistance in animal derived *E. faecalis* and *E. faecium* strains reveal possible selective pressure of antibiotic usage. Therefore, the European Union member states have started gathering and evaluating data on antimicrobial resistance of animal *E. faecalis* and *E. faecium* strains from both farm and slaughterhouse premises (EFSA, 2008).

16.2 *Enterococcus* Biology

Enterococci have been around for a long time. Theodor Escherich described *Enterococcus faecalis* as early as 1886, naming it at the time as

Foodborne Pathogens and Antibiotic Resistance, First Edition. Edited by Om V. Singh.

Micrococcus ovalis, with *E. faecium* following at 1899 (Jay *et al.*, 2005). The *Enterococcus* genus was not distinguished from the *Streptococus* genus for decades. It was not until 1984 that the *Enterococcus* group has been designated through the use of DNA hybridization and 16S rRNA sequencing, with its first members being *E. faecium* and *E. faecalis*. According to 16S rRNA analysis, enterococci are more closely related to *Vagococcus*, *Tetragenococcus*, and *Carnobacterium* than to *Streptococcus* and *Lactococcus* (M'hir *et al.*, 2012). Enterococci exhibit low guanine plus cytosine content in DNA, which is quite common in the related *Clostridium* and *Bacillus* spp. (M'hir *et al.*, 2012). Currently seven groups have been established (Table 16.1).

Enterococci are Gram-positive, catalase-negative, facultative anaerobic non-spore-forming-cocci that can occur as single cells or organized in pairs or chains (Foulquié Moreno *et al.*, 2006). They exhibit high tolerance in adverse environmental conditions. They are more thermophilic than other Gram-positive cocci, with growth occurring in temperatures ranging from 5 to 50°C and survival for up to 30 min at 60°C (Araújo and Ferreira, 2013). According to Van den Berghe *et al.* (2006) the optimum, minimum, and maximum temperatures for growth on brain heart infusion under aerobic conditions was 42.7, 6.5, and 47.8 °C, respectively. Their strong homeostasis mechanisms result in high resistance to pH, salt, metals, and desiccation (Fisher and Phillips, 2009). Their salt tolerance is quite marked since they can grow at 6.5% NaCl. Also they can grow in a broad pH range (Giraffa, 2002; Franz *et al.*, 2003). Enterococci can produce a variety of products such as aromatic compounds (Centeno *et al.*, 1999), enzymes (Sarantinopoulos *et al.*, 2001; Ghrairi *et al.*, 2008) and bacteriocins (enterocins) (Cleveland *et al.*, 2001; Achemchem *et al.*, 2005). The enterocins produced include enterocins A, B, I, L, and P (Fisher and Philips, 2009), which are active against *Listeria* spp., *Clostridium* spp., and *Staphylococcus aureus* (Campos *et al.*, 2006). Bacteriocin production has been studied extensively, with at least 17 different bacteriocins already identified (Fisher and Philips, 2009), since that, except from pathogen control, there is evidence that they can be used on cancer treatment (Nami *et al.*, 2014).

16.3 *Enterococcus* as a Probiotic

The *Enterococcus* strain most researched for its probiotic properties is *E. faecium* SF68 (Lewenstein *et al.*, 1979; Bellomo *et al.*, 1980). *E. faecium* SF68 is highly efficient in controlling human diarrhea probably due to its origin as a

Table 16.1 Members Species of the *Enterococcus* Groups (Klein, 2003; Franz *et al.*, 2011).

Enterococcus group	Member species
E. avium group	*E. avium, E. devriesei, E. gilvus, E. malodoratus, E. pseudoavium. E. raffinosus*
E. cecorum group	*E. cecorum, E columbae*
E. dispar group	*E. dispar, E. asini, E. canintestini, E. hermanniensis, E. pallens*
E. faecalis group	*E. faecalis, E. caccae, E. haemoperoxidus, E. moraviensis, E. silesiacus, E. termitis, E. ureasiticus, E. quebecensis*
E. faecium group	*E. faecium, E. canis, E. durans, E. hirae, E. mundtii, E. phoeniculicola, E. ratti, E. villorum, E. thailandicus*
E. gallinarum group	*E. gallinarum, E. casseliflavus*
E. saccharolyticus group	*E. saccharolyticus, E. acquimarinus, E. camelliae, E. italicus, E. sulfureus*

commensal bacterium of the human gut and its quick reproduction. Also it has been reported to inhibit *E. coli*, *Salmonella* serovars, *Shigella* spp., and *Enterobacter* spp. *in vitro* (Lewenstein *et al.*, 1979; Bellomo *et al.*, 1980; Canganella *et al.*, 1996). Also, *E. faecium* SF68 has been shown to prevent against antibiotic associated diarrhea (Borgia *et al.*, 1982; Wunderlich *et al.*, 1989; D'Souza *et al.*, 2002), and to be effective in the treatment of irritable bowel syndrome (Gade and Thorn, 1989; Kim *et al.*, 2006; Enck *et al.*, 2008). Except from the gut-associated benefits, enterococci have been used so as to lower the blood cholesterol levels (Agerholm-Larsen *et al.*, 2000; Franz *et al.*, 2001, 2011). Also *E. faecalis* containing Symbioflor 1 (SymbioPharm, Herborn, Germany) has been used as an immune regulator for the prevention and treatment of chronic recurrent hypertrophic sinusitis or bronchitis (Habermann *et al.*, 2002).

Concerning the use of *Enterococcus* spp. in animals, their use involves among others *E. faecium* SF68 for the prevention and treatment and diarrhea in pigs. *E. faecium* SF68 can survive stomach passage (Spieler and Männer, 1996) and therefore can easily be administered *per os*. Its efficacy has been shown against colibacillosis in weaning piglets (Underdahl *et al.*, 1982). Also, Taras *et al.* (2006) and Zeyner and Boldt (2006) report reduced mortality and pre- and post-weaning diarrhea after *E. faecium* SF68 or *E. faecium* DSM 10633 usage respectively. Several research papers deal with the possible immunostimulatory effects or the competitive exclusion of enteric pathogens by *E. faecium* (Broom *et al.*, 2006; Scharek *et al.*, 2005, 2007). In poultry, growth promotion and better feed conversion has been observed after *E. faecium* SF68 administration (Samli *et al.*, 2007). The efficacy on growth promotion of probiotic mixtures containing enterococci has been shown among others by Mountzouris *et al.* (2007) and Capcarova *et al.* (2010). Since there is abundant evidence that enterococci favor the colonization of the intestine by lactic acid bacteria, the possible competitive exclusion of pathogenic bacteria, namely *Campylobacter*, can be of tremendous value to food hygiene.

16.4 *Enterococcus* in Food

Enterococci belong to the lactic acid bacteria and they are of interest to Food Science since they contribute to spoilage and are used as starter cultures in fermented products (Franz *et al.*, 2011). For a long time, *Enterococcus* species were thought as harmless to humans and unimportant medically. The recent involvement of enterococci in nosocomial infections and their marked antimicrobial resistance has posed questions toward the safety of use of these bacteria in technological food applications. Since enterococci inhabit the mammalian gastrointestinal tract, they often contaminate foods of animal origin such as meats, fermented and cooked meats, milk, and cheese (Franz *et al.*, 2011).

16.4.1 *Enterococcus* as a Microbial Indicator in Water and Food

The occurrence of enterococci in fecal material has made researchers propose their use as indicators of water quality as early as 1900. Nowadays, enterococci are accepted as intestinal colonizers in many animals ranging from intervertebrates, reptiles, and birds, to man (Aarestrup *et al.*, 2000a; Werner, 2012). *E. faecalis* predominates in human faeces with counts ranging from 10^5 to 10^7 CFU/g, whereas *E. faecium* counts range from 10^4 to 10^5 CFU/g (Chenoweth and Schaberg, 1990; Franz *et al.*, 2011). They can be found in diverse environments such as sewage, soil and water soil, surface waters, ocean water, sewage, and on plants (Franz *et al.*, 2011; Werner, 2012). The main advantages for their use as indicators of fecal contamination of water are i) they do not multiply in water ii) they occur in smaller numbers in feces than *Escherichia coli*, and iii) they are more persistent in water than coliforms. Still, except from *E. faecalis* and *E. faecium*, the fecal origin of the other 28 members of the *Enterococcus* genus remains to be clarified. Although the use of enterococci as a microbial indicator in water is well established, their use in food is less advantageous. Their multiple possible origins in the food-processing environment (Mundt, 1986) along with their resistance in environmental factors, including disinfection, make them less ideal in monitoring safety status of food (Knechtges, 2012), as shown by Birollo *et al.* (2001).

16.4.2 Meat and Meat Products

Occurrence of enterococci in raw meat products is well established since it is reported by several studies (Chingwaru *et al.*, 2003; Sustackova *et al.*, 2004; Novais *et al.*, 2005; McGowan *et al.*, 2006; Messi *et al.*, 2006; Kaszanyitzky *et al.*, 2007; Miranda *et al.*, 2007; Fluckey *et al.*, 2009; Gousia *et al.*, 2011; Vignaroli *et al.*, 2011; Aslam *et al.*, 2012; Cho *et al.*, 2013; Choi *et al.*, 2013; Gundogan *et al.*, 2013; Jahan *et al.*, 2013; Klibi *et al.*, 2013; Jahan and Holley, 2014; Pesavento *et al.*, 2014; Gousia *et al.*, 2015; Guerrero-Ramos *et al.*, 2016). In meat products from Turkey (Kasimoglu-Dogru *et al.*, 2010), Italy (Busani *et al.*, 2004), Greece (Gousia *et al.*, 2015), Spain (Martin *et al.*, 2005), and United States (Hayes *et al.*, 2003) the most prevalent *Enterococcus* spp. was *E. faecium* followed by *E. faecalis*. On the contrary, *E. faecalis* was reported to be more predominant in meat samples from Denmark (Aarestrup *et al.*, 2000b), and Italy (Pesavento *et al.*, 2014). Reports from Germany (Lemcke and Bülte, 2000), and Tunisia (Klibi *et al.*, 2013) show almost equal prevalence of *E. faecalis* and *E. faecium*. Although in some cases the differences in observed prevalence are quite marked, the cause of this phenomenon is probably due to seasonal variations, geographical regions, and different methods of detection and isolation (Gousia *et al.*, 2015).

Heat treatment of meat products can result in selection of enterococci, because of their heat resistance. Therefore, enterococci usually form part of the spoilage flora of cured meat products, such as canned hams and chub-packed luncheon meats (Magnus *et al.*, 1986, 1988). Accordingly they have been isolated from various Spanish, German, and Italian meat products (Marchesino *et al.*, 1992; Aymerich *et al.*, 1996; Casaus *et al.*, 1997; Cintas *et al.*, 1997), with their concentration reaching large numbers, as observed in salami and Landjäger were *Enterococcus* counts ranged from 10^2 to 2.6×10^5 CFU g^{-1} (Teuber *et al.*, 1996). Moreover, they predominate in fermented raw sausages or cooked meat products (Martin *et al.*, 2005; Paramithiotis *et al.*, 2008; Ruiz-Moyano *et al.*, 2009). In a survey of 31 naturally fermented sausages from Spain, the number of enterococci ranged from 1.3 to 4.48 $\log_{10}$ CFU g^{-1}, in comparison to total LAB counts ranging from 7.12 to 9.07 $\log_{10}$ CFU g^{-1} (Hugas *et al.*, 2003). The biochemical activity of enterococci might contribute to aroma development by their glycolytic, proteolytic, and lipolytic activities. Furthermore, metmyoglobin reduction activity has been described for meat enterococci (Hugas *et al.*, 2003).

16.4.3 Cheese

Enterococci are quite common in raw milk (Chingwaru *et al.*, 2003; Zamfir *et al.*, 2006; Kagkli *et al.*, 2007; Franciosi *et al.*, 2009; Giannino *et al.*, 2009; Moraes *et al.*, 2012; Gaaloul *et al.*, 2015; Gundogan *et al.*, 2013; Perin *et al.*, 2014; Hammad *et al.*, 2015). Since the enterococci are resistant to high salt concentrations and generally in adverse environmental conditions, their existence in cheese is quite common (Burdychová, 2009; Psoni *et al.*, 2006; Ortigosa *et al.*, 2008; Franciosi *et al.*, 2009; Giannou *et al.*, 2009; Serio *et al.*, 2010; Ahmadova *et al.*, 2013; Favaro *et al.*, 2014; Hammad *et al.*, 2015; Vrabec *et al.*, 2015). The most predominant enterococci in cheese are *E. faecium*, *E. faecalis* and *E. durans* contributing to ripening and aroma production (Tzanetakis and Litopoulou-Tzanetaki, 1992; Sarantinopoulous *et al.*, 2001; Giraffa, 2002; Randazzo *et al.*, 2008). Especially in in artisanal cheeses, enterococci often predominate at the end of the ripening period (Litopoulou-Tzanetaki, 1990; Bouton *et al.*, 1998; Manolopoulou *et al.*, 2003; Cosentino *et al.*, 2004; Jurkovic *et al.*, 2006; Pimentel *et al.*, 2007; Fuka *et al.*, 2010).

16.4.4 Other Foods

Although not widely searched for, enterococci have been isolated from egg contents (Papadopoulou *et al.*, 1997; Schwaiger *et al.*, 2010; Gousia *et al.*, 2015). Also enterococci are abundant in plant derived fermentation products, mostly researched on food of African or Asian origin (Hamad *et al.*, 1997; Moreno *et al.*, 2002; Onda *et al.*, 2002; Yousif *et al.*, 2005; Oguntoyinbo *et al.*, 2007; Franz *et al.*, 2011). In olive fermentation, enterococci are implicated in oleuropein breakdown, reducing its bitter taste (Asehraou *et al.*, 1992; van den Berg *et al.*, 1993; Lavermicocca *et al.*, 1998; de Castro *et al.*, 2002; Franz *et al.*, 2011). The presence and role of enterococci in fermented vegetable foods has been discussed extensively by M'hir *et al.* (2013). The elucidation of several beneficial properties, such as aroma production, positive contribution to the ripening process, and bacteriocin production exhibited by the existence of enterococci in foodstuffs made them ideal candidates for starter culture organisms.

16.5 Antibiotic Resistance

Enterococci, especially *E. faecalis* and *E. faecium*, are renowned for their genome plasticity. They are able to integrate and utilize Mobile Genetic Elements (MGEs) such as plasmids, transposons, insertion sequences, and prophages (Galloway-Peña *et al.*, 2012; Mikalsen *et al.*, 2015). They easily acquire resistance determinants that they can later transfer to strains of the same species, to other species of the same genus, and to numerous pathogenic or not-pathogenic bacteria, even in absence of selective pressure (Al-Nassir *et al.*, 2008), thus acting as a reservoir of antibiotic resistance-associated genes (Miller *et al.*, 2014; Iweriebor *et al.*, 2015; Jahan *et al.*, 2015). Of both species, *E. faecalis* is less host or origin-specific and is considered a better receiver and donor of MGEs; however, *E. faecium* normally expresses resistance and virulence at higher levels (Pesavento *et al.*, 2014; O'Driscoll and Crank, 2015; Yu *et al.*, 2015).

Two distinct clades of *E. faecium* have been described. Clade A includes strains that cause nosocomial infections in humans, whereas clade B comprises community-related strains (Mikalsen *et al.*, 2015). Genetic differences between the two clades are detected through the entire genome with a characteristic trend for strains of clade A to harbor twice as many genes associated with MGEs, and more readily express MultiDrug-Resistant (MDR) phenotypes (Galloway-Peña *et al.*, 2012; Hollenbeck and Rice, 2012; Miller *et al.*, 2014). Transfer of resistant bacteria or resistance determinants from animals and foods of animal origin to humans is also a point of concern (Hammerum *et al*, 2010; Iweriebor *et al*, 2015). Long-term use of virginiamycin, a streptogramin agent, as a growth promoter in livestock led to the selection of *E. faecium* with cross-resistance to quinupristin/dalfopristin, also a streptogramin but used in human medicine (Soltani *et al.*, 2000). Some researchers report the presence of high-risk clonal complexes of zoonotic enterococci in both humans and animals, whereas others suggest a clear genetic diversity between enterococcal strains in foods or food-producing animals and those within the clinical setting (Klein, 2003; Freitas *et al.*, 2011; Tzavaras *et al.*, 2012; Novais *et al.*, 2013; Jahan *et al.*, 2015). In any case, through direct or indirect routes, resistance can be transferred to human pathogens and complicate treatment (Foulquié Moreno *et al.*, 2006; Iweriebor *et al.*, 2015).

Antibiotic resistance in enterococci is generally produced by at least one of the following mechanisms: target modification, alterations that affect access of the drugs to the target, and enzymatic drug inactivation (Table 16.2). Resistance can also be intrinsic (such as in the

Table 16.2 Distribution of General Mechanisms of Resistance Exhibited by *Enerococcus* spp. to Selected Classes of Antibiotics.

Antibiotic class	Target modification[a]	Altered access of drug	Drug inactivation
β-lactams	+	−	+
Aminoglycosides	+	+	+
Glycopeptides/ lipoglycopeptides	+	−	−
Lipopeptides	+	−	−
Fluoroquinolones	+	+	−
Oxazolidinones	+	−	−
Macrolides/lincosamides	+	+	+[b]
Streptogramins	+	+	+
Pleuromutilins	+	+	−
Tetracyclines/ glycylcyclines	+	+	−
Rifamycins	+	−	−
Sulfonamides/trimethoprim	+	−	−

[a] Includes by-pass of drug activity by activation of alternative biochemical pathways.
[b] Lincosamides only.

cases of certain β-lactams, aminoglycosides and macrolides/ lincosamides) or acquired, through mutations in specific genes or acquisition of exogenous genetic material (Werner, 2012; Hollenbeck and Rice, 2015).

16.5.1 Resistance to β-Lactams

Antibiotics of the broad β-lactam class inhibit the synthesis of peptidoglycan, the basic polymeric structure of the bacterial wall, by binding to transpeptidase and transglycosidase enzymes, collectively called Penicillin Binding Proteins (PBPs) (Waxman and Strominger, 1983). Bactericidal activity to both Gram-positive and Gram-negative bacteria is generally displayed by penicillins, cephalosporins/cephamycins, and carbapenemes (with variations among individual agents), since monobactams are active only against the latter (Johnson and Cunha, 1995).

Enterococci are intrinsically moderately resistant (10–100 times less susceptible than streptococci) or, more properly, tolerant to specific β-lactams (particularly penicillin, ampicillin, or cephalosporins/cephamycins), due to the presence of a species-specific chromosomal gene (*pbp4* in *E. faecalis, pbp5* in *E. faecium*) constitutively encoding a low-affinity PBP4/5 (Ono *et al.*, 2005; Hollenbeck and Rice, 2012; Kristich and Little, 2012) and/or to the superoxide radical-partitioning activity of the enzyme superoxide dismutase (Bizzini *et al.*, 2009). Tolerance renders the otherwise bactericidal β-lactams, bacteriostatic.

Another mechanism that leads to reduced susceptibility to penicillins in *E. faecium* involves activity of an L,D-transpeptidase (Ldt_{fm}) and formation of a restructured peptidoglycan layer (Miller *et al.*, 2014). Acquired, high-level resistance to penicillins can be mediated by mutations in *pbp4/5* genes, followed by production of respective PBPs with even lower affinity, and also by production of β-lactamases, enzymes that differentially inactivate β-lactam antibiotics. Point mutations in *pbp5* (potentially carried on Tn*5382* conjugative transposons) is the main mechanism for high-level penicillin resistance in *E. faecium*, whereas over-production of β-lactamases, encoded in plasmid/transposon-borne *bla* genes, is largely restricted to *E. faecalis*

(Hollenbeck and Rice, 2012; Werner, 2012). In the latter species, alterations in PBP4 confer resistance to both, penicillins and carbapenems (Ono *et al.*, 2005).

Apart from the above, several specific pathways participate in the acquisition of high-level cephalosporin resistance. These include a two-component signal transduction system (CroRS), a transmembrane serine/threonine kinase (IreK) and putative mutations in *rpoB* gene, which encodes the β subunit of the enzyme RNA polymerase (Kristich and Little, 2012; Miller *et al.*, 2014).

Ampicillin can be efficaciously used in the treatment of infections caused by susceptible enterococci. In the case of β-lactamase-producing *E. faecalis*, the addition of a β-lactamase inhibitor (e.g., sulbactam) is recommended. When a bactericidal effect is desirable, the inclusion of an aminoglycoside (gentamicin or streptomycin) increases the cure rates, provided that the pathogen does not present high-level aminoglycoside resistance. In patients with imminent or established renal failure, use of aminoglycosides is prohibited and they are replaced by newer-generation cephalosporins or imipenem or imipenem plus vancomycin (Arias *et al.*, 2010; Hollenbeck and Rice, 2012; Pesavento *et al.*, 2014; O'Driscoll and Crank, 2015).

16.5.2 Resistance to Aminoglycosides

Aminoglycosides are concentration-dependent, bactericidal drugs, inhibitors of protein synthesis. They hinder peptide elongation at the 30S small ribosomal subunit and lead to rapid bacterial death (Nicolau *et al.*, 1995; Kohanski *et al.*, 2007). Enterococci (like most Gram-positive bacteria) are also intrinsically resistant, at least to low aminoglycoside concentrations, primarily due to decreased penetration into the bacterial cell through the cell wall, but moreover due to the frequent occurrence of specific aminoglycoside-modifying enzymes, such as Aac(6')-Ii and Aph(3')-IIIa and the activity of a ribosomal methyltransferase (EfmM) resulting in overall resistance to tobramycin, netilmicin, and kanamycin (Arias *et al.*, 2010). Due to these, only two drugs, streptomycin and gentamicin are used in the treatment of enterococcal infections, always in combination with cell wall/cell membrane-acting agents, such as β-lactams, daptomycin (their synergy relies on β-lactams or daptomycin acting on the bacterial cell wall or cell membrane, respectively, thus facilitating entry of aminoglycosides), or quinupristin/dalfopristin.

High-level resistance to streptomycin and gentamicin, totally precluding clinical use of these agents, can be attained by ribosomal alterations and/or the acquisition of genes encoding aminoglycoside-modifying enzymes, such as the streptomycin-inactivating Ant(6')-Ia and Ant(3") or the gentamicin-inactivating Aac(6') Ie-Aph(2")Ia, Aph(2")-Ib, Aph(2")-Ic, Aph(2")-Id and Aac(6')-Im (Chow *et al.*, 2001; Kobayashi *et al.*, 2001; Hollenbeck and Rice, 2012). The above enzymes, along with Ant(4')-Ia, which confers resistance to tobramycin, amikacin, kanamycin and neomycin are encoded by genes mostly located on MGEs, for instance Tn*4001*/Tn*5281* (Werner, 2012) that can be transferred horizontally.

16.5.3 Resistance to Glycopeptides and Lipoglycopeptides

Indisputably, the most intriguing issue in enterococcal antibiotic resistance concerns vancomycin, a glycopeptide that was initially introduced in human medicine for the treatment of infections caused by methicillin-resistant *Staphylococcus aureus* (MRSA) (Rodvold and McKoneghy, 2014; Biedenbach *et al.*, 2015; Münch *et al.*, 2015). Vancomycin is bacteriostatic against enterococci (in contrast, it is considered bactericidal against *Staphylococcus* spp. and *Streptococcus* spp.) and like β-lactams, it inhibits synthesis of a functional peptidoglycan layer (French, 2006; Brauers *et al.*, 2007). However, vancomycin mechanism of action is different, as it selectively binds to the D-Ala-D-Ala terminus of the pentapeptide precursors, inhibiting the cross-linking between two parallel

N-acetylglucosamine-*N*-acetylmuramic acid polymer chains (Gold, 2001; Werner, 2012). Resistance arises from synthesis of altered, low-affinity pentapeptide side chains with D-Ala-D-Lac or D-Ala-D-Ser termini.

To date, nine resistance phenotypes have been described in vancomycin-resistant *Enterococcus* spp. (VRE), namely VanA, VanB, VanD, VanM (D-Ala-D-Lac modification and high-level resistance), VanC, VanE, VanG, VanL, and VanN (D-Ala-D-Ser modification and low-level resistance), attributable to the presence of nine respective gene clusters (Hollenbeck and Rice, 2012; Miller *et al.*, 2014). VanA is the most common and studied resistance type, mainly exhibited by *E. faecium* and yielding high-level resistance to both, vancomycin and teicoplanin (another glycopeptide antibiotic). At least seven enzymes are involved in expression of VanA-type resistance with two of them (VanA and VanH) being responsible for synthesis of low-affinity precursors, two (VanX and VanY) inactivating the normal precursors, and two more (VanR and VanS) regulating gene expression (the role of another enzyme, VanZ, has not been clarified yet).

The presence of a glycopeptide antibiotic in the milieu activates *vanS* and initiates operon expression. VanB is the second most common operon. A different *vanS* gene ($vanS_B$), is not activated by the presence of teicoplanin and so, VanB-positive strains remain susceptible to this agent (Gold, 2001; Hollenbeck and Rice, 2012; Werner, 2012). Both resistance operons are known to be associated with MGEs, *vanA* with Tn*1546* and *vanB* with Tn*5382*/Tn*1549* (Arthur and Courvalin, 1993; Foulquié Moreno *et al.*, 2006; Mikalsen *et al.*, 2015). The presence of Tn*1546* on plasmids (Garnier *et al.*, 2000) was held responsible for the dissemination of vancomycin resistance from *E. faecium* and the emergence of Vancomycin-Resistant *Staphylococcus aureus* (VRSA) in the early 2000s (Biedenbach *et al.*, 2015).

Concerning other Van phenotypes, VanM also yields high-level resistance in *E. faecium*, VanD is a non-transferable type reported in both *E. faecalis* and *E. faecium*, VanC offers low-level resistance and is intrinsic in *E. gallinarum* and *E. casseliflavus*, while VanE, VanG, VanL, and VanN are less frequent low-level resistance variants (O'Driscoll and Crank, 2015). Of outmost clinical importance are enterococcal strains (predominantly of *E. faecium*) that present ampicillin and vancomycin resistance coupled with high-level aminoglycoside resistance, as they are very pathogenic and difficult to eradicate.

Lipoglycopeptide antibiotics are bactericidal and allegedly more potent than vancomycin against *Enterococcus* spp. (Münch *et al.*, 2015; Nnedu and Pankey, 2015). Telavancin targets both, cell wall and cell membrane disrupting their integrity (Pankey and Ashcraft, 2013). However, this agent lacks activity against VanA strains (O'Driscoll and Crank, 2015). On the contrary, oritavancin also interacts with PBPs and secondary cellular biochemical pathways and kills VanA *E. faecium*, preferably when used in combination with an aminoglycoside (gentamicin) for the suppression of mutant selection (Arias *et al.*, 2010; Münch *et al.*, 2015).

16.5.4 Resistance to Lipopeptides

As a concentration-dependent, bactericidal drug (French, 2006), daptomycin is widely used against VRE, despite lacking a respective approval by the FDA or the EMA (Tran *et al.*, 2013; Trubiano *et al.*, 2013). Daptomycin (chemically a cyclic lipopeptide) oligomerizes within the bacterial cell membrane and forms a channel or pore-like structure, which promotes leakage of intracellular ions (mainly potassium) to the extracellular matrix, resulting in bacterial death (Miller *et al.*, 2014). The precise term for daptomycin resistance is "non-susceptibility," since (in contrast to staphylococci) CLSI and EUCAST clinical resistance breakpoints for enterococci have not yet been set. Non-susceptibility in *E. faecalis* is likely governed by redistribution of cellular cardiolipin content and a decrease in phosphatidylglycerol levels, both molecules being essential for daptomycin activity.

These events are initiated by mutations in the *liaF* gene (mirrored in substitutions in LiaFSR regulatory system) and also in specific genes involved in phospholipid metabolism and biogenesis, namely *gpdD* and *cls* (Hancock and Perego, 2004; Tran *et al.*, 2013; Miller *et al.*, 2014). A similar, albeit not identical, mechanism is responsible for non-susceptibility in *E. faecium*, since it is again mutations in multiple genes encoding enzymes implicated in lipid metabolism (*cfa* and *cls*) and components of regulatory systems, most notably *yycG* (encoding protein YycG, the histidine kinase sensor of the YycFG system) and *liaS* (encoding protein LiaS of the LiaFSR system) that are responsible for increased daptomycin minimum inhibitory concentrations (Szurmant *et al.*, 2005; Hollenbeck and Rice, 2012; Miller *et al.*, 2014). In the latter species, substitutions in components of the phosphoenolopyruvate:sugar phosphotransferase system are also likely to be connected with non-susceptibility. Due to the special mechanism of action of daptomycin, cross-resistance with other classes of antibiotics has not been detected (Kelesidis *et al.*, 2011).

Monotherapy of complicated enterococcal infections with daptomycin in humans is generally reported to be accompanied by an increased incidence of failure and (rarely) non-susceptibility emergence depended on the duration of treatment. Use of high doses and combinational therapies with fifth generation cephalosporins, gentamicin, linezolid, tigecycline, quinupristin/dalfopristin, ampicillin, ampicillin plus gentamicin, or gentamicin plus rifampicin have yielded favorable results and generally prevent mutant selection and amplification (Kelesidis *et al.*, 2011; Shukla *et al.*, 2013; O'Driscoll and Crank, 2015).

16.5.5 Resistance to Fluoroquinolones

Among fluoroquinolones, those belonging to the third (levofloxacin, sparfloxacin) and fourth (gatifloxacin, moxifloxacin) generations have significant activity against Gram-positive bacteria (Hooper, 2002). They act through inhibition of DNA gyrase and topoisomerase IV, two enzymes necessary for bacterial DNA transcription and replication (Hooper and Jacoby, 2015). DNA gyrase inhibition is the main mechanism in Gram-negative bacteria, whereas topoisomerase IV is the main fluoroquinolone target in Gram-positive microorganisms, like enterococci (Werner, 2012). Each of these enzymes is a tetramer, consisting of two pairs of two subunits (GyrA and GyrB for DNA gyrase; ParC and ParE for topoisomerase IV).

The mechanisms of resistance of enterococci to fluoroquinolones include amino acid substitutions (mainly in GyrA and ParC), activity of efflux pumps and protection of DNA gyrase from inhibition (Jacoby, 2005). Mutations in *gyrA* and *parC* genes can cause desensitization of the enzymes to fluoroquinolone activity, especially when they occur in specific areas, called "Quinolone Resistance Determining Regions," or commonly "QRDRs." Mutations in *gyrB* and *parE* can also reduce susceptibility, but are less frequent in clinical strains, whereas sequential mutations in all genes confer continuously increasing levels of resistance (Hooper and Jacoby, 2015).

In *E. faecalis*, *emeA* gene is known to encode a NorA-related efflux pump, responsible for the removal of fluoroquinolone molecules out of the bacterial cell and the increase of minimum inhibitory concentration values (Jonas *et al.*, 2001). Mutations in other, regulatory genes that control the transcription of this efflux pump structural proteins also lead to an increased removal of the antibiotic from the bacterial cell. Finally, in *E. faecalis*, a chromosomally located *qnr* gene encodes a protein (EfsQnr) that protects DNA gyrase from inhibition (Hooper and Jacoby, 2015).

In sharp contrast to Gram-negative bacteria, in Gram-positive microorganisms, including enterococci, plasmid-mediated quinolone resistance (PMQR) is not observed (Strahilevitz *et al.*, 2009). Fluoroquinolones have been occasionally used in the treatment of infections caused mainly by *E. faecalis*, in monotherapy or in combination with other agents, like quinupristin/dalfopristin or ampicillin. However, due to the rapid emergence of resistance and the availability of more potent alternatives, they are not considered first-line drugs against enterococci (Arias *et al.*, 2010).

16.5.6 Resistance to Oxazolidinones

Currently, linezolid is the only antibiotic approved by the FDA for the treatment of complicated VR *E. faecium* infections, such as endocarditis or bacteremia (Diaz *et al.*, 2012; O'Driscoll and Crank, 2015). Its bacteriostatic activity is manifested after binding to the 23S rRNA subunit and discontinuing protein synthesis at a very initial stage (Bozdogan and Appelbaum, 2004; Brauers *et al.*, 2007; Werner, 2012). Although considered relatively rare, resistance in both *E. faecalis* and *E. faecium* strains has been reported. This is a result of mutations in genes encoding 23S rRNA (typically those that concern domain V) and ribosomal proteins (L3 and L4) or of plasmid-mediated acquisition of the *cfr* gene. The latter encodes a methyltrasnferase responsible for modification of the linezolid-binding region of the 23S rRNA and also contributes to resistance to chloramphenicol, florfenicol, lincosamides and streptogramins A (dalfopristin) (Bozdogan and Appelbaum, 2004; Hollenbeck and Rice, 2012; Bourgeois-Nicolaos *et al.*, 2014; Deshpande *et al.*, 2015).

Remarkably, however, another agent of this class, tedizolid, is active against *cfr*-positive strains (O'Driscoll and Crank, 2015). Target modification through the above pathways is the only mechanism of oxazolidinone resistance in enterococci described so far. Just like in the case of daptomycin, resistance to oxazolidinones is highly associated with the duration of therapy and its emergence can be suppressed by the concurrent use of other drugs, such as doxycycline (Arias *et al.*, 2010; Werner, 2012; O'Driscoll and Crank, 2015). Since linezolid is not bactericidal, some researchers advocate its use in severe VRE infections only as an alternative agent, when the combinations of β-lactams plus aminoglycosides and the daptomycin- or quinupristin/ dalfopristin-based combinational regimens cannot be administered (Arias *et al.*, 2010).

16.5.7 Resistance to Macrolides, Lincosamides, Streptogramins and Pleuromutilins

Macrolides (i.e., erythromycin, clarithromycin, azithromycin) and lincosamides (lincomycin, clindamycin) have long been considered as alternative options against Gram-positive pathogens, after β-lactam antibiotics. Nevertheless, their antibacterial spectrum also includes a number of Gram-negative and intracellular bacteria, for example, *Campylobacter* spp. and *Chlamydia* spp. They are bacteriostatic agents that inhibit protein synthesis by binding to the ribosomal 50S subunit. Their use has been constantly declining during the past several years due to the widespread development of resistance (Portillo *et al.*, 2000; Leclercq, 2002).

Macrolide/lincosamide resistance pathways in *Enterococcus* spp. include target modification, efflux of the antibiotic, and drug inactivation. Target modification involves dimethylation of the 23S rRNA subunit of the 50S ribosomal subunit through action of Erm proteins, encoded by *erm* genes. Four classes of *erm* genes have been identified in pathogenic bacteria, namely *erm*(A), *erm*(B), *erm*(C), and *erm*(F) with a plethora of them being carried in plasmids and transposons. Up to date, *erm*(A) and *erm*(B) class genes have been detected in *E. faecalis* and *E. faecium*. Their expression leads to resistance not only to macrolides/lincosamides, but also to streptogramins B (quinupristin), a trait known as "MLS_B phenotype." When the expression of the aforementioned genes is constitutive, MLS_B phenotype infers high-level cross-resistance to all agents belonging to these classes of antibiotics. However, in the case of inducible resistance due to the presence of an antibiotic-inducer, strains can remain susceptible to non-inducer agents, displaying variable MLS_B phenotypes (Portillo *et al.*, 2000; Schmitz *et al.*, 2000; Leclercq, 2002; Miller *et al.*, 2014).

Target modification is the predominant resistance mechanism for macrolides in both *E. faecalis* and *E. faecium* and for lincosamides in *E. faecium* only. In contrast, *E. faecalis* is intrinsically fully resistant to lincosamides as well as to streptogramins A (dalfopristin) and pleuromutilins, due to the presence and expression of the *lsa* gene that encodes an ATP-Binding Cassette (ABC) homologue active transporter, Lsa(A), and yields the so called "LS_AP phenotype" (this must not be confused with the pattern obtainable by expression of the *cfr* gene, which is observed in both *E. faecalis* and *E. faecium* and does not render resistance to pleuromutilins) (Singh *et al.*, 2002; Isnard *et al.*, 2013; Miller *et al.*, 2014). In *E. faecium*, mutations in another, intrinsic and species-specific *lsa*-like gene, *eat*(A), has also been shown to markedly reduce susceptibility to lincomycin/clindamycin, dalfopristin and tiamulin, a pleuromutilin antibiotic (Isnard *et al.*, 2013). Concerning macrolide efflux, a positive correlation between the presence of *mef* genes (also encoding ABC efflux pumps) and low-level erythromycin resistance has moreover been suggested for enterococci (Luna *et al.*, 1999). Finally, expression of the *lnu*(B) gene, at least in *E. faecium*, is responsible for the synthesis of a lincosamide nucleotidyltransferase, a lincosamide-only inactivation factor (Leclercq, 2002).

Quinupristin/dalfopristin is a mixture of streptogramin B (quinopristin) and A (dalfopristin), at a ratio of 30:70, that also inhibits protein synthesis at the ribosomal level (O'Driscoll and Crank, 2015). It is bacteriostatic against *E. faecium* (French, 2006) but inactive (as aforementioned) against *E. faecalis* due to the intrinsic presence of the *lsa* gene (Singh *et al.*, 2002; Hollenbeck and Rice, 2012). A former FDA official approval for its use against VR *E. faecium* infections has been withdrawn (O'Driscoll and Crank, 2015). Apart from already-mentioned mechanisms (*cfr*, *erm*, and *eat*(A) genes expression or mutations), other *E. faecium* ABC efflux pumps can reduce the intracellular concentrations of streptogramins A and B. These efflux pumps are encoded by the *vga*(D) gene (van Duijkeren *et al.*, 2014) and the indigenous *msrC* gene (Portillo *et al.*, 2000) respectively, with the *vga*(D) gene also conferring resistance to pleuromutilins. Dalfopristin inactivation by bacterial acetyltransferases Vat(D) and Vat(E), and quinupristin molecular modification by lyase VgbA (and possibly VgbB) have also been reported (Soltani *et al.*, 2000; Miller *et al.*, 2014).

Interestingly, specific plasmids isolated in *E. faecium* have been shown to simultaneously carry *erm*, *vat*(D), *vat*(E) and *vgb*(A) genes, thus yielding extended resistance to macrolides, lincosamides, and streptogramins A and B (Hollenbeck and Rice, 2012). Off-label use of quinupristin/ dalfopristin against *E. faecium*-related infections is reserved for cases after clinical failure with daptomycin. Combination with other agents (e.g., doxycycline plus rifampicin, high dose ampicillin, imipenem, levofloxacin or gentamicin) should always be preferred (Arias *et al.*, 2010; Hollenbeck and Rice, 2012; O'Driscoll and Crank, 2015).

Use of pleuromutilin antibiotics has been associated mostly with veterinary medicine (tiamulin, valnemulin) and less with treatment of infections in humans (retapamulin). Despite enterococci not being considered target bacteria, resistance has been testified and is generally linked to expression of plasmid/transposon-borne genes (*cfr*, *vga*(D) and *lsa*) or perhaps to mutations in chromosomal genes, such as *rplC* (which encodes ribosomal protein L3) and those encoding 23S rRNA (van Duijkeren *et al.*, 2014).

16.5.8 Resistance to Tetracyclines and Glycylcyclines

Classic tetracyclines (i.e., tetracycline, oxytetracycline, doxycycline, minocycline) are bacteriostatic drugs that exert their activity by binding to the 30S small ribosomal subunit of the susceptible bacteria. Thus, they inhibit the binding of aminoacyl-tRNA to the mRNA-ribosome complex and subsequently, the protein synthesis (Roberts, 1996). Individual agents of this class, such as doxycycline and minocycline, have been sporadically, though successfully, used in the treatment of severe infections caused by strains of *Enterococcus* spp. and tested in pharmacodynamic *in vitro* models, especially in combination with quinupristin/dalfopristin plus rifampicin, linezolid or chloramphenicol (Arias *et al.*, 2010).

Resistance in clinical enterococcal strains is due to both over-expression of efflux pumps and alterations in the ribosomal conformation (Huys *et al.*, 2004; Cauwerts *et al.*, 2007; Miller *et al.*, 2014). The genes responsible are namely *tet*(K), *tet*(L), *tet*(M), *tet*(O), and *tet*(S). Another identified gene, *tet*(U), is no longer considered to be a resistant determinant, but rather a mis-annotated 3' end of a *rep* gene (Caryl *et al.*, 2012). Efflux pumps are encoded by the usually plasmid-carried *tet*(K) and *tet*(L) genes. Their over-expression leads to prime resistance to tetracycline and doxycycline, but not to minocycline (Huys *et al.*, 2004; Schwartz *et al.*, 2009). The triplet of *tet*(M), *tet*(O), and *tet*(S) genes encode proteins that indirectly alter the three-dimensional structure of the ribosome, inhibiting the binding of new drug molecules and yielding resistance also to minocycline (Huys *et al.*, 2004; Miller *et al.*, 2014).

The most frequently encountered tetracycline-resistance gene, *tet*(M) is commonly part of the bacterial core genome but can be carried on Tn*916* or Tn*1545* transposons, in the latter case often in contiguity with the macrolide-resistance gene *erm*(B) and the kanamycin-resistance gene *aph*A-3 (Mikalsen *et al.*, 2015); hence the significant percentage of tetracycline-resistant strains also exhibiting the MLS_B phenotype (Cauwerts *et al.*, 2007).

Tigecycline is a new-generation, glycylcycline antibiotic with a mechanism of action similar to that of tetracyclines and activity against VRE. Although there have been scarce reports on resistance in *E. faecalis* strains, the exact determinants remain unknown (Werner, 2012). Monotherapy of enterococcal infections with tigecycline is discouraged due to lack of sufficient clinical data. However, the absence of reported substantial resistance to this agent makes it a potentially appealing addition to combination regimens (Arias *et al.*, 2010).

16.5.9 Resistance to Rifamycins

Rifampicin (also known as rifampin, especially in the United States) disrupts DNA-dependent RNA synthesis in susceptible bacteria by inhibition of RNA polymerase (Brandis *et al.*, 2015). A major drawback of rifampicin use is the rapid advent of resistance after the onset of treatment (Enne *et al.*, 2004). The mechanism involves mainly mutations in the *rpoB* gene and the subsequent reduction in the binding affinity of rifampicin, in both *E. faecalis* and *E. faecium* (Kristich and Little, 2012; Brandis *et al.*, 2015).

Despite the fact that rifampicin in not regularly used in the treatment of infections caused by enterococci, there have been reports on respectable clinical cure rates, especially when administered in combination with agents from different classes, for example, quinupristin/dalfopristin plus doxycycline for the treatment of *E. faecium* endocarditis or daptomycin plus gentamicin (Arias *et al.*, 2010). Results from *in vitro* studies have shown that daptomycin, along with lipoglycopeptides (telavancin) enhance rifampicin uptake by the bacterial cell in *E. faecium* and this could explain their synergy in activity (Pankey and Ashcraft, 2013).

16.5.10 Resistance to Sulfonamides/Trimethoprim

Although both sulfonamides and trimethoprim are bacteriostatic agents, their combination is known to be synergistic and bactericidal (Batzias *et al.*, 2005), as they act through inhibition of sequential steps in tetrahydrofolic acid biosynthesis beginning with *p*-aminobenzoic acid reaction with pteridine; sulfonamides inhibit dihydropteroate reductase and trimethoprim inhibits dihydrofolate reductase (Wormser *et al.*, 1982). Since the vast majority of bacteria cannot utilize exogenous folic acid, intracellular synthesis is necessary for the maintenance of their normal metabolism. Enterococci, remarkably, are able to absorb folic acid existing in the extracellular space and thus, naturally circumvent sulfonamides/trimethoprim activity in vivo (Zervos and Schaberg, 1985).

16.6 *Enterococcus* Infection

16.6.1 Infection in Humans

16.6.1.1 Bacteremia

Bacteremia is the third most common presentation of enterococcal disease, especially in seriously ill patients, who receive antibiotics (Sood *et al.*, 2008; O'Driscoll and Crank, 2015). Enterococci globally account for approximately 10% of all cases of bacteremia (Rajkumani *et al.*, 2015), since the proportion of infections caused by enterococci has increased over the last years (Nielsen, 2015). They have been reported as the most common primary cause of intra-abdominal infections, endovascular sources, and urinary tract infections (Hamada *et al.*, 2015), while older age, male gender and severe underlying diseases have been associated with the acquisition of bloodstream infections (BSIs) (Billington *et al.*, 2014). VRE bacteremia cases, in particular, are caused by *E. faecium* (Rosa *et al.*, 2014), which has higher rates of antibiotic resistance, and has been associated with higher mortality than *E. faecalis* (Chou *et al.*, 2008; Billington *et al.*, 2014). Although nosocomial enterococcal bacteremias are commonly polymicrobial, 80 to 90% of the cases of VRE bacteremia have been reported as monomicrobial (Cetinkaya *et al.*, 2000).

Additionally, many cases of VRE bacteremia may be transient and clinically irrelevant in patients who have undergone invasive procedures (Mutters *et al.*, 2013). Colonized cancer patients, diabetes mellitus, patients who have undergone a gastrointestinal procedure, acute renal failure, and exposure to vancomycin have been major risk factors for developing VRE BSI (Kara *et al.*, 2015). A significant incidence of VRE *E. faecium* bacteremia has been reported among patients with solid and hematologic malignancies, while it compromises mostly intensive care unit (ICU) patients (Rosa *et al.*, 2014). There is also evidence that enterococci are a significant cause of bacteremia in patients with HIV infection (Aamodt *et al.*, 2015). Mortality rates in patients with VRE BSIs range from 20 to 46% (Balli *et al.*, 2010), although they may reach values higher than 60% for VRE *E. faecium* infections (Rosa *et al.*, 2014) and they are usually associated with severe pre-existing illness (McBride *et al.*, 2010).

Patients with BSI due to VRE are 2.5 times more likely to die than patients with BSI due to vancomycin-susceptible strains (Balli *et al.*, 2010). However, some reports suggest that VRE bacteremia may serve more as a marker of severity of illness than as an independent risk factor for mortality (Orsi and Ciorba, 2013). What is more, mortality in bacteremia with ampicillin-resistant strains is higher than in those patients with ampicillin-sensitive infections (Hamada *et al.*, 2015).

Since enterococci are among the first bacteria to colonize the neonatal gastrointestinal tract, either through oral ingestion, breast milk, or through the mother's vaginal and gastrointestinal flora during delivery (Shantala *et al.*, 2014), they have been reported as rare cause of early onset neonatal sepsis, associated with prematurity, severe underlying diseases, such as necrotizing enterocolitis, prolonged hospital stay, exposure to antibiotics, nasogastric tubes, and multiple intravascular devices (Murray, 1990; Butler, 2006).

16.6.1.2 Endocarditis

Infective endocarditis is a systemic and life-threating disease with high morbidity and mortality rates ranging from 19 to 38% (Damasco *et al.*, 2014). Enterococci are responsible for 5 to 20% of the cases of bacterial endocarditis, they represent the third cause of both native and prosthetic valve infectious endocarditis, following staphylococcal and streptococcal species (McDonald *et al.*, 2005; Bassetti *et al.*, 2013). Additionally, due to the rise in hospital-associated enteroccocal infections, the frequency of the infection caused by these microorganisms is actually increasing (Nigo *et al.*, 2014). Furthermore, enterococcal bacteremia results in endocarditis in 8 to 32% of the cases (McDonald *et al.*, 2005). *E. faecalis* remains the most frequent cause of the infection (Murray, 1990), associated with central venous lines, liver transplantation, and mitral valve infections (O'Driscoll and Crank, 2015), whereas its incidence in literature varies from 17.3 to 47% (Francischetto *et al*, 2014).

Enterococcal endocarditis usually occurs in older patients (Falcone *et al.*, 2015), combined with more comorbidities than the other types of infective endocarditis (Pericas *et al.*, 2015), more often in men than in women (McDonald *et al.*, 2005), occasionally in children and rarely in infants (Murray, 1990), whereas left-sided involvement is the most frequent (McDonald *et al.*, 2005). Common risk factors include genitourinary and biliary portals (Murray, 1990). Affected patients are hospitalized, immunocompromised, those who attend a dialysis center or have other comorbidities (Butler, 2006).

The most common symptoms are the presence of murmur, fever, weight loss, malaise, and generalized aches (O'Driscoll and Crank, 2015). Clinical features, such as heart failure or systemic embolization, have been reported in 46% and 26% of the patients respectively (McDonald *et al.*, 2005). Treatment options are limited due to the emergence of high-level aminoglycosides resistance, vancomycin resistance, and multidrug resistance, especially for *E. faecium* isolates (Nigo *et al.*, 2014; Pericas *et al.*, 2015). Despite major medical advances and new potent antimicrobial agents, the prevalence and mortality have not improved over the last years (Nigo *et al.*, 2014).

16.6.1.3 Intra-abdominal, Pelvic, and Soft-Tissue Infections

The second most frequent enterococcal infections that have been reported are intra-abdominal and pelvic infections or post-surgery wound infections (Sood *et al.*, 2008). Enterococci are isolated regularly along with other enteric bacteria from such infections, including Gram-negative and anaerobic microorganisms, since they are commensals of the gastrointestinal tract and usually act synergistically to their production (Butler, 2006; O'Driscoll and Crank, 2015). They are usually isolated from intra-abdominal abscesses, especially from immunocompromised patients having significant medical co-morbidities (Len *et al.*, 2014).

However, they are responsible for causing monomicrobial peritonitis infections, which occur most often in patients with liver cirrhosis or in patients who undergo chronic peritoneal dialysis (Bibi *et al.*, 2015; O'Driscoll and Crank, 2015). Additionally, they are frequently isolated from cases of postoperative peritonitis, ranging from 17 to 32% of the cases, whereas these patients usually have significantly more surgical complications (Seguin *et al.*, 2012). Moreover, enterococci have also caused acute salpingitis, peripartum maternal infection, such as endometritis with bacteremia (Murray, 1990).

Enterococcal wound and soft-tissue infections are also rising steadily and they are associated with previous antimicrobial therapy, multiple surgical interventions and longer hospitalization (Rajkuman *et al.*, 2014), but there has also been an increasing rate of VRE skin and skin-structure infections during the last decade (Sader *et al.*, 2010). Vertebral osteomyelitis has been reported as a rare infection caused by *Enterococcus spp*, as a result of a hematogenous spread from an infection of the urinary tract (Kow and Ferzandi, 2013).

16.6.1.4 Urinary Tract Infections

Urinary tract infections (UTIs) are some of the most frequently occurring nosocomial infections (Shigemura *et al.*, 2005), and the most common type of clinical disease, caused by enterococci, both within and outside hospital settings (Barros *et al.*, 2009). More often, they are hospital-acquired, associated with urinary tract abnormalities, urinary catheters or prolonged antibiotic treatment (Bitsori *et al.*, 2005), although over the last years there is an increasing isolation rate of enterococci in outpatients as well (Shigemura *et al.*, 2005). Acute uncomplicated cystitis, pyelonephritis (Ronald, 2002), prostatitis, and perinephric abscess have been reported (Murray, 1990). The continuous use of antibiotics as a treatment or prophylaxis induces enterococcal colonization, therefore, enterococcal UTIs are considered to be the consequence of this therapeutic regimen (Bitsori *et al.*, 2005). Enterococci are usually ranked among the top five pathogens for UTIs and the resistance patterns of these organisms can vary significantly among hospitals, countries, and continents (Shigemura *et al.*, 2005).

The increasing rates of VRE (Heintz *et al.*, 2010), that are mostly responsible for colonization of the urinary tract or asymptomatic bacteriuria (Bitsori *et al.*, 2005), are of particular concern since they are associated with increased mortality, health cost, and limited treatment options (Heintz *et al.*, 2010). Moreover, the unnecessary use of antibiotics in patients who are simply colonized with VRE in the urine has become a big problem (Heintz *et al.*, 2010). What is more, a remarkable characteristic of the past 20 years has been the increasing isolation rate of *E. faecalis* (Shigemura *et al.*, 2005). Enterococcal UTIs are more frequent at the ages before 10 and after 60 years, when urinary malformations are more frequent (Barros *et al.*, 2009).

Indicatively in children, the nosocomial cases range from 12 to 15%, whereas the community-acquired episodes range from 4.4 to 5.6% (Bitsori *et al.*, 2005). *Enterococcus* has also been identified as the most frequent uropathogen among elderly male patients and outpatients, related mainly to urinary catheters (Barros *et al.*, 2009). Intensive care units (ICUs) exhibit the highest prevalence of nosocomial infections with an estimated rate of 8-21% for nosocomial UTIs (Bagshaw and Laupland, 2006). Several nosocomial surveillance studies conducted in American and European ICUs have reported enterococci as the third predominant bacterial pathogen of these infections (Bagshaw and Laupland, 2006).

Enterococci appear to be important uropathogens also among patients with diabetes (Ronald, 2002), as well as in UTIs that occur early after kidney transplantation, whereas ampicillin–resistant *E. faecium* has been isolated from renal allograft recipients with recurrent UTIs (Mitra and Alangaden, 2011). Enterococcal UTIs are an important source of bacteremia in adults (Bitsori *et al.*, 2005).

16.6.1.5 Uncommon Infections

Enterococcal meningitis is uncommon, ranging from 0.3 to 4% of the total bacterial cases. Additionally, central nervous system (CNS) infections are rarely caused by VRE (Wang *et al.*, 2014). VR *E. faecium* is a more frequent cause than VR *E. faecalis*, representing 82% and 5% of the infections, respectively (O'Driscoll and Crank, 2015). Patients with VRE CNS infection tend to be older, to have serious underlying diseases, such as CNS non-malignant diseases, hematologic malignancies (Wang *et al.*, 2014), pulmonary and cardiac diseases (O'Driscoll and Crank, 2015). The symptoms include fever, altered mental status, headache, meningeal signs, whereas patients with spontaneous meningitis exhibit higher frequency of bacteremia than patients with postoperative infection (Pintado *et al.*, 2003). Limited therapeutic options are available for the treatment of VR *E. faecium* (Knoll *et al.*, 2013). The mortality rate has been reported from 13 to 33% (Zeana *et al.*, 2001). Enterococci have also been associated with skin and skin structure infections as a part of polymicrobial infections (O'Driscoll and Crank, 2015). Moreover, prosthetic join infections associated to arthroplasty due to enterococci have been reported in patients, who had at least some comorbidity and these infections were usually polymicrobial (Tornero *et al.*, 2014).

Conclusively, although enterococci have low intrinsic virulence, they are capable of causing severe infections, especially hospital-acquired. Moreover, they have emerged as one of the most frequent causes of nosocomial infections in developed countries. Infections are usually present in immunocomprised patients, including variable clinical manifestations. Over the last years, the rate of these infections has significantly risen worldwide and the treatment is of great concern due to the multidrug resistant strains, especially for *E. faecium* multi-resistant isolates.

16.6.2 Infection in Animals

Enterococci in animals are opportunistic pathogens with the infections attributed to a large extent to poor hygienic conditions. Enterococci have been isolated from various infections in animals including bovine mastitis. The most implicated enterococci are *E. faecalis* and *E. faecium*, with *E. durans*, *E. avium*, *E. hirae*, and *E. pseudoavium* also been reported to cause disease in animals (Simjee *et al.*, 2006). In poultry enterococcosis has been reported in various avian species worldwide, with the disease occurring in an acute or subacute/chronic form. It is caused mainly by *E. avium*, *E. durans*, *E. faecalis*, *E. faecium*, *E. hirae*, and *E. faecalis* with the clinical signs including depression, lethargy, ruffled feathers, diarrhea, decrease in egg production, and head tremors. If untreated, most affected birds die (Morishita, 2015). In fish, a clinical syndrome exhibited by loss of appetite, erratic swimming, darkening of body color, eye lesions, external haemorrhagic lesions, and ulcerations have been attributed to *Enterococcus* spp. and *Streptococcus* spp. with the clinical signs of this syndrome varying considerably among species of affected fish (Athanassopoulou and Roberts, 2004).

16.7 *Enterococcus* Epidemiology

16.7.1 Prevalence in Nosocomial Settings

Enterococci are normal inhabitants of the intestinal tract of individuals and animals and of insects and nematodes (Arias and Murray, 2013), but they can also be found on the skin, in the genitourinary tract and in the oral cavity (O'Driscoll and Crank, 2015). Enterococci were implicated in causing infections in individuals about a century ago, when they were isolated from patients with endocarditis and urinary tract infections (UTIs) (Murray, 1990). Over the last years, they have emerged as a major cause of nosocomial and community-acquired infections (Hamada *et al.*, 2015). They were firstly reported as common causes of hospital-acquired infections in the mid-late 1970s, probably related to the increasing use of third-generation cephalosporins, to which enterococci are naturally resistant (Cetinkaya *et al.*, 2000).

Enterococcal infections may be caused by at least 12 species, including *E. avium*, *E. casseliflavus*, *E. durans*, *E. faecalis*, *E. faecium*, *E. gallinarum*, *E. hirae*, *E. malodoratus*, *E. mundtii*, *E. pseudoavium*, *E. raffinosus*, and *E. solitarius* (Sood *et al.*, 2008). However, the most important causes of enterococcal infections in individuals are due to *E. faecalis* and *E. faecium*, which account for up to 80% and 20% of the cases respectively (Orsi and Ciorba, 2013). In the last few years, enterococcal infections have become frequent occurrences in hospital settings, with increasingly common isolates that are resistant to multiple antibiotics (Barros *et al.*, 2009). Moreover, enterococci demonstrate significant resistance to many antimicrobials such as ampicillin, aminoglycosides, cephalosporins, cotrimoxazole, clindamycin and glycopeptides, including vancomycin (Bitsori *et al.*, 2005; Sood *et al.*, 2008; Orsi and Ciorba, 2013). *E. faecalis* is most likely to be susceptible to ampicillin and vancomycin, whereas *E. faecium* has a higher incidence of resistance to multiple antibiotics (Hamada *et al.*, 2015).

Vancomycin resistance was first described in 1988 and VRE have emerged as an increasing common clinical problem worldwide (McBride *et al.*, 2010), especially in developed countries, such as U.S. and European countries (Rajkumani *et al.*, 2015). Their treatment poses a challenge because of their multidrug resistance (Cardenas *et al.*, 2014). Most of the VRE isolates are *E. faecium*, since the resistance to vancomycin is present in about 80% of the *E. faecium* isolates (Falcone *et al.*, 2015). Germany, Greece, England, Ireland, and Portugal have high VRE rates of >10%, whereas in most European countries an increase in ampicillin-resistant enterococci has been observed since the year 2000 (Sidler *et al.*, 2014).

Hospital outbreaks of infection or colonization have been reported with both VanA and VanB isolates (Cetinkaya *et al.*, 2000), although strains with *VanA* resistance predominate globally, unlike *VanB* strains that may be locally prevalent (Woodford and Livermore, 2009). VRE epidemiology differs worldwide, since the reservoirs in the United States are usually present in healthcare units, whereas in Europe healthy populations are the main carriers, due to the use of avoparcin in the agricultural industry during the 1980s (Orsi and Ciorba, 2013), which confers cross-resistance to vancomycin (Sood *et al.*, 2008). The transmission can occur through direct contact with colonized or infected patients or indirectly through healthcare workers, contaminated medical equipment, or environmental surfaces (Orsi and Ciobra, 2013). Risk factors for developing a nosocomial VRE infection also include gastrointestinal colonization, severe underlying diseases, such as diabetes, renal failure or hemodialysis, a long period of hospitalization, especially in surgical or intensive-care units, solid organ and hematopoietic stem cell transplantation, the presence of urinary catheters, and multiple courses of antimicrobials (Arias and Murray, 2013; Rosa *et al.*, 2014).

16.7.2 Foodborne Outbreaks

The first report of foodborne disease caused by enterococci dates back to 1926 where two outbreaks of gastroenteritis were attributed to cheese in which enterococci were abundant. In addition, no other pathogen like *S. aureus* or *Salmonella* was found (Oprea and Zervos, 2007). The symptoms observed were consistent with the clinical presentation of biogenic amine intoxication, an observation that is in accordance with the potency of enterococci to produce biogenic amines in food matrices (Bover-Cid *et al.*, 2001; Economou *et al.*, 2007). Efforts to prove that enterococci can cause disease in healthy volunteers have produced contradictory results (Oprea and Zervos, 2007).

16.7.3 Prevalence in Agriculture and Farm Animals: The Missing Link?

The first report of vancomycin-resistant enterococci was made by Bates *et al.* (1993) who have isolated VRE from farm animals in the United Kingdom. Since vancomycin was never used in animal therapy in the United Kingdom, cross-resistance to avoparcin, a glycopeptide used as a feed additive since the 1970s in Europe was blamed. Since then several sources of vancomycin resistant enterococci have been described, including dogs, cats, horses, wild birds, foxes, wood frogs, ostriches, pigs, pork, broilers, poultry meat, environmental samples, and sewage. *E. faecium* was the most commonly isolated VRE species with only occasional isolation of VR *E. faecalis*. Over the years there is a steady persistence of a low percentage of *E. faecium* exhibiting resistance to glycopeptides in poultry, cattle, and pig.

In 1997, Woodford *et al.* have reported the isolation of quinupristin/dalfopristin resistant *E. faecium* from humans in the UK, in which quinupristin/dalfopristin and other streptogramins were not licensed at the time for human therapy. Resistance to this antibiotic combination was linked to the use of virginiamycin in animals. Streptogramin A resistance encoding genes [*vat*(D), *vat*(E), *vga*(D), and *vat*(H)] were isolated from animals and sewage in most parts of the world (Hammerum, 2012).

The observations of cross-resistance in *Enterococcus* spp. among others, have led authorities to discourage or ban antibiotic usage in animals, unless utterly necessary. Avoparcin use was banned in almost everywhere with only Australia permitting its use for therapeutic reasons. In 2006, the European Union has banned the use of antimicrobials for growth promotion. Still the rates of resistance to vancomycin and quinupristin/dalfopristin of enterococci isolated from animals remain high, although the antibiotics responsible were not in contact any more with commensal enterococci in animals. This phenomenon can be explained by nonspecific selection of resistant strains by other factors like other antimicrobials, metals, disinfectants, or bacterial interaction or fitness (Hammerum, 2012; Garcia-Migura *et al.*, 2014).

Since the genetic background of environmental fitness, including antimicrobial resistance, can be found in plasmids, various environmental factors can exert pressure toward the conservation of antibiotic resistance genes. Two such examples can be shown by the use of tetracycline, tylosin, and copper sulphate in animals, which can be related to persistence of vancomycin resistance (Hammerum, 2012). The gene *erm(b)* encoding resistance to tylosin is located in the same plasmid with *vanA* on *E. faecium* (Aaerstrup, 2000b). Likewise the copper resistance gene *tcrB* was located in a plasmid bearing *vanA* gene (Hasman and Aarestrup, 2005). Therefore the use of substances such as copper sulphate can select toward the persistence of vancomycin genes.

When animal and human *Enterococcus* isolates were compared by pulsed field gel electrophoresis, similar and highly similar pulsotypes were found in vancomycin resistant animal and human isolates (Van Den Bogaard *et al.*, 1997; Hammerum *et al.*, 2004). In later studies, amplified fragment length polymorphism analysis and multilocus sequence typing were used, revealing that the *E. faecium* strains responsible for disease in humans could be grouped in a clonal complex (CC17), whereas the animal strains belonged to other groups.

In contrast, dogs have been shown to possibly be a reservoir to human strains from clinical infections or hospital outbreaks (Damborg *et al.*, 2008; Ghosh *et al.*, 2011). Still there is evidence that enterococci belonging to CC17 can also be found in animals, since CC17 *E. faecium* strains have been isolated from pig, chicken meat, rabbit meat, and veal (Lopez *et al.*, 2009; Freitas *et al.*, 2011). Still the clonal complexes of *E. faecium* prevailing in animals are others than CC17. For *E. faecalis*, similar observations have been made, since outbreak or hospital acquired *E. faecalis* isolates have been categorized in clonal complexes (CC2, CC9, and CC87) similar to those of animals. Pigs have been shown to harbor *E. faecalis* belonging to CC2 (ST6), whereas *E. faecalis* strains belonging to ST16, ST116, ST40, and ST97 have been isolated from both animals (pigs, poultry), animal products (pork and turkey meat), and healthy and hospitalized humans (Ruiz-Garbajosa *et al.*, 2006; Kawalec *et al.*, 2007; Agersø *et al.*, 2008; Freitas *et al.*, 2009; Solheim *et al.*, 2009; Larsen *et al.*, 2010, 2011).

Enterococci of animal origin can colonize the human intestine, as reported by Berchieri (1999) and Sørensen *et al.* (2001). The close proximity of resistant enterococci of animal origin with commensal enterococci can result in transfer of the genetic elements of antimicrobial resistance. In a hospital setting, animal VRE strains can be acquired prior to hospitalization or via an animal food product, and given the opportunity, transfer resistance to the possibly present hospital CC17 *E. faecium* strains (Hammerum, 2012). This can be enhanced by the use of antibiotics. It is clear that though *E. facium* of animal origin usually are not a danger by themselves, they can provide the genetic pool of resistance that can be transferred to human pathogenic strains (Hammerum, 2012). On the contrary the animal-derived *E. faecalis* strains seem to be quite similar to human pathogenic strains and therefore might pose a possible human health hazard (Larsen *et al.*, 2010, 2011).

References

Aamodt, H., et al. (2015). Genetic relatedness and risk factor analysis of ampicillin-resistant and high-level gentamicin-resistant enterococci causing bloodstream infections in Tanzanian children. *BMC Infectious Diseases, 15*, p. 107.

Aarestrup, F.M., et al. (2000a). Comparison of antimicrobial resistance phenotypes and resistance genes in *Enterococcus faecalis* and *Enterococcus faecium* from humans in the community, broilers, and pigs in Denmark. *Diagnostic Microbiology and Infectious Disease, 37*, pp. 127–137.

Aarestrup, F.M., et al. (2000b). Associations between the use of antimicrobial agents for growth promotion and the occurrence of resistance among *Enterococcus faecium* from broilers and pigs in Denmark, Finland, and Norway. *Microbial Drug Resistance, 6*, pp. 63–70.

Achemchem, F., et al. (2005). *Enterococcus faecium* F58, a bacteriocinogenic strain naturally occurring in Jben, a soft, farmhouse goat's cheese made in Morocco. *Journal of Applied Microbiology, 99*, pp. 141–150.

Agerholm-Larsen, L., et al. (2000). The effect of a probiotic milk product on plasma cholesterol: a meta-analysis of short-term intervention studies. *European Journal of Clinical Nutrition, 54*(11), pp. 856–60.

Agersø, Y., et al. (2008). Vancomycin-resistant *Enterococcus faecalis* isolates from a Danish patient and two healthy human volunteers are possibly related to isolates from imported turkey meat. *Journal of Antimicrobial Chemotherapy, 62*, pp. 844–845.

Ahmadova, A., et al. (2013). Evaluation of antimicrobial activity, probiotic properties and safety of wild strain *Enterococcus faecium* AQ71 isolated from azerbaijani motal cheese. *Food Control, 30*(2), pp. 631–41.

Al-Nassir, W.N., et al. (2008). Both oral metronidazole and oral vancomycin promote persistent overgrowth of vancomycin-resistant enterococci during treatment of *Clostridium difficile*-associated disease. *Antimicrobial Agents and Chemotherapy, 52*(7), pp. 2403–2406.

Araújo, T.F. and Ferreira, C.L.L.F. (2013). The genus *Enterococcus* as probiotic: safety concerns. *Brazilian Archives of Biology and Technology, 56*(3), pp. 457–466.

Arias, C. and Murray, B. (2013). The rise of the *Enterococcus:* beyond vancomycin resistance. *Nature Reviews Microbiology, 10*(4), pp. 266–278.

Arias, C.A., Contreras, G.A. and Murray, B.E. (2010). Management of multidrug-resistant enterococcal infections. *Clinical Microbiology and Infection, 16*(6), pp. 555–562.

Arthur, M. and Courvalin, P. (1993). Genetics and mechanisms in glycopeptides resistance in Enterococci. *Antimicrobial Agents and Chemotherapy, 37*(8), pp. 1563–1571.

Asehraou, A., Faid, M. and Jana M. (1992). Physico-chemical properties and the microflora of Moroccan black table olives. *Grasas y Aceites, 43*, pp. 130–133.

Aslam, M., et al. (2012). Characterization of antimicrobial resistance and virulence genes in *Enterococcus* spp. isolated from retail meats in Alberta, Canada. *International Journal of Food Microbiology, 156*(3), pp. 222–230.

Athanassopoulou, F. and Roberts, R.J. (2004). Streptococcal infections of farmed fish. *Journal of the Hellenic Veterinary Medical Society, 55*(2), pp. 136–144.

Aymerich, T., et al. (1996). Biochemical and genetic characterization of enterocin A from *Enterococcus faecium*, a new antilisterial bacteriocin in the pediocin family of bacteriocins. *Applied and Environmental Microbiology, 62*, pp. 1676–1682.

Bagshaw, S. and Laupland, K. (2006). Epidemiology of intensive care unit-acquired urinary tract infections. *Current Opinions in Infectious Diseases, 19*, pp. 67–71.

Balli, E., Venetis, C. and Miyakis, S. (2014). Systematic review and meta-analysis of linezolid versus daptomycin for treatment of vancomycin-resistant enterococcal bacteremia. *Antimicrobial Agents and Chemotherapy, 58*(2), pp. 734–739.

Barros, M., Martinelli, R. and Rocha, H. (2009). Enterococcal urinary tract infections in a University Hospital: Clinical studies. *Brazilian Journal of Infectious Diseases, 13*(4), pp. 294–296.

Bassetti, M., et al. (2013). Gemtamicin once-daily in enterococcal endocarditis. *International Journal of Cardiology, 168*(5), pp. 5033–5034.

Bates, J., Jordens, Z. and Selkon, J.B. (1993). Evidence for an animal origin of vancomycin-resistant enterococci. *Lancet, 342*, pp. 490–491.

Batzias, G.C., Delis, G.A. and Koutsoviti-Papadopoulou, M. (2005). Bioavailability and pharmacokinetics of sulphadiazine, N^4-acetylsulphadiazine and trimethoprim following intravenous and intramuscular administration of a sulphadiazine/ trimethoprim combination in sheep. *Veterinary Research Communications, 29*, pp. 699–712.

Bellomo, G., et al. (1980). A controlled double-blind study of SF68 strain as a new biological preparation for the treatment of diarrhea in pediatrics. *Current Therapeutic Research, 28*, pp. 927–934.

Berchieri, A. (1999). Intestinal colonization of a human subject by vancomycin-resistant *Enterococcus faecium. Clinical Microbiology and Infection, 5*, pp. 97–100.

Bibi, S., et al. (2015). Clinical, laboratory and bacteria profile of spontaneous bacterial peritonitis in chronic liver disease patients. *Journal of the College of Physicians and Surgeons Pakistan, 25*(2), pp. 95–99.

Biedenbach, D.J., et al. (2015). In vitro activity of oritavancin and comparator agents against staphylococci, streptococci, and enterococci from clinical infections in Europe and North America, 2011-2014. *International Journal of Antimicrobial Agents, 46*(6), pp. 674–81.

Billington, E.O., et al. (2014). Incidence, risk factors, and outcomes for *Enterococcus spp. Blood stream infections: A population-based study. International Journal of Infectious Disease, 26*, pp. 76–82.

Birollo, G.A., Reinheimer, J.A. and Vinderola, C.G. (2001). Enterococci vs. non-lactic acid microflora as hygiene indicators for sweetened yoghurt. *Food Microbiology, 18*, pp. 597–604.

Bitsori, M., et al. (2005). Community-acquired enterococcal urinary tract infections. *Pediatric Nephrology, 20*, pp. 1583–1586.

Bizzini, A., et al. (2009). The *Enterococcus faecalis* superoxide dismutase is essential for its tolerance to vancomycin and penicillin. *Journal of Antimicrobial Chemotherapy, 64*(6), pp. 1196–1202.

Borgia, M., et al. (1982). A controlled clinical study on *Streptococcus faecium* preparation for the prevention of side reactions during long-term antibiotic treatments. *Current Therapeutic Research, 31*, pp. 265–271.

Bourgeois-Nicolaos, N., et al. (2014). The emergence of linezolid resistance among enterococci in intestinal microbiota of treated patients is unrelated to individual pharmacokinetic characteristics. *Antimicrobial Agents and Chemotherapy, 58*(5), pp. 2681–2687.

Bouton, Y., Guyot, P. and Grappin, P. (1998). Preliminary characterization of microflora of Comté cheese. *Journal of Applied Microbiology, 85*, pp. 123–131.

Bover-Cid, S., et al. (2001). Amino acid-decarboxylase activity of bacteria isolated from fermented pork sausages. *International Journal of Food Microbiology, 66*(3), pp. 185–189.

Bozdogan, B. and Appelbaum, P.C. (2004). Oxazolidinones: activity, mode of action, and mechanism of resistance. *International Journal of Antimicrobial Agents, 23*(2), pp. 113–119.

Brandis, G., et al. (2015). Comprehensive phenotypic characterization of rifampicin resistance mutations in *Salmonella* provides insight into the evolution or resistance in *Mycobacterium tuberculosis. Journal of Antimicrobial Chemotherapy, 70*, pp. 680–685.

Brauers, J., et al. (2007). Bactericidal activity of daptomycin, vancomycin, teicoplanin and linezolid against *Staphylococcus aureus, Enterococcus faecalis* and *Enterococcus faecium* using human peak free serum drug concentrations. *International Journal of Antimicrobial Agents, 29*, pp. 322–325.

Broom, L.J., et al. (2006). Effects of zinc oxide and *Enterococcus faecium* SF68 dietary supplementation on the performance, intestinal microbiota and immune status of weaned piglets. *Research in Veterinary Science, 80*, pp. 45–54.

Burdychová, R. (2009). Study of enterococci during cheese manufacture and ripening and evaluation of their role in tyramine production. *Acta Universitatis Agriculturae et Silviculturae Mendelianae Brunensis, 57*(5), pp. 49–56.

Busani, L., et al. (2004). Antimicrobial susceptibility of vancomycin-susceptible and -resistant enterococci isolated in Italy from raw meat products, farm animals, and human infections. *International Journal of Food Microbiology, 97*(1), pp. 17–22.

Butler, K. (2006). Enterococcal infection in children. *Seminars in Pediatric Infectious Diseases, 17,* pp. 128–139.

Campos, C.A., et al. (2006). Preliminary characterization of bacteriocins from *Lactococcus lactis, Enterococcus faecium* and *Enterococcus mundtii* strains isolated from turbot (*Psetta maxima*). *Food Research International, 39,* pp. 356–364.

Canganella, F., et al. (1996). Microbiological investigation on probiotic preparations used for animal feeding. *Microbiological Research, 151,* pp. 167–175.

Capcarova, M., et al. (2010). The effect of selected microbial strains on internal milieu of broiler chickens after preoral administration. *Research in Veterinary Science, 91*(1), pp. 132–137.

Capita, R. and Alonso-Calleja, C. (2013). Antibiotic-resistant bacteria: a challenge for the food industry. *Critical Reviews in Food Science and Nutrition, 53*(1), pp. 11–48.

Cardenas, A.M., Andreacchio, K. and Edelstein, P. (2014). Prevalence and detection of mixed-population enterococcal bacteremia. *Journal of Clinical Microbiology, 52*(7), pp. 2604–2608.

Caryl, J.A., et al. (2012). "*tet*(U)" is not a tetracycline resistance determinant. *Antimicrobial Agents and Chemotherapy, 56*(6), pp. 3378–3379.

Casaus, P., (1997). Enterocin B, a new bacteriocin from *Enterococcus faecium* T136 which can act synergistically with enterocin A. *Microbiology, 143,* pp. 2287–2294.

Cauwerts, K., et al. (2007). High prevalence of tetracycline resistance in *Enterococcus* isolates from broilers carrying the *erm(B)* gene. *Avian Pathology, 36*(5), pp. 395–399.

Centeno, J.A., et al. (1999). Effects of the addition of *Enterococcus faecalis* in Cebreiro cheese manufacture. *International Journal of Food Microbiology, 48,* pp. 7–111.

Cetinkaya, Y., Falk, P. and Mayhall, G. (2000). Vancomycin-resistant enterococci. *Clinical Microbiology Reviews, 13*(4), pp. 686–707.

Chenoweth, C. and Schaberg, D. (1990). The epidemiology of enterococci. *European Journal of Clinical Microbiology and Infectious Diseases, 9*(2), pp. 80–89.

Chingwaru, W., Mpuchane, S.F. and Gashe, B.A. (2003). *Enterococcus faecalis* and *Enterococcus faecium* isolates from milk, beef, and chicken and their antibiotic resistance. *Journal of Food Protection, 66,* pp. 931–936.

Cho, J.I., et al. (2013). Prevalence and antimicrobial resistance of *Enterococcus* spp. isolated from meat and fishery production in Korea. *Food Science and Biotechnology, 22*(1), pp. 161–165.

Choi, J. and Woo, G. (2013). Molecular characterization of high-level gentamicin-resistant *Enterococcus faecalis* from chicken meat in Korea. *International Journal of Food Microbiology, 165*(1), pp. 1–6.

Chou, Y.Y., et al. (2008). Vancomycin-resistant enterococcal bacteremia: comparison of clinical features and outcome between *Enterococcus faecium* and *Enterococcus faecalis. Journal of Microbiology, Immunology and Infection, 41*(2), pp. 124–129.

Chow, J.W., et al. (2001). Aminoglycoside reistance genes *aph(2")-Ib* and *aac(6')-Im* detected together in strains of both *Escherichia coli* and *Enterococcus faecium. Antimicrobial Agents and Chemotherapy, 45*(10), pp. 2691–2694.

Cintas, L.M., et al. (1997). Biochemical and genetic characterization of enterocin P, a novel sec-dependent bacteriocin from *Enterococcus faecium* P13 with a broad antimicrobial spectrum. *Applied and Environmental Microbiology, 63,* pp. 4321–4330.

Cleveland, J., et al. Bacteriocins: safe natural antimicrobials for food preservation. *International Journal of Food Microbiology, 71*(1), pp. 1–20.

Cosentino, S., et al. (2004). Genotypic and technological characterization of enterococci isolated from artisanal Fiore Sardo cheese. *The Journal of Dairy Research, 71*, pp. 444–450.

Damasco, P.V., et al. (2014). Infective endocarditis in Rio de Janeiro, Brazil: a 5-year experience at two teaching hospitals. *Infection, 42*, pp. 835–842.

Damborg, P., Sorensen, A.H. and Guardabassi, L. (2008). Monitoring of antimicrobial resistance in healthy dogs: first report of canine ampicillin-resistant *Enterococcus faecium* clonal complex 17. *Veterinary Microbiology, 132*, pp. 190– 196.

De Castro, A., et al. (2002). Utilization of *Enterococcus casseliflavus* and *Lactobacillus pentosus* as starter cultures for Spanish-style green olive fermentation. *Food Microbiology, 19*, pp. 637–644.

Deshpande, L.M., et al. (2015). Detection of a new *cfr*-like gene, *cfr*(B), in *Enterococcus faecium* isolates recovered from human specimens in the United States as part of the SENTRY antimicrobial surveillance program. *Antimicrobial Agents and Chemotherapy, 59*(10), pp. 6256–6261.

Diaz, L., et al. (2012). Transferable plasmid-mediated resistance to linezolid due to *cfr* in a human clinical isolate of *Enterococcus faecalis. Antimicrobial Agents and Chemotherapy, 56*(7), pp. 3917–3922.

D'Souza, A.L.D., et al. (2002). Probiotics in prevention of antibiotic associated diarrhoea: meta analysis. *British Medical Journal, 324*, pp. 1–6.

Economou, V., et al. (2007). Changes in histamine and microbiological analyses in fresh and frozen tuna muscle during temperature abuse. *Food Additives and Contaminants, 24*(8), pp. 820–32.

EFSA (European Food Safety Authority) and ECDC (European Centre for Disease Prevention and Control). (2015). EU Summary Report on antimicrobial resistance in zoonotic and indicator bacteria from humans, animals and food in 2013. *EFSA Journal, 13*(2), pp. 178–, 4036.

EFSA (European Food Safety Authority). (2008). Report from the Task Force on Zoonoses Data Collection including guidance for harmonized monitoring and reporting of antimicrobial resistance in commensal *Escherichia coli* and *Enterococcus* spp. from food animals. *The EFSA Journal, 141*, pp. 1–44.

Enck, P., et al. (2008). A mixture of *Escherichia coli* (DSM 17252) and *Enterococcus faecalis* (DSM 16440) for treatment of the irritable bowel syndrome — a randomized controlled trial with primary care physicians. *Neurogastroenterology and Motility, 20*, pp. 1103–1109.

Enne, V.I., et al. (2004). Rifampicin resistance and its fitness cost in *Enterococcus faecium. Journal of Antimicrobial Chemotherapy, 53*, pp. 203–207.

Falcone, M., Russo, A. and Venditti, M. (2015). Optimizing antibiotic therapy of bacteremia and endocarditis due to staphylococci and enterococci: New insights and evidence from the literature. *Journal of Infection and Chemotherapy, 21*, pp. 330–339.

Favaro, L., et al. (2014). Bacteriocinogenic potential and safety evaluation of non-starter *Enterococcus faecium* strains isolated from home made white brine cheese. *Food Microbiology, 38*, pp. 228–39.

Fisher, K., Phillips, C. (2009). The ecology, epidemiology and virulence of *Enterococcus. Microbiology, 155*(6), pp. 1749–57.

Fluckey, W.M., et al. (2009). Diversity and susceptibility of *Enterococcus* isolated from cattle before and after harvest. *Journal of Food Protection, 72*, pp. 766–774.

Foulquié Moreno, M.R., et al. (2006). The role and application of enterococci in food and health. *International Journal of Food Microbiology, 106*(1), pp. 1–24.

Franciosi, E., et al. (2009). Presence of enterococci in raw cow's milk and "puzzone di moena" cheese. *Journal of Food Processing and Preservation, 3*(2), pp. 204–17.

Francischetto, O., et al. (2014). Healthcare-associated infective endocarditis: a case series in a referral hospital from 2006 to 2011.

Arquivos Brasileiros de Cardiologia, 103(4), pp. 292–298.

Franz, C.M.A.P., et al. (2001). Bile salt hydrolase activity of enterococci isolated from food: screening and quantitative determination. *Journal of Food Protection, 64*, pp. 725–729.

Franz, C.M.A.P., et al. (2003). Enterococci in foods—a conundrum for food safety. *International Journal of Food Microbiology, 88*, pp. 105–122.

Franz, C.M., et al. (2011). Enterococci as probiotics and their implications in food safety. *International Journal of Food Microbiology, 151*(2), pp. 125–40.

Freitas, A.R., et al. (2009). Clonal expansion within clonal complex 2 and spread of vancomycin-resistant plasmids among different genetic lineages of *Enterococcus faecalis* from Portugal. *Journal of Antimicrobial Chemotherapy, 63*, pp. 1104–1111.

Freitas, A.R., et al. (2011). Human and swine hosts share vancomycin-resistant *Enterococcus faecium* CC17 and CC5 and *Enterococcus faecalis* CC2 clonal clusters harboring Tn*1546* on indistinguishable plasmids. *Journal of Clinical Microbiology, 49*(3), pp. 925–931.

French, G.L. (2006). Bactericidal agents in the treatment of MRSA infections – the potential role of daptomycin. *Journal of Antimicrobial Chemotherapy, 58*, pp. 1107–1117.

Fuka, M.M., et al. (2010). Bacterial communities associated with the production of artisanal Istrian cheese. *International Journal of Food Microbiology, 142*, pp. 19–24.

Gaaloul, N., et al. (2015). Isolation and characterization of large spectrum and multiple bacteriocin-producing *Enterococcus faecium* strain from raw bovine milk. *Journal of Applied Microbiology, 118*(2), pp. 343–355.

Gade, J. and Thorn, P. (1989). Paraghurt for patients with irritable bowel syndrome. A controlled clinical investigation from general practice. *Scandinavian Journal of Primary Health Care, 7*, pp. 23–26.

Galloway-Peña, J., et al. (2012). Genomic and SNP analyses demonstrate a distant separation of the hospital and community-associated clades of *Enterococcus faecium*. *PLoS ONE, 7*(1), p. e30187.

Garcia-Migura, L., et al. (2014). Antimicrobial resistance of zoonotic and commensal bacteria in Europe: The missing link between consumption and resistance in veterinary medicine. *Veterinary Microbiology, 170*(1–2), pp. 1–9.

Garnier, F., et al. (2000). Characterization of transposon Tn1549, conferring VanB resistance in *Enterococcus* spp. *Microbiology, 146*(Pt6), pp. 1481–1489.

Ghosh, A., Dowd, S.E. and Zurek, L. (2011). Dogs leaving the ICU carry a very large multi-drug resistant enterococcal population with capacity for biofilm formation and horizontal gene transfer. *PLoS ONE, 6*, p. e22451.

Ghrairi, T., et al. (2008). Purification and characterization of bacteriocins produced by *Enterococcus faecium* from Tunisian Rigouta cheese. *Food Control, 19*, pp. 162–169.

Giannino, M.L., et al. (2009). Study of microbial diversity in raw milk and fresh curd used for Fontina cheese production by culture-independent methods. *International Journal of Food Microbiology, 130*, pp. 188–195.

Giannou, E., et al. (2009). Identification and biopreservation potential of *Enterococcus* spp. isolated from fully ripened graviera, a traditional hard Greek cheese. *Italian Journal of Food Science, 21*(2), pp. 135–47.

Giraffa, G. (2002). Enterococci from foods. *FEMS Microbiology Reviews, 26*, pp. 163–171.

Gold, H.S. (2001). Vancomycin-resistant enterococci: mechanisms and clinical observations. *Clinical Infectious Diseases, 33*, pp. 210–219.

Gousia, P., et al. (2011). Antimicrobial resistance of major foodborne pathogens from major meat products. *Foodborne Pathogens and Disease, 8*(1), pp. 27–38.

Gousia, P., et al. (2015). Vancomycin-resistance phenotypes, vancomycin-resistance genes, and resistance to antibiotics of enterococci isolated from food of animal origin. *Foodborne Pathogens and Disease, 12*(3), pp. 214–220.

Guerrero-Ramos, E., et al. (2016). Antimicrobial resistance and virulence genes in enterococci from wild game meat in Spain. *Food Microbiology, 53,* pp. 156–164.

Gundogan, N., Ataol, O. and Torlak, F.O. (2013). Determination of some virulence factors in *Staphylococcus aureus, Enterococcus faecalis* and *Enterococcus faecium* isolated from meat and milk products. *Journal of Food Safety, 33*(4), pp. 387–393.

Habermann, W., et al. (2002). Reduction of acute recurrence in patients with chronic recurrent hypertrophic sinusitis by treatment with a bacterial immunostimulant (*Enterococcus faecalis* bacteria of human origin). *Arzneimittel-Forschung, 52,* pp. 622–627.

Hamad, S.H., et al. (1997). Characterisation of the bacterial flora of Sudanese sorghum flour and sorghum sourdough. *Journal of Applied Microbiology, 83,* pp. 764–770.

Hamada, Y., et al. (2015). Clinical features of enterococcal bacteremia due to ampicillin-susceptible and ampicillin-resistant enterococci: An eight-year comparison study. *Journal of Infection and Chemotherapy, 21,* pp. 527–530.

Hammad, A.M., Hassan, H.A. and Shimamoto, T. (2015). Prevalence, antibiotic resistance and virulence of *Enterococcus* spp. in Egyptian fresh raw milk cheese. *Food Control, 50,* pp. 815–820.

Hammerum, A.M. (2012). Enterococci of animal origin and their significance for public health. *Clinical Microbiology and Infection, 18*(7), pp. 619–625.

Hammerum, A.M., et al. (2004). A vancomycin-resistant *Enterococcus faecium* isolate from a Danish healthy volunteer, detected 7 years after the ban of avoparcin, is possibly related to pig isolates. *Journal of Antimicrobial Chemotherapy, 53,* pp. 547–549.

Hammerum, A.M., Lester, C.H. and Heuer, O.E. (2010). Antimicrobial-resistant enterococci in animals and meat: a human health hazard? *Foodborne Pathogens and Disease, 7*(10), pp. 1137–1146.

Hancock, L.E. and Perego, M. (2004). Systematic inactivation and phenotypic characterization of two-component signal transduction systems of *Enterococcus faecalis. Journal of Bacteriology, 186*(23), pp. 7951–7958.

Hasman, H. and Aarestrup, F.M. (2005). Relationship between copper, glycopeptide, and macrolide resistance among *Enterococcus faecium* strains isolated from pigs in Denmark between 1997 and 2003. *Antimicrobial Agents and Chemotherapy, 49,* pp. 454–456.

Hayes, J., et al. (2003). Prevalence and antimicrobial resistance of *Enterococcus* species isolated from retail meats. *Applied and Environmental Microbiology, 69*(12), pp. 7153–7160.

Heintz, B.H., Halilovic, J. and Christensen, C.L. (2010). Vancomycin-resistant enterococcal urinary tract infections. *Pharmacotherapy, 30*(11), pp. 1136–1149.

Hollenbeck, B.L. and Rice, L.B. (2012). Intrinsic and acquired resistance mechanisms in *Enterococcus. Virulence, 3*(5), pp. 421–433.

Hooper, D.C. (2002). Fluoroquinolone resistance among Gram-positive cocci. *Lancet Infectious Diseases, 2,* pp. 530–538.

Hooper, D.C. and Jacoby, D. (2015). Mechanisms of drug resistance: quinolone resistance. *Annals of the New York Academy of Sciences, 1354*(1), pp. 12–31.

Hugas, M., Garriga M. and Aymerich MT. (2003). Functionality of enterococci in meat products. *International Journal of Food Microbiology, 88*(2-3), pp. 223–233.

Huys, G., et al. (2004). Prevalence and molecular characterization of tetracycline resistance in *Enterococcus* isolates from food. *Applied and Environmental Microbiology, 70*(3), pp. 1555–1562.

Isnard, C., et al. (2013). Genetic basis for *in vivo* and *in vitro* resistance to lincosamides, streptogramins A, and pleuromutilins (LS_AP phenotype) in *Enterococcus faecium. Antimicrobial Agents and Chemotherapy, 57*(9), pp. 4463–4469.

Iweriebor, B.C., Obi, L.C. and Okoh, A.I. (2015). Virulence and antimicrobial resistance factors of *Enterococcus* spp. isolated from fecal samples from piggery farms in Eastern Cape, South Africa. *BMC Microbiology, 15,* p. 136.

Jacoby, G.A. (2005). Mechanisms of resistance to quinolones. *Clinical Infectious Diseases, 41*(S2), pp. S120–S126.

Jahan, M., Krause, D.O. and Holley, R.A. (2013). Antimicrobial resistance of *Enterococcus* species from meat and fermented meat products isolated by a PCR-based rapid screening method. *International Journal of Food Microbiology, 163*(2–3), pp. 89–95.

Jahan, M. and Holley, R.A. (2014). Incidence of virulence factors in enterococci from raw and fermented meat and biofilm forming capacity at 25 °C and 37 °C. *International Journal of Food Microbiology, 170*, pp. 65–69.

Jahan, M., et al. (2015). Horizontal transfer of antibiotic resistance from *Enterococcus faecium* of fermented meat origin to clinical isolates of *E. faecium* and *Enterococcus faecalis. International Journal of Food Microbiology, 199*, pp. 78–85.

Jay, J.M., Loessner, M.J., and Golden, D.A. (2005). *Modern Food Microbiology*, 7th ed. New York: Springer.

Johnson, D.H. and Cunha, B.A. (1995). Aztreonam. *Medical Clinics of North America, 79*(4), pp. 733–473.

Jonas, B.M., Murray, B.E. and Weinstock, G.M. (2001). Characterization of *emeA*, a *norA* homolog and multidrug resistance efflux pump, in *Enterococcus faecalis. Antimicrobial Agents and Chemotherapy, 45*(12), pp. 3574–3579.

Jurkovic, D., et al. (2006). Molecular identification and diversity of enterococci isolated from Slovak Bryndza cheese. *Journal of General and Applied Microbiology, 52*, pp. 329–337.

Kagkli, D.M., et al. (2007). Contamination of milk by enterococci and coliforms from bovine faeces. *Journal of Applied Microbiology, 103*(5), pp. 1393–405.

Kara, A., et al. (2015). Risk of vancomycin-resistant enterococci bloodstream infection among patient colonized with vancomycin-resistant enterococci. *Brazilian Journal of Infectious Diseases, 19*(1), pp. 58–61.

Kasimoglu-Dogru, A., Gencay, Y.E. and Ayaz, N.D. (2010). Prevalence and antibiotic resistance profiles of *Enterococcus* species in chicken at slaughter level; absence of *vanA* and *vanB* genes in *E. faecalis* and *E. faecium. Research in Veterinary Science, 89*(2), pp. 153–158.

Kaszanyitzky, E.J., et al. (2007). Antimicrobial susceptibility of enterococci strains isolated from slaughter animals on the data of Hungarian resistance monitoring system from 2001 to 2004. *International Journal of Food Microbiology, 115*, pp. 119–123.

Kawalec, P.Z, et al. (2007). Clonal structure of *Enterococcus faecalis* isolated from Polish hospitals: characterization of epidemic clones. *Journal of Clinical Microbiology, 45*, pp. 147–153.

Kelesidis, T., et al. (2011). Daptomycin nonsusceptible enterococci: an emerging challenge for clinicians. *Clinical Infectious Diseases, 52*(2), pp. 228–234.

Kim, Y.G., et al. (2006). The effects of probiotics on symptoms of irritable bowel syndrome. *The Korean Journal of Gastroenterology, 47*, pp. 413–419.

Klein, G. (2003). Taxonomy, ecology and antibiotic resistance of enterococci from food and the gastro-intestinal tract. *International Journal of Food Microbiology, 88*(2–3), pp. 123–131.

Klibi, N., et al. (2013). Species distribution, antibiotic resistance and virulence traits in enterococci from meat in Tunisia. *Meat Science, 93*(3), pp. 675–680.

Knechtges, P.L. (2012). *Food Safety: Theory and Practice.* Burlington: Jones and Bartlett Learning.

Knoll, B.M., Hellmann, M. and Kotton, C.N. (2013). *Enterococcus faecium* meningitis in adults: case series and review of the literature. *Scandinavian Journal of Infectious Diseases, 45*(2), pp. 131–139.

Kobayashi, N., et al. (2001). Distribution of aminoglycoside resistance genes in recent clinical isolates of *Enterococcus faecalis, Enterococcus faecium* and *Enterococcus avium.Epidemiology and Infection, 126*(2), pp. 197–204.

Kohanski, M.A., et al. (2007). A common mechanism of cellular death induced by bactericidal antibiotics. *Cell, 130*, pp. 797–810.

Kow, N., Ferzandi, T. (2013). *Enterococcus* osteomyelitis secondary to pyelonephritis. *International Urogynecology Journal, 24*, pp. 691–692.

Kristich, C.J. and Little, J.L. (2012). Mutations in the β subunit of RNA polymerase alter intrinsic cephalosporin resistance in enterococci. *Antimicrobial Agents and Chemotherapy, 56*(4), pp. 2022–2027.

Larsen, J., et al. (2010). Porcine-origin gentamicin-resistant *Enterococcus faecalis* in humans, Denmark. *Emerging and Infectious Diseases, 16*, pp. 682–684.

Larsen, J., et al. (2011). Porcine and human community reservoirs of *Enterococcus faecalis*, Denmark. *Emerging and Infectious Diseases, 17*, pp. 2395–2397.

Lavermicocca, P., et al. (1998). Characterization of lactic acid bacteria isolated from olive phylloplane and table olive brines. *Italian Journal of Food Science, 10*, pp. 27–39.

Leclercq, R. (2002). Mechanisms of resistance to macrolides and lincosamides: nature of the resistance elements and their clinical implications. *Antimicrobial Resistance, 34*, pp. 482–492.

Len, O., et al. (2014). Daptomycin is safe and effective for the treatment of Gram-positive cocci infections in solid organ transplantation. *Transplant Infectious Disease, 16*, pp. 532–538.

Lewenstein, A., Frigerio, G. and Moroni, M. (1979). Biological properties of SF68, a new approach for the treatment of diarrhoeal diseases. *Current Therapeutic Research, 26*, pp. 967–981.

Litopoulou-Tzanetaki, E. (1990). Changes in numbers and kinds of lactic acid bacteria during ripening of Kefalotyri cheese. *Journal of Food Science, 55*, pp. 111–113.

López M, Sáenz Y, Rojo-Bezares B, Martínez S, del Campo R, Ruiz-Larrea, F., Zarazaga, M. and Torres C. (2009). Detection of *vanA* and *vanB2*-containing enterococci from food samples in Spain, including *Enterococcus faecium* strains of CC17 and the new singleton ST425. *International Journal of Food Microbiology, 133*, pp. 172–178.

Luna, V.A., et al. (1999). A variety of Gram-positive bacteria carry mobile *mef* genes. *Journal of Antimicrobial Chemotherapy, 41*, pp. 19–25.

M'hir, S., et al. (2012). Technological, functional and safety aspects of enterococci in fermented vegetable products: a mini-review. *Annals of Microbiology, 62*(2), pp. 469–481.

Magnus, C.A., Ingledew, W.M. and McCurdy, A.R. (1986). Thermal resistance of streptococci isolated from pasteurized ham. *Canadian Institute of Food Science and Technology Journal, 19*, pp. 62–67.

Magnus, C.A,, McCurdy, A.R. and Ingledew, W.M. (1988). Further studies on the thermal resistance of Streptococcus *faecium* and Streptococcus *faecalis* in pasteurized ham. *Canadian Institute of Food Science and Technology Journal, 21*, pp. 209–212.

Manolopoulou, E., et al. (2003). Evolution of microbial populations during traditional Feta cheese manufacture and ripening. *International Journal of Food Microbiology, 82*, pp. 153–161.

Marchesino, B., et al. (1992). Microbiological events during commercial meat fermentations. *Journal of Applied Bacteriology, 73*, pp. 203–209.

Martin, B., et al. (2005). Genetic diversity and safety aspects of enterococci from slightly fermented sausages. *Journal of Applied Microbiology, 98*, pp. 1177–1190.

McBride, S., Upton, A. and Roberts, S. (2010). Clinical characteristics and outcomes of patients with vancomycin-susceptible *Enterococcus faecalis* and *Enterococcus faecium* bacteraemia – a five-year retrospective review. *European Journal of Clinical Microbiology and Infectious Diseases, 29*, pp. 107–114.

McDonald, J.R., et al. (2005). Enterococcal endocarditis: 107 cases from the international collaboration on endocarditis merged database. *American Journal of Medicine, 118*, pp. 759–766.

McGowan, L.L., et al. (2006). Prevalence and antimicrobial resistance of enterococci isolated from retail fruits, vegetables, and meats. *Journal of Food Protection, 69*(12), pp. 2976–2982.

Messi, P., et al. (2006). Vancomycin-resistant enterococci (VRE) in meat and environmental samples. *International Journal of Food Microbiology, 107*(2), pp. 218–222.

Mikalsen, T., et al. (2015). Investigating the mobilome in clinically important lineages of *Enterococcus faecalis. BMC Genomics, 16*, p. 282.

Miller, W.R., Munita, J.M. and Arias, C.A. (2014). Mechanisms of antibiotic resistance in enterococci. *Expert Review of Anti-Infective Therapy 12*(10), pp. 1221–1236.

Miranda, J.M., et al. (2007). Antimicrobial resistance in *Enterococcus* spp. strains isolated from organic chicken, conventional chicken, and turkey meat: A comparative survey. *Journal of Food Protection, 70*(4), pp. 1021–1024.

Mitra, S. and Alangaden, J.A. (2011). Reccurent urinary tract infections in kidney transplant recipients. *Current Infectious Disease Reports, 13*, pp. 579–587.

Moraes, M., et al. (2012). Bacteriocinogenic and virulence potential of *Enterococcus* isolates obtained from raw milk and cheese. *Journal of Applied Microbiology, 113*, pp. 318–328.

Moreno, M.R.F., et al. (2002). Microbial analysis of Malaysian tempeh, and characterization of two bacteriocins produced by isolates of *Enterococcus faecium. Journal of Applied Microbiology, 92*, pp. 147–157.

Morishita, T.Y. (2015). Overview of Enterococcosis in Poultry. In: S Aiello, M Moses, M Steigerwald, eds., *Merck Veterinary Manual*. Available at: http://www.merckvetmanual.com/mvm/poultry/enterococcosis/overview_of_enterococcosis_in_poultry.html [accessed November 10, 2015].

Mountzouris, K.C., et al. (2007). Evaluation of the efficacy of a probiotic containing *Lactobacillus, Bifidobacterium, Enterococcus*, and *Pediococcus* strains in promoting broiler performance and modulating cecal microflora composition and metabolic activities. *Poultry Science, 86*, pp. 309–317.

Münch, D., et al. (2015). Structural variations of the cell wall precursor Lipid II and their influence on binding and activity of the lipoglycopeptide antibiotic oritavancin. *Antimicrobial Agents and Chemotherapy, 59*(2), pp. 72–781.

Mundt, O.J. (1986). Enterococci. In: P.H.A. Sneath, N.S. Mair, M.E. Sharpe, J.G. Holt, eds., *Bergey's Manual of Systematic Bacteriology, vol. 2*, 4th edition. Baltimore: Williams and Wilkins, pp. 1063–1065.

Murray, B. (1990). The life and times of the *Enterococcus. Clinical Microbiology Reviews, 3*(1), pp. 46–65.

Mutters, N., et al. (2013). Control of the spread of Vancomycin Enterococci in hospitals. *Medicine, 110*(43), pp. 725–731.

Nami, Y., et al. (2014). A newly isolated probiotic *Enterococcus faecalis* strain from vagina microbiota enhances apoptosis of human cancer cells. *Journal of Applied Microbiology, 117*(2), pp. 498–508.

Nicolau, D.P., et al. (1995). Experience with a once-daily aminoglycoside program administered to 2,184 adult patients. *Antimicrobial Agents and Chemotherapy, 39*(3), pp. 650–655.

Nielsen, S.L. (2015). The incidence and prognosis of patients with bacteremia. *Danish Medical Journal, 62*(7), p. B5128.

Nigo, M., et al. (2014). What's new in the treatment of enterococcal endocarditis? *Current Infectious Disease Reports, 16*(10), p. 431.

Nnedu, O.N. and Pankey, G.A. (2015). Update on the emerging role of telavancin in hospital-acquired infections. *Journal of Therapeutics and Clinical Risk Management, 11*, pp. 605–610.

Novais, C., et al. (2005). High occurrence and persistence of antibiotic resistant enterococci in poultry food samples in Portugal. *Journal of Antimicrobial Chemotherapy, 56*, pp. 1139–1143.

Novais, C., et al. (2013). Spread of multidrug-resistant *Enterococcus* to animals and humans: an underestimated role for the pig farm environment. *Journal of Antimicrobial Chemotherapy, 38*, pp. 2746–2754.

O'Driscoll, T. and Crank, C. (2015). Vancomycin-resistant enterococcal infections: epidemiology, clinical manifestations, and optical

management. *Infection and Drug Resistance, 8*, pp. 217–230.
Oguntoyinbo, F.A., et al. (2007). In vitro selection and evaluation of *Bacillus* starter cultures for the production of okpehe, a traditional African fermented condiment. *International Journal of Food Microbiology, 113*, pp. 208–218.
Onda, T., et al. (2002). Widespread distribution of the bacteriocin-producing lactic cocci in miso paste products. *Journal of Applied Microbiology, 92*, pp. 695–705.
Ono, S., Muratani, T. and Matsumoto, T. (2005). Mechanisms of resistance to imipenem and ampicillin in *Enterococcus faecalis. Antimicrobial Agents and Chemotherapy, 49*(7), pp. 2954–2958.
Oprea, S. and Zervos, M. (2007). *Enterococcus* and its association with foodborne illness. In: S. Simjee, ed., *Foodborne Diseases*. New Jersey: Humana Press, pp. 157–174.
Orsi, G.B. and Ciorba, V. (2013). Vancomycin resistant enterococci healthcare associated infections. *Annali di Igiene: Medicina Preventiva e di Comunità, 25*(6), pp. 485–92.
Ortigosa, M., et al. (2008). Sources of enterococci in idiazábal-type cheese. *International Journal of Food Microbiology, 125*(2), pp. 146–52.
Pankey, G.A. and Ashcraft, D.S. (2013). In vitro synergy of telavancin and rifampin against *Enterococcus faecium* resistant to both linezolid and vancomycin. *The Ochsner Journal, 13*, pp. 61–65.
Papadopoulou, C., et al. (1997). Bacterial strains isolated from eggs and their resistance to currently used antibiotics: is there a health hazard for consumers? *Comparative Immunology, Microbiology and Infectious Diseases, 20*(1), pp. 35–40.
Paramithiotis, S., et al. (2008). Identification and characterization of *Enterococcus* spp. in Greek spontaneous sausage fermentation. *Journal of Food Protection, 71*, pp. 1244–1247.
Pericas, J.M., et al. (2015). Enterococcal endocarditis revisited. *Future Microbiology, 10*, pp. 1215–1240.
Perin, L.M., et al. (2014). Virulence, antibiotic resistance and biogenic amines of bacteriocinogenic lactococci and enterococci isolated from goat milk. *International Journal of Food Microbiology, 185*, pp. 121–126.
Pesavento, G, et al. (2014). Prevalence and antibiotic resistance of *Enterococcus* spp. isolated from retail cheese, ready-to-eat salads, ham, and raw meat. *Food Microbiology, 41*, pp. 1–7.
Pimentel, L.L., et al. (2007). Assessment of safety of enterococci isolated throughout traditional Terrincho cheesemaking: virulence factors and antibiotic susceptibility. *Journal of Food Protection, 70*, pp. 2161–2167.
Pintado, V., et al. (2003). Enterococcal meningitis: a clinical study of 39 cases and review of the literature. *Medicine, 82*(5), pp. 346–364.
Portillo, A., et al. (2000). Macrolide resistance genes in *Enterococcus* spp. *Antimicrobial Agents and Chemotherapy, 44*(4), pp. 967–971.
Psoni, L., et al. (2006). Genotypic and phenotypic heterogeneity in *Enterococcus* isolates from batzos, a raw goat milk cheese. *International Journal of Food Microbiology, 109*(1–2), pp. 109–120.
Rajkumari, N., Mathur, P. and Misra, M. (2014). Soft tissue and wound infections due to *Enterococcus spp.* Among hospitalized trauma patients in a developing country. *Journal of Global Infectious Diseases, 6*(4), pp. 189–193.
Rajkumari, N., et al. (2015). Magnitude of enterococcal bacteremia in trauma patients admitted for intensive trauma care: a tertiary care experience from South Asian Country. *Journal of Laboratory Physicians, 7*(1), pp. 38–42.
Randazzo, C.L., et al. (2004). *Lactobacillus casei*, dominant species in naturally fermented Sicilian green olives. *International Journal of Food Microbiology, 90*, pp. 9–14.
Randazzo, C.L., et al. (2008). Effect of wild strains used as starter cultures and adjunct cultures on the volatile compounds of the Pecorino Siciliano cheese. *International Journal of Food Microbiology, 122*, pp. 269–278.
Roberts, M.C. (1996). Tetracycline resistance determinants: mechanisms of action, regulation of expression, genetic mobility, and distribution. *FEMS Microbiology Reviews, 19*, pp. 1–24.

Rodvold, K.A. and McKoneghy, K.W. (2014). Methicillin-resistant *Staphylococcus aureus* therapy: past, present, and future. *Clinical Infectious Diseases, 58*(S1), pp. S20–S27.

Ronald A. (2002). The etiology of urinary tract infection: Traditional and emerging pathogens. *American Journal of Medicine, 113*(1A), pp. 14S–19S.

Rosa R., et al. (2014). Vancomycin-resistant *Enterococcus faecium* bacteremia in a tertiary care hospital: epidemiology, antimicrobial susceptibility, and outcome. *BioMed Research International.* Article ID 958469.

Ruiz-Garbajosa, P., et al. (2006). Multilocus sequencetyping scheme for *Enterococcus faecalis* reveals hospital-adapted genetic complexes in a background of high rates of recombination. *Journal of Clinical Microbiology, 44*, pp. 2220–2228.

Ruiz-Moyano, S., et al. (2009). Safety and functional aspects of preselected enterococci for probiotic use in Iberian dry-fermented sausages. *Journal of Food Science, 74*, pp. M398–M404.

Sader, H., Farrell, D. and Jones, R. (2010). Antimicrobial susceptibility of Gram-positive cocci isolated from skin and skin-structure infections in European medical centers. *International Journal of Antimicrobial Agents, 36*, pp. 28–32.

Samli, H.E., et al. (2010). Effects of *Enterococcus faecium* supplementation and floor type on performance, morphology of erythrocytes and intestinal microbiota in broiler chickens. *British Poultry Science, 51*, pp. 564–568.

Sarantinopoulous, P., et al. (2001). Biochemical properties of enterococci relevant to their technological performance. *International Dairy Journal, 11*, pp. 621–647.

Scharek, L., et al. (2007). Impact of the probiotic bacteria *Enterococcus faecium* NCIMB 10415 (SF68) and *Bacillus cereus* var. toyoi NCIMB 40112 on the development of serum IgG and faecal IgA of sows and their piglets. *Archives of Animal Nutrition, 61*, pp. 223–234.

Scharek, L., et al. (2005). Influence of a probiotic *Enterococcus faecium* strain on development of the immune system of sows and piglets. *Veterinary Immunology and Immunopathology, 105*, pp. 151–161.

Schmitz, F.-J., et al. (2000). Prevalence of macrolide-resistance genes in *Staphylococcus aureus* and *Enterococcus faecium* isolates from 24 European university hospitals. *Journal of Antimicrobial Chemotherapy, 45*, pp. 891–894.

Schwaiger, K., Schmied, E.-M.V. and Bauer, J. (2010). Comparative analysis on antibiotic resistance characteristics of *Listeria* spp. and *Enterococcus* spp isolated from laying hens and eggs in conventional and organic keeping systems in Bavaria, Germany. *Zoonoses Public Health 57*, pp. 171–180.

Schwartz, B.S., et al. (2009). Doxycycline, not minocycline, induces its own resistance in multidrug-resistant community-associated methicillin-resistant *Stahylococcus aureus* clone USA300. *Clinical Infectious Diseases, 48*, pp. 1483–1484.

Seguin, P., et al. (2012). Are enterococci playing a role in postoperative peritonitis in critically ill patients? *European Journal of Clinical Microbiology and Infectious Diseases, 31*, pp. 1479–1485.

Serio, A., et al. (2010). Evaluation of metabolic activities of enterococci isolated from pecorino abruzzese cheese. *International Dairy Journal, 20*(7), pp. 459–64.

Shantala, G.B., et al. (2014). Neonatal septicaemia caused by vancomycin resistant *Enterococcus faecium-* A case report. *Journal of Clinical and Diagnostic Research, 8*(11), pp. DD03–DD04.

Shigemura, K., et al. (2005). Pathogen occurrence and antimicrobial susceptibility of urinary tract infection cases during a 20-year period (1983-2002) at a single institution in Japan. *Japanese Journal of Infectious Diseases, 58*, pp. 303–308.

Shukla, B.S., et al. (2013). Treatment considerations in vancomycin-resistant enterococcal bacteremia: daptomycin or linezolid? A review. *International Journal of Clinical Pharmacology, 35*, pp. 697–703.

Sidler, J., et al. (2014). Enterococci, *Clostridium difficile*, and ESBL-producing bacteria: epidemiology, clinical impact and prevention in ICU patients. *Swiss Medical Weekly, 144*, p. w14009.

Simjee, S., et al. (2006). *Enterococcus*. In: F.M. Aarestrup, ed., *Antimicrobial Resistance in Bacteria of Animal Origin*. Washington, DC: ASM Press, pp. 315–328.

Singh, K.V., Weinstock, G.M. and Murray BE. (2002). An *Enterococcus faecalis* ABC homologue (Lsa) is required for the resistance of this species to clindamycin and quinupristin-dalfopristin. *Antimicrobial Agents and Chemotherapy, 46*(6), pp. 1845–1850.

Solheim, M., et al. (2009). Comparative genomics of *Enterococcus faecalis* from healthy Norwegian infants. *BMC Genomics, 10*, p. 194.

Soltani, M., et al. (2000). Mechanisms of resistance to quinupristin-dalfopristin among isolates of *Enterococcus faecium* from animals, raw meat, and hospital patients in Western Europe. *Antimicrobial Agents and Chemotherapy, 44*(2), pp. 433–436.

Sood, S., et al. (2008). Enterococcal infections and antimicrobial resistance. *Indian Journal of Medical Research, 128*, pp. 111–121.

Sørensen, T.L., et al. (2001). Transient intestinal carriage after ingestion of antibiotic-resistant *Enterococcus faecium* from chicken and pork. *New England Journal of Medicine, 345*, pp. 1161–1166.

Spieler, A. and Männer, K. (1996). Milchsäurebakterein als probiotischer Futterzusatzstoff. *Kraftfutter, 5*, pp. 200–203.

Strahilevitz, J., et al. (2009). Plasmid-mediated quinolone resistance: a multifaceted threat. *Clinical Microbiology Reviews, 22*(4), pp. 664–689.

Sustackova, A., Nápravníková, E. and Schlegelová J. (2004). Antimicrobial resistance of *Enterococcus* spp. isolates from raw beef and meat products. *Folia Microbiologica, 49*(4), pp. 411–417.

Szurmant, H., et al. (2005). YycH regulates the activity of the essential YycFG two-component system in *Bacillis subtilis*. *Journal of Bacteriology, 187*(15), pp. 5419–5426.

Taras, D., et al. (2006). Performance, diarrhoea incidence, and occurrence of *Escherichia coli* virulence genes during long-term administration of a probiotic *Enterococcus faecium* strain to sows and piglets. *Journal of Animal Science, 84*, pp. 608–617.

Teuber, M., Perreten, V. and Wirsching, F. (1996). Antibiotikumresistente Bakterien: Eine neue Dimension in der Lebensmittelmikrobiologie. *Lebensmitteltechnologie, 29*, pp. 182–199.

Tornero, E., et al. (2014). Characteristics of prosthetic joint infections due to *Enterococcus* sp. and predictors of failure: a multi-national study. *Clinical Microbiology and Infection, 20*(11), pp. 1219–1224.

Tran, T.T., et al. (2013). Whole-genome analysis of a daptomycin-susceptible *Enterococcus faecium* strain and its daptomycin-resistant variant arising during therapy. *Antimicrobial Agents and Chemotherapy, 57*(1), pp. 261–268.

Trubiano, J.A., et al. (2013). The prevention and management of infections due to multidrug resistant organisms in haematology patients. *British Journal of Clinical Pharmacology, 79*(2), pp. 195–207.

Tzanetakis, N. and Litopoulou-Tzanetaki, E. (1992). Changes in numbers and kinds of lactic acid bacteria in Feta and Teleme, two Greek cheeses from ewes' milk. *Journal of Dairy Science, 75*, 1389–1393.

Tzavaras, I., et al. (2012). Diversity of *vanA*-type vancomycin-resistant *Enterococcus faecium* isolated from broilers, poultry slaughterers and hospitalized humans in Greece. *Journal of Antimicrobial Chemotherapy, 67*, pp. 1811–1818.

Underdahl, N.R., Torres-Medina, A. and Doster, A.R. (1982). Effect of *Streptococcus faecium* C-68 in control of *Escherichia coli*-induced diarrhea in gnotobiotic pigs. *American Journal of Veterinary Research, 12*, pp. 2227–2232.

Van den Berg, D.J.C., et al. (1993). Isolation, screening and identification of lactic acid bacteria from traditional food fermentation processes and culture collections. *Food Biotechnology, 7*, pp. 189–205.

Van den Berghe, E., De Winter, T. and De Vuyst, L. (2006). Enterocin A production by *Enterococcus faecium* FAIR-E 406 is characterised by a temperature- and pH-dependent switch-off mechanism when growth is limited due to nutrient depletion. *International Journal of Food Microbiology, 107*, pp. 159–170.

Van Den Bogaard, A.E., Jensen, L.B. and Stobberingh, E.E. (1997). Vancomycin-resistant enterococci in turkeys and farmers. *New England Journal of Medicine, 337*, pp. 1558–1559.

Van Duijkeren, E., et al. (2014). Pleuromutilins: use in food-producing animals in the European Union, development of resistance and impact on human and animal health. *Journal of Antimicrobial Chemotherapy, 69*, pp. 2022–2031.

Vignaroli, C., et al. (2011). Multidrug-resistant enterococci in animal meat and faeces and co-transfer of resistance from an *Enterococcus durans* to a human *Enterococcus faecium*. *Current Microbiology, 62*(5), pp. 1438–1447.

Vrabec, M., et al. (2015). Antibiotic resistance and prevalence of *Enterococcus* spp. and *Escherichia coli* isolated from bryndza cheese. *Italian Journal of Animal Science, 14*(4), pp. 609–614.

Wang, J., et al. (2014). Central nervous system infections due to vancomycin-resistant enterococci: case series and review of the literature. *International Journal of Infectious Diseases, 25*, pp. 26–31.

Waxman, D.J. and Strominger, J.L. (1983). Penicillin-binding proteins and the mechanism of action of β-lactam antibiotics. *Annual Review of Biochemistry, 52*, pp. 825–869.

Werner, G. (2012). Current trends of emergence and spread of vancomycin-resistant Enterococci. In: M. Pana, ed., *Antibiotic Resistant Bacteria - A Continuous Challenge in the New Millennium*. Rijeka, Croatia: InTech Press.

Willems, R.J., et al. (2005). Global spread of vancomycin- resistant *Enterococcus faecium* from distinct nosocomial genetic complex. *Emerging and Infectious Diseases, 11*, pp. 821–828.

Woodford, N., et al. (1997). Methicillin-resistant *Staphylococcus aureus* and vancomycin-resistant enterococci. *Lancet, 350*, p. 738.

Woodford N, Livermore D. 2009. Infections caused by Gram-positive bacteria: a review of the global challenge. *Journal of Infection 59*(S1), pp. S4–S16.

Wormser, G.P., Keusch, G.T. and Heel, R.C. (1982). Co-trimoxazole (Trimethopim-sulfamethoxazole) an updated review of its antibacterial a activity and clinical efficacy. *Drugs 24*(6), pp. 459–518.

Wunderlich, P.F., et al. (1989). Double-blind report on the efficacy of lactic acid-producing *Enterococcus* SF68 in the prevention of antibiotic-associated diarrhoea and in the treatment of acute diarrhoea. *The Journal of International Medical Research, 17*, pp. 333–338.

Yousif, N.M., et al. (2005). Molecular characterization, technological properties and safety aspects of enterococci from 'Hussuwa', an African fermented sorghum product. *Journal of Applied Microbiology, 98*, pp. 216–228.

Yu, J., et al. (2015). Molecular characterization and resistant spectrum of *Enterococci* isolated from a haematology unit in China. *Journal of Clinical and Diagnostic Research, 9*(6), pp. DC04–DC07.

Zamfir, M., et al. (2006). Biodiversity of lactic acid bacteria in Romanian dairy products. *Systematic and Applied Microbiology, 29*, pp. 487–495.

Zeana, C., et al. (2001). Vancomycin-resistant *Enterococcus faecium* meningitis successfully managed with linezolid: case report and review of the literature. *Clinical Infectious Diseases, 33*, pp. 477–482.

Zervos, M.J. and Schaberg, D.R. (1985). Reversal of the *in vitro* susceptibility of enterococci to trimethoprim-sulfamethoxazole by folinic acid. *Antimicrobial Agents and Chemotherapy, 28*(3), pp. 446–448.

Zeyner, A. and Boldt, E. (2006). Effects of a probiotic *Enterococcus faecium* strain supplemented from birth to weaning on diarrhoea patterns and performance of piglets. *Journal of Animal Physiology and Animal Nutrition, 90*, pp. 25–31.

17

Antibiotic Resistance in Seafood-Borne Pathogens

Sanath Kumar[1], *Manjusha Lekshmi*[1], *Ammini Parvathi*[2], *Binaya Bhusan Nayak*[1], *and Manuel F. Varela*[3]

[1] *QC Laboratory, Harvest and Post Harvest Technology Division, ICAR-Central Institute of Fisheries Education (CIFE), Seven Bungalows, Versova, Andheri (W), Mumbai, 400061, India*
[2] *CSIR-National Institute of Oceanography (NIO), Regional Centre, Dr. Salim Ali Road, Kochi, 682018, India*
[3] *Eastern New Mexico University, Department of Biology, 1500 S. Ave. K, Station 33, Portales, New Mexico, 88130, USA*

17.1 Human Pathogenic Bacteria in Seafood

The global increase in the consumption of seafood has resulted in the increased mobility of fishery products across the world. For example, nearly half the seafood consumed in the United States is imported from other countries, mainly from Asia. This trend, although desirable as means of meeting the food demands of increasing global population, has also raised concerns on seafood acting as vehicles for human pathogens. In developed countries such as the United States, European nations, and Canada, seafood is responsible for a significant proportion of outbreaks (Iwamoto *et al.*, 2010). Epidemiological data from the United States suggest that a large proportion (76%) of the seafood-borne disease outbreaks is caused by bacterial pathogens, followed by viruses (21%) and parasites (2.6%). Among different seafood types, majority of seafood-borne outbreaks involved molluscan shellfish (45%), followed by fish (39%) and crustaceans (16%) (Iwamoto *et al.*, 2010). *Escherichia coli*, *Salmonella enterica*, *Campylobacter jejuni*, *Listeria monocytogenes*, and vibrios are the major pathogenic bacteria in seafood (Amagliani *et al.*, 2012). Humans, birds and animals are the reservoirs of enteric pathogens, while other highly opportunistic pathogens such as *Listeria monocytogenes* are widely distributed in nature. The presence of enteric bacteria such as *E. coli* and *Salmonella* in seafood is solely due to fecal contamination of the aquatic environment. Others such as vibrios are natural inhabitants of coastal-marine environments and hence could be found in considerable numbers in freshly harvested seafood. *V. cholerae*, *V. parahemolyticus*, and *V. vulnificus* are the prominent human pathogenic vibrios found in seafood.

The problem of seafood contamination with pathogenic microorganisms has become a key food safety issue in developing and highly populous countries of the world, where anthropogenic pollution of aquatic environments is a major problem. The routes of contamination of seafood are several and often very difficult to identify. Contamination of coastal water bodies such as estuaries, creeks and backwaters occurs due to direct discharge of domestic sewage and as a consequence fish and shellfish harvested from such water bodies harbor enteric pathogens. Fish and shellfish harvested from deep sea

Foodborne Pathogens and Antibiotic Resistance, First Edition. Edited by Om V. Singh.

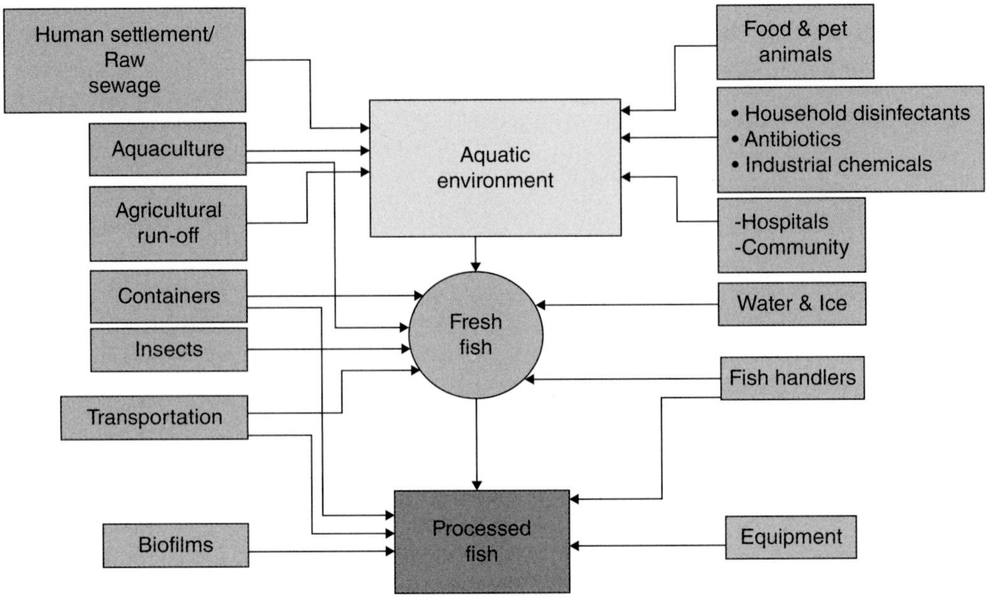

Figure 17.1 Schematic representation of different sources of contamination of fish and shellfish with pathogenic and antibiotic resistant bacteria.

are free from enteric pathogens, although they may harbor the naturally occurring vibrios. However after landing of fresh seafood, introduction of pathogens can occur through several portals of contamination as depicted in Figure 17.1. These also constitute critical control points at which application of suitable pathogen intervention strategies can prevent seafood contamination with pathogens. In the following sections, we discuss some of the important bacterial pathogens that can compromise the safety of seafood, their sources, virulence traits, and the emerging problem of antibiotic resistance in them.

17.1.1 *Escherichia coli* as an Indicator and a Pathogen

Escherichia coli is the most abundant facultative anaerobe present in the intestine of humans and several other warm blooded animals, comprising both non-pathogenic variants with modest ability to cause disease, and pathogenic variants, which are able to cause enteric and/or extra-intestinal diseases in various human and animal hosts (Johnson *et al.*, 2002). *E. coli* is classified into more than 700 serotypes based on the somatic antigen "O," flagellar antigen "H," and capsular antigen "K" (Eisenstein *et al.*, 2000), the majority of which are part of the regular intestinal microflora and are not harmful. A proportion of these are human and animal pathogens, involved in disease conditions like diarrhea, hemorrhagic colitis, dysentery, bladder and kidney infections, septicemia, hemolytic uremic syndrome, meningitis, and so on (Croxen *et al.*, 2013).

The native habitat of *E. coli* is the lower gastrointestinal tract of warm-blooded animals, being present in human and animal feces at an average of approximately 10^9/g (Edberg *et al.*, 2000). Hence its occurrence in food and water samples is considered as an indication of recent fecal contamination, since the organism does not survive well outside the human host for extended periods of time. *E. coli* is considered as the best indicator of fecal contamination and cheaper, precise, sensitive, and user-friendly methods are available for its detection directly from food and water samples.

E. coli is classified into five different pathogroups based on the presence of various virulence genes present on virulence-related plasmids or chromosomal pathogenicity islands, namely enterotoxigenic *E. coli* (ETEC), enteropathogenic *E. coli* (EPEC), enterohemorrhagic or Shiga toxin-producing *E. coli* (EHEC/STEC), enteroinvasive *E. coli* (EIEC), and enteroaggregtive *E. coli* (EAEC) (Nataro and Kaper, 1998). ETEC causes "traveler's diarrhea," characterized by watery diarrhea with low-grade fever, abdominal cramps, and nausea. In extreme conditions, it causes rice-water–like stools, resulting in dehydration. ETEC adheres to small intestinal mucosal cells and produce enterotoxins that are either heat labile (LT) or heat stable (ST), causing diarrhea (Sousa, 2006). The LTs are oligomeric toxins that are structurally and functionally strongly linked to the cholera enterotoxin (CT) expressed by *Vibrio cholerae* (Sixma *et al.*, 1993). On the other hand, STs are small, monomeric toxins with multiple cysteine residues. EPEC causes severe infantile diarrhea by invading the mucosal cells. The symptoms include fever, vomiting, and dehydration. This group of *E. coli* characteristically do not produce enterotoxins, but can form characteristic attaching and effacing lesions on the surface of epithelial cells lining the gastrointestinal tract with the help of pili, and are hence termed as attaching and effacing pathogens (Croxen *et al.*, 2013). The transmission of EPEC from person to person is through the fecal-oral route *via* contaminated food and water. Shiga toxin-producing *E. coli* (STEC) produces Shiga-like toxin 1 (Stx1) and Shiga-like toxin 2 (Stx2). The most common serotype responsible for large-scale, foodborne outbreaks is O157:H7, although numerous other serotypes of STEC are responsible for sporadic foodborne infections involving hemorrhagic colitis, bloody diarrhea and the hemolytic uremic syndrome (HUS) (Sousa 2006). Cattle are the major reservoirs of STEC, and foodborne infections are generally associated with foods of bovine origin. Among different pathotypes of *E. coli*, STEC are more serious pathogens owing to the severity of infections caused by them and the associated morbidities and mortalities.

Studies from India suggest the incidence of STEC in fresh seafood (Kumar *et al.*, 2001; Surendraraj *et al.*, 2010). Isolation of STEC has been reported from mussels, oysters and cockles in France (Gourmelon *et al.*, 2006). The isolation of ETEC in fresh fish and shellfish has been reported from Brazil (Teophilo *et al.*, 2002) and Korea (Koo *et al.*, 2012). In 2004, an outbreak of gastroenteritis associated with the consumption of butterfly shrimp occurred in Nevada, USA (Jain *et al.*, 2008). Examples of foods that can be potential agents of *E. coli* infection include fish products such as sashimi, oysters, mussels, clams, and cockles, which are often consumed raw or without further heat treatment.

17.1.2 *Salmonella enterica*

Foodborne outbreaks of *Salmonella* are reported from all over the world, and *Salmonella* is the second most important foodborne pathogen in terms of number of outbreaks. The natural habitat of *Salmonella* spp. is in the gastrointestinal tract of animals, birds, and reptiles. Infections with the human-restricted serovars *S.* Typhi and *S.* Paratyphi lead to invasive, life-threatening systemic disease known as typhoid fever, whereas non-typhoidal serovars (NTS) cause self-limiting gastroenteritis with intestinal inflammation and diarrhea (Zhang *et al.*, 2003). *Salmonella* contamination of marine environments has been recognized as a serious threat to human health worldwide (Gomez *et al.*, 1997). Seafood harvested from contaminated water acts as a vehicle for transmission of salmonellosis, and the human infection is generally associated with the consumption of raw or undercooked bivalve mollusks (Heinitz *et al.*, 2000). Contamination of seafood at various stages during and after harvest results in the introduction of *Salmonella* into fish and fishery products (Amagliani *et al.*, 2012). *Salmonella* is an environmentally persistent pathogen that is capable of surviving and proliferating in diverse environments including processing plants (Winfield and Groisman, 2003). *Salmonella* can form biofilms on a wide variety of contact surfaces, both inorganic and organic, that are resistant to biocides and disinfectants and can survive for prolonged periods of time in conditions of nutrient deficiency and become constant sources of food contamination.

17.1.3 Pathogenic Vibrios

The genus *Vibrio* consists of more than 35 species, distributed ubiquitously in marine environments throughout the world. As vibrios form part of the indigenous microflora of the marine environment, they are often associated with foodborne infections through consumptions of raw or undercooked seafoods. The most important pathogens of this genus are *Vibrio cholerae, V. parahemolyticus, V. vulnificus, V. alginolyticus,* and to a lesser extent *V. fluvialis, V. metschnikovii,* and *V. anguillarium*. The clinical signs may range from gastroenteritis to wound infection, otitis, and septicemia depending on the bacterial species causing disease. The main source of *Vibrio* is seafood and there are many reports of seafood-associated vibriosis outbreaks from all over the world (Iwamoto *et al.*, 2010).

V. cholerae is of most concern among the vibrios because of its ability to cause cholera. Of more than 200 *V. cholera* serogroups, serotypes O1 and O139 are associated with cholera, characterized by massive acute diarrhea (Rabbani *et al.*, 1999). These serotypes get into the aquatic environment through fecal contamination. There are several outbreaks of cholera linked to seafood. However, organisms of *V. cholerae* serogroups other than O1 and O139 (non-O1 non-O139 serogroups) have been associated with sporadic cases of foodborne outbreaks of gastroenteritis, but have not spread in epidemic form. Choleragenic *V. cholerae* are characterized by their ability to produce cholera toxin that is encoded by the *ctx* gene (Kaper *et al.*, 1995). However, it is not possible to distinguish O1 from non-O1 by general laboratory techniques.

V. parahemolyticus is a halophilic vibrio that occurs naturally in the marine environment. The bacterium inhabits aquatic habitats of varying salinity and is the major causative agent of some of the most serious diseases in fish and shellfish. The human pathogenic *V. parahemolyticus* causes acute gastroenteritis characterized by diarrhea, vomiting, and abdominal cramps through consumption of contaminated raw fish or shellfish. However, not all strains of *V. parahemolyticus* cause illness in humans; in fact, the majority of strains isolated from the environment or seafood are not pathogenic. The pathogenic strains of *V. parahemolyticus* are those that produce Thermostable Direct Hemolysin (TDH) toxin encoded by the *tdh* gene. TDH is an enzyme that lyses human red blood cells on "Wagatsuma" blood agar plates, which is referred to as the Kanagawa phenomenon-positive (KP^+). The majority (90%) of the *V. parahemolyticus* strains isolated from clinical cases are KP^+, whereas only 1 to 2% of the strains of environmental origin are KP^+. Another toxin produced by KP-negative *V. parahemolyticus* strains is the TDH-Related hemolysin (TRH) toxin encoded by the *trh* gene. TRH-positive *V. parahemolyticus* can cause gastroenteritis and are considered potentially pathogenic (Honda *et al.*, 1987).

Vibrios are generally considered to be highly susceptible to most clinically used antimicrobials. However, during the past few decades, antimicrobial resistance has emerged and evolved in many bacterial genera due to the excessive use of antimicrobials in humans, agriculture, and aquaculture (Cabello, 2006). This emerging issue has gained great concern due to increased resistance of pathogenic vibrios toward clinically used antimicrobials (Tang *et al.*, 2002). Tetracycline has been recommended as the antimicrobial of choice and alternative treatments are combinations of expanded-spectrum cephalosporins (e.g., ceftazidime) and doxycycline or a fluoroquinolone alone (Tang *et al.*, 2002). Recent studies suggest that multi-drug resistant *Vibrio* spp. are rampant in seafood which obtain antibiotic resistance *via* plasmids or other mobile genetic elements (Al-Othrubi *et al.*, 2014).

17.1.4 *Listeria monocytogenes*

Listeria monocytogenes is an important foodborne pathogen responsible for several outbreaks each year all over the world. The organism is widely distributed in nature and therefore can easily contaminate food. Infections occur in very young and elderly people, pregnant women and those with a weak immune system (Gahan and Hill, 2014). Most outbreaks have occurred with ready-to-eat, minimally processed foods of animal origin. Two serovars 1/2a and 4b are commonly involved in foodborne listeriosis, although 13 serovars of *L. monocytogenes* are known. Seafood harbors *L. monocytogens* due to postharvest secondary contamination. Contamination can occur in the processing environment from various sources such as the raw material, water, processing equipment, and surfaces. *L. monocytogenes* forms strong biofilms in processing environments, which are highly resistant to cleaning and disinfection. Biofilms can act as constant sources of contamination for raw and finished products. A few reports of seafood-borne *L. monocytogenes* infections suggest the involvement of vacuum-packed, cold-smoked fish, shrimp, smoked mussels, "Gravad" rainbow trout, and imitation crab meat (Rocourt *et al.*, 2000). Seafood products such as lightly salted or marinated products and smoked fish can pose the risk of *L. monocytogenes* infection when consumed without further heat treatment (Huss *et al.*, 2000). Since *L. monocytogenes* can grow at temperatures of up to 3.3°C, RTE seafood products should be stored at <3°C. The United States has imposed zero tolerance limit for *L. monocytogenes* in 25 grams of ready-to-eat seafood products such as smoked fish or crab meat.

17.2 An Overview of Bacterial Antimicrobial Resistance Mechanisms

Bacteria, after exposure to antimicrobial agents, may develop drug-resistance mechanisms or in some cases, may be intrinsically resistant to certain antibiotics. These resistance determinants reduce the antimicrobial activities of the drugs. There are four major molecular mechanisms of bacterial drug resistance:

i) The metabolic degradation of the antimicrobial agent involving some microbial enzymes that convert the drug into an ineffective compound incapable of killing or inhibiting the growth of the bacterium.
ii) Alteration of cellular drug target through various modifications so that the drug no longer effectively binds to its target. Such targets in a bacterial cell include the ribosomal protein subunits, nucleic acids, metabolic enzymes, and the bacterial cell wall components.
iii) Reduced drug permeability that involves the inability of the drug to gain entry into the cytoplasm of the bacterial cell. Such a drug-resistant bacterium might have opted to reduce the expression of a putative drug transport system in the membrane thus reducing cellular drug entry.
iv) Antimicrobial agent efflux pump systems, which operate by extruding drugs from the bacterial cell, thereby reducing the effective cellular drug concentrations (Wright, 2005; Kumar and Varela, 2013).

Bacteria may be single drug- or multidrug-resistant (MDR). Multiple mechanisms are associated with a MDR phenotype, and these are genetically encoded on some mobile genetic elements such as the transposons and plasmids (Gold and Moellering Jr., 1996). Bacteria become MDR when they acquire multiple genetic factors from different sources through various mechanisms of conjugation, transduction or transformation (Jacoby and Archer, 1991). Apart from the acquired resistance, bacteria have chromosomally encoded intrinsic resistance mechanisms such as the *marRAB* locus of *Escherichia coli*, which confers intrinsic resistance to a number of antibiotics such as tetracyclines, chloramphenicol, cephalosporins, penicillins, nalidixic acid, fluoroquinolones, rifampin, organic solvents, and so on (Alekshun and Levy, 1997). The antibiotic resistance mediated by the *marRAB* locus is due to efflux mechanisms, which involves the AcrAB efflux pump encoded on *marRAB* locus (Alekshun and Levy, 1997). In *E. coli, Salmonella enterica,* and *K. pneumoniae*, the *marRAB* locus is constitutively expressed making these bacteria resistant to several antibiotics and organic solvents (Webber *et al.*, 2006). Several modulators of *mar* transcription have been identified, which include benzoate, sodium salicylate, paraquat, dinitrophenol, menadione, and plumbaginetc (Randall and Woodward, 2001).

17.2.1 Antimicrobial Efflux Pumps in Bacteria

An important antimicrobial resistance mechanism involves the active efflux of drugs from the cells of bacteria (Kumar and Varela, 2012; Kumar and Varela, 2013). Active drug efflux reduces the intracellular concentrations of antimicrobial agents of bacteria, conferring resistance to single and multiple drugs and resulting in compromised efficacies of clinical chemotherapy (Levy, 2002; Andersen *et al.*, 2015). In terms of energetics, there are several general categories of active versus passive efflux mechanisms. The first category is called primary active transport, a system that uses the energy stored in the form of ATP and its hydrolysis to transport and extrude antimicrobials across the biological membrane against the drug concentration gradients (Wilkens, 2015). Another active transport mechanism, known as the phosphotransferase system (PTS) or group translocation, uses an enzymatic modification of the drug substrate as it is transported through the membrane against the drug concentration gradient (Parr and Saier, 1992). A third active transport system uses the energy stored in the form of an ion gradient that is generated by cellular respiration to actively efflux antimicrobial agents from bacteria (Poolman and Konings, 1993). The second category is called passive transport, a system which does not require biological energy but is driven by the solute gradient itself down its concentration gradient. These drug and multidrug transport systems would make good targets for modulation in terms of specific inhibitors of the transporter systems or in terms of regulation of gene expression systems, both approaches of which are of intensive focuses presently (Varela *et al.*, 2013).

17.3 Antibiotic Resistant Bacteria in the Aquatic Environment

Contamination of aquatic environment with human pathogenic bacteria in general and antibiotic resistant bacteria in particular, is a critical problem worldwide. Anthropogenic activity due to human settlement is the major source of pollution in developing countries. Release of untreated sewage into rivers, estuaries, and lakes not only results in eutrophication of water, but also has direct influence on the nature and composition of bacterial microflora in these ecosystems. Pathogenic, antibiotic resistant bacteria may be directly introduced into the aquatic environments from terrestrial land run-off. On the other hand, antimicrobial compounds such as disinfectants and antibiotics are directly released into the aquatic environment through domestic sewage, which has direct bearing on the emergence of antibiotic resistant bacteria. Sublethal levels of antibiotics in the aquatic environment lead to the development of antibiotic resistance.

Antibiotics are used in livestock for reasons other than treatment of diseases, such as for growth promotion and prophylaxis. The prolonged use of antibiotics results in the development of resistance in bacteria associated with farm animals. Resistant bacteria enter into the environment through animal manures, abattoir wastes, and sewage. Studies have shown the presence of residues of different antibiotics in streams receiving urban wastewater and populations of antibiotic resistant bacteria. Gradual development of resistance to antibiotics occurs as a consequence of direct exposure to antibiotic residues. Although many such resistant bacteria are not pathogenic to humans, they nevertheless can act as reservoirs of antibiotic resistance genes. Since the genetic factors responsible for antibiotic resistance are generally encoded on mobile elements, transfer of resistant determinants to other bacteria can lead to development of a resistant population within the aquatic environment.

The use of contaminated water for recreation, drinking, and agriculture results in the spread of resistant bacteria in human and animal populations. While prudent use of antibiotics is critical to halt further development of resistance to antibiotics, it is equally important to develop and implement waste management techniques to reduce environmental contamination with antibiotic residues and the resistant bacteria.

17.4 Antimicrobial Resistance in Seafood-Borne Pathogens

The discovery of antibiotics changed the perspective of treating human infectious diseases in the twentieth century. With the advent of microbiological, biochemical, and analytical techniques, numerous antibacterial compounds could be identified, purified, and even synthesized for human use. Over the years, the antibiotics, apart from being used in treating human and animal infections, have also been used in animal husbandry and aquaculture for growth promotion, improvement of feed efficiency, prophylaxis, and treatment of infections (Cabello, 2006). However, unlike in humans where antibiotics are prescribed for a specific duration following diagnosis of bacterial infections, the antibiotic use in non-human animals and aquaculture is arbitrary, often without isolating or identifying the causative agent and many times, unwarranted. As a consequence of unregulated use of antibiotics for purposes other than treatment of infections, humans are exposed to antibiotic resistant pathogens through food and the environment.

The morbidity and mortalities due to various infectious diseases in densely populated, low-income countries are clearly increasing, and their treatment is critically affected by the development of antibiotic resistance among pathogenic bacteria. The use of antibiotics for short-term economic gains can inflict a lasting negative impact on the human health and jeopardize chemotherapeutic applications of antibiotics (Walsh and Amyes, 2004; Kumar and Varela, 2013). The use of antibiotics in food growing facilities is responsible for the presence of resistant bacteria in the food chain. Many food animals and birds are the natural reservoirs of human pathogenic bacteria such as the *Salmonella enterica*, Shiga toxin-producing *Escherichia coli*, and *Campylobacter*.

The use of antimicrobials in food animals naturally leads to the development of resistance in these bacteria, which are easily transmitted to humans *via* food. In addition, since resistant bacteria reside in these food animals, their excreta will contaminate the environment with such bacteria, which in turn will reach the humans via water or food. Contamination of aquatic environment with pathogenic, antibiotic resistant bacteria has direct impact on aquatic food such as fish and shellfish being carriers of such bacteria. Biomagnification of pathogenic bacteria can occur in filter feeding animals such as oysters and clams, which can accumulate pathogens to numbers several fold higher than the surrounding water.

Several studies across the world have isolated antibiotic resistant human pathogens from fresh, processed, and ready-to-eat seafood/seafood products (Table 17.1). The magnitude of human health risk due to antibiotic resistant bacteria has been highlighted by a recent study in which ready-to-eat shrimps from four countries were found to harbor multidrug-resistant bacteria. Forty-two percent of the isolates from shrimp were resistant to antimicrobial agents, which included important human pathogens such as *E. coli*, *Enterococcus* spp., *Salmonella* spp., *Shigella flexneri*, *Staphylococcus* spp., and *Vibrio* spp. The presence of antibiotic resistant

Table 17.1 Incidences of Antibiotic Resistant Human Pathogenic Bacteria in Seafood and Seafood Products.

Seafood type	Source	Country	Resistant pathogen isolated	Resistance profile	Reference
Oysters & mussels	Markets & restaurants	Brazil	*Vibrio parahemolyticus*	AMP, CPD	Rojas *et al.*, 2011
Fresh fish	Retail markets	China	*Aeromonas & Enterobacteriaceae*	TE, SXT, β-lactams	Ye *et al.*, 2013
Shrimp	Retail markets	China	*V. vulnificus*	S, GEN, TOB, CZ	Pan *et al.*, 2013
Fish products	RTE, salted & smoked	Greece	*Staph. aureus* & *Staphylococcus* spp.	MET	Sergelidis *et al.*, 2014
Fresh fish	Retail market	India	*E. coli*	Cephalosporins, carbapenems, CIP	Singh *et al.*, 2014
Fresh seafood	Retail markets	India	*V. cholerae*	CPD, TIC, AMC, CL	Kumar and Lalitha, 2013
Fish & shrimps	Landing centre	India	*A. hydrophila*	TR, TE, NA, RIF, S	Thayumanavan *et al.*, 2003
Oysters, Fish	Fresh	India	*S. enterica* Weltevreden	Multiple antibiotics	Deekshit *et al.*, 2012
Fresh seafood	Fish markets	India	*A. hydrophila*	SMZ, E, TE, TM	Vivekanandhan *et al.*, 2002
Seafood products	Raw and RTE	Iran	*L. monocytogenes*	P, AMP, TE, VA, CIP	Fallah *et al.*, 2013
Seasoned, dried fish fillet	Retai markets	Korea	*E. coli*	CIP, NA, S, SMZ, TE	Ryu *et al.*, 2012
Fresh fish	Markets	Malaysia	*Aeromonas* spp.	Multiple antibiotics	Radu *et al.*, 2003
Oysters, Fish	Fresh seafood	Mexico	*L. monocytogenes*	Multiple antibiotics	Rodas-Suárez *et al.*, 2006
Fresh Fish	Retail markets	Morocco	*S. enterica*	NA, AMP, SM, TE	Bouchrif *et al.*, 2009
Fish & shellfish	Uncooked	Thailand	Multiple enteric pathogens	AMP, third- & fourth-generation cephalosporins	Woodring *et al.*, 2012
Bloody cockles, green mussels &oysters	Retail markets	Thailand	*Campylobacter* spp.	E, NA, CIP	Soonthornchaikul and Garelick, 2009
Fresh fish	Commercial markets	Turkey	*Enterobacteriaceae* & *Vibrionaceae*	TE, AMP, CEP, S	Matyar *et al.*, 2004
Fresh fish	Anchovy, trout & sea bream	Turkey	*Staph. aureus* & *S. enterica*	E, TE, PG, SMZ, GEN	Onmaz *et al.*, 2015
Fish & shellfish	Imported	USA	*S. enterica*	AMP, AMC, C, SXT, TE,	Zhao *et al.*, 2003; Ponce *et al.*, 2008; Khan *et al.*, 2006
Shrimp	Imported	USA	*K. pneumoniae* & *K. oxytoca*	Multiple antibiotics	Nawaz *et al.*, 2012

AMC = Amoxicillin/clavulanate; Amp = ampicillin; C = Chloramphenicol; CEP = Cephalothin; CIP = Ciprofloxacin; CL = Colistin; CPD = Cefpodoxime; CZ = Cefazolin; E = Erythromycin; GEN = Gentamicin; MET = Methicillin; NA = Nalidixic acid; P = Penicillin; PG = Penicillin G; RIF = Rifampicin; S = Streptomycin; SMZ = Sulfamethoxazole; TE = Tetracycline; TIC = Ticarcillin; TOB = Tobramycin; TR = Trimethoprim; VA = Vancomycin.

bacteria in ready-to-eat products constitutes a major human health hazard since the consumers are directly exposed to the risk of infection by these pathogens upon consumption. In a study in the United States, of 105 *Salmonella* strains isolated from imported seafood during 2000–2005, two isolates belonging to serovars Oslo and Bareily were resistant to trimethoprim/sulfamethoxazole, sulfisoxazole, ampicillin, tetracycline, and chloramphenicol, and these were isolated from seafood imported from India and Vietnam (Khan *et al.*, 2006). Both of these isolates harbored class I integrons suggesting that the resistance genes could be easily transferred to other pathogenic bacteria. There is a growing concern on the presence of multiple antibiotic resistant bacteria in seafood from Asian countries, which lack stringent laws to prohibit the misuse of antibiotics, resulting in rapid development and dissemination of resistance. Antibiotic resistant bacteria that emerge in human populations quickly enter the food chain and get disseminated rapidly. From a seafood safety point of view, immediate measures are needed to develop strategies to prevent the spread of antibiotic resistant bacteria through seafood.

17.4.1 Antibiotic Resistant Enterobacteriaceae

Members of the family Enterobacteriaceae, which include important foodborne pathogens, such as *Salmonella enterica* and *Escherichia coli*, are known to cause diverse types of infections ranging from wound infection to meningitis, and are also known agents of nosocomial infections. In the recent years, the chemotherapeutic options for enterobacteria are getting severely narrowed owing to the development of resistance to multiple antibiotics, the most notable among these being resistance to β-lactam group of antibiotics. The β-lactam group of antibiotics, which includes penicillins, cephalosporins, monobactams, and carbapenems constitutes a major component of antibiotic treatment against Enterobacteriaceae.

The mechanism of β-lactam resistance is through the production of β-lactamases, which hydrolyze these antibiotics into ineffective compounds (Bush *et al.*, 1995). Some of the earliest known lactamases are classified as metallo-β-lactamases (MBL) requiring zinc metal ion for their action and these enzymes degrade penicillins, cephalosporins, and carbapenems but not monobactams (Ambler, 1980). Equally important are the AmpC lactamases produced by several Enterobacteriaceae members including *E. coli, Salmonella enterica, Klebsiella pneumoniae, Enterobacter aerogenes,* and *Proteus mirabilis.*

Among the rapidly spreading multidrug-resistant bacteria are the extended spectrum-β-lactamase (ESBL)-producing Enterobacteriaceae. ESBLs can hydrolyze penicillins and broad spectrum β-lactams such as cefotaxime, ceftazidime, ceftriaxone, and monobactams such as aztreonam (Philippon *et al.*, 1989), but not cephamycins and carbapenems, and are inhibited by β-lactamase inhibitors such as clavulanate, sulbactam, and tazobactam (Jacoby and Carreras, 1990). A recent report from the Infectious Diseases Society of America listed ESBL-producing *Klebsiella* spp. and *E. coli* among the most important six drug-resistant microbes to which new therapies are urgently needed (Talbot *et al.*, 2006). Resistance to all cephalosporins, fluoroquinolones, and carbapenems leaves little scope to control these MDR bacteria with available antibiotics.

The prevalence of ESBL-producing gram-negative pathogens is high in Asian countries such as China, Japan, Korea, and India (Hawkey, 2008). Of specific importance are the extended spectrum β-lactase-producing *E. coli, Klebsiella* spp., and carbapenem-resistant *Acinetobacter* (Talbot *et al.*, 2006). ESBLs have originated from the TEM-1, TEM-2, and SHV-1 β-lactamases that confer resistance to penicillins and first generation cephalosporins due to point mutations as a result of the exposure to third generation and extended spectrum cephalosporins (Chaudhary and Aggarwal, 2004). Extended-spectrum β-lactamases (ESBLs), unlike their parent β-lactamases, confer resistance to a variety of cephalosporins such as cefotaxime, ceftriaxone, ceftazidime, and aztreonam but not against carbapenems (Bradford, 2001; Giske *et al.*, 2009).

The plasmid-borne ESBLs predominantly identified among Enterobacteriaceae are CTX-M, TEM, and SHV types (Paterson and Bonomo, 2005). The production of extended spectrum β-lactamases (ESBL) and Amp-C lactamase productions have been implicated in widespread resistance to third generation cephalosporins by many Gram-negative bacteria (Black *et al.*, 2005). Both ESBL and AmpC- β lactamases are plasmid-encoded. The bacteria producing ESBL are generally multi-resistant and complicate the treatment, leaving physicians with very few therapeutic options.

The presence of multi-resistant pathogenic bacteria among human populations means that such bacteria would often find their way into aquatic environment and then to the seafood. The occurrence of antibiotic resistant *E. coli* in seafood in India has been reported (Kumar et al., 2005). High levels of ampicillin- and ciprofloxacin-resistant *E. coli* have been isolated from seafood (Kumaran *et al.*, 2010). The occurrence multiple drug resistant, ESBL-producing enterobacteria in seafood is worrying. A recent study suggests that ESBL-producing enterobacteria harboring multiple ESBL genes and resistant to a wide range of cephalosporins are widely distributed in fresh seafood in India (Kumar *et al.*, 2014). The majority of the isolates from fresh seafoods from retail markets were resistant to cefotaxime, cefpodoxime, and ceftazidime, aztreonam, and piperacillin/tazobactam antibiotics (Kumar *et al.*, 2014). A study from Korea on commercial seafood has reported the incidence of cephalothin- and ampicillin-resistant *E. coli* (Ryu *et al.*, 2012). Miranda and Zemelman (2001) reported multiple antibiotic resistant enterobacteria in commercial fish captured from Concepcion Bay, Chile.

The recent discovery of $bla_{\text{NDM-1}}$ (New Delhi Metallo-β-Lactamase) – encoding *K. pneumoniae* has raised concerns on the occurrence and spread of antibiotic resistant bacteria in India (Yong *et al.*, 2009). $bla_{\text{NDM-1}}$, found on a 140 kb plasmid, is easily transferable and confers resistance to all β-lactams except aztreonam. The NDM-producing *K. pneumoniae*, believed to have originated from India, is now reported from all over the world (Nordmann *et al.*, 2011). Though $bla_{\text{NDM-1}}$ was initially reported in *K. pneumoniae* and *E. coli*, it has been found in other enterobacterial species and also in non-enterobacteria such as *Acinetobacter* spp. suggesting a faster dissemination of the gene in the environment.

NDM-positive bacteria are increasingly being reported from aquatic ecosystems raising concerns of such bacteria getting established in the environment and disseminated *via* foods (Walsh *et al.*, 2011). $bla_{\text{NDM-1}}$-positive *K. pneumoniae* has been reported from Kim Nguu River of Vietnam (Isozumi *et al.*, 2013). $bla_{\text{NDM-1}}$-harboring *Acinetobacter baumannii* has been isolated from the sewage of hospitals in Beijing, China (Zhang *et al.*, 2013). Isolations of $bla_{\text{NDM-1}}$-positive enterobacteria from clinical settings have been reported from Mumbai (Deshpande *et al.*, 2010). However, studies are needed to determine if $bla_{\text{NDM-1}}$ bacteria in the environment are derived from clinical sources or other reservoirs, such as animals and birds also exist. Nevertheless, bla_{NDM}-producing enterobacteria represent an emerging food safety concern.

The quinolone-resistant Enterobacteriaceae have been on the rise worldwide (Castanheira, 2008). Quinolone resistance is conferred by the mutation in gyrase or DNA topoisomerase II gene and also due to plasmid-borne *qnr* genes (Hooper, 2000). Following the discovery of plasmid mediated quinolone resistance in *K. pneumoniae* in the United States, a similar mechanism of resistance has been found in other members of Enterobacteriaceae such as *Citrobacter freundii, Escherichia coli, Enterobacter cloacae, Cronobacter sakazakii, Klebsiella pneumoniae, and K. oxytoca* from India and elsewhere (Nazik *et al.*, 2005; Magesh *et al.*, 2011). The hypothesis that antibiotic resistance genes can evolve in harmless bacteria, which later transfer them to human pathogens is strengthened by a recent study in which an aquatic bacterium *Shewanella algae* harboring the *qnrA* gene was found to be the natural reservoir for the resistance gene (Poirel *et al.*, 2005).

The use of antibiotics in animal husbandry may also contribute to the dissemination of such bacteria into the environment. It is important to identify the sources and routes of contamination of food and water with antibiotic resistant bacteria to control their spread in the environment. Strategies such as scientific management of domestic sewage, regulation of human settlement and pollution along the coast, and development of hygienic market facilities for seafood are needed to ensure quality and safety of seafood.

17.5 Antimicrobials in Aquaculture and their Human Health Consequences

The fast-growing aquaculture sector has a major role in fulfilling nutritional requirements of the growing population of the world. At present, aquacultured fish and shellfish contribute to about 20% of the world's animal protein intake (FAO, 2004). It is expected that by 2030, nearly 60% of food fish will be from aquaculture. The increasing demand for food fish has resulted in the rapid expansion of aquaculture. The commercial aquaculture, which has replaced the traditional aquaculture, is intensive in nature, both in terms of capital input and the aquaculture practices. Intensive stocking, use of protein-rich commercial feeds, growth promoting products, and application of antimicrobials for the control of microbial and parasitic diseases are all integral to intensive aquaculture practices. Microbial diseases, both viral and bacterial, parasitic infections, and changes in physico-chemical conditions, which adversely affect the health of cultured animals, can cause large-scale losses due to mortalities and consequent economic losses. These concerns force the use of antimicrobials, pesticides, bioremediators (both biological and chemical), and probiotics as prophylactic as well as disease control measures (Romero *et al.*, 2012). When antibiotics are used, their residues could be found in pond water and sediments, and the contamination of natural environments can occur through pond effluents.

The environmental fate of antibiotic residues is not well understood. Antibiotics used in aquaculture ponds can remain in sediments for periods ranging from several days to several months and can also disturb the biogeochemical cycles and primary production by inhibiting natural bacteria in water and the sediment. Sublethal levels of such antibiotics in aquaculture ponds or in the environment can lead to the development of resistance, both in pathogenic and non-pathogenic bacteria (DePaola *et al.*, 1995). Examples of such bacteria include *Aeromonas hydrophila*, *Edwardsiella tarda*, *Yersinia ruckeri*, *Citrobacter freundii*, *Lactococcus garviae*, *Photobacterium damselae*, *Vibrio anguillarum*, *Vibrio salmonicida*, *V. harveyi*, *Photobacterium psychrophilum*, and *Pseudomonas fluorescens* (Karunasagar *et al.*, 1994; Sorum, 2006) (summarized in Table 17.2).

Bacteria become resistant to multiple drugs when they acquire resistance mechanisms (genes) through horizontal transfer and mutations, or through increased expression of efflux pumps (Kumar *et al.*, 2013). More importantly, the use of one antibiotic can lead to the development of cross-resistance to multiple other unrelated drugs. For example, multidrug resistant bacteria were isolated from a Salmon farm in Chile, which used only florfenicol (Fernández-Alarcón, 2010). Scientific evidences suggest that some of the resistant forms of human pathogens have indeed acquired the genes from aquaculture environment (Angulo *et al.*, 2000; Neimann *et al.*, 2003). Antibiotics are used in aquaculture by farmers without proper guidance. Often, antibiotic use can be avoided if several good aquaculture practices are followed, such as good pond water quality, low to medium stocking density, water exchange and treatment, and proper diagnosis of the problem. Many diseases of fish and shrimps are due to parasites or

Table 17.2 Antibiotic resistant Bacteria in Aquaculture Farms and Farm-Raised Aquatic Animals.

Source	Country	Resistant bacteria	Resistance profile	Reference
Fish and shellfish	Australia	*Vibrio* spp., *Aeromonas* spp.	AMP, AMX, CN, TE, NA	Akinbowale *et al.*, 2006
Fresh water fish	Bangladesh	*Aeromonas* spp., *Pseudomonas* spp.	Multiple antibiotics	Chowdhury, 1998
Shrimp	Brazil	*Salmonella enterica*	TE, SXT, C	Ribeiro *et al.*, 2010
Salmon farm	Chile	Diverse Gram-negative bacteria	TE	Miranda *et al.*, 2003
Fish	China	*Escherichia coli*	AMP, NA, FFC, TE, SMX	Jiang *et al.*, 2012
Sea cucumber and sea urchin farm	China	*Pseudoalteromonasatlantica*, *Vibrio* spp.	TE, CEP, S	Dang *et al.*, 2006
Fish, water, and sediment	Denmark	*Flavobacterium psychrophilum*, *Aeromonas* spp..	OXA, AMX, TR, TE, FFC	Schmidt *et al.*, 2000
Shrimp *(Penaeus monodon)*	India	*Vibrio harveyi*	Multiple antibiotics	Abraham *et al.*, 1997
Penaeus monodon larvae	India	*Vibrio harveyi*	Multiple antibiotics	Karunasagar *et al.*, 1994
Fish	India	*Aeromonas* spp.	TE, AMX, AMP, NV, PB	Hatha *et al.*, 2005
Cultured *Seriola*	Japan	*Photobacterium damselae* ssp. *piscicida*	SMM, FLU, C, K, OXA	Kawanishi *et al.*, 2006
Fish and seawater	Korea	*Vibrio* spp., *Lactococcus garvieae*, *Photobacterium damsela*	TE	Kim *et al.*, 2004
Shrimp *(Penaeus monodon)*	Philippines	*Vibrio harveyi*	OTC, FZ, OXA	Tendencia and de la Peña, 2001
Freshwater farm	South Africa	*Aeromonas* spp.	TE, AMX, E, NA	Jacobs and Chenia, 2007
Integrated fish farming	Thailand	*Enteroacoccus*, *Acinetobacter*	OXA, SMX	Petersen *et al.*, 2002
Ornamental fish	USA	Gram-positive & -negative bacteria	TE, AMP, AZ, CIP, GEN, SMT	Rose *et al.*, 2013
Shrimp and fish ponds	Vietnam	*Acinetobacter*	Sulfonamides	Hoa *et al.*, 2008
Catfish *(Pangasius)*	Vietnam	*Edwardsiella ictaluri*	S, OTC, TR	Dung *et al.*, 2008

AMP = Ampicillin; AMX = Amoxycillin; AZ = Azithromycin; C = Chloramphenicol; CE = Cephalothin; CN = Cephalexin; CIP = Ciprofloxacin; E = Erythromycin; FFC = Florfenicol; FLU = Flumequine; FZ = furazolidone; GEN = Gentamycin; K = Kanamycin; NA = Nalidixic acid; PO = Polymyxin; NV = Novobiocin; OTC = Oxytetracycline; OXA = oxolinic acid; PB = Polymyxin; S = Streptomycin; S-T = sulfadiazine-trimethoprim; SMM = Sulfamonomethoxine; SMX = Sulfamethoxazole; SMT = Trimethoprim-Sulfamethoxazole; TE = Tetracycline; TR = Trimethoprim.

viruses, and the use of antibiotics in such bacteria not only fails to control mortalities occurring in the pond, but also alters the microbial balance in the pond.

The use of antibiotics in aquaculture has a direct role in the development of resistance against them by microorganisms. The intensive aquaculture relies heavily on antibiotics for prophylaxis and management of disease outbreaks (FAO, 2004; Sapkota *et al.*, 2008). Antibiotics administered through feed leach into the water and sediment and more than half of the antibiotics consumed by animals through feed are again excreted into water in active forms. The consumption of medicated feed again depends on environmental factors in the pond and the physiological conditions of animals.

The Asian countries, which contribute to 90% of aquacultured fish and shellfish, lack stringent laws to control the use of chemicals in aquaculture. The presence of antibiotic residues in fish and shellfish can be responsible for the rejection of shipments resulting in severe economic losses to the exporter. Precise information on the type of antimicrobials used, resistance profiles of bacteria in culture environment, and the fate of antibiotic residues in farm effluents are lacking from major fish growing countries in Asia.

A large number of antibiotics are used in aquaculture that include tetracyclines, chloramphenicol, florfenicol, fluoroquinolones such as enrofloxacin and norfloxacin, gentamicin, sulfonamides, nitrofurans, macrolides, and tiamulin (Holmstrom *et al.*, 2003; Soto-Rodríguez *et al.*, 2006). Apart from direct adverse effects of antibiotic residues such as allergies in consumers, the development and spread of antibiotic resistance is a serious issue associated with the use of antibiotics in aquaculture. The occurrence of multidrug resistant bacteria in farm environment is a health hazard to those who work directly with the fish.

Consumption of fish harboring resistant bacteria can lead to transfer of resistance traits to gut microflora, although scientific studies supporting this hypothesis are lacking. Nevertheless, aquacultured fish and shellfish can potentially act as vehicles for resistant bacteria from one geographical region to the other. Within the farm environment, populations of resistant bacteria are selected exposing cultured animals and farm workers to risk of infection by such bacteria. Studies have shown the presence of antibiotic resistant bacteria in cultured fish, shellfish, and the aquaculture environments as a consequence of antibiotic use. Considering the environmental and human health impact of antibiotics in aquaculture, the use of antibiotics in aquaculture has been banned since 2005.

The World Health Organization (WHO) has classified some antibiotics as critically important for treatment of human infections and prohibited their use in veterinary or aquaculture (WHO, 2011). In the United States, a few antibiotics such as tetracycline, florfenicol, and sulfadimethoxine-ormetoprim are approved for use in aquaculture by the Food and Drug Administration (FDA). A Joint FAO/OIE/WHO expert consultation on antimicrobial use in aquaculture and antimicrobial resistance in 2006 calls for prudent use of antibiotics in aquaculture, its proper monitoring and management in accordance with the WHO global principles, the recommendations of OIE (International office of epizootics), and the Codex code of practice for fish and fishery products (WHO, 2011).

In addition, environment-friendly farming involving best aquaculture practices such as implementation of biosecurity measures, use of disease-free animals for stocking and the use of probiotics, immunostimulants, and vaccines to boost the immunity of animals against outbreak of diseases are being intensely promoted among farming communities. These global initiatives for scientific management of aquaculture farms need to be strictly implemented by the developing economies that produce most of aquacultured fish and shellfish, in order to safeguard the health of consumers and the environment. Since aquaculture is an important livelihood activity in many countries in Asia, scientific management is critical for the sustainability of the aquaculture sector.

17.6 Future Directions

The presence of human pathogenic bacteria compromises the safety and acceptability of seafood and can lead to detentions and rejections causing economic losses to the producer. Therefore, good manufacturing practices and identification and management of possible sources of contamination of fish and shellfish from the farm to the consumer are necessary to ensure the safety of seafood. It is important to monitor seafood for the occurrence of antibiotic-resistant pathogenic bacteria in order to prevent their human transmission and also to prevent implication of seafood as vehicles of antibiotic resistant bacteria. A rational introspection of the use of antibiotics in aquaculture is needed and a strict implementation of the recommendations of WHO is necessary to safeguard the potency of antibiotics for their use in human medicine. The fast-growing seafood sector requires concerted efforts from all stakeholders to produce fish and shellfish that are free from pathogens and chemicals of human health concern.

References

Abraham, T.J., Palaniappan, R. and Dhevendaran K. (1997). Epibiotic infestation of luminous bacteria in penaeid shrimp *Penaeus indicus* (H. Milne Edwards) larvae. *Ind. J. Mar. Sci., 26*, pp. 209–211.

Akinbowale, O.L., Peng, H. and Barton, M.D. (2006). Antimicrobial resistance in bacteria isolated from aquaculture sources in Australia. *J. Appl. Microbiol., 100*, pp. 1103–1113.

Alekshun, M.N. and Levy, S.B. (1997). Regulation of chromosomally mediated multiple antibiotic resistance: the *mar* regulon. *Antimicrob. Agents Chemother., 41*, pp. 2067–2075.

Al-Othrubi, S.M.Y., et al. (2014). Antibiotic resistance of *Vibrio parahaemolyticus* isolated from cockles and shrimp sea food marketed in Selangor, Malaysia. *Clin. Microbiol., 3*, p. 1–7.

Amagliani, G, Brandi, G. and Schiavano, G.F. (2012). Incidence and role of *Salmonella* in seafood safety. *Food Res. Int., 45*, pp. 780–788.

Ambler, R.P. (1980). The structure of β-lactamases. *Philos. Trans. R. Soc. London B. Biol. Sci., 289*, pp. 321–331.

Andersen, J.L., et al. (2015). Multidrug efflux pumps from *Enterobacteriaceae*, *Vibrio cholerae* and *Staphylococcus aureus* bacterial food pathogens. *Int. J. Environ. Res. Public Health, 12*, pp. 1487–547.

Angulo, F., et al. (2000). Origins and consequences of antimicrobial resistant nontyphoidal *Salmonella*: implications for the use of fluoroquinolones in food animals. *Microb. Drug Res., 6*, pp. 77–83.

Black, J.A., Moland, E.S. and Thomson, K.S. (2005). AmpC disk test for detection of plasmid-mediated AmpC beta-lactamases in *Enterobacteriaceae* lacking chromosomal AmpC beta-lactamases. *J. Clin. Microbiol., 43*, pp. 3110–3113.

Bouchrif, B., et al. (2009). Prevalence and antibiotic-resistance of *Salmonella* isolated from food in Morocco. *J. Infect. Develop. Countries, 3*, pp. 35–40.

Bradford, P.A. (2001). Extended-spectrum beta-lactamases in the 21st century: characterization, epidemiology, and detection of this important resistance threat. *Clin Microbiol Rev., 14*, pp. 933–951.

Bush, K., Jacoby, G.A. and Medeiros, A.A. (1995). A functional classification scheme for beta-lactamases and its correlation with molecular structure. *Antimicrob. Agents Chemother., 39*, pp. 1211–1233.

Cabello, F.C. (2006). Heavy use of prophylactic antibiotics in aquaculture: a growing problem for human and animal health and for the environment. *Environ. Microbiol., 8*, pp. 1137–1144.

Castanheira, M., et al. (2008). Rapid emergence of $bla_{CTX\text{-}M}$ among *Enterobacteriaceae* in U.S. Medical Centers: molecular evaluation from the MYSTIC Program (2007). *Microb. Drug Resist., 14*, pp. 211–216.

Chaudhary, U., Aggarwal, R. (2004). Extended spectrum -lactamases (ESBL) - an emerging threat to clinical therapeutics. *Indian J. Med. Microbiol., 22*, pp. 75–80.

Chowdhury, M.B.R. (1998). Involvement of Aeromonads and Pseudomonads in diseases of farmed fish in Bangladesh. *Fish Pathol., 33*, pp. 247–254.

Croxen, M.A., et al. (2013). Recent advances in understanding enteric pathogenic *Escherichia coli. Clin. Microbiol. Rev., 26*, pp. 822–880.

Dang, H., et al. (2006). Concurrence and *cat* and *tet* genes in multiple antibiotic-resistant bacteria isolated from a sea cucumber and sea urchin mariculture farm in China. *Microb. Ecol., 52*, pp. 634–643.

Deekshit, V.K., et al. (2012). Detection of class 1 integrons in *Salmonella* Weltevreden and silent antibiotic resistance genes in some seafood-associated nontyphoidal isolates of *Salmonella* in south-west coast of India. *J. Appl. Microbiol., 112*, pp. 1113–1122.

DePaola, A., Peeler, J.T. and Rodrick, G.E. (1995). Effect of oxytetracycline-medicated feed on antibiotic resistance of gram-negative bacteria in catfish ponds. *Appl. Environ. Microbiol., 61*, p. 3513.

Deshpande, P., et al. (2010) New Delhi Metallo-beta lactamase (NDM-1) in *Enterobacteriaceae*: Treatment options with carbapenems compromised. *J. Assoc. Physicians India, 58*, pp. 147–149.

Dung, T.T., et al. (2008). Antimicrobial susceptibility pattern of *Edwardsiella ictaluri* isolates from natural outbreaks of bacillary necrosis of *Pangasianodon hypophthalmus* in Vietnam. *Microb. Drug Res., 14*, pp. 311–316.

Edberg, S.C., et al. (2000). *Escherichia coli*: the best biological drinking water indicator for public health protection. *J. Appl. Microbiol., 88*, pp. 106S–116S.

Eisenstein, B. and Zaleznik, D. (2000). *Enterobacteriaceae*. In: D. Mandell, ed., *Bennett's Principles and practice of infectious diseases*, 5th ed. London: Churchill Livingstone, pp. 2294–2310.

Fallah, A.A., Saei-Dejkordi, S.S. and Mahzounieh, M. (2013). Occurrence and antibiotic resistance profiles of *Listeria monocytogenes* isolated from seafood products and market and processing environments in Iran. *Food Control., 34*, pp. 630–636.

FAO. (2004). The state of world fisheries and aquaculture: Rome Italy, Food and Agriculture Organization of the United States, Fisheries Department, pp. 1–153.

Fernández-Alarcón, C., et al. (2010). Detection of the *floR* gene in a diversity of florfenicol resistant Gram-negative bacilli from freshwater salmon farms in Chile. *Zoonoses Public Health, 57*, pp. 181–188.

Gahan, C.G.M. and Hill, C. (2014). *Listeria monocytogenes*:survival and adaptation in the gastrointestinal tract. *Front. Cell. Infect. Microbiol., 4*, pp. 9.

Giske, C.G., et al. (2009). Redefining extended-spectrum beta-lactamases: balancing science and clinical need. *J. Antimicrob. Chemother., 63*, pp. 1–4.

Gold, H.S. and Moellering, R.C. Jr. (1996). Antimicrobial-drug resistance. *N. Engl. J. Med., 335*, pp. 1445–1453.

Gomez, T.M., et al. (1997). Foodborne salmonellosis. *World Health Stat Q., 50*, pp. 81–89.

Gourmelon, M., et al. (2006). First isolation of Shiga toxin 1d producing *Escherichia coli* variant strains in shellfish from coastal areas in France. *J. Appl. Microbiol. 100*, pp. 85–97.

Hoa, P.T.P., et al. (2008). Detection of the *sul1*, *sul2*, and *sul3* genes in sulfonamide-resistant bacteria from wastewater and shrimp ponds of north Vietnam. *Sci. Total Environ., 405*, pp. 377–384.

Hatha, M., et al. (2005). Antibiotic resistance pattern of motile aeromonads from farm raised fresh water fish. *Int. J. Food Microbiol., 98*, pp. 131–134.

Hawkey, P.M. (2008). Prevalence and clonality of extended-spectrum beta-lactamases in Asia. *Clin. Microbiol. Infect., 14*, pp. 159–165.

Heinitz, M.L., et al. (2000). Incidence of *Salmonella* in fish and seafood. *J. Food Prot., 63*, pp. 579–592.

Holmström, K., et al. (2003). Antibiotic use in shrimp farming and implications for environmental impacts and human health. *Int. J. Food Sci. Technol., 38*, pp. 255–266.

Honda, S., et al. (1987). Gastroenteritis due to Kanagawa negative *Vibrio parahaemolyticus. Lancet, 1*, pp. 331–332.

Hooper, D.C., et al. (1987). Mechanisms of action of and resistance to ciprofloxacin. *The Am. J. Med., 82*, pp. 12–20.

Huss, H.H., Jørgensen, L.V. and Vogel, B.F. (2000). Control options for *Listeria monocytogenes* in seafoods. *Int. J. Food Microbiol., 62*, pp. 267–274.

Isozumi, R., et al. (2012) $bla_{(NDM-1)}$-positive *Klebsiella pneumoniae* from environment, *Vietnam. Emerg. Infect. Dis., 18*, pp. 1383–1385.

Iwamoto, M., et al. (2010). Epidemiology of seafood-associated infections in the United States. *Clin. Microbiol. Rev., 23*, pp. 399–411.

Jacobs, L. and Chenia, H.Y. (2007). Characterization of integrons and tetracycline resistance determinants in *Aeromonas* spp. isolated from South African aquaculture systems. *Int J Food Microbiol., 114*, pp. 295–306.

Jacoby, G.A. and Archer, G.L. (1991). New mechanisms of bacterial resistance to antimicrobial agents. *N. Engl. J. Med., 324*, pp. 601–612.

Jacoby, G.A. and Carreras, I. (1990). Activities of β-lactam antibiotics against *Escherichia coli* strains producing extended-spectrum β-lactamases. *Antimicrob. Agents Chemother., 34*, pp. 858–862.

Jain, S., et al. (2008). An outbreak of enterotoxigenic *Escherichia coli* associated with sushi restaurants in Nevada, 2004. *Clin. Infect. Dis., 47*, pp. 1–7.

Jiang, H.X., et al. (2012). Prevalence and characteristics of β-lactamase and plasmid-mediated quinolone resistance genes in *Escherichia coli* isolated from farmed fish in China. *J. Antimicrob. Chemother., 67*, pp. 2350–2353.

Johnson, J.R. and Russo, T.A. (2002). Extraintestinal pathogenic *Escherichia coli*: “the other bad *E. coli*”. *J. Lab. Clin. Med., 139*, pp. 155–162.

Karunasagar, I., et al. (1994). Mass mortality of *Penaeus monodon* larvae due to antibiotic-resistant *Vibrio harveyi* infection. *Aquaculture, 128*, pp. 203–209.

Kaper, J.B., Morris, J.G. and Levine, M.M. (1995). *Cholera. Clin. Microbiol. Rev., 8*, pp. 48–86.

Kawanishi, M., et al. (2006). Drug resistance and random amplified polymorphic DNA analysis of *Photobacterium damselae* ssp. *piscicida* isolates from cultured Seriola (yellowtail, amberjack and kingfish) in Japan. *Lett. Appl. Microbiol., 42*, pp. 648–653.

Khan, A.A., et al. (2006). Characterization of class 1 integron resistance gene cassettes in *Salmonella enterica* serovars Oslo and Bareily from imported seafood. *J. Antimicrob. Chemother., 58*, pp. 1308–1310.

Kim, S.R., Nonaka, L. and Suzuki, S. (2004). Occurrence of tetracycline resistance genes *tet*(M) and *tet*(S) in bacteria from marine aquaculture sites. *FEMS Microbiol. Lett., 237*, pp. 147–156.

Koo, H.J., et al. (2012). Phylogenetic group distribution and prevalence of virulence genes in *Escherichia coli* isolates from food samples in South Korea. *World J. Microbiol. Biotechnol., 28*, pp. 1813–1816.

Kumar, R. and Lalitha, K.V. (2013). Prevalence and molecular characterization of *Vibrio cholerae* O1, non-O1 and non-O139 in tropical seafood in Cochin, India. *Foodborne Pathog. Dis., 10*, pp. 278–283.

Kumar, S, et al. (2013). Bacterial antimicrobial efflux pumps of the MFS and MATE transporter families: A review. *Recent Res. Devel. Antimicrob. Agents Chemother., 7*, pp. 1–21.

Kumar, S., et al. (2001). Detection of Shiga-toxigenic *Escherichia coli* (STEC) in fresh seafood and meat marketed in Mangalore, India by PCR. *Lett. Appl. Microbiol., 33*, pp. 334–338.

Kumar, S., et al. (2005). Prevalence and antibiotic resistance of *Escherichia coli* in tropical seafood. *World J. Microbiol. Biotechnol., 21,* pp. 619–623.

Kumar, S., et al. (2014). Occurrence of multiple antibiotic resistant extended spectrum-β-lactamase (ESBL) producing enterobacteria in fresh seafood. *J Vet. Sci. Technol., 5,* p. 3.

Kumar, S. and Varela, M.F. (2012). Biochemistry of bacterial multidrug efflux pumps. *Int. J. Mol. Sci., 13,* pp. 4484–4495.

Kumar, S. and Varela, M.F. (2013). Molecular mechanisms of bacterial resistance to antimicrobial agents. In: A. Méndez-Vilas, ed., *Microbial pathogens and strategies for combating them: science, technology and education.* Badajoz, Spain: Formatex.

Kumaran, S., et al. (2010). Antibiotic resistant *Escherichia coli* strains from seafood and its susceptibility to seaweed extracts. *Asian Pac. J. Trop. Med., 3,* pp. 977–981.

Levy, S.B. (2002). Active efflux, a common mechanism for biocide and antibiotic resistance. *Symp Ser. Soc. Appl. Microbiol., 31,* pp. 65S–71S.

Magesh, H., et al. (2011). Identification of plasmid-mediated quinolone resistance genes *qnrA1, qnrB1,* and *aac(6')*-1b-cr in a multiple drug-resistant isolate of *Klebsiella pneumoniae* from Chennai. *Ind. J. Med. Microbiol., 29,* pp. 262–268.

Matyar, F., et al. (2004). Prevalence and resistance to antibiotics in Gram negative bacteria isolated from retain fish in Turkey. *Ann. Microbiol., 54,* pp. 151–160.

Miranda, C.D., et al. (2003). Diversity of tetracycline resistance genes in bacteria from Chilean salmon farms. *Antimicrob. Agents Chemother., 47,* pp. 883–888.

Miranda, C.D. and Zemelman, R. (2001). Antibiotic resistant bacteria in fish from the Concepción Bay, Chile. *Mar. Pollut. Bull., 42,* pp. 1096–1102.

Nataro, J.P. and Kaper, J.B. (1998). Diarrheagenic *E. coli. Clin. Microbiol. Rev., 11,* pp. 142–201.

Nawaz, M., et al. (2012). Isolation and characterization of multidrug-resistant *Klebsiella* spp. isolated from shrimp imported from Thailand. *Int. J. Food Microbiol., 155,* pp. 179–184.

Nazik, H., Ongen, B. and Kuvat, N. (2005). Investigation of plasmid-mediated quinolone resistance among isolates obtained in a Turkish intensive care unit. *Jpn. J. Infect. Dis., 61,* pp. 310–312.

Neimann, J., et al. (2003). A case-control study of risk factors for sporadic *Campylobacter* infections in Denmark. *Epidemiol. Infect., 130,* pp. 353–366.

Nordmann, P., et al. (2011). Does broad spectrum β-lactam resistance due to NDM-1 herald the end of the antibiotic era for treatment of infections caused by Gram-negative bacteria? *J. Antimicrob. Chemother., 66,* pp. 689–692.

Onmaz, N.E., et al. (2015). Occurrence and antimicrobial resistance of *Staphylococcus aureus* and *Salmonella* spp. in retail fish samples in Turkey. *Mar. Pollut. Bull., 90,* pp. 242–246.

Pan, J., et al. (2013). Molecular characterization and antibiotic susceptibility of *Vibrio vulnificus* in retail shrimps in Hangzhou, People's Republic of China. *J. Food Prot., 76,* pp. 2063–2068.

Parr, T.R., Jr. and Saier, M.H., Jr. (1992). The bacterial phosphotransferase system as a potential vehicle for the entry of novel antibiotics. *Res. Microbiol., 143,* pp. 443–447.

Paterson, D.L. and Bonomo, R.A. (2005). Extended-spectrum beta-lactamases: a clinical update. *Clin Microbiol Rev., 18,* pp. 657–686.

Petersen, A., et al. (2002). Impact of integrated fish farming on antimicrobial resistance in a pond environment. *Appl. Environ. Microbiol., 68,* pp. 6036–6042.

Philippon, A., Labia, R., and Jacoby, G. (1989). Extended-spectrum beta-lactamases. *Antimicrob. Agents Chemother., 33,* pp. 1131–6.

Poirel, L., et al. (2005). Vibrionaceae as possible sources of Qnr-like quinolone resistance determinants. *J. Antimicrob. Chemother., 56,* pp. 1118–1121.

Ponce, E., et al. (2008). Prevalence and characterization of *Salmonella enterica* serovar Weltevreden from imported seafood. *Food. Microbiol., 25,* pp. 29–35.

Poolman, B. and Konings, W.N. (1993). Secondary solute transport in bacteria. *Biochim. Biophys. Acta., 1183,* pp. 5–39.

Rabbani G.H. and Greenough, W.B. (1999). Food as a vehicle of transmission of cholera. *J. Diarrhoeal Dis. Res., 17,* pp. 1–9.

Radu, S., et al. (2003). Prevalence and resistance to antibiotics for *Aeromonas* species from retail fish in Malaysia. *Int. J. Food Microbiol., 81,* pp. 261–266.

Randall, L.P. and Woodward, M.J. (2001). Multiple antibiotic resistance (*mar*) locus in *Salmonella enterica* serovar Typhimurium DT104. *Appl. Environ. Microbiol., 67,* pp. 1190–1197.

Ribeiro, R.V., et al. (2010). Incidence and antimicrobial resistance of enteropathogens isolated from an integrated aquaculture system. *Lett. Appl. Microbiol., 51,* pp. 611–618.

Rocourt, J., Jacquet, C. and Reilly A. (2000). Epidemiology of human listeriosis and seafoods. *Int. J. Food Microbiol., 62,* pp. 197–209.

Rodas-Suárez, O.R., et al. (2006). Occurrence and antibiotic sensitivity of *Listeria monocytogenes* strains isolated from oysters, fish, and estuarine water. *Appl. Environ. Microbiol., 72,* pp. 7410–7412.

Rojas, M.V., et al. (2011). Characterization of *Vibrio parahaemolyticus* isolated from oysters and mussels in São Paulo, Brazil. *Rev. Inst. Med. Trop. Sao Paulo, 53,* pp. 201–205.

Romero, J., Feijoo, C.J., and Navarrete, P. (2012). Antibiotics in Aquaculture-use, abuse and alternatives. In: E. Carvalho, ed., *Health and environment in aquaculture* ISBN: 978-953-51-0497-1, InTech.

Rose, S., et al. (2013). Imported ornamental fish are colonized with antibiotic-resistant bacteria. *J. Fish Dis., 36,* pp. 533–542.

Ryu, S.H., et al. (2012). Antimicrobial resistance and resistance genes in *Escherichia coli* strains isolated from commercial fish and seafood. *Int. J. Food Microbiol., 152,* pp. 14–18.

Sapkota, A., et al. (2008). Aquaculture practices and potential human health risks: Current knowledge and future priorities. *Environ. Int., 34,* pp. 1215–1226.

Schmidt, A.S., et al. (2000). Occurrence of antimicrobial resistance in fish-pathogenic and environmental bacteria associated with four danish rainbow trout farms. *Appl. Environ. Microbiol., 66,* pp. 4908–4915.

Sergelidis, D., et al. (2014). Isolation of methicillin-resistant *Staphylococcus* spp. from ready-to-eat fish products. *Lett. Appl. Microbiol., 59,* pp. 500–506.

Singh, A.S., et al. (2014). Isolation of *Escherichia coli* harboring $bla_{\mathrm{NDM\text{-}5}}$ from fresh fish in India. *J. Microbiol. Immunol. Infect., pii:* S1684–1182(14)00234-5. doi: 10.1016/j.jmii.2014.11.004 .

Sixma, T.K., et al. (1993). Refined structure of *Escherichia coli* heat labile enterotoxin, a close relative of cholera toxin. *J. Mol. Biol., 230,* pp. 890–918.

Soonthornchaikul, N. and Garelick, H. (2009). Antimicrobial resistance of *Campylobacter* species isolated from edible bivalve molluscs purchased from Bangkok markets, Thailand. *Foodborne Pathog. Dis., 6,* pp. 947–51.

Sorum, H. (2006). Antimicrobial drug resistance in fish pathogens. In: F. Aarestrup, ed., *Antimicrobial resistance in bacteria of animal origin.* Washington, DC: ASM Press, pp. 213–238.

Soto-Rodríguez. S., Armenta. M., Gomez-Gil. B. (2006) Effects of enrofloxacin and florfenicol on survival and bacterial population in an experimental infection with luminescent *Vibrio campbellii* in shrimp larvae of *Litopenaeus vannamei. Aquaculture, 255,* pp. 48–54.

Sousa, C.P. (2006). The versatile strategies of *Escherichia coli* pathotypes: A Mini Review. *J. Venom. Anim. Toxins incl. Trop. Dis., 12,* p. 3.

Surendraraj, A., Thampuran, N. and Joseph, T. (2010). Molecular screening, isolation, and characterization of enterohemorrhagic *Escherichia coli* O157:H7 from retail shrimp. *J. Food Prot., 73,* pp. 97–103.

Talbot, G.H., et al. (2006). Bad bugs need drugs: an update on the development pipeline from the Antimicrobial Availability Task Force of the Infectious Diseases Society of America. *Clin. Infect. Dis., 42*, pp. 657–668.

Tang, H.J., et al. (2002). In vitro and invivo activities of newer fluoroquinolones against *Vibrio vulnificus. Antimicrob. Agents Chemother., 46*, pp. 3580–3584.

Tendencia, E.A. and de la Peña, L.D. (2001). Antibiotic resistance of bacteria from shrimp ponds. *Aquaculture, 195*, pp. 193–204.

Teophilo, G.N., et al. (2002). *Escherichia coli* isolated from seafood: Toxicity and plasmid profiles. *Int. Microbiol., 5*, pp. 11–14.

Thayumanavan, T., et al. (2003). Incidence of haemolysin-positive and drug-resistant *Aeromonas hydrophila* in freshly caught finfish and prawn collected from major commercial fishes of coastal South India. *FEMS Immunol. Med. Microbiol., 36*, pp. 41–45.

Varela, M.F., Kumar, S. He, G. (2013). Potential for inhibition of bacterial efflux pumps in multidrug-resistant *Vibrio cholerae. Indian J. Med. Res., 138*, pp. 285–287.

Vivekanandhan, G., et al. (2002). Antibiotic resistance of *Aeromonas hydrophila* isolated from marketed fish and prawn of South India. *Int. J. Food Microbiol., 76*, pp. 165–168.

Walsh, F.M. and Amyes, S.G. (2004). Microbiology and drug resistance mechanisms of fully resistant pathogens. *Curr. Opin. Microbiol., 7*, pp. 439–444.

Walsh, T.R., et al. (2011). Dissemination of NDM-1 positive bacteria in the New Delhi environment and its implications for human health: an environmental point prevalence study. *Lancet Infect. Dis., 11*, pp. 355–362.

Webber, M.A., et al. (2006). Overexpression of *marA, soxS and acrB* in veterinary isolates of *Salmonella enterica* rarely correlates with cyclohexane tolerance. *J. Antimicrob. Chemother., 57*, pp. 673–679.

WHO (World Health Organization). (2011). *Critically important antimicrobials for human medicine. WHO Advisory Group on Integrated Surveillance of Antimicrobial Resistance (AGISAR). 3rd revision, 2011.* Geneva: WHO.

Wilkens, S. (2015). Structure and mechanism of ABC transporters. *F1000 Prime Rep., 7*, p. 14.

Winfield, M.D. and Groisman, E.A, (2003). Role of non-host environments in the lifestyles of *Salmonella* and *Escherichia coli. Appl. Environ. Microbiol., 69*, pp. 3687–3694.

Woodring, J., et al. (2012). Prevalence and antimicrobial susceptibilities of *Vibrio, Salmonella,* and *Aeromonas* isolates from various uncooked seafoods in Thailand. *J. Food Prot., 75*, pp. 41–47.

Wright, G.D. (2005). Bacterial resistance to antibiotics: enzymatic degradation and modification. *Adv. Drug Del. Rev., 57*, pp. 1451–1470.

Ye, L., et al. (2013). Antibiotic-resistant bacteria associated with retail aquaculture products from Guangzhou, China. *J. Food. Prot. 76*, pp. 295–301.

Yong, D., et al. (2009). Characterization of a new metallo-beta-lactamase gene, *bla*(NDM-1), and a novel erythromycin esterase gene carried on a unique genetic structure in *Klebsiella pneumoniae* sequence type 14 from India. *Antimicrob. Agents Chemother., 53*, pp. 5046–5054.

Zhang, C., et al. (2013). Higher isolation of NDM-1 producing *Acinetobacter baumannii* from the sewage of the hospitals in Beijing. *PLoS One, 8*, p. e64857.

Zhang, S., et al. (2003) Molecular pathogenesis of *Salmonella enterica* serotype Typhimurium-induced diarrhea. *Infect. Immun. 71*, pp. 1–12.

Zhao, S., et al. (2003). Antimicrobial-resistant *Salmonella* serovars isolated from imported foods. *Int. J. Food Microbiol., 84*, pp. 87–92.

18

Antimicrobial Resistance of *Campylobacter* sp.

Tareq M. Osaili[1] *and Akram R. Alaboudi*[2]

[1] *Dept. of Nutrition and Food Technology, Faculty of Agriculture, Jordan Univ. of Science and Technology, Irbid, Jordan*
[2] *Dept. of Veterinary Pathology and Public Health, Jordan Univ. of Science and Technology, Irbid, Jordan*

18.1 Introduction

Campylobacters are gram-negative, slim, curved rods bacteria with a polar flagellum located at one or both ends of the cell makes the bacteria highly motile. *Campylobacter* spp. are thermophilic growing at optimum at 42°C and microaerophilic requiring oxygen concentrations of 3–15% and carbon dioxide concentrations of 3–5%. The thermophilic characteristic of the microorganism is believed to be an adaptation to the bird intestines and the environment of animal. *Campylobacter* spp. have a small genome of 1.6–1.7 Mbp, and that may explain the complex nutritional requirements of this microorganism in laboratory culture (Griffiths and Park, 1990).

Campylobacter spp. are the most frequently recognized bacterial cause of human gastroenteritis worldwide (Silva *et al.*, 2011). *Campylobacter* spp. causes a foodborne infection called campylobacteriosis. In the United States, it is estimated 8,463 cases of human campylobacteriosis occur each year (Scallan *et al.*, 2011). In the European Union, more than 214,000 human cases were reported in 2012 (EFSA and ECDC, 2014). *Campylobacter* spp. are members of the epsilonproteobacterial group containing two clinical important species; *C. jejuni* and *C. coli*. The species that are most commonly associated with human infection are *C. jejuni, which* is most commonly associated with human infection (80–90%), whereas *C. coli* is responsible for 5–10% of human infection (EFSA and ECDC, 2013).

Campylobacter spp. are part of the normal intestinal flora of wild and domesticated animals and birds. Poultry is considered the most significant reservoir worldwide. Post-slaughter processing of poultry does not reduce the level of colonization of poultry and may increase of cross-contamination of previously uncontaminated carcasses (Sack *et al.*, 2001). Poultry and poultry meat products are the main source of human infection. Ingestion of contaminated milk, meat, and water are known as an important source of human infection. Also, pet animals (dogs and cats), wild birds, and other animals are sources of infection (Sheppard *et al.*, 2009). Cross-contamination of foods from contaminated raw food may also represent a significant source of infection.

C. jejuni causes acute gastroenteritis, diarrhea, fever, and abdominal pain. It is implicated in extra-intestinal infections that include reactive arthritis, septicemia, bacteremia, meningitis, and hemolytic uremic syndrome (Foley, 2012; Ge *et al.*, 2013). Usually, *Campylobacter* infections are short duration, mild, and self-limiting disease lasting only a few days, and the treatment with antibiotics is commonly not

Foodborne Pathogens and Antibiotic Resistance, First Edition. Edited by Om V. Singh.

required. However, antibiotic therapy is necessary under specific clinical situations; however, in severe cases, infection in immune-suppressed patients and long-lasting infections (Guyard-Nicodème *et al.*, 2015).

Macrolides, such as erythromycin and azithromycin, are now generally considered to be the optimal drug for treating campylobacteriosis. Fluoroquinolones (ciprofloxacin) and tetracycline have been used as an alternative therapy of clinical campylobacteriosis (Allos, 2001). Aminoglycoside (gentamicin) treatment is used in very severe cases of *Campylobacter* infections, that is, bacteremia (Corcoran *et al.*, 2006; Alfredson and Korolik 2007; Kurinic *et al.*, 2007). However, therapy complicated as antimicrobial resistance in *Campylobacter* isolates from human infections has become increasingly common after the approval of some antibiotics in poultry farming (Engberg *et al.*, 2001) especially resistance to fluoroquinolones, macrolides, tetracycline, and chloramphenicol (Cody *et al.*, 2010; EFSA and ECDC, 2011; Silva, 2011).

18.2 Antimicrobial Resistance

Antimicrobials include all substance that kill microorganisms or inhibits their growth in a clinical setting, materials, and on surfaces. Antimicrobials include antibiotics that are natural substances produced by fungi or bacteria and used to treat bacterial infections in humans and animals. There are also synthetic antibacterial drugs or chemotherapeutics that are purely chemically derived products. "Antibiotic" is often used as synonym of the general term "antimicrobial agent." Antimicrobial resistance is the ability of a microorganism to resist the growth inhibitory or killing effect of an antimicrobial above the normal sensitivity of the specific bacterial strain (Butaye *et al.*, 2015).

The availability and use of antibiotics for treating infectious diseases effectively improved the health and life expectancy of humans and the health of animals. But, the abuse of antibiotics has resulted in antimicrobial resistance in bacteria to the therapeutic agents used.

Bacterial resistance for many different drugs is being widely reported particularly in developing countries due to drug misuse and absence of strict regulation controlling their uses. The cause of this high percent of resistance among human isolates to most of antibiotics is because of the approval for the use of those drugs in agriculture. Antimicrobial resistance in *Campylobacter* may be acquired by different mechanisms in particularly genetic manipulation. The general level of resistance among animal or human isolate in any particular geographical location depends on the regulations concerning the control on the wise and strict uses of antimicrobials in medicine and agriculture.

In recent times, concern of *Campylobacter* infections has increased because of the frequent isolation of antimicrobial resistant strains in animals and food (Smole-Možina *et al.*, 2011). The increase of antimicrobial resistant of *Campylobacter* spp is due to several factors including:

i) the inappropriate usage of antibiotics in the treatment of infections in humans, especially in the cases where infections tends to be self-limiting;
ii) the excessive use of antibiotic in veterinary medicine for the treatment and prevention of infections;
iii) the use of antibiotics as growth promoters, although they have been banned in food animals in several countries (Maćkiw *et al.*, 2012; Wei *et al.*, 2016). Antimicrobials treatment of *Campylobacter* infections is not usually recommended except in severe cases where macrolides and fluoroquinolones are the preferred options for treatment. However, fluoroquinolone resistance of *Campylobacter* spp. has been observed and increased in human infections (Nachamkin *et al.*, 2002).

The antimicrobial resistant bacteria and genes may be transferred to human via the foods (Verraes *et al.*, 2013). Many studies have supposed of a link between the antibiotics abuse during agricultural production and antimicrobial resistance of human pathogens in which food is one of the possible transfer routes (Mayrhofer *et al.*, 2006; Carattoli, 2008; Srinivasan *et al.*, 2008; Depoorter *et al.*, 2012).

18.3 Consequences of Foodborne Antimicrobial Resistance on Humans

Antimicrobial resistant bacteria may be ingested by human and present public health risk. Infections with resistant bacteria to more than one class or type of antibiotics have become more frequent in health care settings. Therefore, antimicrobial options for treating infections are very limited, and in some situations, no effective antimicrobial found or more expensive or more toxic antimicrobials are needed. Emerging of antimicrobial resistance of foodborne pathogens increases the risk of infections by increasing the number of hospitalizations and death in humans. In the United States, CDC estimates that more than 2 million people are seriously infected every year with bacteria that are resistant to one or more of the antimicrobial that are designed to treat those infections with at least 23,000 dying as a result. The antimicrobial resistance of microorganisms is not only a problem for infected humans; they also threaten the community. These microorganisms will persist and possibly infect others and extending the resistance problem (CDC, 2013).

Antimicrobial resistance genes can be presented in food either as DNA fragments or enclosed in bacteria and bacteriophages. These genes may be involved in public health risk as pathogenic bacteria can pick up antimicrobial resistance genes from the gene pool and probably transfer them to other pathogenic bacteria (Verraes *et al.*, 2013).

18.4 Antimicrobial Resistance Mechanisms

Bacteria respond to antibacterial drugs by showing different forms of resistance. As antibacterial drugs are being in use widely, so did the level and complexity of resistance mechanism. To understand the antimicrobial resistance mechanisms, we need to know how antimicrobials attack the microbial cells. Different mechanisms of action of antimicrobials have been proved: i) interference with bacterial cell wall synthesis, like β-lactamase (penicillin) and glycopeptides (vancomycine) (McManus, 1997); ii) protein synthesis inhibition either by binding to 50S ribosomal subunit (macrolides or chloramphenicol) or binding to bacterial isoleucyl-t RNA synthetase (mupirocin) (Drlica and Zhao, 1997; McManus, 1997); iii) interference with nucleic acid synthesis by inhibiting DNA synthesis (fluoroquinolones) or RNA synthesis (rifampin) (Drlica and Zhao, 1997); iv) inhibition of metabolic pathways (sulfonamides) (Tenover, 2006) and v) disruption of bacterial membrane structure (polymyxins) (Storm, 1997; Carpenter, 2004).

Some bacteria are innately resistant to one or more classes of antimicrobials (Tenover, 2006). Bacteria can also become resistant either through mutation and selection (vertical evolution) or by acquiring genetic information that encodes resistance from other bacteria through transformation, conjugation or transduction (Horizontal evolution) (Drlica and Zhao, 1997). Spontaneous mutation may take different forms like modifying or altering binding site, production of enzymes that inactivate drug, altering membrane protein channel for drug entry, and up-regulation pump that expels drug from the cell (Tenover, 2006). Acquiring resistance from other resistant bacteria may take the following forms: organism acquire gene encoding enzymes such as β-lactamase that destroy the drug before reaching the organism, acquire efflux pump that extrudes antibacterial drug before it reaches the target site, acquiring genes for metabolic pathway that produces new cell wall lacking the binding site, and by limiting the access of antimicrobial drug to intracellular site. Some bacteria may acquire more than one mechanism and thus have multidrug resistance by being resistance to three or more drug classes (Drlica and Zhao, 1997).

18.5 Antimicrobial Susceptibility Testing of *Campylobacter*

The antimicrobial resistance tests of *C. jejuni* are detailed in an excellent review by Ge *et al.* (2013). The antimicrobial resistance tests can be grouped in two major methods: standardized susceptibility testing method (phenotypic test methods) and genetic test methods.

18.5.1 Standardized Susceptibility Testing Methods

These testing methods have been published as the standards of the Clinical and Laboratory Standards Institute (CLSI, 2015) in the United States and the European Committee on Antimicrobial Susceptibility Testing (EUCAST, 2012 and 2016) in Europe.

The standardized methods include disk diffusion, agar dilution, and broth microdilution methods. Reference strain *C. jejuni* ATCC 33560 is usually used as the standard.

In the antimicrobial susceptibility tests, the antimicrobial activity against the test microorganism is measured by determining minimum inhibitory concentration (MIC) or inhibition zone diameter. The CLSI and EUCAST established guidelines for standardized antimicrobial susceptibility testing and developed and published interpretation criteria for the organisms as resistant, intermediate, or susceptible.

18.5.1.1 Susceptibility Testing by Disk Diffusion Method

Antimicrobial susceptibility testing usually uses the disc diffusion method because it is simple, convenience, not expensive and ability of testing many antimicrobial agents in one set of experiment. Although reproducibility of the results has not been achieved of the disk diffusion method in campylobacters, the CLSI has approved the variation of this method as a way to screen the resistance of *Campylobacter* isolates to ciprofloxacin and erythromycin (CLSI, 2010).

The growth media used in the disk diffusion method of CLSI or EUCAST is either autoclaved, liquid, and lukewarm Mueller-Hinton agar with sheep blood (5%) or Mueller-Hinton agar with defibrinated horse blood (5%), respectively, in sterile Petri dishes. After solidification of the agar, the plates are dried at room temperature in a laminar airflow bench. An adjusted inoculum of fresh culture of *Campylobacter* isolate at concentration of 0.5 MacFarland standard is streaked on the agar plate. Then, the antimicrobial disks or strips are applied on the surface of the agar and plates are incubated in microaerophilic conditions (10% CO_2, 5% O_2, and 85% N_2) at 37°C for 48 h or 42°C for 24 h. This method uses growth up to the edge of the disk or strip as an indicator of microbial resistance (no zone of inhibition) (Ge *et al.*, 2013; CLSI, 2015; EUCAST, 2016).

18.5.1.2 Susceptibility Testing by Agar Dilution Method

Plate dilution on Mueller-Hinton agar with 5% defibrinated sheep blood is used to test the antimicrobial susceptibility of *Campylobacter* isolates. The agar is mixed with twofold serial dilutions of antimicrobial agent and poured into plates. After solidification, each agar plate is inoculated with the bacterial inoculum adjusted to 0.5 McFarland turbidity standard. Plates are then incubated at 37°C for 48 h or 42°C for 24 h under microaerophilic atmosphere. Plates without antibiotics are used as control.

The results are expressed as MIC (minimum inhibition concentration), which is the lowest antimicrobial concentration that no visible growth could be observed (Ge *et al.*, 2013; Rossi *et al.*, 2015).

18.5.1.3 Microbroth Dilution Method

In this method, a cation-adjusted Mueller-Hinton broth with 2.5–5% lysed horse blood is mixed with twofold dilutions of the antimicrobial agents and transferred to microtitre plates. After that, a standardized suspension of bacteria is inoculated in the broth and incubated at 37 °C for 48 h or at 42 °C for 24 h under microaerophilic atmosphere. The results are expressed as MIC (ClSI, 2015; EUCAST, 2016; Ge *et al.*, 2013; Rossi *et al.*, 2015).

18.5.2 Genetic Test Methods

Genetic methods are based on the detection of a resistance gene or its product in the bacteria in order to detect antimicrobial resistance (Kahlmeter, 2012). This method provides rapid and direct identification of *Campylobacter* resistance. Fluoroquinolones-resistant genes can be determined by PCR amplification of the quinolone-resistance-determining region (QRDR) in the DNA gyrase gene (*gyrA*). A single point mutation (usually leading to Thr-86-Ile) in the QRDR is sufficient to identify high-level resistance to fluoroquinolones. Thr-86-Ile mutations are identified by the mismatch amplification mutation assay PCR (MAMA-PCR) (Maćkiw *et al.*, 2102; Ge *et al.*, 2013).

Macrolide resistance can be detected as a point mutation of the 23S RNA gene in the region corresponding to nucleotides 2074 and 2075 (Lehtopolku *et al.*, 2011). Tetracycline resistance can be detected by PCR amplification of the *tet* (O) gene located generally on the plasmid or sometimes on the chromosome (Gibreel *et al.*, 2004). Because chromosomal target mutations comprise the majority of *Campylobacter* resistance to fluoroquinolones and macrolides, DNA sequencing of the previously mentioned target genes is sufficient for detecting the mutations (Gibreel and Taylor, 2006).

18.6 *Campylobacter* Antimicrobials Resistance: Global Overview

The development of *Campylobacter* antimicrobial resistance against fluroqinilones involves both animals and humans isolates. For example, in the Netherlands in 2009, approximately 55% of the human *Campylobacter* isolates were resistant against ciprofloxacin compared to 35% in the period 2002–2005 (Silva *et al.*, 2011). The resistance of *C. jejuni* and *C. coli* to fluroqinilone increased significantly over the last 15 years, with a rate of 75 to 88% in Spain and Thailand, respectively (Nachamkin *et al.*, 2000; Engberg, 2001). In 2008, a study has shown that the antimicrobial resistance of *C. jejuni* against fluroqinilones in poultry ranged from 16.7% in Denmark and up to 100% in Hungary (Geenen *et al.*, 2011). High resistance for poultry local isolates (Al-Natour *et al.*, 2016) to enrofloxacin (100%) was found; however, it does not match low resistance (7%) reported in Sweden (Lindmark *et al.*, 2004) or 29.5% in Japan (Haruna *et al.*, 2012).

Rapid emergence of fluoroquinoles-resistant *Campylobacter* worldwide may be partially attributed to the enhanced fitness of this pathogen and due to the broad use of fluoroquinoles in veterinary medicine, especially in poultry (Geenen *et al.*, 2011).

Generally, macrolides are the drug of choice for treatment of campylobacteriosis. Development of macrolides resistance during therapy has not originated from human studies; however, numerous studies linked it to the veterinary use of antibiotics (macrolides-lincosamide group) (Silva *et al.*, 2011). Most of the data about azithromycin shows the same or less resistance level compared with erythromycin (Engberg, 2001).

Worldwide, the resistance to aminoglycosides (gentamicin) is generally very low because it is rarely used in poultry industry (Deckert *et al.*, 2010). However, gentamicin is used locally in veterinary medicine practices and therefore, a local study on human isolates reported a resistance rate of 84% (Jaradat *et al.*, 2015). In a study conducted in Tanzania on *C. jejuni* isolated from humans, the resistant rate of azithromycin, erythromycin, tetracycline, norfloxacin and other antibiotics ranged from 44 to 89% (Komba *et al.*, 2015).

A study in Iran on *Campylobacter* isolated from different sources (different animals, river water, and sewage), showed that 75% of *C. jejuni* isolates were susceptible to tobramycin (Eghbali *et al.*, 2011). By the same meanings, 10% of the human isolates were susceptible to tobramycine, this rate is close to 4% reported for local layer isolates (Al-Natour *et al.*, 2016).

C. jejuni in most circumstances is resistant to most cephalosporins (i.e., cefotaxime, ceftazidime, and cefpirome), although we can find other species of *Campylobacter* are susceptible to cephalosporins (Blaser and Engberg, 2008). In a study conducted in Nigeria on *Campylobacter* isolates from stools of infants 72% of the isolates were susceptible to nitrofurantoin (Adekunle and Onilude, 2015). These findings are almost close to the result of local human isolates (Jaradat *et al.*, 2015) that 76% of the isolates were susceptible to nitrofurantoin.

The resistance rates of *Campylobacter* spp. toward imipenem (carbapenems) remained under 1%, they recommended to use it only in extra intestinal infections to prevent development of resistance (Silva *et al.*, 2011). In a study conducted in Finland during the period of 1995–2000, of the 376 *C. jejuni* human isolates only 2% of the isolates were resistant to erythromycin and azithromycin or cefotaxime but there no resistant was found to imipenem (Hakanen *et al.*, 2003). In a local study (Jaradat *et al.*, 2015), 50 human isolates (100%) were susceptible to imipenem and this is compatible with the results of a study conducted in Kuwait on human *C. jejuni* isolates during 2002-2010, where all the isolates were uniformly susceptible to imipenem (Albert *et al.*, 2013). This is not surprising in the way that imipenem is not used in veterinary medicine as it is quite expensive in comparison to other drugs. Most *C. jejuni* usually produces β-lactamases therefore it is traditionally not treated with β-lactams (ampicillin). Imipenem contains β-lactamase inhibitors as the last enzyme being traditionally produced by *Campylobacter*.

Broad-spectrum and inexpensive tetracycline is commonly in use in poultry, large animal, and human medicines. Therefore, high resistance rates have been reported in Jordan (Al-Natour *et al.*, 2016), UK (Piddock *et al.*, 2008), the United States (Son *et al.*, 2007), Japan (Itoh *et al.*, 1984), and Australia (Pratt and Korolik, 2005). High resistance figures were also recorded in the United States for conventionally or organic grown chicken or in Canada (Deckert *et al.*, 2010). In a local study on broiler isolates, we found that 45% of the isolates were totally sensitive to the doxycycline (Osaili *et al.*, 2012), which is again close to (56%) reported for local human study (Jaradat *et al.*, 2015).

One of the lowest resistance pattern recorded locally was toward florfenicol (phenicol) (6.5%); this trend is in parallel line with very low resistance percentages (0.0–2.6) reported for chloramphenicol (Deckert *et al.*, 2010), and no resistance to florfenicol (Zhao *et al.*, 2010). In another study on 38 broiler isolates, 21 isolates (55%) showed multiple drug resistance and 17 isolates were totally sensitive to the tested antimicrobials (erythromycin, gentamycin, norfloxacin ciprofloxacin, doxycycline tetracycline, tilmicosin, amoxicillin, chlortetracycline, and trimethoprim). Of the 21 multi-drug resistant isolates, 2 showed resistant to only two antibiotics (amoxicillin and trimethoprim), 3 were resistant to nine antimicrobials and the other 16 isolates were totally resistant to all tested drugs (Osaili *et al.*, 2012). In addition to quinolones, isolates are commonly resistant to macrolides, tetracycline, and ampicillin. Results have showed that *CmeG* gene functions as a multidrug efflux transporter contributing to antibiotic resistance and oxidative defense in campylobacters (Jeon *et al.*, 2011). The general percentage of multi-resistant isolates has been reported up to 15–43% (Hänninen and Hannula, 2007; Bester and Essack, 2008).

For unclear reasons, *C. coli* strains isolated from chickens and other animals showed resistance to certain groups of antibiotics (macrolides and fluroqinilones) more than *C. jejuni* strains (D'lima *et al.*, 2007; Kim *et al.*, 2008). Significant differences among countries and species are commonly reported and almost all of the conducted studies showed high resistance to erythromycin in *C. coli* than *C. jejuni* (EUCAST, 2012).

Reports presented in this respect that present a good example on the misuse of antibiotics in developing countries could be explained here. Most of the literatures considered erythromycin (macrolides) and ciprofloxacin (fluroqinilones) as the drugs of choice and these are not compatible with the results of the local human study, because 100% of the isolates were resistant to erythromycin and 44% were resistant to ciprofloxacin (Jaradat *et al.*, 2015). In another local study on 96 *C. jejuni* isolates from layer farms (Al-Natour *et al.*, 2016), the results showed high level of resistance (100%) toward the two drugs of choices (erythromycin and ciprofloxacin) and relatively high level of resistance (84%) to gentamicin in comparison to 56% of resistance for local human isolates (Jaradat *et al.*, 2015). In this respect it is worth to mention the exceptionally high resistance among layer isolates because of their long life span in comparison to broiler chicken and being so have more chance of being exposed to antimicrobials.

18.7 Antimicrobial Resistance of *Campylobacter* Isolates From the Middle East Region

Limited data is available on *Campylobacter* resistance from Middle Eastern countries. In 1998, a study conducted in Lebanon tested the antimicrobial susceptibility of different *Campylobacter* isolates from human, chicken caeca, and raw chicken carcasses. The results showed that most isolates are highly to moderately susceptible to gentamicin (97%), amoxicillin clavulanate (95%), clindamycin (77%), chloramphenicol (77%), and ampicillin (69%). Lower susceptibility was observed against tetracyclines (49%), erythromycin (47%), ciprofloxacin (39%), and norfloxacin (36%) (Talhouk *et al.*, 1998). In Jordan, a study conducted of human isolates found that 77% of the isolates exhibit resistance against cotrimoxazole (Battikhi, 2002). In another study in Jordan on 38 broiler isolates, 21 isolates (65%) showed multiple drugs resistance and the remaining 17 isolates were sensitive to the tested antimicrobials (erythromycin, gentamicin, norfloxacin, ciprofloxacin, doxycycline, tetracycline, tilmicosin, amoxicillin, chlortetracycline, and trimethoprim) (Osaili *et al.*, 2012). Other studies (Al-Natour *et al.*, 2016; Jaradat *et al.*, 2015) conducted in Jordan on the antimicrobial resistance of *C. jejuni* isolated from human and chicken are represented in Tables 18.1 and 18.2. Also in a study done in Kuwait by Albert *et al.* (2005) found that 53% of the isolates were resistance to ciprofloxacin. In a study performed in Iraq by a Harvard study team on human isolates, the results indicated that all of the isolates were resistant to cephalothin, trimethoprim, and sulphamethoxazole and all of the isolates were sensitive to chloramphenicol and gentamicin (Moore *et al.*, 2006). In Saudi Arabia, a study on *C. jejuni* and *C. coli* human isolates showed that resistant to tetracycline and erythromycin was observed in 7.3% and 32.7% of those isolates, respectively (Moore *et al.*, 2006).

18.8 Strategies to Prevent Future Emergences of Bacterial Resistance

Campylobacter and its antimicrobial resistance represent a challenge of global dimensions to human and veterinary medicines with the prospect of the therapeutic failure for life-saving treatments. It is important to minimize this resistance by including and promoting appropriate and judicious use of antimicrobials in both humans and animals so that the use of those drugs should be limited to cases where they are necessary to ensure animal or human health (USFDA, 2004; DANMAP, 2009; NARMS, 2011; EUCAST, 2012). Resistance of *Campylobacter* infection is not well monitored and no surveillance program exists in most developing countries. The need for additional studies on the use of antimicrobials in humans and animals are needed and the empiric therapy must be based on locally assessed susceptibility profiles

Table 18.1 Antimicrobial Susceptibility of 50 Diarrheic and Non-Diarrheic Local Human *C. jejuni* Isolates.

Antimicrobial	No (%) Susceptible	No (%) Intermediate	No (%) Resistant
Gentamicin	15 (30%)	7 (14%)	28 (56%)
Imipenem	48 (96%)	2 (4%)	0 (0%)
Cefotaxime	2 (4%)	0 (0%)	48 (96%)
Erythromycin	0 (0%)	0 (0%)	50 (100%)
Azithromycin	6 (12%)	0 (0%)	44 (88%)
Tobramycin	5 (10%)	7 (14%)	38 (76%)
Nitrofurnation	38 (76%)	7 (14%)	5 (10%)
Ciprofloxacin	14 (28%)	14 (28%)	22 (44%)
Norfloxacin	3 (6%)	8 (16%)	39 (78%)
Doxycycline	19 (38%)	3 (6%)	28 (56%)

Table 18.2 Antimicrobial Susceptibility of 96 Local Layer Chicken *C. jejuni* Isolates.

Antibiotic	Resistant (R) %	Intermediate (I) %	Susceptible (S) %
Ampicillin	100	0	0
Ampicillin/ sublactam	100	0	0
Pipracillin/ tazobactam	0	0	100
Cefazolin	100	0	0
Cefoxitin	1	0	99
Ceftazidime	0	1	99
Ceftriaxone	61.5	38.5	0
Cefepime	55	0	45
Imipenem	0	0	100
Gentamicin	84	14.6	1
Tobramycin	0	95	4.5
Ciprofloxacin	35.4	0	64.6
Levofloxacin	100	0	0
Nitrofurantoin	15.6	16.7	67.7
Trimethoprim/ sulfamethoxazole	88.5	0	11.5

(Engberg, 2006). The common misuse and unrestricted use of antimicrobials in both veterinary and human medicines in developing countries necessitates an effective monitoring program for antimicrobial resistance of foodborne pathogens and may be considered as an urgent need for these countries (Moore *et al.*, 2006).

The periodical investigation of pathogens to antimicrobials is an important issue in dealing with animal or human bacterial isolates

(Aarestrup, 1999). *Campylobacter* resistance increases year after year and this may also be due to empirical therapy for diarrheal disease without performing antimicrobial susceptibility test or because the overuse of antimicrobials, which increases the risk of resistance, not only for *Campylobacter* but also for other bacteria. The high resistance may indicate abuse handling of antibiotics and thus necessity a strict control on antimicrobial usage in animal industry.

It is somewhat difficult to compare between different studies because different techniques or variable breakpoints values were used. Therefore, again an extensive study is needed to establish an internationally recommended method. Several comparisons among drug susceptibility testing methods have tested antimicrobial susceptibility in *Campylobacter* spp. with contradicting results (Ge *et al.*, 2002; McDermott *et al.*, 2005; Luangtongkum *et al.*, 2007; Miflin *et al.*, 2007; Lehtopolku *et al.*, 2012). These apparent disagreements in determining the resistance values depend on the investigated antimicrobial agent(s) and the used cutoff values (Ge *et al.*, 2002; Moore *et al.*, 2006; Luangtongkum *et al.*, 2007).

We should also remember that low levels of resistance compromise therapy may not be detected by traditional susceptibility testing techniques (Tenover *et al.*, 2004).

References

Aarestrup, F.M. (1999). Association between consumption of antimicrobial agents in animal husbandry and occurrence of resistance bacteria among food animals. *International Journal of Antimicrobial Agents, 12,* pp. 279–285.

Adekunle, O.C. and Onilude, A.A. (2015). Antimicrobial resistance and plasmid profiles of *Campylobacter* Species from infants presenting with diarrhoea in Osun State, Nigeria. *Open Journal of Medical Microbiology, 5,* pp. 17–21.

Al-Natour, M.Q., Alaboudi, A.R., Osaili, T.M. and Obaidat, M.M. (2016). Resistance of *Campylobacter jejuni* isolated from layer farms in northern Jordan using microbroth dilution and disc diffusion techniques. *Journal of Food Science 81,* pp. M1749–M1753.

Albert, M.J. (2013). In *vitro* susceptibility of *Campylobacter jejuni* from Kuwait to tigecycline antimicrobial agents. *Indian Journal of Medical Research, 137,* pp. 187–190.

Albert, M.J., et al. (2005). Ciprofloxacin resistance and its molecular mechanism in *Campylobacter* spp. isolated in Kuwait. *Microbial and Drug Resistant, 11,* pp. 266–270.

Alfredson, D.A. and Korolik, V. (2007). Antibiotic resistance and resistance mechanisms in *Campylobacter jejuni* and *Campylobacter coli. FEMS Microbiology Letters, 277,* pp. 123–132.

Allos, B.M. (2001). *Campylobacter jejuni* infections: update on emerging issues and trends. *Clinical Infectious Diseases, 32,* pp. 1201–1206.

Battikhi, M.N. (2002). Epidemiological study on Jordanian patients suffering from diarrhoea. *New Microbiology, 25,* pp. 405–412.

Bester, L.A. and Essack, S.Y. (2008). Prevalence of antibiotic resistance in *Campylobacter* isolates from commercial poultry suppliers in KwaZulu-Natal, South Africa. *Journal of Antimicrobial Chemotherapy, 62,* pp. 1298–300.

Blaser, M.J. and Engberg, J. (2008). "Clinical Aspects of *Campylobacter jejuni* and *Campylobacter coli* Infections." In: I. Nachamkin, C.M. Szymanski, M.J. Blaser, eds., *Campylobacter.* Washington, DC: ASM Press, p. 99–121.

Butaye, P., Argudín, M.Á. and Threlfall, J. (2015). Introduction to Antimicrobial-Resistant Foodborne Pathogens. In: C.-Y. Chen, X. Yan, C.R. Jackson, eds., *Antimicrobial Resistance and Food Safety: Methods and Techniques.* New York: Elsevier, pp. 2–15.

Carattoli, A. (2008). Animal reservoirs for extended spectrum beta-lactamase producers. *Clinical Microbiology and Infection, 14,* pp. 117–123.

Carpenter, C.F. and Chambers, H.F. (2004). Daptomycin: another novel agent for treating infection due to drug-resistant gram-positive pathogens. *Clinical and Infectious Diseases, 38*, pp. 994–1000.

CDC. (2013). Antimicrobial resistance threats in the United States, 2013. Department of Health and Human Services. Centers for Disease Control and Prevention. Available at: http://www.cdc.gov/drugresistance/pdf/ar-threats-2013-508.pdf [Accessed 10 November 2015].

CLSI. (2015). Methods for Antimicrobial Dilution and Disc Susceptibility Testing of Infrequently Isolated or Fastidious Bacteria. 3rd Edition (M45). Wayne, PA.

CLSI. (2010). Methods for Antimicrobial Dilution and Disk Susceptibility Testing of Infrequently Isolated or Fastidious Bacteria; Approved Guideline—2nd Edition (M45-A2). Wayne, PA.

Cody, A.J., et al. (2010). Ciprofloxacin-resistant campylobacteriosis in the UK. *Lancet, 376*, p. 1987.

Corcoran, D., et al. (2006). An investigation of the molecular mechanisms contributing to high-level erythromycin resistance in *Campylobacter. International Journal of Antimicrobial Agents, 27*, pp. 40–45.

DANMAP (Danish Programme for surveillance of antimicrobial consumption and resistance in bacteria from animals, food and humans). (2009). Use of antimicrobial agents and occurrence of antimicrobial resistance in bacteria from food animals, foods and humans in Denmark. Available at: http://www.danmap.org/pdfFiles/Danmap_2009 [Accessed 10 November 2015].

Deckert, A., et al. (2010). Prevalence and antimicrobial resistance in *Campylobacter* spp. isolated from retail chicken in two health units in Ontario. *Journal of Food Protection, 73*, pp. 1317–1324.

Depoorter, P., et al. (2012). Assessment of human exposure to 3rd generation cephalosporin resistant *E. coli* (CREC) through consumption of broiler meat in Belgium. *International Journal of Food Microbiology, 159*, pp. 30–38.

D'lima, C.B., et al. (2007). Clonal population structure and specific genotypes of multidrug-resistant *Campylobacter coli* from turkeys. *Applied and Environmental Microbiology, 73*, pp. 2156–2164.

Drlica, K. and Zhao, X. (1997). DNA gyrase, topoisomerase iv and the 4- quinolones. *Microbiology and Molecular Biology Reviews, 61*, pp. 377–392.

EFSA (European Food Safety Authority), ECDC (European Centre for Disease Prevention and Control). (2014). The European Union summary report on trends and sources of zoonoses, zoonotic agents and food-borne outbreaks in 2012. *EFSA J., 12*, p. 3547.

EFSA (European Food Safety Authority), ECDC (European Centre for Disease Prevention and Control). (2013). The European Union summary report on trends and sources of zoonoses, zoonotic agents and food-borne outbreaks in 2011. *EFSA J., 11*, p. 3129.

EFSA (European Food Safety Authority), ECDC (European Center for Disease Prevention and Control). (2011). The European Union summary report on antimicrobial resistance in zoonotic and indicator bacteria from humans, animals and food in the European Union in 2009. *EFSA J. 9*, p. 2154.

Eghbali, M., Baserisalehi, M. and Bahador, N. (2011). Antimicrobial susceptibility of thermophilic *Campylobacter* spp. isolated from environmental samples in tonekabon. *International Journal of Molecular and Clinical Microbiology, 1*, pp. 21–24.

Engberg, J., et al. (2001). Quinolone and macrolide resistance in *Campylobacter jejuni* and *C. coli:* resistance mechanisms and trends in human isolates. *Emerging Infectious Diseases, 7*, pp. 24–34.

Engberg, J., et al. (2006). Antimicrobial resistance in *Campylobacter*. In: F.M. Aerestrup, ed., *Antimicrobial Resistance in Bacteria of Animal Origin*. Washington, DC: ASM Press, pp. 269–292.

EUCAST. (2016). Breakpoint tables for interpretation of MICs and zone diameters.

Version 6.0, 2016. Available at: http://www.eucast.org/fileadmin/src/media/PDFs/EUCAST_files/Breakpoint_tables/v_6.0_Breakpoint_table.pdf. Accessed DATE: 8 June 2016.

EUCAST. (2012). Clinical breakpoints, epidemiological cut-off (ECOFF) values and EUCAST disk diffusion methodology for *Campylobacter jejuni* and *Campylobacter coli*. Available at: http://www.eucast.org/fileadmin/src/media/PDFs/EUCAST_files/Consultation/Campylobacter_wide_consultation_August_2012.pdf. [Accessed 10 November 2015].

Foley, S. (2012). *Campylobacter jejuni*. In: K.A. Lampel, S. Al-Khaldi, S.M. Cahill, eds., *Bad bug book: Foodborne pathogenic microorganisms and natural toxins handbook*, 2nd ed. Washington DC: FDA.

Ge, B., et al. (2002). Comparison of the E test and agar dilution method for *in vitro* antimicrobial susceptibility testing of *Campylobacter*. *Journal of Antimicrobial and Chemotherapy, 50,* pp. 487–494.

Ge, B., et al. (2013). Antimicrobial resistance in *Campylobacter*: susceptibility testing methods and resistance trends. *Journal of Microbiological Methods, 95,* pp. 57–67.

Geenen, P.L., et al. (2011). Risk profile on antimicrobial resistance transmissible from food animals to humans. Bilthoven, The Netherlands. Available at: www.rivm.nl/bibliotheek/rapporten/330334001.pdf [Accessed 10 November 2015].

Gibreel, A. and Taylor, D.E. (2006). Macrolide resistance in *Campylobacter jejuni* and *Campylobacter coli*. *Journal of Antimicrobials and Chemotherapy, 58,* pp. 243–255.

Gibreel, A., et al. (2004). Incidence of antibiotic resistance in *Campylobacter jejuni* isolated in Alberta, Canada, from 1999 to 2002, with special reference to tet (O)-mediated tetracycline resistance. *Antimicrobial Agents and Chemotherapy 48,* pp. 3442–3450.

Griffiths, P.L. and Park, R.W.A. (1990). Campylobacters associated with human diarrhoeal disease. *Journal of Applied Bacteriology, 69,* pp. 281–301.

Guyard-Nicodème, M., et al. (2015). Prevalence and characterization of *Campylobacter jejuni* from chicken meat sold in French retail outlets. *International Journal of Food Microbiology, 203,* pp. 8–14.

Hakanen, J.A., et al. (2003). Multidrug resistance in *Campylobacter jejuni* strains collected from Finnish patients during 1995–2000. *Journal Antimicrobial and Chemotherapy, 52,* pp. 1035–1039.

Hänninen, M.L. and Hannula, M. (2007). Spontaneous mutation frequency and emergence of ciprofloxacin resistance in *Campylobacter jejuni* and *Campylobacter coli*. *Journal Antimicrobial and Chemotherapy, 60,* pp. 1251–1257.

Haruna, M., et al. (2012). Prevalence and susceptibility of Campylobacter in broiler flocks in Japan. *Zoonoses Public Health, 59,* pp. 241–245.

Itoh, T., et al. (1984). Antimicrobial susceptibility of *Campylobacter jejuni* and *Campylobacter coli* isolated from human and animals in Japan. *Journal of the Japanese Association for Infectious Diseases, 58,* pp. 1206–1212.

Jaradat, G., Abu basha, E. and Alaboudi, A. (2015). *Campylobacter jejuni* as diarrheal causative agent in Northen Jordan, M.Sc. thesis, University of Science and Technology, Irbid, Jordan.

Jeon, B., et al. (2011). Contribution of CmeG to antibiotic and oxidative stress resistance in *Campylobacter jejuni*. *Journal Antimicrobial and Chemotherapy, 66,* pp. 79–85.

Kahlmeter, G. (2012). EUCAST in 2012: aggregate MIC distributions and ECOFFs/*Campylobacter* spp. and AST. Available at: http://www.fda.gov/AnimalVeterinary/SafetyHealth/AntimicrobialResistance/NationalAntimicrobialResistanceMonitoringSystem/ ucm334828.htm [Accessed 10 November 2015].

Kim, J.S., Kim J.W. and Kathariou S. (2008). Differential effect of temperature on natural transformation to erythromycin and nalidixic acid resistance in *Campylobacter coli*.

Applied and Environmental Microbiology, 74, pp. 6121–6125.

Komba, E.V., et al. (2015). Prevalence, antimicrobial resistance and risk Factors for thermophilic *Campylobacter* infections in symptomatic and asymptomatic humans in Tanzania. *Zoonoses and Public Health, 62,* pp. 557–568.

Kurinic, M., et al. (2007). Mechanism of erythromycin resistance of *Campylobacter* spp. isolated from food, animals and humans. *International Journal of Food Microbiology, 120,* pp. 186–190.

Lehtopolku, M., et al. (2011). Ribosomal mutations as the main cause of macrolide resistance in *Campylobacter jejuni* and *Campylobacter coli*. *Antimicrobial Agents and Chemotherapy, 55,* pp. 5939–5941.

Lehtopolku, M., et al. (2012). Inaccuracy of the disk diffusion method compared with the agar dilution method for susceptibility testing of *Campylobacter* spp. *Journal of Clinical Microbiology, 50,* pp. 52–56.

Lindmark, H., et al. (2004). Genetic characterization and antibiotic resistance of *Campylobacter jejuni* isolated from meats, water, and humans in Sweden. *Journal of Clinical Microbiology, 42,* pp. 700–706.

Luangtongkum, T., et al. (2007). Comparison of Antimicrobial Susceptibility Testing of *Campylobacter* spp. by the Agar Dilution and the Agar Disk Diffusion Methods. *Journal of Clinical Microbiology, 45,* pp. 590–594.

Maćkiw, E., et al. (2102). Antibiotic resistance in *Campylobacter jejuni* and *Campylobacter coli* isolated from food in Poland. *Food Control, 23,* pp. 297–301.

Mayrhofer, S., et al. (2006). Antimicrobial resistance in commensal *Escherichia coli* isolated from muscle foods as related to the veterinary use of antimicrobial agents in food-producing animals in Austria. *Microbial Drug Resistant, 12,* pp. 278–283.

McDermott, P.F., et al. (2005). Broth microdilution susceptibility testing of *Campylobacter jejuni* and the determination of quality control ranges for fourteen antimicrobial agents. *Journal of Clinical Microbiology, 43,* pp. 6136–6138.

McManus, M.C. (1997). Mechanism of bacterial resistance to antimicrobial agents. *American Journal of Health-System Pharmacy, 541,* pp. 1420–1433.

Miflin, J.K., Templeton, J.M. and Blackall, P.J. (2007). Antibiotic resistance in *Campylobacter jejuni* and *Campylobacter coli* isolated from poultry in the South-East Queensland region. *Journal of Antimicrobial Chemotherapy, 59,* pp. 775–778.

Moore, J. E., et al. (2006). The epidemiology of antibiotic resistance in *Campylobacter. Microbes and Infection, 8,* pp. 1955–1966.

Nachamkin, I., Engberg, J. and Aarestrup, F.M. (2000). Diagnosis and Antimicrobial Susceptibility of *Campylobacter* Species. In: I. Nachamkin and M.J. Blaser, eds., *Campylobacter, 2nd edition*. Washington, DC: ASM Press, pp. 45–66.

Nachamkin, I., Ung, H. and Li, M. (2002). Increasing fluoroquinolone resistance in *Campylobacter jejuni*, Pennsylvania, USA, 1982–2001. *Emerging Infectious Diseases, 8,* pp. 1501–1503.

NARMS (National Antimicrobial Resistance Monitoring System - CDC) (2011). National Antimicrobial Resistance Monitoring System - Enteric Bacteria Annual Report. Available at: www.cdc.gov/narms/ [Accessed 10 November 2015].

Osaili, T.M., Alaboudi, A.R. and Al- Akhras, R.R. (2012). Prevalence and antimicrobial susceptibility of *Campylobacter* spp. in live and dressed chicken in Jordan. *Foodborne Pathogens and Disease, 9,* pp. 54–58.

Piddock, L. J., Griggs, D., Johnson, M. M., Ricci, V., Elviss, N.C., et al. (2008). Persistence of *Campylobacter* species, strain types, antibiotic resistance and mechanisms of tetracycline resistance in poultry flocks treated with chlortetracycline. *Journal of Antimicrobial Chemotherapy, 62,* pp. 303–315.

Pratt, A. and Korolik, V. (2005). Tetracycline resistance of Australian *Campylobacter jejuni* and *Campylobacter coli* isolates. *Journal of Antimicrobial Chemotherapy, 55,* pp. 452–460.

Rossi, M., et al. (2015). Antimicrobial Resistance and *Campylobacter jejuni* and *C. coli*. In: C.-Y. Chen, X. Yan, C.R. Jackson, eds., *Antimicrobial Resistance and Food Safety: Methods and Techniques*, New York: Elsevier, pp. 55–69.

Sack, D.A., et al. (2001). Antimicrobial resistance in shigellosis, cholera and campylobacteriosis. World Health Organization /CDS/DRS/2001.8 pp. 1–51. WHO, Geneva, Switzerland. Available at: http://www.who.int/drugresistance/Antimicrobial_resistance_in_shigellosis_cholera_and_cam.pdf. [Accessed 10 November 2015].

Scallan, E., et al. (2011). Foodborne illness acquired in the United States—major pathogens. *Emerging Infectious Diseases, 17*, pp. 7–15.

Sheppard, S.K., et al. (2009). *Campylobacter* genotyping to determine the source of human infection. *Clinical Infectious Diseases, 48*, pp. 1072–1078.

Silva, J., et al. (2011). *Campylobacter* spp. as a Food-borne Pathogen: A Review. *Frontiers in Microbiology, 2*, p. 200.

Smole-Možina, S., et al. (2011). *Campylobacter* and its multi-resistance in the food chain. *Trends in Food Science &Technology, 22*, pp. 91–98.

Son, I., et al. (2007). Prevalence of Arcobacter and *Campylobacter* on broiler carcasses during processing. *International Journal of Food Microbiology, 113*, pp. 16–22.

Srinivasan, V., et al. (2008). Distribution of tetracycline and streptomycin resistance genes and class 1 integrons in *Enterobacteriaceae* isolated from dairy and nondairy farm soils. *Microbial Ecology, 55*, pp. 184–193.

Storm, D.R., Rosenthal, K.S. and Swanson, P.E. (1997). Polymxin and related peptide antibiotics. *Annual Review of Biochemistry, 46*, pp. 723–763.

Talhouk, R.S., et al. (1998). Prevalence, antimicrobial susceptibility and molecular characterization of *Campylobacter* isolates recovered from humans and poultry in Lebanon. *Journal of Medical Libanais, 46*, pp. 310–316.

Tenover, F.C., Weigel, L.M. and Appelbaum, P.C. (2004). Vancomycin resistant *Staphylococcus aureus* isolate from a patient in Pennsylvania. *Antimicrobial Agents and Chemotherapy, 48*, pp. 275–280.

Tenover, F.C. (2006). Mechanisms of antimicrobial resistance in bacteria. *American Journal of Medicine, 119*(6 Suppl. 1), pp. S3–S10.

U.S. Food and Drug Administration. (2004). Antimicrobial Resistance & Drug Development FDA/IDSA/ISAP, Working Group Presentations and Transcripts, April *15–16*, 2004.

Verraes, C., et al. (2013). Antimicrobial Resistance in the Food Chain: A Review. *International Journal Environmental Research and Public Health, 10*, pp. 2643–2669.

Wei, B., et al. (2016). Prevalence and antimicrobial resistance of *Campylobacter* spp. isolated from retail chicken and duck meat in South Korea. *Food Control, 62*, pp. 63–68.

Zhao, S., et al. (2010). Antimicrobial resistance of *Campylobacter* isolates from retail meat in the United States between 2002 and 2007. *Applied and Environmental Microbiology, 76*, pp. 7949–7956.

19

Prevalence and Antibiogram of Pathogenic Foodborne *Escherichia coli* and *Salmonella* spp. in Developing African Countries

Adeyanju Gladys Taiwo (DVM, MVPH)

CAHS (COCIN Animal Health Services) Vwang, Jos South, Nigeria

19.1 Introduction

Advancement in technology has greatly improved the standard of living for the population in some countries, whereas for others, it has left them barely able to find a roof over their heads, let alone a decent hygienic meal because of the competition arising from the populace. "Although population growth has slowed for the world as a whole, this report reminds us that some developing countries, especially in Africa, are still rapidly growing" as said by the Under-Secretary-General for Economic and Social Affairs, Wu Hongboan. The UN projections of an increase in the African population growth is a warning that resources meant for consumption or to save and invest for the future will be scarce, not to mention health care and education (NTA, 2013).

Causes of food poisoning resulting in diseases are by microorganisms (bacteria, fungi, mould, yeasts and viruses) or physical (plastics, glass, elastic bands, wood chips or band-aids) and chemical hazards(cleaning chemicals or foods with naturally occurring toxins) (Victorian Government Health Information, 2009).

Foodborne diseases due to pathogenic organisms are common in developing countries of Africa such as the Sub-Saharan, West,and East Africa (FAO, 2005). Food contaminated by harmful bacteria, viruses, parasites, or chemical substances is responsible for more than 200 diseases and cancers. The most prevalent foodborne zoonotic diseases are cholera, typhoid, shigellosis, acute aflatoxicosis, and konzo (an acute paralysis caused by consumption of high cyanide cassava) (WHO, 2015). On a global outlook, WHO has estimated a global record of 31 foodborne hazards in 2015, causing 32 diseases being 11 diarrhoeal disease agents (1 virus, 7 bacteria, 3 protozoan), 7 invasive infectious disease agents (1 virus, 5 bacteria, 1 protozoan) 10 helminths, and 3 chemicals as shown in Table 19.1.

Population boom has been seen in most African countries, whereas their economy has not improved. People scavenge for meals as even menial jobs are difficult to come by. Also, standards are being compromised due to various reasons such as negligence, poor supervision/unsupervision, absence of policy structure, unavailability of resources necessary for standardization, or simply ignorance. Most cases of foodborne diseases especially in the rural areas are unrecognized, whereas fewer cases are counted due to poverty and lack of resources for food control management and services. The problems resulting from foodborne diseases are either biological or chemical in nature.

Foodborne Pathogens and Antibiotic Resistance, First Edition. Edited by Om V. Singh.

Table 19.1 Foodborne Hazards in WHO Global Estimates.

Diarrheal Disease Agents	*Campylobacter* spp., *Cryptosporidium* spp., *Entamoeba histolytica*, Enteropathogenic *E. coli* (EPEC), Enterotoxigenic *E. coli* (ETEC), *Giardia* spp., Norovirus, *Salmonella enterica* (non-invasive infections) non-typhoidal, *Shigella* spp., Shiga toxin-producing *E. coli* (STEC), *Vibrio cholerae*
Invasive Infectious Disease Agents	*Brucella* spp., *Hepatitis A* virus, *Listeria* spp., *Mycobacterium bovis*, *Salmonella enterica* (invasive infections) non-typhoidal, *Salmonella enterica* Paratyphi A, *Salmonella enterica* Typhi
Helminths	*Ascaris* spp., *Echinococcus multilocularis*, *Echinococcus granulosus*, *Clonorchis sinensis*, *Fasciola* spp., Intestinal flukes, *Opisthorchis* spp., *Paragonimus* spp., *Taenia solium*, *Toxoplasma gondii*, *Trichinella* spp.
Chemicals	Aflatoxin, Cassava cyanide, Dioxin

19.2 Factors that Play a Role in the Epidemiology of Foodborne Diseases

19.2.1 Microbial Adaptation

This is due to the therapeutic use and misuse of antimicrobial agents both in human and animal population (GAO, 2004). The indiscriminate self-prescription/medication by people when they or their animals are ill instead of visiting the hospital because of insufficient funds has seen a gradual but distructive incline favoring survival of bacteria strains that are resistant to agents of medications, hence a classification as multipledrug-reistant (MDR), extensively drug-resistant (XDR), and pandrug-resistant (PDR) (Magiorakos *et al.*, 2011).

19.2.2 Health Sector

Most government health care centers in rural areas are unders staffed, whereas others have specific days in a week in which a specialist can attend to patients though for some limited hours within the day. Patients come early in the morning to join a queue waiting to visit with a specialist. Others within the long waiting number of patients who could not be attended to are requested to come the following week. The hygienic standard of some of the hospitals and primary health centers is such that disease and infections are spread to patients who come for treatment or care. Water is not readily made available for cleaning; there are also no disinfectants and sanitizers. Access to drugs is also limited, whereas at certain times it is near its expiration date by the time the drugs get into the interior villages. Immunosuppressive treatments have therefore increased. To bridge this gap, several health agencies have arisen. An example is the "DOCTORS WITHOUT BORDERS," which established mobile clinics in South Africa (MSF, 2013).

19.2.3 Demographic Changes, Technology, and Development

Economic and technical developments have seen the introduction of new foods, especially fastfoods. However, ignorance and negligence on the part of food handlers, together with an increase in mass catering, are important factors in foodborne illness. Personal hygiene, dietary habits, the environmental condition, access to social conveniences, availabilty of portable water, cultural beliefs, and rituals are some interrelated factors that lead to foodborne illnesses (FAO, 2005; Musa and Akande, 2013). The proportion of the population susceptible to foodborne problems is increasing especially in Africa. Globilization and the need to make endsmeet has seen the migration of people across borders where they are introduced to new foods and dietary habits, and therefore, high risks of foodborne diseases (Van de Venter, 2000; FAO, 2005).

19.2.4 Civil Unrest

The practice for treatment is empirically based on clinical diagnosis along with the availability of antibiotics sold over the counter without necessary prescription as seen to be common among displaced people moving en masse from areas of communal conflicts or terrorist attacks to safer places. Poor people are typically at a greater risk of becoming infected but they are also less likely to be able to access health care, including genuine antimicrobials. The urban poor, most arising from such civil unrest and residing in camps for displaced persons, often live in cramped, overcrowded accommodations, which facilitates the dissemination of not only airborne diseases but also foodborne diseases.

19.3 Food Poisoning and Food Vending

The African region is one of the world's most food-insecure regions and is burdened by malnutrition. Unsafe food has been linked to the deaths of an estimated 2 million people annually. Infants, young children, pregnant women, the elderly, and those with an underlying illness are particularly vulnerable.

There are multiple sources of food contamination as shown in Figure 19.1.

Chemical poisoning in food has been reported in several countries such as Kenya, Angola, Nigeria, Senegal, and South Africa, among others. More than 100,000 cases of cholera in 22 countries such as Tanzania, Nigeria, Mozambique, Namibia, andSouth Africa resultedin over 1700 deaths as reported in 2014 (WHO, 2012; FSI, 2015). As a trade and in a bid to provide support and care for families, street food vendors are seen to populate most areas where people generally trade in goods and services. In a typical West African country such as Nigeria, a rich variety of vending food is easily prepared either at home, in stalls, or by the roadside and made available for purchase and consumption (Chukuezi, 2010).

The vendors can be seen along the streets hawking their supplies, carting it, or having shacks where they sell. Workers and students who are hungry purchase food from these food vendors not minding the hygiene of the food nor of the utensils been used. Personal hygiene is the most common source of food contamination arising from the vendors not wearing aprons or covering their heads while cooking and or serving, not caring for dirty and long nails, blowing of air into polythene bags in a bid to open it before use, handling of money while serving food and so on, as noted by Chukuezi (2010).

For example, some food vendors who hawk or cart already cooked meals can be seen with a bucket of water in which they add a little soap and use it all day in washing plates, cups, or cutlery. For sponge, most use cellulose foam. Fresh water is rarely made use of within the day for washing or rinsing of individual plates, cups or curtleries. Some vendors use cellulose foam in cleaning the plate and dish meals to other hungry buyers because of unavailability of water. Biofilms are produced over time on contact surfaces and serve as source of cross-contamination (Kariyawasam and Asooriya, 2006; Van Houdt and Michiels, 2010; Ishola and Adeyanju, 2014). Also, inadequate or complete lack of power supply necessitates that leftover food is not refrigerated and is reheated and sold the next day to unsuspecting buyers so as not to incur financial waste. Drinks sold are either company made or home brewed. Home-brewed drinks are packaged either in polythene bags or recycled plastic bottle (often picked by the roadside or at functions and are washed with water and at times detergent, if affordable). Access to portable water is impossible for some people, even in their residential areas. Because they need to trade, they use well water, stream water, or water purchased from water hawkers to enabled them to brew their produce for sell. Sources of pathogenic organisms from food vendors are not limited to brewed drinks or already cooked meals but include fruit salads in take-away packs, fresh fruits, nuts, baked, or fried foods, indigenous

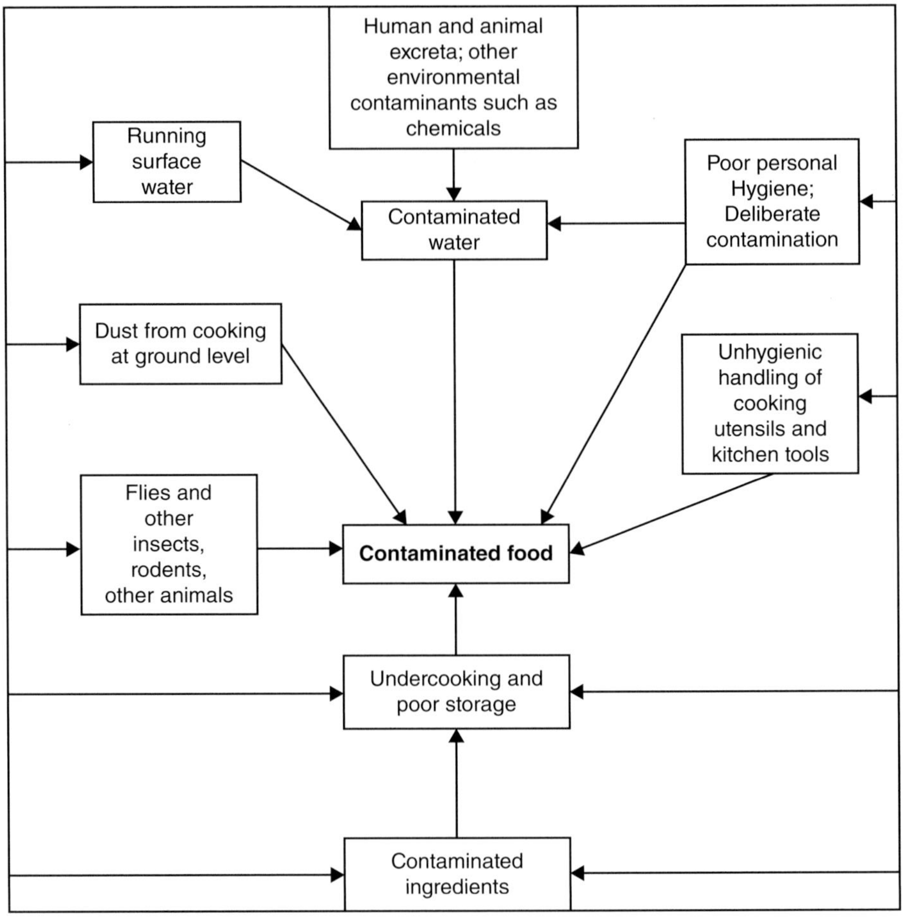

Figure 19.1 Multiple sources of food contamination.

snacks, fermented milk (Nono), grilled meat and fish, and so on (Musa and Akande, 2003; Reubenand Owuna, 2013; Ogah *et al.*, 2015).

In most African countries, meat sources includes a variety of birds, reptiles, wild animals, bats, frogs, fish, pigs, goat, sheep, donkeys, camels, cattle, and dogs, depending on traditional or cultural preference. Fish, beef, mutton/chevon, cheval, pork, chicken, duck, guinea fowl, quail, turkey, and dog meat are the most common meats. As an example, frozen poultry meat in some West African countries comes from the European Union (EU). Countries supplied are mainly Benin, Ghana, Nigeria, Senegal, Togo, and Ivory Coast (USDA, 2002). Most fresh meat is obtained from government-built abattoir facilities and from proliferative cooperative backyard-slaughter slabs found in residential areas. Abattoirs share similar sanitary problems with markets. These problems include but are not limited to institutional regulations, enforcement and monitoring,and above all, poor supervision by ill-trained, ill-equipped, and corrupt officials with no regulated practices by the supervisors.

19.4 Foodborne Colibacillosis and Salmonellosis

Colibacillosis and Salmonellosis have for decades been described as the leading causesof foodborne illnesses worldwide (Panisello *et al.*, 2000).

Salmonella is the most incriminated pathogenic microorganisms of bacterial food poisoning followed by *Escherichia coli*, especially Shiga Toxin-producing *Escherichia coli* (STEC) for which raw or undercooked foodstuffs get contaminated either during primary production (e.g., slaughtering) or further processing and handling (e.g., cross-contamination during processing and human-to-food contamination via food handlers). Food vendors play a very important role in the transmission of these illnessess through several means as reported by Musa and Akande, who noted that the attitude of the vendors regarding personal hygiene such as proper care of skin lesions to avoid flies transfering pathogens from the lesions onto food or utensils, covering of hair during sales of food, well-kept finger nails, among other cleanliness habits,as well as their attitude to the sales of food such as proper re-heating, covering of food, cleaning of serving tables, washing of utensils with clean water and so on, play a great role in transmission of foodborne illnesses (Musa and Akande, 2003; Ogah *et al.*, 2015). This is the situation observed in several developing countries (FAO, 2005). Consumers have little opportunity to assess the safety of the food they eat in some African regions. Microorganism can colonize to form biofilms on solid surfaces when adequate nutrients, minerals, and organic matter are present in the processing environment. Free-living bacteria are more susceptible to sanitizers and antimicrobial cleaning agents than bacteria in biofilms. Biofilms are difficult to remove from food contact surfaces and equipment by normal cleaning, thus posing a food safety risk (Kariyawasam and Asooriya, 2006). As a prerequisite, food vendors should be required to undergo a medical examination before selling food for the public especially in schools. Antimicrobial use in animals does select resistant foodborne pathogens that may be transmitted to humans as food contaminants thereby the occurences of related human ilnesses arising from antibiotic resistance (Angulo *et al.*, 2004; GAO, 2004), most of which are of food-animal origin. *Escherichia coli* has increased in humans (GAO, 2004), whereas antibiotic-resistant *Salmonella typhimurium* DT 104 have moved from animals to humans through foods of animal origin (Ribot *et al.*, 2002). As noted by Clark, a breakdown of sources of *Escherichia coli*bacteria are as follows:

- Food: 69%
- Water: 18%
- Animals or their environment: 8%
- Person-to-person: 6%

Some simple steps one can take to reduce risk of *Escherichia coli* infection include:

- Wash hands thoroughly before and after eating and after going to the bathroom
- Wash all fruits and vegetables before eating by skinning them if possible and washing them before eating
- Check with your local department of health to find out which restaurants in your area have had recent problems with sanitation
- Avoid allowing raw meats to come into contact with other foods while cooking
- Do not allow children to share bath water with anyone who has diarrhea or symptoms of stomach flu
- Wash hands thoroughly after any contact with farm animals
- Care should be taken when changing diapers of children with diarrhea
- Proper cooking and re-heating of foods
- Avoid drinking any non-chlorinated water (Clark, 2015).

A critical step is cataloging foodborne illnesses for future referencing and studies as it relates to antibiotic resistance over time.

19.5 Antibiotic Resistance

The relative or complete lack of effect of an antimicrobial agent is known as antibiotic resistance. A major problem is the familiarity with medicines and free access to unregulated drugs leads to self-medication. The quality of some drugs are counterfeit because of an inability to regulate the drugs. Rigorous standards and strict laws govern the quality and distribution of medicines. In countries with high occurences of infectious diseases, standards are expensive to develop and maintain; therefore, these countries adopt pharmacopeias from abroad. However, implementing quality assurance standards written in these books may be difficult or impossible in underequipped laboratories, and enforcing laws has been a challenge in many parts of Africa. Due to this, local producers import the active ingredient and most of the formulation ingredients or excipients they employ but the drug quality for some of them are sub-standard with lower-than-stated-doses which promotes antimicrobial resistance (Okeke *et al.*, 2007). A formal surveillance system for antibiotic resistance is lacking in many countries in the developing world.

A summary of the general causes of antibiotic resistance (Dyner, 2009):

- Exposure to sub-optimal levels of antimicrobials
- Exposure to broad-spectrum antibiotics
- Inactivating enzymes
- Exposure to microbes carrying resistant genes
- Lack of hygiene in clinical environments
- Use of antibiotic in food/agriculture

19.6 Reasons for Resistance Against Specific Antibiotics

Some specific antibiotics and the factors resulting in their resistance (Dyner, 2009):

- Beta-lactams: β-lactamases and alteration of PBPs
- Aminoglycosides: Enzyme modification of the drug, ineffective transport and altered ribosomal binding.
- Macrolides: Decreased permeability of cell wall to drug, alteration in the 5oS ribosomal binding site, inactivation by enzymatic hydrolysis and efflux pump
- Sulfonamides: Overproduction of PABA
- Trimethoprim: Plasmid mediated dehydrofolate reductases
- Vancomycin: Low frequency of resistance
- B-Lactamases inhibitors: Hyperproducers of β-lactamase, new β-lactamses resistant to inhibitors, chromosomal cephalosporinases
- Carbapenems: porin mutations, efflux pump overproduction, and zinc metalloenzymes and othe β-lactamases
- Fluoroquinolones: Decreased number of porins in target cells, efflux mechanisms, and mutations in DNA gyrase and topoisomerase

Because bacteria have a rapidly changing nature in their pattern of resistance (Krietsch, 2013), the consequences of these resistant organisms as regards human health include an increased frequency of more serious infections, treatment failure, and use of more expensive drugs (Angulo *et al.*, 2004; Cóilín and Richard, 2013). *Salmonella* and *Escherichia coli* have been documented as being multiresistant to nearly all commonly available antibiotic.

19.7 Antibiotic Resistance of *Salmonella*

Outbreaks of multidrug-resistant *S. typhimurium* definitive type 104 (DT 104) have showed resistance to chloramphenicol, ampicillin, streptomycin, sulfonamides, chloramphenicol, and tetracycline (Herikstad *et al.*, 1997). Fluoroquinolones,which have been known to be active against a wide range of bacteria including enteric bacteria,are gradually losing this ability. A decreased susceptibility to fluoroquinolones have increased among *Salmonella* spp from food animals (Aarestrup *et al.*, 2003). Quinolone-resistant strains of *Salmonella* have been reported in other continents with multidrug-resistant *S.* Typhimurium definitive type 104 (DT 104) being widespread. Although fluoroquinolone class of drugs is not recommended in children, resistance to quinolones in adults have seen a gradual increase (Herikstad *et al.*, 1997), which is worrying. The biggest current issue in drug-resistant *Salmonella* is cephalosporin resistance as researced by CDC (Centersfor

Disease Control and Prevention) wherein *Salmonella typhimurium* and *Salmonella heidelberg* are at risk of acquiring CTX-M resistance.

Instead of CTX-M (a form of the beta-lactamase enzymes) breaking down cephalosporins, it confers resistance to the bacteria that produces this enzyme. Currently, about 5% of *Salmonella* strains are resistant to cephalosporins, mostly in cases of *Salmonella* Heidelberg and *Salmonella* Newport. The three most important front-line antibiotics used in treatment of Salmonellosis are ceftriaxone, ciprofloxacin, and azithromycin (Krietsch, 2013). Drug-resistant *Salmonella* continues to pose a public health threat, particularly as resistance spreads across classes of drugs, necessitating the use of more expensive drugs, which makes treatment less effective, and in worse-case scenarios, leaves infections untreatable.

19.8 Antibiotic Resistance of *Escherichia coli*

Escherichia coli acquires antimicrobial resistance faster than other conventional bacteria due. Although antibiotics are not recommended for treatment of *Escherichiacoli* O157:H7 infections in humans, there is evidence that bacterial isolates are resistant to some antibiotics (Aibinu *et al.*, 2007). Enteroaggregative *E. coli* (EAEC) and enterotoxigenic fluoroquinolone–resistant *Escherichiacoli* strains often show resistance to other drugs such as ampicilin, tetracycline, chloramphenicol, trimethoprin, sulphamethoxazole, and gentamicin (Komp *et al.*, 2003), with a significant increase in fluoroquinolones-resistant *Escherichia coli* in many countries over the last few decades (Viroy *et al.*, 2005).

The wide spread use of ciprofloxacin in broilers accounted for varying degree of resistance as noted by Miranda (2008), who recorded the resistance of the drug against *Escherichia coli* being higher in broilers than in pigs. Multiple antimicrobial resistance exhibited by *E. coli* O157:H7 strains from fermented milk is an indication of possible antibiotic abuse as recorded by Reuben and Owuna (2013) who observed a high level of tetracycline resistance, which might be the result of the antibiotic being readily available for the routine prophylaxis and treatment among cattle as well as in promoting growth (Olatoye, 2010). In general, due to an increase in resistance against multiple antibiotics resulting in increase resistance pattern, it becomes necessary to increase the period of withdrawal of an antibiotic after therapy.

19.9 How to Combat Foodborne Diseases and Antibiotic Resistance

- Use narrow antibiotic coverage when possible
- Restrict antimicrobial use
- Track resistance pattern
- Routine testing of *Salmonella* and *E. coli* should be carried out and report made to the U.S. CDC for further actions.
- Routine surveillance of foodborne diseases
- Improve the social and environmental conditions in the country
- Continuing education should be made compulsory for healthcare workers
- Routine screening of food vendors

Acknowledgments

Dr. H. Mychael for highly proficient proofreading as well as editing this book and Dr. (Mrs) O.O. Ishola for her ever-present support.

References

Aarestrup, F.M., et al. (2003). Is it time to change fluoroquinolone breakpoints for *Salmonella* spp. *Antimicrob Agents Chemother.*, *47*(2), pp. 827–829

Aibinu, I.E., Peters, R.F., Amisu, K.O., Adesida, S.A., Ojo, M.O. and Tolu, O. (2007). Multidrug Resistance in *E. coli* O157 Strains and the Public Health Implication. *Journal for Animal Science, 3*(3), pp. 22–33.

Angulo F.J., Nargund, V.N. and Chiller, T.C. (2004). Evidence of an Association Between Use of Anti-microbial Agents in Food Animals and Anti-microbial Resistance Among BacteriaIsolated from Humans and the Human Health Consequences of Such Resistance. *J.Vet. Med. B Infect. Dis. Vet. Public Health, 51,* 374–379.

Nunan, C. and Young, R. (2013). Antibiotic resistance - the impact of intensive farming on human health. A report for the Alliance to Save Our Antibiotics. Available at: https://www.ciwf.org.uk/media/3758854/Antibiotics-Alliance-briefing-10-March-2013.pdf (Accessed June 7, 2016).

Chukuezi, C. (2010). Food Safety and Hyienic Practices of Street Food Vendors in Owerri, Nigeria. *Studies in Sociology of Science, 1*(1), pp. 50–57. Available at: http://www.cscanada.net/index.php/sss/article/view/j.sss.1923018420100101.005 [Accessed June 7, 2016].

Clark, M. (2015). Foodborne illness: Common bacteria and viruses that cause food poisoning. Available at: http://www.foodborneillness.com/ecoli_food_poisoning/ [Accessed June 7, 2016.].

Dyner, L. (2009). Antibiotic Resistance. Available at: http://peds.stanford.edu/Tools/documents/AntibioticResistanceLLD.pdf [Accessed June 7, 2016].

FAO. (2005). Practical Actions to Promote Food Safety. FINAL REPORTFAO/WHO Regional Conference on Food Safety for Africa. October 2005. Harare, Zimbabwe. Available at: ftp://ftp.fao.org/es/esn/foodsafetyforum/caf/draftreport_en.pdf http://www.fao.org/docrep/meeting/010/a0215e/A0215E00.htm#TOC [Accessed June 7, 2016].

Food Safety International (FSI). (2015). Africa: Food/Water Borne Illness Outbreaks. Available at: http://regionalnews.safefoodinternational.org/page/Africa%3A+Food%2FWater+Borne+Illness+Outbreaks http://regionalnews.safefoodinternational.org/page/Africa%3A+Food%2FWater+Borne+Illness+Outbreaks [Accessed June 7, 2016.]. http://www.foodsafetynews.com/2015/12/who-releases-first-global-estimates-of-foodborne-disease/#.VqyBTFKXSWw

Herikstad, H., et al. (1997). Emerging quinolone-resistant *Salmonella* in the United States. *Emerg Infect Dis., 3,* pp. 371–372.

Ishola, O.O. and Adeyanju, G.T. (2014). Frozen Retail Poultry Meat Contact Surfaces as Sources of *Salmonella* and *Escherichia coli* Contamination in Ibadan, Oyo State, Nigeria. *American Journal of Infectious Diseases and Microbiology, 2,* pp. 81–85.

Kariyawasam, M.G.I.U. and Asooriya, S.P.J. (2006). A preliminary study on the prevalence of biofilms on in selected Sri Lankan fish and shrimp processing factories: set back in current cleaning process. *Sri Lankan J.Aquat. Sci., 11,* pp. 75–83.

Komp, L.P., Karlsson, A. and Hughes, D. (2003). Mutation rate and evolution of fluoroquinolone resistance in Escherichia coli isolates from patient with urinary tract infections. *Antimicrob Agents Chemother., 47,* pp. 3222–3232.

Krietsch, B. (2013). Food Safety News. Emerging Pathogens: Antibiotic Resistance Slowly Growing in Salmonella. Available at: http://www.foodsafetynews.com/2013/09/antibiotic-resistance-slowly-growing-in-salmonella/#.V1ca7iHXldg [Accessed June 7, 2016].

Magiorakos A.-P., et al. (2011). Multidrug-resistant, extensively drug-resistant and pandrug-resistant bacteria: an international expert proposal for interim standard definitions for acquired resistance. *Clin Microbiol Infect., 18,* pp. 268–281.

Miranda, J.M., et al. (2008). Evolution of resistance in poultry intestinal Escherichia coli during three commonly used antimicrobial therapeutic treatments in poultry. *Poult Sci., 87,* pp. 1643–1648.

MSF. (2013). New Roots for rural health Challenging unequal access in South Africa. The Musina Model of Care 2013. Available at: http://www.msf.org/sites/msf.org/files/musina_new-roots-for-rural-health_south-africa.pdf [Accessed June 8, 2016].

Musa, O.I. and Akande, T.M. (2003). Food Hygiene Practices of Food Vendors in Secondary Schools in llorin. *The Nigerian Postgraduate Medical Journal, 10*(3). Available at: https://www.unilorin.edu.ng/publications/omusa/Food%20Hygiene%20Practices%20of%20Food%20Vendors%20in%20Secondary%20Schools%20.pdf [Accessed June 8, 2016].

National Transfer Accounts Bulletin. (2013). *Population Change and Economic Growth in Africa, 6*. Available at: http://www.ntaccounts.org/doc/repository/NTAbulletin6.pdf [Accessed June 8, 2016].

Ogah, J.O., Adekunle, O.C. and Adegoke, A.A. (2015). Prevalence of Salmonellosis among Food Handlers and the Health Implications on the Food Consumers in Lagos State, *Nigeria. J Med Microb Diagn., 4*, p. 180.

Okeke, N.I., et al. (2007). Perspective: Growing Problem of Multidrug-Resistant Enteric Pathogens in Africa. Emerging Infectious Diseases,13(11). Available at: http://wwwnc.cdc.gov/eid/article/13/11/pdfs/07-0674.pdf [Accessed June 8, 2016].

Olatoye, I.O. (2010). The incidence of antibiotics susceptibility of Escherichia coli0157:H7 from beef in Ibadan Municipal, Nigeria. *Afr J Biotechnol., 9*(8), p. 1196.

Panisello, P.J., et al. (2000).Application of foodbornedisease outbreak data in the development and maintenance ofHACCP systems. *Int J Food Microbiol., 59*, pp. 221–234.

Reuben R.C. and Owuna G. (2013). Antimicrobial, Resistance Patterns Of *Escherichia Coli* O157:H7 From Nigerian Fermented Milk Samples In Nasarawa State, Nigeria. *International Journal of Pharmaceutical Science Invention, 2*(3), pp. 38–44.

Ribot, E.M., et al. (2002). *Salmonellaenterica* serotype Typhimurium DT104 Isolated from Humans, United States, 1985, 1990, and 1995. *EmergingInfectious Diseases, 8*(4), pp. 387–391.

USDA, GAIN report #NI2025:1-6. (2002). Nigeria poultry products. In: PoultryUpdate, 2002. Washington, DC: USDA Foreign Agricultural Service.

U.S. Government Accountability Office (GAO). (2004). Federal Agencies Need to Better Focus Efforts to Address Risk to Humans from Antibiotic Use in Animals. Available at: http://www.gao.gov/new.items/d04490.pdf [Accessed June 8, 2016].

Van Houdt, R. and Michiels, C.W. (2010). Biofilm formation and the food industry, a focus on the bacterial outer surface. Review article. Journal of Applied Microbiology ISSN 1364-5072. Available at: http://www.flandersfood.com/sites/default/files/Biofilm%20formation%20and%20the%20food%20industry,%20a%20focus%20on%20the%20bacterial%20outer%20surface.pdf [Accessed June 8, 2016].

Van de Venter, T. (2000). Emerging food-borne diseases: a global responsibility. Available at: http://www.fao.org/3/a-x7133m/x7133m01.pdf [Accessed June 8, 2016].

Victorian Government Health Information (2009). Department of Health Safety. Food safety. What causes food contamination? Available at: http://dofoodsafely.health.vic.gov.au/topics/topic3/page1.php [Accessed date June 8, 2016].

Viroy, M., et al. (2005). Longitudinal trends in antimicrobial susceptibilities across long term care facilities emergence of fluoroquinolone resistance. *Infect Control Hosp Epidemiol., 26*, pp. 56–62.

World Health Organization (WHO). Regional Office for Africa Brazzaville. (2012). Manual for Integrated Foodborne Disease Surveillance in the WHO African Region. Available at: http://apps.who.int/iris/bitstream/10665/170262/1/foodborne-disease-manual.pdf [Accessed June 8, 2016].

World Health Organization (WHO) regional office for Africa (2015). Overview: Food Safety. Available at: http://www.afro.who.int/en/food-safety/overview.html http://www.afro.who.int/en/clusters-a-programmes/hpr/food-safety-and-nutrition-fan/overview.html [Accessed June 8, 2016].

20

Evolution and Prevalence of Multidrug Resistance Among Foodborne Pathogens

Sinosh Skariyachan[1,2], *Anagha S. Setlur*[1,2], *and Sujay Y. Naik*[1,2]

[1] *Department of Biotechnology, Dayananda Sagar College of Engineering, Bangalore, Karnataka, India*
[2] *Visvesvaraya Technological University, Belagavi, Karnataka, India*

20.1 Introduction

Antibiotic resistance or multi-drug resistance (MDR) is defined as the ability of bacteria to survive exposure to a definite concentration of a drug or an antimicrobial agent. The rapid advent of drug resistant bacteria in the recent years is occurring globally, jeopardizing the efficiency of antibiotics used commonly for the treatment of numerous diseases (Ventola, 2015). This has been attributed to the misuse of the medications as well as lack of advances in new drug developments (Gould and Bal, 2013). From the late 1960s to the early 1980s, the pharmaceutical industry developed several new antibiotics to resolve the resistance crisis, but subsequently, the antibiotic pipeline began to dry up and fewer new drugs were introduced (Sengupta *et al.*, 2013). As a consequence, years after the first patients were treated with antibiotics; bacterial infections have once again become the foremost peril (Spellberg and Gilbert, 2014). The major causes of the evolution of bacterial multi-drug resistance in 2015-2016 include overuse, widespread use in agriculture, lack of new drug developments, misuse, and other regulatory difficulties.

20.2 Major Causes of the Evolution of Bacterial Drug Resistances

20.2.1 Overuse

The overuse of antibiotics distinctly leads to an evolution of resistance (Read and Woods, 2014). Certain epidemiological reports have claimed a direct statistical relationship between the consumption of antibiotic and the emergence of resistance (Goossens, 2009). An analysis of the IMS Health Midas database implied that in 2010, 22.0 standard units (a unit equaling one dose, i.e., one pill, capsule, or ampoule) of drugs were prescribed per person in the United States (Van Boeckel *et al.*, 2014).

20.2.2 Widespread Use in Agriculture

Utilization of antibiotics for the improvement of livestock has been a practice prevalent from many decades (Gross, 2013). These drugs enter the body of animals, the meat of some, are consumed by humans. This transmission results in the development of resistant bacteria in humans, especially the MDR (multi-drug resistance) in

Foodborne Pathogens and Antibiotic Resistance, First Edition. Edited by Om V. Singh.

food borne pathogens and gut microbes (Bartlett *et al.,* 2013). The environmental microbiome will also be drastically affected, in turn triggering adverse health effects.

20.2.3 Lack of New Drug Developments

Since antibiotics are used for a moderately brief period of time and has limited use in chronic ailments, the development of new drugs has been a setback and one of the important causes for instigating microbial resistance to existing drugs (Wright, 2014). Additionally, the introduction of new drugs is expensive and takes years to come into human trials. The upshot is the use of cheap, generally operative drugs, which is beneficial for the public but however, has the downside of causing severe microbial resistance (Piddock, 2012).

20.2.4 Misuse of Antibiotics

Studies have shown that misuse of antibiotics or flawed prescription where 30 to 50% of antibiotics consumption all over the world is faulty, another contributor to promote bacterial growth and resistance (Luyt *et al.*, 2014). Such medications have uncertain therapeutic advantages and endanger patients to potential hitches of antibiotic therapy (Lushniak, 2014). Sub-inhibitory antibiotic concentrations can encourage the development of antibiotic resistance by causing genetic alterations, namely, changes in gene expression, HGT, and mutagenesis (Viswanathan, 2014).

20.2.5 Other Regulatory Difficulties

Obtaining regulatory approvals for a new drug has always been an obstacle since it is time consuming. This leads to prolonged usage of existing drugs, which gravitates toward development of bacterial resistance. This can be an indirect factor for MDR in pathogens.

The emergences of multi-drug resistant pathogens have also adversely been affecting the pharma and health care sectors. The public and private sectors find it problematic to develop new antibiotics for a specific disease. For instance, the foodborne pathogen, *Staphylococcus aureus,* has become an MDR (MRSA-Methicillin-Resistant *Staphylococcus aureus* and VRSA-Vancomycin-Resistant *Staphylococcus aureus*) spawning critical issues to the pharmaceutical companies for the introduction of explicit drugs for *Staphylococcal enterotoxicosis* (Thompson *et al.*, 2013). In 2013, the CDC (Centers for Disease Control and Prevention) declared that the human race is now in the "post-antibiotic era" and in 2014, the World Health Organization (WHO) cautioned that the antibiotic resistance crisis is gradually becoming dire. MDR bacteria have been declared a considerable threat to the global economy as well as to the diverse healthcare sectors (Michael *et al.*, 2014).

Foodborne illnesses are on the rise in several parts of the world and are evolving as a grave public health hazard. The CDC reports that 76 million foodborne infections, including 5,000 deaths and 325,000 hospitalizations, occur in the United States every year. Foodborne Diseases Active Surveillance Network (FoodNet) studies demonstrated that *Salmonella, Campylobacter, Shigella, Cryptosporidium,* and Shiga toxin *Escherichia coli* (STEC) O157 persist to be the prominent causes of death (Crim *et al.*, 2015). They are developing into multi-drug resistant pathogens and treatment for these diseases is a complex issue (Nyachuba, 2010). While MDR outbreaks have been compiled since the 1970s, the majority (34 to 58%) has been fallen since 2000 and dairy products caused a quarter of the outbreaks (14 outbreaks) (Ferri *et al.*, 2015). The major outbreaks of MDRs in the recent years include *Salmonella, Escherichia coli,* and *Listeria* infections. These bacteria accounted for approximately 91% of multi-state outbreaks in 2015 (CDC, 2015). *Salmonella* caused the most hospitalizations and could be traced to have been transmitted from eggs and chickens (Yeni *et al.*, 2015). The presence of *Listeria monocytogenes* in several food samples such as cheese plants was discovered at a recent survey conducted between 2008 and 2012 in Chile, which is suggestive of its emerging pathogenicity (Saludes *et al.*, 2015).

Entero-hemorrhagic *Escherichia coli* (EHEC) have proven capability to induce acute intestinal disease and the hemolytic uremic syndrome (HUS). Cattle are the main pool for EHEC, where the bacteria can survive for years (Karch *et al.*, 2015). There have been reports of *Campylobacter* infections as well, that are commonly spread through poultry meat where *Campylobacter jejuni,* another emerging MDR pathogen, demonstrated 20.8% isolation in several food samples when compared with other food pathogens (Dan *et al.*, 2015). According to WHO, almost one-third (30%) of the deaths due to foodborne infections and intoxications occurs in children below the age of five years. One in every 10 people in the world infects post-ingesting contaminated food. Of these, 420,000 people face death, along with 125,000 children under the age of five years. Reports are also available where foodborne illnesses have taken a toll on different parts of the world such as Africa (Salmonellosis), the United States (Campylobacteriosis and Enterohemolytic *E. coli*), eastern Mediterranean region (Campylobacteriosis, Salmonellosis), and European and Southeast Asian regions (WHO, 2015). Almost all the capable infecting agents (e.g., bacteria, fungi, virus, and parasite) have high levels of MDR with greater mortality and morbidity and therefore they are also known as "super bugs" (Tanwar *et al.*, 2014).

Since the development of MDR drugs to treat food borne and other infections pose a difficulty, efforts are being made to formulate and design new drugs to combat MDR pathogens. An assessment of how the main resistance patterns and related infections can be managed in the absence of more conclusive information is necessary (Viale *et al.*, 2015). Some studies have shown a drastic reduction in the MDR *Salmonella* and *Escherichia coli* O157: H7, when treated with antimicrobial interventions that are presently in use in several U.S. beef processing plants (Doyle, 2015). Global counter measures direct unions and countries in authorizing and devising new regulations.

Antibiotic administration to farm animals for promoting their growth has also been restricted by the European Union. Since the expenditures to develop an MDR drug are high, research is also being focused on the development of organic compounds to combat severe microbial resistance globally. The worldwide economy is also being supported for, to regulate the production of safe and healthy foods devoid of harmful pathogens. Further, large-scale measures for the management of antibiotics at farm level are being carried out to decrease a large percentage of MDR development (Koluman and Dikici, 2013). For instance, multi-drug resistant tuberculosis is a leading public health concern and introducing a new drug for MDR-TB costs 200 times more than what it takes to develop a normal drug. Thus, a global financial and investment strategy is being formulated to provide a long-term technical assistance for TB (Hwang and Keshavjee, 2014).

20.3 Food Poisoning and Foodborne Illness—An Overview

20.3.1 Major Food Sources and Food Products

The major part of food materials are acquired from plants and animals, which are an enormous source of diverse microorganisms (Brenner *et al.*, 2005). The vital sources of foodborne pathogens in transmitting the hazardous diseases are:

- *Campylobacter jejuni*
  This organism spreads through raw and undercooked poultry and other meat, raw milk, untreated water, red meat, and ground beef (Bull *et al.*, 2006).

- *Clostridium botulinum*
These species spread via improperly prepared home-canned foods; honey (should not be fed to children less than 12 months old), beans, sweet corn, beets, asparagus, spinach, chard, tomatoes, apricots, peaches, meat, and poultry products (Dodds, 1993).

- *Clostridium perfringens*
These are isolated from frozen foods, fruits, vegetables, home-prepared foods, raw meat, poultry, and fishes (Brynestad, 2002).

- *Escherichia coli*
The sources for this organism include beef, especially undercooked or raw hamburger, raw milk, unpasteurized juices, and ciders (Hancock *et al.*, 1998).

- *Listeria monocytogenes*
This bacteria mainly spreads through unpasteurized dairy products, including soft cheeses, sliced deli meats, smoked fish, hot dogs, salads (i.e., egg, ham, seafood, and chicken salads), raw meat and vegetables (Kathariou, 2002).

- *Salmonella typhi* and *Salmonella paratyphi*
Raw and undercooked eggs, undercooked poultry and meat, fresh fruits, vegetables, unpasteurized dairy products such as ice creams and cheeses, eggs, hash, sausages, cured meats, sandwiches, and chili (Altekruse *et al.*, 1997).

- *Staphylococcus aureus*
These bacteria mainly spreads through cooked foods high in protein (e.g., cooked ham, salads, bakery products, dairy products) that are held too long at room temperature, custard, creams, bakery products, poultry products, fish and fish products, salad dressings and sauces (Bennett, 2001).

- *Shigella dysenteriae*
The main source of these organisms include salads, unclean water, any food handled by someone who is infected with the bacterium, potato, tuna, shrimps, macaroni, chicken, vegetables, milk, dairy products, and meat (Martin *et al.*, 1986).

- *Vibrio cholerae* and *Vibrio vulniferans*
These bacteria are spreading through raw or undercooked seafood, particularly shellfish, oysters, fishes, shrimps and crustaceans (Pires *et al.*, 2012).

- *Yersinia enterocolitica*
This organism has been isolated from various food sources such as beef, pork, liquid egg, soft cheese, raw and pasteurized milk, fish, shrimps, crabs, turkey, and raw oysters (Bozcal *et al.*, 2015).

20.4 Factors that Influence the Growth of Foodborne Pathogens in Food Products

The microorganisms utilize the food substrates as nutrition for their growth and survival. The primary reason for the spoilage of food items is the increase in the number of microorganisms and consumption of nutrients thereby causing enzymatic reactions and alterations in the food composition resulting in noxious odors that occur due to the breakdown of some of the constituents of foods. The microorganisms use the food as its carbon, nitrogen, and protein source and multiply incessantly (Skariyachan and Abhilash, 2012). This growth of microbes can be either due to intrinsic or extrinsic factors.

20.4.1 Intrinsic Factors

The parameters that affect the microbial growth from within the food items are called intrinsic factors. These include:

- **Hydrogen Ion Concentration (pH)**: The growth of pathogens in food is dependent upon the pH of the food substrates. If the pH of the food is less than 7, acidophiles grow, neutrophiles at pH7 and alkalophiles such as proteolytic bacteria at pH greater than 7 (Doelle, 2014).
- **Water activity (moisture content)**: It is the ratio of vapor pressure of the solution to the vapor pressure of the solvent. Bacteria require

more water activity than yeasts or moulds and gram-positive bacteria need lesser water than gram-negative bacteria. For instance, *S. aureus* grows at water activity of 0.86 and *Clostridium* at 0.94 (Troller, 2012).

- **Redox Potential (Eh)**: As the oxidation-reduction potential increases for a food item, the growth of the microbes increases and thus changes the composition of the foods (Banwart, 2014).
- **Composition of nutrients**: The presence of carbon sources, such as esters, alcohols, celluloses, hemicelluloses, amino acids, and peptides for protein sources, will boost the growth of pathogenic bacteria.
- **Anti-nutritional factors (ANF)**:These inhibitory substances present in foods reduce the availability of one or more nutrients, leading to growth of bacteria that can survive under extreme conditions. ANFs further may inhibit few bacteria and allow the growth of other pathogens. For example, Avidin, present in egg whites, binds to Biotin making it biologically unavailable (Sathe, 2013).
- **Biological structure of foods**:The more complex the structure of a food is, the harder it is for the microbes to enter and vice-versa. Egg has a complex hard shell, providing the inner nutritional content protection from bacterial attacks.

20.4.2 Extrinsic Factors

The parameters that are substrate-independent and depend mainly on the food storage environments are known as intrinsic factors. These include:

- **Relative humidity (RH)**:When foods with low water activity are placed in high relative humid environments, the food takes up more moisture until equilibrium is attained. Further, RH is inversely proportional to temperature (Jay, 2012).
- **Atmospheric gases**: Presence of oxygen, carbon dioxide, ozone, and so on is very important in places where food items are deposited. These gases control the microorganisms in foods. Modified Atmosphere Packaged foods (MAP) make use of these gases.
- **Temperature**: The temperature of the foods determines the type of bacteria that grow in it. Psychrophiles, such as *Enterococcus, Lactobacillus, Shewanella, Pseudomonas, Micrococcus,* and so on breed at temperatures of 20 to 30 °C. Mesophiles (20 to 45 °C) include *Listeria, Eshcerichia, Salmonella, Shigella,* and so on, whereas thermophiles (55 to 65 °C), include *Bacillus, Clostridium,* and so on (Ratkowsky *et al.*, 1982).
- **Other microbes**: The bacteria residing in food can undergo certain negative reactions. The presence of one bacterium may inhibit the emergence of another due to the secondary metabolites produced by it. These include bacteriocins, organic acids, harmful hydrogen peroxides, and so on (Vuyst *et al.*, 1994).

20.5 Food Poisoning and Foodborne Infections

Food poisoning refers to the disease caused by microorganisms transmitted through food, either by direct ingestion of the germ or by the release of toxins that harm the stomach and intestines (Hennekinne *et al.*, 2015). Foodborne diseases caused due to poisoning can be categorized into two:

- **Food intoxication**: It is caused by the formation of bacterial toxins in foods (Hayes, 2013). The three major types include:
  - Staphylococcal enterotoxicosis: *Staphylococcus aureus* intoxication
  - Botulism: *Clostridium botulinum* intoxication
  - Gastroenteritis: *Clostridium perfringens* and *Bacillus cereus* intoxication
- **Foodborne infection**: These are caused by the direct attack of the host through intestinal tract (Swaminathan and Feng, 1994). The types of food infections are:
  - Salmonellosis: caused by enterotoxin and cytotoxin of *Salmonella* spp.

- *Clostridium perfringens gastroenteritis:* It iscaused by enterotoxin release during sporulation of *Clostridium perfringens* type A to E in the intestinal tracts.
- *Bacillus cereus* gastroenteritis: It is caused by an exotoxin release of *Bacillus cereus* in the intestinal tract.
- Enteropathogenic *E. coli* infection: It is caused by various serotypes of *E.coli* such as invasive type and enterotoxigenic type.
- Other types: Yersiniosis, Listeriosis, Campylobacteriosis, Vibrioparahaemolyticus infections, and so on.

20.6 An Illustration of Major Foodborne Gastroenteritis

20.6.1 Salmonellosis

It is also known as enteric fever caused by two major organisms, *Salmonella typhi* that causes typhoid, and *salmonella paratyphi* (type A and B) that induces paratyphoid. These are gram-negative, rod-shaped bacterium whose incubation time after it enters the body is 8 hours, after which it starts multiplying, triggering the disease (de Jong *et al.*, 2014). These bacteria can be cultured in a selective medium called SS Agar (Salmonella-Shigella Agar, where dark-centered colonies indicate presence of *Salmonella*) and can grow at temperatures of 37 °C and alkaline pH (Rahaman *et al.*, 1975). The transmission of the disease occurs through contaminated foods and spread through vectors such as flies. The important symptoms include fever, malaise, chills, abdominal pain, nausea, drowsiness and green colored-foul smelling stools (Jacobsen *et al.*, 2011).

20.6.2 Shigellosis

It is also known as bacillary dysentery caused primarily by *Shigella dysenteriae*. These are gram-negative, non-spore forming rods, which initiates the infection in the body after 1 to 7 days of entry into the host. As mentioned afore, *Shigella* can be grown in S-S Agar where the colonies appear smooth and transparent. This bacterium can also be cultured in MacConkey's Agar (selective media). It is a mesophile and grows at slightly alkaline pH of 7.2 to 7.4 (Echeverria *et al.*, 1991). An endotoxin called "Shiga Toxin," a pentamer rich in carbohydrates, binds to RBC and creates a channel that leads to immense loss of water. The main after-effects of shigellosis include Reiter's disease (urinary tract disease in males) and HUS (hemolytic uremic syndrome, in females). The symptoms include severe dysentery, abdominal cramps, chills, diarrhea, watery/bloody stools, nausea, and body pain (Wilson *et al.*, 2011).

20.6.3 Listeriosis

This foodborne disease spreads through contaminated vegetables and fruits. It is caused by *Listeria monocytogenes* (a gram-positive, non-spore forming rods), a notorious pathogen that affects humans, cattle, and fishes. The period of disease initiation post entry is 7 to 8 days.

Listeria fails to grow in basic medium, but grows in LPM medium as they are fastidious microorganisms (Lee and McClain, 1986). It grows at temperatures of 30 to 37°C and at alkaline pH. The key features of this disease includes infection in pregnant women, neonatal infections, affects the CNS causing "Meningitis encephalitis" and severe gastroenteritis (Mylonakis *et al.*, 2002). *Listeria monocytogenes* is ubiquitous in the environment and its central route of transmission is through ingestion of contaminated foods. The symptoms are stiff neck, loss of body balance, and confusion. It is noted that there is no body pain and these symptoms are not easily identifiable in the early stages of the infection (Lovett *et al.*, 1987).

20.6.4 Staphylococcal enterotoxicosis

Staphylococcus aureus causes a pyogenic infection and symptoms appear after 24 hours. It is gram-positive facultative anaerobic cocci that attack by producing toxins. It survives at 35 to 37 °C and at high alkaline conditions. A selective media, BTBL Agar (Bromo Thymol Blue agar) and Mannitol Salt Agar can be used for the growth of this bacterium. Large yellow-colored disc-like colonies of *S. aureus* will be observed (Oussalah *et al.*, 2007). Enterotoxins of A, B, C1, C2, D, and E are produced by this bacteria where A and B are the most predominant toxin. The molecular weight is around 26000 to 30000 Da. These organisms are commonly present in dairy products, wine, meat, poultry, and so on. The common symptoms are body discomfort, salvation, nausea, weak pulse, headache, chills, abdominal cramps, shallow respiration, and fever (Gould *et al.*, 2012).

20.6.5 Clostridial Infections

Clostridium is gram-positive anaerobic, spore-forming, hyper-thermophilic bacteria. It survives at temperatures of 80°C and at alkaline pH (7 to 7.2). The factors that cause the growth of these bacteria are spore germination, anaerobic growth conditions, highly proteinaceous food items, moisture content, and high redox potential

20.6.6 Botulism

It is caused by *Clostridium botulinum,* where the symptoms characterize after 12 to 36 hours. *C. botulinum* grows in anaerobic medium such as Robinson Cooked Meat Agar and differential media like Blood Agar. Strong beta-hemolytic colonies are produced in blood agar, where a yellow zone of clearance is observed as a result of RBC lysis (Hospital *et al.*, 2016). Botulinum toxin type A to E consists of two chains of 100 kDa each, which are the main virulent factors. It can be classified into three groups (Dohrmann *et al.*, 2015). Group 1 (type A toxin), Group 2 (type E, B, and F toxins), and Group 3 (type C and D toxins). The symptoms are digestive disturbance, dizziness, diarrhea, fatigue, fever, chills, headache, nausea, and constipation.

20.6.7 Gas Gangrene

It is caused by *Clostridium perfringens,* where the symptoms materialize within 6 to 24 hours. *C. perfringens* grows in TSC Agar (Thiosulphate Cycloserin) and are identified as dark, black-colored colonies. Enterotoxins (CPE proteins) such as alpha, beta, delta, epsilon, or iota containing 319 amino acids are the main virulent factors. These toxins bind to either claudin or a 50-kDa eukaryotic membrane receptor that leads to the formation of 90 kDa CPE-containing complex in host cells. The symptoms include formation of gas gangrenes that are smelly, nausea, fever, and so on (Uzal *et al.*, 2015).

20.6.8 *Vibrio parahemolytics* Infection

Vibrio is a gram-negative, comma-shaped, non-spore–forming organism. The incubation period of the disease is 2 to 48 hours. These bacteria can grow in TCBS Agar (Thiosulphate Chlorine Bile Salt Agar) as well as BTBL agar medium. Alpha and beta-hemolytic colonies are produced by most of the strains. *V. cholerae* produces golden-yellow colonies and *V. parahemolyticus,* green-colored colonies. A pathogenicity island (PAI) is formed, which contains the Thermostable Direct Hemolysin gene (*TDH* gene) that codes for cholera toxin and thus is transmitted through contaminated food and water. The symptoms include loss of water, diarrhea, chills, headache, fever, and prostrations (Vezzulli *et al.*, 2013).

20.6.9 Yersiniosis

Yesiniosis is caused by gram-negative, non-spore–forming bacteria called *Yersinia enterocolitica.* The incubation period of the disease is 24 to 36 hours.

The major consequences of this disease are pseudo-appendicitis (a false appendicitis is formed and reactive arthritis (long-lasting nerve endings and joint pain (auto-immune condition))). This organism spreads through beef, pork, liquid egg, soft cheese, raw, and pasteurized milk.

The indications include abdominal pain, fever, dehydration, nausea, malaise, and vomiting (Harakeh *et al.*, 2012).

20.6.10 Campylobacteriosis

It is caused by a gram-negative spirochete called *Campylobacter jejuni and Campylobacter coli.* The incubation time for this disease is 2 to 10 days. Most of the strains produce powerful enterotoxins that are heat labile and thus appropriate cooking of foods can prevent this foodborne disease. The symptoms are abdominal pain and cramps, diarrhea, bloody stools, headache, and fever (Campana *et al.*, 2012).

20.7 Major Types of Antibiotics Used to Treat Foodborne Infections

Although some antibiotic applications are quite ineffective against major foodborne infections, since the bacteria have become multi-drug resistant, extreme-drug resistant, and pan-drug resistant, some diseases are cured via antibiotics. The major drugs used for the treatment include:

20.7.1 Salmonellosis

In cases where antibiotics are necessary, trimethoprim/sulfamethoxazole (TMP-SMX), amoxicillin, or ampicillins are the top choices. Ceftriaxone, cefotaxime, or fluoroquinolones are the useful options for antimicrobial-resistant strains, though fluoroquinolones are not favored for people younger than 18 years of age. For those with an infection in a specific tissue or organ (invasive disease), treatment with an enlarged-spectrum cephalosporin is suggested, until it is known if the bacteria is susceptible to one or more of the commonly used antibiotics listed (Gopinath *et al.*, 2012).

20.7.2 Shigellosis

The following antibiotics are used to treat *Shigella* mediated infections:

- Beta-lactams: Ampicillin, amoxicillin, third-generation cephalosporins (cefixime, ceftriaxone), and pivmecillinam (not available in the United States)
- Quinolones: Nalidixic acid, ciprofloxacin, norfloxacin, and ofloxacin
- Macrolides: Azithromycin
- Others: sulfonamides, tetracycline, cotrimoxazole, and furazolidone (Das *et al.*, 2013).

Most clinical infections with *S. sonnei* are self-limited (48–72 h) and may not require antibiotic therapy. If an ampicillin and TMP-SMX resistant strain is isolated or if the susceptibility is unknown, parenteral ceftriaxone sodium, fluoroquinolone (e.g., ciprofloxacin, ofloxacin), azithromycin dihydrate or oral cefixime are the drugs of choice (Xiao *et al.*, 2012).

20.7.3 Clostridial Infections

- **Botulism***:* Since toxins are the major cause of this poisoning, antitoxins are commonly used for the treatment. HBAT (Heptavalent Botulinum Anti-Toxin) is the drug of choice (Hill *et al.*, 2013). Antitoxins are specified for naturally occurring non-infant botulism. Equine-derived antitoxins that stimulate immediate immunity against *Clostridium botulinum* toxins A, B, C, D, E, F, and G are also preferred. The recommended antibiotics are Penicillin G, Chloramphenicol, and Clindamycin for botulism wounds (Supuran, 2012).
- **Gas gangrene***:* Recent studies have suggested that protein synthesis inhibitors such as clindamycins are the drug of choices. Chloramphenicol, rifampicin, and tetracyclines may be more effective since they impede the synthesis of exotoxins, which cause gas gangrenes. In spite of increasing clindamycin resistance among anaerobes, cases of clindamycin-resistant *C. perfringens* are exceptional (Au *et al.*, 2013). A combination of clindamycin and metronidazole is used for patients allergic to penicillin drugs. A combination of penicillin and metronidazole may be incompatible and is not recommended. Newer and innovative antibiotics such as daptomycin,

linezolid, and tigecycline have not been studied in patients with gas gangrene, but are still considered as options. Drugs like penicillin-G, amoxicillin, ticarcillin, piperacillin, cefazolin, cefoxitin, cefotetan, third-generation cephalosporins, macrolides, metronidazole, imipenem, and so on maybe used but some bacteria are resistant to these antibiotics (Fejes *et al.*, 2013).

20.7.4 Campylobacter Gastroenteritis

This infection is a self-limited infection, which can usually be cured easily. In the most brutal cases of this disease, antibiotics are administered. Macrolide antibiotics such as erythromycin, clarithromycin, or azithromycin are the most effective agents for *Campylobacter jejuni*. Fluoroquinolone antibiotics (ciprofloxacin, levofloxacin, gatifloxacin, or moxifloxacin) can also be used, but resistance to this class has been at a rise, due to the use of this class of antimicrobials in poultry feed (Qin *et al.*, 2012).

20.7.5 Yersiniosis

Y. enterocolitica is usually susceptible in vitro to aminoglycosides, chloramphenicol, tetracycline, trimethoprim-sulfamethoxazole (TMP-SMZ), piperacillin, ciprofloxacin, and third-generation cephalosporins. The isolates are often resistant to penicillin, ampicillin, and first-generation cephalosporins, as the organism produces beta-lactamase. The clinical failure with cefotaxime has been reported recently. Resistance to macrolides and fluoroquinolones is also intermittently reported. Clinically, *Y. enterocolitica* infection responds well to aminoglycosides, TMP-SMZ, ciprofloxacin, and doxycycline (Rosner *et al.*, 2013).

20.7.6 *Vibrio cholerae* Infections

This disease involves dehydration and thus rehydration via fluid intake is the most approved treatment. However, antibiotics like doxycycline, tetracycline, furazolidone, trimethoprim-sulphamethoxyzole, ciprofloxacine, ampicillin, and erythromycin are recommended for treatment (Singh *et al.*, 2015).

20.7.7 *Staphylococcus aureus* Infections

The antibiotics commonly prescribed to treat this infection include certain cephalosporins, nafcillin, and other related antibiotics such as sulfa drugs and vancomycin (Hasan *et al.*, 2016). Vancomycin is required to treat severe poisoning because many strains of these bacteria have become resistant to other traditional medicines. But vancomycin and some other antibiotics must be given intravenously. This organism has become resistant to methicillin and vancomycin and is therefore known as MRSA (methicillin-resistant *Staphylococcus aureus*) and VRSA (vancomycin-resistant *Staphylococcus aureus*) respectively (Mandal *et al.*, 2015).

20.7.8 Listeriosis

This infection is self-limited but needs medical care in case of severe listeriosis. Ampicillin is generally the preferred agent, but penicillin is also sometimes suggested. Invasive listeriosis occurs during pregnancy and infects the newborn and thus combinations of antibiotics are used in such cases (Spitzer *et al.*, 1986).

20.8 Mechanisms of Evolution of Antibiotic Resistance in Food Products

Evolution of multi-drug resistance is an emerging menace. Albeit several new drugs have been developed commercially, the resistance of food borne pathogens to these drugs still persist due to sustained drug exposure (Loeffler and Stevens, 2003). The antimicrobial drugs normally act on the microbes by impeding a metabolic pathway such as nucleotide synthesis in turn causing the hindering of RNA/DNA synthesis and thereby protein synthesis ensued by rupture of the cell membrane and by competing with the substrate of any enzyme involved in the synthesis of the cell wall (Chethana *et al.*, 2013). MDR in bacteria present in food products may be bred by either of the following mechanisms. Firstly, the bacteria may have multiple genes, which each code for resistance to a single drug. Secondly, the overexpression of genes, which code for multidrug efflux pumps, might also induce multiple drug resistance (Nikaido, 2009). Hence, microorganisms have evolved numerous mechanisms to overcome the effectiveness of drugs. The commonly encountered mechanisms of resistance can be categorized into two (Tanwar *et al.*, 2014):

- Intrinsic resistance: The bacterial resistance which originates from inside the host cell. The generalized intrinsic mechanisms are shown in Figure 20.1.
- Extrinsic resistance: The bacterial resistance which occurs from the external environment of the host cell.

The main intrinsic resistance mechanisms are explained in the following sections.

20.8.1 Chromosomal Mutations

Mutations are any small changes in the DNA of the host through addition, deletion, or substitution of a single or several base pairs. In a bacterial DNA, a mutation might produce significant changes in the binding site of an antibiotic, which may prevent the binding of antibiotic to the site entirely. Reports revealed that the bacterial protein LexA (SOS gene repressor) maybe responsible for the acquisition of resistance due

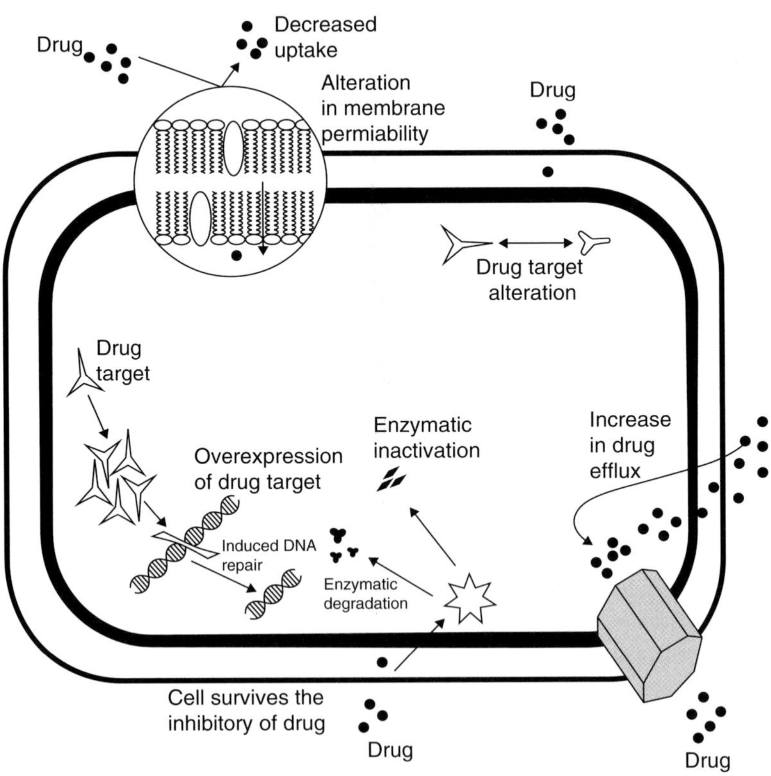

Figure 20.1 Generalized mechanism of multi-drug resistance representing the intrinsic mechanisms of resistance: Chromosomal mutations, inactivation of drug by enzymes, and action of efflux pumps.

to mutations, making the bacteria such as *Salmonella* resistant to quinolones and rifampicin (Walsh *et al.*, 2012). DNA destruction encourages auto-proteolytic activity of LexA. During this mechanism, the transcription of genes such as Pol II, Pol IV, and Pol V (three non-essential DNA polymerases) occurs, which are necessary for mutations post DNA damage (Cirz *et al.*, 2005). The bacteria with mutations that can survive for prolonged periods will multiply and pass on the mutated genes to their offspring, thereby initiating an evolution of antibiotic resistance. Although the mutations allow the bacteria to be resistant, they also have few drawbacks. For instance, a ribosomal mutation may alter the binding site of an antibiotic and defend the host cell but will also dwindle the synthesis of proteins (Andersson and Hughes, 2012). Further, a study also compared the antibiotic resistant strains of *Escherichia coli* and *Salmonella* spp. to their antibiotics and a decrease in the growth rates were observed (Levin *et al.*, 2000). The mechanism is shown in Figure 20.2.

20.8.2 Inactivation of the Drugs by Enzymes

This is a regular mechanism for antibiotics of natural origin such as aminoglycosides, which are deactivated by acetylation, adenylation, or enzymatic phosphorylation. Beta-lactams are also inactivated by enzymatic hydrolysis in the periplasm. Genes that code for inactivating enzymes generate resistance on plasmids as additional genetic components (Davies and Wright, 1997). Aminoglycoside-inactivating enzymes can add acetyl, adenyl, or phosphoryl groups to the drug to inactivate them (Wright, 1999). It is also reported that the source of these enzymes are the antibiotic producing host bacteria, which produce the antibiotic to protect themselves (Lin *et al.*, 2015). Another important example is beta-lactamases, which can cleave

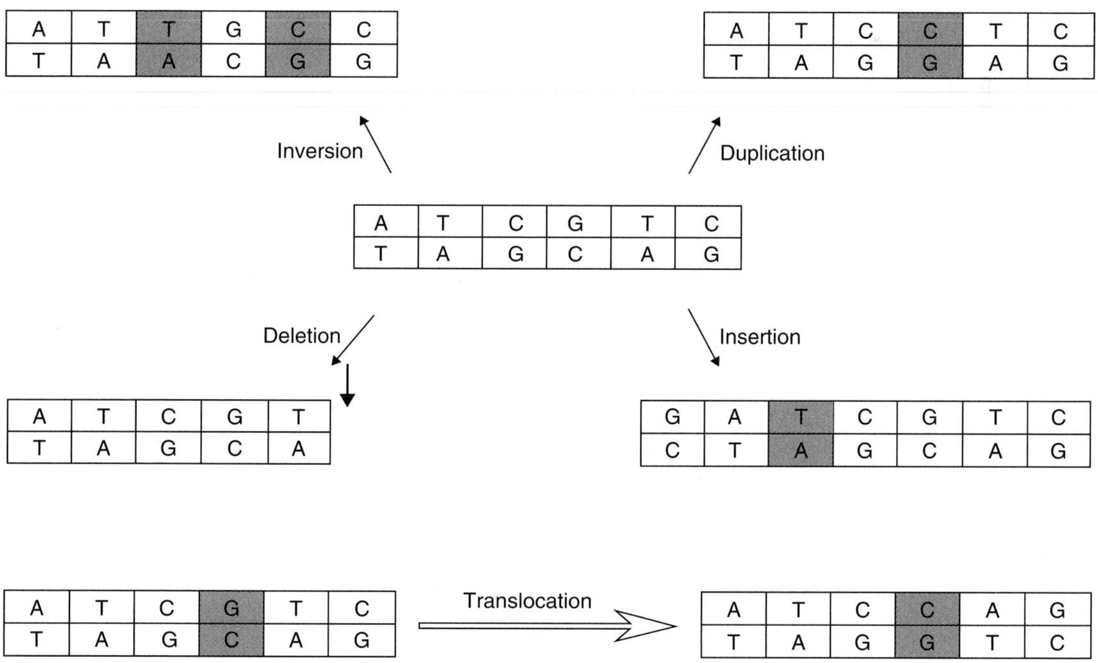

Figure 20.2 Chromosomal mutations indicating inversion, duplication, deletion, insertion, and translocation responsible for causing antibiotic resistance in host bacterial cells.

the beta-lactam family of antibiotics and cause resistance. *Staphylococcus aureus,* a notorious foodborne pathogen has developed an immense resistance initiated by a beta-lactamase coded for it by a plasmid. Albeit this difficulty was resolved by the development of Methicillin and related compounds that defy the hydrolysis via enzymes, another enzyme called TEM beta-lactamase, was described in gram-negative bacteria, which induce resistance (Rodvold *et al.*, 2014).

20.8.3 Resistance of Bacteria Through Efflux Pumps

The efflux pumps are carrier proteins involved in the removal of toxic substrates (including pertinent antibiotics) from the interior of the cells to the exterior. These proteins exist in both gram-negative and gram-positive bacteria (Bambeke *et al.*, 2000). These pumps can be specific for one substrate or may transport a collection of structurally unrelated compounds (antibiotics of various classes) and such pumps are usually linked with multiple drug resistance (MDR). The efflux pump genes are a part of operon found on plasmids, with a regulatory gene controlling its expression. Over-expression (resulting from mutations) of these genes may lead to resistance to substrates (antibiotics). For instance, increased expression of acrAB gene in *E. coli* can cause resistance to some *E. coli*-related antibiotics (Thanassi *et al.*, 1997). This mechanism is considered to be an intrinsic resistance, which permits the survival of the host in a hostile environment like in the presence of drugs (Li *et al.*, 1994). This is observed in food borne pathogens such as *Campylobacter jejuni* (CmeABC7.8), *Escherichia coli* (AcrAB-TolC, AcrEF-TolC, EmrB, EmrD9), *Salmonella* (AcrB11), and *Staphylococcus aureus* (NorA12). These efflux fluoroquinolones and thus an evolutionary mechanism of resistance are developed within a cell (Wang *et al.*, 2001; Webber and Piddock, 2001; Pomposiello and Demble, 2000). The mechanism of efflux pumps is shown in Figure 20.3.

20.8.4 Horizontal Gene Transfer

The external resistance mechanisms include the three major types of horizontal gene transfers (HGT). This term, also known as lateral gene transfer (LGT), refers to the transfer of genes between organisms, which is one of the major causes of evolution of many organisms (Rossi *et al.*, 2014). The mechanisms are conjugation, transformation, and transduction.

20.8.4.1 Conjugation

Most of the bacteria have plasmids that are small, circular, double-stranded, extra-chromosomal DNA that possess a multitude of attributes. These are present separately from the bacterial chromosome and carry specialized genes responsible for providing antibiotic resistance. Bacterial conjugation is the transfer of genetic material (plasmids) between bacterial cells by direct cell-to-cell contact or by a bridge-like connection (pili) between two cells. During this mechanism, the donor cell provides a conjugative or mobilizable genetic element that is most often a plasmid or a transposon (jumping genes). Most conjugative plasmids have a system ensuring that the recipient cell does not already contain a similar element. The donor cell first produces a pilus. This pilus attaches to the recipient cell and brings both the cells close together. The mobile plasmid is nicked and a single strand of DNA is transferred to the recipient cell. Both the donor and recipient cells at this time synthesize a complementary DNA strand to produce the original double-stranded circular plasmid and thus both the cells become viable donors. The donor cell, which transfers the gene is called F^+, since they contain the F plasmid (an episome with a length of 100 kb that can integrate itself into the bacterial chromosome by homologous recombination), whereas the recipient cell is known as F^- plasmid. The initiation of conjugation occurs when the enzyme relaxase creates a nick in the plasmid to be transferred. The nicked strand, or T-strand, is then unwound from the unbroken strand and transferred to the recipient cell in a 5' to 3'

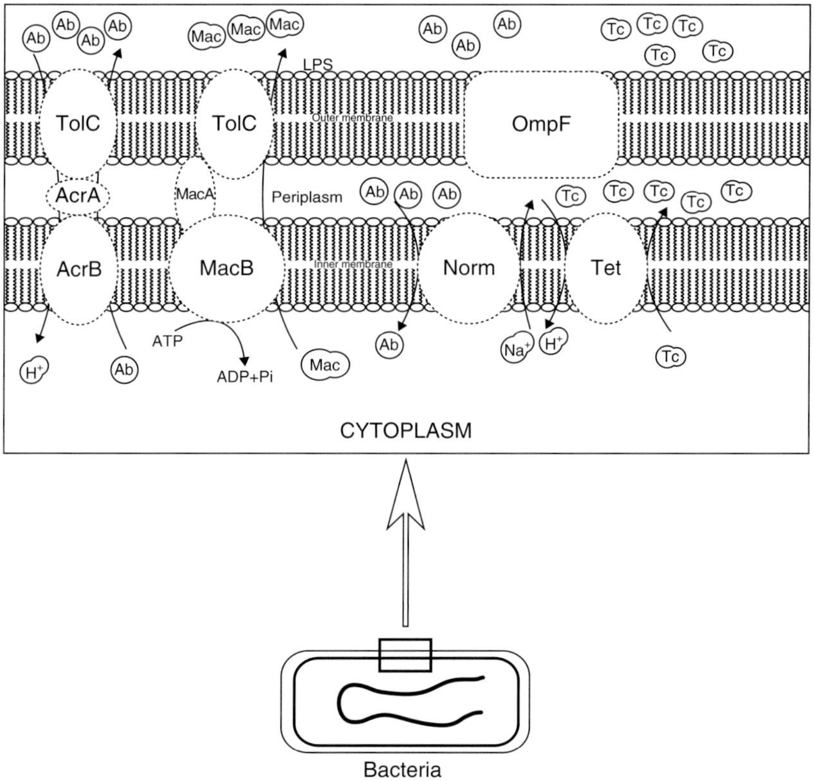

Figure 20.3 Action of Efflux pumps in antibiotic resistance. The substrates of AcrAB (e.g., aminoglycosides) are recognized by a portion of AcrB, which faces the cytoplasm while others like tetracycline migrate through the inner membranes to AcrB binding pockets. The *E. coli* Tet flux is the secondary transporters that use a proton motive force to transport the genes. NorM is a transporter that exploits the gradient of sodium ions. MacAB is a primary transporter that uses the energy from ATP hydrolysis to drive the efflux pumps. Single transport systems such as Tet and NorM and transport porins such as OmpF across the membranes.

terminus direction. The remaining strand is replicated either independent of the conjugative action or in combination with conjugation. Thus, the resistant genes transferred leads to an evolution of antibiotic resistant bacteria (Clewell, 2013). The mechanism of conjugation is displayed in Figure 20.4.

20.8.4.2 Transformation

The genetic alteration of a cell resulting from the direct uptake, incorporation, and expression of exogenous genetic material from its surroundings is called transformation. This mechanism occurs naturally in some species of bacteria, but it can also be affected by artificial means in other cells. For transformation to occur, bacteria must be in a state of competence, which may happen as a time-limited response to environmental conditions such as starvation and cell density. There are two forms of transformation and competence: natural and artificial. Natural transformation is a bacterial adaptation for DNA transfer that depends on the expression of numerous bacterial genes whose products appear to be devised to carry out this process. The capacity for natural transformation occurs in a number of prokaryotes, and therefore around 67 prokaryotic species are known to undergo this process. Competence is triggered by high cell density and nutritional limitation. For instance, in

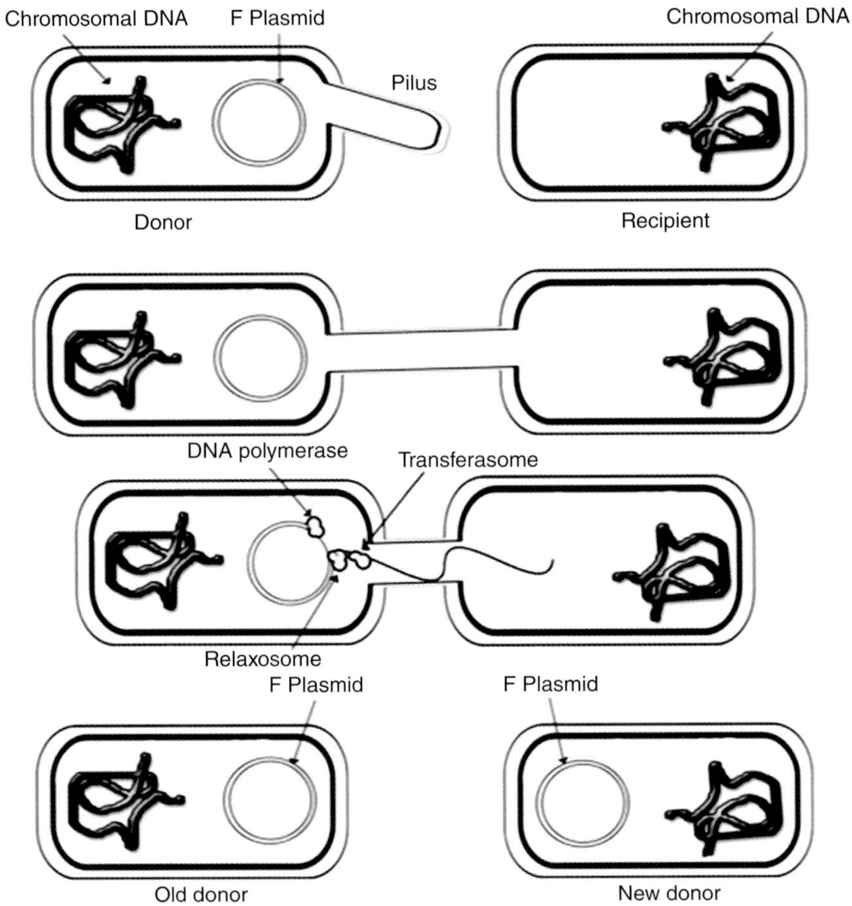

Figure 20.4 Mechanism of bacterial conjugation that leads to antibiotic resistance. A conjugation pilus is formed in the donor bacterium, which attaches to the recipient bacterium. The antibiotic gene present on the plasmid is transferred via the conjugation bridge, making the recipient, same as the original donor after replication of the plasmid.

Streptococcus mutants, transformation occurs at high cell density and is related with biofilm formation, a phenotypic method of transferring antibiotic resistance.

When bacterial cells destruct naturally, their cells break apart and the cell material, including pieces of the chromosome, gets spilled out into the environment. If there is another bacterial cell in the vicinity that is competent for transformation it can take up that free DNA (by development of pilus or secretion system) and either incorporate it into their genome or degrade it. This method can give the bacterial cell a new attribute or an ability that make it better adapted to extreme environments such as drug resistance development. Transformation can also occur in artificial laboratory conditions where competence can be deliberately induced within a cell.

The cells are incubated in solution containing divalent cations (commonly calcium chloride) under cold conditions, before being exposed to a heat shock. This causes pores to be produced in the cell membrane of the bacteria, through which the DNA uptake occurs via chemically induced competence. Moreover, the divalent positive cations attract the negative phosphate groups present on the DNA, which is to be taken up by the cell. This adherence is ensued by entry of the DNA of interest into the cell due to

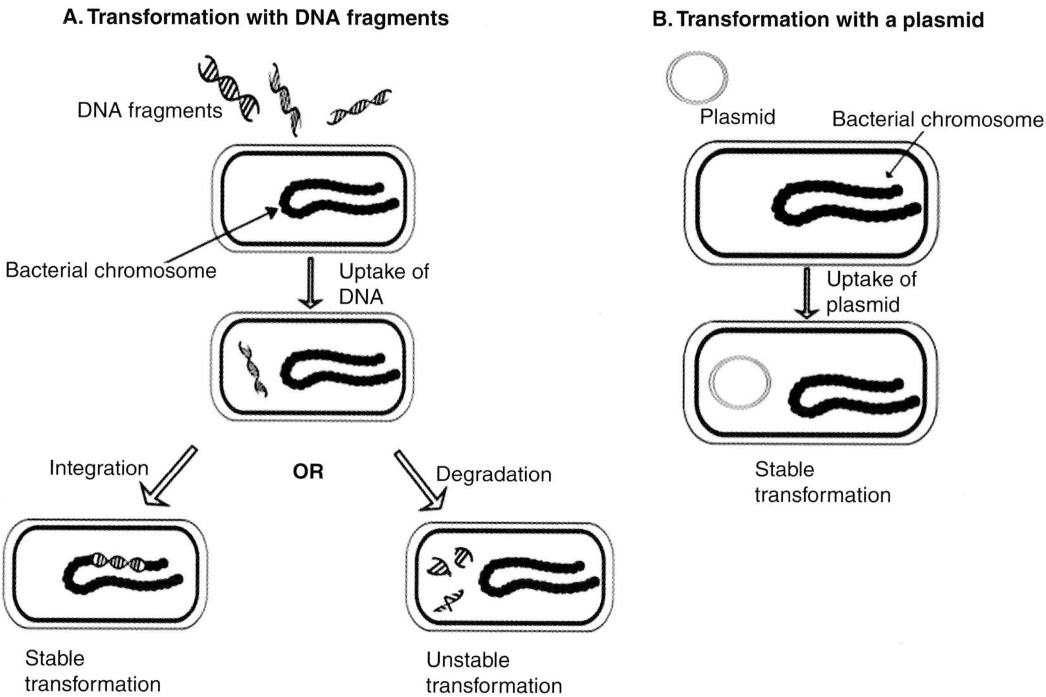

Figure 20.5 Mechanism of bacterial transformation. (A) Transformation with DNA fragments, indicating the DNA uptake (antibiotic resistant gene) from the environment followed by integration (stable transformation) or degradation (unstable transformation). (B) Transformation with a plasmid, leading to stable transformation.

application of sudden heat shock. The pores are closed by treatment with ice. Electroporation, is also a method of inducing artificial transformation (Chen and Dubnau, 2004). The mechanism of transformation is shown in Figure 20.5.

20.8.4.3 Transduction

The process by which DNA is transferred from one bacterium to another via a virus (bacteriophages) or a viral vector is called transduction. This mechanism does not involve any physical contact between the donor cell and the recipient cell as is with conjugation. Transduction is essential since it explains one mechanism by which antibiotic drugs become ineffective due to the transfer of antibiotic resistant genes between bacteria. When bacteriophages infect a bacterial cell, their normal mode of reproduction is to release their genetic elements into the host cell, reverse transcribe the RNA into DNA, and integrate along with the chromosomal DNA of the host. An enzyme called integrase is the main mediator of this process. The multiplication of the virus occurs by two mechanisms: lytic and lysogenic cycles. Lytic cycles occur soon after the integration of viral and bacterial genes, whereas the lysogenic cycles occur after a period of dormancy of the viral DNA within the host genes.

The new viral particles so formed break open the host cell, destroying it, and enter a new host bacterium to transfer the genetic material. During this mechanism, the transfer of antibiotic resistant genes from the previous bacterium to the new host cell occurs, thereby instigating an evolution of resistant bacteria (Zinder, 1955). Transduction can occur in two ways: generalized transduction and specialized transduction. In generalized transduction, a bacterial genemay be transferred to another via a phage, but no viral DNA will be transferred. This may occur either by recombination or by headful packaging (Wilson *et al.*, 1979).

In specialized transduction, the genes that get relocated depend upon where the phage genome is located on the chromosome. This occurs when the prophage excises imprecisely from the chromosome so that bacterial genes lying adjacent to the prophage are included in the excised DNA. This is then packaged into a new virus particle, which then delivers the DNA into a new bacterium. This is most commonly observed in *E. coli*. The mechanism of transduction is illustrated in Figure 20.6.

20.9 Evolution of XDR and PDR Bacteria

The term multi-drug resistance, literally means "resistant to many drugs," however, in most gram-positive and gram-negative bacteria, MDR refers to resistance against three or more antimicrobial classes (Falagas *et al.*, 2006).

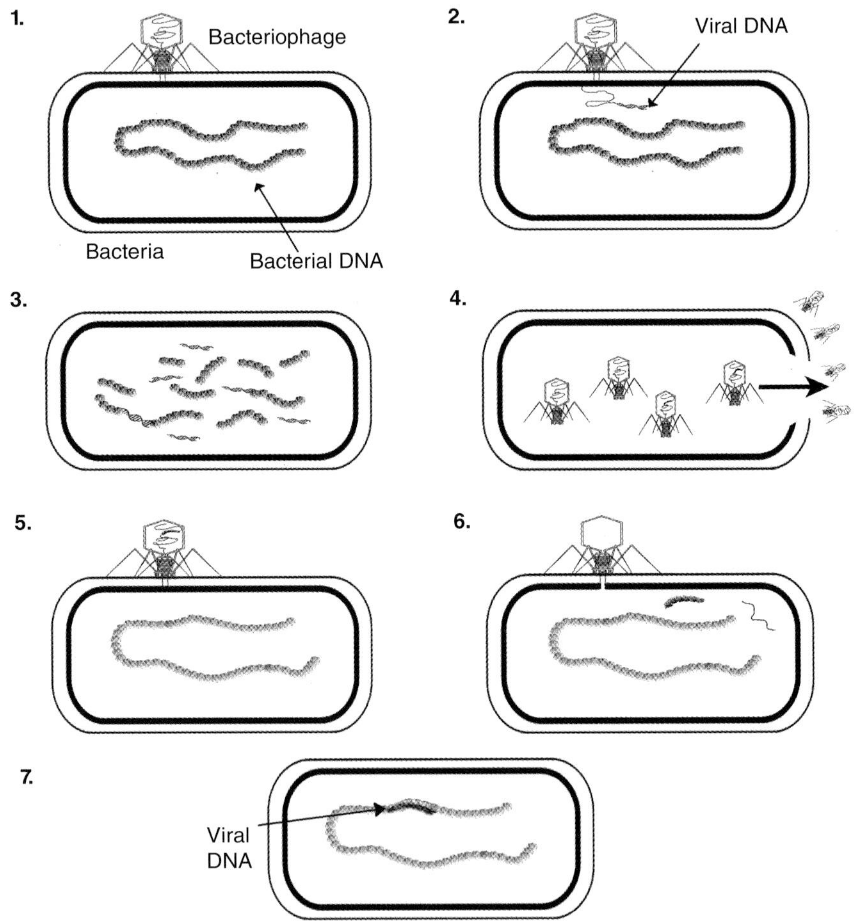

Figure 20.6 Mechanism of bacterial transduction for the transfer of antibiotic resistant genes from one bacterium to another (1) The bacteriophage attaches to the surface of the host bacterial cell. (2) Viral genome is introduced into the host cell. (3) Integration of the viral gene with the bacterial genome. (4) Formation of new viral particles and its entry into the environment after destroying the host cell. (5) New phages infect another host bacterium. (6) The viral genome containing the resistant genes from the previous bacterium is incorporated into the new host. (7) Integration of the resistant genes from the previous to the new host.

Further, this term can also be applicable for resistance against one key antimicrobial (Siegel *et al.*, 2007). Recently, studies have indicated an increased in the resistance of bacteria toward a number of drugs due to various afore mentioned reasons. Their susceptibility toward antibiotics has decreased and thus the needs for more number of drugs to treat several pathogenic foodborne infections are a necessity. This further leads to the emergence of extreme drug resistance or extensive drug resistance (XDR), which are resistant to first-line antibiotics such as rifampicin and isoniazid, plus a fluoroquinone and also resistant to one of the three major injectable second-line drugs such as amikacin, kanamycin, and capreomycin with regard to tuberculosis. *Clostridium* can grow and survive at temperatures above 100°C and are resistant to aminoglycosides, quinolones, and phenolic disinfectants. Therefore, these species are categorized as XDRs. Moreover, the evolution of resistant bacteria does not end at extreme resistance. From the Greek prefix "pan," the term PDR or pan drug resistance has emerged. Certain bacteria have resistance toward all antimicrobial agents. This includes the resistance toward all commercially available antibiotics and the routinely tested antimicrobials. For example, *Staphylococcus aureus, Enterococcus* spp, *Acinetobacter,* and so on, are now an emerging PDR, resistant to almost all available antibiotics (Walsh and Toleman, 2012).

20.10 Need for Caution and WHO/FDA Stands Toward the Development of MDR Pathogens in Foods

The emergences of drug-resistant pathogens are turning out to be a critical issue, especially in case of food products. Food is the source of living and the presence of pathogens in food are adversely harming humans as well as other animals. This problem is generally tough to prevent, but it can however be controlled (Freire-Moran *et al.*, 2012). The ecological balance may be disrupted and there is an urgent need for the development of strong antibiotics and drugs to combat these resistant bacteria. Although, steps are being taken with this regard, the fatality rates are still on the rise and thus, introduction of new drugs are not the only solution. There is an urgent need to address the lack of effective treatments to meet the increasing public health burden caused by multidrug-resistant bacteria, in particular against gram-negative bacteria (Bush *et al.*, 2011). Various strategies and solutions for solving this issue need to be formulated and executed.

The U.S. FDA (Food and Drug Administration) aims at reducing the use of antibiotics in livestock (Kuehn, 2014). This could cut down the transmission of resistant pathogens to food. As the U.S. FDA sets antimicrobial drug-residue limits in animal products (Baynes *et al.*, 2016), this information can be further used to aid the agency to establish better tactics for antibiotic use. This knowledge can also be used to help comprehend the contribution of different antimicrobials used in the food-production environment to the spread of multi-drug resistance and augmented virulence of food borne pathogens, and can be used to assist risk evaluations. The U.S. FDA is also considered a pivotal partner in the construction of new drugs to battle the strong and resistant microorganisms. Financial support toward the areas affected chiefly by MDR pathogens are being implemented by the WHO (World Health Organization) (Bush, 2015). The main goals of the CDC (Centers for Disease Control and Prevention) toward resolving this crisis include slowing the emergence of resistant bacteria and preventing the spread of infections caused by these bacteria, strengthening the national surveillance efforts to combat foodborne diseases, advancement and rapid development of innovative diagnostic tests for characterizing the resistant bacteria, development of new antibiotics for the treatments, and improving the international collaboration and capacities for antibiotic resistance (House, 2014). Therefore, a global approach toward solving this menace of drug resistance has already been put to effect.

20.11 Possible Solutions and Recommendations for Prevention

As the emergence of MDR become a threat to the food sectors, there is high scope for receiving suggestions and advice for the prevention or control of the issues arises due to MDR. The following points can be considered for the prevention or control of MDR pathogens:

- Using the appropriate antimicrobial for an infection, for example, no antibiotics for viral diseases and infections (House, 2014)
- Identifying the causative organism wherever possible
- Selecting an antibiotic that targets the specific organism and a specific disease-causing protein, rather than relying on a broad-spectrum drug (Bennett *et al.*, 2014)
- Completing an entire duration of antibiotic treatment recommended by the physician
- Using the precise dosage for disease treatment
- Utilizing combination therapies of antibiotics- for combating MDR pathogens (Worthington and Melander, 2013)

Furthermore, knowledge and awareness of MDR pathogens and its diseases must be spread in order to make the people discern the problems and the methods to fight it. Most importantly, maintaining a personal hygiene is the crucial step in decreasing the development of such microorganisms (Lee *et al.*, 2013). Providing monetary support to the countries that face dangerous health adversities may be a possible solution for the development of new and innovative drugs for subduing the activity of resistant bacteria (Davtyan *et al.*, 2015).

20.12 Conclusion

The world today is riddled by numerous harmful diseases, with foodborne diseases being the most troublesome. The emergence of multiple drug–resistant bacteria is one of the primary causes for instigating such diseases. The evolution of these dangerous bacteria has led to an increase in the fatality rates and the humans as well as other animals are at an immense risk. Therefore, an understanding of the various MDR pathogens in food, the causes of appearance, mechanism of evolution, and the methods to combat them, is necessary. The overuse and misuse of drugs, lack of development of new antibiotics, and usage of drugs in agriculture are the main causes of emergence of drug resistant pathogens. Moreover, the transfer of the drug resistant genes from one bacterium to another via several mechanisms of horizontal gene transfer (transformation, conjugation, and transduction) and chromosomal mutations are studied to comprehend the engendering of resistant foodborne pathogens. Although global measures are being taken worldwide by the WHO and the U.S. FDA to limit the drug resistance, certain health measures need to be explored individually to exterminate this menace. MDR pathogens have now transformed to XDR and PDR, which are an additional threat to the society. The major outbreaks of foodborne diseases due to resistant bacteria indicate that high caution is required. Additionally, the evolution of resistant bacteria may also lead to the development of new diseases, which may prove detrimental to the society. Thus, the prevalence of drug resistance is a vital issue that must be addressed instantly.

References

Altekruse, S.F., Cohen, M.L. and Swerdlow, D.L. (1997). Emerging foodborne diseases. *Emerging Infectious Diseases*, *3*, p. 285.

Andersson, D.I. and Hughes, D. (2012). Evolution of antibiotic resistance at non-lethal drug concentrations. *Drug Resistance Updates, 15,* pp. 162–172.

Au, T.T., et al. (2013). Clostridial hepatic gas gangrene with massive intravascular haemolysis and acute renal failure: A case report and literature review. *Hong Kong Journal of Emergency Medicine, 20,* p. 292.

Bambeke, V.F., Balzi, E. and Tulkens, P.M. (2000). Antibiotic efflux pumps. *Biochemical Pharmacology, 60,* pp. 457–470.

Banwart, G. (2014). *Basic Food Microbiology.* New York: Springer Science & Business Media.

Bartlett, J.G., Gilbert, D.N. and Spellberg, B. (2013). Seven ways to preserve the miracle of antibiotics. *Clin Infect Dis., 56,* pp. 1445–1450.

Baynes, R.E., et al. (2016). Health Concerns and Management of Select Veterinary Drug Residues. *Food and Chemical Toxicology, 88,* pp. 112–122.

Bennett, J.E., Dolin, R. and Blaser, M.J. (2014). *Mandell, Douglas, and Bennett's Principles and Practice of Infectious Diseases.* New York: Elsevier Health Sciences.

Bennett, R.W. (2001). *Staphylococcus aureus.* New York: John Wiley and Sons, Inc., pp. 201–220.

Bozcal, E., et al. (2015). Isolation of pathogenic *Yersinia enterocolitica* strains from different sources in Izmir region, Turkey. *Folia Microbiologica, 60,* 523–529.,

Brenner, D.J., Staley, J.T. and Krieg, N.R. (2005). Classification of procaryotic organisms and the concept of bacterial speciation. In: *Bergey's Manual of Systematic Bacteriology.* New York: Springer, pp. 27–32.

Brynestad, S. and Granum, P.E. (2002). Clostridium perfringens and foodborne infections. *International Journal of Food Microbiology, 74,* pp. 195–202.

Bull, S.A., et al. (2006). Sources of *Campylobacter spp.* colonizing housed broiler flocks during rearing. *Applied and Environmental Microbiology, 72,* pp. 645–652.

Bush, K., et al. (2011). Tackling antibiotic resistance. *Nature Reviews Microbiology, 9,* pp. 894–896.

Bush, K. (2015). Investigational Agents for the Treatment of Gram-Negative Bacterial Infections: A Reality Check. *ACS Infectious Diseases, 1,* pp. 509–511.

Campana, R., et al. (2012). Antagonistic activity of Lactobacillus acidophilus ATCC 4356 on the growth and adhesion/invasion characteristics of human Campylobacter jejuni. *Current Microbiology, 64,* pp. 371–378.

Centers for Disease Control and Prevention (CDC). (2015). Multistate foodborne disease outbreaks sicken thousands, highlight need for action. *Morbid Mortal Weekly Rep., 404,* pp. 639–3286.

Chen, I. and Dubnau, D. (2004). DNA uptake during bacterial transformation. *Nature Reviews Microbiology, 2,* pp. 241–249.

Chethana, G.S., et al. (2013). Review on multidrug resistant bacteria and its implication in medical sciences. *Journal of Biological Scientific Opinion, 1,* pp. 32–37.

Cirz, R.T., et al. (2005). Inhibition of Mutation and Combating the Evolution of Antibiotic Resistance. *PLoS Biol., 3,* p. e176.

Clewell, D.B., ed. (2013). Bacterial conjugation. New York: Springer Science & Business Media.

Crim, S.M., et al. (2015). Preliminary Incidence and Trends of Infection with Pathogens Transmitted Commonly Through Food—Foodborne Diseases Active Surveillance Network, 10 US Sites, 2006–2014. *MMWR. Morbidity and Mortality Weekly Report, 64,* pp. 495–499.

Dan, S.D., et al. (2015). Antibiotic susceptibility and prevalence of foodborne pathogens in poultry meat in Romania. *The Journal of Infection in Developing Countries, 9,* pp. 35–41.

Das, J.K., et al. (2013) Antibiotics for the treatment of Cholera, Shigella and Cryptosporidium in children. *BMC Public Health, 13*(Suppl 3), p. S10.

Davies, J. and Wright, G.D. (1997). Bacterial resistance to aminoglycoside antibiotics. *Trends Microbiol., 5,* pp. 234–240.

Davtyan, K., et al. (2015). Key role of tuberculosis services funding mechanisms in tuberculosis control and elimination. *European Respiratory Journal, 45,* pp. 289–291.

de Jong, A., et al. (2014). Antimicrobial susceptibility of Salmonella isolates from healthy pigs and chickens (2008–2011). *Veterinary Microbiology, 171,* pp. 298–306.

Dodds, K.L. (1993). *Clostridium botulinum* in foods, in A.H.W. Hauschid and K.L. Dodds (eds) *Clostridium botulinum:* Ecology and Control in Foods. New York: Marcel Dekker, pp. 53–68.

Doelle HW (2014) Bacterial metabolism. Academic Press, London, 59–p.

Dohrmann, A.B., et al. (2015). Clostridium cluster I and their pathogenic members in a full-scale operating biogas plant. *Applied Microbiology and Biotechnology, 99,* pp. 3585–3598.

Doyle, M.E. (2015). Multidrug-Resistant Pathogens in the Food Supply. *Foodborne Pathogens and Disease, 12,* pp. 261–279.

Echeverria, P., Sethabutr, O. Pitarangsi, C. (1991). Microbiology and diagnosis of infections with Shigella and enteroinvasive Escherichia coli. *Review of Infectious Diseases, 13*(Supplement 4), pp. S220–S225.

Falagas, M.E., Koletsi, P.K. and Bliziotis, I.A. (2006). The diversity of definitions of multidrug-resistant (MDR) and pandrug-resistant (PDR) *Acinetobacter baumannii* and *Pseudomonas aeruginosa. J Med Microbiol., 55,* pp. 1619–1629.

Fejes, I., Dégi, R. and Végh, M. (2013). Clostridium septicum gas gangrene in the orbit: a case report. *Infection, 41,* pp. 267–270.

Ferri, M., et al. (2015). Antimicrobial Resistance: A Global Emerging Threat to Public Health Systems. *Critical Reviews in Food Science and Nutrition,* doi: 10.1080/10408398.2015.1077192.

Freire-Moran, L., et al., and the ECDC-EMA Working Group. (2011). Critical shortage of new antibiotics in development against multidrug-resistant bacteria—time to react is now. *Drug Resistance Updates, 14,* pp. 118–124.

Goossens, H. (2009). Antibiotic consumption and link to resistance. *Clin Microbiol Infec., 15,* pp. 12–15.

Gopinath, S., Carden, S. and Monack, D. (2012). Shedding light on Salmonella carriers. *Trends in Microbiology, 20,* pp. 320–327.

Gross, M. (2013) Antibiotics in crisis. *Curr Biol., 23,* pp. R1063–R1065.

Gould, I.M. and Bal, A.M. (2013). New antibiotic agents in the pipeline and how they can overcome microbial resistance. *Virulence, 4,* pp. 185–191.

Gould, I.M., et al. (2012). New insights into meticillin-resistant *Staphylococcus aureus* (MRSA) pathogenesis, treatment and resistance. *International Journal of Antimicrobial Agents, 39,* pp. 96–104.

Hancock, D.D., et al. (1998). Multiple sources of *Escherichia coli* O157 in feedlots and dairy farms in the northwestern USA. *Preventive Veterinary Medicine, 35,* pp. 11–19.

Harakeh, S., et al. (2012). Highly resistant *Yersinia enterocolitica* isolated from dairy based foods in Lebanon. *The International Arabic Journal of Antimicrobial Agents, 2,* doi:10.3823/70.

Hasan, M.A., et al. (2016). Identification of putative drug targets in Vancomycin-resistant Staphylococcus aureus (VRSA) using computer aided protein data analysis. *Gene, 575,* pp. 132–143.

Hayes, R. (2013). *Food Microbiology and Hygiene.* New York; Springer Science & Business Media.

Hennekinne, J.A., et al. (2015). European Food Poisoning Outbreaks Involving Meat and Meat-based Products. *Procedia Food Science, 5,* pp. 93–96.

Hill, S.E., et al. (2013). Foodborne botulism treated with heptavalent botulism antitoxin. *Annals of Pharmacotherapy, 47,* p. e12.

Hospital, X.F., et al. (2016). A study on the toxigenesis by *Clostridium botulinum* in nitrate and nitrite-reduced dry fermented sausages. *International Journal of Food Microbiology, 218,* pp. 66–70.

House, W. (2014), National strategy for combating antibiotic-resistant bacteria. Available at: http://www.cdc.gov/drugresistance/federal-engagement-in-ar/national-strategy/index.html [Accessed on December 1, 2015].

Hwang, T.J. and Keshavjee, S. (2014). Global Financing and Long-Term Technical Assistance for Multidrug-Resistant Tuberculosis: Scaling Up Access to Treatment. *PLoS Med. 11*, p. e1001738.

Jacobsen, A., et al. (2011). The Salmonella enterica pan-genome. *Microbial Ecology, 62*, pp. 487–504.

Jay, J.M. (2012). *Modern food microbiology*. Springer.

Karch, H., et al. (2015). Enterohemorrhagic *E. coli* (EHEC): Environmental-Vehicle-Human Interface. In: *Zoonoses-Infections Affecting Humans and Animals*. Netherlands: Springer, pp. 235–248.

Kathariou, S. (2002). *Listeria monocytogenes* virulence and pathogenicity, a food safety perspective. *Journal of Food Protection, 65*, pp. 1811–1829.

Koluman, A. and Dikici A. (2013). Antimicrobial resistance of emerging foodborne pathogens: status quo and global trends. *Critical Reviews in Microbiology, 39*, pp. 57–69.

Kuehn, B.M. (2014). FDA moves to curb antibiotic use in livestock. *JAMA, 311*, pp. 347–348.

Lee, C.R., et al. (2013). Strategies to minimize antibiotic resistance. *International Journal of Environmental Research and Public Health, 10*, pp. 4274–4305.

Lee, W.H. and McClain, D. (1986). Improved Listeria monocytogenes selective agar. *Applied and Environmental Microbiology, 52*, pp. 1215–1217.

Levin, B.R., Perrot, V. and Walker, N. (2000). Compensatory mutations, antibiotic resistance and the population genetics of adaptive evolution in bacteria. *Genetics, 154*, pp. 985–997.

Lin, J., et al. (2015). Mechanisms of antibiotic resistance. *Frontiers in Microbiology, 6*.

Loeffler, J. and Stevens, D.A. (2003). Antifungal drug resistance. *Clinical Infectious Diseases, 36*, pp. S31–S41.

Lovett, J., Francis, D.W. and Hunt, J.M. (1987). *Listeria monocytogenes* in raw milk: detection, incidence, and pathogenicity. *Journal of Food Protection, 50*, pp. 188–192.

Lushniak, B.D. (2014). Antibiotic resistance: a public health crisis. *Public Health Rep., 129*, pp. 314–316.

Luyt, C.E., et al. (2014). Antibiotic stewardship in the intensive care unit. *Crit Care., 18*, p. 480.

Mandal, S.M., Ghosh, A.K. and Pati, B.R. (2015). Dissemination of antibiotic resistance in methicillin-resistant *Staphylococcus aureus* and vancomycin-resistant *S. aureus* strains isolated from hospital effluents. *American Journal of Infection Control, 43*, pp. e87–e88.

Martin, D.L., et al. (1986). Contaminated produce—a common source for two outbreaks of Shigella gastroenteritis. *American Journal of Epidemiology, 124*, pp. 299–305.

Michael, C.A., Dominey-Howes, D. and Labbate, M. (2014). The antibiotic resistance crisis: causes, consequences, and management. *Front Public Health, 2*, p. 145.

Mylonakis, E., et al. (2002). Listeriosis during pregnancy: a case series and review of 222 cases. *Medicine, 81*, pp. 260–269.

Nikaido, H. (2009). Multidrug resistance in bacteria. *Annu Rev Biochem., 78*, pp. 119–146.

Nyachuba, D.G. (2010). Foodborne illness: is it on the rise? *Nutrition Reviews, 68*, pp. 257–269.

Oussalah, M., et al. (2007). Inhibitory effects of selected plant essential oils on the growth of four pathogenic bacteria: *E. coli* O157: H7, *Salmonella typhimurium*, *Staphylococcus aureus* and *Listeria monocytogenes*. *Food Control, 18*, pp. 414–420.

Piddock, L. (2012). The crisis of no new antibiotics—what is the way forward? *Lancet Infect Dis., 12*, pp. 249–253.

Pires, S.M., et al. (2012). Attributing human foodborne illness to food sources and water in Latin America and the Caribbean using data from outbreak investigations. *International Journal of Food Microbiology, 152*, pp. 129–138.

Pomposiello, P.J. and Demple, B. (2000). Identification of SoxS-regulated genes in *Salmonella enterica* serovar *typhimurium*. *Journal of Bacteriology, 182,* pp. 23–29.

Qin, S., et al. (2012). Identification of a novel genomic island conferring resistance to multiple aminoglycoside antibiotics in Campylobacter coli. *Antimicrobial Agents and Chemotherapy, 56,* pp. 5332–5339.

Ratkowsky, D.A., et al. (1982). Relationship between temperature and growth rate of bacterial cultures. *Journal of Bacteriology, 149,* pp. 1–5.

Rahaman, M.M., Huq, I. and Dey, C.R. (1975). Superiority of MacConkey's agar over Salmonella-Shigella agar for isolation of *Shigella dysenteriae* type 1. *Journal of Infectious Diseases, 131,* pp. 700–703.

Read, A.F. and Woods, R.J. (2014). Antibiotic resistance management. *Evol Med Public Health., 2014,* p. 147.

Rodvold, K.A. and McConeghy, K.W. (2014). Methicillin-resistant *Staphylococcus aureus* therapy: past, present, and future. *Clinical Infectious Diseases, 58,* pp. S20–S27.

Rosner, B.M., et al. (2013). Clinical aspects and self-reported symptoms of sequelae of Yersinia enterocolitica infections in a population-based study, Germany 2009–2010. *BMC Infectious Diseases, 13,* p. 236.

Rossi, F., et al. (2014). Horizontal gene transfer among microorganisms in food: current knowledge and future perspectives. *Food Microbiology, 42,* pp. 232–243.

Saludes, M., Troncoso, M. and Figueroa, G. (2015). Presence of *Listeria monocytogenes* in Chilean food matrices. *Food Control, 50,* pp. 331–335.

Sathe, S.K. (2013). Chemistry and Implications of Antinutritional Factors in Dry Beans and Pulses. *Dry Beans and Pulses Production, Processing and Nutrition,* pp. 359–377.

Sengupta, S., Chattopadhyay, M.K. and Grossart, H.P. (2013). The multifaceted roles of antibiotics and antibiotic resistance in nature. *Front Microbiol., 4,* p. 47.

Siegel, J.D., et al. (2007). Management of multidrug-resistant organisms in health care settings, 2006. *Am J Infect Control., 35*(suppl 2), pp. 165–193.

Singh, P., et al. (2015). Prevalence of multidrug resistant altered *Vibrio cholerae* O1 isolates among Diarrhoeal patients in Delhi during 2008–2012. *Prevalence, 5,* pp. 622–628.

Skariyachan, S. and Abhilash, M. (2012). *Introduction to Food Biotechnology.* New Dehli: CBS Publishers and Distributers. Spellberg, B. and Gilbert, D.N. (2014). The future of antibiotics and resistance: a tribute to a career of leadership by John Bartlett. *Clin Infect Dis., 59,* pp. S71–S75.

Spitzer, P.G., Hammer, S.M. and Karchmer, A.W. (1986). Treatment of Listeria monocytogenes infection with trimethoprim-sulfamethoxazole: case report and review of the literature. *Review of Infectious Diseases, 8,* pp. 427–430.

Swaminathan, B. and Feng, P. (1994). Rapid detection of food-borne pathogenic bacteria. *Annual Reviews in Microbiology, 48,* pp. 401–426.

T Supuran, C. (2012). Inhibition of bacterial carbonic anhydrases and zinc proteases: from orphan targets to innovative new antibiotic drugs. *Current Medicinal Chemistry, 19,* pp. 831–844.

Tanwar, J., et al. (2014). Multidrug resistance: an emerging crisis. *Interdisciplinary Perspectives on Infectious Diseases.*

Thanassi, D.G., Cheng, L.W. and Nikaido, H. (1997). Active efflux of bile salts by *Escherichia coli. Journal of Bacteriology, 179,* pp. 2512–2518.

Thompson, J.M., et al. (2013). Antibiotic resistant *Staphylococcus aureus* in hospital wastewaters and sewage treatment plants with special reference to methicillin-resistant *Staphylococcus aureus* (MRSA). *Journal of Applied Microbiology, 114,* pp. 44–54.

Troller, J. (2012). *Water activity and food.* New York: Elsevier.

Uzal, F.A., et al. (2015). Animal models to study the pathogenesis of human and animal

Clostridium perfringens infections. *Veterinary Microbiology, 179,* pp. 23–33.
Van Boeckel, T.P., Gandra, S. and Ashok, A. (2014). Global antibiotic consumption 2000 to 2010: an analysis of national pharmaceutical sales data. *Lancet Infect Dis., 14,* pp. 742–750.
Ventola, C.L. (2015). The Antibiotic Resistance Crisis: Part 1: Causes and Threats. *Pharmacy and Therapeutics, 40,* pp. 277–283.
Vezzulli, L., Colwell, R.R. and Pruzzo, C. (2013). Ocean warming and spread of pathogenic vibrios in the aquatic environment. *Microbial Ecology, 65,* pp. 817–825.
Viale, P., et al. (2015). Treatment of MDR-Gram negative infections in the 21st century: a never ending threat for clinicians. *Current Opinion in Pharmacology, 24,* pp. 30–37.
Viswanathan VK (2014) Off-label abuse of antibiotics by bacteria. *Gut Microbes, 5,* pp. 3–4.
Vuyst, L.D. and Vandamme, E.J. (1994). *Bacteriocins of lactic acid bacteria: microbiology, genetics and applications.* New York: Springer, doi: 10.1007/978-1-4615-2668-1.
Walsh, T.R. and Toleman, M.A. (2012). The emergence of pan-resistant Gram-negative pathogens merits a rapid global political response. *Journal of Antimicrobial Chemotherapy, 67,* pp. 1–3.
Wang, H., et al. (2001). Genetic characterization of highly fluoroquinolone-resistant clinical Escherichia coli strains from China: role of acrR mutations. *Antimicrobial Agents and Chemotherapy, 45,* pp. 1515–1521.
Webber, M.A. and Piddock, L.J.V. (2001). Absence of mutations in marRAB or soxRS in acrB-overexpressing fluoroquinolone-resistant clinical and veterinary isolates of Escherichia coli. *Antimicrobial Agents and Chemotherapy, 45,* pp. 1550–1552.
WHO (World Health Organization). (2016). WHO'S First Ever Global Estimates of Foodborne Diseases Find Children Under 5 Account for Almost One Third Of Deaths. *Saudi Med J., 37,* pp. 109–110.
Wilson, B.A., et al. (2011). *Bacterial pathogenesis: a molecular approach,* 3rd ed. Washington, DC: American Society for Microbiology.
Wilson, G.G., et al. (1979). High-frequency generalised transduction by bacteriophage T4. *Nature, 20,* pp. 80–82.
Wright, G.D. (2014). Something new: revisiting natural products in antibiotic drug discovery. *Can J Microbiol., 60,* pp. 147–154.
Worthington, R.J. and Melander, C. (2013). Combination approaches to combat multidrug-resistant bacteria. *Trends in Biotechnology, 31,* pp. 177–184.
Wright, G.D. (1999). Aminoglycoside-modifying enzymes. *Curr. Opin. Microbiol., 2,* pp. 499–503.
Xiao, G.G., et al. (2012). A school outbreak of Shigellasonnei infection in China: clinical features, antibiotic susceptibility and molecular epidemiology. *Indian Pediatrics, 49,* pp. 287–290.
Yeni, F., et al. (2015). Most Common Foodborne Pathogens and Mycotoxins on Fresh Produce: A Review of Recent Outbreaks. *Critical Reviews in Food Science and Nutrition. 56,* pp. 1532–1544.
Zinder, N.D. (1955). Bacterial transduction. *Journal of Cellular and Comparative Physiology, 45* pp. 4523–4549.

Index

Foodborne Pathogens and Antibiotic Resistance, First Edition. Edited by Om V. Singh.

m

Food Science and Technology from Wiley Blackwell

Decontamination of Fresh and Minimally Processed Produce	Gomez-Lopez	9780813823843
Progress in Food Preservation	Bhat	9780470655856
Food Safety for the 21st Century: Managing HACCP and Food Safety throughout the Global Supply Chain	Wallace	9781405189118
The Microbiology of Safe Food, 2nd edition	Forsythe	9781405140058

SENSORY SCIENCE, CONSUMER RESEARCH & NEW PRODUCT DEVELOPMENT

Olive Oil Sensory Science	Monteleone	9781118332528
Quantitative Sensory Analysis: Psychophysics, Models and Intelligent Design	Lawless	9780470673461
Product Innovation Toolbox: A Field Guide to Consumer Understanding and Research	Beckley	9780813823973
Sensory and Consumer Research in Food Product Design and Dev, 2nd Ed	Moskowitz	9780813813660
Sensory Evaluation: A Practical Handbook	Kemp	9781405162104
Statistical Methods for Food Science	Bower	9781405167642
Concept Research in Food Product Design and Development	Moskowitz	9780813824246
Sensory and Consumer Research in Food Product Design and Development	Moskowitz	9780813816326

FOOD INDUSTRY SUSTAINABILITY & WASTE MANAGEMENT

Food and Agricultural Wastewater Utilization and Treatment, 2nd Edition	Liu	9781118353974
Sustainable Food Processing	Tiwari	9780470672235
Food and Industrial Bioproducts and Bioprocessing	Dunford	9780813821054
Handbook of Sustainability for the Food Sciences	Morawicki	9780813817354
Sustainability in the Food Industry	Baldwin	9780813808468
Lean Manufacturing in the Food Industry	Dudbridge	9780813810072

FOOD LAWS & REGULATIONS

Guide to US Food Laws and Regulations, 2nd Edition	Curtis	9781118227787
Food and Drink - Good Manufacturing Practice: A Guide to its Responsible Management (GMP6), 6th Edition	Manning	9781118318201
The BRC Global Standard for Food Safety: A Guide to a Successful Audit, 2nd Edition	Kill	9780470670651
Food Labeling Compliance Review, 4th edition	Summers	9780813821818

DAIRY FOODS

Lactic Acid Bacteria: Biodiversity and Taxonomy	Holzapfel	9781444333831
From Milk By-Products to Milk Ingredients: Upgrading the Cycle	de Boer	9780470672228
Milk and Dairy Products as Functional Foods	Kanekanian	9781444336832
Milk and Dairy Products in Human Nutrition: Production, Composition and Health	Park	9780470674185
Manufacturing Yogurt and Fermented Milks, 2nd Edition	Chandan	9781119967088
Sustainable Dairy Production	de Jong	9780470655849
Advances in Dairy Ingredients	Smithers	9780813823959
Membrane Processing: Dairy and Beverage Applications	Tamime	9781444333374
Analytical Methods for Food and Dairy Powders	Schuck	9780470655986
Dairy Ingredients for Food Processing	Chandan	9780813817460
Processed Cheeses and Analogues	Tamime	9781405186421
Technology of Cheesemaking, 2nd edition	Law	9781405182980

SEAFOOD, MEAT AND POULTRY

Seafood Processing: Technology, Quality and Safety	Boziaris	9781118346211
Should We Eat Meat? Evolution and Consequences of Modern Carnivory	Smil	9781118278727
Handbook of Meat, Poultry and Seafood Quality, second edition	Nollet	9780470958322
The Seafood Industry: Species, Products, Processing, and Safety , 2nd Edition	Granata	9780813802589
Organic Meat Production and Processing	Ricke	9780813821269
Handbook of Seafood Quality, Safety and Health Effects	Alasalvar	9781405180702

BAKERY & CEREALS

Oats Nutrition and Technology	Chu	9781118354117
Cereals and Pulses: Nutraceutical Properties and Health Benefits	Yu	9780813818399
Whole Grains and Health	Marquart	9780813807775
Gluten-Free Food Science and Technology	Gallagher	9781405159159
Baked Products - Science,Technology and Practice	Cauvain	9781405127028

BEVERAGES & FERMENTED FOODS/BEVERAGES

Encyclopedia of Brewing	Boulton	9781405167444
Sweet, Reinforced and Fortified Wines: Grape Biochemistry, Technology and Vinification	Mencarelli	9780470672242
Technology of Bottled Water, 3rd edition	Dege	9781405199322
Wine Flavour Chemistry, 2nd edition	Bakker	9781444330427
Wine Quality: Tasting and Selection	Grainger	9781405113663

PACKAGING

Handbook of Paper and Paperboard Packaging Technology, 2nd Edition	Kirwan	9780470670668
Food and Beverage Packaging Technology, 2nd edition	Coles	9780813812748
Food and Package Engineering	Morris	9780813814797
Modified Atmosphere Packaging for Fresh-Cut Fruits and Vegetables	Brody	9780813812748

For further details and ordering information, please visit www.wiley.com/go/food